KV-388-172

The Physics and Chemistry of Minerals and Rocks

edited by

R. G. J. Strens
School of Physics,
University of Newcastle upon Tyne

A Wiley–Interscience Publication

JOHN WILEY & SONS
London · New York · Sydney · Toronto

Library of Congress Cataloging in Publication Data:

NATO Advanced Study Institute, Newcastle upon Tyne, 1974.
The physics and chemistry of minerals and rocks.

Proceedings of the NATO Advanced Study Institute on Petrophysics, held at the University of Newcastle upon Tyne Apr. 22–26, 1974.
'A Wiley–Interscience publication.'
1. Mineralogy—Testing—Congresses. 2. Rocks—Testing—Congresses. 3. Geochemistry—Congresses.
I. Strens, R. G. J. II. Title.
QE351.N37 1974 549 75–6930

ISBN 0 471 83368 1

Photosetting by Thomson Press (India) Limited, New Delhi and printed in Great Britain at The Pitman Press, Bath

Foreword by the Director of the NATO Institute

The remarkable development in the 1920's of the theory of the internal constitution of stars, associated particularly with the name of Sir Arthur Eddington, was possible because he was able to assume that the equation of state within these bodies was the gas law. Our understanding of the internal constitution of planets has lagged behind because no such simple equation of state is available. For the Earth the empirical data of seismology provided a relationship between elastic constants, density, and pressure, and empirical relationships between compressibility and pressure have been used to determine the internal constitution of other terrestrial planets, though with some dubious assumptions about their composition and internal temperatures. The internal constitutions of the major planets were put on a more satisfactory basis theoretically by Wildt and important progress was made in applying the quantum theory of hydrogen as the basis of the discussion.

The development of high pressure techniques by P. W. Bridgeman in the 1920's and 30's gave promise that laboratory experiments could provide some of the essential data. In the last decade very great progress has been made, both in static and dynamic methods of reaching exceedingly high pressures and generating the experimental techniques of physical measurements in these conditions. We have now entered a period in which there is a fascinating interplay between laboratory experimental results, quantum mechanical theories of the behaviour of materials at high pressure, and the increasing geophysical data on the deep interiors of planets. It was, therefore, thought timely to hold the NATO Advanced Study Institute on this subject and it is hoped that the Proceedings will be of great help to students entering this field and will be a catalyst for further collaborative work between the disciplines of high pressure physics and geophysics.

I am sure that participants will wish me to thank NATO for their generous financial support, and Mr. W. F. Mavor and Mrs. Marion Turner for their hard work, which contributed much to the success of the Institute.

S. K. RUNCORN F.R.S.

Preface

Although invitations to contribute were phrased in general terms, the papers in this volume have three main themes: transport properties, including electrical and thermal properties and convection; the dependence of the macroscopic properties of rocks and minerals on their microstructure; and the reactions, properties and structure of minerals, particularly those containing transition metal ions, at high pressure.

The meeting showed that our understanding of the response of rocks to stress, whether by elastic or plastic deformation, or by fracture, is advancing rapidly. The rates of such very large scale processes as mantle convection and continental drift are found to depend on the nature and concentration of dislocations, and atomic diffusion rates. In a few years, these advances may lead to the development of quantitative rather than qualitative theories of rock deformation and mantle convection.

Work on the determination of crystal structure at high pressure and at high temperature, and on the computer modelling of the response of crystal structures to changes in pressure, temperature and composition should greatly increase our understanding of mineral properties and reactions under these conditions, and ultrahigh pressure phase equilibrium studies are now reaching beyond the 400–650 km transition zone, towards lower-mantle conditions. Spectroscopic and other work suggests that major changes in mantle properties will result from changes in the bonding of transition metal ions (especially Fe^{2+}) at high pressure.

I should like to thank Marion Turner for her cheerful and efficient handling of the secretarial work involved in preparing the scientific programme and editing thc proceedings.

R. G. J. STRENS

Contributors

R. M. ABU-EID — *Department of Earth and Planetary Sciences, MIT, Cambridge, Mass. 02139, USA.*

S. AKIMOTO — *Institute for Solid State Physics, University of Tokyo, Roppongi, Minato-ku, Tokyo 106, Japan.*

D. BAMFORD — *Department of Geology with Geophysics, University of Birmingham, PO Box 363, Birmingham B15 2TT.*

W. A. BASSETT — *Department of Geological Sciences, University of Rochester, NY 14627, USA.*

G. D. BORLEY — *Department of Geology, Royal School of Mines, Prince Consort Road, London SW7, UK.*

A. M. BOULLIER — *Institut des Sciences de la Nature, 44037 Nantes, France.*

M. BUKOWINSKI — *Institute of Geophysics, University of California, Los Angeles, California 90024, USA.*

R. G. BURNS — *Department of Earth and Planetary Sciences, MIT, Cambridge, Mass. 02139, USA.*

B. R. CLARK — *Department of Geology and Mineralogy, University of Michigan, Ann Arbor, Michigan 48104, USA.*

L. S. COLLETT — *Geological Survey of Canada, 601 Booth Street, Ottawa, Ontario, Canada K1A OE8.*

M. J. DEMPSEY — *School of Physics, University of Newcastle, Newcastle upon Tyne NE1 7RU, UK.*

H. G. DRICKAMER — *School of Chemical Sciences, University of Illinois, Urbana, Ill. 61801, USA.*

Al. DUBA — *Lawrence Radiation Laboratory, University of California, Livermore, Calif. 94550, USA.*

W. B. DURHAM — *Department of Earth and Planetary Sciences, MIT. Cambridge, Mass. 02139, USA.*

J. T. ENGELDER — *Lamont-Doherty Geological Observatory of Columbia University, Palisades, New York 10964, USA.*

C. W. FRANK — *Sandia Laboratories, Albuquerque, New Mexico 87115, USA.*

K. FUCHS — *Geophysikalisches Institut der Universitat, Hertzestrasse 16, D-75, Karlsrube 21, Germany.*

C. GATEAU — *Bureau de Recherches Geologiques et Miniers, BP 6009, 45018 Orleans-Cedex, France.*

K. A. Goettel — *Department of Earth and Planetary Sciences, MIT, Cambridge, Mass. 02139, USA.*

C. Goetze — *Department of Earth and Planetary Sciences, MIT, Cambridge, Mass. 02139, USA.*

Y. Gueguen — *Institut des Sciences de la Nature, Faculte des Sciences, 44037 Nantes, France.*

B. M. Hamil — *Department of Geology and Geological Engineering, Michigan Technological University, Houghton, Michigan 49931, USA.*

R. M. Hazen — *Department of Geological Sciences, Harvard University, 20 Oxford Street, Cambridge, Mass. 02138, USA.*

H. C. Heard — *University of California, Lawrence Livermore Laboratory, Livermore, Calif. 94550, USA.*

A. Hirn — *Institut de Physique du Globe, Universite 6, 11 Quai St. Bernard, F-75230 Paris, France.*

F. E. Huggins — *Geophysical Laboratory, 2801 Upton Street, Washington DC 20008, USA.*

T. J. Katsube — *Geological Survey of Canada, 601 Booth Street, Ottawa, Ontario, Canada K1A OE8.*

W. C. Kelly — *Geology and Mineralogy, University of Michigan, Ann Arbor, Michigan 48104, USA.*

R. Kind — *Geophysikalisches Institut der Universitat, Hertzestrasse 16, D-75, Karlsruhe 21, Germany.*

G. Kinsland — *Department of Geological Sciences, University of Rochester, Rochester, NY 14627, USA.*

O. J. Kleppa — *The James Franck Institute, University of Chicago, 5640 Ellis Avenue, Chicago, Ill. 60637, USA.*

L. Knopoff — *Institute of Geophysics and Planetary Physics, University of California, Los Angeles, 90024, USA.*

D. L. Kohlstedt — *Department of Earth and Planetary Sciences, MIT, Cambridge, Mass. 02139, USA.*

K. Langer — *Mineralogisch-petrologisches Institut der Universität Bonn, 53 Bonn, Poppelsdorfer Schloss, Germany.*

A. D. Law — *University Chemical Laboratory, University of Kent, Canterbury CR2 7NH, Kent, UK.*

G. W. Lorimer — *Department of Metallurgy, University of Manchester, Manchester M13 9PL. UK.*

H. Mao — *Geophysical Laboratory, 2801 Upton Street, Washington DC 20008, USA.*

Y. Matsui — *Institute for Thermal Spring Research, Okayama University, Misasa, Tottori-Ken 682-02, Japan.*

L. Merrill — *High Pressure Data Center, Brigham Young University, Provo, Utah 84602, USA.*

L.-C. Ming — *Department of Geological Sciences, University of Rochester, Rochester, NY 14627, USA.*

R. Nashner — *Institute of Geophysics and Planetary Physics, University of California, Los Angeles, Calif. 90024, USA.*

G. R. Olhoeft — *Geophysics Laboratory, Department of Physics, University of Toronto, Toronto, Ontario M5S 1A7, Canada.*

A. J. Piwinskii — *University of California, Lawrence Livermore Laboratory, Livermore, Calif. 94550, USA.*

C. T. Prewitt — *Department of Earth and Space Sciences, State University of New York, Stony Brook, New York 11790, USA.*

J. M. Prevosteau — *Bureau de Recherches Geologiques et Miniers, BP 6009, 45018 Orleans-Cedex, France.*

C. Prodehl — *Geophysical Institute, University of Karlsruhe, Hertzstr. 16, D-75 Karlsruhe 21, Germany.*

R. R. Reeber — *Department of Mineralogy and Petrology, University of Cambridge, Downing Place, Cambridge, UK.*

D. Richter — *Department of Earth and Planetary Sciences, MIT, Cambridge, Mass. 02139, USA.*

S. K. Runcorn — *School of Physics, University of Newcastle, Newcastle upon Tyne NE1 7RU, UK.*

L. Rybach — *Institut fur Geophysik ETHZ, Postfach 266, CH-8049 Zurich, Switzerland.*

R. N. Schock — *University of California, Lawrence Livermore Laboratory, Livermore, Calif. 94550, USA.*

R. D. Shannon — *Experimental Station, Central Research Department, E.I. du Pont de Nemours, Wilmington, Delaware 19398, USA.*

G. H. Shaw — *Department of Physical Sciences, Florida International University, Tamiami Trail, Miami, Florida 33144, USA.*

G. Simmons — *Department of Earth and Planetary Sciences, MIT, Cambridge, Mass. 02139, USA.*

G. Smith — *Department of Solid State Physics, Australian National University, Canberra ACT 2600, Australia.*

S. Sriruang — *Division of Geotechnics, Royal Thai Irrigation Dept. Bangkok 4, Thailand.*

L. Steinmetz — *Geophysikalisches Institut der Universität, Hertzestrasse 16, D-75, Karlsruhe 21, Germany.*

R. G. J. Strens — *School of Physics, University of Newcastle, Newcastle upon Tyne NE1 7RU, UK.*

Y. Syono — *Institute for Iron, Steel and Other Metals, Tohoku University, Katahiracho, Sendai 980, Japan.*

H. G. Tolland — *Department of Chemical Engineering, University of Newcastle, Newcastle upon Tyne NE1 7RU, UK.*

F. Vaillant — *Bureau de Recherches Geologiques et Miniers, BP 6009, 45018 Orleans-Cedex, France.*

N. Warren — *Institute of Geophysics and Planetary Physics, University of California, Los Angeles, CA 90024, USA.*

S. WHITE — *Department of Geology, Imperial College, Prince Consort Road, London SW7, UK.*

F. W. WRAY — *Department of Geodesy and Geophysics, University of Cambridge, Madingley Road, Cambridge, UK.*

Contents

PART II

PART I

The Plastic Deformation of Minerals

G. W. Lorimer

Department of Metallurgy, Faculty of Science,
University of Manchester, Manchester M13 9PL

The plastic deformation of a crystalline solid can occur in a number of distinguishable and independent ways while crystallinity and continuity are retained (Ashby, 1972). A stress which exceeds the theoretical yield stress of a perfect crystal causes plastic flow. This form of plastic deformation can be called defect-less flow, and is distinct from all other forms of plastic deformation which require some structural defect to be present. Glide motion of dislocations (line defects), and at high temperatures ($T > 0.5\ T_m$, where T_m is the absolute melting temperature) dislocation climb, are two other separate and distinguishable mechanisms of plastic deformation. At high temperatures the migration of point defects can give rise to plastic deformation through matrix diffusion—Nabarro–Herring creep—or grain boundary diffusion—Coble creep. Twinning is the last in this list of mechanisms, but, unlike the others, it can provide only limited plastic deformation.

There are other mechanisms in addition to those noted above which can give rise to bulk flow of a mass of material, e.g. fine scale fracture coupled with some form of lubrication which characterizes cataclastic flow, *but these are not basic processes of plastic deformation of crystalline solids for the continuity of the crystal is not maintained.*

Which of the above deformation mechanisms predominates in a polycrystalline solid depends on the physical properties of that solid (e.g. elastic moduli) its microstructure (e.g. grain size) and on the temperature, pressure and applied shear stress. Ashby (1972) has popularized the concept of the 'deformation mechanism map' on which the realms of the different deformation mechanisms are plotted in stress–temperature space. The map is divided into fields and within one field a particular deformation mechanism predominates. Figure 1, from a paper by Stocker and Ashby (1973), is a deformation map for olivine Fo_{85}–Fo_{95} with a grain size of one millimetre and under a confining pressure of 10 kb. Superimposed on the map are contours of constant strain rate. At any temperature and applied shear stress the dominant deformation mechanism can be defined. The accuracy of this form of deformation map depends critically on the accuracy of the rate equations for each of the deformation processes.

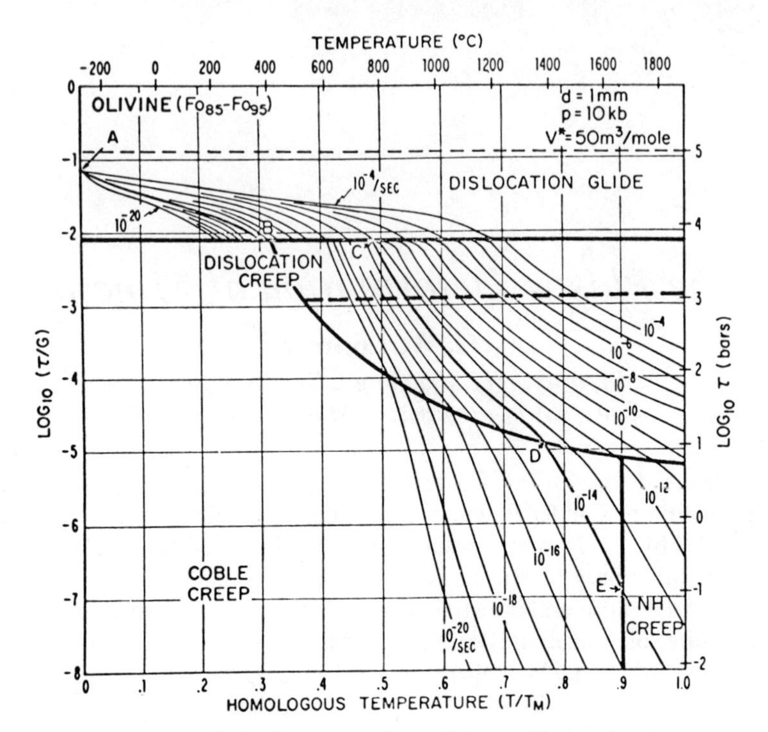

Figure 1. Deformation map for olivine with strain rate contours superimposed (Reproduced with permission from R. L. Stocker and M. F. Ashby, *Rev. Geophys. Space Phys.*, **11**, 391 (1973) copyright by American Geophysical Union).

While these equations are known to a reasonable degree of accuracy for metallic systems, and the experimental constants in the equations can be obtained from laboratory experiments, in oxides and minerals the lack of extensive experimental data means that the deformation maps are, at best, qualitative. Nevertheless they are an elegant method of indicating the relationships between the various mechanisms of plastic deformation as a function of the temperature and applied shear stress.

Two of the mechanisms of plastic deformation—dislocation glide and dislocation climb, combined in dislocation creep—are discussed below. These two mechanisms have been selected because optical and electron optical petrographic investigations of many deformed minerals indicate that these two mechanisms predominate during the plastic deformation of minerals in the earth's crust.*

Dislocation glide is the mechanism which enables plastic deformation to occur at stresses a small fraction of that required to provide defect-free flow.

*Stocker and Ashby (1973) and Weertman (1970) have proposed that dislocation creep is also the predominant deformation mechanism in the upper mantle.

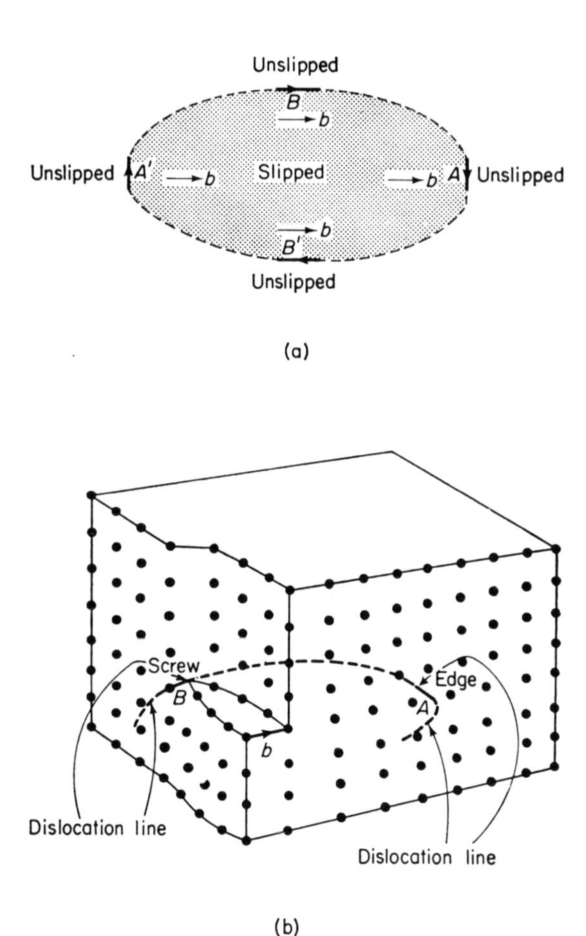

Figure 2. Geometry of a dislocation loop (a) and a portion of this loop in a primitive lattice (b). The Burgers vector, slip plane, pure edge and pure screw components are indicated. (Reproduced from Peter Gay, *The Crystalline State*, p. 289, Oliver and Boyd, Edinburgh, 1972, by permission).

The dislocation line delineates the border between portions of a crystal which have deformed (slipped) under an applied shear stress and those regions which have not (Figure 2(a)). Deformation proceeds by the expansion of the dislocation loop in the slip plane. When the loop intersects the surface a step is produced; this unit of deformation is characteristic of the dislocation: the Burgers vector, **b** While the crystal above the slip plane has been sheared by a unit **b** relative to the crystal below the slip plane, the shear has occurred by the stepwise breaking and remaking of atomic bonds as the dislocation loop has expanded: this process can proceed at a much smaller stress than that required to shear all of the atomic bonds in the slip plane simultaneously—the theoretical shear stress. The lattice geometry around the dislocation line changes continuously around the dislocation loop: where the displacement vector **b** is perpendicular to the dislocation line the dislocation is termed a pure edge dislocation, and where **b** is parallel to the dislocation line the dislocation is pure screw (Figure 2(b)). The screw and edge proportions of the dislocation line vary continuously around the dislocation loop. The detailed geometry of dislocations in simple crystal lattices has been described by several authors (e.g. Read, 1953; Hull, 1965; Weertman and Weertman, 1964).

The stress required to move a straight dislocation line in an otherwise perfect crystal is called the Peierls–Nabarro stress, τ_p. Detailed calculations of τ_p are difficult (Kelly, 1966) but the qualitative results show that

$$\tau_p \propto \exp(-W)$$

where W is the width of the dislocation. The dislocation width is defined as the width of the region, measured perpendicular to the dislocation line and in the slip plane, within which the relative displacement of the atoms above and below the slip plane is greater than half of its maximum value (Kelly, 1966).

The width of a dislocation varies inversely as τ_{max}/G, the ratio of the theoretical shear strength to the shear modulus. This ratio is critically dependent on the type of bonding in the crystals; in metals where bonding is multipolar and non-directional the ratio τ_{max}/G is small, the dislocation width is large and dislocation glide occurs under small shear stresses at low temperatures. In ionic and covalent bonded crystals (cf. minerals) bonding is highly polarized and directional, the ratio τ_{max}/G is large, the dislocations are narrow and dislocation motion is extremely difficult.

An additional obstacle to dislocation motion in compounds with complex crystal structures is that a single Burgers vector cannot be used to characterize the motion of each ion type in the structure; the ions have to undertake a coordinated movement to retain the crystal structure during the passage of a dislocation. This process, known as synchroshear, was first described by Kronberg (1967) to explain the motion of dislocations in alumina. Synchroshear can occur only if atom migration can take place at the dislocation core, i.e. if the temperature is sufficiently high that diffusion can occur at a sensible rate.

In metals at temperatures $> 0.5\ T_m$ a second mechanism of dislocation motion occurs: dislocation climb. Vacant sites are a thermodynamically stable component of a lattice because of the large entropy of mixing associated with their random distribution, and in metals the equilibrium vacancy concentration near the melting point is approximately 10^{-5}.* At temperatures $> 0.5\ T_m$ there is a sufficient vacancy concentration to allow climb of dislocations to proceed; vacant lattice sites are absorbed at the dislocation core and the dislocation 'climbs' out of its slip plane. The geometry of dislocation climb in a simple primitive lattice is shown in Figure 3.

Non-conservative climb of dislocations enables the dislocations to move

*By rapid quenching the vacancies can be retained and vacancy agglomerations, e.g. voids and loops, can then be observed using transmission electron microscopy.

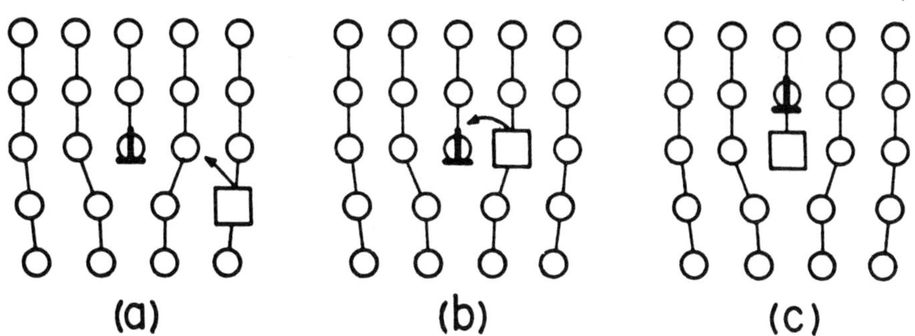

Figure 3. Geometry of dislocation climb in a simple primitive lattice.

in three dimensions in a crystal and adopt minimum energy configurations. During dislocation climb dislocations of opposite sign may interact and/or annihilate one another, dislocations are absorbed into grain boundaries and lost to free surfaces and dislocations may knit themselves into low energy arrays or sub-grain boundaries.

Under an applied shear stress at an elevated temperature dislocation multiplication and dislocation glide occur concurrently with dislocation climb. At low strain rates the dynamic balance between deformation and annealing, between dislocation production and annihilation, is described as dislocation climb-controlled solid state creep. The dislocation configurations which are produced during creep deformation depend on a number of parameters: T/T_m, strain rate, crystal structure and stacking fault energy. In many pure metals and alloys (cf. Honeycombe, 1968) the dislocation distributions have been characterized by examination of specimens at room temperature after creep deformation. From these studies three main theories of creep have emerged: the dislocation climb theory, the dislocation jog theory and the work hardening/recovery-creep theory (Lagneborg, 1972). In an effort to determine which, if if any, of these theories is correct, Henderson–Brown and Hale (1974) have carried out in-situ creep experiments on an Al–1%Mg alloy using high voltage electron microscopy (HVEM). These experiments have shown that in specimens thicker than 5 μm the dislocation distributions produced by creep testing thin foils in the electron microscope are similar to those observed in bulk specimens under similar stress/temperature conditions. During the creep experiments in the HVEM the detailed motion of individual dislocations was monitored. Figure 4 is a series of micrographs taken from a video tape recording of one of these experiments. A dislocation (dotted) can be seen to move towards the subgrain boundary, its motion is slowed as it moves through the boundary and it then accelerates when the boundary is cleared. The variation in the velocity of this dislocation as a function of its distance from the sub-boundary is shown in Figure 5.* The results of these investigations indicate that the

*These results were kindly provided by Dr. K. F. Hale. Copies of the film showing dislocation motion in Al–Mg alloys during creep can be obtained from Dr. K. F. Hale at the Division of Inorganic and Metallic Structure, National Physical Laboratory, Teddington, Middlesex, TW11 OLW.

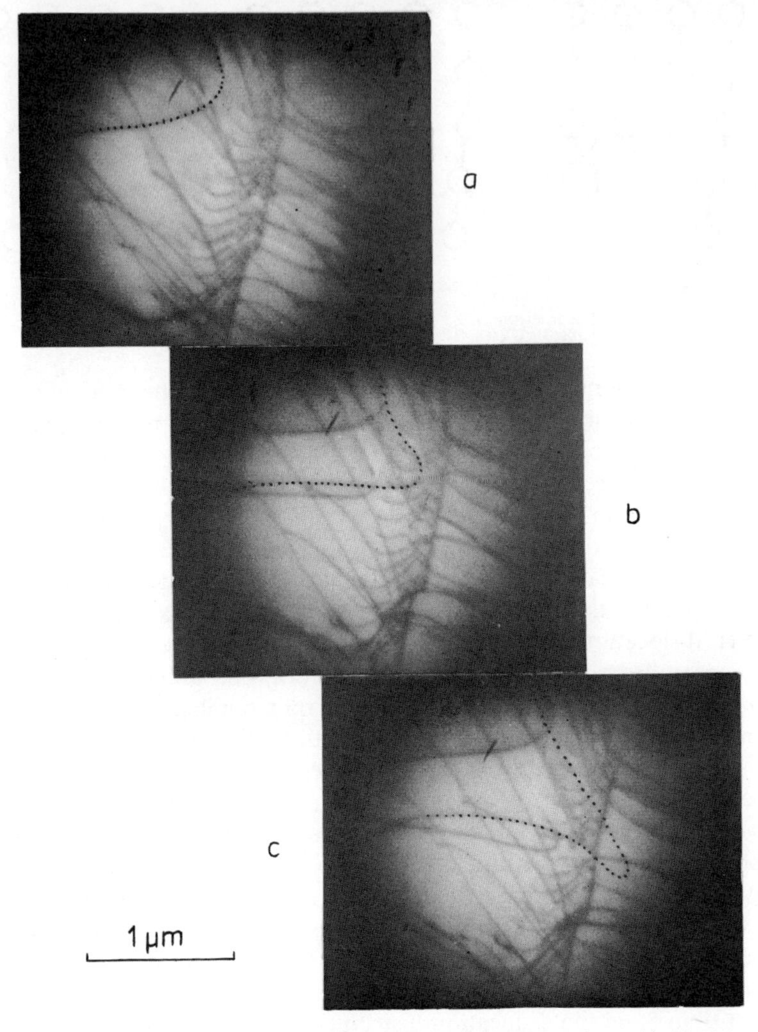

Figure 4. HVEM micrographs of in-situ deformation of an Al–1% Mg alloy under a stress of 7.6 MN/m^2 at 300 °C. The moving dislocation is dotted. (a) 0 seconds, (b) 58 seconds (c) 106 seconds. The stress axis is vertical. (Reproduced from M. Henderson–Brown and K. F. Hale in P. R. Swann, C. J. Humphreys and M. J. Goringe (ed.) *Proceedings of the Third International Conference on High Voltage Electron Microscopy*, p. 206, Academic Press, London, 1974. (Reproduced by permission of the National Physical Laboratory. Crown copyright reserved)

dislocation/sub-boundary interaction is of paramount importance and that the work hardening/recovery theory of creep (McLean and Hale, 1961; McLean, 1966; Lagneborg, 1969) is confirmed.

At the present time it is not possible to carry out in-situ deformation experiments on minerals in the high voltage electron microscope. However ion-thinned

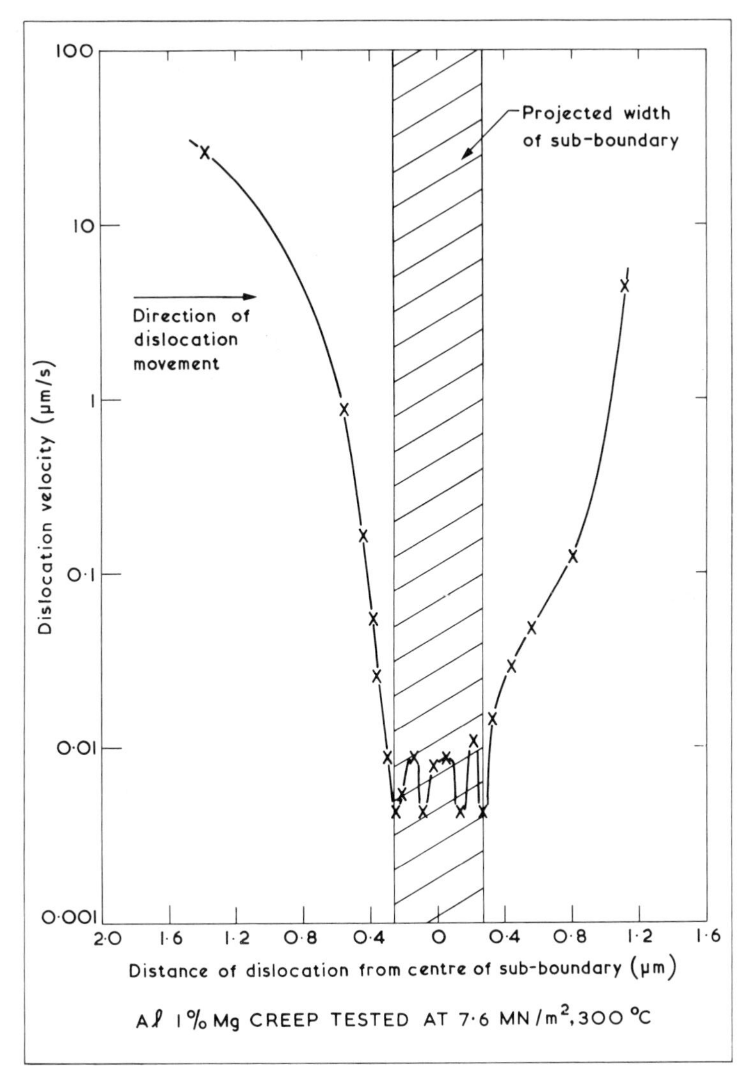

Figure 5. Variation in velocity of dislocation in Figure 4 passing through sub-boundary network. (Reproduced from M. Henderson–Brown and K. F. Hale in P. R. Swann, C. J. Humphreys and M. J. Goringe (ed.), *Proceedings of the Third International Conference on High Voltage Electron Microscopy*, p. 206, Academic Press, London, 1974. (Reproduced by permission of the National Physical Laboratory. Crown Copyright reserved.)

specimens of naturally and artificially deformed minerals have been examined by conventional and high voltage electron microscopy. White has reported microstructures in naturally deformed quartzite (White, 1971, 1973, 1974) and Boland *et al.* (1971) have observed dislocation distributions in olivine (dunite) which are consistent with dislocation climb-controlled solid state creep. An

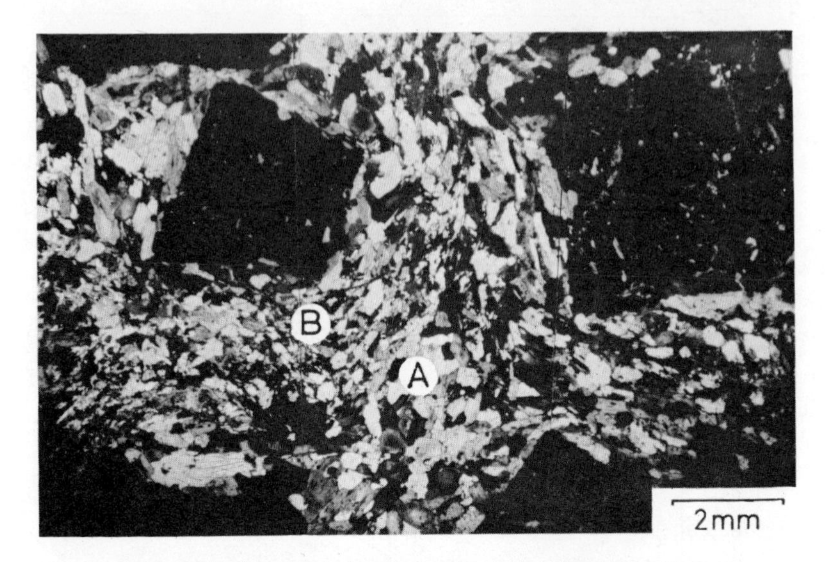

Figure 6. Optical micrograph (crossed polars) showing large, dark garnets and a fine-grained omphacite matrix. The omphacite has 'flowed' around the garnets (Reproduced from P. E. Champness, W. S. Fyfe and G. W. Lorimer, *Contr. Miner. Petrol.*, **43**, 91 (1974) by permission).

ideal 'model' specimen in which the deformation features in the optical microscope can be correlated with the dislocation distribution, as seen in the electron microscope, is the eclogite shown in Figure 6. The eclogite consists of garnet (dark) grains embedded in a pyroxene (omphacite) matrix. At this magnification the omphacite appears to have 'flowed' around the garnets. All of the omphacite grains show some signs of deformation with patchy, or less commonly, undulose extinction. The omphacite grains which are furthest away from the garnets, for example those near A in Figure 6, show less signs of deformation and are larger than those which are adjacent to the garnets. Those omphacite grains which are near the corners of the garnets, i.e. at stress concentrations such as area B in Figure 6 where extensive flow has occurred, appear the most severely deformed in the optical microscope. An electron micrograph from an area such as A in Figure 6 is shown in Figure 7. The dislocation density is low, about 10^9 lines cm^{-2}, and most of the dislocations are in subgrain boundaries. Figure 8 is a HVEM* micrograph from an area such as B in Figure 6 where the grain size is small and the deformation has been concentrated. There is a high dislocation density in the region, approximately 10^{12} lines cm^{-2}, with some of the dislocations in extensive subgrain networks and others as

*The combination of ion-thinned samples and 1000 kV (HVEM) microscopy has proved to be particularly useful for examining dislocation sub-structures and phase distributions in minerals. Not only is the useable specimen thickness approximately ten times that at 100 kV but radiation damage of minerals such as the feldspars is reduced (Lorimer and Champness, 1974).

Figure 7. Microstructure of the omphacite from a region such as A in Figure 6 which contains a low density of dislocations in a stable sub-boundary array. HVEM 1000 kV. (Reproduced from P. E. Champness, W. S. Fyfe and G. W. Lorimer, *Contr. Miner. Petrol.*, **43**, 91 (1974) by permission).

individuals within these networks. Some regions, e.g. at F, contain planar arrays of deformation, a primary deformation feature (dislocations on their original slip plane).

The optical and electron optical structures observed in the eclogite indicate that the omphacite microstructure has been produced by the combination of deformation and annealing processes. While both recovery—the removal of point defects and rearrangement of dislocations (polygonization) without a decrease in the dislocation density—and recrystallization—the reduction in dislocation density and migration of grain boundaries to produce equiaxed defect-free grains—processes can occur by post-deformation annealing, the

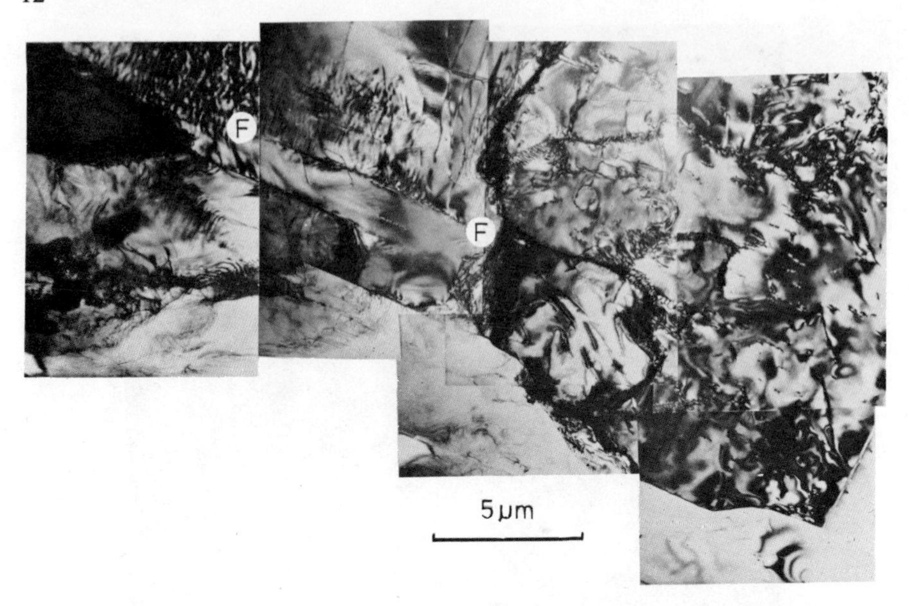

Figure 8. Microstructure of the omphacite from a region such as B in Figure 6 which contains a high density of dislocations. Many sub-boundaries, planar dislocation arrays and individual dislocations can be seen. HVEM 1000 kV. (Reproduced from P. E. Champness, W. S. Fyfe and G. W. Lorimer, *Contr. Miner. Petrol.*, 43, 91 (1974) by permission).

presence of recrystallized grains, Figure 6, which contain a subgrain structure and individual, tangled dislocations indicate that the microstructure was formed by a process of dynamic recovery and recrystallization. In the regions where the maximum 'flow' of material has occurred, i.e. where the strain rate was the greatest, sufficient energy has been stored by the deformation process to initiate recrystallization. These new grains, in turn, have been deformed and some of the new dislocations produced have climbed into subgrain boundary arrays (Figure 8).

A recent high voltage electron microscopic investigation of a naturally-deformed sodic plagioclase (Lorimer *et al.*, 1974) has revealed a dislocation substructure similar to that observed in the pyroxene phase of the eclogite. The plagioclase was from a fine-grained biotite–muscovite–epidote gneiss of the Eastern Zillertal Alps (North Tyrol, Austria). Figure 9 is a HVEM micrograph of the sample showing a high dislocation density, approximately 10^{12} lines cm^{-2}, and an extensive sub-boundary network. The lamellar feature is a twin which itself contains many smaller twins and subgrain boundaries. The sample is exsolved on a fine (20–40 nm) scale.

A quick sum can be done to put the high dislocation density into perspective. If it is assumed that the thickness of the sample is approximately 1 μm and that an area 1 μm^2 on the micrograph in Figure 9 contains 10 dislocations, then the number of dislocation lines in one cubic centimetre of the sample is 10^{13}. If

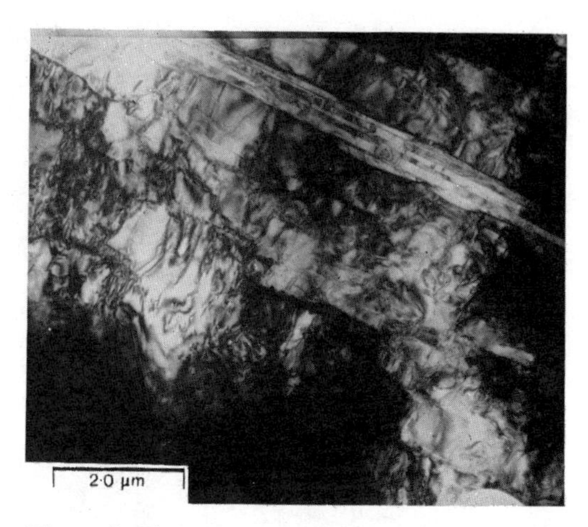

Figure 9. Electron micrograph of deformed plagioclase showing an extensive subgrain structure and numerous individual dislocations. The feature in the top right of the micrograph is a twin which has been sheared and which contains sub-boundaries. The sample is exsolved with a wavelength of 20 to 40 nm. HVEM 1000 kV. (Reproduced from G. W. Lorimer and P. E. Champness in P. R. Swann, C. J. Humphreys and M. J. Goringe (ed.), *Proceedings of the Third International Conference on High Voltage Electron Microscopy*, p. 301, Academic Press London, 1974. Reproduced by permission of the Royal Microscopical Society).

the dislocation lines have the shortest possible length, i.e. are parallel to the electron beam (an obvious underestimation of total length, but this is roughly balanced by the underestimation of specimen thickness of 1 μm), then the total length of dislocation line in the specimen is

$$10^{13}\,\frac{\text{lines}}{\text{cm}^3} \times 10^{-4}\,\frac{\text{cm}}{\text{line}} = 10^9 \text{ cm of dislocation line cm}^{-3}$$

or 10,000 kilometres of dislocation line in every cubic centimetre!

Most of the electron microscopic investigations of deformation structures in minerals have been concentrated on the silicates. However some preliminary investigations have been carried out on the deformation of sulphides by Dr. T. M. Banks at the Physical Metallurgy Centre of the British Steel Corporation in Rotherham. Sulphides—MnS, FeS and mixed sulphides—are present in most steels; their number, size and shape have a profound effect on the mechanical properties and machinability of wrought steel products. Figure 10(a) and 10(b) are of αMnS inclusions which have been rolled (in steel) at 900 °C and 1200 °C, respectively. The inclusion rolled at 900 °C contains a large number of free dislocations as well as an extensive sub-grain network—a partially recovered

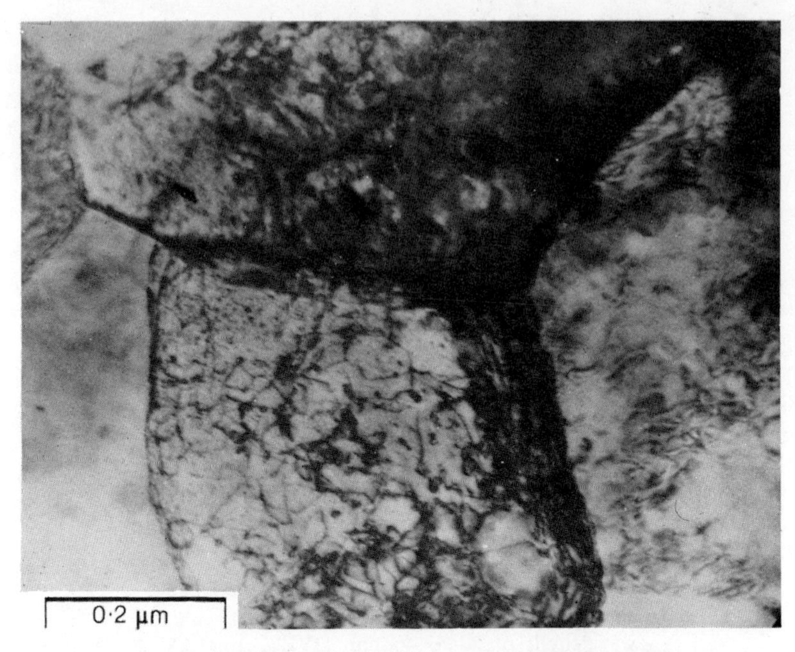

Figure 10(a). αMnS inclusion rolled (in steel) at 900 °C exhibiting a partially recovered dislocation structure. HVEM 1000 kV (courtesy T. M. Banks, unpublished work).

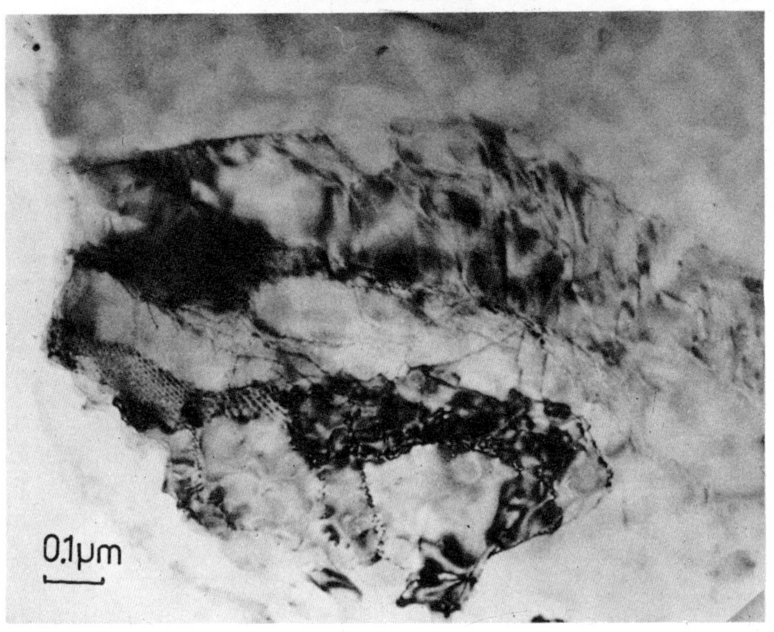

Figure 10(b). αMnS inclusion rolled (in steel) at 1200 °C exhibiting a totally recovered and partially recrystallized dislocation structure. HVEM 1000 kV. (courtesy T. M. Banks, unpublished work).

dislocations structure—while in the inclusion rolled at 1200 °C all of the dislocations are in sub-grain boundaries, a fully recovered and partially recrystallized dislocation structure. The simple structure of MnS and the high temperature of deformation, 0.8 T_m at 1200 °C, enables rapid dislocation climb and boundary migration to occur; the deformation process is one of hot working—simultaneous deformation and recrystallization at elevated temperature.

The dislocation distributions which are observed in naturally-deformed silicates are often similar to those found in metals which have undergone deformation by hot working or steady state creep. The process of hot working is associated with high temperatures and high strain rates in the range 0.5 to 500 sec^{-1}, (cf. the MnS above), while solid state creep, controlled by the climb of dislocations, is observed at slower strain rates. At the strain rates which can be expected to operate in geological conditions, typically 10^{-12} to 10^{-16} sec^{-1}, and the low temperatures encountered the deformation process is best described as dislocation-climb-controlled solid state creep (cf. the eclogite and plagioclase feldspar above).

The dislocation distributions in silicate minerals which have been deformed in the laboratory are usually very different from the distributions observed in naturally-deformed specimens. Barber and Wenk (1973), Boland (1974) and McLaren *et al.* (1967, 1969, 1970) in studies of laboratory deformed limestones, olivines and quartz, respectively, have observed planar arrays of dislocations—primary deformation structures produced by dislocation glide—and not the complex three-dimensional networks characteristic of naturally-deformed silicates. The reason for the difference in the dislocation distributions is the high strain rates used during the laboratory experiments as compared with the natural deformation conditions. As can be seen from the deformation map shown in Figure 1 above, the effect of increasing the strain rate at a given temperature can be to change the predominant deformation mechanism from dislocation climb to dislocation glide.

Another parameter which must be considered when discussing the plastic deformation of silicates is the effect of a small amount (< 0.1 per cent) of water in the structure (*not* at the grain boundaries) on the mechanics of dislocation motion. Griggs (1967) and Griggs and Black (1965) have shown that the addition of a small amount of water to otherwise pure quartz increases the plasticity (reduces the strength) dramatically. Transmission electron microscopy of these 'water-weakened' samples (McLaren *et al.*, 1967) revealed primary dislocation substructures consistent with deformation by dislocation glide. The water-weakening of quartz was attributed by Griggs (1967) to a hydrolysis reaction at the dislocation core. Formation of a hydrolyzed bridge between silicon ions by the reaction

$$\mathrm{Si{-}O{-}Si} + H_2O \longrightarrow \mathrm{Si{-}OH.HO{-}Si}$$

produces a weak hydrogen bond between the Si—OH groups. It was proposed that dislocation glide takes place by the exchange of hydrolysed bridges at the dislocation core. Although the detailed research on water weakening has been carried out on quartz, all silicates contain Si—O—M bridges and hydrolysis of these could promote dislocation glide.

The concept of hydrolytic weakening was developed to explain easy glide of dislocations, but it is possible that a similar process can result in dislocation climb at low temperatures ($0.3 < T < 0.5\ T_m$). The absorption of vacancies at the core of a dislocation in a simple structure can be easily visualised to produce dislocation climb (Figure 3 above). However the absorption of a vacancy at a dislocation core in a mineral must also be associated with a synchroshear-type of ion shuttle for dislocation climb to proceed. The concept of a vacant lattice site in a mineral is in itself far from simple. At temperatures above 0 K vacancies are a thermodynamically stable constituent of a mineral lattice, because of entropy considerations. However the difference in binding energies of the constituent ions, the necessity to maintain charge balance and the presence of stoichiometric structural voids leads to complications.

Dislocation distributions are observed in naturally deformed silicates which could only be formed by dislocation climb, but it is clear that the detailed mechanism of dislocation climb in minerals has yet to be resolved.

ACKNOWLEDGEMENTS

The author would like to thank his wife, Dr. P. E. Champness, for her assistance with the preparation of the manuscript. The high voltage electron microscopy was carried out at the Swinden Laboratories of the British Steel Corporation on an instrument made available by the Science Research Council.

REFERENCES

Ashby, M. F. (1972). *Acta Met.*, **20**, 887.
Boland, J. N., A. C. McLaren and B. E. Hobbs (1971). *Contr. Miner. Petrol.*, **30**, 53.
Champness, P. E., W. S. Fyfe and G. W. Lorimer (1974). *Contr. Miner. Petrol.*, **43**, 91.
Gay, P., *The Crystalline State* (1972). Oliver and Boyd, Edinburgh.
Griggs, D. (1967). *Geophys. J. R. Astr. Soc.*, **14**, 19.
Henderson-Brown, M. and K. F. Hale (1974). In *Proceedings of the Third International Conference on High Voltage Electron Microscopy*, ed. P. R. Swann, C. J. Humphreys and M. J. Goringe, Academic Press, London, p. 206.
Honeycombe, R. K. (1968). *The Plastic Deformation of Metals*, Edward Arnold, London.
Hull, D. (1965). *Introduction to Dislocations*, Pergamon Press, Oxford.
Kelly, A. (1966). *Strong Solids*, Clarendon, Oxford.
Kronberg, M. L. (1957). *Acta Met.*, **5**, 507.
Lagneborg, R. (1969). *Metall. Sci. J.*, **3**, 161.
Lorimer, G. W., H.-U. Nissen and P. E. Champness, in press. *Schweiz. Mineral. Petrogr. Mitt.*
Lorimer, G. W. and P. E. Champness, (1974). In *Proceedings of the Third International Conference on High Voltage Electron Microscopy*, ed. P. R. Swann, C. J. Humphreys and M. J. Goringe, Academic Press, London, p. 301.
McLean, D. (1966). *Rep. Prog. Phy.*, **99**, 1.
McLean, D. and K. F. Hale, (1961). *Report on a Symposium on Structural Processes in Creep*, organized by the Iron and Steel Institute and the Institute of Metals, p. 19.

Read, W. T. (1953). *Dislocations in Crystals*, McGraw–Hill, New York.
Stocker, R. L. and M. F. Ashby (1973). *Rev. Geophys. Space Phys.*, **11**, 391.
Weertman, J. (1970). *Rev. Geophys. Space Phys.*, **8**, 195.
Weertman, J. and J. Weertman (1964). *Elementary Dislocation Theory*, Macmillan, New York.
White, S. (1971). *Nature Physical Science*, **234**, 175.
White, S. (1973). *J. Mat. Sci.*, **8**, 490.
White, S. (1974). In *Proceedings of the Third International Conference on High Voltage Electron Microscopy*, ed. P. R. Swann, C. J. Humphreys and M. J. Goringe, Academic Press, London, p. 317.

Evidence of Superplasticity in Mantle Peridotites

Y. Gueguen and A. M. Boullier
Laboratoire de Tectonophysique, Nantes

1 INTRODUCTION

The peridotite nodules from kimberlites are divided into four main textural groups principally on the basis of the texture and the preferred orientation of olivine and also that of enstatite in some cases, (Boullier and Nicolas, 1973, 1975): the coarse granular and coarse tabular ones represent the granular type of Boyd and Nixon (1972) and are undeformed; the porphyroclastic and mosaic ones represent the sheared type of the same authors and show an increasing deformation. Finally, one mosaic fluidal subtype is defined and is the subject of this paper.

The geochemical studies of Boyd and Nixon (1973) and MacGregor (1975) show that the nodules of this subtype are the deepest ones: they originated at a depth of 200 km and a temperature of 1400 °C.

Boullier and Nicolas (1973) first attributed the fluidal texture to an intensive and nonpenetrative shearing superimposed on an older mosaic texture. In a later paper, the same authors (Boullier and Nicolas, 1975) proposed that this deformation took place in the superplastic field on the basis of analogy with superplastic textures in metals. The present paper is intended to show from observations at optical and electron microscope scales, that all the criteria required for superplasticity are present here and that this superplastic process is the last stage of the tectonic evolution of the kimberlites.

2 OBSERVATIONS

2.1 The Sample

The nodule is a garnet–lherzolite made of 65 per cent olivine, 28 per cent orthopyroxene, 3.5 per cent clinopyroxene and 3.5 per cent garnet. The contribution to flow of the last two minerals can be neglected and the rock considered as a two-phase sample. The nodule shows a well developed foliation defined by mineral flattening. However, the lineation is weak in this plane.

2.2 The Thin Section

The observation with the optical polarizing microscope was made on a section cut perpendicular to the foliation and parallel to the lineation. Olivine forms an equant mosaic of small grains (0.07 mm) (see Figure 1). However, as observed in many peridotites from nodules of basalts and from massifs, some elongated porphyroclasts remain (Figure 2), but the size of their subgrains is the same as the mosaic one. Some euhedral and undeformed tablets of olivine also appear at the edges of the porphyroclasts or inside them. The preferred orientation of the porphyroclasts is the one usually observed in highly deformed peridotites: [100] parallel to the lineation and [010] perpendicular to the foliation. The orientation of the olivine grains forming the mosaic has been studied (Boullier and Nicolas, 1973). It shows that local domains

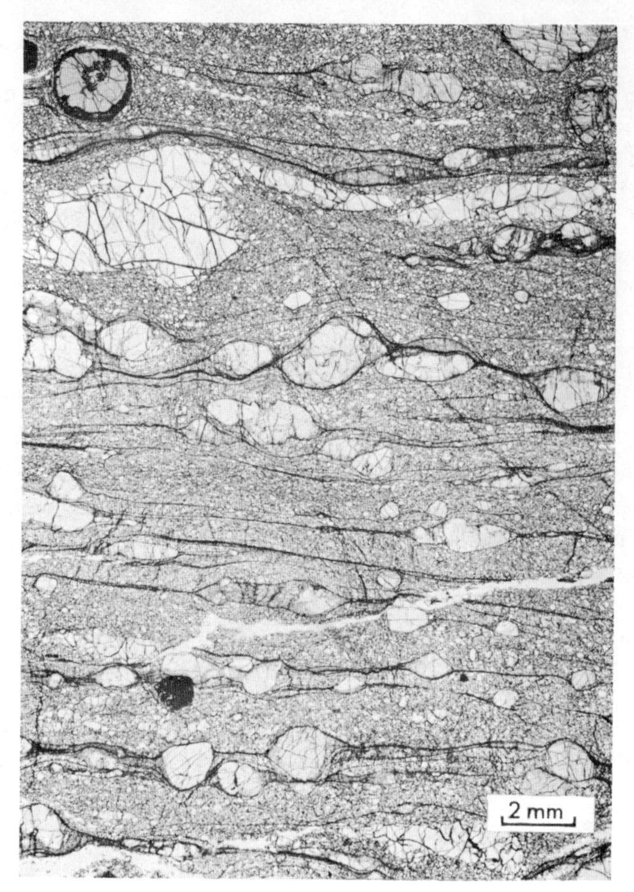

Figure 1. Thin section described in the text. Plane polarized light.

Figure 2. Olivine porphyroclast recrystallizing. Crossed Nicols.

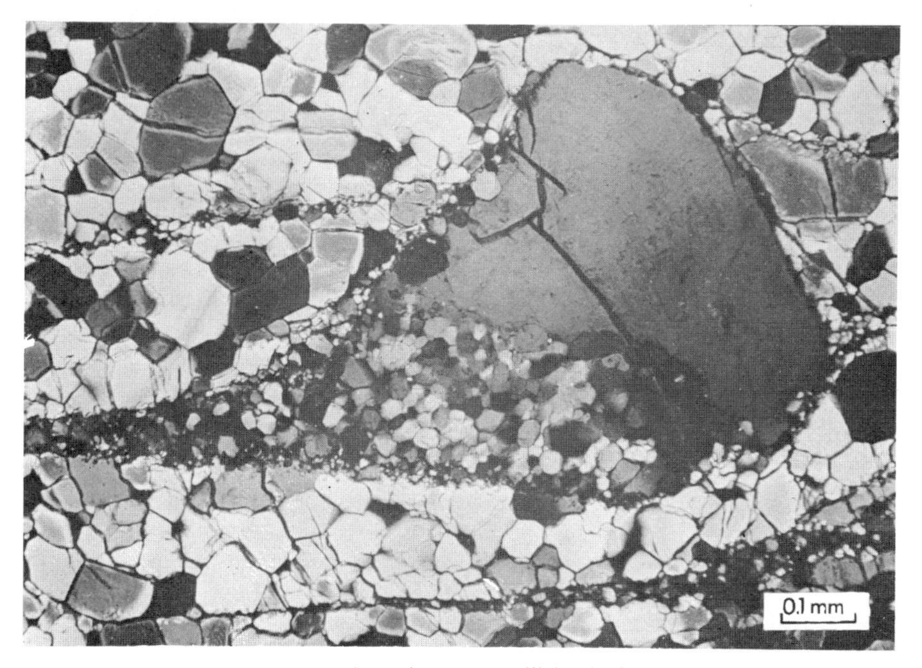

Figure 3. Orthopyroxene porphyroclast recrystallizing in fine grains forming stripes. Crossed Nicols.

have a strong preferred orientation differing one from the other. The total preferred orientation is weak. This suggests that the domains are derived from porphyroclasts which have recrystallized by the increasing misorientation process (Poirier and Nicolas, in preparation), transforming subgrains into grains beyond 10–15 per cent of rotation. This process could also explain the similarity in size of porphyroclast subgrains and grains forming the mosaic.

Orthopyroxene porphyroclasts are deformed and have an undulatory extinction. They are recrystallized on their edges in very fine grains (Figure 3) which also form stripes 0.3 to 0.01 mm thick, through the olivine mosaic, giving a fluidal aspect to the rock. The structure of the fine-grained equiaxed orthopyroxenes in the stripes is also a mosaic one with triple junctions at 120° (see Figure 3). The average size of those grains is about 10 μm. The stripes are

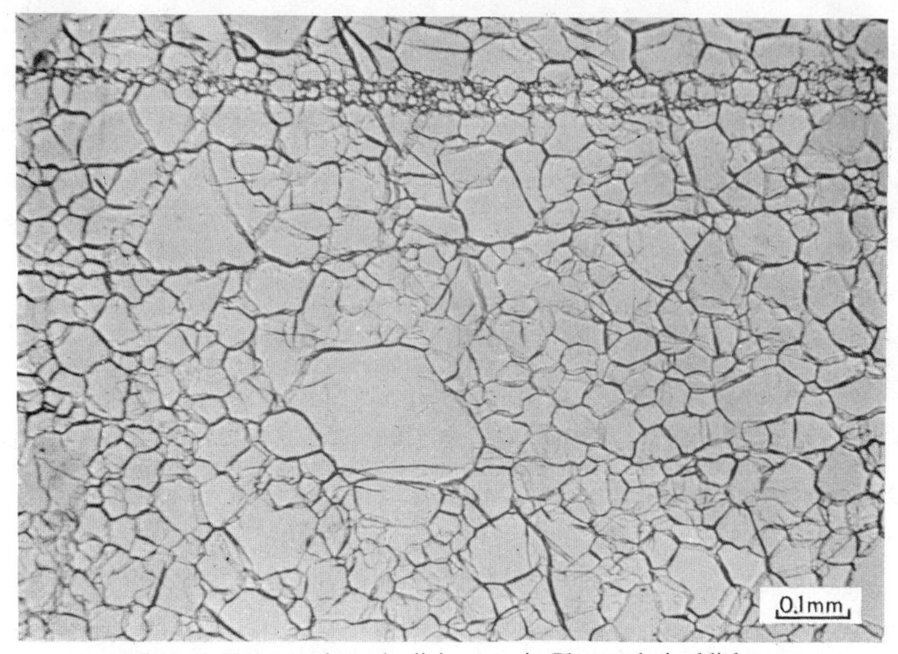

Figure 4. Fracture through olivine mosaic. Plane polarized light.

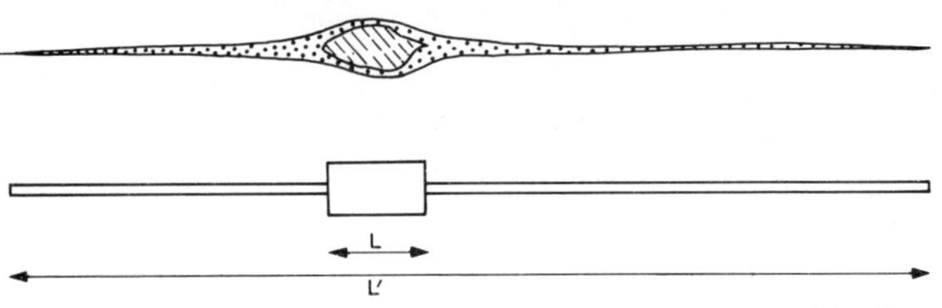

Figure 5. Percentage of deformation in orthopyroxene: $\varepsilon = \dfrac{L' - L}{L}$

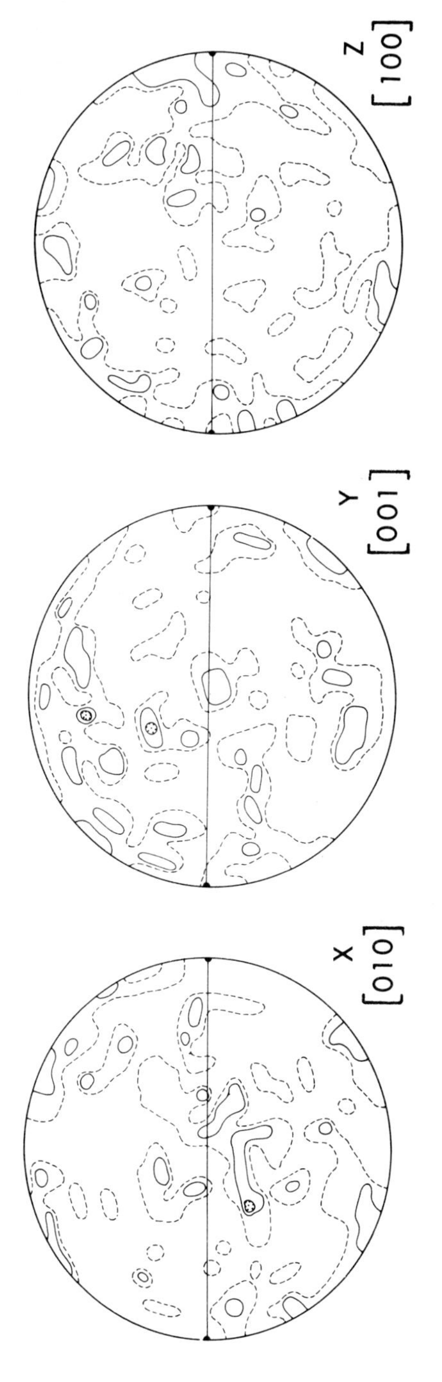

Figure 6. Fine grained orthopyroxenes preferred orientation (100 grains).

parallel to the porphyroclasts elongation, pass round the obstacles such as garnets or other pyroxenes, and end in the olivine in a discontinuity line (Figure 4). The distance between two stripes varies from 0.1 mm to 2 mm. Their length is at least 25 mm on average since they commonly go through the thin section. We calculated an average percentage of deformation, using the length of the porphyroclasts and the length of the stripes formed by the small orthopyroxenes (Figure 5). The result is:

$$\varepsilon = \frac{L' - L}{L} = 8.40 \text{ (rational deformation 2.20)}$$

The preferred orientation of the orthopyroxene porphyroclasts is strong: [001] parallel to the lineation and [100] perpendicular to the foliation. The fine grained orthopyroxenes which form the stripes have no fabric at all (Figure 6).

2.3 Observations with the Transmission Electron Microscope

Thin sections of the same sample were prepared for Transmission Electron Microscopy, using the ion-bombardment technique: 1 μm thin sections were studied at 1 MeV and 2000 Å thin sections at 100 kV. The results on the dislocation densities and microstructures are summarized in table I, as well as the results concerning the preferred orientations.

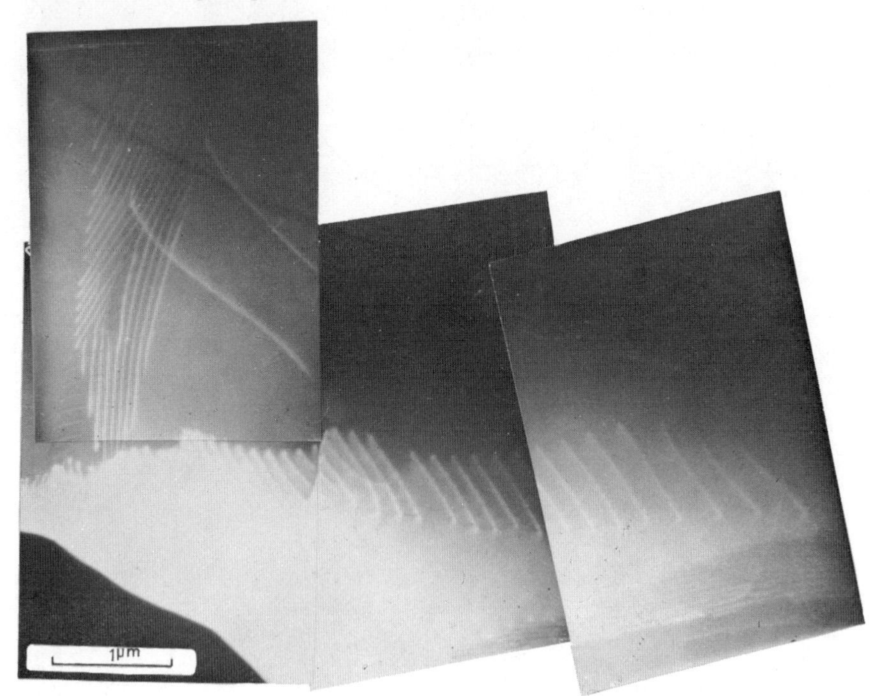

Figure 7. Dislocations in an olivine grain from the olivine mosaic (1MeV). Dark field.

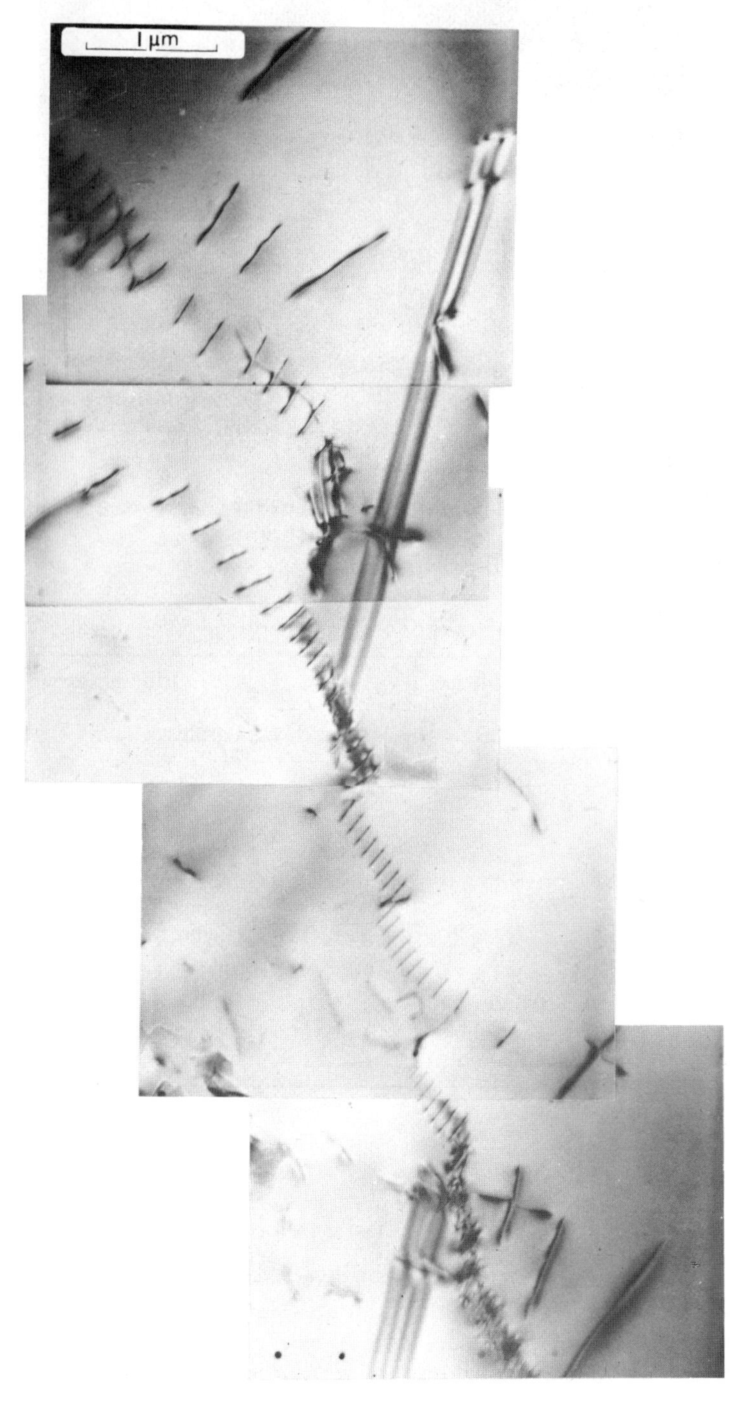

Figure 8. Dislocations and stacking faults in orthopyroxene porphyroclast (100 keV). Bright field.

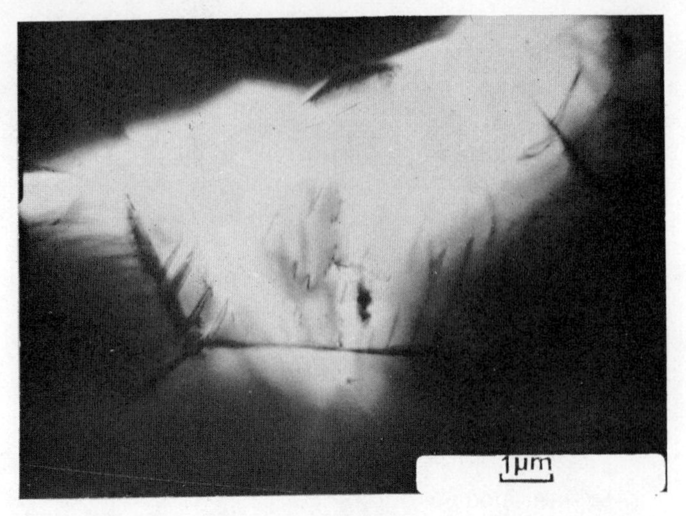

Figure 9. Dislocations in a recrystallized orthopyroxene grain (100 keV). Bright field.

Table 1. Characteristics of olivine and orthopyroxene crystals

	Olivine		Orthopyroxene	
	Porphyroclasts	Mosaic	Porphyroclasts	Mosaic
Dislocation density (cm^{-2})	5×10^7 [a]	10^7	10^8	5×10^7
Microstructure	Cells formed[a]	Cells formed (Figure 7)	Cells incompletely formed (Figure 8)	No cells (Figure 9)
Preferred orientations	[100] parallel to the lineation [010] perpendicular to the foliation	Locally strong	[001]parallel to the lineation [100] perpendicular to the foliation	No preferred orientation (Figure 6)

[a]Gueguen and Boullier, in preparation

3 INTERPRETATION

3.1 Evidence of an Unusual Type of Plastic Deformation

The above observations indicate that a very strong deformation took place in these peridotites at a high temperature. Several facts demonstrate that this deformation is not of a usual kind:

—Orthopyroxene recrystallization is rarely observed in peridotites from nodules as well as from massifs. Annealing experiments (Goetze and

Kohlstedt, 1973) demonstrate that defects are not mobile in orthopyroxene even at 1350 °C. Since recrystallization can occur only when defects are mobile, and since their mobility increases with T/T_m we conclude that in orthopyroxene, it requires an unusually high value of T/T_m.

—The large percentage (Figure 5 above) of deformation observed in orthopyroxene suggests that the strain-rate was faster than the average plate tectonics strain-rate 10^{-14} s^{-1}.

—However the stress was not very large, if we consider the moderate dislocation density ($\rho = 10^7$ cm^{-2}) in olivine mosaic. Unfortunately it does not seem possible to define its quantitative value with accuracy. There are two possible ways of evaluating it, but neither is satisfactory. The first is to use the dislocation density which is related to the internal stress:

$$\sigma_i \sim \mu b \rho^{1/2}$$

where μ is the shear modulus, **b** is the Burgers vector, and ρ is the free dislocation density. Not only is ρ_i not exactly equal to the applied stress, but the dislocation density, after deformation, is no longer the same as during deformation. An undetermined amount of post-deformation recovery takes place in natural deformations as well as in experimental ones. It is obviously very difficult to quantify this recovery. The second way to evaluate the stress is to use the subgrain size-stress relationship:

$$\frac{\sigma}{\mu} = \frac{L_0}{L}$$

where L is the subgrain size and L_0 a constant. We then have to assume that the subgrains observed on a much smaller scale in high voltage electron microscopy (cells) have a completely different behaviour from the optical visible subgrains considered here, as pointed out by Green and Radcliffe (1972). If this hypothesis is right, knowing L_0 and L provides us with an indirect measure of σ. But L, in a given rock, always varies by a factor of ten. In the studied sample the average value of L is 0.07 mm. L_0 is not accurately known. Weertmann (1970) used $L_0 = 5 \times 10^{-5}$ mm but two data points from experimental deformation of olivine (Raleigh and Kirby, 1970) suggest that $L_0 = 4 \times 10^{-4}$ mm. There are clearly not enough data to get a representative value for L_0. So it does not seem possible to give any significant value for σ. The deformation is controlled by orthopyroxene since enstatite stripes end in fractures through the olivine mosaic (Figure 4 above). This means that olivine could not keep up with the deformation and that it responded by fracturing. However it is well known that orthopyroxene is much less ductile than olivine.

It is difficult to account for all these facts by any usual plastic flow mechanism. It is especially difficult to understand how the strain rate can be (geologically) fast when deformation is controlled by the less ductile mineral. So a completely different flow mechanism is required. We suggest that superplasticity could be this mechanism.

3.2 Characteristics of Superplasticity

Superplasticity is characterized by an unusual tensile extensibility: $\varepsilon \simeq 1000$ per cent. Two kinds of superplasticity must be distinguished. One occurs when P.T. conditions correspond to a phase change, the other occurs when polycrystals are fine-grained. We are concerned here with the second kind of superplasticity. So far it has been studied only in alloys: two phases are required in order to keep the grain-size very small (a few microns). A detailed review of these studies is given in Suery (1974). Superplastic behaviour can appear above 0.3 T_m for stresses below a certain limit: for instance $\sigma < 10^{-3}$ μ in lead (Ashby and Verall, 1973), where μ is the shear modulus. Correspondingly the strain-rate must not exceed a certain limit either: for instance $\dot{\varepsilon} < 10^{-4}$ in lead. These limits depend on the grain size: the larger the grain-size is, the smaller the limit. Above this limit, deformation is controlled by dislocation climb. During superplastic flow, grain-boundary sliding is dominant. Consequently, dislocation density is not high, no cells are formed and the fabric is destroyed by grain-rotation. All of these are observed here in the orthopyroxene grains as shown by the comparison of Table 1 and Table 2.

Two theoretical models which explain superplasticity have been advanced:

(a) Hayden *et al.* (1972) suggested that grain-boundary sliding was dominant but that the rate-controlling effect was dislocation climb within the grains. Their result shows that the strain-rate $\dot{\varepsilon}$ is:

$$\dot{\varepsilon}_1 = \frac{K}{d}\dot{\varepsilon}_{\text{d.c.}} \tag{1}$$

where $\dot{\varepsilon}_{\text{d.c.}}$ stands for the strain-rate produced by the usual high temperature dislocation creep, d is the grain-size and K is a constant. They found $(K/d) \simeq 50$, so that the superplastic strain-rate is fifty times the plastic strain-rate. They have shown that, in some cases, superplasticity can take place in structures where the grain-size is not small, but where grains are very elongated. The thickness of the grains may then be almost equal to the size of the cells and recrystallization takes place at the edges of the elongated grains, the subgrains being transformed into grains. The new grains thus produced are very small since their size is that of the cells. So as soon as they are produced, they display a superplastic behaviour. The strain-rate is then:

$$\dot{\varepsilon}_2 \simeq \left(\frac{K}{d} + 1\right)\dot{\varepsilon}_{\text{d.c.}} \tag{2}$$

In this case, the superplastic flow is controlled by the production rate of small grains.

(b) A different model is given by Ashby and Verall (1973). Grain-boundary sliding is the dominant mechanism, but it is diffusion-accommodated. Their model is consistent with equiaxed grains, no cell formation, destruction of fabric and low values of strain-rate. The result is:

Table 2. Comparison of plastic and superplastic flow

	Grain-size	Preferred orientation	Dislocation microstructure	Dislocation density	$\left(\frac{T}{T_m}\right)$	Stress	Strain-rate
Plastic flow at high temperatures (dislocation creep)	No limits. Usually from a few microns to a few millimetres	Preferred orientation developed	Cells are formed	Can be as high as 10^{11} cm^{-2}	> 0.3	$\sigma \leqslant$ breakdown stress which is $10^{-3}\mu$[a] or thereabouts	Can be as high as 10^{-1}
Superplastic flow	$d < 10$ microns	No preferred orientation. The fabric is destroyed by grain rotation	No Cells	Moderate ($\leqslant 10^8$ cm^{-2})	> 0.3	$\sigma \leqslant$ limit depending on $\frac{T}{T_m}$ and d In lead at $0.5\, T_m$ this limit is $10^{-4}\, \mu$ if $d = 8$ microns[b]	$\dot{\varepsilon} <$ limit depending on $\frac{T}{T_m}$ and d In lead at $0.5\, T_m$ this limit is 10^{-8} if $d = 8$ microns[b]

[a] Stocker and Ashby (1973)
[b] Ashby and Verall (1973)

$$\dot{\varepsilon}_3 \simeq 100 \frac{\Omega}{kTd^2} \sigma D \tag{3}$$

where Ω is the atomic volume, k the Boltzmann constant, σ the stress, D a self-diffusion coefficient, d the grain-size, and T the temperature (K). In this case, the strain-rate is proportional to the stress, which means that the medium is Newtonian.

3.3 Superplasticity as the Third Stage of the Tectonic History of these Nodules

Using our observations and the above theoretical and experimental results it is possible to reconstruct the following schematic development of the deformation undergone by these nodules (figure 10).

(1) The rock was flowing in the mantle. The flow was controlled as usual by the more ductile mineral, olivine. The strain-rate $\dot{\varepsilon}$ was

$$\dot{\varepsilon}_4 = \frac{\sqrt{3^{n+1}}}{2} AD \frac{\mu \mathbf{b}}{kT} \left(\frac{\sigma}{\mu}\right)^n \tag{4}$$

where A and n are the Dorn constants: $A = 1.2 \times 10^4$ and $n = 4.2$ for olivine (Stocker and Ashby, 1973). D is a self-diffusion coefficient for

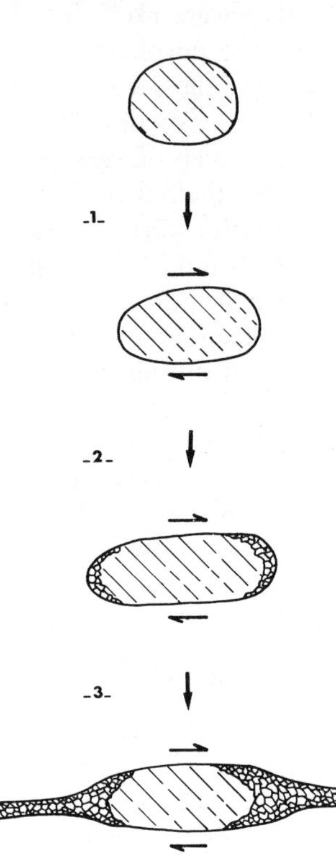

Figure 10. Interpretation of the deformation process in an orthopyroxene grain.

olivine: $D = 3 \times 10^4 \exp - (135 \text{ kcal mole}^{-1}/kT) \text{ cm}^2 \text{ s}^{-1}$ (Goetze and Kohlstedt, 1973) **b** is the Burgers vector, and the other symbols have the same meaning as above.

(2) For some reason, T/T_m increased and recrystallization took place in orthopyroxene. Cells were formed (figure 8 above) and became new small grains (figure 9 above). At this stage, olivine was already almost completely recrystallized. The flow was controlled either by olivine (Equation (4)), or by the small orthopyroxene grains produced (Equation (2)). The lack of data on diffusion and creep in orthopyroxene prevent us from comparing (2) and (4) quantitatively. But since the strain-rate is usually slower in orthopyroxene than in olivine, and since here, in equation (2). there is an 'amplification factor' K/d, we suggest that both $\dot{\varepsilon}_2$ and $\dot{\varepsilon}_4$ could well have the same magnitude. Since $\dot{\varepsilon}_2$ is relative to orthopyroxene and $\dot{\varepsilon}_4$ to olivine, $\dot{\varepsilon}$ would be about the same in both. The existence of this second stage is demonstrated by the fact that deformation of the fine-grained orthopyroxene stripes could take place only when some orthopyroxene recrystallization had occurred. If the stripes had been produced first by elongation and then by recrystallization, they would have displayed a fabric, but they do not. However the stripes were not formed right at the beginning of orthopyroxene recrystallization. If they were, then the small grains would have gone into the stripes as soon as they were produced and there would be none left around the orthopyroxene porphyroclasts, which is not the case. If, as suggested, $\dot{\varepsilon}_2 \simeq \dot{\varepsilon}_4$ during the second transition stage, the flow is not truly superplastic but many small orthopyroxene grains are produced.

(3) When the numer of these grains was large enough, grain-boundary sliding within them was no longer controlled by the production of these grains (Equation (2)) but by the diffusion in them, according to the Ashby–Verall model (Equation (3)). Then $\dot{\varepsilon} \simeq \dot{\varepsilon}_3$ and the flow is truly superplastic. Stripes were formed and fractures appeared in olivine which could not keep up with the deformation. During the second stage $\dot{\varepsilon} \simeq \dot{\varepsilon}_2 \simeq \dot{\varepsilon}_4$ but during the third one $\dot{\varepsilon} \simeq \dot{\varepsilon}_3 \gg \dot{\varepsilon}_2$ or $\dot{\varepsilon}_4$. We can compare $\dot{\varepsilon}_3$ and $\dot{\varepsilon}_2$ since the unknown diffusion coefficient in orthopyroxene which appears in (2) and (3) is eliminated in this calculation. Using the value of K given by Hayden *et al.* (1972) and the values of Dorn parameters given by Stocker and Ashby (1973) the result is $\dot{\varepsilon}_2 \ll \dot{\varepsilon}_3$ if $\sigma \ll 10^{-4}\ \mu$, that is $\dot{\varepsilon} \simeq \dot{\varepsilon}_3$ for stresses lower than 80 bars. The process was then interrupted and the nodule was brought up to the surface.

3.4 The Interpretation Related to a Diapiric Origin

The above interpretation explains all the observations made on these nodules. It is also in agreement with the model developed by Green and Gueguen (1974) to explain the origin of kimberlites. These authors suggested that an upwelling was initiated at a depth of about 300 km. Then a diapir formed and rose up to the point where the solidus was reached. At this point a magma was produced which then brought the nodules up to the surface. In this model, the plastic deformation took place in the diapir, during the upwelling. The nodule described is located at the top of the paleogeotherm, that is at the point where T/T_m is maximum. Thus that stage (2) in this model corresponds to the beginning of the upwelling. Since T/T_m is much higher than usual, recrystallization can take place in orthopyroxenes. This stage (2) leads naturally into stage (3). The stresses are in agreement with the condition stated above, $\sigma \ll 10^{-4}\ \mu$, since in this diapir model, the stresses would be in the region of 100 bars. Finally this model can provide us with a lower limit of the strain-rate which is geologically fast as expected: the strain $\varepsilon = 220$ per cent was developed in less than 3 million years since this is the time upwelling takes according to the Green–Gueguen model. Then $\dot{\varepsilon} > 2.20/3\text{m.y}$ that is $\dot{\varepsilon} > 2 \times 10^{-14}$. Thus even with a moderate stress, the strain-rate was geologically fast (but metallurgically slow).

4. CONCLUSION

Superplasticity occurs in peridotites only in special conditions. It seems that these conditions, present in the case of some nodules from kimberlites, are exceptional in the mantle. Superplasticity has never been described in nodules from basalts (Mercier and Nicolas, 1975) and even in the case of nodules from kimberlites, it has been described only for some nodules, located at the top of the paleogeotherms (Boullier and Nicolas, 1973, 1975).

However the appearance of superplasticity has important consequences, the first being that the usual situation is reversed: the generally more plastic olivine does not control the plastic flow in this case. This emphasizes how important it is from a plasticity point of view to consider that the mantle is not made of a continuum of olivine, but, at a first approach, of two phases which are olivine and orthopyroxene.

ACKNOWLEDGMENTS

We are grateful to A. Nicolas for discussions and suggestions and we thank the Laboratoire de Microscopie Electronique de Toulouse and the Laboratoire de Métallurgie Physique de Poitiers for the use of their electron microscopes.

REFERENCES

Ashby, M. F. and Verall, R. A. (1973). *Acta Metallurgica*, **21**, 149–163.
Boullier, A. M. and Nicolas, A. (1973). In *Lesotho Kimberlites*, ed. P. H. Nixon, pp. 57–66.
Boullier, A. M. and Nicolas, A., (1975). *Physics and Chemistry of the Earth*, **9**, 462–476. Pergamon Press.
Boyd, F. R. and Nixon, P. H. (1972). *Carnegie Institution, Annual Report Year Book 71*, pp. 362–373.
Boyd, F. R. and Nixon, P. H. (1973). Extended Abstracts of Papers, *International Conference on Kimberlites*, pp. 47–50.
Goetze, C. and Kohlstedt, D. L. (1973). *J.G.R.*, **78**, No. 28, 5961–5971.

Green, H. W., II and Gueguen, Y. (1974). *Nature*, **249**, 617–620.
Green, H. W., II and Radcliffe S. V. (1972). *Earth and Planetary Science Letters*, **15**, 239–247.
Gueguen, Y. in preparation.
Hayden, H. W., Floreen, S. and Goodell, P. D. (1972). *Metallurgical Transactions*, **3**, 833–842.
MacGregor, I. D. (1975). *Physics and Chemistry of the Earth*, **9**, 455–466. Pergamon Press.
Mercier, J. C. and Nicolas, A. (1975). *J. Petrology*, **16**, 454–487.
Poirier, J. P. and Nicolas, A., in preparation.
Raleigh, C. B. and Kirby, S. H. (1970). *Mineralogical Society of America, Special Paper 3*, pp. 113–121.
Stocker, R. L. and Ashby, M. F. (1973). *Review of Geophysics and Space Physics*, **2**, No. 2, 391–426.
Suery, M. (1974). *Thesis*, University of Metz, 162 pp.
Weertman, J. (1970). *Review of Geophysics and Space Physics*, **8**, No. 1, 145–168.

Experimental Deformation of Single Crystal Olivine with Application to Flow in the Mantle

D. L. Kohlstedt, C. Goetze, and W. B. Durham

Massachusetts Institute of Technology
Cambridge, Massachusetts 02139, U.S.A.

INTRODUCTION

Experimental and theoretical investigations of the high-temperature steady-state creep behaviour of olivine and olivine-rich rocks have rapidly multiplied during the past five years in response to active interest in upper-mantle flow associated with large-scale plate motions. (For a review, see Weertman and Weertman, 1974.) The main obstacle in applying the laboratory creep data to earth problems is the necessity to extrapolate to geological conditions, particularly the low strain rates and the high pressures. Deformation studies of a wide variety of materials, including metals, oxides, salts, and ice, have demonstrated that steady-state creep in solids obeys a flow law of the form

$$\dot{\varepsilon} = Kf(\sigma) \exp\left[-(Q + PV)/RT\right] \quad (1)$$

where $\dot{\varepsilon}$ is the creep rate, $\sigma = \sigma_1 - \sigma_3$ is the differential stress, $f(\sigma)$ is an empirical function of σ, K is a constant, Q is the activation energy for self-diffusion of the slowest moving species, V is the corresponding activation volume, P is the hydrostatic pressure, T is the temperature, and R is the gas constant. (See, for example, Sherby and Burke, 1967, and Kirby and Raleigh, 1973.) If the activation energy and activation volume are known from laboratory measurements at mantle temperatures and pressures, then extrapolation in T and P via Equation (1) is straightforward. Extrapolation in stress is more complex. The term $f(\sigma)$, which ideally would be calculated from first principles, changes its functional form in going from a 'low' to an 'intermediate' stress regime and from an 'intermediate' to a 'high' stress regime. In practice, $f(\sigma)$ and the divisions between 'low', 'intermediate', and 'high' stress regimes are only known confidently when experimentally determined. Thus, we have recently extended the empirical flow law for dry olivine, which has been previously reported for the 1 to 10 kb differential stress range, to the geophysically more interesting region near 100 bars (Kohlstedt and Goetze, 1974).

In the present paper, we first review published flow law results for dry olivine and plot a flow law accurate to within one order of magnitude in strain rate in the range 50–15,000 bars and 800–1,600 °C near one bar hydrostatic pressure. This flow law is compared with models for dislocation controlled creep. Second, new creep data are presented for olivine single crystals which are oriented to favour climb at the expense of glide. These data provide a lower limit for the polycrystalline flow law. Third, two possible indicators of paleostress are considered, dislocation density and grain size. Laboratory data confirm a one-to-one correspondence between dislocation density and applied stress. Laboratory and field data indicate that grain size is also a predictable function of stress. Finally, the empirical flow law is tested against geological observation. Using a published geotherm, the strain rate appropriate for the upper mantle and the thickness of the lithosphere are calculated for differential stresses of 100 and 200 bars.

EXPERIMENTAL FLOW LAW

Published Results

Investigations of the high-temperature, steady-state creep behaviour of mantle-related materials include experiments on natural olivine-rich polycrystalline aggregates and on natural olivine single crystals (Carter and Ave'Lallemant, 1970; Raleigh and Kirby, 1970; Blacic, 1972; Phakey, Dollinger, and Christie, 1970; Kirby and Raleigh, 1973; Post and Griggs, 1973; Kohlstedt and Goetze, 1974). Data from the polycrystalline samples have the advantage of better simulating the constraints imposed on grains deforming in the earth. Unfortunately, the presence of phases such as pyroxene lowers the solidus temperature of these rocks to approximately 1400 °C. To deform to several per cent strain at this temperature on a laboratory time scale, a stress of 1 kb or greater is required.

Table 1. Activation energy and applicable temperature range for high-temperature creep and dislocation climb

Activation energy	Temperature range	Experiment	Reference
130 kcal/mol	725–1325 °C	Polycrystal creep	Post and Griggs, 1973
120 ± 17	1100–1350	Polycrystal creep	Carter and Ave'Lallemant, 1970
135 ± 30	1290–1450	Loop collapse	Goetze and Kohlstedt, 1973
140 ± 30	1200–1400	Dislocation climb	Goetze and Kohlstedt, 1973
126 ± 2	1400–1650	Single crystal creep	Kohlstedt and Goetze, 1974
125 ± 15	1400–1600	Single crystal creep	Present study

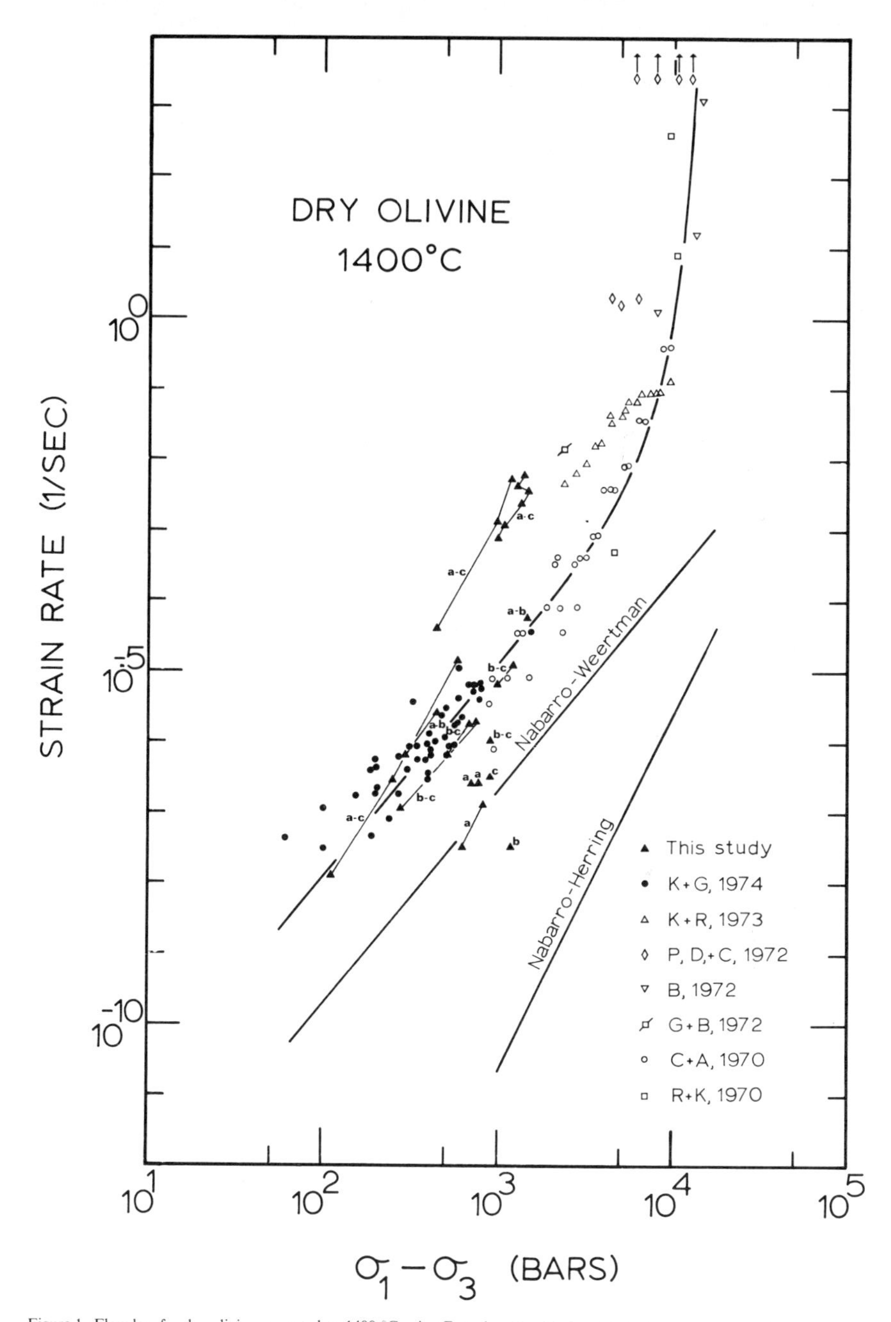

Figure 1. Flow law for dry olivine corrected to 1400 °C using Equation (1) with Q = 125 kcal/mol. Polycrystalline data from Raleigh and Kirby (1970), Carter and Ave'Lallemant (1970), Goetze and Brace (1972), Blacic (1972), and Kirby and Raleigh (1973) are included. Single crystal results from Phakey, Dollinger and Christie (1972), Kohlstedt and Goetze (1974), and the present study (Table 2) are also included. A best fit curve assuming $\dot{\varepsilon} \propto \sigma^3$ at low stresses ($\sigma < 2$ kb) is sketched. Theoretical curves for Nabarro (1967) creep as modified by Weertman (1968) and for Nabarro (1948)–Herring (1950) creep modified for a stress-grain size relation (Goetze, 1974) are included. The orientation of the crystals deformed in the present study are as follows: a, b, and c indicate that σ_1 was parallel to [100], [010], and [001] respectively; a–b, b–c, and a–c indicate that σ_1 was midway between [100] and [010], [010] and [001], [100] and [001], respectively. Solid triangles connected by thin lines are data points from a single sample.

To extend the creep data on mantle-related materials to lower stresses, we have deformed single crystals of olivine (Kohlstedt and Goetze, 1974; present study) in the temperature range 1150 to 1650 °C. These stress versus strain rate data are plotted in Figure 1 together with creep data from polycrystalline samples. All data have been adjusted to a common temperature of 1400 °C using the flow law given in Equation (1) with an activation energy of 125 kcal/mol. The activation energies obtained from single crystal and polycrystal deformation experiments and from studies of dislocation climb in olivine are listed in Table 1. They show that an activation energy of 125 ± 10 kcal/mol is appropriate for high-temperature dislocation creep in olivine. Thus the empirical relationship between stress and strain rate shown in Figure 1 can be extrapolated with some confidence over the temperature range 800–1600 °C.

For $100 < \sigma < 2000$ bars, $f(\sigma) = \sigma^n$ with $n = 3$ is a reasonable fit to the data of Figure 1. Differential stress tests at constant temperature on single crystal samples varify that $n = 3 \pm 1$ in this stress range. An $f(\sigma) = \sigma^3$ is theoretically predicted for steady-state creep controlled by diffusion-limited dislocation glide (Weertman and Weertman, 1965) and by dislocation climb (e.g. Barrett and Nix, 1965; Nabarro, 1967; Weertman, 1968). (See Weertman, 1974, for a review.) The magnitude of the observed creep rate, however, is higher by a factor of 10 to 100 than these theories predict.

The lower stress ($\sigma < 1$ kb) creep data for olivine can be compared with dislocation creep data in polycrystalline MgO (Langdon and Pask, 1970) and Al_2O_3 (Cannon and Sherby, 1973) if the data are plotted as $\dot{\varepsilon}/D$ versus σ/μ to eliminate material parameters, where D is the lattice diffusion constant of the slowest moving species and μ is the shear modulus. Such a comparison has been presented by Kohlstedt and Goetze (1974, Figure 9). The creep data from these oxides are in agreement both in magnitude and slope with the olivine creep data and thus support the flow law shown in Figure 1 at the lower stresses.

New Data

Our previous single crystal creep experiments (Kohlstedt and Goetze, 1974) have been extended to study the influence of crystal orientation on the flow law. The results of these runs are summarized in Table 2. The experimental deformation apparatus was the same as that described previously (Kohlstedt and Goetze, 1974) except that molybdenum anvils were used for runs 7304–2 and 7304–3 and tungsten anvils for the remainder. The buffering gas controlling the oxygen fugacity was a 3:1 H_2:CO_2 mixture; with this gas mixture the interior of the crystals remained a clear olivine colour after heating to 1600 °C for several hours. The crystals were oriented so that both ends of a 4 mm long sample were within 4° of the desired orientation; low angle tilt boundaries produced misorientation of up to 4° between the ends of a given sample.

Three independent slip systems with Burgers vectors [100], [010], and [001] have been identified for olivine by transmission electron microscopy studies

Table 2. Experimental flow law data for oriented dry olivine single crystals

Specimen	Orientation	ε (%)	T (°C)	$\sigma_1 - \sigma_3$ (bars)	$\dot{\varepsilon}$ (10^{-5}/sec)	$-\log_{10} \dot{\varepsilon}_{1400}$
7403–2	b–c	4.8	1500	910	0.75	6.0
7403–3	a	1.8	1600	624	0.15	7.5
				832	0.60	6.9
7403–4	a	9	1600	700	1.25	6.6
7403–5	a	2	1600	777	1.22	6.6
7403–6	b	2.6	1600	1196	0.17	7.5
7403–7	c	2	1600	918	1.7	6.5
7404–1	a–c	25	1600	112	0.06	7.9
				250	1.4	6.5(5)
				585	68	4.8(5)
7405–1	a–c	5.2	1400	448	4	4.4
			1250	1008	3	2.9
			1150	1232	0.7	2.3
7406–1	a–c	8.3	1150	1290	0.60	2.4
				1415	0.87	2.2
				990	0.11	3.1
				1100	0.17	2.9
				1347	0.35	2.6
				1470	0.52	2.4(5)
7406–2	b–c	21	1600	280	0.56	6.9(5)
				742	9.4	5.7
7406–3	a–b	18	1600	297	3.5	6.2
				444	12.2	5.6
7406–4	a–b	0.15	1250	1464	0.15	4.2(5)
7406–5	b–c	15	1400	1020	0.65	5.2
				1220	1.24	4.9
			1600	510	3.25	6.2
				694	9.12	5.7(5)

(Phakey, Dollinger, and Christie, 1972; Goetze and Kohlstedt, 1973).* Extension along the principal axes of olivine is thus not possible by glide alone. Nevertheless, polycrystalline olivine can be plastically deformed to large strains both in the earth and in the laboratory. Other mechanisms, such as dislocation climb, must be active in olivine deformation. To test the role of climb, single crystals of olivine were deformed with [100], [010], or [001] oriented parallel to the load axis, σ_1. The results are plotted on Figure 1 as a, b and c, respectively. The data are approximately two orders of magnitude lower in strain rate than predicted by the experiments on arbitrarily oriented crystals (Kohlstedt and Goetze, 1974), and they are in fair agreement with the theoretical flow law derived by Nabarro (1967) as modified by Weertman (1968) for creep by dislocation climb alone. These a, b and c data provide a lower limit to the flow law for polycrystalline olivine. Though climb is slower than glide, it does

*A very low density of $\langle 101 \rangle$ Burgers vectors has also been established for olivine (Goetze and Kohlstedt, 1973; present study). If the slip planes associated with these $\langle 101 \rangle$'s are $\{101\}$, then one additional slip system is available. If the slip plane is (010), then no new systems are added.

proceed at high enough rates to contribute significantly to flow in olivine polycrystals. Glide, however, appears to account for most of the strain observed in samples where both glide and climb can operate.

The strain rates for the crystal orientations labelled a–b and b–c (σ_1 oriented midway between [100] and [010] and midway between [010] and [001], respectively) agree well with the creep data for arbitrarily oriented single crystals. Differential tests indicate that Q and n for these crystals are also 125 ± 15 kcal and 3 ± 1. Crystals oriented with σ_1 midway between [100] and [001] (labelled a–c) show similar strain rates at 1600 °C, 100 to 500 bars, but show higher strain rates at 1150 to 1400 °C, 500 to 1500 bars, than other crystal orientations. The most frequently observed glide systems—(100) [001] and (001) [100]—are both most favourably oriented for these a–c crystals. It is not presently known whether the high strain rates reflect changes in temperature or stress sensitivity.

Discussion

That the experimental deformation of olivine is consistent with a dislocation creep mechanism is demonstrated by the following:

(1) Transmission electron microscopy comparisons of the undeformed and the deformed crystals verify that dislocations nucleate and move. An equilibrium density of dislocations proportional to the square of the applied stress is observed (Kohlstedt and Goetze, 1974).

(2) The recovery kinetics are rapid enough to accommodate the necessary dislocation motion (Goetze and Kohlstedt, 1973).

(3) The stress sensitivity of the strain rate, Figure 1, is typical of creep models involving high-temperature dislocation mechanisms (Weertman and Weertman, 1974).

The theoretical stress-strain rate relation predicted for Nabarro–Herring (i.e. diffusional) creep (Nabarro, 1948; Herring, 1950) is sketched in Figure 1 above. The grain size, d, is taken to be inversely proportional to the stress squared, $d \propto 1/\sigma^2$ (see section on Paleostress Indicators; Goetze, 1974). Nabarro–Herring creep appears to contribute little to the observed creep rates, and it is least important at low stresses where the grain size is largest.

Insufficient data are available to discuss quantitatively Coble (grain-boundary diffusional) creep (Coble, 1963). Qualitatively, the $d \propto 1/\sigma^2$ relation proposed by Goetze (1974) indicates that Coble creep is likely to be most important at stresses near 10 kb where the grain size approaches a few microns (Figure 4 below). The presence of water or other fluid phase in the grain boundaries would enhance Coble creep.

At stresses above about 2 kb, the observed creep rate of olivine is faster than predicted from extrapolation of a third power law. A breakdown of power-law creep is commonly observed in metals and is thought to be associated with the presence of high vacancy concentrations which accelerate dislocation climb (Weertman, 1957; Barrett and Nix, 1965). The excess vacancies are generated by dislocation intersection processes. In olivine the dislocation density at 10 kb is approaching $10^{10}/cm^2$ (Figure 3 below). The empirical relation

$$\dot{\varepsilon} = AD (\sinh \beta\sigma)^n,$$

where A, n and β are constants independent of stress, has been proposed for high stress deformation (Garofalo, 1963). Weertman (1974) has suggested that power law creep should break down near $\sigma/\mu = 10^{-3}$; that is, for olivine $\sigma \simeq 7$ kb. The data are inadequate to test this functional relation for olivine.

In Figure 2 the activation energy of 125 kcal/mol is used to shift the flow law (the heavy line drawn through the data in Figure 1) to temperatures between 800 and 1600 °C in 100 °C intervals. Before we apply this flow law to problems concerning creep in the mantle, we should consider the following questions:

(1) Will a zone of partial melt or the presence of water substantially increase the strain rates reported here? While the work of Arzi (1972) on partial melt in granites would suggest that no large change in strain rate occurs when a few per cent of melt is introduced, that of Murrell and Chakravarty

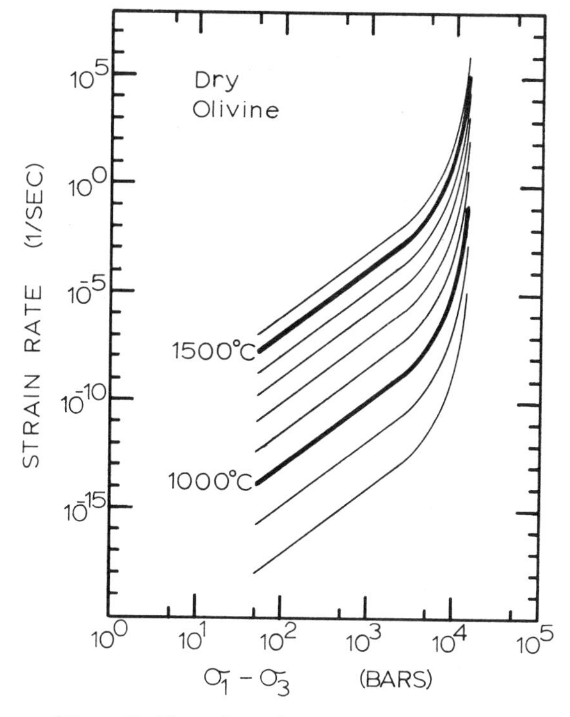

Figure 2. Flow law for dry olivine obtained from Figure 1 adjusted to temperatures in the range 800 to 1600 °C in 100 °C intervals with an activation energy of 125 kcal/mol using Equation (1).

(1973) shows a substantial increase in strain rate when melt is introduced. The role of partial melt is therefore unclear. Carter and Ave'Lallemant (1970) and Blacic (1972) present deformation results for several dunite samples and for one olivine single crystal which are believed to contain water. For a strain rate of 10^{-14}/sec and a temperature of 1250 °C, Carter and Ave'Lallemant's data predict a flow stress almost 3 orders of magnitude lower for 'wet' dunite than for 'dry' dunite. Carter and Ave' Lallemant suggest that above 1000 °C, the dry dunite results yield more reasonable stresses and viscosities when extrapolated to earth conditions. Our calculations (see the section on Application of Experimental Flow Law) based on the dry olivine flow law of Figure 2 support their conclusion

(2) What is the effect of hydrostatic pressure, P, on the experimental flow law results?
Increasing pressure acts to decrease the creep rate at a fixed stress and temperature, as is seen from Equation (1), by making diffusion more difficult. No experimental data for the activation volume for diffusion of any species in olivine are available. Estimates of the activation volume based on pressure-melting relation and molar volume arguments range between 10 and 40 cm^3/mole (Stocker and Ashby, 1973; Goetze and Brace, 1972). Uncertainties in the strain rate due to uncertainties in the activation volume become comparable with the scatter in the experimental flow law near 100 km.

(3) Does the mechanism controlling plastic flow in laboratory experiments also operate in the earth? Two arguments suggest that it does. First, as illustrated in the Applications section, the experimental flow law of Figure 1 is consistent with geophysical observations. Second, dislocation microstructures observed in natural olivines are comparable in dislocation density, Burgers vectors, and sub-boundary formation, with those observed in laboratory-deformed olivines (see, for example, Green and Radcliffe, 1972a, 1972b). It should be noted that increasing pressure (depth) is likely to affect Nabarro–Herring creep through an Arrhenius term similar to that for dislocation creep in Equation (1). The relative influence of pressure on Coble creep is not presently known.

PALEOSTRESS INDICATORS

Dislocation Density

Theory predicts (see Weertman, 1968) and experiments varify (Barrett and Nix, 1965; McLean and Hale, 1961; Bilde-Sörensen, 1972; Poirier, 1972; Kohlstedt and Goetze, 1974) that the 'free' dislocation density, ρ, during steady-state creep is proportional to the square of the applied stress,

$$\sigma_1 - \sigma_3 = \alpha\mu\mathbf{b}\rho^{1/2} \tag{2}$$

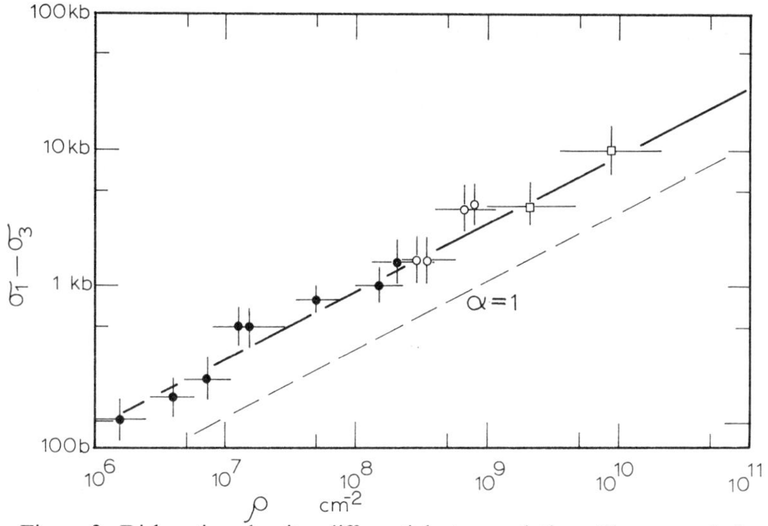

Figure 3. Dislocation density–differential stress relation. The open circles are data from Phakey, Dollinger and Christie (1972), the squares from Raleigh and Kirby (1973), and the closed circles from Kohlstedt and Goetze (1974). A best fit line is shown as is the line theoretically predicted by Equation (2) with $\alpha = 1$.

where **b** is the active Burgers vector and α is a constant of order unity. Dislocations aligned in tilt boundaries are not included in ρ because their long-range stress fields effectively cancel. A plot of dislocation density versus applied stress for experimentally deformed olivine in presented in Figure 3. These data represent creep tests performed over a 600 °C range in temperature, suggesting that the relation between $\sigma_1 - \sigma_3$ and ρ shown in Figure 3 is insensitive to temperature. Equation (2) with $\alpha = 1$ is plotted as a dashed line in Figure 3. The experimental data have the slope predicted by Equation (2); they are best fitted with $\alpha = 3$.

Although the data of Figure 3 are for dry olivine, similar results are anticipated when water is present. The parameters μ and **b** in Equation 2 should not be strongly affected by water content.

Transmission electron microscopy of crystals deformed in our laboratory indicate that for a given stress the equilibrium dislocation density in olivine is reached at strains of less than 2 per cent. If the deformation stress is removed, the dislocation–dislocation interaction forces will drive the dislocations into low strain-energy sub-boundaries. The rates of the dislocation recovery processes in olivine, which have been studied by Goetze and Kohlstedt (1973), are quite fast at temperatures above 1000 °C; for example, the free dislocation density will decrease by a factor of 2 in 1 hour at 1100 °C, with a corresponding increase in sub-boundary density. Thus, care must be exercised in applying the data in Figure 3 to naturally deformed olivines.

Raleigh and Kirby (1970) have suggested that in olivine subgrain size is inversely proportional to the applied stress. Weertman (1968) demonstrates that subgrain size-stress data for metallic systems are consistent with this inverse proportionality. If the dislocation substructure is in equilibrium with a given stress, then the stress predicted from the free dislocation density should be in agreement with that predicted from the subgrain size. Insufficient data are presently available to test this point. When the deforming stress is removed, the sub-boundary density recovers to a lower density but at a rate slow compared to that of the free dislocation density (Goetze and Kohlstedt, 1973). Thus the existence of numerous sub-boundaries and few free dislocations should be indicative of a crystal which has recovered statically.

Goetze (1974) has suggested a third paleostress indicator, grain size. The experimental observations which demonstrate that a systematic relation between grain size and stress exists are summarized for olivine in the following paragraphs.

Grain Size

Observations supporting a dependence of grain size on stress are the following:

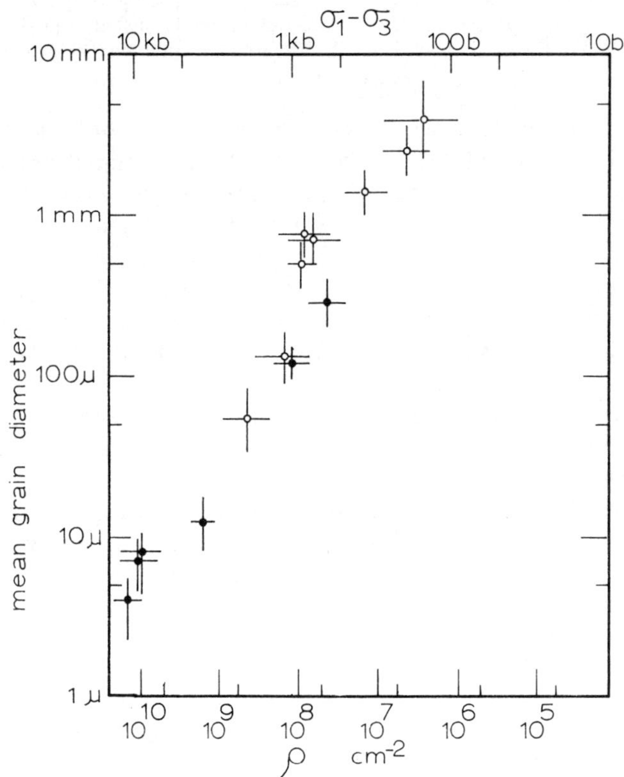

Figure 4. Mean grain size versus dislocation density (differential stress) for olivine. Open circles are observations on field samples and closed circles on experimentally deformed specimens (see Table 3).

Table 3. Dislocation density or differential stress and grain size for the fourteen naturally and six artificially deformed olivines listed.

Locality	ρ	$\sigma_1 - \sigma_3$	Grain size	References	Comments
Salt Lake Crater, Hawaii	$1–7 \times 10^6$		4 mm	Green and Radcliffe, 1972a	Curvy boundaries
Nunavak Island, Alaska	1×10^6		3 mm		Cumulate texture
Hualalai, Hawaii	5×10^6		2.5 mm	Green and Radcliffe, 1972a	
Ronda Massif, Spain	1.7×10^7		1.4 mm		
Lanzo Massif # MPL–6	2×10^6		1.0 mm	Nicolas *et al.*, 1972	Soap-bubble grain
Lanzo Massif # PBW–3	2.5×10^6		600 μm	Nicolas *et al.*, 1972	Soap-bubble grain structure
Ivrea Zone, Trivera, Italy	$\sim 10^8$		750 μm		Slightly coldworked?
Jackson County, North Carolina	7×10^7		700 μm		Slightly coldworked?
Lunar Crater, Nevada	$1–2 \times 10^8$		120 μm	Green and Radcliffe, 1972a	Partially recrystallized
Mt. Albert, Quebec	2.5×10^8		50–100 μm	Goetze and Kohlstedt, 1973	Partially recrystallized
Vourinos Ophiolite Complex, Greece	$3–6 \times 10^8$		30–100 μm	Green and Radcliffe, 1972a	Partially recrystallized
Twin Sisters Range, Washington	1×10^8		300–700 μm		Partially recrystallized
Anita Bay, New Zealand	Complex		Complex	Boland *et al.*, 1971	
Thaba Putsoa, Lesotho	Complex		Complex	Boyd and Nixon, 1973 Boullier and Nicolas, 1973	
Experimental					
N–25		10.5 kb	~ 4 μm	Raleigh and Kirby, 1970	All partially recrystallized
N–26		3.5–4 kb	10–15 μm	Also Green and Radcliffe, 1972a, b	
N–27		9.5 kb	4–12 μm		
G–28		~ 1 kb	120 μm	Ave'Lallemant and Carter, 1970	Completely recrystallized
N–154		~ 600 b	200–400 μm		
GB–20		9 kb	< 10 μm	Blacic, 1972	Partially recrystallized

(1) For a given stress, coarse-grained aggregates syntectonically recrystallize to a finer grain size, while fine grain aggregates undergo substantial grain growth.
(2) Kamb (1972) has demonstrated that a stable grain size is reached in polycrystalline ice deformed to large (20 to 40 per cent) strains.

Transmission electron microscopy and optical microscopy observations, either by us or by others, on twelve natural peridotite, dunite, and lherzolite samples and six experimentally deformed samples form the data base which we used to test for a systematic stress or dislocation density–grain size relation. Table 3 summarizes the data and lists the localities of the samples with appropriate references. The dislocation density–grain size data are plotted in Figure 4. The stress scale in Figure 4 is taken from Figure 3. The results from two natural samples, Anita Bay dunite and a sheared kimberlite nodule, could not reasonably be plotted in Figure 4, because in these specimens small grains and large grains have widely different dislocation densities. Goetze (1974) presents evidence indicating that a correlation between grain size and stress exists for quartz; he has also outlined the physical basis for the observed inverse square relation. The agreement in Figure 4 between data from laboratory and naturally deformed specimens which span many orders of magnitude in strain rate and several hundreds of degrees in temperature supports the idea that the grain size is predominantly controlled by the deforming stress.

Although verification of a well-defined correlation between grain size and stress requires substantially more experimental and theoretical evidence, a systematic trend is suggested by Figure 4.

APPLICATION OF EXPERIMENTAL FLOW LAW

It is generally accepted (Raleigh and Kirby, 1970; Carter and Ave'Lallemant, 1970; Goetze and Brace, 1972; Kirby and Raleigh, 1973) that the flow of olivine controls the creep rate in the first few hundred kilometres of the upper mantle. The ultimate test, therefore, of whether Figure 1 above correctly predicts the behaviour of the mantle must come from localities in the mantle in which temperature, strain rate, and stress are independently known from geophysical data. Post and Griggs (1973) have shown from the Fennoscandian uplift data that a flow law with $n = 3$ best fits the observations. However, to date there has been no analysis of the uplift using an assumed geotherm and a nonlinear flow law to compute the actual distribution of stress and strain rate. Such a computation is beyond the scope of this paper. The data of Figure 2 above predicts stresses of tens of bars causing a strain near 10^{-4} in a few thousand years at upper mantle temperatures as required by the uplift data.

Another test of the applicability of the experimental flow law results to the earth is possible in South Africa. Boyd and Nixon (1973) have reported a paleogeotherm to 170 km from specimens brought to the surface in the Lesotho kimberlites. Goetze (1974) has recently shown that the 'perturbed' portion of the geotherm obtained from sheared specimens is altered due to shear heating

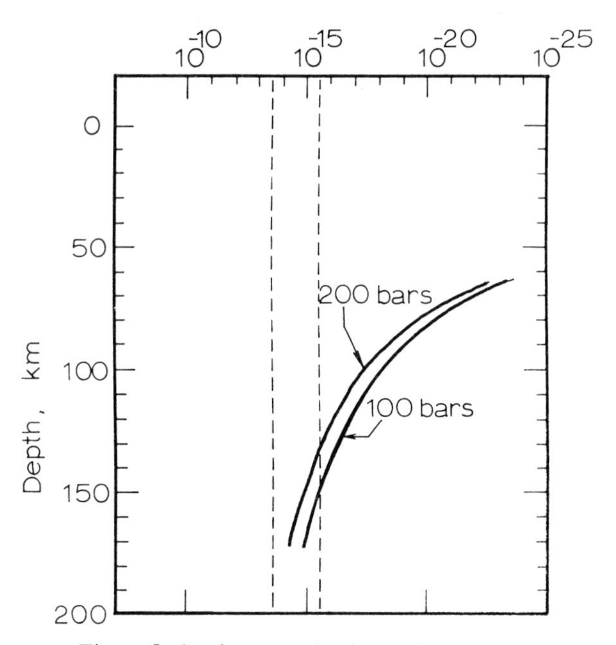

Figure 5. Strain rate–depth profile calculated from the flow law of Figure 2 for differential stresses of 100 and 200 bars. Boyd and Nixon's (1973) geotherm and a pressure correction discussed in the text were assumed.

of the specimens during eruption. Both the unsheared lherzolite specimens and the unrecrystallized grains of the sheared lherzolites have grain sizes in the vicinity 2–5 mm, indicating (Figure 4) a differential stress of approximately 100–200 bars. These samples come from the depth range 120–280 km. If a displacement rate of 1 to 5 cm/year is distributed as a uniform shear over 100 to 700 km, the local strain rate would lie between $10^{-13\frac{1}{2}}$–$10^{-15\frac{1}{2}}$/sec. We take these values to bracket the true strain rate at the base of the African plate under Lesotho. In Figure 5 we plot the strain rate resulting from a differential stress of 100 and 200 bars as a function of depth according to Figure 2 above. The effect of pressure is included by assuming an activation volume of 15 cm^3/mol which is numerically the same as tying the flow law to the melting point of olivine (Kirby and Raleigh, 1973; Raleigh and Kirby, 1970; Carter and Ave' Lallemant, 1970; Weertman, 1970). The uncertainty near 150 km is probably one to two decades in strain rate. Within the experimental accuracy of the flow law, the predicted strain rate does approach the observed strain rate in the depth range 150 to 200 km (Figure 5).

It is interesting to note that Figure 5 predicts a lithospheric thickness of 140–160 km. Boyd and Nixon's (1973) geotherm is one of the coldest known or proposed. Solving the same problem using Clarke and Ringwood's (1964) continental geotherm gives a lithospheric thickness of 110 km; their oceanic geotherm yields 60 km. It is hoped that more data will become available, in time, to test further the consistency of field and laboratory measurements.

CONCLUSIONS

(1) The flow law for dry olivine is experimentally determined for the differential stress range 100 to 10,000 bars. Below 2,000 bars, the data are consistent with a third power stress law; above 2,000 bars, the flow law becomes increasingly sensitive to stress with increasing stress.
(2) An activation energy of 125 ± 10 kcal/mol is observed for creep and for dislocation climb in olivine over the temperature range 725–1650 °C.
(3) Creep rates measures for olivine single crystals oriented with [100], [010], or [001] parallel to σ_1 are approximately two orders of magnitude slower than predicted by the experimental flow law and provide a lower limit to the polycrystalline flow law.
(4) The dislocation density in experimentally deformed olivine crystals reaches equilibrium at a strain of less than 1 to 2 per cent and is proportional to the square of the applied stress.
(5) The grain size of experimentally and naturally deformed olivine grains varies roughly as the inverse of the dislocation density.
(6) Strain rates predicted by the proposed flow law agree to within 1 to 2 decades accuracy with observed strain rates under the South African plate.

ACKNOWLEDGEMENTS

The authors gratefully acknowledge support from NSF Grant No, GA 36280.

REFERENCES

Arzi, A. (1972). *Trans. Am. Geophys. Union*, **53**, 513.
Barrett, C. R., and W. D. Nix (1965). *Acta. Met.*, **13**, 1247–1258.
Bilde-Sörensen, J. B. (1972). *J. Amer. Ceram. Soc.*, **55**, 606–609.
Blacic, J. D. (1972). In *Fracture and Flow of Rocks*, eds. H. C. Heard, I. Y. Borg, N. L. Carter, C. B. Raleigh, *Geophys. Monogr. Ser.*, Vol. 16, pp. 109–115.
Boland, J. N., A. C. McLaren, B. E. Hobbs (1971). *Contrib, Mineral. Petrol.*, **30**, 53–63.
Boullier, A.-M., and A. Nicolas (1973). In *Lesotho Kimberlites*, ed. P. H. Nixon, pp. 57–66.
Boyd, F. R., and P. H. Nixon (1973). In *Annual Report of the Director*, Geophysical Laboratory, 1972–1973, pp. 431–445.
Cannon, W. R., and O. D. Sherby (1973). *J. Amer. Ceram. Soc.*, **56**, 151–160.
Carter, N. L., and H. G. Ave'Lallemant (1970). *Geol. Soc. Amer Bull.*, **81**, 2181–2202.
Clarke, S. P., and A. E. Ringwood (1964). *Rev. Geophys.*, **2**, 35–88.
Garofalo, F. (1963). *Trans. AIME*, **227**, 351–356.
Goetze, C., Manuscript in preparation.

Goetze, C., and W. F. Brace (1972). *Tectonophysics*, **13**, 583–600.
Goetze, C., and D. L. Kohlstedt (1972). *J. Geophys. Res.*, **78**, 5961–5971.
Herring, C. (1950). *J. Appl. Phys.*, **21**, 437–445.
Green, H. W., and S. V. Radcliffe (1972). In *Flow and Fracture of Rocks*, eds. H. C. Heard, I. Y. Borg, N. L. Carter, and C. B. Raleigh, *Geophys. Monogr. Ser.*, Vol. 16, pp. 139–156.
Green, H. W., and S. V. Radcliffe (1972b). *Earth & Planet. Sci. Lett.*, **15**, 239–247.
Kamb, B. (1972). In *Flow and Fracture of Rocks*, eds. H. C. Heard, I. Y. Borg, N. L. Carter, and C. B. Raleigh, *Geophys. Monogr. Ser.*, Vol. 16, pp. 211–242.
Kirby, S. H., and C. B. Raleigh (1973). *Tectonophysics*, **19**, 165–194.
Kohlstedt, D. L., and C. Goetze (1974). *J. Geophys. Res.*, **79**, 2045–2051.
Langdon, T. G. and J. A. Pask (1970) *Acta. Met.*, **18**, 505–510
McLean, D., and K. F. Hale (1961). In *Structural Processes in Creep*, p. 19. Iron and Steel Institute, London.
Murrell, S. A. F., and S. Chakravarty (1973). *Geophys. J. R. astr. Soc.*, **34**, 211–250.
Nabarro, F. R. N. (1948). In *Report of the Conference on the Strength of Solids*, Bristol, The Physical Society of London, pp. 75–90.
Nabarro, F. R. N. (1967). *Phil. Mag.* **16**, 231–237.
Nicolas, A., J. L. Bouchez, and F. Boudier (1972). *Tectonophysics*, **14**, 143–171.
Phakey, P., G. Dollinger, and I. Christie, (1972). In: Flow and Fracture of rocks (eds. H. C. Heard, I. Y. Borg, N. L. Carter and C. B. Raleigh), *Geophys. Monogr. Ser.*, Vol. 16, pp. 117–138.
Poirier, J. P. (1972). *Phil. Mag.*, **26**, 713–725.
Post, R. L., and D. T. Griggs (1973). *Science*, **181**, 1242–1244.
Raleigh, C. B., and S. H. Kirby (1970). *Mineral. Soc. Amer. Spec. Paper 3*, 113–121.
Sherby, O. D., and P. M. Burke (1967). *Progress in Materials Science*, Vol. 13, eds. B. Chalmers and W. Hume-Rothery, pp. 325–390, Oxford: Pergamon Press.
Stocker, R. L., and M. F. Ashby (1973). *Rev. Geophys. Space Phys.*, **11**, 391–426.
Weertman, J. (1957). *J. Appl. Phys.*, **28**, 362–364.
Weertman, J. (1968). *Trans. Amer. Soc. Metals*, **61**, 681–694.
Weertman, J. (1970). *Rev. Geophys. Space Phys.*, **8**, 145–168.
Weertman, J. (1974). In *Dorn Memorial Symposium*, eds. J. C. M. Li and A. K. Mukherjee, New York: Plenum Press, in press.
Weertman, J., and J. R. Weertman (1965). *Physical Metallurgy*, ed. R. W. Cahn, pp. 793–819, Amsterdam: North-Holland.
Weertman, J., and J. R. Weertman (1974). *Annual Review of Earth and Planetary Science*, **3**, in press.

Experimental Deformation of Common Sulphide Minerals

B. R. Clark and W. C. Kelly
Department of Geology and Mineralogy. The University of Michigan, Ann Arbor, Michigan 48104, U.S.A.

INTRODUCTION

The motion of dislocations is ultimately responsible for a variety of plastic deformation features that are visible on the scale seen in the optical microscope. These features record the mechanisms that are most active during deformation. In some cases, the mechanisms themselves are dominant under particular confining pressure and temperature (P–T) ranges, and the observable features can then be used to deduce the physical conditions when deformation took place. The sulphides are a unique group of minerals widely distributed in nature, which can be deformed plastically at the low temperatures and pressures corresponding to conditions of relatively shallow (1–10 km) burial in the earth's crust. Under these conditions, most silicates are brittle, and although calcite develops deformation twins, the twinning process is not indicative of specific P–T conditions, at least when exhibited on an optical scale.

The common sulphide ore minerals—pyrite, sphalerite, chalcopyrite, galena, and pyrrhotite—display a variety of deformation features, some of which are sensitive to pressure and particularly to temperature conditions. Thus it may be possible to use the appearance of some of the common features formed in sulphides to set limits on P–T conditions under which rocks at shallow depths were deformed.

Under surficial or extremely shallow conditions (< 1 km), the sulphides deform largely as brittle materials, as do most other rock-forming minerals. Here the strength and the appearance of the sulphides after deformation are controlled by a process of a pervasive brittle microfracturing commonly called *cataclasis.* In laboratory experiments the cataclastically deformed rocks commonly reach a fracture or yield strength, then maintain an approximately constant differential stress as they are further strained. The yield strength is highly sensitive to confining pressure because an increased confining pressure increases the stress normal to a fracture surface, thus inhibiting sliding along

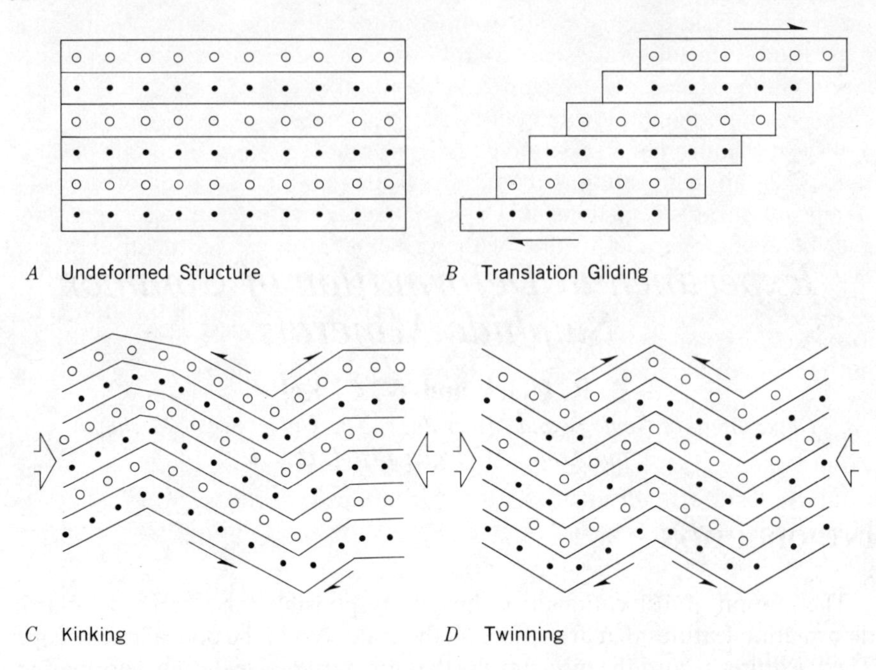

Figure 1. Changes in internal structure of a hypothetical crystal due to different plastic deformation mechanisms.

the fracture. As pressure is increased, competing mechanisms that are insensitive to confining pressures become dominant. The yield stress is also sensitive to temperature although the cataclastic mechanism itself is probably not. Instead, an increased temperature increases the ease with which competing plastic mechanisms become activated and ultimately dominant. Consequently, increases in the depth of burial, i.e., pressure and temperature, decrease the importance of cataclasis as a deformation mechanism. In the sulphides, the plastic mechanisms that replace cataclasis are translation gliding and twin gliding.

Translation gliding (Figure 1B) can be simply illustrated as a result of edge dislocations propagating and moving through a crystal, producing an offset of the portion of the crystal on one side of the glide plane with respect to the rest of the crystal on the opposite side of the plane. The individual slip planes commonly appear as *slip lines* where they intersect a surface. In many cases, the slip lines are observable after etching as rows of etch pits corresponding to intersections with the surface of edge dislocations lying in the slip plane. Translation gliding takes place only along specific sets of crystallographic planes, usually the close-packed planes, and only in specific direction within those planes. A shear stress is required to drive the dislocations along the glide planes, so allowable glide planes at approximately 45° between the maximum and minimum principal compressive stresses are most likely to be activated. The offset takes place with no fracturing or frictional sliding on the glide plane, and is thus not very sensitive to the confining pressure. However, translation gliding is sensitive to temperature and takes place at much lower shear stresses at high than at low temperatures.

If the mechanism of translation gliding is limited to planes in certain orientations, then an aggregate may contain crystals in which no glide plane has a significant shear stress along it. For example, a single allowable glide plane may be aligned parallel to one of the principal stresses. For deformation to occur, the plane must somehow become rotated into an orientation such that a shear stress is applied to it. A common mechanism for such a rotation is the formation of a *kink band* (Figure 1C) in which a segment of the crystal is kinked into a new orientation, then can deform further by gliding along the now favourably oriented glide plane. Kink boundaries vary from diffuse to sharp, depending on the degree to which the lattice is bent. Interlimb angles are highly variable as well. Because of the geometry of kinking, the kink boundaries regularly form at high angles to the maximum compressive stress; thus the orientation of kinks becomes an indicator of stress orientations at the time of deformation.

Deformation twinning (Figure 1D) develops from more complicated dislocations. However, these twins are also caused by shear stresses applied parallel to allowable twin planes in a crystal. In many cases, layers of atoms within a twinned segment move to a new position which produces a mirror image of the untwinned lattice, with the twin plane being the mirror plane. The twin boundaries are consistently sharp since the change in orientation takes place across one or two planes, and the twins may be very thin. The importance of twinning as a deformation mechanism is limited by the fact that a fixed amount of internal rotation is produced by twinning, and thus there is a maximum shape change that can take place in this way. Again the need for a high resolved shear stress along a twin plane, combined with the crystallographic control of allowable planes, limits the orientations of possible twins. Therefore, the twins can also be used to determine orientations of principal stresses at the time of deformation.

One or more of the features resulting from these potential mechanisms appears in naturally deformed sulphides if (1) the mechanism is a dominant one under the P–T conditions of deformation *and* (2) later recrystallization has not entirely removed the evidence.

We shall first review the deformation features produced experimentally and the dependence of the features on P–T conditions, then discuss the kinds of information that might be recorded by the minerals, and finally comment briefly on the effects of annealing and recrystallization.

DEFORMATION FEATURES IN SULPHIDES

The work reported here includes recent contributions from workers at a number of institutions including Michigan. Early deformation experiments on sulphide minerals were conducted by Mügge (1898, 1920), and Buerger (1928) at room temperature and high, but unknown confining pressures. They made very important discoveries about the deformation mechanisms active at room temperatures. Recently, the pioneering work of Siemes and his students at the Technische Höchschule at Aachen provided data on sulphide behaviour at room temperature under accurately controlled high pressures. Together with studies at high temperatures by Roscoe at McGill University, Atkinson at Imperial College, and the present authors, this work has given us a fairly complete understanding of the behaviour of these minerals in the laboratory over a variety of environmental conditions. Rather than review the work of each worker individually, we will refer to them and to others at appropriate places in the text.

The minerals of greatest interest from a tectonic standpoint are of course the most common sulphides: pyrite, sphalerite, chalcopyrite, galena, and pyrrhotite. From the diversity of crystal structures in the group, we might expect a diversity of behaviour, and this assuredly is the case. Essentially all the laboratory studies have been on aggregates of natural sulphides, so in addition to the experimental errors inherent in strength or ductility measurements, the starting materials are neither pure nor perfectly homogeneous. A relatively small temperature and pressure range was normally covered in most experiments, because the sulphides are generally chemically reactive enough either to recrystallize, or at least to transform to new phases under conditions of high grade metamorphism. Consequently, 500 °C and 5 kb are the approximate limits of most sulphide studies at present. Laboratory strain rates from 10^{-4} to 10^{-6} sec^{-1} have usually been used, thus requiring extrapolations of both laboratory strengths and deformation mechanisms to the strain rates near 10^{-13} sec^{-1} normally assumed for natural deformation. Each of the features found in the experimental runs has been described from naturally occurring ore deposits, indicating that the sulphides may contain valuable tectonic information.

Pyrite (FeS_2)

A series of deformation experiments by Adams (1910), Mügge (1920), Newhouse and Flaherty (1930), Bridgman (1937), and Graf and Skinner (1970) all showed that pyrite was extremely strong and brittle, in some experiments to temperatures as high as 650 °C and confining pressures to 25 kb (Graf and Skinner, 1970). The only recognizable mechanism was cataclasis, but this in itself might be useful tectonically if a significantly high percentage of the microfractures that form during cataclasis had a preferred orientation relative to the principal stress directions. However, Graf and Skinner pointed out that their experimental runs had a variety of fractures formed both during and after the deformational stage. Since the two groups were difficult to separate, it seems likely that the use of fractures to derive directions of principal stresses would be difficult if not impossible.

The experimental results are in agreement with the conclusions from field studies of deformed pyrite (Stanton, 1959; Davis, 1972) in which highly fractured grains of pyrite have commonly been interpreted as resulting from tectonic deformation. Because of the enormous range of temperatures and pressures over which pyrite is brittle, it does not allow any estimate of the depth at which deformation occurred.

Such nonductile behaviour in pyrite might be predicted from the crystal structure of the AX_2 sulphides. In pyrite, the iron atoms lie in a face-centered cubic arrangement, with pairs of sulphur atoms located with a centre of gravity half way along each of the cube edges. Pairs of sulphur atoms in adjacent planes lie in different orientations, so that even though the structure can be imagined as analogous to face-centered cubic structures in metals, the sulphur atoms preclude the possibility that the {111} planes will be planes of easy glide.

Sphalerite (ZnS)

Sphalerite also has a face-centred cubic structure, but in this case the {111} planes are alternately close packed planes of zinc and sulphur atoms and it might be predicted that {111} planes could act as translation or twin glide

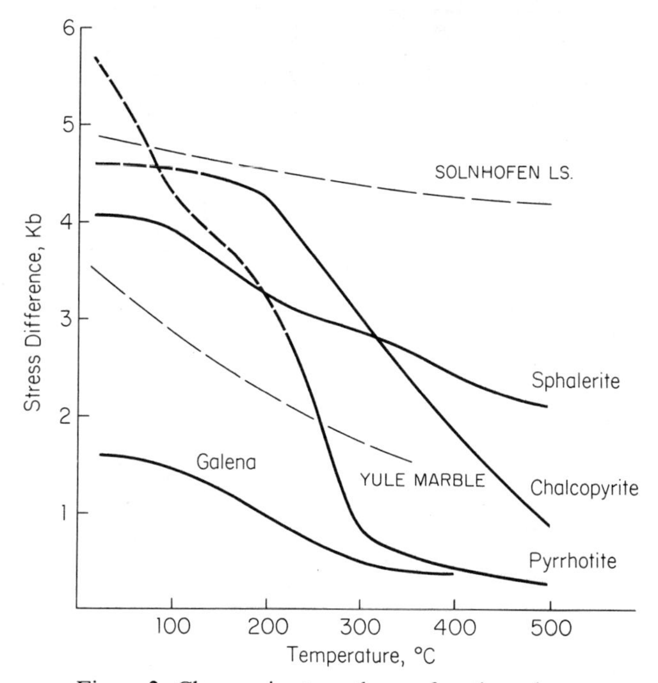

Figure 2. Changes in strength as a function of temperature for sulfide ores and selected other aggregates. Strain rate 7×10^{-5} S^{-1}, confining pressure 1 kb. After Clark and Kelly (1973). (Portions reproduced by permission from *Economic Geology* 1973, vol. 68, p. 351.)

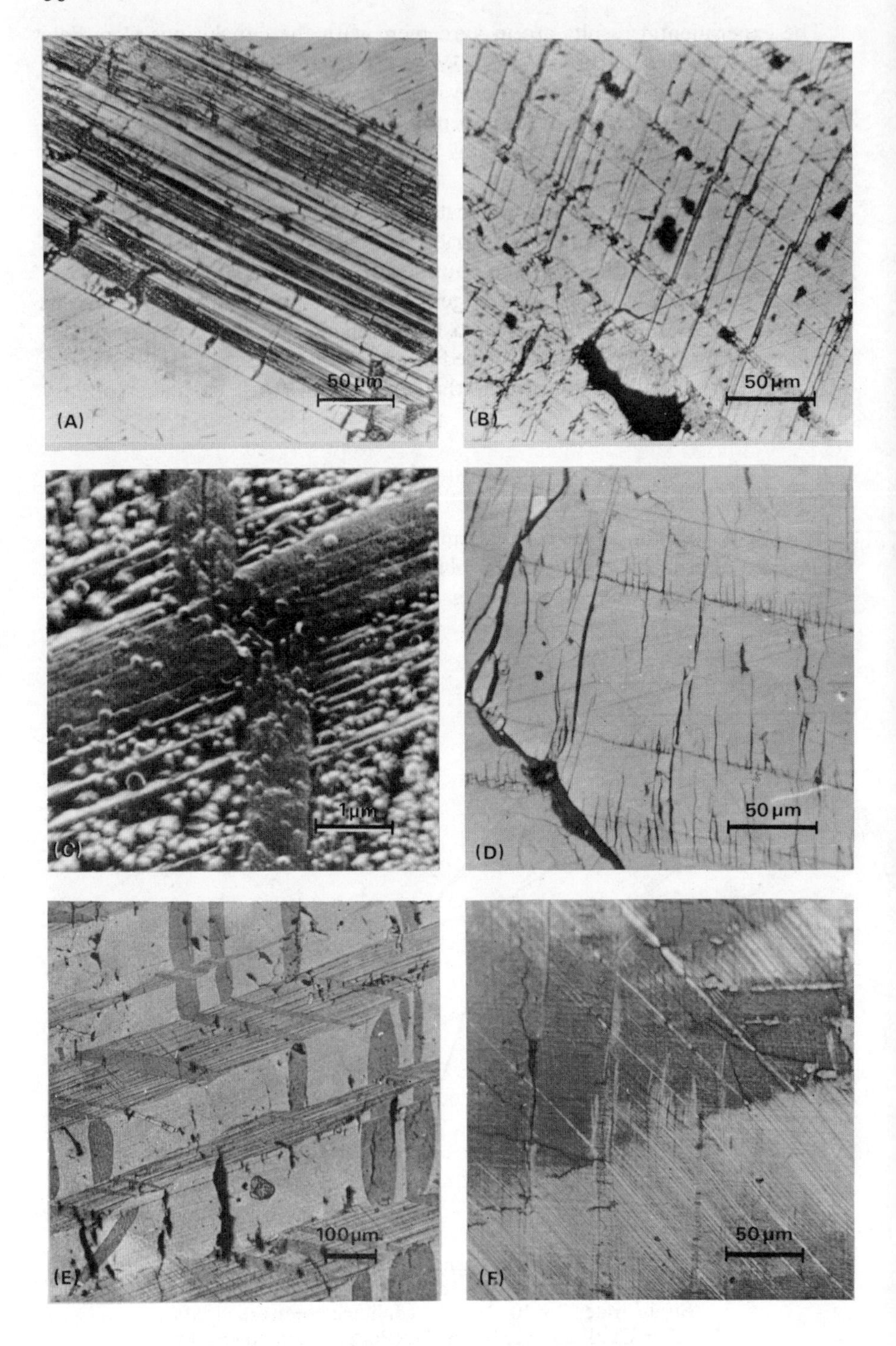
50 μm
(A)
50 μm
(B)
1 μm
(C)
50 μm
(D)
100 μm
(E)
50 μm
(F)

planes, or both. Early studies at room temperature by Veit (1922) and Buerger (1928) produced striations on {111} planes that Veit interpreted as translation glide planes, and Buerger interpreted as twins. More recent studies by Clark and Kelly (1973) established both twinning and translation gliding as active mechanisms under experimental strain rates to temperatures as high as 500 °C. The mechanisms do not appear to change significantly as a function of pressure or temperature with the exception of some cataclastic behaviour at the equivalent of very shallow depths of burial. The strength of sphalerite is not highly sensitive to temperatures so that it becomes one of the stronger sulphide minerals at high temperature. Strengths are on the order of 4 kb at room temperature and 2.5 kb at 500 °C (Figure 2).

Optical features produced by experimental deformation include sets of abundant, fine, closely spaced polysynthetic twins that can be easily distinguished from sets of broad growth (?) twins commonly found in undeformed or recrystallized starting materials Figure 3A. Both types of twins appear to use {111} planes as twin planes. The presence of growth (?) twins in one orientation usually truncates deformation twins using other {111} planes. However, deformation twins may actually try to cross each other, leading to elongate holes on zones of crushed material at the intersections (Figure 3B,C). The amount of slip or translation gliding that has occurred is normally very difficult to estimate, because features that look like fine slip lines in polished section might also be thin twins. We were able to establish that translation gliding occurred at all temperatures to 500 °C and pressures to 2000 bars by recognizing bending and occasional kinking of the sets of twin planes, which requires some mechanism other than twinning (Clark and Kelly, 1973).

Sphalerite deformation twins have commonly been reported in field studies (Richards, 1966; Stanton, 1959; Vokes, 1963) and again correlate well with experimentally produced twins. Although the broad range of environments in

Figure 3. Optical and scanning electron microscope (SEM) photomicrographs of deformation features in sphalerite and chalcopyrite. Sphalerite etched with hydriodic acid, chalcopyrite with solution of hydrogen peroxide and ammonium hydroxide. Maximum compression north-south in figures. Deformation conditions are given as confining pressure in bars/temperature in °C/ total strain in percent.

(A) Fine polysynthetic deformation twins in sphalerite contained within a single broad growth twin. 2000 b/24 °C/10 %.

(B) Intersecting sets of sphalerite deformation twins. Note that the northeast trending set is offset by the northwest trending set and is thus older. 2000 b/200 °C/10%.

(C) SEM view of intersecting sphalerite deformation twins. At the intersection the later twinning is not permitted within the earlier twin, and the zone where they cross becomes granulated. 2000 b/350 °C/10 %. (Reproduced by permission from *Economic Geology* 1973, vol. 68, p. 344.)

(D) Microfracturing originating at fine twins in chalcopyrite. 1000 b/24 °C/6.5%.

(E) Sets of {112} deformation twins confined to {110} inversion twins in chalcopyrite. Untwinned region in original crystal apparently not properly oriented for deformation twins to form. 1000 b/200 °C/23 %.

(F) A rare example of minor kinking of deformation twins in chalcopyrite, implying that some translation gliding is active. 2000 b/300 °C/23 %.

which deformation can produce twins limits their value as a $P-T$ indicator, they could be used with care to determine principal stress directions using methods developed by Turner (1953) for calcite twins.

Chalcopyrite ($CuFeS_2$)

Chalcopyrite is essentially isostructural with sphalerite except that with both copper and iron in the structure, the unit cell is tetragonal, and the $\{112\}$ planes are equivalent to the $\{111\}$ planes in sphalerite. The c cell dimension in chalcopyrite is almost exactly twice (1.97 times) the a cell dimension. Mügge (1920) and Buerger (1928) showed that the $\{112\}$ plane was a slip plane in chalcopyrite at room temperature, but neither found any twin gliding. In more recent experiments, abundant twins have been produced in the laboratory at temperatures from 100 ° to 500 °C (Roscoe, 1973; Atkinson, 1974; Clark and Kelly, in preparation). At low temperatures and pressures, chalcopyrite is considerably more brittle than sphalerite, and at least some cataclasis affects the samples deformed experimentally even at temperatures as high as 400 to 500 °C (Figure 3D). In general, with increasing pressure and temperature, the dominant mode of deformation changes from cataclasis to deformation twinning with minor translation gliding under both conditions. Strengths drop from nearly 4 kb at room temperature to near 1 kb at 500 °C (Figure 2).

Deformation twins can be distinguished optically from growth or inversion twins in chalcopyrite as in sphalerite on the basis of their very fine polysynthetic appearance. In addition, the twins of non-deformation origin appear to use the $\{110\}$ and $\{102\}$ planes as twin planes almost exclusively. Such pre-existing twins normally terminate the later deformation twins, leading to individual bands (early twins) containing *en echelon* sets of deformation twins while adjacent bands (untwinned material) may contain no deformation twins at all (Figure 3E). The details of twin interactions may become very complicated, but careful study of the relative ages of several sets of twins of different origins may reveal a well-documented history of tectonic and metamorphic events that affected natural chalcopyrites.

Evidence for translation gliding is found at all experimental pressures and temperatures, and it may play a significant role in chalcopyrite deformation (Lang, 1968). At room temperature, experiments on single crystals produced sets of $\{112\}$ slip lines together with cataclasis. At high temperatures, the deformation twins were commonly slightly curved or kinked, which in the absence of significant cataclasis, implies that translation gliding was active under these conditions as well (Figure 3F).

There is good reason to believe that deformation twins in natural chalcopyrite are much more abundant than commonly reported in the literature. A few good descriptions of field examples do exist (Ramdohr, 1928; Richards, 1966; Vokes, 1963), but in most studies, etching techniques are not used on chalcopyrite. Thus deformation twins are not readily observed even in cases where chalcopyrite is shown by textural relations to have been mobilized. The authors even found it necessary to reject blocks of natural chalcopyrite from both the Kidd Creek and Icon mines in eastern Canada for use in some experimental deformation studies because of pervasive sets of deformation twins.

Galena (PbS)

Galena displays a somewhat different type of behaviour from any of the preceding sulphides. It is by far both the weakest and the most ductile of the common sulphides, a fact that might be predictable from its simple cubic NaCl-type structure. Translation gliding is the most important ductile deformation mechanism, and brittle behaviour in galena is almost completely lacking except at room temperature. Early studies of deformation of galena in the laboratory by Mügge (1898) and Buerger (1928) established the slip system $\{001\},\langle 110\rangle$ from the study of slip lines formed on the surfaces of the deformed specimens. A detailed study of deformation of galena at room temperature and confining pressures to 5000 bars was published by Lyall and Paterson (1966), and high temperature deformation studies have been reported by Salmon *et al.* (1974) and by Atkinson (1974). These studies showed that the strength of galena was normally less than that of any of the other sulphides (Figure 2), and that under a wide variety of conditions to 400 °C and 5000 bars galena deformed almost exclusively by translation gliding. Because the set of $\{001\}$ planes is not sufficient to allow galena in some orientations to be deformed into any arbitary shape required by surrounding grains, an additional means for allowing deformation must be utilized in at least some galena grains. At low temperatures the new mechanism is likely to be local fracturing, and cataclastic textures in galena are produced under these conditions. At high temperature, the galena is much more likely to deform by kinking, and then by slip on the newly reoriented $\{001\}$ planes in the kinked region. Deformation twins were only produced when galena was deformed at extremely high strain rates such as during impact (Lyall, 1966).

The most useful microscopic feature formed by the deformation of galena is the pattern of kink bands and associated features. The kink band boundaries consistently form at high angles to the maximum compressive stress. In addition, the cleavage is so perfect along $\{001\}$ that any sort of irregularities in the geometry of kinked segments which might cause local tensile stresses either in the kink or in the surrounding matrix, open up the cleavages and leave obvious holes in the grains (Figure 4A). In two dimensions, the holes are long, flat elliptical cracks with the longest axis parallel to the maximum compression, and forced apart perpendicular to the maximum compression. At deformation temperatures above 300 °C, the opened cleavages were less common, and at least some 'syntectonic annealing' accompanied the high temperature deformation runs. At that stage it is presumed that the annealing began to be a more efficient way to remove local stress differences than tensile fracturing.

In general it was not possible to observe the slip lines themselves in standard optical microscopy of polished sections. The commonly used etchant of hydro-

50 µm
(A)
1µm
(B)
50 µm
(C)
100 µm
D)
50 µm
(E)
50 µm
(F)

chloric acid and thiourea produced erratic etch pit patterns, and in only a few cases did the etch pits mark planes that were likely to have been active slip planes during deformation. In these cases, the planes were consistently the {001} planes. However, for samples deformed at low temperatures, Salmon *et al.* (1974) reported one technique that did show slip planes quite effectively. Galena specimens were cleaved rather than cut and polished, then the cleavage surface was shadowed with gold and viewed in the scanning electron microscope. The great depth of field of this microscope allowed large areas of the irregular cleavage surface to be examined in focus. Cleavage surfaces along which the sample broke were probably old incipient cleavage fractures that had been potential surfaces of breakage before the sample was deformed. During deformation, that incipient fracture surface recorded the translation gliding along slip planes as a series of step-like features offsetting the surface. When broken open, the surface appears like those shown in Figure 4B. In some samples, the $\langle 110 \rangle$ nature of the slip direction can clearly be deduced from the offset of the crystal on two perpendicular cleavage faces. At high temperatures, this fracturing technique is not effective, even though there is abundant evidence from kinking that translation gliding was still active. The reason for this is probably that the planes along which fracturing takes place are new cleavage fractures, not inherited from the pre-deformation stage. Since translation gliding has little effect on the internal structure of the crystal (Figure 1B), the new fracture cuts across it along the shortest path, oblivious to the previous offset of the crystal by slip.

Recrystallization has a profound effect on the final texture of deformed galena, even at temperatures as low as 200 or 300 °C (Stanton and Gorman,

Figure 4. Optical and scanning electron microscope (SEM) photomicrographs of deformation features in galena and pyrrhotite. Galena samples etched with hydrochloric acid–thiourea solution. Maximum compression north–south in figures of experimental samples. Deformation conditions are given as confining pressure in bars/temperature in °C/total strain in per cent.

(A) Kink bands and associated opened cleavage fractures in deformed galena. The opened fractures show the degree of crystal rotation within the kink band. 500 b/24 °C/13%.
(B) SEM view of slip lines produced on faces of cleaved galena sample. Note indication of $\langle 110 \rangle$ slip direction from appearance of slip lines on two perpendicular {100} planes. 1000 b/24 °C/9 %.
(C) Naturally deformed and annealed specimen of galena from Star Mine, Coeur d'Alene district, Idaho, showing retention of opened cleavages through recrystallization.
(D) Kink bands in pyrrhotite formed near magnetite grain. These kinks also developed open fractures. 1000 b/250 °C/12 %.
(E) Lensatic twins formed in pyrrhotite during deformation. Such twins appear to be strongly favoured by high confining pressures and differential stresses, but may not be true deformation twins (see text). 1000 b/400 °C/12 %. (Reproduced by permission from *Economic Geology* 1973, vol. 68, p. 342.)
(F) Kink bands (trending northwest) and twins (trending north-south) intersecting in pyrrhotite. Despite offset of twin, the kinks are interpreted to have formed first. 1000b/500 °C/12 %. (Reproduced by permission from *Economic Geology* 1973, vol. 68, p. 342.)

1968). We have recently completed a study of annealing textures in galena (in preparation) and recognize a variety of features that indicate a post-deformational recrystalization phase. In particular, the character of the kink boundaries changes from broad, poorly defined zones between segments with different orientations in a grain, to sharp lines of demarcation, that eventually surround the kinked region and produce elongate grains. The new elongate grains appear to be moderately stable features that may play a large role in the formation of textures such as gneissic galena (Ramdohr, 1928; 1969). In addition, especially in galena deformed at high temperature, a mosaic of subgrains forms and eventually evolves into an equidimensional granular texture. Given enough time, it is likely that these equidimensional grains will overtake the elongate grains and most evidence of kinking will be erased. However, it is interesting to note that even in ores such as the Coeur d'Alene galenas, in which the equigranular texture has been well developed naturally, the open cleavages that accompanied an earlier stage of kinking have not been eliminated (Figure 4C).

Pyrrhotite ($Fe_{1-x}S$)

Pyrrhotite displays the most interesting deformation features of any of the common sulphides. In addition, it suffers enormous changes in strength as a function of deformation temperatures in the laboratory (Figure 2). The phase relations for pyrrhotite are quite complex (Desborough and Carpenter, 1965), and relate to a variety of possible superlattices of either hexagonal or monoclinic symmetry depending upon the precise number of iron vacancies present and their degree of ordering. The fundamental crystal structure is the NiAs structure in which the metal atoms and sulphur atoms alternate in close packed hexagonal layers.

Buerger's (1928) studies of pyrrhotite showed deformation largely by cataclasis, but he reported faint striations parallel to (0001) which he cited as evidence of translation gliding at room temperature. He reported no twinning. More recent studies by Graf and Skinner (1971) showed that at very high pressures (to 11.5 kb) and high temperatures (to 400 °C) a series of twins formed during deformation. In low temperature runs from the Graf and Skinner studies, Ypma and Schull (personal communication, 1972) showed that a large number of prominent kink bands had formed. Our studies have shown that pyrrhotite undergoes a significant change in deformation mechanisms as a function of pressure and temperature (Clark and Kelly, 1973). At low P–T conditions in the laboratory, it deforms largely by cataclasis. As the P–T is increased, observable amounts of translation gliding, and abundant kink bands, accompany cataclasis. Temperatures above approximately 250 °C bring about large numbers of twins and a major drop in strength.

Kinks can be seen in polished sections of deformed pyrrhotite as an irregular bending of the lattice that is readily apparent when the polars are nearly crossed (Figure 4D). No etching is necessary since pyrrhotite is anisotropic. As in galena, some kink boundaries are sharp but many are gradational. Furthermore, the kinking operation tends to open fractures across the kinks similar to the opened cleavages in galena, especially at low confining pressures. Kinking is particularly limited in pyrrhotite because there is only one set of active translation glide planes, the (0001) planes. Pyrrhotite thus remains a strong mineral even though the kinking is abundantly developed in some grains.

Twinning on $\{10\bar{1}2\}$ first appears in deformed pyrrhotite at temperatures of 250–300 °C (Figure 4E,F). The onset of twinning correlates closely with a substantial drop in strength of pyrrhotite aggregates. We originally interpreted the twins as deformation twins (Clark and Kelly, 1973), on the basis of this close correlation of strength reduction with the first appearance of the twins and the orientation of twins at high angles, but a number of arguments can be made for alternative interpretations. First, the twins appear as broad lensatic features rather than narrow polysynthetic bands. Second, sets of twins developing on different planes in the $\{10\bar{1}2\}$ family seem to have the same crystallographic orientation, such that intersecting twins seem simply to merge with each other and not to interfere. Third, it was possible to form a few randomly oriented lensatic twins by bringing a sample to temperatures above 300 °C under hydrostatic pressure alone, although samples heated to 300 ° under atmospheric pressure produced no twins. A reasonable alternate explanation would be that the twins are in fact inversion twins, and that at temperatures of 250 °–300 °C and above and at high confining pressure pyrrhotite transforms to a high pressure polymorph which permits other translation glide systems to become active. A high temperature phase (Desborough and Carpenter, 1965) is already known to exist at temperatures slightly above 300 °C, but the phase change involves only a change in the symmetry of the superstructure, and it seems intuitively unlikely that such a change could have a major effect on the permissible slip systems. It seems more likely that a high pressure phase exists (see also Taylor and Mao, 1970) but has not been recognized at these P–T conditions. Clearly this problem requires some future study.

One important non-tectonic feature is the abundance of release fractures accompanying major temperature changes, at least at laboratory rates. This may be due to (1) volumetric changes associated with inversion, or (2) anisotropic coefficients of thermal expansion and/or elasticity of the grains, with the result that high residual stresses developed during the runs. We suspect the first explanation, because release fractures seem to be especially common in the twinned samples. It is important to recognize that fractures alone are not necessarily an indication of deformation in pyrrhotite, although with care an investigator might be able to separate fracture due to temperature changes from those due to cataclastic deformation. Kinks and twins have been recognized in nature (Vokes, 1963; Stanton, 1959) and are likely to be considerably more valuable for tectonic interpretation.

TECTONIC APPLICATIONS

The original purpose of this series of experiments on the common sulphides was to gain an understanding of the types of deformation mechanisms that are active at high temperatures and pressures in the laboratory, and to correlate these as well as possible with mechanisms that can be recognized from naturally deformed sulphide ores. The obvious practical application of the experimental data is to use naturally observed deformation features to interpret the tectonic history of ore deposits, and thus to help answer important questions about the relative timing between ore emplacement and the deformation and metamorphism of the host rock. But a further extremely important application of the experimental results is in the interpretation of the tectonic history of mountain belts in which ores may be present in only trace or minor amounts. If they were emplaced early in the history of the belt they could act as sensitive recorders of shallow or late tectonic events. Because of their ductile behaviour at such low temperatures and pressures, the sulphides are likely to contain information about deformation that may have left the silicate minerals unaffected. All of the features described from the experimental studies have been recognized in natural samples of ores, suggesting that despite the enormous extrapolation required in correlating laboratory and natural strain rates, many of the laboratory mechanisms are the rate-limiting, or at least important, mechanisms in the field. Little application of the laboratory data has yet been made to field studies, but work underway at The University of Michigan is very encouraging.

Several types of information from the laboratory studies appear to be worthy of attempts to apply them to naturally deformed rocks. In particular, the strength differences as a function of temperature, and to a lesser extent confining pressure, may produce a sequence of 'relative mobilities' that is an indication of the conditions under which the ores were deformed. Furthermore, the presence or absence of specific deformation features in one or more of the sulphide minerals, particularly such features as twins in pyrrhotite or kinks that have not recrystallized in galena, may set limits on the P–T conditions during and after deformation. Finally, the orientations of particular features have been shown to be closely correlated with principal stress and/or strain directions in the experimentally deformed sulphides, and should be very useful in determining regional stress patterns in tectonic areas as well.

The strengths of all of the sulphides decrease with an increase in temperature, and can be expected to be fairly independent of confining pressure as long as deformation is largely plastic. Based on the highly plastic behaviour of nearly all sulphides even at low confining pressures and at the rapid experimental strain rates, it is reasonable to postulate that at natural strain rates the sulphides will be plastic at virtually all depths. Hence, the strengths of these minerals should be a function of temperature, and differences of strength in natural ores would probably be recognized as differences in the degree to which the ores had become mobilized during deformation (see review by Vokes, 1969). Several other factors will tend to reduce the absolute values of strength, including grain size, the actual strain rates, the presence of reactive fluids, and the degree of syntectonic recrystallization. Thus the sulphide strengths measured in the laboratory probably represent maximum strength values. Furthermore, sphalerite, galena, and chalcopyrite all decrease in strength fairly regularly with increasing temperature, and although the experiments show that the strength of chalcopyrite decreases more rapidly with temperature than that of sphalerite, the temperature at which the one mineral becomes stronger than the other must be dependent on the several factors mentioned above. However, in pyrrhotite, the strength drop that occurs between 200 ° and 300 °C in the laboratory is so large that it is quite likely that in ores deformed at high temperatures, pyrrhotite will appear weaker than sphalerite and perhaps chalcopyrite as well, while if deformed at low temperatures, pyrrhotite will appear as the strongest of the three. Independent evidence of this temperature might be gained from the presence or absence of the lensatic twins that accompany high-temperature deformation of pyrrhotite. As an example, Hewett and Soloman (1964) have reported a sequence of increasing mobility for the Mount Isa ores of pyrite—sphalerite—pyrrhotite—galena. From the experimental results, this would suggest that temperatures at the time of deformation were high enough to allow pyrrhotite to behave as a weak mineral. An independent check should be whether twins are present in the pyrrhotite, but they did not describe the microscopic structures.

The appearance of specific deformation features such as kinks or twins in limited temperature ranges may be of some value in a few of the ore minerals. Like calcite, both sphalerite and chalcopyrite develop deformation twins over nearly the entire P–T range examined in the laboratory. A possible exception may be the room temperature deformation of chalcopyrite where no twins form, but we have collected no data on the effect of strain rate on the production of deformation twins, and it may be that twins form at even lower temperatures at natural strain rates.

The two most promising possibilities for using deformation features for limiting tectonic environments appear to be the presence of kink bands in galena and the presence of twins in pyrrhotite. The kink bands in galena recrystallize at temperatures greater than about 200 °C in the laboratory to form individual elongate grains. Their previous history as kink bands can be deduced from the deflections of old cleavages across the fractures in either the formerly kinked region or the surrounding galena, or occasionally extending into both and deflected at the old kink boundary. The presence of kink bands in the unrecrystallized state must limit post-kinking temperatures to less than 200 °C, although of course the precise temperatures cannot be determined.

The second valuable change in deformation mechanisms occurs in pyrrhotite around 250 °–300 °C, when twins begin to form. Although kinks still form at temperatures greater than 300 °C, we have not been able to produce twins at temperatures below 250 °C, and they do not become common below 300 °C. It remains a question whether the mere presence of twins is enough to establish the fact that deformation has affected the pyrrhotite. In one hydrostatic run we were able to produce a few twins, indicating that deforming stresses were not necessary. However, the twins from the deformation runs did form in orientations of relatively high resolved shear stress in the samples during deformation. This suggests that the shear stresses did have a significant control over the formation of the twins, even if they are not the classical polysynthetic deformation twins. From our studies it appears that a combination of kinks and twins in deformed pyrrhotite points to a high temperature during deformation (greater than 300 ° at laboratory strain rates), whereas the presence of kinks but absence of twins would suggest that the pyrrhotite was deformed at low temperatures.

In comparing the behaviour of the deformation mechanisms in the sulphides with those in the metals, it is interesting to note that in two of the sulphides, chalcopyrite and pyrrhotite, twinning appears to be favoured by high temperatures and presumably low strain rates, just the opposite of the great majority of the metals. This suggests that the mechanisms by which twins are produced in some sulphides are more sensitive to temperature than the translation glide mechanisms. In terms of the commonly accepted flow laws, this indicates that the activation energies for twinning are higher than those for slip (self diffusion) and implies that there may be temperature-sensitive factors controlling the complicated dislocation motion that produces twinning, which are not related to diffusion of vacancies in the lattice. To our knowledge, this problem has not been investigated in minerals.

The third major application of the laboratory deformation data to field problems is in relating the geometry of the deformation features to the principal stress directions (Salmon *et al.*, 1974). Carter and Raleigh (1969) have reviewed the techniques by which both twins and kinks can be used to determine the principal stress directions, based on the principles that both features require high shear stresses to form, and that the greatest shear stress is always in an orientation 45 ° between the maximum and minimum principal stress. For twins, the maximum compressive stress can be determined uniquely by noting the most common orientation of twin planes and the sense of rotation of the twinned segment of a grain relative to the untwinned segment; the twinned segment tends to be rotated toward the minimum stress from the maximum stress. The technique was developed by Turner (1953) for calcite twins, and has been applied in many studies of calcite-bearing rocks since that time. In the sulphides, the opaque nature of the minerals may require that the technique be modified in order to locate the stress axes precisely. However, approximate orientations can be determined quite rapidly. Principal stresses can also be determined from kink bands (Griggs *et al.*, 1960; Raleigh, 1968). The same principles apply to kinks as to twins: the orientation of the kink plane, the kink direction, and the sense of movement in the kink, when measured for a statistically significant number of kinks are sufficient to determine the principal stress directions. The sulphide minerals that contain the best examples of kink bands are galena and pyrrhotite. Again the opacity of these minerals makes the complete determination of the necessary orientations difficult, but rapid approximate determinations are quite easy to do.

One feature whose orientations may be extremely easy to find, and may be as useful for determining principal stress directions as the twins or kinks, are opened cleavage fractures. Since the sulphides are most useful under conditions of shallow burial, they can be expected to form a number of open fractures like those produced in all of the laboratory materials and found in such natural ores as the Coeur d'Alene galena. The best examples are associated with kinks in galena and pyrrhotite, and twins in sphalerite. The opened fractures appear to form experimentally in the plane containing the maximum and intermediate compressive stresses and perpendicular to the minimum compressive stress. The precise orientations of a statistical sample of these fractures can be measured using a reflecting universal stage.

Problems of Recrystallization

Our preliminary studies of annealing of sulphide ores indicate that many of the features imposed on the ore minerals during deformation can be removed by allowing extensive recrystallization to occur after deformation. The work of Siemes and his colleagues (Siemes, 1964; Saynisch, 1970; Bayer and Siemes, 1971) suggests that information can still be obtained from the ores based on their crystallographic fabric patterns even after some recrystallization. Stanton and Gorman (1968) have shown that recrystallization in the form of triple junction equilibration begins at temperatures as low as 100 °C in galena and sphalerite, and 200 °C in chalcopyrite. However, a number of features in galena are retained to considerably higher temperatures in our laboratory annealing studies, and it appears likely that the deformation features described above will be useful over a range of P–T conditions where nucleation and new grain growth remain limited. At higher P–T conditions, crystallographic fabric data from the sulphides can be used to complement the information provided by fabrics and deformation features present in the silicates.

Conclusions

The experimental studies of the sulphide minerals suggest that they can play a major role in the analysis of the tectonic history of rocks in which they are found. The information recorded in the sulphides is probably different than that recorded in common silicate minerals, because the sulphides are weak and ductile under P–T conditions in which the silicates are still strong and brittle. Thus, not only are the sulphides more likely to form easily interpreted plastic deformation features under conditions of shallow burial, but they should also be considerably more sensitive to small stress differences than the silicates.

The obvious direct application of sulphide deformation studies is to the interpretation of tectonic histories of ore deposits, and beyond that to the establishment of times of emplacement of ores in tectonic belts. While metamorphosed and deformed sulphide deposits have been postulated from textural evidence in mountain belts around the world (Vokes, 1969), there is now some firm basis for beginning to establish directions of principal stresses and, to some degree, P–T conditions associated with periods of sulphide deformation.

A second important application of the sulphide studies is their potential use for analysing shallow or late tectonic activity in the mountain belts themselves. Care must be taken to establish that the ores were emplaced before the tectonic event, but if the ores have recorded the event, they then contain a considerable amount of information about the stresses and environment of deformation. Many of the sulphide minerals are considerably more common and widely distributed than the relatively few mining centres in the world imply. Although most deposits are small and not economically exploitable, the presence of minor amounts of the sulphides can be enough to allow their use as tectonic indicators. The use of sulphides in mountain belts for tectonic interpretation is probably limited to areas in which the metamorphic grade is low, or deformation after metamorphism has occurred, but these are precisely the conditions under which the silicates yield little information.

The problems of natural annealing and the very slow strain rates in nature limits the significance of the laboratory measurements. In general, we conclude that the laboratory strength values are maximum values to be expected in nature, and ductilities are minima. Thus, brittle $P-T$ conditions in the laboratory probably extend to greater equivalent depths than they do in nature. These limitations notwithstanding, a careful application of the knowledge of deformation features and the behaviour of sulphides under laboratory conditions can lead to a significant increase in the amount of information available to workers attempting to interpret tectonic histories from field and petrographic studies.

ACKNOWLEDGEMENTS

These studies were supported by the National Science Foundation, Grant # GA-25164. We are indebted to many graduate students who played a part in accumulating the data for sulphides and who helped in development of instrumentation, especially Bette Salmon, Floyd Price, and David Chapman. Dr. Donald Peacor provided needed crystallographic expertise. Illustrations were prepared by Derwin Bell. Lastly, we particularly want to thank our friends in the mining industry who provided us with starting materials.

REFERENCES

Adams, F. D. (1910). *Jour. Geology*, **18**, 489–535.
Atkinson, B. K. (1974). *Inst. Mining and Met. Trans., Sec. B.*, **83**, 19–28.
Bayer, H., and Siemes, H. (1971). *Mineral. Deposita*, **6**, 225–244.
Bridgman, P. W. (1937). *Proc. Am. Acad. Arts Sci.*, **71**, 387–460.
Buerger, M. J. (1928). *Am. Mineralogist*, **13**, 1–17, 35–51.
Carter, N. L. and Raleigh, C. B. (1969). *Bull. Geol. Soc. America*, **80**, 1231–1246.
Clark, B. R. and Kelly, W. C. (1973). *Econ. Geol.*, **68**, 332–352.
Davis, G. A. (1972). *Econ. Geol.*, **67**, 634–655.
Desborough, G. A., and Carpenter, R. H. (1965). *Econ. Geol.*, **60**, 1431–1450.
Graf, J. L., Jr., and Skinner, B. J. (1970). *Econ. Geol.*, **65**, 206–215.
Griggs, D. T., Turner, F. J., and Heard, H. C. (1960). *Geol. Soc. Amer. Mem.*, **79**, 39–104.
Hewett, R. L., and Solomon, P. J. (1964). *Intl. Geol. Congr., 22nd, India*, **5**, 571–595.

Lang, H. (1966). *Thesis*, Rhein-Westf. Tech. Hoch. Aachen.
Lyall, K. D. (1966). *Am. Mineral.*, **51**, 243–247.
Lyall, K. D., and Paterson, M. S. (1966). *Acta Metallurgica*, **14**, 371–383.
Mügge, O. (1898). *Neues Jahrb. Mineral. Geol. Palaontol., Abh.*, **1**, 71–158.
Mügge, O. (1920). *Neus Jahrb. Mineral. Geol. Palaontol.*, 24–25.
Newhouse, W. H., and Flaherty, G. F. (1930). *Econ. Geol.*, **25**, 600–620.
Raleigh, C. B. (1968). *Jour. Geophys. Res.*, **73**, 5391–5406.
Ramdohr, P. (1928). *Neues Jahrb. Mineral. Geol. Palaontol., Abt. A*, **57**, 1013–1068.
Ramdohr, P. (1969). *The Ore Minerals and their Intergrowths*, Oxford, Pergamon, 1174 pp.
Richards, S. M. (1966). *Austral. CSIRO- Minerag. Invest., Tech. Pap.* **5**, pp. 1–24.
Roscoe, W. E. (1973). *Ph.D. Thesis*, McGill Univ., Montreal, Canada.
Salmon, B. C., Clark, B. R., and Kelly, W. C. (1974). *Econ. Geol.*, **69**, 1–16.
Saynisch, H. J. (1970). In Paulitsch, P., ed., *Experimental and Natural Rock Deformation*, Berlin, Springer–Verlag, pp. 209–252.
Siemes, H. (1964). *Neus Jahrb, Mineral. Geol. Palaontol., Abh.* **102**, 1–30.
Stanton, R. L. (1959). *Can. Mining Met. Bull.*, **52**, 357–368.
Stanton, R. L., and Gorman, H., (1968). *Econ. Geol.*, **63**, 907–923.
Taylor, L. A. and Mao, H. K. (1970). *Science*, **170**, 850–851.
Turner, F. J. (1953). *Am. Jour. Sci.*, **251**, 276–298.
Veit, K. (1922). *Neues Jahrb. Mineral. Geol. Palaontol., Beilage-Band*, **45**, 121–148.
Vokes, F. M. (1963). *Norg. Geol. Undersokelse*, **222**, 1–126.
Vokes, F. M. (1969). *Earth-Sci. Rev.*, **6**, 99–143.

New Applications of the Diamond Anvil Pressure Cell: (I) Effect of Pressure on Ultimate Strength in Crystalline Materials

W. A. Bassett and G. Kinsland
Department of Geological Sciences
University of Rochester
Rochester, N.Y. 14627

MEASUREMENT OF ELASTIC STRAIN

A polycrystalline sample can be compressed between diamond anvils in a pressure cell that permits diffraction observations on the sample when an X-ray

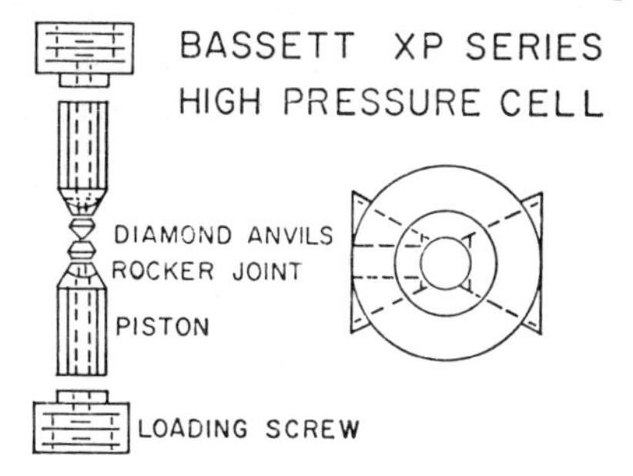

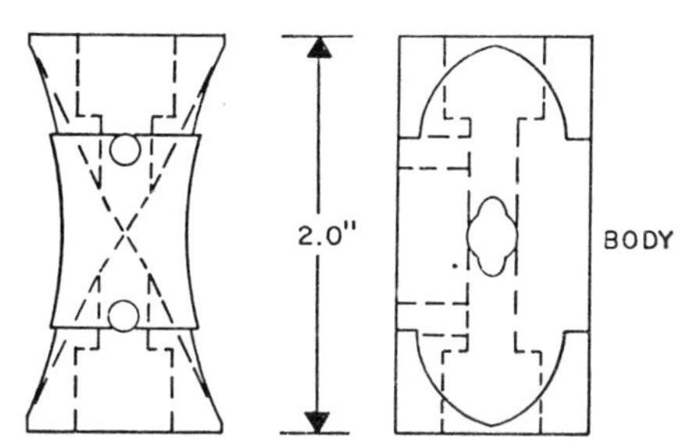

Figure 1. Diagram of the diamond anvil press modified to permit passage of an X-ray beam perpendicular to the loading axis.

beam traverses the sample perpendicular to the compression axis (Figure 1). The diffraction rings recorded on a flat film are ellipses due to the elastic strain in the crystallites. Ultimate strength as a function of pressure can be calculated from the observed strains. The method has been applied to NaCl and MgO.

The modification of the diamond cell used permits a large conical dispersion of X-rays (Figure 1). As load is applied to the anvils, the sample undergoes elastic strain and also plastic strain as manifested by extrusion of the sample from between the anvils.

If the elastic strain in the polycrystalline sample is isotropic, a diffraction ring, e.g. the 200 in MgO, is a circle. If, however, the sample is uniaxially strained, the ring becomes an ellipse, since the 200 atomic layers in some of the crystallites are more closely spaced than the 200 atomic layers in other crystallites depending upon their orientations with respect to the compression axis. Thus, d_{200} measured parallel to the compression axis is smaller than the d_{200} perpendicular to the compression axis. The change in d_{200} calculated from the diffraction rings provides a means for measuring the elastic strain in the sample. This strain can be separated into an isotropic component, due to pressure, and an uniaxial component, due to uniaxial stress. The magnitude of this uniaxial stress can be calculated from the measured strain using the elastic constants of the sample. Since plastic as well as elastic strain occurs in the sample, the sample is considered to be work hardened. It is assumed that the calculated uniaxial stress is the maximum that the sample can support and therefore equals the ultimate strength. Pressure can be calculated from the isotropic strain using the sample's elastic constants. The values for ultimate strength and pressure can then be plotted (Kinsland, 1974).

ULTIMATE STRENGTH VERSUS PRESSURE FOR NaCl AND MgO

Stresses for NaCl calculated from the strains using the Hildebrand equation

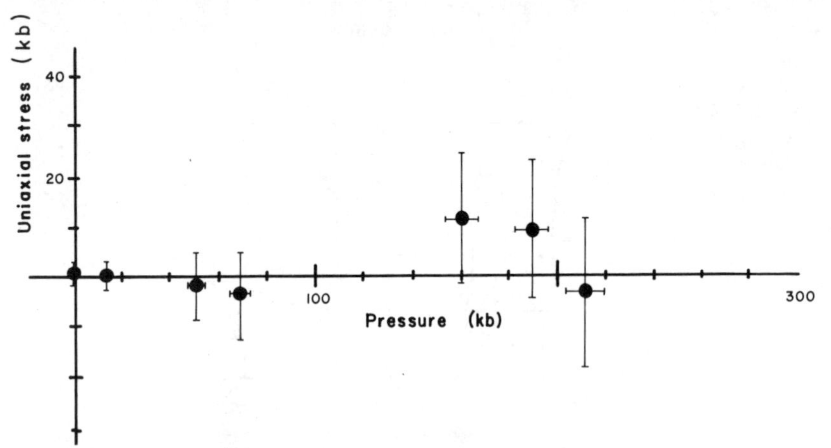

Figure 2. The uniaxial stress, interpreted as ultimate strength, plotted versus pressure for NaCl at 25 °C.

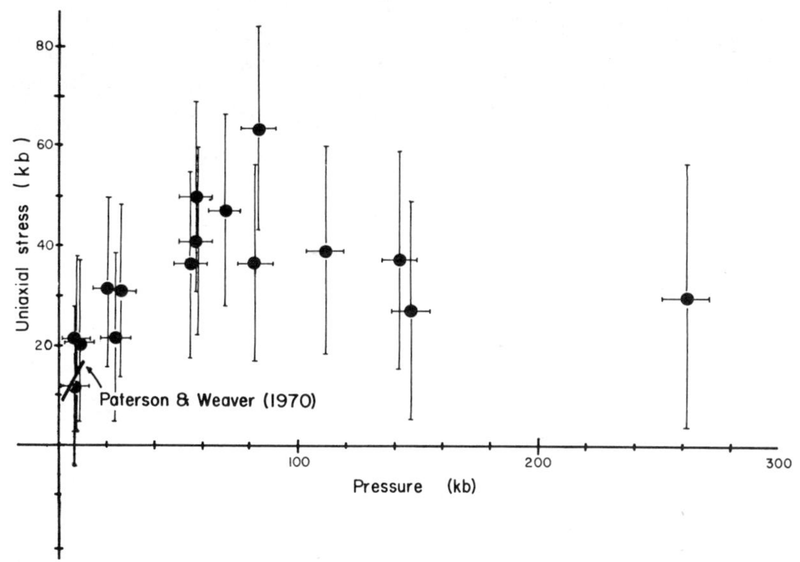

Figure 3. The uniaxial stress, interpreted as ultimate strength, plotted versus pressure for MgO at 25 °C. The data of Paterson and Weaver (1970) are included for comparison.

(Weaver *et al.* 1971) were used to plot the ultimate strength versus pressure at 25 °C (Figure 2). It should be noted that the ultimate strength of NaCl does not depart from zero by more than experimental error up to 200 kbar confining pressure. It is concluded, therefore, that the ultimate strength of NaCl lies below the detectability limit of the X-ray methods employed. It follows that NaCl is an excellent pressure transmitting medium to a pressure of 200 kbar when an isotropic stress (hydrostatic pressure) to within 5 kbar is desired.

A plot of ultimate strength versus pressure at 25 °C for MgO does depart significantly from zero over the range of pressure from zero to 260 kbar (Figure 3). The ultimate strength increases with pressure up to approximately 80 kbar and remains constant or declines slightly at pressures above that up to 260 kbar. The break in the plot is interpreted to occur at the point where sufficient crystallographic slip planes are activated to satisfy the von Mises criterion and the sample becomes truly ductile.

REFERENCES

Kinsland, G. L. (1974). *Ph.D. Thesis*, Department of Geological Sciences, University of Rochester.

Paterson, M. S. and C. W. Weaver (1970). *J. Amer. Ceramic Soc.*, **53**, 463–471.

Weaver, J. S., T. Takahashi, and W. A. Bassett (1971). In *Accurate Characterization of the High-Pressure Environment* Ed. E. C. Lloyd, National Bureau of Standards Publication 326, pp. 189–199.

The Role of Dislocation Processes During Tectonic Deformations, with Particular Reference to Quartz

S. White,

Departments of Geology and Metallurgy, Royal School of Mines, Imperial College, London S.W. 7

1 INTRODUCTION

Recent electron microscopy studies of both tectonically and laboratory deformed minerals and rocks have indicated that dislocation processes are important tectonic deformation mechanisms (see Green and Ratcliffe, 1972; McLaren and Hobbs, 1972; White, 1973a, 1974a,b; White and Treagus, 1974; Ardell, Christie and Tullis, 1974; Kohlstedt and Goetz, this volume). It will be shown that the mechanical behaviour of a mineral is dependent upon the ease with which recovery can occur. This in turn influences the type of intragranular strain features which develop, the recrystallization process and ultimately the fabric. The last feature will affect the mechanical and physical properties of rocks such as seismic, magnetic and electrical anisotropies, flow strength and possibly subsequent microfracture patterns. Thus an understanding of the role of dislocation processes during tectonic deformations is of prime importance.

This contribution will deal with quartz in particular, giving a summary of the dislocation structures in quartz grains deformed in a variety of tectonic environments and a description of the dislocation processes involved in recrystallization. The defects in plagioclase, carbonates, pyroxene and olivine will be compared briefly with those in quartz. Finally the environments in which dislocation processes are important will be outlined and possible relationships between flow rate and stress discussed.

2 DISLOCATION SUBSTRUCTURES IN DEFORMED QUARTZ

The defects in naturally deformed quartz grains have been described by White, 1971, 1973a,b,c, 1974a,c; McLaren and Hobbs, 1972; White and Treagus, 1974; and in experimentally deformed quartz by McLaren and Retchford,

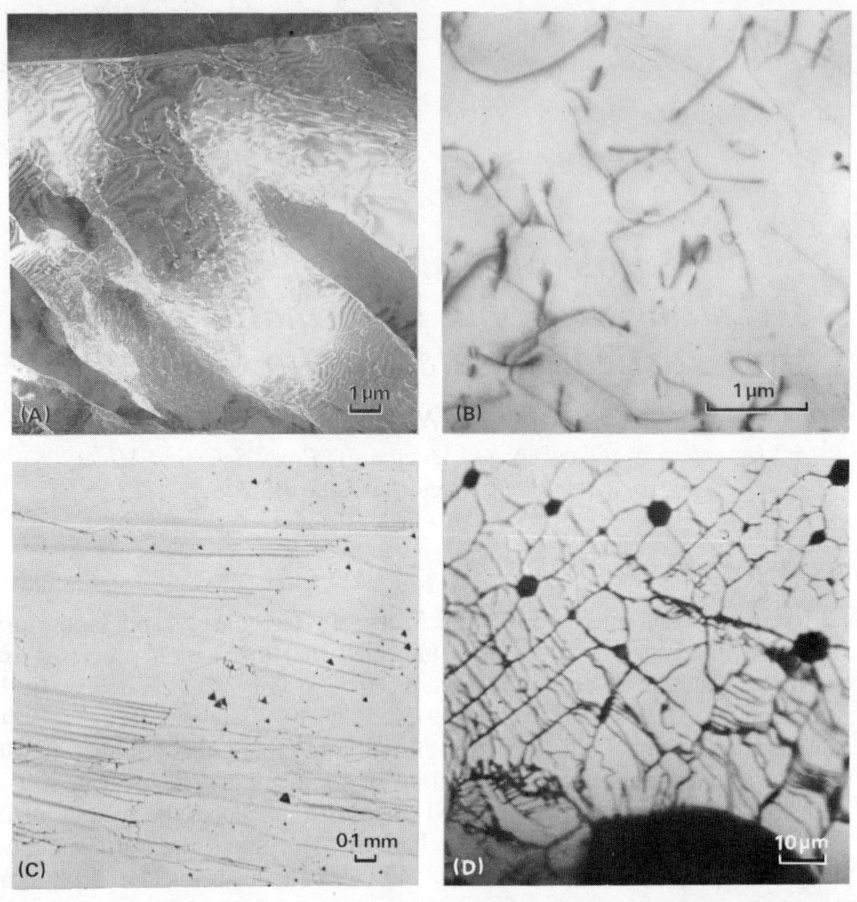

Figure 1. Typical dislocation structures in underformed and tectonically deformed quartz. (A) A dark field electron micrograph showing a typical sub-grain structure in tectonically deformed quartz. 1000 kV. (B) An electron micrograph of an irregular three dimensional arrangement of dislocations inside a sub-grain. 1000 kV. (C) An optical micrograph of etched-out dislocations, dark lines, marking out growth cells in undeformed single crystal quartz. (D) An optical micrograph of an etched out dislocation network and bubbles along a healed fracture in single crystal quartz.

1969; McLaren *et al.*, 1970; Hobbs, McLaren and Paterson, 1972; Ardell, Christie and Tullis, 1974.

Sub-grain structures containing an irregular three dimensional array of unbound, internal dislocations are very common in naturally deformed quartz (Figures 1 A and B). That these are due to deformation can be deduced from comparisons of defects in undeformed single crystals from hydrothermal veins and in experimentally deformed quartz. The single crystals are almost free from dislocations, whose densities are so low that they are difficult to detect in an electron microscope and other techniques must be used. Etching and X-ray topography are the most profitable. Typical dislocation structures in such crystals are shown in Figures 1 C and D. Walls of dislocations do form but they mark out growth cells (Lang and Miuscov, 1967; McLaren, Osborne and Saunders, 1971). Most individual dislocations occur along steps in twin boundaries and along twin intersections (Phakey, 1969) and radiating from inclusions (Nielsen and Foster, 1960) where they are created by stress due to lattice mismatch and differences in the thermal expansion coefficients of the quartz and inclusion.

High densities of dislocations without evidence of recovery, particularly sub-grain formation, are typical of structures in quartz that has been experimentally deformed at strain rates of 10^{-5} to 10^{-7} sec^{-1} and temperatures in the range 600 °C to 900 °C (see above references). Higher strain rates ($> 10^{-5}$ sec^{-1}) and temperatures below 500 °C favour more brittle deformation modes such as deformation twinning (McLaren *et al.*, 1967) and cataclasism (Borg *et al.*, 1960). Recrystallization during experimental deformation occurs without a preceding recovery phase (Ardell, Christie and Tullis, 1974).

The above structures differ from those found in metamorphic quartzites that have been deformed under 'normal' tectonic conditions, i.e. strain rates of around 10^{-15} and differential stresses of less than 500 bar. As shown in Figures 1 A and B the individual grains in these rocks show ample evidence of recovery and the dislocation substructures are typical of crept and hot worked metals and ceramics (discussed by White, 1974b). High densities of dislocations (Figure 4 B below) with less evidence of recovery plus deformation twins, fractured and ground quartz and glass are found in some high stress, high strain rate environments (see also White, 1973c). Their dislocation substructures are similar to those in experimentally deformed quartz and in cold worked metals and ceramics. The difference between deformation structures in tectonically and experimentally deformed quartz suggests that experimentally determined flow laws should not be extrapolated to slower strain rate tectonic conditions and may only be applicable to limited high strain rate natural environments. However, this does not mean that experimental studies are of limited or little geological value. They provide vital data on dislocation dynamics, the effect of variations in deformation parameters on deformation mechanisms, operative slip systems and mechanical behaviour, all of which are required if natural deformation processes are to be understood.

2.1 Relationship between Dislocation Substructures and Optical Strain Features

The plasticity of quartz during tectonism has often been commented upon and attributed to dislocation processes (e.g. Spry, p. 58, 1969). The reason for this is the presence of intracrystalline strain features; namely undulatory extinction, deformation bands and lamellae and sub-grains which can be seen in a petrological microscope. These were recognized as being identical to features produced by creep in ceramics and metals (White, 1971, 1974b) and were shown to be the product of dislocation substructures (White, 1973a;

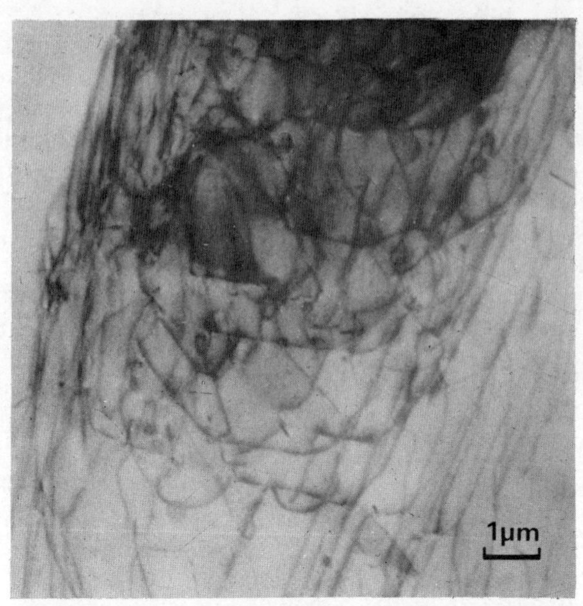

Figure 2. An electron micrograph of a band with high density of dislocations which causes banded undulatory extinction. 1000 kV.

White and Treagus, 1974). Undulatory extinction results when there is a predominance of dislocations of one sign or when groups of narrowly spaced sub-grain walls approximately parallel to prism planes have a similar sense of tilt. Both substructures reflect the presence of bending moments during tectonism. Banded undulatory extinction forms when the dislocations are arranged in bands of varying density (Figure 2) or when sub-grains of opposite tilt are interleaved. Deformation bands are a group of small elongate sub-grains contained within a large elongate sub-grain. In some cases, what appears optically as a sub-grain may in fact be a single sub-grain, but more often it contains smaller, slightly misorientated (less than 1 °) sub-grains, i.e. a sub-grain in sub-grain structure. Most deformation lamellae in quartz tectonites are narrow sub-grains or the decorated walls of a sub-grain. However, in rapidly deformed quartz they are deformation twins and shear fractures which are decorated by glass or debris along the slip planes (White, 1973a,c), and are similar to many lamellae produced in quartz experimentally. Other lamellae in laboratory deformed quartz are narrow zones of high dislocation densities and are morphologically similar to the structure in Figure 2 but are narrower (Hobbs, McLaren and Paterson, 1972). Lamellae related to sub-grain structures are rarely produced during dynamic laboratory deformations which is further evidence of the differences between experimentally and naturally deformed specimens. White (1973c) has stressed that lamellae in tectonites cannot be used to determine paleo-stress orientations or slip planes without confirming by electron microscopy that they have been produced by slip. Thus some conclusions derived from the assumption that all lamellae are slip planes may be invalid.

2.2 Development of Dislocation Substructures and Optical Strain Features

White (1974b) showed that intracrystalline optical strain features as well as recovered dislocation structures are typical of those produced by dislocation creep in metals and ceramics, and used this to outline the progressive formation of dislocation substructures and accompanying optical effects in naturally deformed quartz. He suggested that they develop during the initial stages of deformation. High densities of dislocations are created during the first increments of plastic strain and cause an initial phase of rapid strain hardening (primary creep). Recovery is initiated when a critical dislocation density, which is dependent upon both the temperature and stress, is reached and transient creep commences. Generally the first sub-grains to form are elongated and narrow, with most dislocation walls approximately parallel to prism or basal planes. They widen and may give way to equidimensional sub-grains, especially at high temperatures which also favour the rapid attainment of steady state creep conditions. A commonly encountered structure is the coexistence of equidimensional and elongate sub-grains. Steady state creep which is the dynamic equilibrium between strain hardening processes (namely dislocation generation and entanglement) and recovery (disentanglement, dislocation annihilation and the arrangement of dislocations in structures without associated long range stress fields) results when a stable dislocation structure has developed. The steady state sub-grain size and the internal unbound dislocation density within the sub-grains are stress, and perhaps temperature, dependent (see Section 2.4). The only subsequent alteration in the substructure is an increase in the misorientation between adjacent sub-grains with increasing strain until eventually recrystallization occurs.

Undulatory extinction forms in the initial period of deformation and is followed by deformation bands and lamellae and finally a well delineated sub-grain structure. Undulatory extinction will always tend to be present because of the unbound dislocations in the sub-grain interiors and the bending common in tectonic deformations. All of the above features develop in the initial stages of strain and will remain until recrystallization occurs, hence their abundance in quartz tectonites.

The presence or absence of certain intracrystalline effects may prove to be indicative of the stage of creep deformation reached by a given quartzite. For example it is unlikely that a quartzite in which only undulatory extinction and bands have developed has been subjected to steady state dislocation creep whereas a well developed sub-grain structure with sub-grains of a smaller size should indicate a steady state deformation.

2.3 Dynamic Recrystallization.

There are currently three known processes by which new grains form and

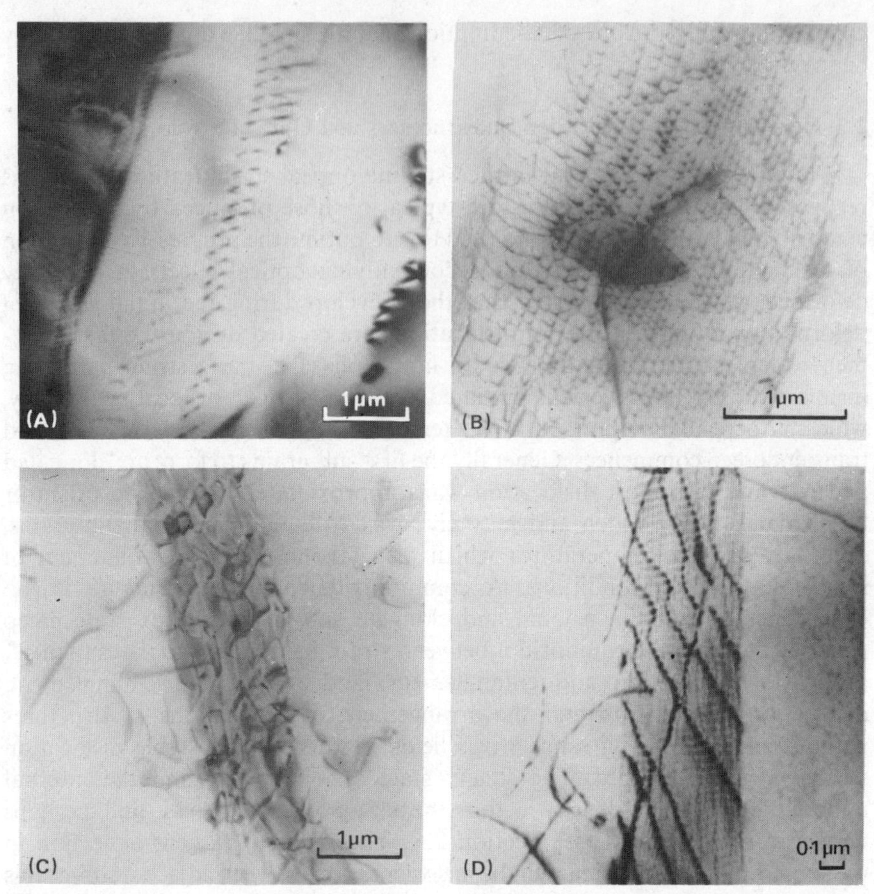

Figure 3. Electron micrographs showing the typical structures of boundaries in quartz. (A) A low angle tilt boundary with a misorientation of less than 1 °, along a prism plane. 1000 kV. (B) A low angle twist boundary. 1000 kV. (C) A complex tilt-twist low angle boundary typical of those causing misorientations of between 1 ° and 10 ° in deformed quartz. 1000 kV. (D) A high angle boundary causing misorientations greater than 10 °. A fine meshwork of structural dislocations can be seen. 1000 kV.

initiate dynamic recrystallization in quartzites (White, 1973b; Wilson 1973; Ardell, Christie and Tullis 1974). They are sub-grain rotation, formation of viable strain free nuclei in highly deformed areas, and grain boundary bulging.

Dislocations are continually produced within sub-grains during steady state deformation. Many are annihilated during reactions with other dislocations but some enter the walls of the sub-grains and become incorporated into them, increasing the complexity of the constituent dislocation network and the misorientation across the walls. The sub-grain boundary gradually changes from a sessile low angle boundary into a high angle one (Figure 3) with a structure identical to a grain boundary (Bollman, p. 187, 1970) that has the ability to move into adjacent grains and possibly trigger off grain boundary bulging. Thus the sub-grains gradually misorientate until they become recognizable separate entities, i.e. small recrystallized grains. Sub-grains misorientate either individually or in clusters as in crept ceramics (see Figure 1 and 3 in Streb and Reppich, 1974; Reppich, 1971). It is not clear why there is this difference but it may be that higher stresses favour the former process. Apparent clustering can occur by the growth or coalescence of sub-grains bounded by high angle boundaries and the subsequent development of a second generation of sub-grains with initial low angle boundaries in their interiors. These will also misorientate and can lead to a further phase of recrystallization.

In the second process, strain free nuclei develop in or adjacent to grain boundaries of quartz grains with high densities of unrecovered dislocations or in highly misorientated areas of the grains such as deformation or kink bands. New grains attributable to this process have been observed in experimentally deformed quartz (Ardell, Christie and Tullis, 1974) and also occur in quartz nodules in cataclasites along the Outer Hebrides Thrust, N. Uist (Figure 4). In the latter case elongate sub-grains and high densities of dislocations formed in the parent grains. New grains, with orientations differing from their hosts, were much smaller than the sub-grains formed in the grain boundary areas (Figure 4), and were deformed by the continuing deformation. A similar process is observed in plagioclase (see Section 3). The exact mechanism by which the nuclei develop has not been observed. This process may be limited to high strain rate, non-steady state deformations in which the rate of dislocation generation exceeds that for recovery and during which the dislocation density continually increases and eventually triggers recrystallization.

Grain boundary bulging occurs when a grain boundary moves into a neighbouring grain (Figure 4) leaving in its wake a strain-free bulge which becomes the nucleus for a new grain (see Figures 2 and 3 in Bailey and Hirsch, 1962). Certain grain boundary relationships appear to be more prone to bulging than others, giving rise to a preferred orientation of the nuclei. As noted above, it is feasible for bulging to occur during creep in conjunction with sub-grain rotation.

The new grains formed by sub-grain rotation will initially have a crystallographic orientation very similar to that of their parents, whereas the remaining processes develop a new fabric which will be controlled by the old host grains. The fabric formed prior to recrystallization will depend upon temperature and stress which may influence the operative slip systems and the total strain until recrystallization occurs (see also Tullis, Christie and Griggs, 1973). A simplistic outline of fabric development in a quartzite can be given if it is assumed that basal slip is dominant and that grains are spherical and without an initial fabric. As the grains deform and elongate they will rotate until their basal plane becomes approximately parallel to the tensile axis σ_1). As a result, c-axes of the deformed grains will form a single maximum perpendicular to the foliation plane if the strain ellipsoid for the deformation is pancake shaped. A cigar shaped ellipsoid should result in a great circle of

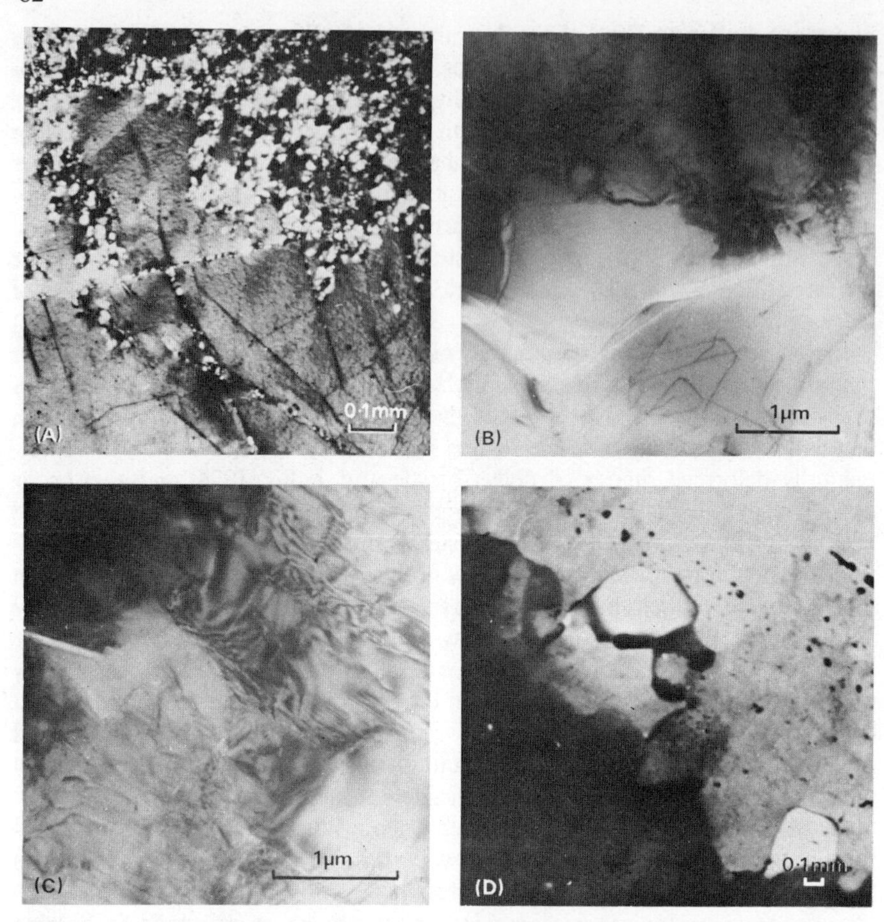

Figure 4. Recrystallization nuclei in quartz and andesine. (A) An optical micrograph of partially recrystallized grains in a quartzite module in a cataclasite. Note the absence of a well developed sub-grain structure in the old grain. (B) An electron micrograph across the contact between old and new grains. A small strain free nucleus can be seen growing into a deformed old grain (dark area) containing high densities of dislocations. Dislocations develop in the new grains away from the contact. 1000 kV. (C) Strain free nuclei developing in deformed andesine. 1000 kV. (D) An optical micrograph illustrating the development of new grains by grain boundary bulging in a deformed quartzite. The light areas are new grains

c-axes perpendicular to the lineation. Thus σ_2 will influence the fabric as demonstrated by Kern and Braun (1973) with experimentally deformed rock salt. Normally the active slip conditions are far more complex but nevertheless the fabric can be determined by the Taylor approach (see Wilson, 1973) as illustrated by the work of Bhattacharyya and Pasayat (1968) and Siemes (1973). The net result is the development of a fabric by glide, the intensity of which may be related to the amount of strain before recrystallization. Thus we find that ribboned quartz often exhibits a simple sharp c-axes fabric (Wilson and Glass, 1974). The basic glide fabric will be preserved if sub-grain rotation is the operative mechanism, or replaced if one of the other mechanisms is responsible for recrystallization: neither type will produce a random fabric. Processes other than sub-grain rotation will produce a host control nucleation (Hobbs, 1968) which has been reported in natural quartzites (Ransom, 1971; Wilson, 1973).

Continuing deformation after recrystallization will modify the new fabrics. They will be randomized by grain boundary sliding and rotation, which are favoured by a fine grain size, whereas grain growth of favourably orientated grains will produce a different fabric (see Wilson, 1973). Glide and elongation of the new grains may sharpen or alter the fabric, depending upon whether the operative slip systems do or do not change. Further deformation with the generation of dislocations in the new grains will inhibit grain growth and will produce mylonites in high stress environments.

2.4 Effect of Deformation Parameters on Substructures

Three main parameters, stress, temperature and 'lattice water', are known to affect the dislocation structures, and therefore the intracrystalline optical strain features in quartz.

The effects of stress can be seen by comparing the defects in mylonitized quartz with those in normal quartz tectonites. Smaller equant sub-grains (White, 1974a) and narrower elongate sub-grains (compare Figures 3 and 7 in White, 1973a) form in mylonites. The internal dislocation densities within mylonites can be higher than in their unmylonitized equivalents, but there are however many exceptions. Similar trends occur in crept and hot-worked metals and ceramics where it has been found that the relationship between stress, σ, and average sub-grain diameter, d, is

$$d \propto B\left(\frac{\sigma}{\mu}\right)^{-m}$$

where μ is the shear modulus and B and m are constants, and the reported values of m range from 0.3 to 1 (Barrett and Nix, 1965; Jonas, Sellars and Tegart, 1969; Lagneborg, 1972). The density of dislocations, ρ, within the sub-grains increases with stress, particularly the internal stress which for low stress environments equals the applied stress,

$$\rho = A\left(\frac{\sigma}{\mu}\right)^{n}$$

where A and n are constants; n has a theoretical value of 2 or 3 (Barrett and Nix, 1965; Weertman and Weertman, 1965; Lagneborg, 1972). However, most experimental determinations have produced a value of about 1 although values as low as 0.33 have been reported (Orlova, Tobolova and Cadek, 1972; Orlova, Pahutova and Cadek, 1972; Orlova and Cadek, 1973; Poirer, 1972). These relationships only apply to steady state deformation. It may prove

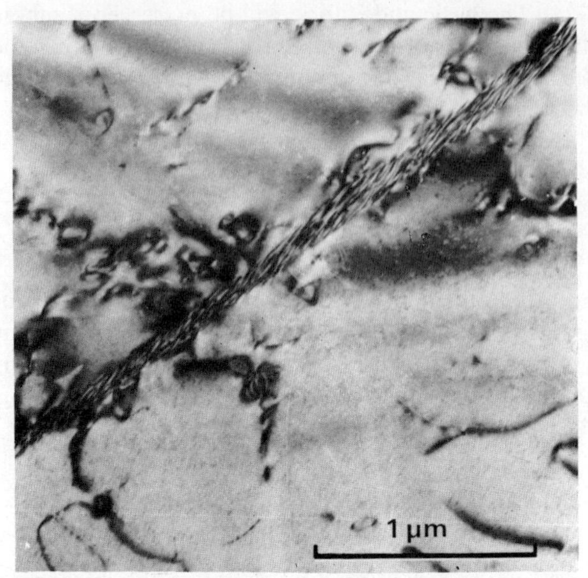

Figure 5. An electron micrograph of a sub-grain wall and dislocations in a quartzite with an annealed texture (compare Figure 1a, White 1973a). 1000 kV.

possible to use sub-grain sizes and dislocation densities in quartz to estimate the magnitude of tectonic stresses. Raleigh and Kirby (1970) and Goetze and Kohlstedt (1973) have recently attempted to determine mantle stresses using this technique. However, the discrepancy between theory and experiment and the inconsistencies in experimental results must be resolved first, and so must the disagreement over the influence of temperature at constant stress on sub-grain size and dislocation density. Streb and Reppich (1973) and Orlova *et al.* (1972a) found that the sub-grain size increased and the internal dislocation density decreased with temperature. However, other researchers (see Challenger and Moteff, 1973) have not detected any influence of temperature. Two additional potential sources of inaccuracy in tectonites are the possible production of a sub-grain structure with low density of internal dislocations, firstly by hydrostatic stresses during post tectonic annealing (Figure 5) and secondly during Nabarro–Herring creep (Passmore, Duff and Vasilos, 1966). The last point shows that dislocation substructures will develop during a deformation in which dislocation processes are not the main mechanism.

Temperature effects can be studied by taking quartz deformed under normal conditions from various metamorphic grades. Sub-grains tend to be equidimensional in high grade rocks and elongate in low grade quartzites, especially if the low grade deformation occurred during retrogression (see White, 1973a, 1974b). The densities of unbound dislocations may decrease as temperature increases but this is not clear. Both observations are consistent with metallurgical observations. The effects of temperature are complicated by an increase in the 'lattice water' content of quartz which aids recovery and has an effect rather similar to an increase in temperature (White, 1974a). Hence equidimensional sub-grains can form in low temperature wet environments (Figure 5; White, 1973).

3. DISLOCATION SUBSTRUCTURES IN PLAGIOCLASE

Deformed plagioclases show a marked difference when they are compared with quartz (see White, 1974c). Sub-grains are rare. They have been reported in albites (Lorimer, Champness and Spooner, 1972) but are not present to any extent in oligoclases, andesines and labradorites (White, 1974c and d). The dislocations exhibit little evidence of recovery being predominantly pure edge and especially pure screw. Albite twins are ubiquitous with pericline deformation twins restricted to the most deformed areas. Although the twins inhibit dislocation movement, the absence of recovery in relatively twin free areas indicates that another factor exists. It has been suggested (White, 1974c) that this is the presence of the $I\bar{1}$ and $I\bar{1}^*$ superlattices (McLaren and Marshall, 1974) which exist in low temperature intermediate plagioclases. Super-dislocations, i.e. two unit dislocations separated by a zone of disorder (an anti-phase boundary), form in superlattices. They climb more slowly than single dislocations and inhibit recovery. However, recovery will be feasible during high temperature deformations above 500 °C when the $I\bar{1}$ or $I\bar{1}^*$ superlattice is wholly or partially disordered. Recovery may be possible in albite which does not form a superlattice. Recrystallization is not preceded by a marked recovery phase and occurs by the formation of strain free nuclei (Figure 4) in the most deformed areas of parent grains. Chemical changes may accompany recrystallization, particularly if this has occurred during retrogression, and may aid nucleation.

4. THE DISLOCATION SUBSTRUCTURES IN OTHER MINERALS

Several electron microscopy studies have been made of olivine (Boland, McLaren and Hobbs, 1971; Green and Radcliffe, 1972; Olsen and Birkeland, 1973; Boland, 1974; Guegen and Boullier, this volume; Kohlstedt and Goetze, this volume). Tectonically deformed olivine exhibits similar dislocation substructures to those described in section 2 for quartz. There are few data available for carbonates. Preliminary results however, have shown that at low temperatures calcite deforms by twinning and dislocation processes with twinning decreasing and dislocation processes and recovery increasing at higher temperatures (Barber and Wenk, 1973; White, 1974d). The dislocations in magnesian limestones have associated stacking faults and in non-stoichiometric dolomites there are large densities of small strain centres. Recovery appears to be inhibited in both cases (White, 1974d). Orthopyroxenes (Green and Radcliffe, 1972; Kohlstedt and Vander Sande, 1973) and clinopyroxenes have partial dislocations separated by a stacking fault (Figure 6) which also inhibit recovery.

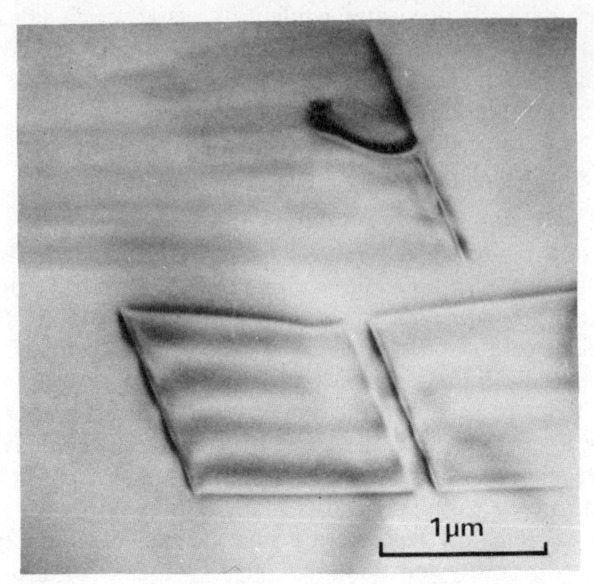

Figure 6. An electron micrograph of partial dislocations separated by stacking faults in clinopyroxene. 1000 kV.

5. THE EXTENT OF DISLOCATION CREEP PROCESSES IN CRUSTAL DEFORMATION

Dislocation structures indicative of creep or hot working are found in the constituent grains of quartzites which have been deformed at low strain rates by normal tectonic stresses from greenschist through to and including granulite facies metamorphic grades (White, 1973a, 1974b). Diffusion processes may occur in low temperature wet environments especially in impure quartzites and may also account for the elongation of quartz grains in slates. Pressure shadows and overgrowths are good indications that such processes have been operative. However, dislocation mechanisms cannot be excluded unless it can be shown that the clastic, or host, grains are free from dislocations or at least free from intracrystalline strain features. The quartz in fibrous pressure shadows contain few dislocations but abundant Brazil twins (Figure 7) which are defects characteristics of solution or diffusion growth. The elongation of quartz grains in many slates is produced by overgrowths which contain abundant small inclusions of phyllosilicates (see Figure 6, White, 1974a). The host grain is free, or almost free, of dislocations. Both features are probably the products of water assisted diffusive processes, which include pressure solution and Riecke process in low temperature environments. Similar features in high temperature rocks are more likely to result from solid state grain boundary diffusion (Coble creep) or lattice diffusion (Nabarro–Herring creep). Recent reports of stress induced geochemical changes in rocks (Berglund and Ekstrom, 1973; Parker, 1973; Stephenson, 1974) indicate that extensive solid state dif-

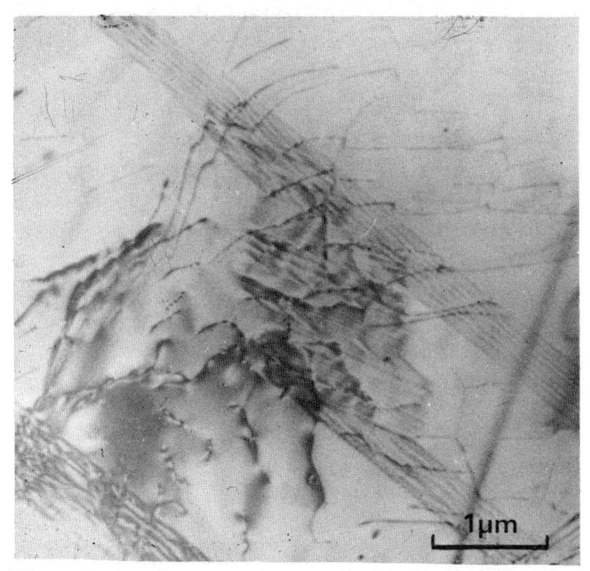

Figure 7. An electron micrograph showing Brazil twin development in a fibrous quartz overgrowth in a low grade quartzite. 1000 kV.

fusion can occur in high grade rocks. All diffusive processes should preserve the existent crystallographic fabric (Paterson, 1973), i.e. the diagenetic fabric will be preserved in low grade rocks. To summarize, present indications for medium and coarse grained quartzites are that low temperature wet environments favour water-assisted diffusive processes and at high temperatures solid state diffusion in the form of either Coble or Nabarro–Herring creep is predominant. Dislocation creep processes are dominant in the intermediate temperature range, especially from mid-greenschist to upper amphibolite grades, but also occur to some extent at lower and higher grades.

6. RELATIONSHIP BETWEEN CREEP RATE AND STRESS

Diffusion processes give a linear relationship between steady state strain rate $\dot{\varepsilon}$, and stress, σ, of the type

$$\dot{\varepsilon} = \frac{B_{nh}\sigma D_l \Omega}{d^2 kT} \text{ for Nabarro–Herring}$$

creep or

$$\dot{\varepsilon} = \frac{B_c \sigma D_{gb} \omega \Omega}{d^3 kT} \text{ for Coble creep.}$$

where B_{nh} and B_c are structure constants, D_l and D_{gb} are the lattice and grain boundary diffusivities, respectively, of the slowest moving ions (which in silicates are most likely to be oxygens), Ω the atomic volume which is pressure dependent, ω the grain boundary width, d the average grain diameter, k Boltzmann's constant, T temperature in K; thus in both cases

$$\dot{\varepsilon} \propto \sigma$$

Water-assisted diffusive processes, including pressure solution, are akin to Coble creep (Elliot, 1973). The strain rate of impure quartzites at low temperature and all quartzites at high temperatures should be linearly proportional to the stress.

There are several theories for steady state dislocation creep and hot working (see Jonas, Sellars and Tegart, 1969; Lagneborg, 1972). They can be divided into three main mechanisms: dislocation climb, dislocation jog, and recovery–strain hardening. All give a relationship of the type

$$\dot{\varepsilon} \propto \left(\frac{\sigma^n}{\mu^{n-1}} \right)$$

where n ranges in value from 3 to 5 depending upon the particular theory. There has been a tendency in the geological literature to concentrate on climb theories (Weertman, 1970; Raleigh and Kirby, 1970; Heard, 1972; Kirby and Raleigh, 1973), but there is no unique dislocation creep theory that unifies all experimental observations; the more studies reported, the greater the controversy and modification of existing theories. Recently Stang, Nix and Barrett (1973) were unable to reconcile their observations on the high temperature creep of an iron–silicon alloy with any current creep theory. A part of the problem may lie in the existence of unknown deformation mechanisms (Ashby, 1972). Another source of confusion has been that many individual theories in each of the three categories were formulated without attention to the actual dislocation structures formed during creep. Jonas *et al.* (1969) and Lagneborg (1972) critically reviewed dislocation creep mechanisms and both concluded that the recovery–strain hardening theories were the most feasible. The essence of these is that a balance exists between strain hardening and strain softening during steady state creep. They can explain most observed phenomena, e.g. the build-up of dislocations during primary creep, associate transient creep with the initiation of recovery, predict constant dislocation structures during secondary creep and finally take into account the sub-grain structures, effects of alloying and inclusions, changes in stoichiometry and stacking fault energy. The occurrence of this mechanism has been confirmed recently (Henderson–Brown and Hale, 1974) during *in-situ* creep experiments in a high voltage electron microscope. It is the only process that actually has been observed during dislocation creep.

Grain boundary sliding is another mechanism capable of producing a steady state flow with $\dot{\varepsilon} \propto \sigma^n$ where n is 1 if the sliding is by diffusion processes (Gifkins, 1968) or 2 if by dislocation glide in grain boundaries (Langdon, 1970). It is particularly common in fine grained materials and can produce superplasticity

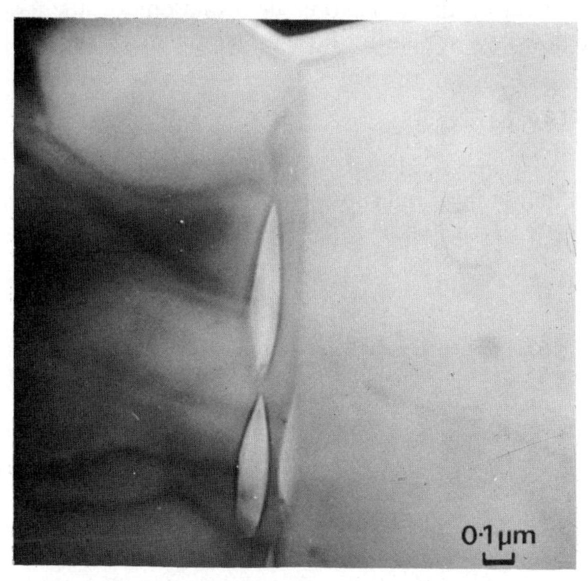

Figure 8. An electron micrograph showing void (bubble) development along a grain boundary in a fine grained quartz mylonite. 1000 kV.

during creep (Gifkins, 1967). Fine grained mylonites have grain boundaries containing voids or bubbles (Figure 8), which must weaken the boundaries, especially if the pore fluid pressure is high. Grain boundary sliding may be an important deformation mechanism in all mylonites and in any fine grained quartzites which have weak grain boundaries.

The above discussion also applies to olivine. The plasticity of both of these minerals reflects the ease with which dislocation recovery (softening) processes occur within them. Minerals such as plagioclase and pyroxenes have dislocation configurations which cannot recover easily and consequently are more brittle. The effect that non-stoichiometry has upon dislocation structures in carbonates is interesting, as it indicates that their rheological properties may be controlled by P_{CO_2}. Will the oxygen partial pressure affect the mechanical behaviour of silicates? This is possible. The diffusion creep behaviour of both Fe_2O_3 and doped Al_2O_3 and MgO are dependent upon P_{o2} (Pascoe and Hay, 1973; Gordon, 1973; Hollenberg and Gordon, 1973).

Furthermore, it must be remembered that present creep theories were formulated for simple metals and pure ceramics, detailed studies of more complex materials and polyphase materials may produce surprises. There are indications that these may have higher values of the stress exponent n for dislocation creep (cit. Davies *et al.*, 1973). Similar n values have been found for carbonate and sulphide rock samples (Hard and Raleigh, 1972; Rutter, 1974; Atkinson, 1972). Therefore, are we being realistic when we attempt to evaluate the rheological properties of rocks during tectonic deformations on the assumption that they will have flow laws similar to those for simple materials?

ACKNOWLEDGEMENTS

The author was supported by a Royal Commission for the Exhibition of 1851 Research Fellowship, and is presently a Royal Society/Mr. and Mrs. John Jaffé Donation Research Fellow.

REFERENCES

Atkinson, B. K. (1972). *Ph.D. Thesis*, Univ. of London.
Ardell, A. J.,Christie, J. M. and Tullis, J. A. (1974). *Cryst. Lattice Defects*, (in press)
Ashby, M. F. (1972). *Acta Met.*, **20**, 887–897.
Bailey, P. and Hirsch, P. B. (1962). *Proc. Roy. Soc.*, **267A**, 11–30.
Barber, D. J. and Wenk, H. R. (1973). *J. Mater. Sci.*, **8**, 500–508.
Barrett, C. R. and Nix, W. D. (1965). *Acta Met.*, **13**, 1247–1258.
Berglund, S. and Ekstrom, T. K. (1974). *Lithos*, **7**, 1–6.
Bhattacharyya, D. S. and Pasayat, S. (1968). *Tectonophysics*, **5**, 303–314.
Boland, J. N. (1964). In Swann, P. R. *et al.* (ed.) *High Voltage Electron Microscopy*, Academic Press, London, 312–316.
Boland, J. N., McLaren, A. C., and Hobbs, B. E. (1971). *Contr. Miner. and Petrol.*, **30**, 53–63.
Bollman, W. (1970). *Crystal Defects and Crystalline Interfaces*. Springer–Verlag, Berlin.
Borg, I., Friedman, M. Handin, J. and Higgs, D. V. (1960). In Griggs, D. and Handin, J. (ed.) *Rock Deformation*. Geol. Soc. Amer. Memoir, **79**, 133–141.
Challenger, K. D. and Moteff, J. (1973). *Met. Trans.*, **4**, 749–755 (1973).
Davies, P. W., Nelmes, G., Williams, K. R. and Wilshire, B., *Metal. Sci. J.*, **7**, 87–92.
Elliott, D. (1973). *Geol. Soc. Amer. Bull.*, **84**, 2645–2654.
Gifkins, R. C. (1967). *J. Inst. Metals*, **95**, 373–377.
Gifkins, R. C. (1968). *J. Amer. Cer. Soc.*, **51**, 69–72.
Goetze, C. and Kohlstedt, D. L. (1973). *J. Geophys. Res.*, **78**, 5961–5971.
Gordon, R. S. (1973). *J. Amer. Cer. Soc.*, **56**, 147–152.
Green, H. W. and Radcliffe, S. V. (1972). In Heard, H. C. *et al.* (ed.) *Fracture and Flow of Rocks*. Amer. Geophysical Union, Washington D.C., 139–156.
Heard, H. C. (1972). In Heard, H. C. *et al.* (ed.) *Fracture and Flow of Rocks*. Amer. Geophysical Union, Washington D.C., 191–209.
Heard, H. C. and Raleigh, C. B. (1972). *Geol. Soc. Amer. Bull.*, **83**, 935–956.
Henderson-Brown, M. and Hale, K. F. (1974). In Swann, P. R., *et al.* (ed.) *High Voltage Electron Microscopy*. Academic Press, London. 206–210.
Hobbs, B. E. (1968). *Tectonophysics*, **6**, 353–401.
Hobbs, B. E., McLaren, A. C. and Paterson, M. S. (1972). In Heard, H. C. *et al.* ed.) *Fracture and Flow of Rocks*. Amer. Geophysical Union, Washington D.C., 29–53.
Hollenberg, G. W. and Gordon, R. S. (1973). *J. Amer. Cer. Soc.*, **56**, 140–147.
Jonas, J. J., Sellars, C. M., and Tegart, W. J. McG. (1969). *Met. Revs.*, **14**, 1–24.
Kern, H. and Brown, G. (1973). *Contrib. Mineral. and Petrol.*, **40**, 169–182.
Kohlstedt, D. L. and Vander Sande, J. B. (1973). *Contrib. Mineral. and Petrol.*, **42**, 81–92.
Kirby, S. H. and Raleigh, C. B. (1973). *Tectonophysics*, **19**, 165–194.
Lagneborg, R. (1972). *Internat. Met. Revs.*, **17**, 130–146.
Lang, A. R. and Miuscov, V. F. (1969). In Sheftal (ed.) *Growth of Crystals*. Consultants Bureau, New York.
Langdon, T. G. (1970). *Philos. Mag.*, **22**, 689–700.

Lorimer, G. W., Champness, P. E. and Spooner, E. T. C. (1972). *Nature Phys. Sci.*, **239**, 108.
McLaren, A. C., Retchford, J. A., Griggs, D. T. and Christie, J. M. (1967). *Phys. Stat. Sol.*, **19**, 631–644.
McLaren, A. C. and Retchford, J. A. (1969). *Phys. Stat. Sol.*, **33**, 657–668.
McLaren, A. C., Turner, R. G., Boland, J. N. and Hobbs, B. E. (1970). *Contrib. Mineral. and Petrol.*, **29**, 104–115.
McLaren, A. C., Osborne, C. F. and Saunders, L. A. (1971). *Phys. Stat. Sol.* **A4**, 235–247.
McLaren, A. C. and Hobbs, B. E. (1972). In Heard, H. C. *et al.* (ed.) *Fracture and Flow of Rocks*. Amer. Geophysical Union, Washington D. C. 55–66.
McLaren, A. C. and Marshall, D. B. (1974). *Contrib. Mineral. and Petrol.*, **44**, 237–249.
Nielsen, J. W. and Foster, F. G. (1960). *Amer. Mineral.*, **45**, 299–310.
Olsen, A. and Birkeland, T. (1973). *Contrib. Mineral. and Petrol.*, **42**, 147–157.
Orlová, A., Patuhova, M. and Cadek, J. (1972). *Philos. Mag.*, **25**, 865–877.
Orlová, A., Tabolova, Z. and Cadek, J. (1972). *Philos. Mag.*, **26**, 1263–1274.
Orlová, A. and Cadek, J. (1973). *Philos. Mag.*, **28**, 891–899.
Parker, R. B. (1974). *Geol. Soc. Amer. Bull.*, **85**, 11–14.
Pascoe, R. T. and Hay, K. A. (1973). *Philos. Mag.*, **27**, 897–914.
Passmore, E. M., Duff, R. H. and Vasilos, T. (1966). *J. Amer. Cer. Soc.*, **49**, 596–600.
Paterson, M. S. (1973). *Revs. Geophys. and Space Phys.*, **11**, 355–389.
Phakey, P. P. (1969). *Phys. Stat. Sol.*, **34**, 105–119.
Poirer, J. P. (1972). *Philos. Mag.*, **26**, 713–725.
Raleigh, B. C. and Kirby, S. H. (1970). *Mineral. Soc. Amer. Spec. Paper*, **3**, 113–121.
Ransom, D. M. (1971). *Mineral. Mag.*, **38**, 83–88.
Reppich, B. (1971). *J. Mater. Sci.*, **6**, 267–269.
Rutter, E. H. (1974). *Tectonophysics*, **22**, 311–334.
Seimes, H. (1974). *Contrib. Mineral. and Petrol.*, **43**, 149–157.
Spry, A. (1969). *Metamorphic Textures*. Pergamon, Oxford.
Stephansson, O. (1974). *Tectonophysics*, **22**, 233–251.
Strang, R. G., Nix, W. D. and Barrett, C. R. (1973). *Met. Trans.*, **4**, 1695–1699.
Streb, G. and Reppich B. (1973). *Phys. Stat. Sol.*, **A16**, 493–505.
Tullis, J., Christie, J. M., and Griggs, D. T., (1973). *Geol. Soc. Amer. Bull.*, **84**, 297–314.
Weertman, J. (1970). *Rev. Geophys.*, **8**, 145–168.
Weertman, J. and Weertman, J. R. (1970). In Cahn, R. W. (ed.) *Physical Metallurgy*, North Holland, Amsterdam, 810–813.
White, S. (1971). *Nature Phys. Sci.*, **234**, 175–177.
White, S. (1973a). *J. Mater. Sci.*, **8**, 490–499.
White, S. (1973b). *Nature*, **244**, 276–278.
White, S. (1973c). *Nature Phys. Sci.*, **245**, 26–28.
White, S. (1974a). In Swann, P. R. *et al.* (ed.) *High Voltage Electron Microscopy*. Academic Press, London, 317–322.
White, S. (1974b). *N. Jh. Miner.* In the press.
White, S. (1974c). Submitted to *Contrib. Mineral and Petrol.*
White S. (1974d). *Proc. Eight. Inter. Electron Mic. Conf., Canberra*, **1**, 482–483.
White, S. and Treagus, J. E. (1974). *N. Jb. Mineral.* In the press.
Wilson, C. J. L. (1973). *Tectonophysics*, **19**, 39–81.
Wilson, C. J. L. and Glass, J. (1974). *Geol. Soc. Amer. Bull.* In the press.

Deformation Textures in Nodules from Kimberlites

G. D. Borley
Department of Geology, Imperial College of Science and Technology, London

INTRODUCTION

Nodules that represent fragments of the earth's mantle from depths down to 200 km are brought to the surface in kimberlite pipes. The commonest types of nodule are: lherzolites (olivine + enstatite (or bronzite) + diopside ± pyrope garnet); harzburgites (olivine + enstatite (or bronzite) ± garnet); eclogites (garnet + Na/Al-rich clinopyroxene); clino- and orthopyroxenites. Many of these nodules are serpentinised and some contain late-formed or secondary phlogopite.

Petrographically the most important and interesting feature of the lherzolite nodules is the evidence they provide of deformation processes in the mantle, and examination of some eighty lherzolite nodules from Lesotho and S. African kimberlites suggests they fall into two broad categories showing different degrees of deformation.

Nodules in the first category are generally of medium to coarse-grained rocks (grain size from 1 to > 5 mm) which may or may not show foliation. Olivine and orthopyroxene show variable strain effects: undulose extinction, microscopic fractures, kink and deformation bands, deformation lamellae, slip-planes, and sub-grain development. Syntectonic and post-tectonic recovery is indicated by recrystallization and annealing. This group includes nodules that contain calcic-diopsides that re-equilibrated at temperatures between 900 °C and 1100 °C and at depths between 100 km and 150 km (Boyd and Nixon 1973a, b and Boyd 1973).

Nodules in the second category are of variable grain size and, again, they may or may not be foliated. They appear to have been deformed under more intense shearing-stresses than nodules in the first category, and deformation seems to have proceeded primarily through a combination of slip (especially in olivine) and sub-grain formation. Evidence of progressive stages of deformation can be discerned as well as recovery. Nodules in this group are known to contain increasingly sub-calcic diopsides that re-equilibrated at temperatures between 1150 °C and 1400 °C and at depths between 150 km and 180 km (ibid).

A number of pyroxenite nodules examined by the author do not fit easily into either of these two broad categories, but too few have so far been studied to permit a classification scheme to be put forward. But it is clear that some clinopyroxenites have been sheared, with consequent exsolution of orthopyroxene, or have deformed through the formation of deformation twin lamellae and have undergone low temperature re-equilibration rather than the high temperature re-equilibration typical of pyroxenes from the sheared lherzolite nodules of category 2 above (Borley and Suddaby, 1975).

Detailed descriptions of textures of lherzolite nodules in the two main categories are not numerous. The earliest and perhaps most complete, if not systematic, description of nodules from kimberlites was that of Williams (1932). Recently Boullier and Nicholas (1973) classified the lherzolite nodules from Lesotho kimberlites on the basis of their microscopic textures and fabrics as: 'coarse-grained' without lineation or foliation; 'tabular' olivine or enstatite textures, medium-grained; 'porphyroclastic', with large deformed porphyroclasts and undeformed smaller neoblasts; 'mosaic', with a single generation of small olivine neoblasts about 0.3 mm grain size). Nodules with the 'mosaic' texture are those referred to by Boyd and Nixon (1973), Johnson (1973), and others as highly-sheared varieties. The paper by Boullier and Nicholas includes detailed descriptions of textures of individual nodules and it therefore provides a valuable reference point for other petrographic observations. More recently Gueguen and Boullier (this volume) have re-examined some of the more highly-sheared nodules and have suggested that certain textures might have been caused by the superplastic flow of enstatite.

The nomenclature used in this paper differs slightly from that of Boullier and Nicholas (1973); the differences will be indicated in the discussion and are summarized in Table 1 below at the end of the petrography section. In general the nomenclature used follows that of Spry (1969) for metamorphic rocks, and is intended to show the gradational nature of deformation, particularly of nodules in category 2, and the textural affinities that deformed mantle nodules have with rocks deformed under crustal metamorphic conditions.

PETROGRAPHY

Category 1 Lherzolite Nodules:

These medium to coarse-grained nodules can be loosely divided into (a) Granular; (b) Granular with well-developed olivine and/or pyroxene subgrains; (c) Porphyroclastic-granular. The three divisions are not sharply defined and (c) is transitional to the more deformed porphyroclastic nodules of category 2.

Typical of (a) is a garnet–lherzolite from Bulfontein Mine. Both olivine

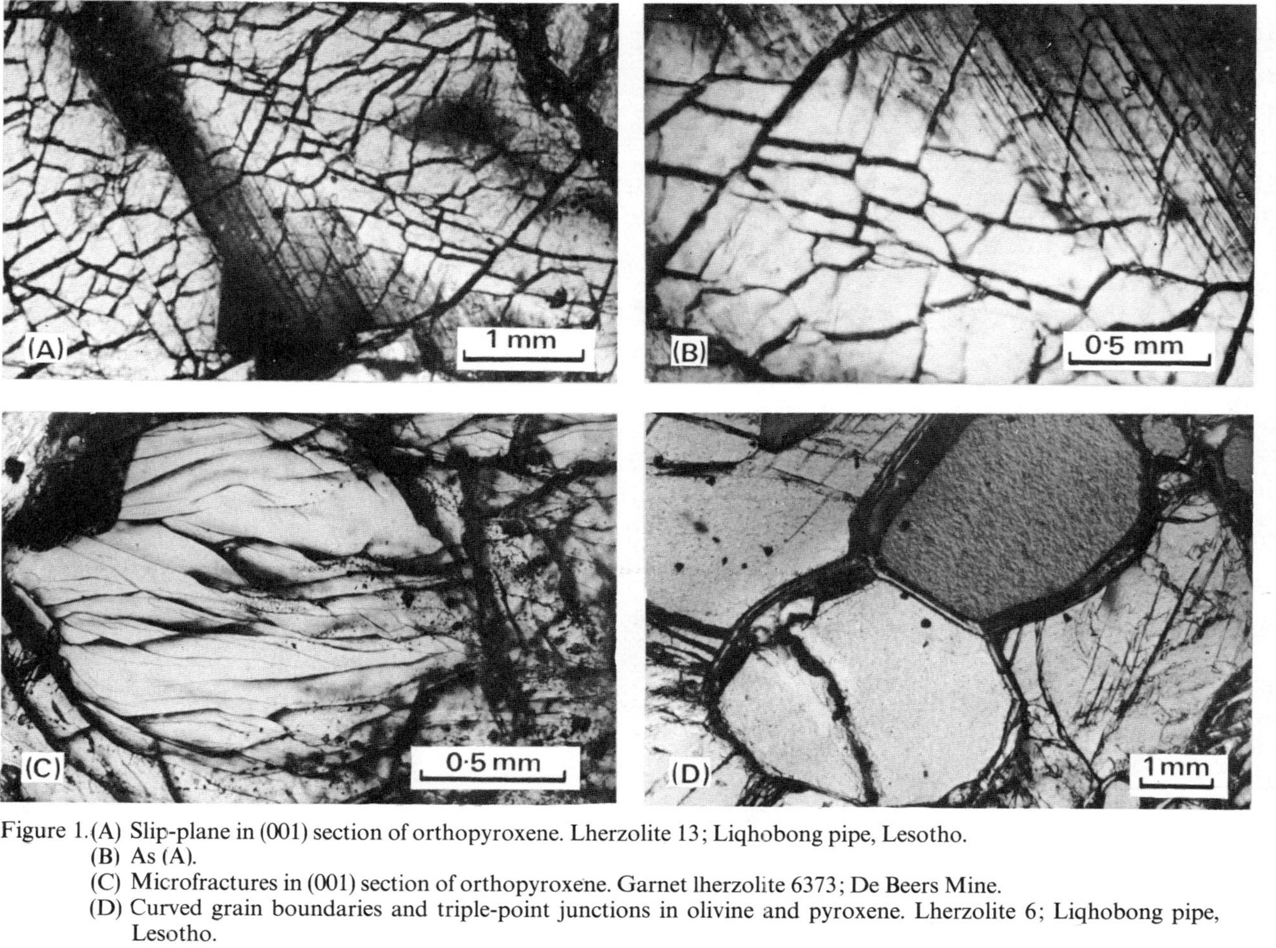

Figure 1. (A) Slip-plane in (001) section of orthopyroxene. Lherzolite 13; Liqhobong pipe, Lesotho.
(B) As (A).
(C) Microfractures in (001) section of orthopyroxene. Garnet lherzolite 6373; De Beers Mine.
(D) Curved grain boundaries and triple-point junctions in olivine and pyroxene. Lherzolite 6; Liqhobong pipe, Lesotho.

and pyroxene show strain effects with poor sub-grain development and there is evidence of recrystallization of some sub-grains. As in other nodules in both categories, recrystallization of sub-grains is accompanied by decreases in microscopic strain effects. In many of the granular nodules where sub-grains have developed from large crystals of olivine and pyroxene, the sub-grains are of highly irregular shape, following fractures that develop along the edges of deformation bands, especially in olivine. But in some examples under (b) above, the shapes of the sub-grains are more regular, being blocky, prismatic, or elongate-tabular. A well-developed slip-plane (probably (100)) may be seen, though rarely, in the orthopyroxene. This slip-plane is shown in Figures 1A and B; it is usually discontinuous and develops in only small parts of large (001) sections. Also present in (001) sections of orthopyroxene are numerous micro-fractures that intersect the main basal cleavages (Figure 1C). In addition to the strain effects that can be seen in individual minerals, these granular nodules frequently show crushing at the edges of large crystals, particularly of olivine, and some recrystallization of the resulting tabulae or prisms to strain-free grains has occurred.

A common feature of these nodules is that coarse, though irregularly sized and shaped, grains of olivine and pyroxene show evidence of annealing, with smooth and slightly curving grain boundaries. Triple point junctions are also seen but the angles at the triple points may deviate from 120 °C (Figure 1D). Textures of this type are seen in lherzolites from Liqhobong pipe (Lesotho), De Beers, and Bulfontein pipes.

Some nodules with textures transitional between the granular nodules of category 1 and the more deformed and variable types of category 2 can be described as porphyroclastic-granular. Good examples are two coarse-grained garnet–lherzolites from Bulfontein Mine, both of which show crushing of olivine grains, especially near grain boundaries. The sub-grains produced are generally either tabular or prismatic and some recrystallization and movement of the sub-grains along grain boundaries appears to have taken place.

Two minerals that are frequently present in these granular nodules, as well as in the nodules of category 2, have not so far been mentioned. The first is pyrope garnet which usually occurs as rounded grains that show little microscopic evidence of strain. Chemically, however, the garnet may show partial breakdown to green spinel, amphibole (possibly a richteritic variety), and phlogopite. The second mineral is phlogopite which, apart from being formed by breakdown of garnet, may be present in clusters of grains or as an interstitial phase in grains as small as 0.1 mm. The phlogopite generally shows strain effects, especially kinking and the development of conjoined deformation bands (Figure 2A). As the origin of phlogopite in these nodules is still debated, it might be worth commenting that the highly deformed nature of much of the phlogopite, even when present as a late interstitial phase, suggests that it may have formed in the mantle; accepting, of course, that the deformation of the nodules generally took place in the mantle and not whilst they were transported to the earth's surface in the kimberlite magma.

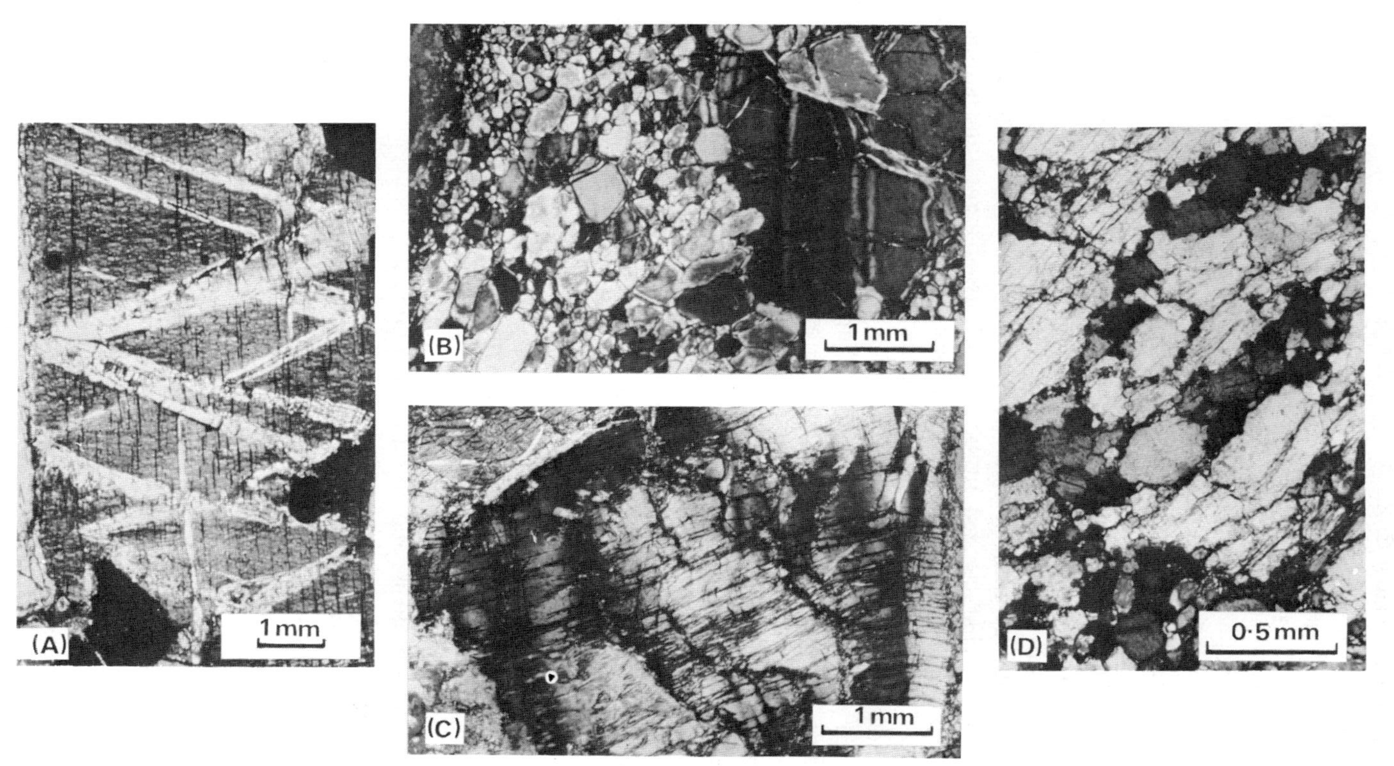

Figure 2. (A) Conjoined deformation bands in phlogopite. Lherzolite 6372; De Beers Mine.
(B) Strained and recrystallized olivine. Lherzolite 6376; De Beers Mine.
(C) Kink and deformation bands in orthopyroxene. Lherzolite 6376; De Beers Mine.
(D) Sub-grains in orthopyroxene. Nicols partly crossed. Lherzolite 6344; Bulfontein Mine.

Category 2 Lherzolite Nodules:

This group contains nodules that are texturally more complex than those previously described, and it includes some nodules classified as 'porphyroclastic' and those classified as 'mosaic' by Boullier and Nicholas (1973). As in the previous group of nodules individual minerals such as olivine and pyroxene show variable strain effects and recrystallization of sub-grains. But, in addition, increased shearing stress has resulted in a more extreme deformation; firstly of olivine and then, later, of enstatite and diopside. Several textural types have been recognized: (a) Porphyroclastic with granoblastic olivine; characterized by porphyroclasts of garnet, strained olivine, and pyroxene, set in a granoblastic olivine matrix of grain size 0.1 mm to 0.3 mm. (b) As (a) but with well-developed enstatite sub-grain mosaics. (c) Porphyroclastic with granoblastic-polygonal olivine; characterized by porphyroclasts of garnet and strained pyroxene set in a matrix of strain-free granoblastic-polygonal olivine of grain size about 0.1 mm. (d) Foliated porphyroclastic with granoblastic-polygonal olivine and enstatite; similar to type (c) but where increased stress appears to have been taken-up by enstatite and diopside rather that by the granoblastic olivine grains. The increased stress might have been due to a continuation of the shearing episode that produced nodules of type (c) or by a later shearing episode.

Textural types (c) and (d) are those referred to by Boullier and Nicholas (1973) as 'mosaic', and type (d) is referred to as indicating the superplastic flow of enstatite by Gueguen and Boullier (this volume). The term mosaic is rejected here in favour of granoblastic-polygonal to indicate the nature of the olivine texture. Mosaic is used in this paper to indicate a crystal sub-grain structure '... in which small adjacent blocks differ slightly in orientation. ...' (Spry 1969), the differences in orientation being demonstrated optically by small variations in birefringence of the grains when the nicols are only partly crossed.

As with the granular nodules of category 1, the sub-divisions in category 2 are not sharp and the textures are gradational, with a reduction in the number of porphyroclasts of olivine and enstatite from (a) to (d).

A number of examples of type (a) nodules have been examined. In all of then the deformed porphyroclasts of olivine are typified by wide deformation bands and lamellae, and there are signs of recrystallisation of sub-grains (Figure 2B). Many of the larger tabular sub-grains of olivine do in fact appear to have been derived by the fracture of olivine porphyroclasts along the boundaries of adjacent deformation bands. In some of these nodules the outlines of the original, coarse, olivine crystals can still be discerned enclosing relic olivine porphyroclasts and sub-grains, but in other nodules the olivine porphyroclasts merge gradually via detached, variably sized, sub-grains into the fine-grained granoblastic olivine matrix. Enstatite porphyroclasts show kinking, bending of cleavages and other evidence of plastic deformation (Figure 2C), and poor sub-grain development.

Textures similar to these are also present in nodules of type (b) but, in addi-

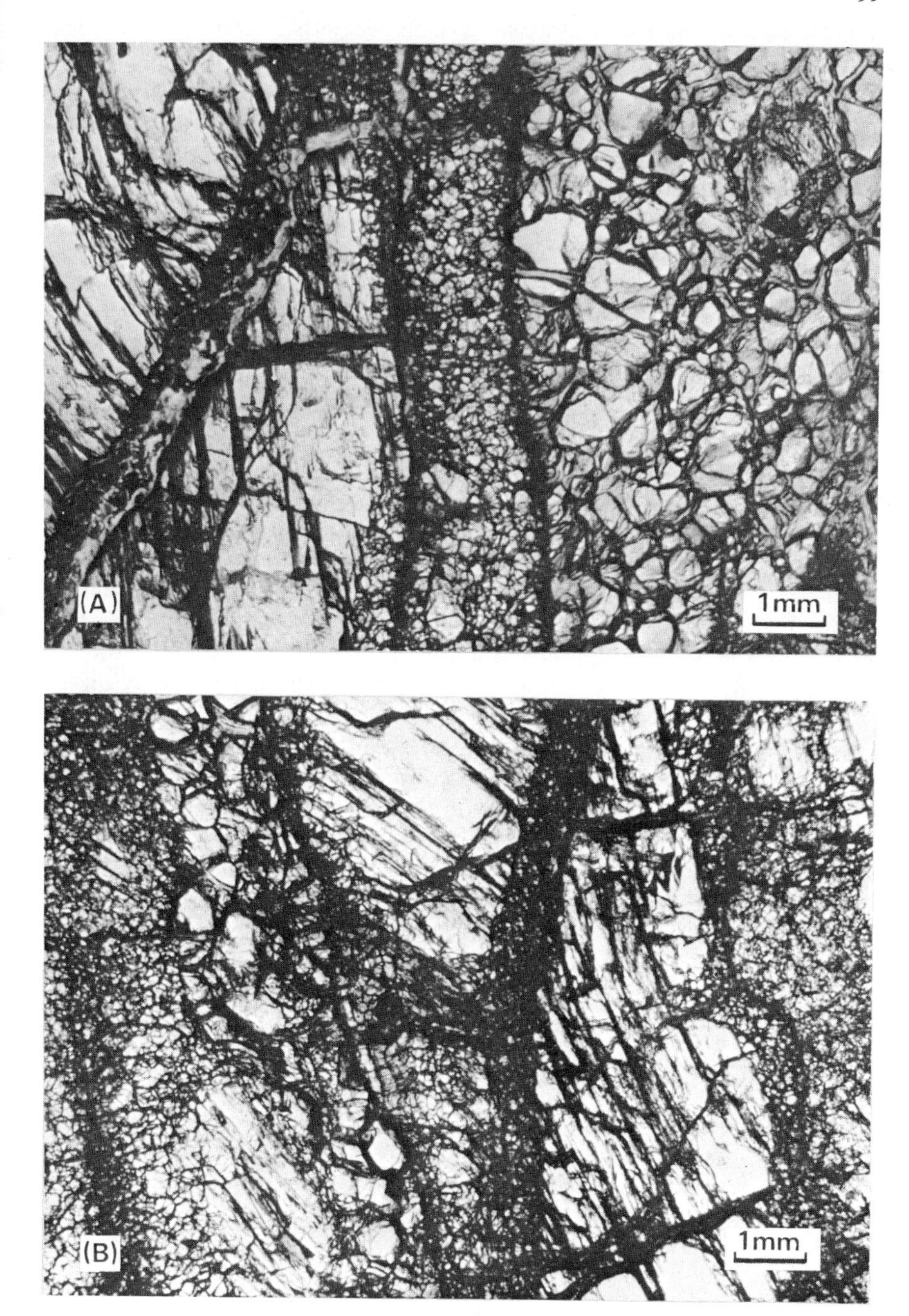

Figure 3. (A) Crushed orthopyroxene, on left, with outer area of fine-grains. Matrix olivine on right. Lherzolite 6364; Jagersfontein Mine.
(B) As (A) but without olivine matrix.

tion, enstatite shows a considerable development of sub-grain mosaics. This texture is seen particularly well in a garnet lherzolite from Bulfontein Mine (Figure 2D). The sub-grains (< 0.5 mm in size), looked at with partly crossed nicols, have the wavy, indefinite boundaries seen sometimes in syntectonically recrystallized minerals, and some of the sub-grains themselves show signs of strain. A feature of this nodule from Bulfontein Mine is that serpentine has, in some areas, totally replaced the olivine grains of the matrix so that the serpentine itself has a granular polygonal texture.

Nodules of type (c) differ from those of (a) and (b) in that olivine is no longer, or only rarely, present as a porphyroclast and is present only as a fine-grained (0.1 to 0.2 mm grain size) granoblastic matrix. The texture is classified as granoblastic polygonal to indicate the considerable degree of polygonization of the olivine, even though many of the olivine grains are now irregularly shaped due to replacement by serpentine.

Perhaps the most striking of the category 2 nodules are those of type (d) which give the appearance of having suffered a late shearing whose effects have been superimposed on those of the earlier deformation. A particularly good example is a garnet–lherzolite from Jagersfontein Mine, originally described by Williams (1932), although similar examples have been described by Boullier and Nicholas (1973) from Lesotho. In the Jagersfontein lherzolite the effects of the late shearing episode appear largely to be restricted to the porphyroclasts of enstatite and, to a lesser extent, diopside. The granoblastic olivine matrix does not appear to have been affected (Figure 3), except in so far as it seems probable that during shearing the olivine grains may have moved by grain boundary sliding (Stocker and Ashby, 1973). The main effects of the shearing-stress are to be seen in the granulation and drawing-out into lozenge- and spindle-shaped crush-zones of the original porphyroclasts of enstatite and diopside (Figure 3). Within some of these zones of deformation, small relic porphyroclasts of pyroxene merge outwards into diffuse, irregularly shaped sub-grains and, finally, in the case of enstatite only, into outer areas of polygonized sub-grains (Figure 4). These polygonal grains of enstatite are, on average 10 to 25 μm or less in size. Faint lines of fine-grained enstatite (which give the lherzolite its foliated appearance) may also run from the relic porphyroclasts into the olivine matrix. Gueguen and Boullier (this volume) consider the enstatite to have undergone superplastic flow, a phenomenon described during the deformation of some fine-grained metals but not previously suggested as taking place in silicates. Unlike the olivine grains of the matrix the fine polygonal enstatite grains are not, of course, separated by serpentine. Both the polygonal olivine and enstatite grains are strain-free. Pyrope garnet in the Jagersfontein lherzolite remained mechanically stable during all stages of deformation, but it clearly became chemically unstable and is frequently pseudomorphed by breakdown products that include small grains of phlogopite.

A summary of textures seen in the various nodules is given in Table 1, which also attempts to compare the nomenclature used with that of Boullier and Nicholas (1973).

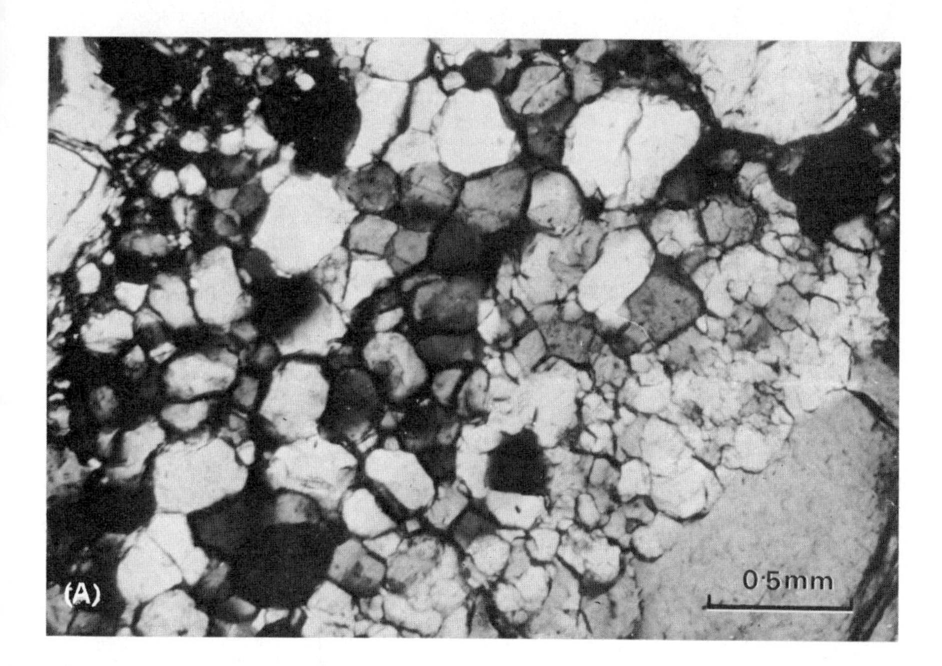

Figure 4. (A) Sub-grains of enstatite around relic porphyroclast. Lherzolite 6364; Jagersfontein Mine.
(B) Closer view of polygonal sub-grains of (A). Lherzolite 6364; Jagerfontein Mine.

Table 1. Summary of nodule textures and nomenclature comparison

Author's nomenclature	Boullier and Nicholas (1973) (based mainly on olivine texture/fabric)	Other comments
Category 1 nodules		
(a) Granular		
(b) Granular olivine with pyroxene sub-grains	Coarse-grained (> 6mm) no foliation or lineation	Annealing textures common. contain calcic-diopsides re-equilibrated between temperatures 900 °C to 1100 °C
(Both medium to coarse-grained)		
—	Tabular textures in olivine/pyroxene	
Porphyroclastic-granular Granulation particularly at grain boundaries	Porphyroclastic olivine with undeformed olivine neoblasts	Transitional to more deformed nodules
Category 2 nodules		
(a) Porphyroclastic with granoblastic olivine	Ditto	↓
(b) As (a) but with well-developed sub-grains of enstatite		Gradational textures. Reduction in number of olivine and enstatite porphyroclasts. Contain increasingly sub-calcic diopsides. Re-equilibrated between temperatures. 1150 °C to 1400 °C
(c) Porphyroclastic with single-generation of granoblastic-polygonal olivine	Mosaic texture to single generation of olivine neoblasts (0.3mm grain size)	
(d) Foliated porphyroclastic with granoblastic-polygonal olivine and enstatite	Ditto (Enstatite texture indicative of super-plastic flow—Gueguen and Boullier, this volume)	

ACKNOWLEDGEMENTS

The Royal Society is thanked for providing a travel grant to visit S. Africa and Lesotho to collect material. The author would also like to thank Prof. A. Nicholas, Anne-Marie Boullier and Yves Gueguen of Nantes University for discussions on nomenclature, and for showing her material in their collection.

REFERENCES

Borley, G. D. and Suddaby, P. (1975). *Mineral. Mag.*, **40**, 6–12.

Boullier, A.-M. and Nicholas, A. (1973). In: *Lesotho Kimberlites*, Lesotho National Dev. Corpn.

Boyd, F. R. (1973). *Geochim. Cosmoch. Acta*, **37**.

Boyd, F. R. and Nixon, P. (1973a). *Ext. Abstr. Inter. Conf. on Kimberlites*, Cape Town Univ.

Boyd, F. R. and Nixon, P. (1973b). In: *Lesotho Kimberlites*, Lesotho National Dev. Corpn.

Gueguen, Y. and Boullier, A.-M. (1976). This volume.

Johnson, J. (1973). *Ext. Abstr. Inter. Conf. on Kimberlites*, Cape Town Univ.

Spry, A. (1969). *Metamorphic Textures*, Pergamon Press.

Stocker, R. L. and Ashby, M. F. (1973). *Rev. Geophys. Space Physics*, **11**.

Williams, A. F. (1932). *Genesis of the Diamond* (2 vols.), Benn.

Microcracks in Rocks

G. Simmons and D. Richter
Department of Earth and Planetary Sciences
Massachusetts Institute of Technology
Cambridge, Massachusetts 02139

INTRODUCTION

Microcracks are very common, if not ubiquitous, in both terrestrial and lunar igneous rocks. They occur in most petrographic thin sections, in laboratory specimens used for the measurement of physical properties, and in rocks *in situ.* They are an important feature of most rocks. Yet they have received very little attention from petrographers (with the important exception of cracks in shocked rocks), perhaps because of the widespread but erroneous belief that all the cracks seen in thin sections were produced when the section was made. Our purpose in writing this article is to describe the petrographic characteristics of cracks, to demonstrate that many different kinds of cracks can be recognized on the basis of objective criteria, and that several types of cracks were produced by different and uniquely identifiable geologic processes.

The presence of cracks in *shocked* rocks has been recognized for some time. The petrographic criteria that are diagnostic of impact processes are summarized by Chao (1967a and b). That the rocks from several terrestrial impact craters contain microfractures was shown by several authors (for examples of the Ries crater see: Chao, 1968; Stöffler, 1965, 1966; von Engelhardt and Stöffler, 1968—for examples from several Canadian craters see: Robertson, Dence, and Vos, 1968; Dence, 1968). DeCarli (1968) and Short (1968) produced microfractures in rocks with shock waves in the laboratory. Short (1966 and 1968) described the microcracks produced in igneous rocks by nuclear explosions. Indeed the degree of fracturing was used by Chao (1967a), Dence (1968), and von Engelhardt and Stöffler (1968) as an important parameter in their qualitative scales of shock metamorphism. The presence of microcracks in lunar rocks has been known since the initial examinations of the Apollo 11 rocks by LSPET (1969) and has been described by many authors (e.g. Chao *et al.*, 1970; von Engelhardt *et al.*, 1970; Carter *et al.*, 1971).

The published literature on the detailed analysis of the origin of microcracks in rocks is very small. In two classic papers on a related subject, Rosenfeld and

Chase (1961) and Rosenfeld (1969) developed the concept that the piezobirefringent halos that surround certain inclusions (solid as well as liquid) in minerals could be used to estimate the pressure–temperature history of the rock. Those papers contain an excellent introduction to the general subject of stress fields around totally enclosed grains which is not repeated here. DeVore (1969) extended their results to calculate the principal thermal stresses in a restrained mineral and in a two-crystal system, and attributed the formation of microcracks to the thermal stress exceeding the brittle fracture strength.

We believe that the microcracks in rocks are caused by any process that raises *local* stresses above *local* strength. Simmons, Wang, Richter, and Todd (1973) calculated strains on the basis of a welded contact model and included specifically the effects due to changes of both pressure and temperature for different relative orientations of several mineral pairs.

A knowledge of microcracks, their characteristics and origin, is significant for both geologists and geophysicists. First, important clues to the tectonic history of many igneous and metamorphic rocks now appear to have been recorded by the microcracks. Thus, one aim of our work on cracks is to provide the basic data for the interpretation of the crack-record. Second, such physical properties of rocks as velocity of elastic waves, compressibility, and electrical conductivity depend strongly on the characteristics of cracks. Improved methods of characterizing the cracks may allow us to estimate better values of physical properties of rocks *in situ* on the basis of data obtained in the laboratory. Hence another aim of our research on cracks is to provide better methods with which to characterize the cracks in rocks. Third, the recognition that cracks in rocks *in situ* do heal is important in such diverse disciplines as petrology (for study of the sequence of tectonic processes), seismology (for earthquake prediction), and plate tectonics (for estimating and for interpreting the various physical properties of descending slabs as a function of time).

DEFINITIONS

Before describing the petrographic characteristics of microcracks, we wish to define certain terms that we have found useful. Abbreviations are shown in parentheses.

Microcrack: An opening that occurs in rocks and has one or two dimensions much smaller than the third. For flat microcracks, one dimension is much less than the other two and the width to length ratio, termed crack aspect ratio, must be less than 10^{-2} and is typically 10^{-3} to 10^{-5}. The length may be as great as metres, but typically is of the order of 100 μm or less. Specifically excluded are such features as joints, faults, vugs, and all large-scale openings in rocks.

Grain boundary crack (GBC): A microcrack associated with a grain boundary. A *coincident* GBC is one that coincides with the grain boundary. A *non-coincident* GBC is one that does *not* coincide with the grain boundary.

Intragrain crack: A microcrack that lies completely within a single grain and does not reach any grain boundary.

Intergrain crack: A microcrack that extends from a grain boundary into the two adjacent grains but does not extend to a second grain boundary.

Multigrain crack (MGC): A microcrack that crosses several grains and hence several grain boundaries.

THE ART OF OBSERVING MICROCRACKS IN ROCKS

Some cracks seen in ordinary petrographic thin sections are caused by the procedures commonly used to make thin sections. Such induced cracks can be quite useful. For example, many of the cleavage cracks so common in the thin sections of certain minerals and useful for identification and recognition of the particular mineral are probably caused by stressing the thin slices of rock both thermally and mechanically when preparing the thin section. However, because we are interested in studying the cracks that were present in the rock *in situ*, we would obviously like to avoid introducing new cracks.

For the study of cracks, we use rock sections 60 to 150 microns thick, i.e. two to five times as thick as the standard petrographic thin section. They are termed crack sections. The thicker section has two main advantages for the study of microcracks. First, it is stronger than the standard thin section and additional cracks are less likely to be produced by the sectioning process. Second, the extra thickness provides a third dimension and we are able to study cracks in three dimensions. Because the optical microscope has a finite depth of field, we can see clearly the features at one level in the crack section even though many other features are present above and below the focal plane. Thus the thick section and the limited depth of field of the microscope allows us to 'follow' the cracks through the section and therefore study them in three dimensions. The crack section is important because certain features of cracks simply cannot be recognized in a two-dimensional section!

Preparation of Crack Sections

The preparation of crack sections is rather simple but two precautions should be observed: (1) the rock slice should be kept near room temperature and (2) excess or unevenly distributed stress on the section should be avoided. Our present procedure for making crack sections is the following:

1. Saw a slice of rock approximately 20 × 40 × 3 mm on a diamond saw. We use a Highland Park model M4 with a paper towel placed in the tray to catch the slice.
2. Grind one face on a hand lap with silicon carbide powder. Grit numbers 240 and 400 are used successively. Only gentle pressures are used.
3. Cement the ground face to a microscope slide with an epoxy that cures at room temperature. We use Tra–Con 2101 and allow it to cure overnight. It is available from the Tra–Con Company, 55 North Street, Medford, Massachusetts.
4. Remove the excess rock by first sawing it in a thin section machine (Ingram

Laboratories Model 103) with vacuum holder and then grinding it on a hand lap with silicon carbide grit numbers 240, 400, and 700 successively. For this grinding procedure, we recommend using a metal backing piece for the glass slides and applying very gentle pressures. (As one gains experience in making crack sections, the metal backing can be omitted.)

5. Place high refractive index oil on the rock slice. Add a standard cover glass.

Crack Decoration

Although many cracks, such as non-coincident grain boundary cracks and cleavage cracks are visible in thin sections with no special treatment most cracks must be decorated in order to see them. Gardner and Pincus (1968) used fluorescent dye penetrants, but in our initial work, we decorated microcracks with carbon particles (Baldridge and Simmons, 1971; and Simmons, Todd, and Baldridge, 1974). Briefly, the cracks were decorated with carbon particles precipitated from furfuryl alcohol. That procedure was (and still is) very important because with it we were able to demonstrate that the cracks we see in sections were really present in the rocks *in situ*. We injected the furfuryl alcohol into the rock at the outcrop and then removed the sample by coring.

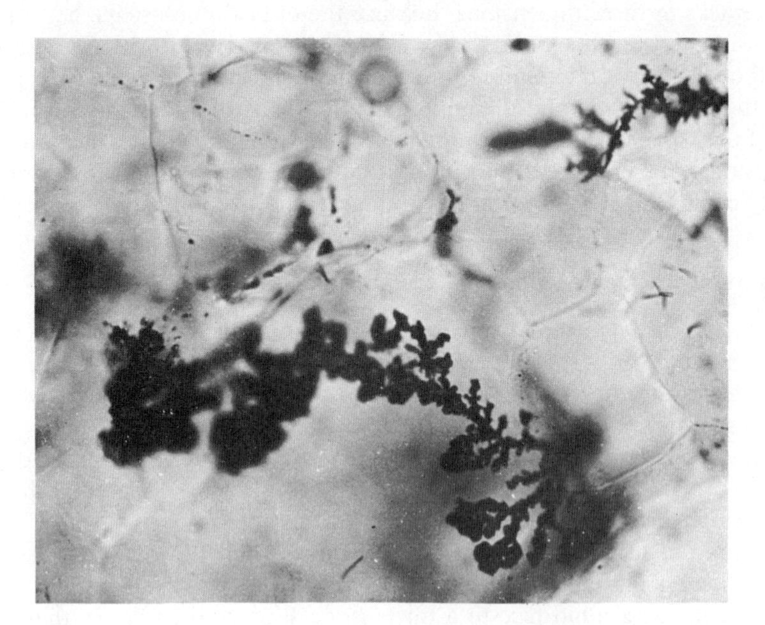

Figure 1. Photomicrograph of the Chelmsford granite with copper deposited in the microcracks. Plane polarized light. Width of field 0.25 mm. The copper extends above and below the field of view. Sometimes the copper is thin enough to be translucent which implies a thickness of about 100 Å.

In a second method of decorating microcracks, we deposit copper electrochemically in the cracks. We use the following technique:

1. Saw the rock into a slab approximately 20 × 40 × 1 mm.
2. Coat one side with silver-conducting epoxy that cures at room temperature (Tra–Con 2902) and at the same time attach an electrical lead to the conducting epoxy.
3. Cure the epoxy overnight.
4. Cover the conducting epoxy, exposed lead, and the remainder of the same side of the slab with non-conducting epoxy (Tra–Con 2101).
5. Cure the epoxy overnight.
6. Place the assembled slice in a bath of $CuSO_4$. Place an electrode of metallic copper in the bath. Connect a 6-volt dc power supply to the electrodes. The current will be less than 1 mA initially, but will increase with time as copper is deposited in the cracks.
7. Remove each day the excess copper that forms on the exposed face of the rock.
8. After one week, remove sample and prepare crack section.

We have used a Cu-electrode in a $CuSO_4$ bath and an Ag-electrode in a cyanide bath. An example of cracks decorated with copper is shown in Figure 1. One limitation to this technique at present is that the cracks do not fill comple-

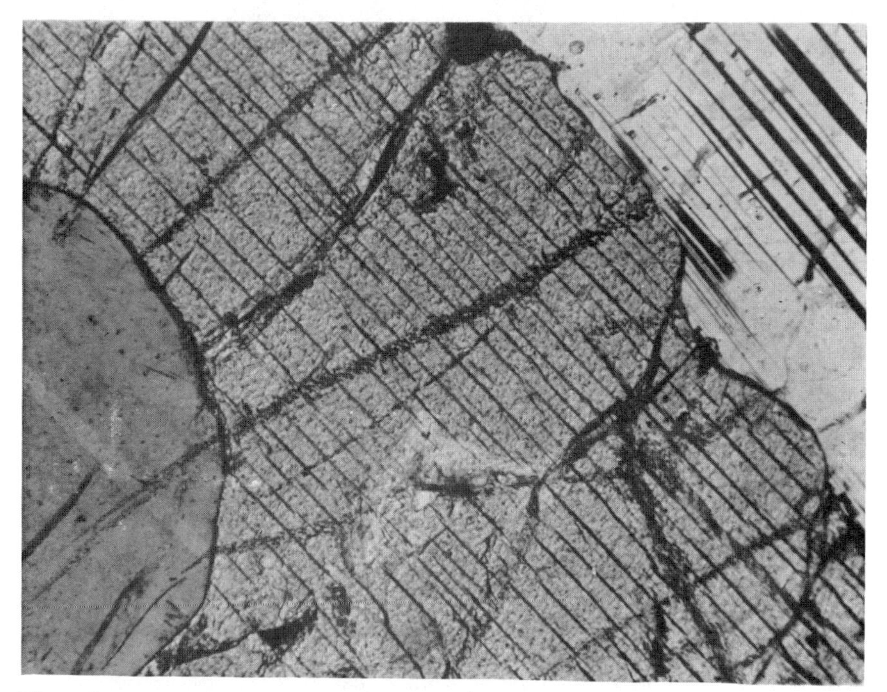

Figure 2. Naturally decorated cracks in the Mt. Tripyramid gabbro. Crossed nicols. Width of field is 0.9 mm. The cracks in the pyroxene and plagioclase grains are marked with small solid opaque phases.

tely and hence one cannot use the technique to examine the extent of all the cracks in a specimen.

A third method and the one that we now use routinely is very simple. We allow any liquids in the cracks to evaporate into the atmosphere, and then apply an index oil to the section. Capillary forces are sufficient to distribute the oil throughout the crack network. An oil with refractive index above the mineral indices is most satisfactory.

And lastly, the cracks in many rocks contain natural decoration which may be very small opaque grains, small quantities of alteration minerals, traces of weathering products, crystallographically oriented rods and tubes, and sometimes faint discoloration of undetermined origin. Examples of naturally decorated cracks are shown in Figures 2–4.

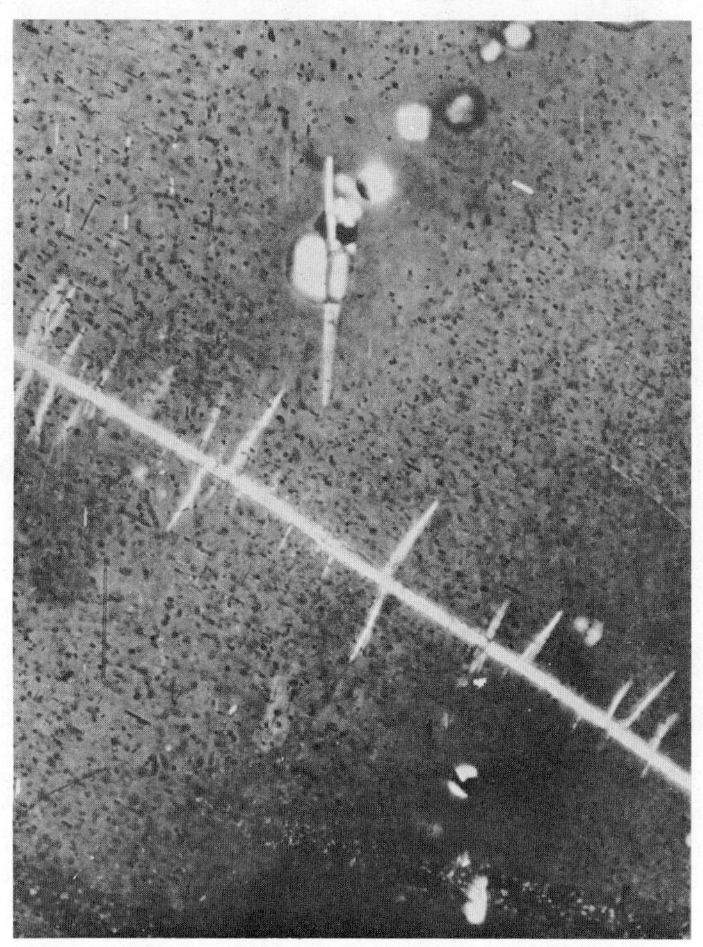

Figure 3. Naturally decorated cracks in plagioclase in the Mt. Tripyramid (NH) gabbro. Crossed nicols. Width of field 0.2 mm. The birefringent material marks the site of a formerly open microcrack.

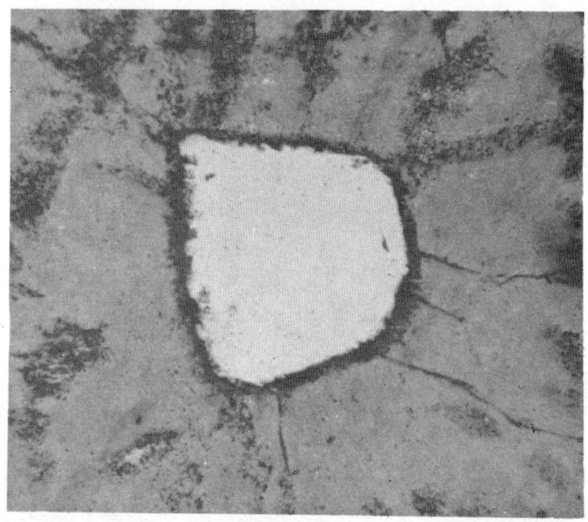

Figure 4. Naturally decorated cracks in a xenolith from a dike in the Spanish Peaks (Colo.) region. Crossed nicols. Width of field 0.6 mm. Both coincident grain boundary and radial cracks have been decorated with sub-microscopic grains of an alteration product. The quartz grain is completely surrounded by feldspar.

ORIGIN AND NATURE OF MICROCRACKS IN ROCKS

Cracks in rocks are produced when the *local* stress exceeds the *local* strength. We suggest that different processes produce cracks with differing and recognizable characteristics. Knowledge of the characteristics allows one to infer the process. On the basis of our present set of data, there appears to be a unique relationship between each process and the characteristics of the cracks produced by the process. Calibration of the characteristics allows one to infer the intensity of the process, though we have not yet calibrated all of the crack characteristics of each process. At the present time, we distinguish the following types of cracks:

d*P*d*T* cracks
Stress induced cracks (SIC)
Radial cracks about totally enclosed grains (RDC)
Concentric cracks about totally enclosed grains (CNC)
Tube cracks (TBC)
Thermal cycling cracks (TCC)
Thermal gradient cracks (TGC)
Shock induced cracks (SHIC)
Cleavage cracks (CLC)
Thin section cracks (TSC)
Cracks of unknown origin (UC)

In this section, we discuss the various kinds of cracks and their characteristics.

1. dPdT cracks

The cracks termed 'dPdT cracks' are strong functions of the relative orientations of pairs of mineral grains and the *tensor* properties of the minerals. Many (possibly all) of the cracks occur between grains or start at grain boundaries; they are grain boundary cracks. A model for their formation has been developed (Simmons, Wang, Richter, and Todd, 1973; Simmons, Richter, Todd, and Wang, 1973) that includes the effects of linear thermal expansion, linear compressibility, and grain boundary strength. The basic hypothesis included in the model is that a grain boundary crack is produced when the *local* linear strain in

Figure 5. Virgin Laramie syenite, crossed nicols. Width of field 0.6 mm. Many cracks originate at the grain boundary. Some are perpendicular to the grain boundary. All grains shown in this photomicrograph are plagioclase. Along the curved boundary perpendicular cracks are restricted to a small range in orientation. The central part of the curved boundary is shown in Figure 6.

Figure 6. dPdT cracks in the virgin Laramie syenite, plane polarized light. Width of field 0.2 mm. Both grains are plagioclase. The grain boundary dips about 45 ° with respect to the plane of the thin section. The cracks normal to the grain boundary, about 10 μm long, are examples of non-coincident GBC.

the vicinity of the grain boundary exceeds the *average* linear strain of the rock. The local strain is calculated on the basis of infinitesimal elasticity theory and the average strain is the Voigt–Reuss–Hill average.

dPdT cracks are those normally found in intrusive igneous rocks. We assume that the grain boundaries in such rocks are completely fused at some time after crystallization. Grain boundary cracks are produced in the rocks during their ascent to the surface of the earth. Examples of dPdT cracks are shown in photomicrographs (Figures 5 and 6) of a pyroxene syenite from the Laramie (Wyo.) anorthosite complex.

The Laramie syenite sample is a virgin sample, a phrase used to describe a sample that has been neither heated nor stressed, either during or after collection. The natural cracks that existed before we collected the rock have not been modified in any way. For instance, heating a sample at atmospheric pressure produces cracks in the specimen (Todd *et al.*, 1972). Application of non-hydrostatic stress also produces cracks in rocks (Hardy, 1972). Our sample of Laramie syenite has not been exposed to such processes. The petrographic characteristics and geologic setting of this rock have been described by Klugman (1966). Its general character can be seen in Figure 5. Plagioclase forms more than 40 per cent of our specimen. Hence, many grain boundaries consist of feldspar–feldspar contacts. Many cracks are associated with the grain boundaries. Some are coincident with the grain boundaries but in these two photographs, those cracks that do not coincide with grain boundaries are easier to see than the coincident grain boundary cracks. For the present discussion, we concentrate on the non-coincident grain boundary cracks. Two dozen such cracks are apparent in Figure 5. The prominent set that occurs along the curved grain boundary near the right-centre of the field is shown enlarged in Figure 6. Each crack in that set is 5–10 μm long, planar, and perpendicular (within a few degrees) to the grain boundary.

Cracked grains can occur within non-cracked grains of the same mineral (Figure 7). Such a relation fits nicely with the model of Simmons *et al.* (1973) for the formation of d*P*d*T* cracks and is caused by the different crystallographic orientations of the two grains with respect to the common grain boundary. This photomicrograph (Figure 7) demonstrates quite clearly that cracks associated with a grain boundary can exist in one grain and be absent in the adjacent. grain. Although many of the cracks extend completely across the cracked grain, note that each one (that crosses the grain) changes direction; the change of direction implies that two cracks grew inward from different boundaries and

Figure 7. d*P*d*T* cracks in virgin Laramie syenite, plane polarized light. Width of field 0.5 mm. This photomicrograph shows a feldspar grain enclosed in another feldspar grain. The enclosed grain has several grain boundary cracks and some extend completely across the grain. The enclosing grain has very few cracks and none is associated with the boundary of the enclosed grain.

united in the interior of the grain. We suggest that the distance over which a non-coincident grain boundary crack will propagate into the interior of a grain is finite. In other areas of the same thin section of the Laramie syenite, we see thin grains with parallel sides in which the grain boundary cracks cross the entire grain but do *not* change direction.

2. Stress induced cracks (SIC)

Cracks caused by non-hydrostatic stress, termed stress cracks, are related chiefly to the principal directions of the non-hydrostatic stress and seem to be largely independent of relative crystallographic orientations of the various minerals once the crack begins to grow; many are likely to cross several grains. Examples of stress cracks are shown in Figures 8 and 9 for a sample of Westerly granite in which cracks were intentionally produced in the laboratory. We loaded uniaxially a cylinder (2.5 cm diameter by 7 cm length) of Westerly granite to about 80 per cent of the fracture strength of the rock but removed the stress before the rock failed completely. A zone of macroscopic cracks developed

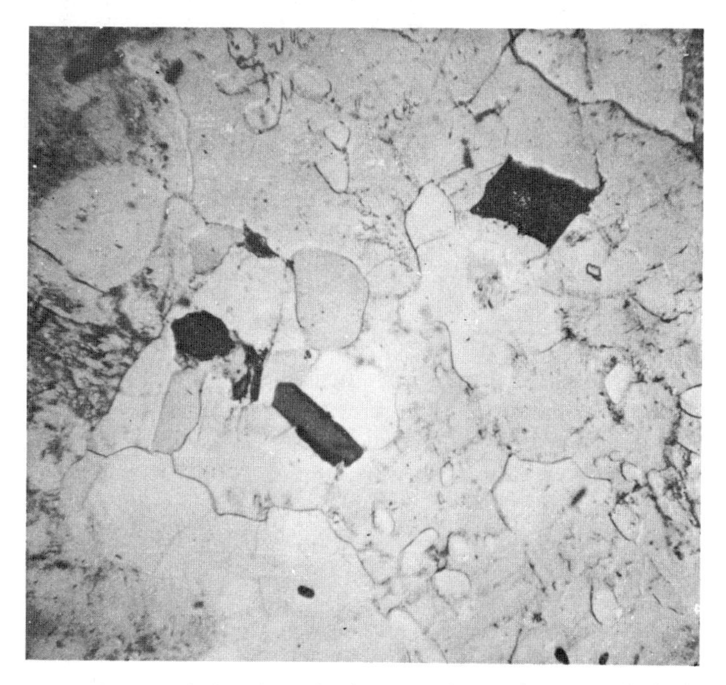

Figure 8. Stress-induced cracks in Westerly granite, crossed nicols. Width of field 2.0 mm. Grains of quartz (light) enclosed in a large feldspar grain (slightly darker). Note that each quartz grain has some dark boundary and some light boundary. The dark portions of the grain boundaries are parallel coincident grain boundary cracks. This peculiar distribution of cracks was caused apparently by the mismatch of elastic properties along the boundary.

Figure 9. Stress-induced crack in Westerly granite, crossed nicols. Width of field 0.7 mm. The large crack that extends diagonally across the entire field was produced by uniaxial stress. The grey, roundish grain is quartz. It is largely surrounded by feldspar. The adjacent darker crystal is muscovite.

at an angle of 40–50 ° to the axis of compression. We cut a single petrographic thin section through the zone perpendicular to the cylinder axis which was used for photomicrographs of Figures 8 and 9. Especially interesting in Figure 8 is the presence of *coincident* grain boundary cracks between quartz and feldspar. Such cracks are not seen in the virgin sample and we infer that these cracks were produced by a mismatch of elastic properties between pairs of grains that amplified the average stress for the whole specimen. Their presence along parts only of a given boundary between two grains is consistent with our general model. But note especially the coincident grain boundary cracks that are continuous along the interfaces between several grains. We term such cracks multigrain boundary microcracks and believe that they are a characteristic feature of some stress induced cracks.

Not all cracks are confined to grain boundaries. Many extend through several grains. For example, note in Figure 9 the large crack that extends diagonally across the entire photomicrograph (and about five times farther). This crack was also produced by uniaxial stress on the Westerly granite sample. Clearly shown in Figure 9 is the role of grain boundary cracks in 'deflecting' growing cracks. A coincident grain boundary crack between the roundish quartz grain and the feldspar crystal was (probably) present already when the large crack was produced.

3. Cracks associated with completely enclosed grains

Mismatch of the *volumetric* properties between a host grain and a totally enclosed grain can produce cracks. If the total volumetric strain of the enclosed grain is less than the volumetric strain of the host, then concentric cracks occur either within the host or along the grain boundary. If the crack is a grain bound-

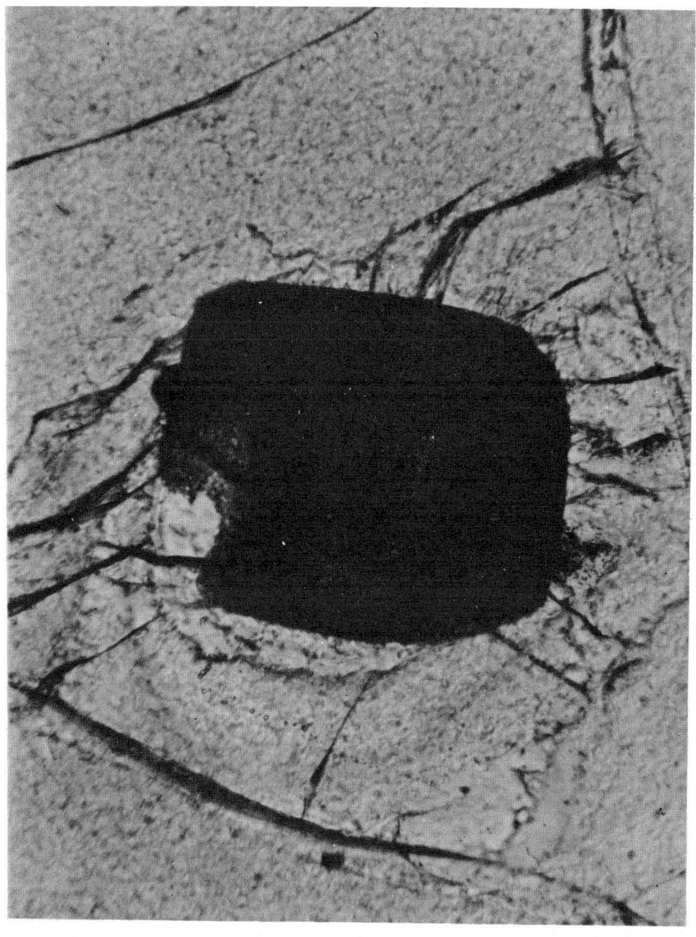

Figure 10. Spinel grain enclosed in olivine. Plane polarized light: Width of field 0.6 mm. The brown spinel is almost opaque in thin section. Differential volume expansion of the two grains caused the cracks. The sample is a virgin peridotite bomb from Kilbourne Hole, New Mexico.

ary crack, then it is a coincident GBC and may extend along the entire boundary. On the other hand, if the volumetric strain of the enclosed grain exceeds that of the host by an amount sufficient to overcome the strength of the host, then radiating, non-coincident grain boundary cracks occur.

An example of radial cracks associated with a spinel grain completely enclosed in an olivine grain is shown in Figure 10. We suggest that those cracks formed in response to the differential volumetric strains of the spinel and olivine due to changes of both pressure and temperature. Similar crack patterns are present around other spinel grains in this rock, a peridotite bomb from Kilbourne Hole, New Mexico. The general features of this area were described by Carter (1970).

A mathematical model for the formation of cracks associated with totally enclosed grains has not yet been developed. However, we note that Rosenfeld (1969) and Rosenfeld and Chase (1961) obtained theoretical expressions for the stress field produced in totally enclosed grains due to change of P and T. Extension of their mathematical formulation to account for crack formation, at least for a first order theory, does not appear to be difficult.

4. Tube cracks

Tubes with two dimensions that are 1–3 μm and the third dimension 100–1000 μm appear to be abundant in igneous rocks. The tubes may be hollow or filled with a solid phase and have circular to elliptical to rectangular cross sections. Some tubes appear to be crystallographically oriented but most tubes studied to date by us have no regular crystallographic orientation. An example of open tube cracks in the Chelmsford (Mass.) granite is shown in Figure 11, a photomicrograph of a thick section (150 μm). These tubes are easily seen in crack-sections but their true nature would not have been apparent from ordinary thin sections. Perhaps the failure of other investigators to recognize such tubes is attributable to the widespread use of thin sections of standard thickness in which open tubes appear as 1–3 μm sized holes and would be easily overlooked.

The open tubes in the Chelmsford granite occur in muscovite grains and are distinct from rutile needles. Not all muscovite grains contain tubes but in some, they are locally quite abundant. They are roughly circular in cross section with diameters of 1–3 μm and lengths of about 100 μm. Many tubes terminate at one end on a grain boundary and a few connect inclusions with grain boundaries, as shown in Figure 11. They are intertwined with each other and have random orientations in the mica; they are definitely not controlled crystallographically.

The tubes in the Chelmsford granite are hollow as can be seen in Figure 12, two photomicrographs obtained with the focal plane of the microscope located at two different depths in the crack section. The finite depth of field of the optical microscope, 2–3 μm at high magnification (P50 objective, 10 × ocular), let us 'follow' the tubes from the upper slide surface down through the entire slice. In the preparation of the crack section, the rock slice was ground to about

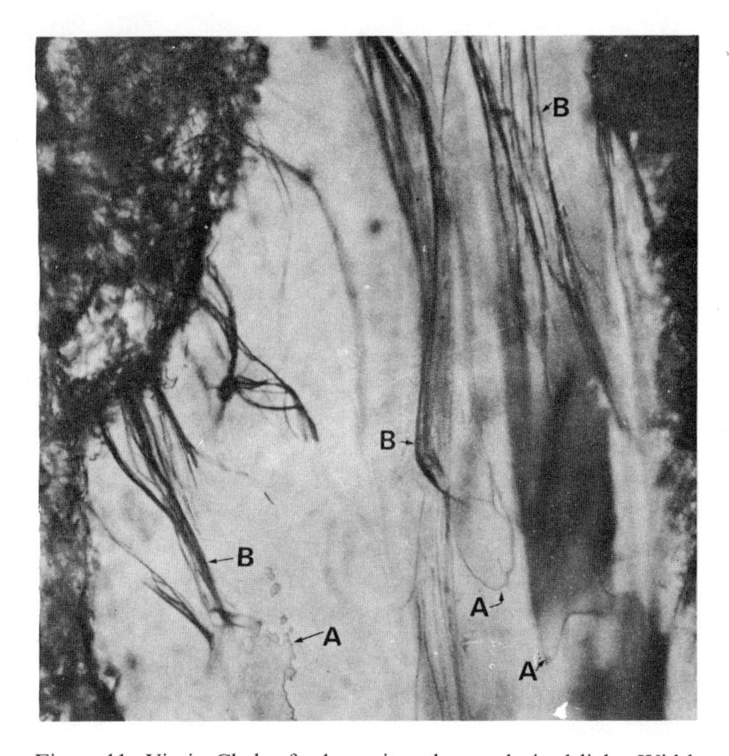

Figure 11. Virgin Chelmsford granite, plane polarized light. Width of field 0.6 mm. Section is 170 microns thick. The features seen in this photomicrograph are 40 microns below the upper surface of the section. The clear area is a single crystal muscovite grain. It contains several flat cracks (some are marked A) and many tubes (a few are marked B). The region just below the inclusion that is slightly left of centre is shown enlarged in Figure 12.

150 μm thickness, the fluids contained in the cracks were allowed to evaporate at room conditions, and then a drop of refractive index oil was placed on the rock slice. Capillary forces drew the oil into the cracks, along with some air. Thus, the microscopist can see empty portions of tubes such as shown in Figure 12, that are spatially continuous with other portions that are filled with index oil, such as shown in the lower left hand corner of the bottom half of Figure 12. Examination with the scanning electron microscope of similar tubes in other samples of Chelmsford granite confirms that these tubes are hollow.

The study of tubes in crack sections provides a basis for recognizing tubes in ordinary thin sections. Shown in Figure 13 are several sets of tube cracks in a grain of feldspar in the Laramie syenite. Both surfaces of the thin section intersect the tubes and therefore the total length of these tubes is unknown.

Although some tubes are hollow, such as those we have described in the

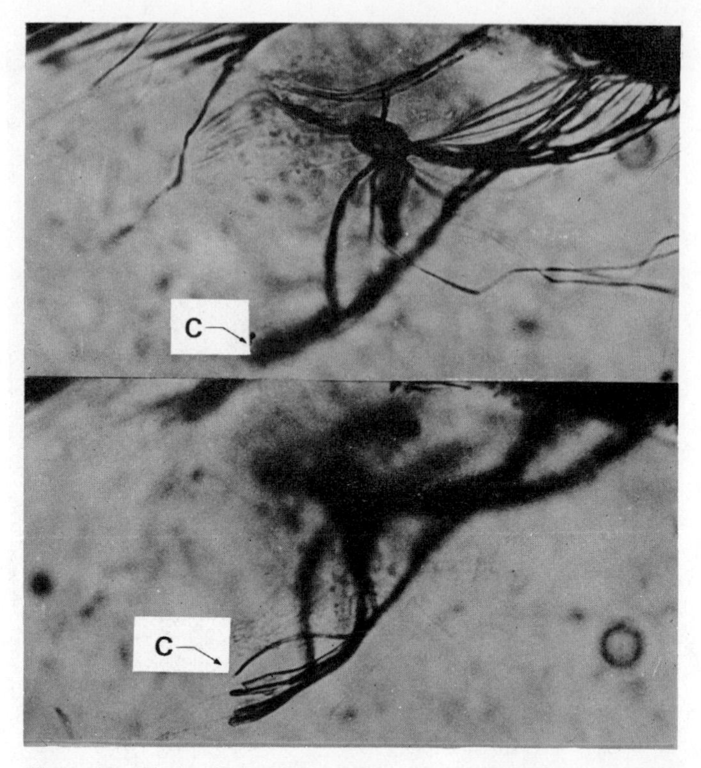

Figure 12. Virgin Chelmsford granite, plane polarized light. Width of field 0.3 mm. In the upper figure, a dozen tubes can be seen that are attached to the spherical 'inclusion'. The lower figure, obtained with the focal plane 35 microns below the position for the upper figure, shows 5 tubes at location C in which the fluid terminates and the physical tubes continue.

Chelmsford granite, others are filled with a solid phase. For example, in the Frederick (Md.) diabase, centimetre-sized phenocrysts of laboradorite that occupy no more than 0.01% by volume of the rock contain crystallographically oriented tubes filled with pyroxene. Similar tubes were observed by Walter *et al.* (1971) in lunar sample 12021; they suggested that the metallic iron which fills those tubes had been deposited from a vapour. The tubes of iron resemble closely the pyroxene-filled tubes that we observe in the Frederick diabase. In many cases (such as shown in Figure 14), one cannot distinguish with the optical microscope alone whether tubes are filled with a solid phase or they are empty and further work with the scanning electron microscope is needed.

Observations of the geometrical relations of tubes may be useful in deducing the origin of tubes. Single tubes appear to be rather common in terrestrial and lunar igneous rocks, particularly if we include those tubes that are filled with a solid phase. Multiple tubes may be arranged in a plane with rather regular spacing between the individual tubes (as in the Laramie syenite, Figure 13)

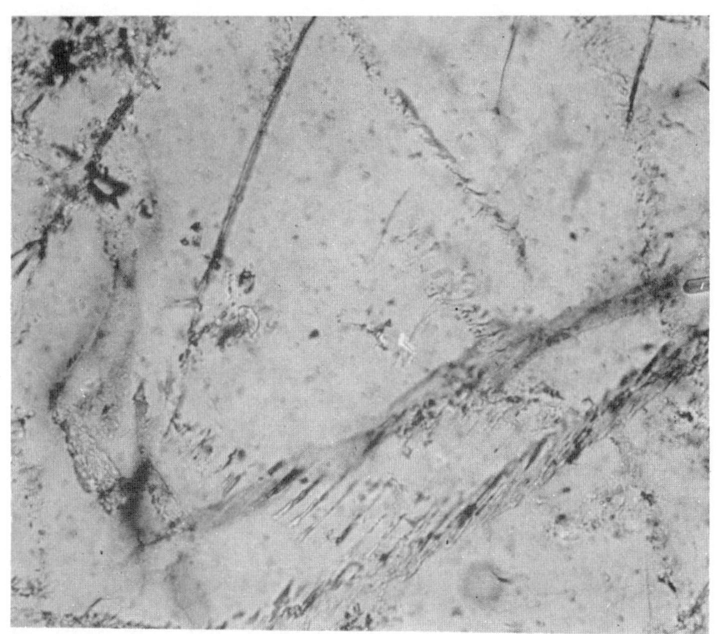

Figure 13. Tube cracks in plagioclase, crossed nicols. Field of view 0.25 mm. These tubes occur in virgin pyroxene syenite from the Laramie (Wyo.) anorthosite complex.

Figure 14. Tubes in plagioclase phenocrysts in Frederick (MD) diabase, plane polarized light. Width of field 0.3 mm. Note the absence of crystallographic control.

or they may be intertwined (as in Chelmsford granite, Figure 11). The tubes themselves may become discontinuous beads (as occur in the Mt. Tripyramid, N. H. gabbro but not illustrated here).

Several different origins for tube cracks appear to be possible. First, some TBC's may be simple solution features produced by either late stage magmatic fluids or by natural ground waters. Such an origin seems attractive for the tubes in the Chelmsford granite in view of the link between inclusions and grain boundaries formed by tubes. Second, some TBC's may be due to the etching of dislocations by ground water. And third, some tubes may be an intermediate stage in the healing of flat cracks. A rather complete spectrum of features that are spatially continuous can be traced from flat, open microcracks through tubes that lie in the plane of the microcrack, through tubes that are themselves discontinuous trains, through planes of inclusions, to material that

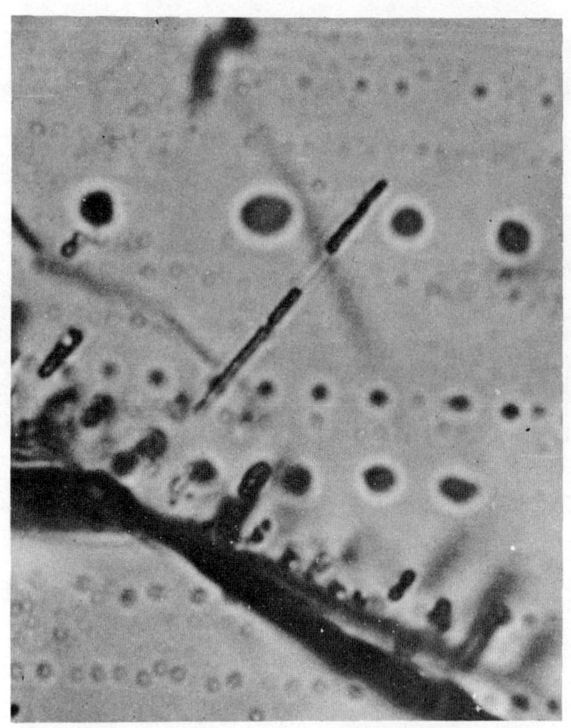

Figure 15. Lunar sample 70017,131. Plane polarized light. Width of field 0.1 mm. This photomicrograph shows a set of parallel tubes in plagioclase along a pyroxene–plagioclase grain boundary. The tubes are partially filled with pyroxene. Note the long tube at centre right; the tube is clearly continuous between the blebs of pyroxene and is hollow. (The fuzzy spots are bubbles in the epoxy cementing the rock to the glass slide.)

has a new, fresh appearence. Although we have not observed the complete spectrum for a single crack, sufficient sets of pairs of the features have been seen that we believe the spectrum is established.

We note specifically that all tubes need not have a common origin. We expect that ground water plays an important role in the formation and subsequent modification of cracks in terrestrial rocks. Water is clearly unimportant in the creation and modification of cracks in lunar rocks, yet such tubes as shown in Figure 15 are very common in lunar samples.

5. *Thermal cycling cracks*

The thermal cycling of rocks produces grain boundary microcracks that are predominantly coincident GBC. We have previously used this process to create cracks in terrestrial rocks so that we could study their effects on physical properties (Wang *et al.*, 1971; Todd *et al.*, 1972 and 1973). We were originally surprised that the properties of rocks containing thermal cycling cracks differed significantly from the properties of rocks with other types of cracks. However, our data on the velocity of compressional waves and static compressibility, summarized in Figure 16, now allow us to recognize the presence of cracks that have been produced in a rock by thermal cycling.

Several different observations indicate that the thermal cycling cracks are

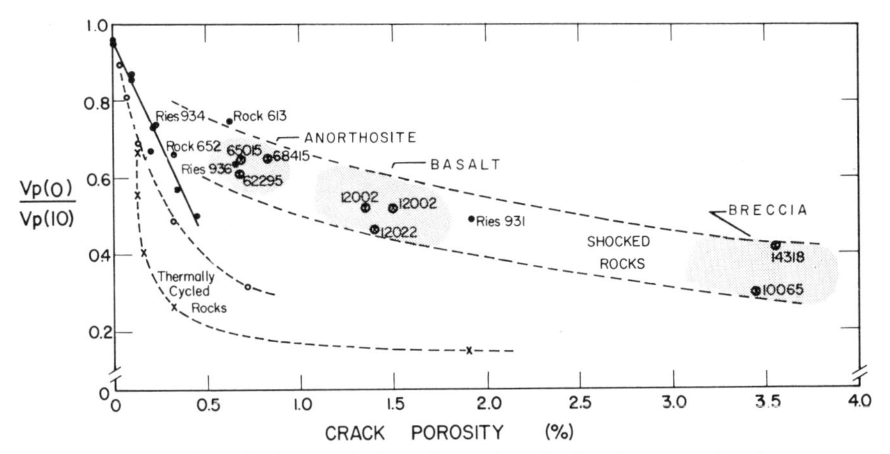

Figure 16. Effect of microcracks in rocks on the velocity of compressional waves. The ratio $V_p(0)/V_p(10)$ is the ratio of the velocity at 1 bar confining pressure to the velocity of a crack-free sample. The solid line represents typical terrestrial igneous rocks. Thermally cycled igneous rocks and shocked rocks separate into two distinct and widely separated fields. The three Ries rocks are granitic samples from the Ries Crater in Germany. The lunar samples (indicated by five digit numbers) separate into distinct groupings within the zone of shocked rocks. Modified from T. Todd, D. Richter, G. Simmons and H. Wang, *Proc. Fourth Lunar Sci. Conf., Vol.* 3, *Physical Properties*. p. 2658. (Reprinted with permission from Pergamon Press, 1973)

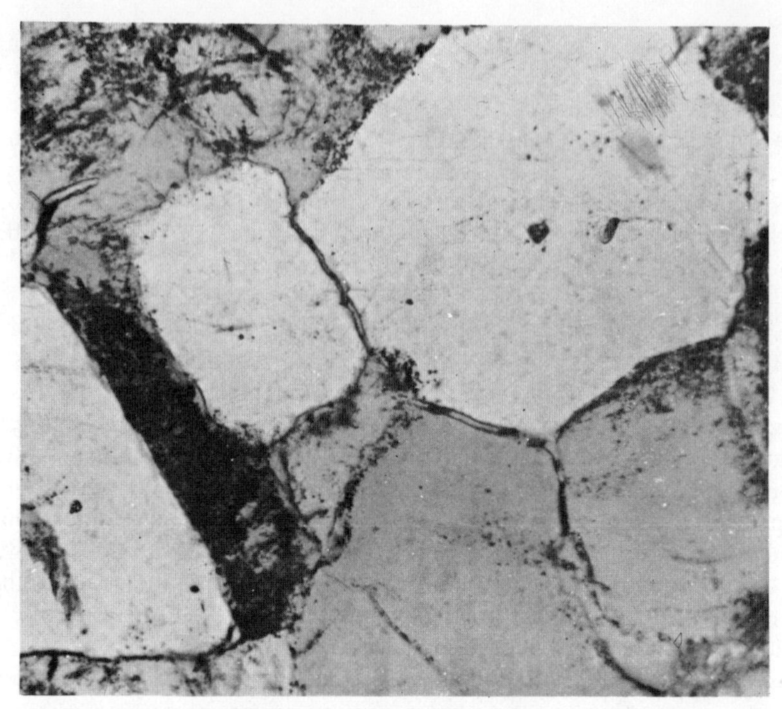

Figure 17. Thermally cycled Westerly granite, plane polarized light. Width of field 0.9 mm. The sample was heated slowly to approximately 950 °C and then cooled before making the thin section. Most grains are separated from adjacent grains by wide grain boundary cracks.

significantly different from other types. First, the data of Figure 16 show that their behaviour with pressure differs from that of d*P*d*T* cracks and shock cracks. Second, Simmons *et al.* (1973) showed that the set of all pairs of crystallographic orientations of two grains that form grain boundary cracks for a given ΔT (at constant ΔP) differs greatly from the set calculated for a given ΔP (at constant ΔT). They have obtained preliminary computer runs for albite–albite and albite–pyroxene pairs. Third, the appearance of thermal cycling cracks in thin sections of a suite of samples of the same rock (exposed to progressively higher maximum temperatures) is quite distinct. See Figure 17, a photomicrograph of Westerly granite that had been thermally cycled to about 950 °C at a heating rate less than 2 °C/min. Note that each individual grain is completely surrounded by a coincident GBC and that non-coincident GBC's are rare.

6. Thermal gradient cracks

The stresses caused by high thermal gradients produce microcracks in rocks. For example, the large and irreversible effects on V_p and V_s observed by Ide (1937) and attributed erroneously by him to cracks produced by high temperatures were in reality caused by cracks produced by thermal gradients. We have shown that temperatures below 350 to 450 °C, *if changed slowly*, produce rather few cracks (Todd *et al.*, 1973; Richter and Simmons, 1974) but a rapid change of temperature produces many cracks. Indeed, Warren and Latham (1970) used thermal gradients across specimens to produce microcracks. We have not yet examined either crack sections or thin sections of samples exposed to high gradients.

7. *Shock cracks.*

Shock waves leave an imprint on rocks that may be examined either with a petrographic microscope or through measurements of bulk physical properties. In Figure 16, we show examples of the effects of shock waves on V_p and on crack porosity. Simmons, Siegfried, Richter, and Schatz (1974) and Siegfried *et al.* (1974) showed that the crack porosity of shocked rock is related to the peak shock pressure. Hence, the measurement of shock induced crack porosity can be used to estimate the intensity of the process.

The petrographic characteristics of shocked rocks have been described by many authors (for example, see French and Short, 1968; Das Ries, 1969; and the Proceedings of the several Lunar Science Conferences). Microcracks have

Figure 18. Shocked Stone Mountain granite, crossed nicols. Width of field 1.3 mm. The sample was located adjacent to a 5 cm. shot hole. The cracks are approximately parallel and were produced by the shock waves.

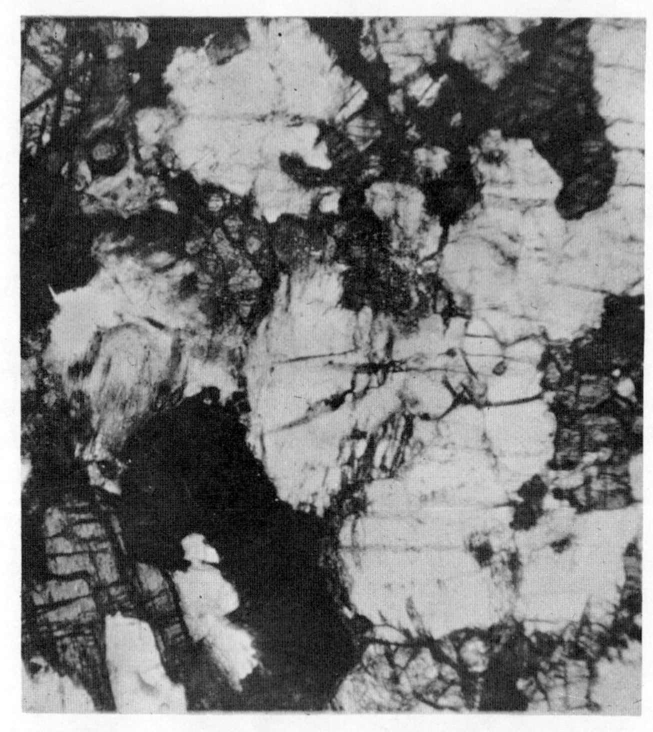

Figure 19. Ries diorite 931, crossed nicols. Width of field 2 mm. Shock cracks in all grains are approximately parallel. Most cracks extend completely across a single grain. The sample is from zone 1 (of Stöffler, 1966.)

been noted often in the petrographic descriptions and were correlated semi-quantitatively with shock pressures by Short (1966) for rocks in the vicinity of the Hardhat underground nuclear explosion. Hörz (1969) related the density of cracks in an experimentally impacted granite block to the calculated pressure. Our own petrographic observations of shock induced cracks are similar to those already described in the literature of shock metamorphism. Shown in Figures 18 and 19 are examples of shocked Stone Mountain (Ga.) granite and a diorite from the Ries meteorite impact crater. The Stone Mountain sample was located adjacent to a 5-centimetre shot hole. The cracks were introduced into the sample during the routine quarrying operations. The Ries diorite, collected from zone 1 of Stöffler (1971), was described by Todd *et al.* (1973). The characteristic petrographic features of shock cracks displayed in Figures 18 and 19 are (1) the presence of many multigrain cracks and (2) the approximate parallelism of many subsets of cracks. The distinct field of shocked rocks in the plot of normalized velocity of compressional waves–crack porosity (Figure 16) provides the basis of another and apparently unique technique for the identification of shock induced cracks in rocks.

8. Cleavage cracks.

Cleavage cracks (CLC) are those cracks parallel to cleavage planes in minerals. Some CLC's are introduced when a thin section is made. They are obviously quite useful to the petrographer and help in the visual recognition of pyroxene, amphibole, etc. More interesting though, from our viewpoint, are the CLC's that were demonstrably present in the rock *in situ*. Some CLC's have been decorated naturally while *in situ* by tubes growing outward from the CLC, by alteration products forming along the CLC's, and by other minerals (particularly opaques) depositing along the crack. An example is shown in Figure 20, a photomicrograph of a peridotite bomb from Kilbourne Hole, New Mexico, in which cleavage cracks in the pyroxene were decorated *in situ* by unidentified alteration products, proving conclusively that the CLC's were present before the thin section was prepared.

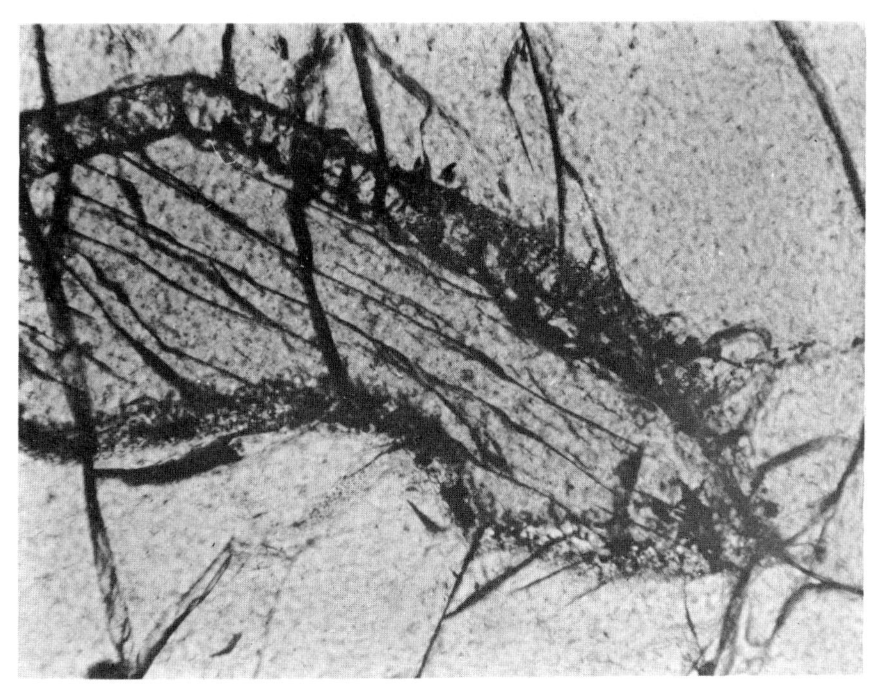

Figure 20. Peridotite bomb, Kilbourne Hole, New Mexico, plane polarized light. Width of field 0.9 mm. The kidney shaped crystal is pyroxene. In the thin section, we see that the long parallel cleavage cracks are marked by alteration material. Note that along one grain boundary is a peculiar set of 'scalloped' cracks; we do not know the origin of them.

9. Thin section cracks

Most thin sections of rocks contain microcracks that were produced when the section was made. Every petrographer has seen them. They are not parti-

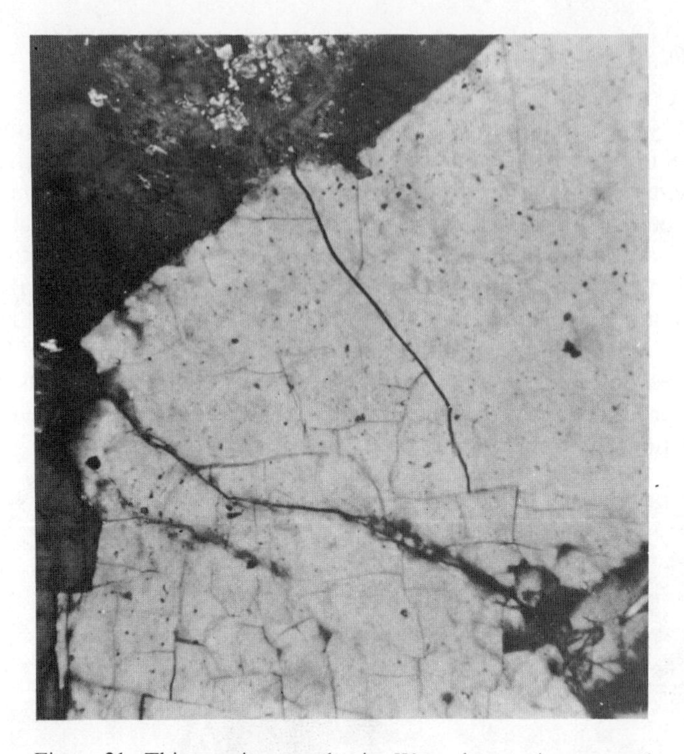

Figure 21. Thin section cracks in Westerly granite, crossed nicols. Width of field 0.5 mm. The large light grey crystal is quartz which contains numerous cracks. The crystal is somewhat too thin for a thin section and the many cracks were introduced during the grinding process.

cularly troublesome for general petrography but neither are they very useful. In Figure 21, we show a typical example. Such cracks (TSC) *are* troublesome though for the petrographer who is interested in cracks that were present in the sample while *in situ*. To avoid TSC's, we make crack sections (described earlier in this manuscript). After one has worked with the crack sections, one can recognize many cracks as being TSC's and can begin to use ordinary thin sections with some confidence.

10. Cracks of unknown origin

Lest the reader think that we have neatly catagorized all the cracks that we have observed, we include three examples of the many cracks of unknown origin present in this section. Shown in Figure 20 are cracks associated with a grain boundary between pyroxene and olivine in a peridotite bomb. Such scalloped boundaries are not rare. In Figure 22, we sketch helicoidal cracks in amphibole (gedrite) crystals found in metamorphosed volcanic rocks by Stout (1974). And a third example is shown in Figure 23, a typical photomicrograph

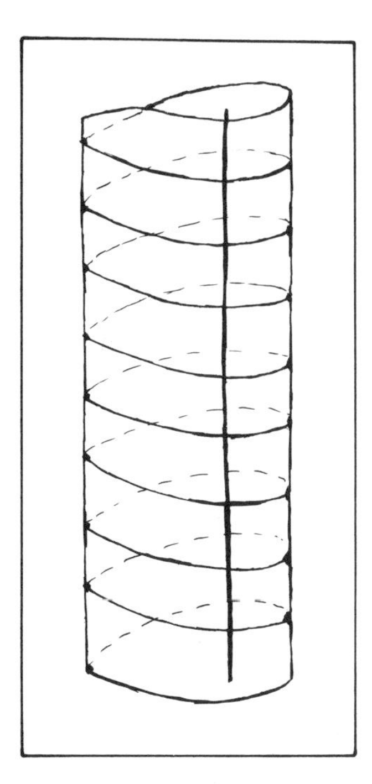

Figure 22. Sketch of helicoidal crack in amphibole (gedrite) from metamorphosed volcanic rock near Keene, N.H., described by Stout (1974). The outer surface has been etched by ground water. With the microscope, one can see that the crack extends from the surface to the axis.

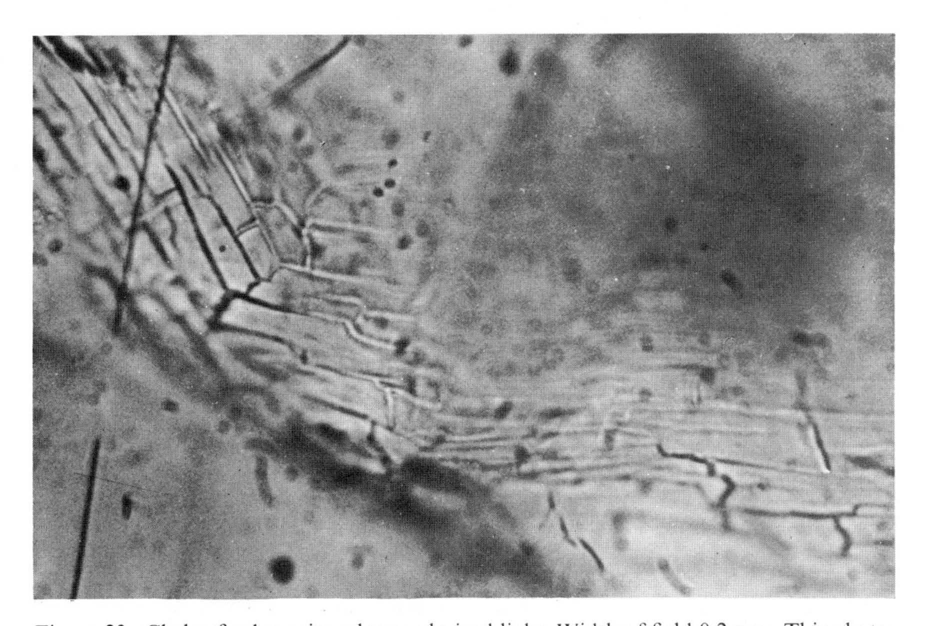

Figure 23. Chelmsford granite, plane polarized light. Width of field 0.2 mm. This photomicrograph is taken from the same 'crack section' in which the tubes of Figure 7 and Figure 8 were seen. Here a quartz–quartz grain boundary shows a brick-like mosaic of cracks of unknown origin. Note the rutile needle that extends across left side of the photo.

of many such structures *on* grain boundary surfaces that resemble the mosaic patterns of brick walls. We do not have satisfactory explanations for the origin of any of these cracks. We expect to find other such features as we continue to examine cracks.

THE HEALING OF MICROCRACKS IN ROCKS

The cracks in any given rock can be annealed with pressure and temperature acting over geologic time. The zeroth order theory would be based on simple classical chemical reaction kinetics. We write

$$\ln c/c_0 = K(t - t_0)$$

where K is a constant, c is concentration of microcracks, t is time, and the subscript denotes initial conditions. The effect of temperature is included in K through an Arrhenius type equation

$$K = A \exp(-Q/RT)$$

where A is another constant, Q is the activation energy, R is the gas constant. Both A and Q are to be determined experimentally.

Experimental data on the annealing of microcracks in rocks does not appear to have been measured yet. But the results of studies on other materials (Table 1) show clearly that healing of artificial cracks can be produced in the laboratory.

Even though the annealing of microcracks in rocks has not yet been demonstrated in the laboratory, abundant evidence exists with which to show that microcracks *have* annealed in the geologic past. In our terminology, healed cracks are the sites of former (open) cracks. They can be recognized in a variety of ways. We have traced open cracks into bubble planes and then into completely healed material (see Figure 24). We have observed a similar relation with tubes. We have seen displacement of twin lamellae along healed cracks. A spectacular example of healed cracks occurs in the Milford, Mass. granite. The general petrography and geologic setting of this granite were described by Shaw (1967). Shown in Figures 25 and 26 is a very long healed crack which

Table 1. Existing studies on microcrack annealing

Material	Form[a]	Experimental technique	Reference
Al_2O_3	SC	heating	Heuer and Roberts (1966)
Al_2O_3	PC	heating	Kirchner *et al.* (1969)
ZnO	PC	heating	Lange and Gupta (1970)
NaCl	SC	compression loading	Grdina and Nevorov (1967)
NaCl and LiF	SC	compression loading	Forty and Forwood (1963)
Glass	Amorphous	compression loading	Wiederhorn and Townsend (1970)

[a]SC = single crystal
PC = polycrystal

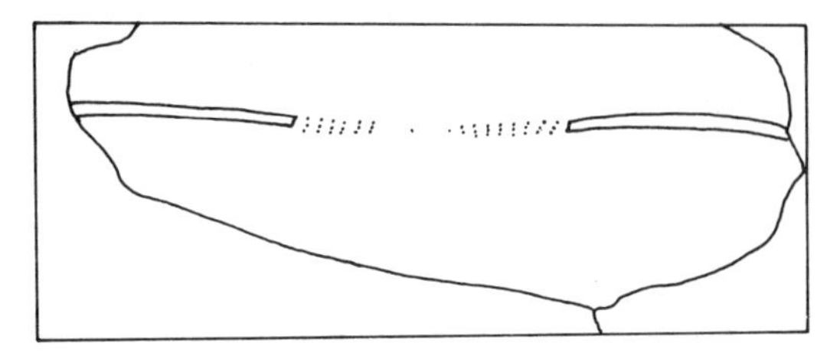

Figure 24. Sketch of an open crack that is spatially continuous with a bubble plane and 'fresh' material. This set of features is interpreted as a healed crack.

can be traced across several feldspar grains and includes bubble planes, coincident grain boundaries, solid phase inclusions, and displaced twin lamellae. An equally striking example is shown in Figure 27 where one can see offset twin

Figure 25. Milford granite, crossed nicols. Width of field 1.5 mm. A long healed crack extends diagonally across the entire field of view. The site of the former crack is marked by fine grained recrystallized material, secondary minerals, bubble planes, and displacement of twin lamellae in plagioclase. See Figure 26 for an enlarged view of the central feldspar grain.

Figure 26. Milford granite, crossed nicols. Width of field 0.5 mm. This photomicrograph is an enlargement of the plagioclase crystal shown in Figure 25. The twin lamellae are sharply offset in one part of the crystal and appear 'dragged' in another part.

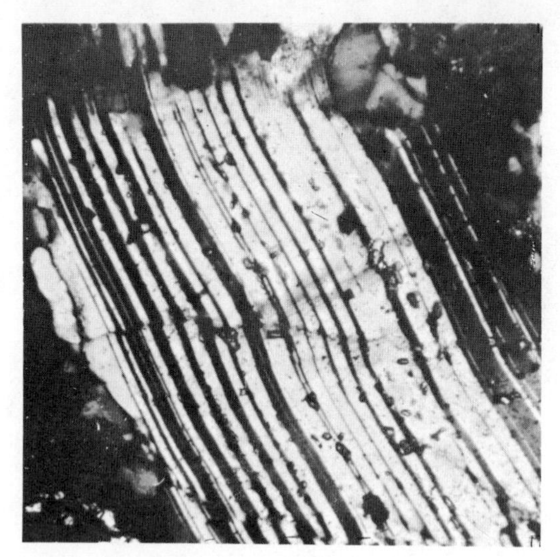

Figure 27. Milford granite, crossed nicols. Width of field 0.5 mm. This photomicrograph shows another example of a plagioclase crystal in which twin lamellae were offset by cracks. The cracks have healed and are now marked by planes of minute bubbles—some of which are barely visible in this photograph.

Figure 28. Schist from vicinity of Seabrook, N. H., crossed nicols. Width of field 0.2 mm. The sample was collected near the Scotland Road fault zone. The stringers of recrystallized quartz which cut across the large quartz crystals mark the sites of former stress induced cracks similar to the one shown in Figure 9.

lamellae, 'drag' along the microcrack and (with the optical microscope) bubble planes, all spatially continuous with completely healed material.

Healed cracks are commonly marked by linear trains of very small grains. For example, in Figure 28, a photomicrograph of a schist from Seabrook, N. H., stringers of very fine grained recrystallized quartz are interpreted as marking the site of healed cracks. From the characteristics of stress induced cracks, we infer that the healed cracks seen in Figure 25–28 were originally SIC's.

Another example of healed cracks is shown in Figure 29, a photomicrograph

Figure 29. Cumberlandite, Cumberland, R. 1., plane polarized light. Width of field 1.5 mm. The large black grain is magnetite which is surrounded by several grains of olivine (light grey). Note the *multiple* sets of cracks concentric with the magnetite grain. These concentric cracks are decorated naturally with small magnetite grains.

of cumberlandite. This rock, collected near Cumberland, R. I., was described by Emerson (1917) and consists almost entirely of magnetite and olivine. The concentric cracks seen in Figure 29 are now healed and are marked by very small grains of magnetite.

Sometimes bubble planes mark the site of a healed crack (see Tuttle, 1949). In Figure 30, a photomicrograph of Westerly granite, we show an example of a healed-d*P*d*T* crack in two quartz grains. As a petrographer gains experience in recognizing healed cracks, he finds them everywhere. We have seen many healed cracks in granite, gabbro, diabase, peridotite bombs from Kilbourne Hole and from the Moses Rock (Utah) diatreme, eclogite inclusions in South African kimberlites, and lunar samples.

CONCLUSIONS

Microcracks produced in rocks by various geologic processes can be recognized and can be characterized uniquely with a combination of techniques. We use the optical microscope and a plot of the normalized velocity of compressional waves versus zero-pressure crack porosity, *at present*, to classify cracks. The different types of cracks present in a given igneous rock can be used to infer the processes that have operated on the rock.

Figure 30. Westerly granite, crossed nicols. Width of field 0.3 mm. The egg-shaped 0.1 mm grains are quartz and are enclosed in feldspar. Note the train of bubbles that extends across the two quartz grains. The train of bubbles lies on a plane that dips about 60 ° to the plane of the section. It probably represents a previous grain boundary crack.

ACKNOWLEDGEMENTS

We have benefitted greatly from discussions with Scott Baldridge, Terry Todd, Herbert Wang, Robert Siegfried, and Michael Feves. Robert Ruppert helped develop the technique for decorating cracks with copper and silver. We appreciate greatly the high quality of photographic work provided by Robert Stevens which has contributed significantly to our progress in understanding microcracks in rocks. Financial support was provided by the National Aeronautics and Space Administration contract NGR-22-009-540 and by the National Science Foundation contract GA-40756.

REFERENCES

Baldridge, S. and G. Simmons (1971). *ABS., Trans. Amer. Geophys. Union*, **52**, 342.

Carter, J. L. (1970). *Geol. Soc. Amer. Bull.*, **81**, 2021–2034.

Carter, N. L., L. A. Fernandez, H. G. Ave'Lallemant and I. S. Leung (1971). *Proc. 2nd Lunar Sci. Conf., Vol. 1, Geochim. et Cosmochim. Acta, Suppl. 2*, 775–795.

Chao, E. C. T. (1967a). In *Researches in Geochemistry, 2.* ed. P. H. Abelson, Wiley, New York, pp. 204–233.

Chao, E. C. T. (1967b). *Science*, **156**, 192–202.

Chao, E. C. T. (1968). In *Shock Metamorphism of Natural Materials*, edited by B. M. French and N. M. Short, Mono Book Corp. Baltimore, pp. 135–158.

Chao, E. C. T., O. B. James, J. A. Minkin, J. A. Boreman, E. D. Jackson, and C. B. Raleigh (1970). *Science*, **167**, 644–647.

Das Ries (1969). *Geologica Bavarica*, Nr. *61*, Bayer. Geologischen Landesamt., Munchen.

De Carli, P. S. (1968). In *Shock Metamorphism of Natural Materials* edited by B. M. French and N. M. Short, Mono Book Corp., Baltimore, pp. 129–134.

Dence, M. R. (1968). In *Shock Metamorphism of Natural Materials*, edited by B. M. French and N. M. Short, Mono Book Corp., Baltimore, pp. 169–184.

DeVore, G. W. (1969). *Contributions to Geology*, **8**, No. 1, 21–36.

Emerson, B. K. (1917). Geology of Massachusetts and Rhode Island, *U.S. Geol. Survey, Bull.* **597**, 289 pp.

v. Engelhardt, W. and D. Stöffler (1968). In *Shock Metamorphism of Natural Materials* edited by B. M. French and N. M. Short, Mono Book Corp., Baltimore, pp. 159–168.

v. Engelhardt, W., J. Arndt, W. F. Muller, and D. Stöffler, (1970). *Proc. of the Apollo 11 Lunar Sci. Conf., Vol. 1, Geochim et Cosmochim Acta, Suppl.* 1, 363–384.

Forty, A. J., and C. T. Forwood, (1963). *Trans. Brit. Cer. Soc.*, **62**, 715–724.

French, B. M., and N. M. Short (ed.), (1968). *Shock Metamorphism of Natural Materials*, Mono Book Corp., Baltimore, 644 pp.

Gardner, R. D. and H. J. Pincus (1968). *Int. J. Rock Mech. Min. Sci.*, **5**, 155–158.

Grdina, Y. V. and V. V. Neverov (1967). *Soviet Physics—Crystallography*, **12**, 421–424.

Hardy, Jr. H. R. (1972). in *Acoustic Emission*, ASTM Special Publication **505**, 41–83.

Heuer, A. H. and J. P. Roberts (1966). *Proc. Brit. Cer. Soc.*, **6**, 17–27.

Hörz, F. (1969). *Contrib. Mineral. and Petrology*, **21**, 365–377.

Ide, J. M. (1937). *J. Geol.*, **45**, 689–716.

Kirchner, H. P., W. R. Beussen, R. M. Gruver, D. R. Platts, and R. E. Walker (1970). Summary Report, Contract N00019-70-C-0418, Naval Air Systems.

Klugman, M. A. (1966). *Mt. Geol.* **3**, 75–84.

Lange, F. F. and T. K. Gupta (1970). *J. Amer. Cer. Soc.*, **53**, 54–55.

LSPET (1969). Preliminary examination of lunar samples from Apollo 11, *Science*, **165**, 1211–1227.

Richter, D. and G. Simmons (1974). *Int. J. Rock Mech. Min. Sci.*, **11**, 403–411.

Robertson, P. B., M. R. Dence, and M. A. Vos (1968). In *Shock Metamorphism of Natural Materials*, edited by B. M. French and N. M. Short, Mono Book Corp., Baltimore, pp. 433–452.

Rosenfeld, J. L. (1969). *Amer. J. Sci.*, **267**, 317–351.

Rosenfeld, J. L., and A. B. Chase (1961). *Amer. J. Sci.*, **259**, 519–541.

Shaw, C. E. (1967). *Ph.D. Thesis*, Brown University, 121 pp.

Short, N. M. (1966). *J. Geophys. Res.*, **71** 1195–1215.

Short, N. M. (1968). In *Shock Metamorphism of Natural Materials* edited by B. M. French and N. M. Short, Mono Book Corp. Baltimore, pp. 219–241.

Siegfried, R. W., G. Simmons, J. Schatz and D. Richter (1974). *Trans. Amer. Geophys. Union*, **55**, ABS. only, 418.

Simmons, G., H. Wang, D. Richter and T. Todd (1973). *Trans. Amer. Geophys. Union*, **54**, ABS. only, p. 451.

Simmons, G., D. Richter, T. Todd and H. Wang (1973). *Geol. Soc. Amer. Abstracts with Programs*, **5**, p. 810.

Simmons, G., R. Siegfried, D. Richter and J. Schatz (1974). ABS, *Fifth Lunar Science Conf.*, pp. 709–711.

Simmons, G., T. Todd and S. Baldridge (1975). *Amer. J. Sci.*, **275**, 318–345.

Stöffler, D. (1965). *N. Jb. Miner. Mh.*, **9–11**, 350–354.
Stöffler, D. (1966). *Contrib. Mineral. and Petrology*, **12**, 15–24.
Stöffler, D. (1971). *J. Geophys. Res.*, **76**, 5541–5551.
Stout, J. H. (1974). *Science*, **185**, 251–253.
Todd, T., H. Wang, W. S. Baldridge, and G. Simmons (1972). *Proc. of the Third Lunar Sci. Conf., Vol. 3, Geochim, et Cosmochim. Acta, Suppl. 3*, 2577–2586, MIT Press, Cambridge.
Todd, T., D. Richter, G. Simmons, and H. Wang, (1973). *Proc. Fourth Lunar Sci. Conf., Vol. 3, Geochim. et Cosmochim. Acta, Suppl.* 4, 2639–2662.
Tuttle, O. F. (1949). *J. Geol.*, **57**, 331–356.
Walter, L. S., B. M. French, K. F. J. Heinrich, P. D. Lowman, A. S. Doan, and I. Adler (1971), *Proc. 2nd Lunar Sci. Conf., Vol. 1, Geochim. et Cosmochim. Acta, Suppl., 2*, 343–358.
Wang, H., T. Todd, D. Weidner, and G. Simmons (1971). *Proc. 2nd Lunar Sci. Conf., Vol. 3, Geochim, et Cosmochim. Acta, Suppl. 2*, 2327–2336.
Warren, N. W. and G. V. Latham (1970). *J. Geophys. Res.*, **75**, 4455–4464.
Wiederhorn, S. M. and P. R. Townsend (1970). *J. Amer. Cer. Soc*, **53**, 486–489.

Effect of Scratch Hardness on Frictional Wear and Stick-Slip of Westerly Granite and Cheshire Quartzite

J. T. Engelder

*Lamont-Doherty Geological Observatory**,
Palisades, N.Y. 10964

INTRODUCTION

Stick-slip is a reasonable laboratory analogue for an earthquake in the upper crust (Brace and Byerlee, 1966). A stick-slip mechanism for earthquakes is appealing because the stress drop calculated for earthquakes is much closer to the stress drop for stick-slip than for the fracture of intact rock (Byerlee and Brace, 1968). Many earthquakes are associated with pre-existing fault zones such as the San Andreas fault of California (Eaton *et al.*, 1970) and so are associated with rocks which already have sliding surfaces. In addition the dynamic behaviour of stick-slip is like that of earthquakes; the particle velocity of stick-slip is 2 to 50 cm/sec and the rupture velocity is several km/sec (Johnson *et al.*, 1973).

A mechanism of stick-slip of rock is related to the frictional wear of rock by brittle fracture (Byerlee, 1967). Byerlee (1967, 1970) proposes that asperities on surfaces in contact become locked and sliding occurs only when the asperities shear off. The static frictional force may be equated with the shear strength of the interlocked asperities.

Shearing of asperities is the mechanism of stick-slip on highly polished Westerly granite (Engelder, 1974). The initial frictional wear of highly polished Westerly granite consists of wear tracks (mainly grooves) shaped with the outline of a carrot with a tip pointing in the direction of slip of the surface containing the groove and a blunt end into which an asperity has indented (Figure 1). The groove lengths are less than or equal to the slip distance during one discrete slip or stick-slip event. The grooves are considerably shorter than total slip on a Westerly granite surface which has been subjected to several stick-slip events. Because of the shape and length of wear grooves, Engelder (1974) suggests that the grooves are generated by the ploughing of asperities

*Contribution no. 2291

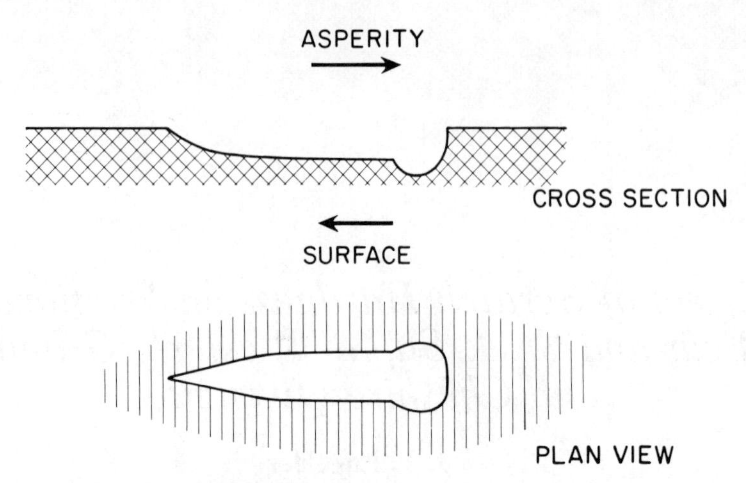

Figure 1. Schematic of an ideal wear groove in feldspar. Arrows indicate sense of slip.

that dig deeper as sliding progresses during a stick-slip event. The ploughing results in an increase in frictional resistance and impedes further sliding. Once sliding has stopped, shearing stress increases until the shear strength of the asperities is exceeded. An instability occurs with the shearing of the asperities and the surfaces slip once more.

This paper presents further observations on the mechanism of stick-slip on highly polished rock surfaces. In particular, attention is focused on the relation between scratch hardness of minerals on the sliding surface and stick-slip. The minimum normal stress at which stick-slip occurs is influenced by a difference in scratch hardness of various minerals on the sliding surface.

EXPERIMENTAL STUDIES

In order to study the mechanism of stick-slip, 1.2 cm diameter by 2.5 cm long cylindrical specimens were slid in compression on surfaces inclined at 35° to the cylindrical axis. Experimental confining pressures on the dry samples were between 0.1 and 2.9 kb with an average sliding displacement rate of 10^{-3} cm/sec. The sliding surfaces were saw cut and then hand lapped against each other to assure a good fit between surfaces. A final polish was achieved with 0.3 μm Linde A polishing compound.

The specimens used for this study are Westerly granite and Cheshire quartzite. Westerly granite consists primarily of quartz, plagioclase, and K-feldspar grains between 0.5 and 2.0 mm in diameter. Other types of grains comprise less than 2 per cent of the sliding surface area. Cheshire quartzite is monomineralic and has grain diameters up to 4.0 mm. The relative contrast in scratch hardness among minerals on the surface of granite versus the lack of contrast in scratch hardness for quartzite prompted the choice of granite and quartzite for this study.

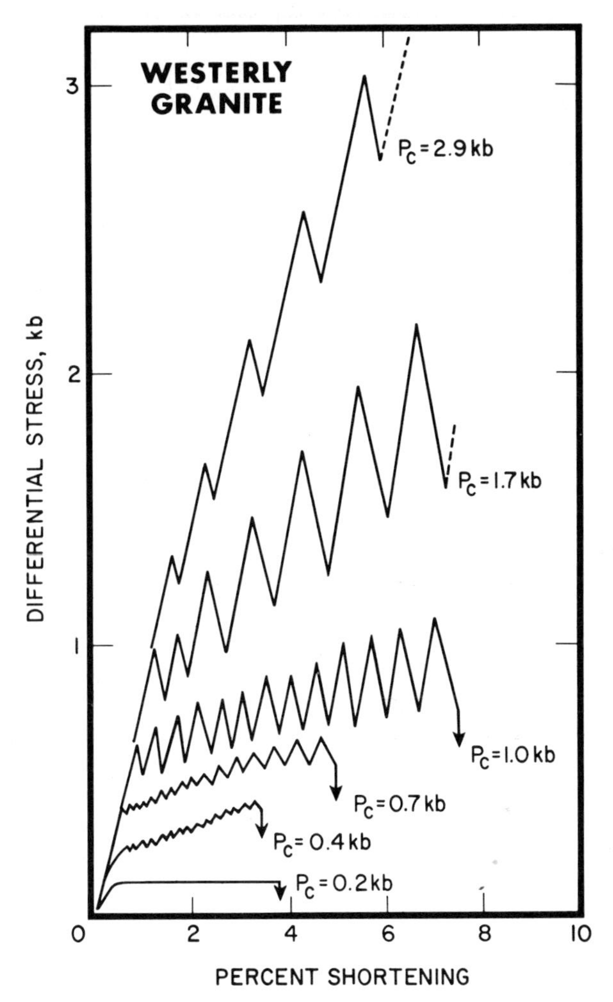

Figure 2. Differential stress versus shortening curve for the sliding of Westerly granite at various confining pressures.

Confining pressure affects the type of frictional sliding of both highly polished Westerly granite and Cheshire quartzite. For Westerly granite stable sliding occurs at confining pressures below 0.3 kb whereas stick-slip occurs above 0.3 kb (Figure 2). For each stick-slip experiment a slight to very marked increase in differential stress is necessary to initiate each successive slip. In contrast to the granite, highly polished Cheshire quartzite goes through a stable sliding to stick-slip transition at a much higher confining pressure (2.5 ± 0.2 kb).

To substantiate ideas on why the stable sliding to stick-slip transition is a function of lithology, Westerly granite and Cheshire quartzite were slid on each other at 1 and 2 kb confining pressure. Dilithologic experiments were prepared

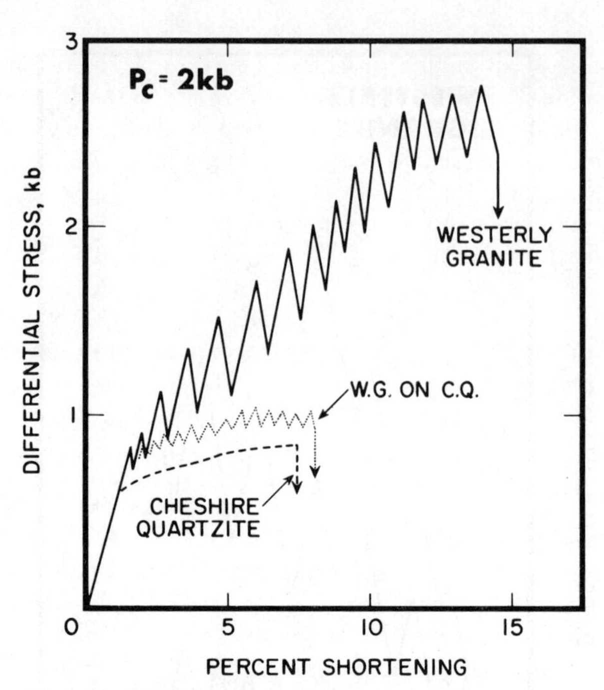

Figure 3. Differential stress versus shortening curves for various specimens at 2 kb confining pressure.

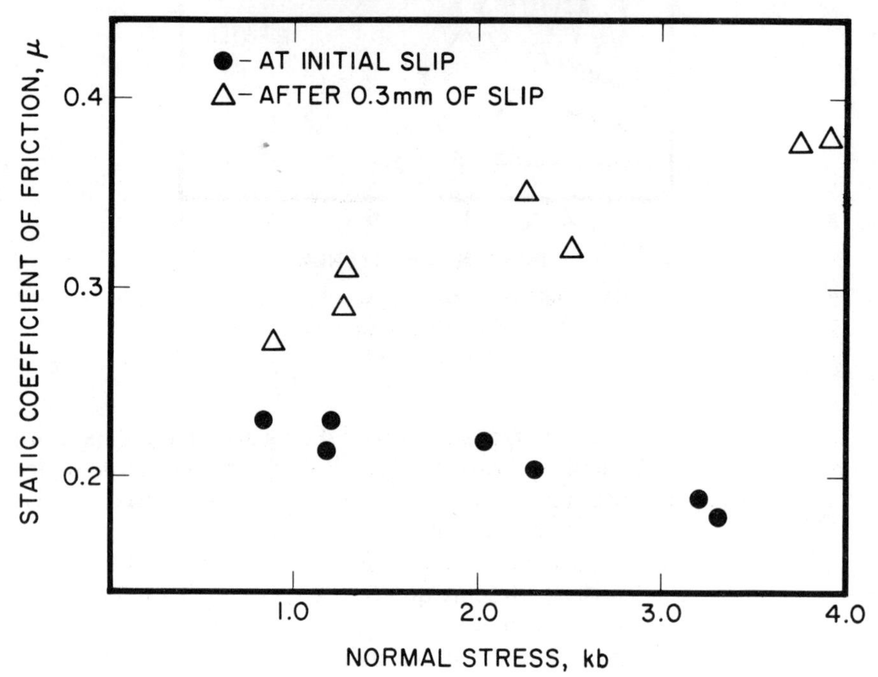

Figure 4. Static coefficient of friction of Westerly granite at various normal stresses.

by lapping a saw cut sliding surface of quartzite against a saw cut sliding surface of granite. The effect of sliding the quartzite on the granite is illustrated in Figure 3. At both 1.0 and 2.0 kb confining pressure Cheshire quartzite slides stably on itself but when slid on Westerly granite stick-slip occurs. The differential stress necessary to initiate slip of Cheshire quartzite on Westerly granite is much less than for Westerly granite on itself.

The static coefficient of friction (μ) at the initiation of slip of Westerly granite on itself follows a slightly decreasing trend with the increase of normal stress (Figure 4). In contrast μ for Westerly granite after 0.3 mm of slip on surfaces that were initially polished increases with confining pressure. As a result of these trends the static coefficient of friction of highly polished Westerly granite varies more per unit displacement at higher confining pressure.

FRICTIONAL WEAR

Engelder (1974) shows that frictional wear on highly polished Westerly granite accompanies stick-slip and occurs only for sliding above 0.3 kb confining pressure. The same relation between wear and stick-slip occurs with the sliding of Cheshire quartzite except that wear and stick-slip occur only at confining pressures above 2.5 ± 0.2 kb.

Figure 5. Reflected-light micrograph of Westerly granite. Sliding surface displaced 2 mm at 2.0 kb confining pressure.

The initial frictional wear of highly polished surfaces of Westerly granite and Cheshire quartzite consists of carrot-shaped wear tracks consisting of grooves, scratches and clumps of gouge. Tracks are continuously generated on the polished surface until the polish is worn off and the surface becomes covered with a layer of gouge. Initial wear tracks are distributed evenly over the Cheshire quartzite surface. In contrast the polished Westerly granite does not wear evenly; a larger number of wear tracks form on feldspar grains. As

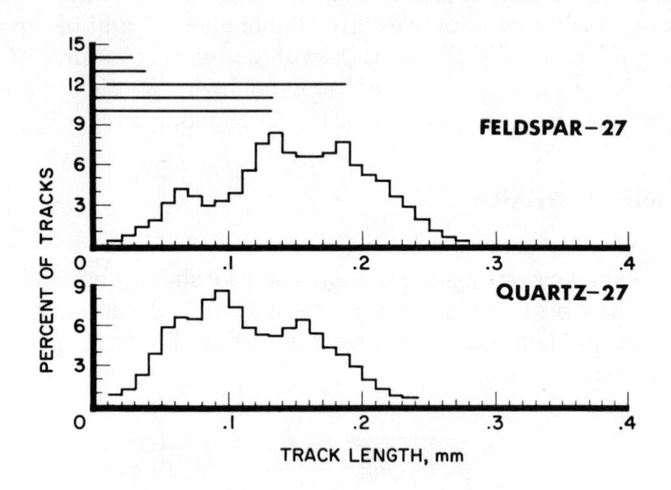

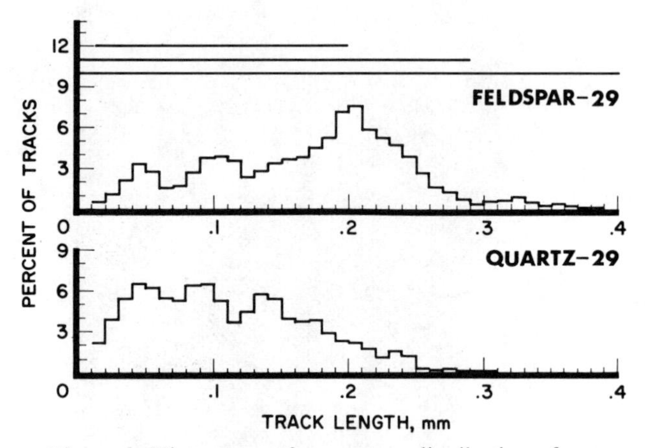

Figure 6. Histograms of percentage distribution of wear track lengths in feldspar and quartz grains on Westerly granite sliding surfaces for experiments 27 and 29 at 2.0 kb confining pressure. Number and length of individual slip lengths are indicated by horizontal lines above histograms. The top horizontal line is the first slip and the bottom the last. At least 300 tracks are included for each histogram.

a result quartz grains are accented in reflected light as those grains with a high reflectivity and few tracks (Figure 5).

Initial measurements of the lengths of the carrot-shaped wear tracks on Westerly granite indicate that most are less than or equal to one slip during a stick-slip experiment (Engelder, 1974). Furthermore, wear tracks on feldspar are slightly longer than those on quartz grains (Figure 6). On either mineral the length of most tracks is less than or equal to the longest slip during a series of stick-slips. The broad distribution of track lengths for the experiments

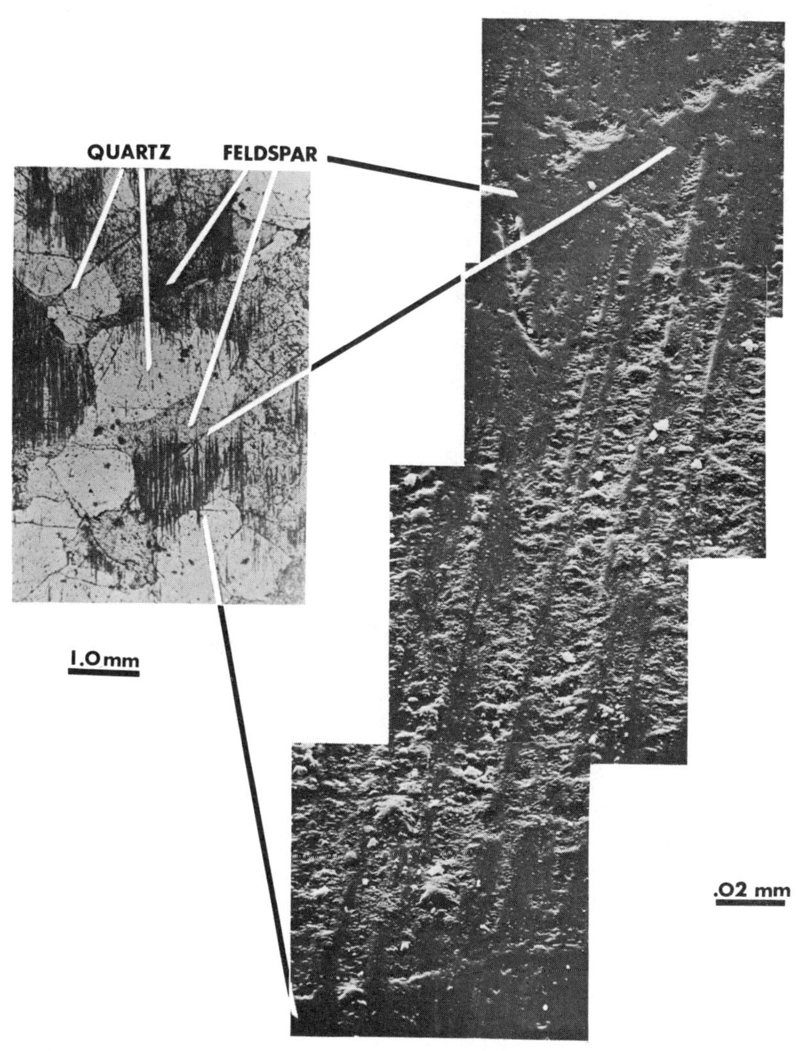

Figure 7. Reflected-light and S.E.M. micrograph of grooves in feldspar. 0.005–0.01 mm white particles in and between grooves are quartz fragments. Surface containing grooves moved toward top of micrograph.

illustrated in Figure 6 is due both to a variety of slip lengths resulting in the total slip of the stick-slip experiment and a variety of track lengths for individual stick-slip events. In addition some tracks are longer than the longest slip length. Wear tracks on the granite are not as numerous when Westerly granite slides on Cheshire quartzite at equivalent confining pressures. In this situation tracks are also less than the length of one slip during stick-slip.

A Cambridge Scanning Electron Microscope equipped for the energy dispersion analysis of X-rays is used to resolve the fine features of the wear tracks. The E.D.A.X. device permits identification of individual crystals on the sliding surface and particles associated with the abraded surface.

Three types of frictional wear tracks may be observed on the Westerly granite surface. They result from the sliding of a quartz asperity on feldspar, a quartz asperity on quartz or feldspar asperity on feldspar, and a feldspar asperity on a quartz. The identity of the asperity is revealed by the presence or absence of foreign particles within a wear track. For example, when a quartz asperity slides on feldspar, a groove is generated which contains quartz particles less than 5 μm in diameter (Figure 7). Grooves in feldspar have a rough bottom

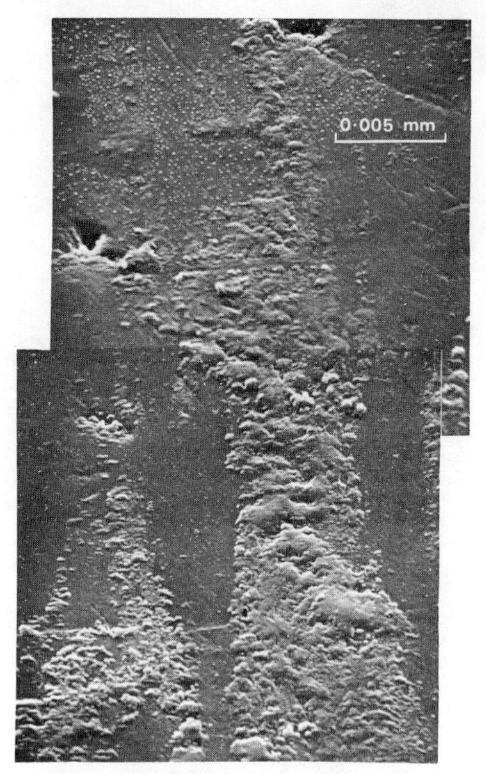

Figure 8. S.E.M. micrograph of a track of feldspar gouge smeared on a quartz grain. Surface on which track formed moved toward top of micrograph.

Figure 9. S.E.M. micrograph of scratches on a quartz grain caused by quartz asperities. Surface containing scratches moved toward top of micrograph.

which is also filled with loose feldspar fragments, many of which exhibit cleavage faces. Most of the grooves have a width to depth ratio which may be as large as 5:1. Grooves in feldspar reach their maximum width and depth after the asperity has travelled less than half of the length of the groove.

The effect of feldspar asperities sliding on quartz are observed by identifying feldspar on quartz surfaces (Figure 8). In this situation the feldspar seems to coat the quartz with a streak which is interpreted as remains of a feldspar asperity as it is worn off during sliding. Apparently the 'softer' feldspar adheres to the quartz and is sheared much like graphite from a pencil which adheres to paper as a pencil point is pushed over the surface.

Quartz surfaces have scratches caused by a series of release fractures which propagated as a sliding asperity presses into the quartz surface (Figure 9). In Engelder (1974) wear tracks on both feldspar and quartz were called grooves based on reflected light microscopy. To be more genetically correct the wear tracks on quartz will be called scratches because ploughing did not occur. The asperities causing these scratches are quartz because the scratches consist entirely of quartz particles; feldspar asperities are observed to wear off and are detected as a streak of feldspar on the surface of the quartz. This type of wear track is the only one found on the surface of Cheshire quartzite.

The situation for feldspar asperities on feldspar is much like that for the quartz asperities on quartz. Scratches are generated with associated release fractures and no foreign particles.

DISCUSSION

The scratch hardness of minerals on sliding surfaces affects the frictional properties of rocks. This is illustrated by considering the effect of feldspar among quartz grains. Highly polished quartz surfaces (Cheshire Quartzite) do not stick-slip below 2.5 kb confining pressure whereas highly polished quartz-feldspar surfaces (Westerly Granite) stick-slip at confining pressures down to 0.3 kb. Because frictional wear seems to be necessary for stick-slip on highly polished quartz surfaces, stick-slip will occur at confining pressures below 2.5 kb only in the presence of minerals with scratch hardnesses less than quartz. Feldspar has a lower scratch hardness than quartz and its greater susceptibility to wear promotes stick-slip of Westerly granite at confining pressures where Cheshire quartzite slides stably.

At less than 2.9-kb confining pressure Westerly granite has a larger static friction than Cheshire quartzite. Granite has three major types of wear contributing to its friction whereas quartzite has only one type of frictional wear. Quartz asperities sliding on feldspar are responsible for the increase in friction of granite relative to quartzite for the following reasons: (1) quartz asperities on quartz grains of Westerly granite should exert the same frictional force as asperities on grains in the quartzite, (2) feldspar asperities on quartz wear off and thus contribute little to the frictional strength of granite, and (3) feldspar surfaces wear much faster than adjacent quartz surfaces. The last point indicates that a large contribution to the frictional force of the highly polished granite is due to the ploughing of asperities into feldspar grains.

Speculation is necessary to explain why wear tracks occur on quartz grains of Westerly granite at confining pressures for which Cheshire quartzite slid stably and no wear tracks were generated. Edges of grains are chipped and rounded on highly polished Westerly granite and thus have been sheared during slip. Sheared grain boundaries are the location of asperities. Presumably a quartz asperity of a certain height is necessary in order to scratch quartz grains on the opposite surface. The elastic stiffnesses of plagioclase and K-feldspar vary more with crystallographic orientation than do the stiffnesses of quartz. When a normal load is applied to adjacent grains of feldspar and quartz these grains are more likely than adjacent grains of quartz to deform elastically in such a manner that a step appears at the grain boundary. Although quartz is elastically anisotropic, much larger normal loads are required before a step is generated which is equivalent in height to that expected for the boundary between a quartz and feldspar. Thus much higher normal loads are required before Cheshire quartzite starts to wear and stick-slip. Slight grain boundary asperities may also result from differential wear of quartz and feldspar during sample preparation.

The longer wear tracks in feldspar grains are believed to be due to the lower scratch hardness of feldspar. The idea is that asperities survive for a longer scratch distance because feldspar into which the asperity is ploughing is not as likely as quartz to reach the strength necessary to shear an asperity. Some feldspar grooves are longer than the distance of the longest slip during stick-slip. This is an indication that some asperities in feldspar do not shear at the initiation of slip but continue to plough through a second slip.

Asperities are more likely to shear at the initiation of a stick-slip event than during sliding. During static contact the asperities penetrate further into the grain containing the groove, thus creating a larger area of contact between the groove and the asperity. This larger area of contact results in the need for a larger shearing force for further ploughing or scratching and, so, effectively increases the strength of the material into which the asperity is penetrating. Further penetration during static contact is the result of time dependent creep of brittle materials under a point load (Scholz and Engelder, 1974).

The frictional properties of sliding surfaces are a function of the amount of wear on the sliding surface. This is illustrated by the difference between μ at the initiation of slip and μ after 0.3 mm of slip (Figure 4): μ changes more during 0.3 mm of slip at higher confining pressures. A plot of number of grooves generated per stick-slip event on Westerly granite versus confining pressure indicates that more grooves are generated per stick-slip event at higher confining pressure (Engelder, 1974). Thus the greatest change in μ for 0.3 mm of displacement and the greatest wear rate occur at higher confining pressures. The rate of change in μ is attributed to the rate of wear. Loose wear particles which are generated during wear interlock and increase the frictional strength of surfaces relative to the same surface prior to wear.

Cheshire quartzite on Westerly granite stick-slips at confining pressures lower than the Cheshire quartzite stable sliding to stick-slip transition. This is due to granite grain boundary asperities which scratch the quartzite. Once the quartzite is scratched quartz wear particles scratch quartz and groove feldspar on the granite. The lower stress drops for quartzite on granite relative to granite on granite is due to a lower rate of wear in the former case. A lower rate of wear means that fewer asperities lock the surfaces and so the frictional strength at the initiation of stick-slip is lower.

SUMMARY

The shape and length of wear tracks indicate that the frictional instability for slip on Westerly granite and Cheshire quartzite is due to the shearing of asperities which have penetrated the sliding surface. The larger magnitude of the frictional instability on Westerly granite relative to Cheshire quartzite is due to feldspar on the sliding surface. The feldspar has a lower scratch hardness which allows more asperity penetration per unit slip. Asperity penetration and subsequent wear seem to be necessary for stick-slip and occur at lower normal stresses on Westerly granite as compared to Cheshire quartzite.

REFERENCES

Brace, W. F. and Byerlee, J. D. (1966). *Science*, **153**, 990–992.
Byerlee, J. D. (1967). *Jour. Geop. Res.*, **72**, 3639–3648.
Byerlee, J. D. (1970). *Tectonophysics*, **9**, 475–486.
Byerlee, J. D. and Brace, W. F. (1968). *Jour. Geop. Res.*, **73**, 6031–6037.
Eaton, J. P., Lee, W. H. K. and Pakiser, L. C. (1970). *Tectonophysics*, **9**, 259–282.
Engelder, J. T. (1974). *Jour. Geop. Res.*, **79**, 4387–4392.
Johnson, T., Wu, F., and Scholz, C. (1973). *Science*, **179**, 278–280.
Scholz, C. H. and Engelder, J. T. (1974). *EOS*, **55**, 428.

A Study of Rock Fracture Induced by Dynamic Tensile Stress and its Application to Fracture Mechanics

B. M. Hamil

Department of Geology and Geological Engineering
Michigan Technological University
Houghton, Michigan 49931

and

S. Sriruang

Division of Geotechnics
Royal Thai Irrigation Department
Bangkok 4, Thailand

INTRODUCTION

The failure mechanisms of rocks have been studied under many different stress conditions. Because rock failure is a slow process in nature, much research has been done under static load conditions. Rock breakage requires rapid loadings, and most breakage studies are performed under dynamic stress conditions.

The purpose of this work was to develop a simple tensile testing method allowing us to study the preferred trajectory of fracture in rocks, document the features observed, and use the results to explain the mechanisms of tensile crack propagation in rocks.

Under point-loading tests, failure occurs with equal probability in all directions in isotropic rocks, but there is a preferred direction of failure in anisotropic rock (McWilliams, 1966). This direction coincides with the direction of minimum tensile strength and minimum sonic velocity. Handin *et al.* (1969) concluded that the state of pre-strain also controlled the orientations of the fractures. At low rates of loading, transgranular defects are the predominant factors influencing the fracture of Charcoal gray granite, but at high rates of loading, both transgranular and intergranular defects are important (Willard and McWilliams, 1969).

From their experiments with Westerly Granite under compression, Friedman *et al.* (1968) observed that more than 75 per cent of the length of shear fracture is parallel to cleavage planes in the feldspars or the grain boundaries or to both, but they did not distinguish transgranular from intergranular cracks. Wawersik and Brace (1971) reported that more than 50 per cent of all cracks in Westerly Granite are transgranular, and concluded that fracture patterns in rock depend not only upon stress or deformation but also on time and unknown factors.

Moavenzadeh *et al.* (1966) Rad and Moavenzadeh (1968), and Rad (1971) studied rock fracture by using a three point bending apparatus. They reported that the specific surface energy is a property of the material and does not depend on sample geometry or orientation of pre-existing flaws. The specific surface energy increases with deflection rate for granite, but remains essentially unchanged for marble. The same is true under thermal treatment. They believe that the weakening of the rock is due to the increase in amounts of intergranular and intragranular cracks upon thermal treatment.

Petersen and Shockey (1971) studied dynamic tensile failure in novaculite through flat-plate impact experiments using a gas gun. They reported that spall cracks normal to the impact direction and other cracks whose planes are more or less parallel to the impact direction are the main cracks formed. From the study of fracture surfaces, they also found that the fractures are mostly intergranular.

In the study of dynamic rock fracture under impulse loads using the Hopkinson split bar method, Attewell (1963) reported that strength increases with rate of loading but the differences are less pronounced for hard rocks which have higher static Young's modulus. Weber and DeMontille (1970) found little difference between the static and dynamic properties of rocks. Fracture initiated at the same stress under both static and dynamic loads (Hakalehto, 1970). The dynamic strength depends on the testing system and especially on the duration of loadings.

Dyke, Conn and Dagan (1969) studied dynamic failure due to high pressure, short duration loadings. They found that the particle distribution characteristics resulting from explosions are dependent on the material, and that the failure stress is a function of sample size.

Bieniawski (1967) studied crack propagation under dynamic tensile stress, caused by impact, using high speed photography. Our experimental results differ from his, but the basic mechanisms for crack propagation are the same.

EXPERIMENTAL PROCEDURE

The study of rock fracture under dynamic tensile stress requires that a single main crack be induced by the applied stress. The method must provide enough energy to break the rock yet keep the damage to specimens to a minimum, as severe disruption of mineral grains along the crack makes reassembly of the broken parts impossible. The exploding wire method used in this research meets these requirements.

The Exploding Wire Apparatus

Professor Been of M.T.U. suggested the use of exploding wires in this work. The apparatus was designed and fabricated by Professor Paul H Lewis of the Department of Electrical Engineering, M.T.U. for a study of the physics of exploding wires. It consists of a charging circuit and capacitor bank, and a remote unit which contains control circuitry for the charging and 'firing' opera-operation. A meter on the remote unit monitors the voltage level of condenser bank, and a 'slow' discharge circuit is available to remove any residual charge or to provide discharge in event of a malfunction of the firing circuit (Lewis, 1963). A sample holder and 'explosion container' were constructed and attached to the main unit (Sriruang, 1972, p. 6).

Specimen Preparation

Thirteen types of rock including sandstones, shales, siltstones, limestones, dolomites, conglomerate, gypsum, granite, gabbro, quartz-diorite, diabase, slate, and schist were subjected to detailed studies, the most extensive experiments being done on limestones, granite, quartz-diorite and gabbro.

Rock cores $2\frac{3}{4}$ inches in diameter were obtained by drilling with a three-inch diameter laboratory core drill through large samples. Exploration diamond drill cores provided samples of other diameters. The standard sample was a 2.5 cm thick right circular cylinder cut from core with a diamond saw. This thickness allows the use of an exploding wire length which provides the optimum explosion energy within the limits of the capability of the apparatus (Lewis, personal communication, 1971). Sample ends were ground parallel to within 1 mm using 200 and 400 grit silicon carbide on a cast iron lap using water as a lubricant. All specimens were dried with compressed air and stored at room temperature.

Twelve sample geometries were used in the study (Figure 1). Most of the basic information was obtained from samples using geometry I. Changes in sample geometry change stress conditions within the sample and allow control of the direction of fracture propagation.

A single slot for the initiation of cracking was cut with a diamond saw blade along the length of all specimens, its depth varying from 3/8 inch in small cores to 1/2 inch in large ones. Slots were cut in a single operation orienting the direction of the cut along a radius and parallel to the axis of the rock cylinder so that slots have equal depth along the length of the specimen. The diamond saw gives a blunt-edged slot, the advantages of which will be explained later.

Sample geometries VII through XI contain circular holes. These were drilled by a single peak tungsten carbide masonry drill of 3/16 inch diameter, the smallest size commercially available. Properly operated, the drill can make clean accurate holes up to 3 inches deep. The drill is driven by a slow-speed drill press. Cooling the bit with cold water after the initial crater is made helps prolong bit life. Cutting edges of the bit were kept sharp on a grinder.

Specimens were set on top of a flat slab during drilling. Both specimen and

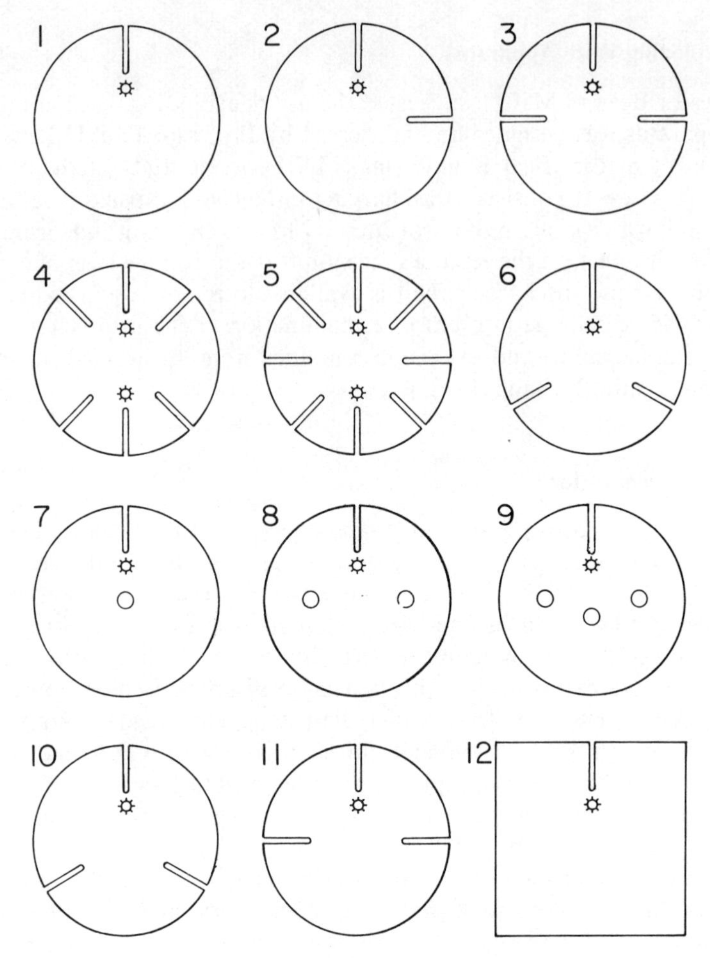

Figure 1. Twelve types of sample geometry used in exploding wire rock fracturing. Charging slots are indicated by*.

support were clamped to the drill press table to prevent any movement due to vibration during drilling. The under support prevented bursting of the bottom of the hole during the final stages of drilling. Light pressure and slow bit speed prevented enlarging of the hole. Petrographic examination showed that grains and grain boundaries surrounding holes were not damaged by the tangential stress of drilling.

The positions of slots and holes in specimens were selected arbitrarily with respect to microstructures in specimens. One purpose of the research is to study the influence of any defects in the specimen on crack propagation. The results of work on sample geometries containing holes are covered elsewhere (Sriruang, 1972).

Methods of Study

Charging slots were positioned where induced cracks, once initiated, would propagate under the influence of the applied stress field, mineral grains and their orientations, and defects present in specimens. These defects, such as pre-existing cracks, veins, joints, bedding planes, phenocrysts, etc., caused local variations of stress fields and therefore variation in crack trajectory and morphology. Selection of samples was important for this reason. Defects in rock should be macroscopic in scale but their abundance should not be so large that induced cracks become too complex for effective study. Thus specimens that contain abundant joints or fine opened cracks were not used in the tests because the pre-existing cracks might be mistaken for induced cracks. All defects were mapped before samples were tested.

The prepared samples were placed between electrodes made of copper bus wire in such a way that the fine copper filament joining the electrodes (the exploding wire) touched the bottom of the charging slot. Specimens were held in by the electrodes with the aid of an epoxy plate which prevented movement of the electrodes during explosions. The tops of the charging slots were then covered with a thin plate of high density rock to confine the explosion pressure yet allow free deformation of the specimen in the direction of applied stress.

The equipment was operated following procedures set up by the builders (Lewis, 1963). The initial explosion was triggered with 2000 volts. If no cracking occurred the voltage was increased in 500 volt increments until crack initiation was observed. Often initial cracks were too fine to be seen with the unaided eye. Low power microscope observation of water-wet rock surfaces often served to identify cracking. Any cracks observed were recorded immediately either by sketching or photography or both. Following this, the specimen was thoroughly dried with compressed air.

The propagation of the initial crack or cracks was studied by repeating the explosion in the slot. Voltage used was reduced for subsequent explosions below the 'critical voltage' required for crack initiation. It was confirmed that less energy is required to propagate than to initiate a crack, as reported by Kutter and Fairhurst (1971). No specimens tested could sustain an explosion of greater than 5000 volts without failing.

Crack propagation can be recorded by sketching or following the methods proposed by Gardner and Pincus (1968) and D'Andrea and Condon (1971). Gardner and Pincus studied cracks on rock surfaces by applying a fluorescent dye penetrant on the polished surfaces and taking photographs of the surfaces under ultraviolet light. D'Andrea and Condon modified this technique by using double-exposed photographs of the rock surfaces at the same positions. The first exposure was taken on the dried surface and the second exposure was taken after the surface was treated with a fluorescent dye penetrant. Thus the surface structures and the cracks could be compared. The latter is more accurate but time consuming, and the interpretation of results from phtographs may be erroneous if the method is used several times with the same specimen. According to the second method, neither the specimen nor the camera can be moved before the second exposure is done.

Two kinds of dye penetrants were used in this study to outline fractures:

'Spot-check'* which is red acrylic plastic which can be seen in ordinary light and 'Zyglo'*, a fluorescent dye which is detectable with long wave ultraviolet light. Spotcheck was used extensively and successfully in this research.

As the induced crack was propagated by repeated explosion, sample geometry determined the eventual fate of the crack and the sample. If the sample failed, the broken parts were reassembled using casein glue (Elmer's glue) after study of the fractured surfaces. Reassembled specimens were then sectioned perpendicular to the crack plane for further study. Unfailed samples had their fractures and slots filled with Elmer's glue for reinforcement before sectioning, as casein glue has higher heat resistance than epoxy resin.

Thin sections were prepared from both faces of test specimens, after which the specimens were still thick enough to be broken further if necessary. Results from this study indicate that thin sections are best made from the outside of fractured specimens because more mineral grain breakage occurs along the fractures in the interior of specimens and the fractures are larger in the interior than at the surface.

Thin sections of normal thickness (about 0.003 cm) were made and left uncovered for further treatment. Spotcheck was then applied to the finished but uncovered thin sections and left for 1 minute or more depending on how much penetration of dye was required. Spotcheck reacts with Lakeside cement which is used to cement the rock to the glass slide, and with Elmer's glue. It penetrates opened cracks and even microcracks. The excess dye was wiped from the slide with clean tissue. The opened cracks and those filled with Lakeside cement and Elmer's glue appeared red. No reactions between Spotcheck and mineral grains were found. After removing the coating dye, high contrast between the stained cracks and rock particles can be observed. The thin sections were dried gently with compressed air. After the fractures were stained with Spotcheck, the thin sections were stained again for mineral identification by using the methods of Laniz, Stevens and Norman (1964), and Lanuron (1966). Plagioclase is stained red, K-feldspar yellow, and quartz colourless. Staining must be done before applying immersion oil and covering the thin sections.

Immersion oil was applied on the thin sections instead of 'Permount' which is normally used to provide the light path between the thin section surface and the cover glass. 'Permount' should not be used because it dissolves the dye present in the cracks. The edges of the cover glass were sealed with Elmer's glue after the excess immersion oil was removed from the glass slide to prevent the immersion oil from drying. Elmer's glue is soluble in dilute acetic acid and can be removed at any time.

RESULTS AND ANALYSIS OF DATA

General Statement

Rock is often assumed isotropic and homogeneous for engineering purposes,

*Products of Magnaflux Corporation, Chicago, Illinois.

but it may be anisotropic and non-homogeneous in reality. Bedded sedimentary rocks can be treated as layered composite materials, while igneous rocks behave as crystalline materials. Sandstones are similar to granular materials, and conglomerates are equivalent to concrete. There is no exact boundary between rocks and other materials.

The results of this research have shown that the direction of crack propagation may or may not be pre-determined by the nature of the rock. Thus particular rules or theories in fracture mechanics are used to explain the mechanisms that cause the cracks to propagate in different ways. Because the mathematical approaches to these theories are very complex and require many data, we prefer verbal and graphical description to mathematical modelling at this stage.

From the study of fracture surfaces, it was found that although rocks exhibit gross brittle failure, some minerals show both brittle and semi-brittle failure. Some microstructures that occurred during the propagation of transgranular cracks can be used to indicate the directions of local crack propagation, and sometimes the macroscopic directions of crack propagation.

Fracture Surface Morphology

A Study of Fracture Surface Structures and Related Mechanisms at Low Magnifications

The fracture surfaces of broken specimens were studied under a binocular microscope using inclined light and magnifications of 3.5 × and 10 ×. The classification of fracture marks proposed by Murgatroyd (1942) and later expanded by Gash (1971) has been adoped here:

Rib marks: Rib marks are approximately semi-circular or arcuate ridges, or distinct changes in plane, which are usually continuous. They are concave to the origin of the main fracture and hence indicate the direction of fracture propagation (Figure 2). They can be subdivided according to the geometrical structure into (1) arcs of approximately similar curvature, marking a change in the plane of the main facture; concave, but not concentric, to the origin of the fracture; and (2) arcuate or completely semi-circular, continuous ridges which are concentric to the origin of frature.

Hackle marks: These features are commonly observed on the fracture surfaces of metals and less commonly on rock and glass. Hackle marks are curved striations, consisting of grooves and ridges which are generally discontinuous. There is no 'slickensiding' or any similar phenomenon. The direction of propagation is invariably as shown in Figure 2 with the hackle marks spreading out as the fracture develops.

Rib and hackle marks display different morphology and are not usually observed on the same fracture surface. They have been found to co-exist on

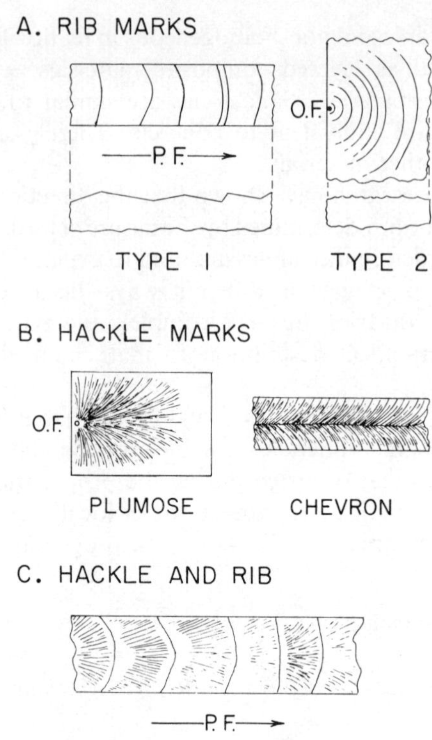

Figure 2. Diagrammatic representation of classes of fracture surface markings. (A) Rib marks, types 1 and 2. (B) Hackle marks, 1 = plumose structure, 2 = chevron structure. (C) Combination of hackle and rib marks. PF = propagation direction of fracture. (Reproduced from P. J. S. Gash, *Tectonophysics*, **12**, 351 (1971) by permission of Elsevier Scientific Publishing Company).

fracture surfaces in glass and rock. The form of the rib marks depends on the stress conditions and not on any preferred lines of rupture (Murgatroyd, 1942). The spacing of the rib marks is affected by fracture velocity; the slower the fracture, the more frequent the marks. Rib marks increase in density with distance from the origin until the surface becomes very rough and the crack may branch. Hackle marks vary considerable in size according to the speed of the fracture. As the crack slows down the hackle marks disappear.

Steps are steeply inclined offsets of the main fracture along other small fractures (Lutton, 1970).

Fracture Surfaces Showing Predominantly Trans-granular Fracture.

The fracture surfaces of a granite specimen having sample geometry type I were studied. The characteristic structures of tensile fracture surfaces mentioned previously could be seen on a relatively flat surface. In igneous rocks, in general, the orientations of mineral grains are random, and cohesion between mineral grains is very strong. The energy applied to break the rocks must overcome the strength of the individual grains and cohesive force between the grain boundaries.

The strength of each mineral grain having well-developed cleavages varies depending on several factors. Cleavage microcracks occur on well-developed crystallographic planes and the number of planes susceptible to cleavage in a polycrystalline material will be a function of orientation. The magnitude of stress and strain play an important part in the formation of microcracks, and these quantities will generally vary with orientation within the material. In a specified grain, having different cleavage planes, the one most nearly normal to the tensile axis will have the greatest stress and strain across it and will be the one most likely to cleave. In a simple tensile stress condition, the normal stress across a cleavage plane will vary with the angle between the normal to the cleavage plane and the tensile axis. As the tensile stress is increased, the first cleavage planes to have a critical stress across them form cleavage cracks which will be perpendicular to the tensile axis. As the tensile stress increases further, additional planes inclined at larger angles to the tensile axis will be exposed to the critical stress and cracks will appear on these planes. The process continues until the tensile stress reaches its maximum value. Only those planes that are perpendicular to the tensile axis will have a pure normal stress across them, all others will have a combination of tensile and shear stress acting on the cleavage plane.

Variations in elastic constants with crystallographic direction, the presence of additional phases with different elastic constants, and the stress fields associated with dislocation distribution will all result in inhomogeneities in the stress across a plane in the material. By using the root-mean-square value of normal and shear stress as a criterion for the formation of microcracks, Kaechele (1967) concluded that the conditions for microcrack formation depend on the cosine of the angle between the crack plane normal and the tensile axis, rather than the cosine squared relationship that results from considering only normal stress. Brace and Walsh (1962), among others, measured the specific surface energy of minerals. They found that the specific surface energy of a cleavage plane in a particular mineral is different from that of other planes. The best cleavage of minerals may be the plane of minimum surface energy. For examples: quartz has $(10\bar{1}1)$ and $(\bar{1}011)$ as its best and good cleavages respectively, all members of the feldspar group are characterized by the cleavages (001) perfect and (010) good, making an angle of nearly 90°. Micas have perfect (001) cleavage which governs completely the orientation of flakes. Although the strength of each cleavage plane in a crystal varies according to its crystallographic orientation, the cleavages may possibly break when the magnitude of stresses and the orientations of mineral grains with respect to the direction of the applied tensile stress are considered together. If the normal to a cleavage plane makes some angle with the tensile stress, shear component of the applied stress will occur along the cleavage plane. Shear stress is a major cause of plastic deformation on cleavage planes and the formation of steps on well-developed cleavages of crystals.

Because the study of fracture morphology was limited in the direction perpendicular to the fracture surfaces, only mineral grains whose cleavage planes parallel the surfaces were investigated. Using magnifications of 3.5 × and 10 ×, we were able to study the structures on fracture surfaces and their orientations with respect to the macroscopic direction of crack propagation. The cleavage fractures are characterized by their extreme flatness and high reflectivity.

Besides rib marks, which were often found in quartz crystals, other structures indicating tensile fracture were studied on the cleavage planes oriented more or less perpendicular to the tensile axis. On these flat surfaces of the crystals the most prominent structures were steps in a plumose arrangement.

In crystals a fracture surface may not be planar but instead may consist of *en echelon* segments of the cleavage plane connected by short steps (Lutton, 1971), as within a large single crystal, or in a single grain of a polycrystal, transgranular cleavage fractures propagate on more than one level, forming steps on the cleavage face. In polycrystalline materials large cleavage steps (differences in level of propagation) form when the crack crosses a grain boundary because the (crystallographic) cleavage planes in adjacent grains usually do not have a common line of intersection. Steps formed at the boundary increase in height with increasing values of the angle between the two lines formed on the grain boundary by the intersection of the respective cleavage planes and the grain boundary (Lange, 1967). The step patterns are not affected by the values of the angle between the normals of the two respective cleavage planes when the cleavages and grain boundary are parallel. When the planes make large angles with the boundary, transgranular fracture is not likely to take place and the crack front often propagates along the grain boundary instead of crossing it. When the angle is zero, no steps are formed at the boundary and no segmentation of the crack front can be seen. As the angle increases, the height of the steps increases and segmentation of the crack front may occur closer to the boundary. For large angles (more than 16 °) crack reinitiation occurs at the boundary and the fracture surface is very irregular. The reinitiation stress for transgranular cleavage increases with the angle between cleavage and grain boundary, but changes very little as the angle between cleavages increases. The above argument is valid for brittle materials that exhibit cleavage but is limited to some extent for materials exhibiting conchoidal fractures, e.g. SiO_2, Al_2O_3, etc. (Lange, 1967). If the cleavage planes in a particular grain are poorly oriented for cleavage the crack front may circumvent the grain and the grain eventually fails by tearing (Tetelman and McEvilly, 1967, p. 102).

Lange observed that during transgranular cleavage there were two modes of energy absorption: (1) the formation of more surface area due to the stepped surface and the undercut cracks associated with the steps, and (2) the plastic deformation associated with the steps. The energy absorbed during fracture increased with increasing angle between cleavage and grain boundary. If the work of fracture is much larger than the surface free energy of a brittle crystal, some dislocations must be generated in the process of cracking (Kelly *et al.*, 1967).

Figure 3 shows a sketch of cleavage fracture on a (001) cleavage plane of a fledspar crystal. The steps indicate the macroscopic direction of crack propagation. The broken edges of cleavage flakes forming steps tend to follow the (110) cleavage plane and lie more or less perpendicular to the macroscopic direction of crack propagation. The crystal shows Carlsbad and albite twinning of which the former is more susceptible to plastic deformation.

Figure 4 shows cleavage fracture of two feldspar crystals. The left crystal was broken along the (001) cleavage plane and the right one along the (010) plane. The macroscopic direction of crack propagation is perpendicular to the edge of the specimen. Steps in *en echelon* arrangement do not seem to develop on the cleavage fracture of the left crystal that shows albite twinning. The right crystal dips gently about 10 °–20 ° to the left. Steps in plumose arrangement can be seen on this crystal. Thus the two cleavages show different modes of fracture.

Figure 5 shows rib marks on a quartz crystal (circled). Microcrack propagates through the crystal to the top of the photograph while the macroscopic direction of crack propagation is from left to right.

Figure 6 shows a sketch of rib marks on fracture surface of the quartz crystal shown in Figure 5 (not to scale). It can be seen that the rib marks radiate from different point sources. They are interfered with by stress waves from neighbouring sources. At the 'intersection' of the stress waves, rib marks are arranged in such a way that the crests of one pattern are located at the troughs of the neighbouring patterns regardless of the directions of microcrack propagation. The stress waves from different sources pass each other like alternating current. The stress waves of each pattern are preserved even in the boundary of the pattern moving in the opposite direction. It seems that the quartz crystal was subjected to compressive stress waves in the direction perpendicular to the plane of the sheet diagram and the microcracks propagated in the plane of the diagram from their origins where the shock pulse initiated. Tensile microfracturing possibly occurred due to the interaction of incident and reflected waves. Rib marks start away from the origin at about the 45 ° reflection point. (For more detailed explanation, see Gash, 1971). The frequencies of the rib marks decrease as the fracture velocities increase away from their origins. The rib marks die out where the fracture surface becomes extremely rough or deceleration of the crack velocity occurs.

Formation of oriented microfractures by interference of stress waves result in a plane or zone of weakness in advance of the main fracture. Thus the main fracture may change its direction as it approaches this weak plane and propagate through it. It was found that sometimes rib marks radiate from their origins in the direction opposed to that of the macroscopic crack propagation. Similar behaviour was found in fine-grained polycrystals (Teleman and McEvily, 1967, p. 103). No steps were found on the faces of single crystals of feldspar having no cleavages and conchoidal fractures were found in the

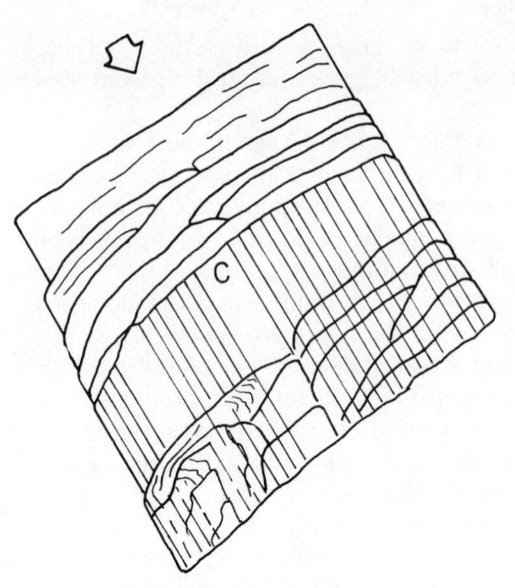

Figure 3. Sketch of cleavage steps in an en-echelon arrangement on a feldspar crystal. C = Carlsbad twinning. A = Albite twinning. Arrow indicates the macroscopic direction of fracture propagation.

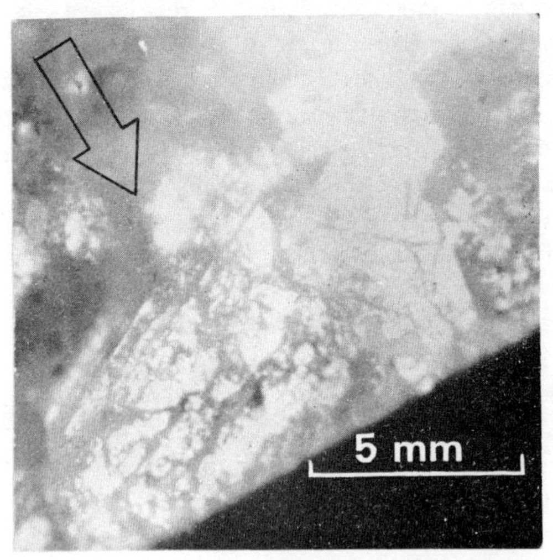

Figure 4. Cleavage fracture of two adjacent feldspar crystals showing different modes of fracture. The left crystal is broken along the (001) cleavage direction, and the right on along the (010) direction. Steps in en-echelon arrangement can be seen on the right crystal. The arrow indicates the macroscopic direction of crack propagation.

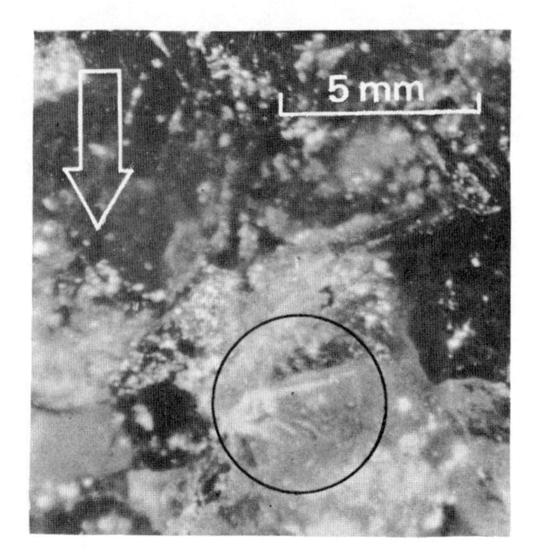

Figure 5. Rib marks on a quartz crystal (circled). Arrow indicates the macroscopic direction of crack propagation.

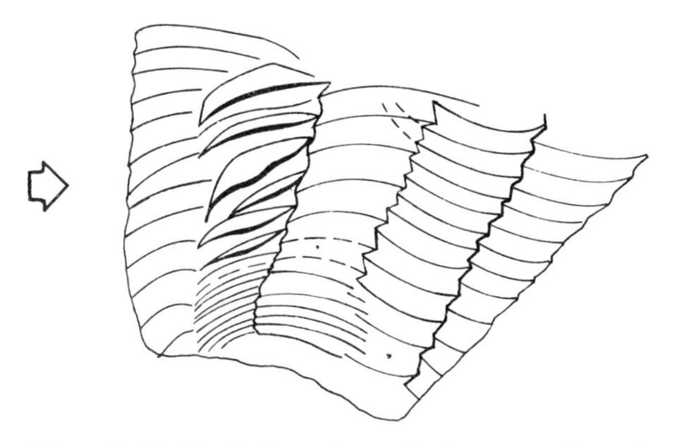

Figure 6. Sketch of rib marks on the fracture surface of the quartz crystal shown in Figure 5 (not to scale). Rib marks radiating from different point sources interfere with stress waves from neighbouring sources. Arrow indicates the macroscopic direction of crack propagation.

broken crystals. Normally the crack propagates along the grain boundary of such feldspar single crystals.

Fracture Surfaces Showing Predominantly Intergranular Fracture

A dolomite specimen of sample geometry I was used with the bedding plane making an angle of about 30 ° with the plane of the slot. The specimen was

Figure 7. Steps in block form in a dolomite crystal. The steps increase in height as the crack propagates in the macroscopic direction of crack propagation indicated by the arrow.

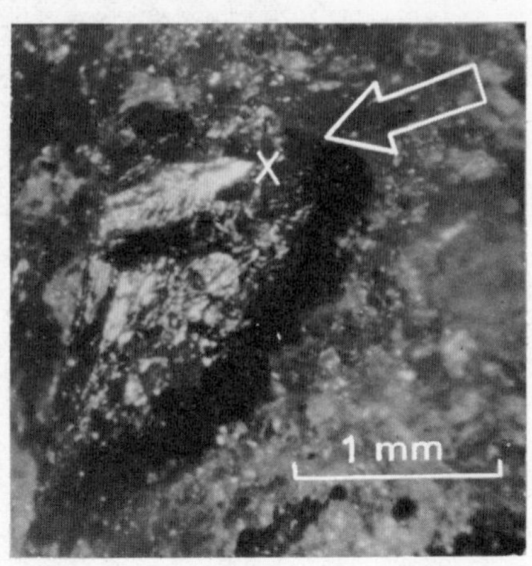

Figure 8. Steps in plumose arrangement in a dolomite crystal. The × indicates the point of crack initiation in the crystal. The arrow indicates the direction of macroscopic crack propagation.

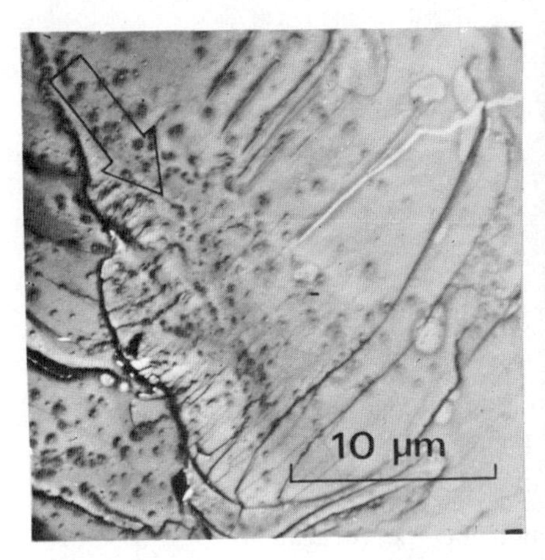

Figure 9. Electron micrograph of the replica of the fracture surface of a feldspar crystal. Dispersion of spherical solid inclusions (dark spots) is not uniform. Cleavage steps are present as hair lines. Artifacts and cracks in the replica may be seen. The arrow indicates the direction of propagation of the crack front.

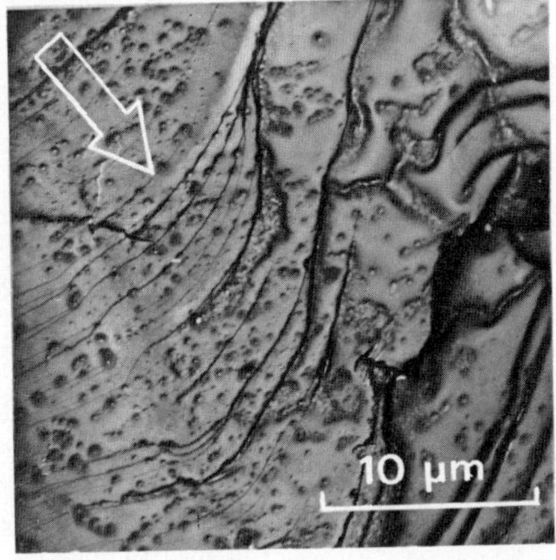

Figure 10. Electron micrograph of the replica of a fracture surface of a feldspar crystal showing coalescence of cleavage steps. The smaller cleavage steps coalesce to larger steps as the crack front moves. Cleavage steps passed through solid inclusions. The arrow indicates the direction of propagation of the crack front.

broken to rupture by a single explosion at 4000 volts. The fracture propagated more or less parallel to the axis of the slot, in the direction perpendicular to the tensile axis.

The failure of the specimens was predetermined since, for anisotropic rocks containing definite planes of weakness making some angles from 60° to 90° with the tensile axis, failure occurred across the planes of weakness which were bedding planes in this case (Youash, 1969). Fracture on a plane parallel to the bedding plane is accomplished by the lowest energy absorption rate while cracking on a plane perpendicular to the bedding plane but in a direction parallel to it requires substantially more energy (Hoagland and Hahn, 1971). In our experiment, the fracture was on a plane making an angle of 30° with the bedding but in a direction parallel to it, requiring less energy. The fracture surface of the broken dolomite specimen was very much less rough than those of granite and gabbro. The rock consists of fine-grained quartz, larger-grained dolomite, and quartz cement. (See Figures 7, 9, 14, 15 and 16), the composition varying from one bedding plane to another. The quartz cement made the rock so hard that drilling was very difficult. Conchoidal fractures containing rib marks were found in large quartz grains. Again, some rib marks indicated that the direction of microcrack propagation opposed that of the macroscopic crack. Steps with plumose arrangement were found in dolomite grains as shown in Figures 7 and 8.

Figure 7 shows steps due to block cracking of a dolomite crystal. The shapes of the steps are very much controlled by cleavages. The steps moved from a lower level to a higher one following the direction of macroscopic crack propagation.

Figure 8 shows steps in plumose arrangement in a dolomite crystal, which was subjected to higher plastic deformation than the one in Figure 9. The mark X shows where crack reinitiation occurs.

Kelly *et al.* (1967) proposed a criterion to predict the failure of a crystal. If the ratio of the largest tensile stress to the largest shear stress close to the tip of a crack in the crystal is greater than the ratio of the ideal cleavage stress to the ideal shear stress, a completely brittle fracture can occur. Otherwise, the crystal must break with some plastic flow, as in the dolomite crystals in Figures 7 and 8.

Fracture Surface Structures and Related Mechanisms at High Magnifications

In general, rock exhibits brittle failure under short duration loadings but some minerals composing the rock display plastic deformation during the fracture process. Those minerals that do not show large plastic deformation will be responsible for the brittle failure of rock. Quartz, for instance, remains brittle and elastic up to fracture under normal environmental conditions but it can be plastically deformed under compression at atmospheric pressure at temperature at 550 °C and above (Baëta and Ashbee, 1967). If the mechanical properties of other minerals present in a rock are known, a better understanding of the failure mechanisms of the rock can be obtained.

Rib marks and steps, which are characteristics of transgranular crack propagation, were found on the fracture surfaces of the specimens cracked by dynamic tensile stress. These structures show that many fracture mechanisms are involved during the process of their formation.

During transgranular cleavage, steps formed at the grain boundary increase in height with increasing angles between cleavages and grain boundaries (Lange, 1967). If the angle is large, reinitiation of the crack front will occur at the grain boundary, and the grain boundary will act as a crack barrier. The stress at the grain boundary will thus increase with the angle between cleavage and grain boundary. The reinitiation stress may cause dislocation of mineral grains at the grain boundary and tearing of the grains ahead of the crack front. Much energy is dissipated at the grain boundary. The energy to form steps less than 1 micrometre in height in a brittle material is small compared to the intrinsic surface energy of the material but is relatively large if the steps are larger than 1 micrometre (Lange, 1967; Lange and Lambe, 1968).

Because most of the dislocation activity is concentrated around the grain boundaries, much plastic deformation occurs in these areas (Evans, 1970). When a crack meets a grain boundary, the crack front will change its direction to the positions where energy consumed by dislocation motion is the smallest, to form an irregular crack front. Thus the stress at the pinning points will be greater than the overall stress at the crack front.

Dislocations are line defects. The atoms close to the dislocations are displaced from their positions in the perfect lattice. When minerals are plastically deformed, dislocations move when the shear stresses acting on them exceed the resistance due to the lattice and the other dislocations so that the dislocation line represents a boundary between the unslipped and slipped areas of the slip planes. There are two types of dislocations: edge and screw dislocations. Normally they occur together forming dislocation loops. With an edge, the displacement produce compressive strain above the dislocation line and tensile strains at the bottom of the dislocation line. For screw dislocations the displacements produce a helical strain field. Pure screw dislocations do not have a tensile stress field and hence do not cause cleavage crack formation (Tetelman and McEvilly, 1967, p. 240). Normally the edge dislocations move faster than the screw dislocations, as the stress required to move screw dislocations is greater than that to move edge dislocations (Evans, 1970). Even more, the stress to form slip bands is very much greater than that to move dislocations. Slip planes may or may not coincide with cleavage planes. When a slip plane intersects an obstacle dislocations are piled up. Microcracks are formed when the concentrated shear stresses at the tip of blocked edge-dislocation band are equal to the theoretical cohesive stress of the cleavage. If the applied stress is increased, the microcracks will increase in length and may lead to fracture. It has been reported that pile-up formation is essentially complete in an extremely short time if the dislocation velocity is high (Rosenfield and Kanninen, 1970). If the stress concentrations which would be generated by equilibrium pile-ups are much larger than the lattice can sustain, the pile-up formation may not be complete.

Dislocations in mica can be formed in nature during the period of crystallization associated with high temperature and pressure environments. During the process of cooling to the normal atmospheric pressure and temperature, mica crystals may be subjected to differential stresses due to thermal and other conditions. Thus dislocations can be formed during this process also (Caslavsky and Vedam, 1970). Most of the dislocations in mica are formed in the basal planes, but some emerge out of the basal planes. If it is required to study dislocations in mica induced by shear stress from an experiment, pre-existing dislocations should be studied prior to the performance of the experiment. Quartz crystals under plastic deformation from compression at atmospheric pressure and temperatures of 550 °C and above, show slip on the system (0001) $[\bar{1}\bar{1}20]$ (Baëta and Ashbee, 1967). At temperatures above 700 °C additional slip systems are formed.

Crystals often contain defects such as cavities, or solid and liquid inclusions, which influence the properties of the crystals and the propagation of cracks. They may be dispersed, or they may form linear arrangements controlled by the crystallographic directions and/or cleavages of the crystal, which may either strengthen or weaken the crystal. Forwood (1968) found that when a crack enters a composite region, a retardation of the crack occurs associated with the formation of cleavage steps and then, after a short penetrating distance, the crack velocity reaccelerated due to the action of constant force. Tipnis and Cook (1966) proposed that ductile shear fracture initiates at non-metallic inclusions because of the difference in plasticity within the inclusions; failure of the inclusion-matrix bond; or brittle fracture of inclusions. Cavitations usually occurs at an inclusion-matrix interface in a plastically deformed matrix. There is a size below which the fracture is inversely proportional to the square root of the particle size, and above this size the applied stress causes cavitation without plastic strain. The critical particle size is 200–300 Å (Tanaka *et al.*, 1970).

When a crack front intersects a defect there will be an interaction between the two and cleavage steps are formed. Lange (1970) proposed that a crack front possesses a line energy, which is an expression of the fracture energy. The fracture energy should depend not only on the amount of energy required to form new surface area but should also depend on the spacing between the dispersed defects. By using the theory of line energy he obtained an expression which indicates that the fracture toughness of a brittle material increases with the decreasing spacings of the defects. This argument is in agreement with those of Hasselman and Fulrath (1966) and Forwood (1968).

Figure 9 above shows an electron micrograph of the replica of fracture surface of a feldspar crystal from a granite specimen. The crack propagates from the top to the bottom of the figure. The cleavage steps can be seen. The dark spots are solid inclusions in the feldspar crystal. The sizes of the spherical inclusions are very much less than 1 μm. The dispersion of the particles is not uniform, and although the particles are very small, their effect on the fracture toughness of the crystal can be recognised. The density of small steps increases with the decrease in spacings of the particles, demonstrating the increase in strength of the crystal where the inclusions are closely spaced. Fewer steps are located where the spacings of the particles are relatively large. Although Lange (1970), Forwood (1968), and Hasselman and Fulrath (1966) did not find these phenomena, these structures support their theory that fracture toughness of a brittle material increases with decreasing spacings of the particles.

Figure 10 above shows another electron micrograph of the replica of the fracture surface of a feldspar from the same granite specimen. The crack propagates approximately from the top to the bottom of the figure. The smaller cleavage steps coalesced with the larger steps to form an *en echelon* arrangement as the crack propagated. The coalescence pattern can be explained by the greater drag exerted on the crack front by the larger steps relative to the smaller steps (Lange and Lambe, 1968). Many cleavage steps pass through the solid inclusions that are well dispersed in the crystal. The replica seems to have deformed along the cleavage steps on the right half of the photograph.

Willard and Hjelmstad (1971), Petersen *et al.* (1971) and Hoagland and Hahn (1971) found several different kinds of structures on the fracture surfaces none of which were similar to those depicted in Figures 9 and 10.

The Model Relating Sample Geometry to Fracture Pattern

From an observation of fracture patterns resulting from the experiments we found that the directions of crack propagation were not random, but were controlled by some definite fracture mechanisms. The simplest form of the complex fracture patterns can be seen in the Sample Geometry II, in which the crack path is very similar to the circular failure planes normally found in soils and some rocks. It is believed that the circular failure surface of a slope has an origin which governs the movements of the slope. The origin is located at an equal distance from the failure plane, and rotation of the slope is about that point. Thus, we use this basic concept of the theory of circular failure in developing our model to explain the fracture mechanisms. Since the procedure normally used in soil mechanics to locate the centre of failure circle requires much information, we developed our own procedure to fit the experimental results. The simplest model for circular crack patterns is shown in Figure 11. Our model is similar to circular failure surfaces of soils and rocks that contain tension cracks whose depths indicate the level of the initiation of failure.

The basic procedure of the graphical method (Table 1) is applicable to other sample geometries provided that a fulcrum can be properly located and the drawing can be done in steps, the nomenclature used being shown in Table 2. Once the theoretical model for each sample geometry is drawn, the directions of crack propagation can be predicted, the directions of maximum stress and maximum energy release can be determined, and the mechanisms that

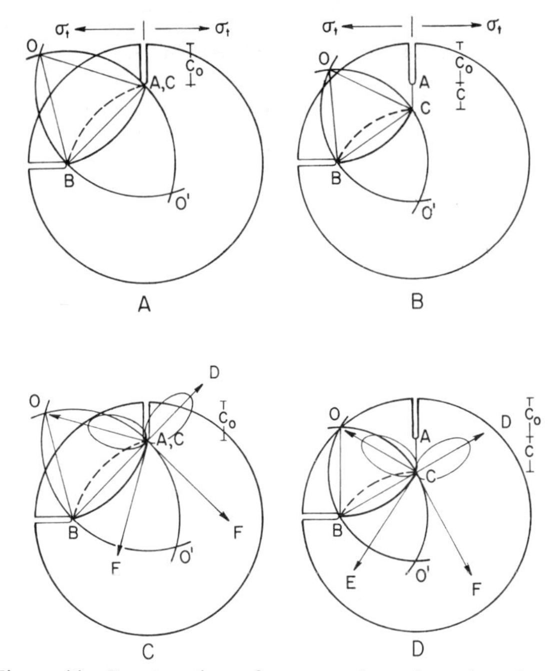

Figure 11. Construction of stress and crack path, using the graphical circular failure model. For details of the construction procedure, see Table 1. For nomenclature see Table 2.

Table 1. Procedure to construct the circular failure model

(1) Draw the sample geometry.

(2) Draw a line joining the tips of the charged and uncharged slots.

(3) With the tip of each slot as a centre, draw an arc of a circle whose radius is equal to the distance between the tips of the slots, cutting each other at O and O′.

(4) With the centre O, draw an arc joining the tips of the slots. Thus, the arc represents the circular path of the crack moving from the tip of the charging slot about the centre O and terminating at the tip of the uncharged slot which acts as a fulcrum. In order to model an alternate propogation of a crack, from A to B and from B to A, the triangle AOB must be equiangular.

(5) Similarly, an arc of the circle O is drawn joining the tips of the slots.

(6) Join OB and OA.

(7) Draw lines AE and AF perpendicular to AO and AB at A respectively.

(8) Extend BA to D.

(9) Draw elliptical loops passing A and symmetrical about AO and AD. (See Figure 11C). These represent isochromatic loops seen in photoelastic studies and serve to indicate maximum stress directions.

If the crack propagates to a critical point C where a change of the direction of crack propagation or branching occurs, the above procedure is applicable to point C regardless of Point A. (See Figure 11B and D).

Table 2. Description of the terminology used in the graphical stress model

(1)	Point A represents the tip of the initial Griffith crack or the charging slot, and C_0 the initial crack length.
(2)	Point B represents the tip of the uncharged slot or any thing that is expected to cause the change of a crack path.
(3)	Point C represents the tip of a running crack at a critical point where change of the direction of crack propagation can be observed, and C is the increased crack length from its initial value.
(4)	Point O represents the origin of maximum energy release (OMER) and the origin of the circular path.
(5)	Point O represents the origin of maximum energy conservation (OMEC) and the mirror image of the OMER.
(6)	The line OA or OC represents the direction of maximum stress pointing towards the OMER at a critical point. (Designated as $\vec{O}$).
(7)	The line OB serves the same purpose as OA or OC for the crack to propagate from B to A or C.
(8)	AE and CE represent the directions of maximum energy release perpendicular to AO and CO respectively. (Designated as $\vec{E}$).
(9)	AD and CD represent the directions of maximum stress on the farther side of the crack. (Designated as $\vec{D}$).
(10)	AF and CF represent the directions of maximum energy release on the farther side of the crack, and are perpendicular to AD and CD respectively. (Designated as $\vec{F}$).
(11)	Arcs AB and CB represent the circular paths of crack propagation controlled by the OMER.
(12)	AB and CB represent the lines of symmetry between the OMER and OMEC. (Designated as $\vec{M}$).
(13)	Ellipses representing isochromatic loops of shear stresses analogous to those seen in photoelastic studies of stress field at crack tips, are added to O and D merely for rapid recognition of the directions of maximum stress. (It should be kept in mind that, in reality, shear stresses are zero at free surfaces and cannot cut across the crack plane.)
(14)	The broken arc AB or CB represents the mirror image of the circular path that is controlled by the OMER.
(15)	σ is the maximum tensile stress acting normal to the plane of the initial crack.

control crack propagation can be explained. It was found that almost all the fracture patterns developed in the sample geometries used in this research were made understandable when the graphical method and theoretical model were applied.

Application of the Model to Experimental Results

Fracture Patterns Associated with Sample Stress Fields

The simple stress field developed in the specimens of Geometry I and XII

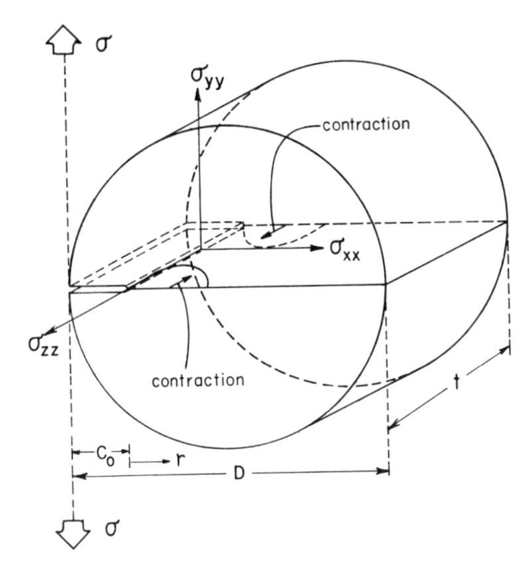

Figure 12. Schematic diagram of the stress field in a simple tensile test using Sample Geometry I. σ = tensile stress, σ_{xx}, σ_{yy}, and σ_{zz}, are stresses in the x, y, and z directions respectively. C_0 = initial crack length, r = distance from the tip of the slot, d = the diameter of the specimen, t = the thickness of the specimen.

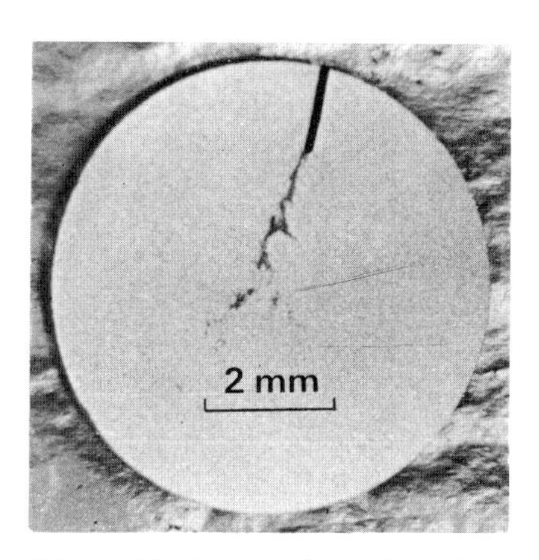

Figure 13. A typical crack pattern developed in Sample Geometry I. The rock is diabase. The induced crack propagates approximately perpendicular to the tensile axis. Crack branching can be seen on the lower end of the main crack. The crack has been stained.

is shown in Figure 12. The dominant stress is tensile acting perpendicular to the axis of the charging slot. The crack propagates essentially perpendicular to the tensile axis. The model developed in the previous section cannot be applied because there is only one slot.

When the specimen is subjected to explosions, tensile stress σ_{yy} will be created ahead of the crack tip causing the mineral grains to elongate elastically along σ_{yy} and contraction will occur normal to it. The contraction is increased with increasing σ_{yy}. The tensile stress is highest at the tip of the slot and decreases with increasing distance away from the tip. The tensile stress (σ_{zz}) which is highest at the centre of the specimen, will counter the force due to contraction. The tensile stress σ_{zz} decreases with increasing distance away from the tip of the slot. The stress σ_{xx} exists to resist the stress due to bending moment and contraction opposed to its direction. It increases from zero at the tip of the slot up to a maximum and then decreases to a constant level with increasing distance r.

The interior of the specimen is subjected to higher stresses than the surface. If the stresses are increased to critical values, the crack will form in the interior first and extend to the surface of the specimen. Jaeger and Hoskins (1966) found the same phenomenon in the Brazilian test. Under dynamic stress conditions, the induced interior crack can be detected in vitreous materials or in a mono-mineral rock by carefully controlling the energy from explosions. The interior crack will be larger than that at the surface. The crack may not start exactly at the half thickness of the specimen due to the anisotropy of the rocks. Since the tip of the slot may be larger than one mineral grain, it is difficult to predict whether the crack will initiate along the grain boundaries or across the mineral grains.

Compressive stress waves from explosions had little influence on macroscopic direction of crack propagation since no spall cracks due to reflective tensile stress waves were found in the specimens, but they could possibly induce microcracks in elastic brittle mineral grains such as quartz. If this is true, random orientation of microcracks should be found anywhere in the specimen. Thin sections of the specimen made prior to the formation of macrocracks did not show such random orientation of microcracks in quartz grains. Therefore, the influence of the compressive stress waves on crack propagation can be neglected. The presence of rib marks in quartz grains mentioned previously must have been caused by a local compressive pulse due to the movement of mineral grains against each other during crack propagation.

Initiation of a crack occurs when the stress conditions at the crack tip satisfied the Griffith criterion: $\sigma = \sigma_{\text{in}} = \sqrt{(2E\gamma/\pi C)}$ where σ is the critical value of the applied uniaxial tensile stress normal to the plane of the crack; E the modulus of elasticity; γ the specific fracture surface energy; C the half-crack length. The initial fracture stress depends to some extent on the geometry of the crack tip. When the radius of curvature of the crack tip (ρ) is less than three times the lattice spacings of the material (a_0), the stress for unstable crack propagation is given by the Griffith criterion. When ρ is more than three times the spacing, the fracture stress is given by

$$\sigma_{\text{in}} \approx \sqrt{\left[\frac{2E\gamma}{\pi C}\left(\frac{\rho}{3a_0}\right)\right]}.$$

Thus unstable crack propagation occurs when the above two equations are satisfied. The atomic bonds at the crack tip are broken, and the crack propagates.

On the atomistic and microscopic levels, mineral grains are not subjected to equal stress fields because of the random orientation of the mineral grains. The crack tip stress fields and the fracture behaviour of the crystals depend on the orientation of the crack with respect to the parent lattice (positions and arrangements of atoms in the crystals) (Chang, 1970; Kaechele, 1967). Theoretical values for the tensile strength of mineral crystals are of the order of 10^5 kg/cm^2, depending on the general inhomogeneity of rocks, their mineral composition, the inherent imperfections of the crystal lattices, and the testing methods (Attewell, 1963). Taking the above information into account, crack initiation will not occur at the same time along the tip of the slot. If the lateral propagation of an induced crack near the tip of the slot occurs, the mineral grains ahead of the crack may be broken along a definite plane, which may not be observed when a thin section cut normal to the crack plane at this point is studied, but this is unusual. Cracks usually start along grain boundaries in sedimentary rocks, but both along grain boundaries and across mineral grains in igneous rocks.

On a macroscopic level, an induced crack usually starts from the curved tip of a slot, and propagates nearly parallel to the axis of the slot, which is perpendicular to the tensile stress. If the induced crack does not start from the tip of the slot but along its length, it is expected that the initiation of the crack is strongly controlled by some microstructure inside the specimen. Slabbing of the specimen is recommened to check the controlling microstructure and its orientation with respect to the axis of the slot. At 'critical voltage' the macroscopic induced crack may be seen on only one side of the specimen. This indicates that the initiation of the induced crack is not uniform. If the crack can be seen on both sides, its length and direction with respect to the axis of the slot may be different on each side.

It was our purpose to study the influence of the nature of rocks on rock fracture, so we did not at first try to control the direction of crack propagation. It is possible to force the crack to propagate in a straight line by cutting a groove on each face of the specimen from the tip of the slot with a diamond saw. However, it is also possible to produce predetermined fracture patterns which are a direct function of the geometry of the specimens.

The Griffith criterion indicates the initial propagation of static cracks in a perfectly brittle material under plane strain conditions. It does not apply to moving cracks because the system accumulates kinetic energy from whatever source initiates the fracture unless the crack is moving very slowly.

When the Griffith criterion is satisfied, instability of the system should occur. As soon as the crack begins growing, the energy demand for further growth increases more than the increase of the stored energy because the crack-driving force (the strain–energy release rate) increases with increasing crack length. The crack growth is checked unless the external load and thus the stored energy is increased again until the stored elastic energy and the energy dissipated are equal. This process goes on continuously until at some stage the driving force finally exceeds the restraining force, and total instability occurs. The initial stage of crack propagation will be stable while the later stage will be unstable. In most cases under static stress conditions the first stage is short in distance and long in time, while the second stage is long in distance and extremely short in time. Under dynamic tensile stress, the propagation of a crack will be unstable (Glucklich, 1971). A crack cannot run rapidly through a solid unless the rate of release of elastic strain energy with the growth of the crack is at least equal to the work done in producing new areas of crack surface (Kelly *et al*., 1967). Fracture propagation is stable, as long as the relationship between the half length (c) of the propagating crack and the applied stress exists and the condition $\sigma > \sigma_{in}$ is maintained (Bieniawski, 1967). Fracture propagation is unstable when the unique relationship between c and σ ceases to exist, that is, when other quantities, e.g. the crack growth velocity, also play a role and fracture propagation cannot be controlled any more by the applied load.

In a simple tension test, when a brittle specimen containing a symmetrical control slot is fractured, the fracture propagates theoretically straight and symmetrically perpendicular to the tensile axis. The direction of propagation may deviate from its theoretical path if there is a lack of symmetry in either the specimen or stress field ahead of the crack. It is common that a crack deviates from its theoretical path when the crack velocity is high. Unstable crack propagation can be expected when a crack propagates at a high velocity. A straight crack path can be controlled by using a double-cantilever beam specimen, or by grooving the surface of the specimen (Berry, 1963).

Using our techniques and sample geometries, it is not very difficult to make a fracture deviate from its natural path or to change the fracture patterns. In a simple tension test, the ideal path of a crack in a brittle material is the line extending from the crack tip perpendicular to the tensile axis. Although we did not intend to control the direction of crack propagation while we did experiments with specimens of Geometry I, it was often found that a crack deviated from its straight path.

Cotterell (1965) defined two classes of fractures:

(I) Fractures that when they deviate from the ideal path have a tendency to return to the original path. The normal stress acting across the prolongation of the crack path is a maximum ahead of the crack. Even if an artificial crack is introduced inclined to the loads the fracture path turns towards the line of symmetry.

(II) Fractures that have tendency to continue to deviate from the ideal path. Although there is symmetry, the fracture deviates from this line because the normal stress across the line of symmetry ahead of the crack is less than the normal stress across the line of local symmetry. In other words, the fractures will propagate along the line of local symmetry because it is a highly probable path for macroscopic growth.

In a real solid the fracture will deviate from the ideal path under the influence of flaws and microscopic anisotropy. If the deviation is accumulative there will be a low probability for any one path (Cotterell, 1965). Thus in rocks whose anisotropic properties are very strong the deviation of the crack propagation from the ideal path is rather common. Class I fractures were normally found in rock specimens of Geometry I when microstructures did not occur in the fracture path. When tests were performed on specimens of the other geometries in which the stress fields were more complex, or when the influence of microstructures was considered, class II fractures were likely to occur. However, it may be possible for class II fractures to develop into class I fractures depending on some factors which will be discussed later. The initial paths of class II fractures, in general, belonged to those of class I fractures, but not always.

Cotterell (1970) proposed a method to determine the stability of fracture path in the compact tension test. The general conclusion is that if a small excursion from the ideal path is made because of some local inhomogeneity the path is stable if the next most probable path is directed back towards the ideal one. The terms stable and unstable paths are used hereafter in this paper. They should not be confused with the terms stable and unstable cracks.

Fracture Patterns Developed in Sample Geometry I.

Each specimen of geometry I contains only one slot, which behaves as an initial crack. The depth of the slot is considered to be the initial crack length (C_0). When the specimen is subjected to the stress from explosions, it will be under compressive stress waves radiating from the slot into the specimen and reflecting from the free boundary in the form of tensile stress waves, and tensile stress. The tensile stress acts on the rock materials in the direction perpendicular to the axis of the slot.

It was found that by using a diamond saw blade to cut the slot, mineral grains along the crack length and at the crack tip were not broken. No visible microcracks were induced during the process of cutting the slot.

A typical crack propagation in a specimen of Geometry I is shown in Figure 13 above. The crack initiates in a diabase specimen from the tip of the slot. The crack path is stable until at about the centre of the specimen branching of the crack occurs. Before branching, the deviation of the crack from its ideal path can be observed. Subsidiary cracks can be seen along its path. Crack branching did not always occur; this figure shows that crack branching is possible in specimens of Geometry I. Unstable crack paths were found in other specimens but usually when defects were involved.

By carefully controlling the amount of explosion voltage close to the critical value for a given rock type, and making successive thin sections, we were able to study the propagation of fine running cracks just before the macroscopic cracks could be observed. It was found that, at some distance away from the tip of the slot, some induced cracks were discontinuous, even though most of the induced cracks were continuous. Branching and coalescence of micro-cracks were common. The sizes of the running cracks varied from place to place. It was found later that the induced cracks that arrested in the specimens showed these phenomena no matter whether the macroscopic cracks occurred or not. These discontinuous cracks may join together if the velocities of the main cracks are high and local stress fields are strong enough.

It is believed that the formation of discontinuous cracks is associated with the presence of Griffith cracks ahead of the main cracks. In rocks, Griffith cracks are always expected, especially in crystalline rocks. They may be present in mineral grains or at the grain boundaries. These microcracks may be open or closed. Cleavages, twinning planes, and deformation lamellae, etc., may be considered as Griffith cracks because they can contribute to discontinuities in mineral grains under suitable stress conditions. Anthony and Congleton (1968) called the Griffith cracks ahead of the main cracks, 'advance cracks', which term is adopted here to distinguish them from the other Griffith cracks that are available in the rock specimens but not associated with crack propagation.

It should be noted that these 'advance cracks' sometimes are very fine and it is very difficult to distinguish them from other Griffith cracks under a microscope without special lighting or staining. The technique of staining the fractures in thin sections used in this research is of great assistance. It was observed that 'advance cracks' were opened under suitable stress conditions and, after staining, they appeared as red or faint pink, depending on their sizes.

Congleton and Petch (1967) first considered this influence of the 'advance crack' on the propagation of the main crack. They found that there are some interactions between the main crack and the 'advance cracks'. The main crack can be arrested locally by running to the 'advance crack'. The stress ahead of the latter will be increased as the former approaches the latter and perhaps will be sufficient to induce propagation of the 'advance crack'. Thus, some rather complex cracking can be expected at this stage.

When the distance r between the tip of the main crack and the centre of the advance crack, whose length is $2c$, is less than c, the advance crack will either be very quickly unloaded as the main crack sweeps past it or it will be absorbed by the main crack (Anthony and Congleton, 1968). On the other hand, if r is longer than approximately $2c$, the 'advance crack' can be accelerated enough for its tip to be moving at a velocity comparable to that of the main-crack tip before coalescence occurs and branching can develop. Thus the stress on the 'advance crack' should remain sufficiently high to maintain extension. The path of the advance crack depends on the local symmetry but, normally, it is along the path of the main crack. If the deviation of the path of the advance crack from that of the main crack is accumulative, the advance crack will not be able to join the main crack. It should be kept in mind that the extension of the advance crack will not occur unless the stress at the tip of the advance crack satisfied the Griffith criterion.

Although Congleton and Petch (1967) did not state that discontinuous propagation of the induced cracks is possible, their theory implies that this phenomenon can occur. It is possible that the discontinuous cracks may be induced by the irregular crack fronts. Thus the induced cracks may be discontinuous in one direction but continuous in the other. However, it is not possible to observe three dimensional propagation of a running crack in an opaque specimen.

The results of the experiments with sedimentary rocks and their metamorphic derivatives showed that the induced cracks propagated mostly along the grain boundaries and less commonly across mineral grains. Petersen and Shockey (1971) obtained similar results. Thus the behaviour of crack propagation in sedimentary rocks is very much different from that in crystalline rocks, especially those of igneous origin, in which the transgranular cracks are dominant. In igneous rocks the degree of crystallization, interlocking, and random orientations of mineral grains are major factors causing the influence of the nature of the rocks on crack propagation. It should be noted that the lineations of igneous rocks also play an important role in controlling the crack propagation but their influence was not systematically studied in this research. In sedimentary rocks, stratification, orientation of mineral grains, degree of packing and cementation, particle shapes and gradation, the difference in strength between rock particles and the matrix or cementing materials control the propagation of the induced cracks. In metamorphic rocks the factors that control crack propagation are those found in igneous and sedimentary rocks.

The common factors that are applicable to all types of rocks are the strength of individual grains and their orientations with respect to the tensile stresses. In general, the fracture patterns are controlled by the stress fields, and perhaps microstructures.

In this section we shall discuss the mechanisms that control crack propagation in sedimentary rocks. Although it has been reported by many authors that in sedimentary rocks, cracks propagate mostly along the grain boundaries, very few of them explained the mechanisms. Some authors believe that the influence of the grain boundaries is the same in any type of rock. It is often said that cracks propagate along grain boundaries because it is possible that high stress concentrations develop and weak planes might be present there. After a search of the literature in rock mechanics it seemed that no one has said that grain boundaries may have less influence than other factors in controlling crack propagation. Wawersik and Brace (1971) and Willard and McWilliams (1969a, b) referred to the problem but offered no explanation. Ko, Haas and Clark (1970) studied the failure criteria for granular rocks. They found that the failure criteria of composite materials are complicated functions of the elastic moduli of matrix, inclusion and composite, and the volume ratios of matrix and inclusion. Whether the materials break by maximum shear, or tensile stress, depending on the type of material, such failure will always be initiated at the grain boundaries. High stress concentrations along the grain boundaries are responsible for the frequent failures observed along grain boundaries. They proposed a theory to predict whether the type of failure of their rock model will be intergranular or intragranular by using strength ratio criteria. When the ratio of uniaxial compressive strength to tensile strength of an individual mineral is less than the ratio of the maximum tensile stress inside the grain boundary to that outside, according to their theory, the failure will be transgranular, if the reverse, failure is intergranular. In other words, the intergranular and intragranular failures are determined by the physical properties of the composite.

Oka and Majima (1970) studied the mechanism of fracturing rock particles. They concluded that the strain energy required to break rock particles is proportional to the third power of particle size and the second power of tensile strength, and is inversely proportional to the value of Young's modulus. Thus the sizes of individual particles indirectly control the crack propagation.

Although it is well known that structural defects in rocks have some influences on rock properties, not much work has been published to show the influences of these microstructures on crack propagation. Willard and McWilliams (1969a) defined defects as either open or closed cracks or as loci at which cracks most likely would occur with sufficient tensile or shear stress to cause failure. Willard and Hjelmstad (1971) have already studied the influence of cavities (voids, partitions or partition micropores) on the fracture character and tensile strength of dacite. They concluded that micropores play a different physical role to concentrate stress than voids in the fracturing of dacite. The influence of voids was studied as part of the research using sample geometries with holes (Sriruang, 1972).

In Figure 14 the induced crack initiates from the tip of the slot along the boundary between the dolomite metacryst and dolomite matrix. Then it penetrates along a cleavage plane in the dolomite cryst. Crack initiation at this location is not common. Perhaps it might be induced by the crack in the interior of the specimen. In Figure 15 the crack initiates from the tip of the slot along the contact between the dolomite and quartz groundmass. It becomes a transgranular crack in a dolomite metacryst. It should be noted that the location of the initiation of the crack was predetermined. The crack initiated as predicted. Crack branching in dolomite is shown in Figure 16. It was found that the crack branching was caused by stress concentration occurring between two dolomite metacrysts which are very much larger than the groundmass. Subsidiary cracks were formed as the branches approached the metacrysts. The subsidiary cracks moved along the grain boundaries between the metacrysts and the groundmass rather than through them. The reasons that the subsidiary cracks did not propagate through the dolomite metacrysts can be explained by the theories of Ko *et al.* (1970), and Oka and Majima (1970). Coalescence

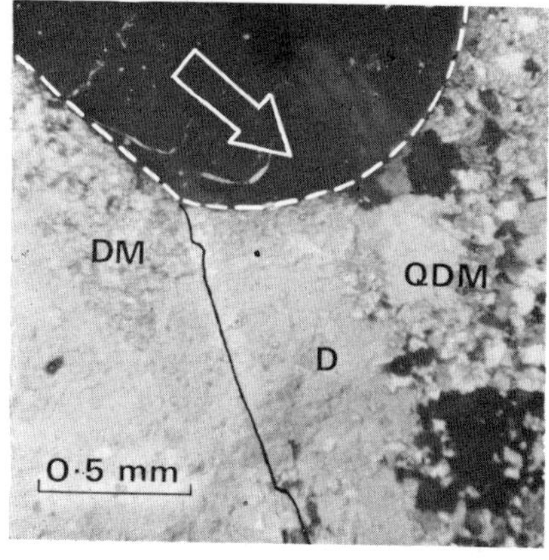

Figure 14. Crack initiation in dolomite shown in thin section under crossed polars. The crack initiates at the contact between the dolomite matrix (DM), and a dolomite metacryst (D). It then travels through the metacryst. QDM = quartz–dolomite matrix. The arrow indicates the axis and tip of the charging slot.

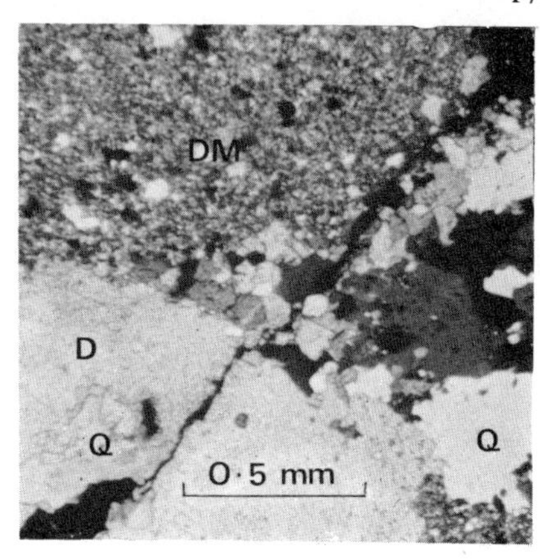

Figure 15. Predetermined crack path in dolomite seen in thin section under crossed polars. The crack initiates at the contact between dolomite and the quartz groundmass. It becomes transgranular when it passes through a dolomite metacryst. DM = dolomite matrix. D = dolomite metacryst. Q = quartz metacryst.

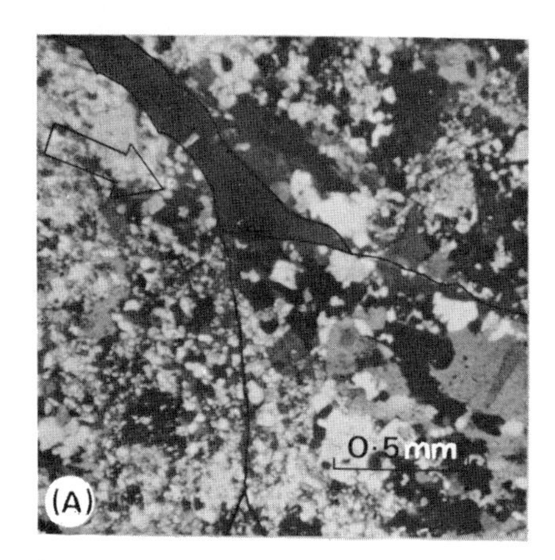

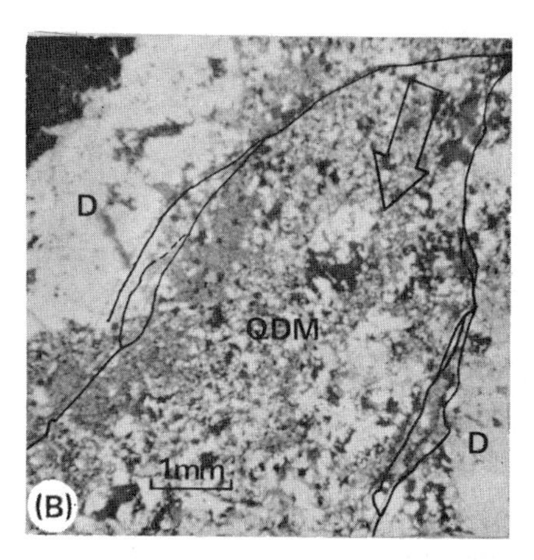

Figure 16. Crack branching in dolomite induced by microstructures as seen in thin section under crossed polars. (A) Crack at branching. (B) Secondary cracks after branching. Crack branching was induced by two dolomite metacrysts. The arrows indicate the direction parallel to the charging slot axis, and the direction of crack travel. D = dolomite metacryst. QDM = quartz–dolomite matrix.

of the subsidiary cracks can be seen after the cracks passed the metacrysts. The mechanisms of crack branching will be described in the next section of the analysis of the results. Dolomite behaves as a crystalline rock.

The distinct character of crack propagation in a granite and a conglomerate is shown in Figure 17 below. The cracks are intergranular in the conglomerate but transgranular in the granite. The sizes of rock particles in conglomerate should be noticed.

Fracture Patterns Associated with Complex Stress Fields

Each specimen geometry contains at least one slot which behaves as an initial crack. The geometry of the initial crack, especially the sharpness of the tip, has some influence on crack propagation. In this research, crack branching was observed within the specimens whose sizes were limited (less than 3 inches in diameter). Selection of the geometry of the initial crack is also important. It has been reported that a very blunt notch will be less stable than a sharp one because the stress level that it will create will be higher and despite relaxation, this level will be sufficient to drive the crack for ahead. With the sharp notch the initial stress will be low and subsequent relaxation will, after a short growth, cause the driving force to sink below the necessary level for propagation (Glucklich, 1971). Thus cracks from blunt notches reach their terminal velocity at a crack length much shorter than cracks from sharp notches (Bradley and Kobayashi, 1972). Anthony *et al.* (1970) reported in the results of their experiments with glass that branching velocities of cracks running from blunt notches were lower than the terminal crack velocity of the glass. This is the reason why blunt notches were used in our experiments. We were able to observe crack branching, sometimes at a distance less than one inch ahead of the tip of the slot. This is the advantage of blunt notches over sharp ones, although we did not do any experiment with sharp notches.

When a brittle crack extends under constant stress, it accelerates and eventually branches. Assuming that a crack propagates in a direction normal to the maximum tensile stress, Yoffe (1951) showed that branching of the crack occurs when its velocity approaches a critical value depending on the properties of the material such as Poisson's ratio. The maximum crack velocity, usually called the terminal crack velocity, is equal to the Rayleigh Wave speed. Generally, at a lower velocity the crack extends in a straight line but at high velocities of crack propagation the symmetrical stress distribution varies and instead of a maximum tensile stress located on the line of the crack, a maximum inclined to the plane of the running crack occurs. The rotation rate and magnitude of the maximum stress increase with fracture velocity and length (Hanson and Sanford, 1970). The lack of a sharp maximum in the normal stress field ahead of the crack tip will cause the crack to oscillate and branch before reaching a critical value in order to release the stored energy. Finkel *et al.* (1970) demonstrated the influence of the fracture conditions on the parameters of the Rayleigh Waves generated on the surfaces of a propagating crack. They found that Rayleigh Waves are radiated only with variation of the crack velocity. As the stored energy is increased, the amplitude of the Rayleigh Wave increased depending on the character of the fracture. Sih (1970) showed the Rayleigh Wave speed is a function of Poisson's ratio.

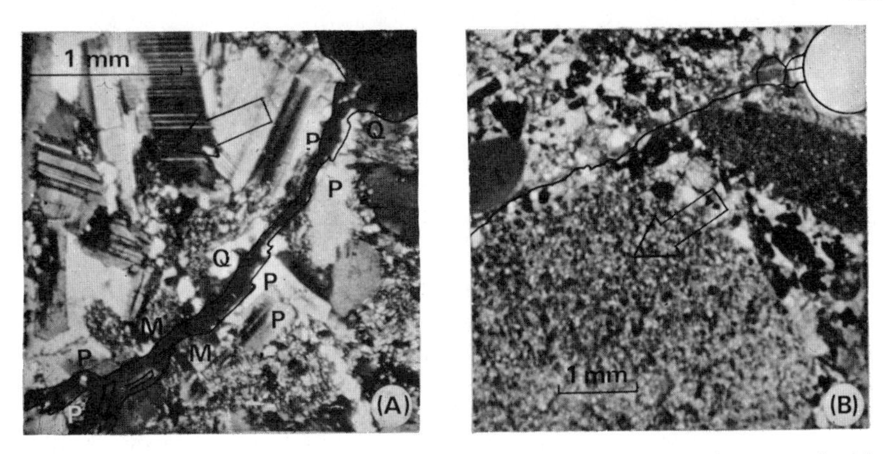

Figure 17. Comparison of crack paths in igneous and sedimentary rocks as seen in thin section under crossed polars. (A) Crack path in granite. (B) Crack path in conglomerate. Cracks are mostly transgranular in the granite and intergranular in the conglomerate. Q = quartz. P = plagioclase. M = mica. Arrows indicate the axes of the charging slots.

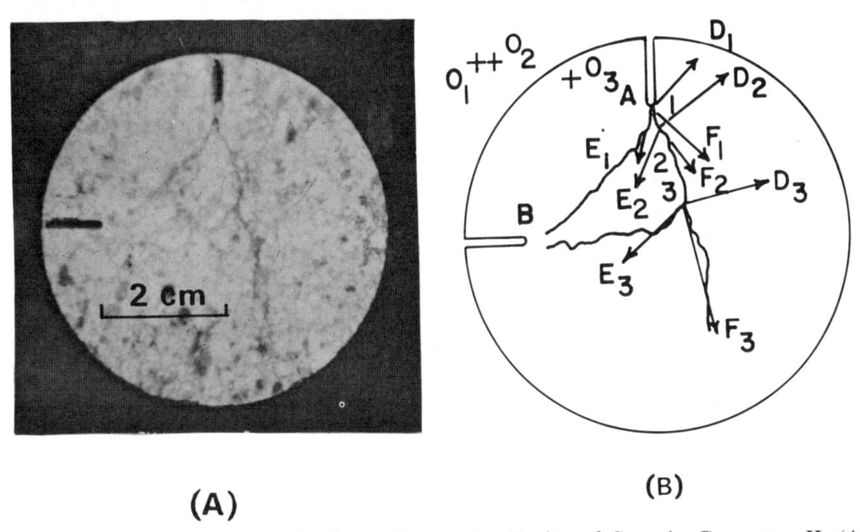

Figure 18. Fracture pattern developed in quartz diorite of Sample Geometry II. (A) Fracture path in the rock specimen. (B) The graphical stress model applied to the fracture pattern. For nomenclature, see Table 2.

Winkler *et al.* (1970) and Sanders (1972) reported that crack velocities are not necessarily limited to the maximum velocity of waves which, under normal loading conditions, are limited to the order of the Rayleigh Wave velocity, or some fraction of an elastic wave velocity. Baëta and Ashbee (1967) found that fractures in quartz crystals compressed to failure, travelled at supersonic velocities. Thus, it is possible that, in rocks, branching of cracks can occur in quartz crystals after the cracks propagate through the quartz crystals.

Congleton and Petch (1967) showed that it is not necessary that branching of a crack occurs at a constant critical velocity. The higher the stress on the crack, the shorter the distance the crack travels before the branch forms. This would be because of faster acceleration of the crack velocity to the critical value. The higher the stress on the crack, the lower the branching velocity. It is not necessary that when the crack velocity is high, branching will occur. They proposed that the mechanism of branching depends on the extension of some advance cracks ahead of the initial running crack. According to their model, the condition necessary for crack branching is that the critical stress intensity factor K near the tip of the running crack is constant and equal to $\pi^{1/2}\sigma_f C_b^{1/2}$, where σ_f is the fracture stress and C_b is the length of the main crack branching. Anthony and Congleton (1968) proposed that branching would probably occur when the ratio of the stress intensity factor at branching, K_b, to the fracture toughness, K_c, is limited to greater than 2. If the ratio is less than 2, the advance cracks do not continue to propagate.

If the ratio becomes very large, a lot of microcracks form ahead of the running crack and branching is not likely to occur. Anthony *et al.* (1970) and Bullen *et al.* (1970) applied both branching criteria to their experiments and obtained good results. However, these criteria have not been used in the study of rock fracture.

When a crack propagates in a direction normal to the maximum tensile stress, the energy released is a maximum. When the crack velocity is high, the stress normal to the line of the crack is a minimum and thus the energy released is a minimum. When the energy released is zero, the crack propagation will be self-maintaining along any direction. At high velocities of crack propagation, due to redistribution of the stress at the crack tip, a maximum stress makes some angle with the line of the crack on either side. Thus, when the crack propagates in the direction perpendicular to the maximum stress that is inclined to the line of the crack, the energy release in this direction will be a maximum. Crack branching favours this condition because there are two such directions when the crack velocity is high (Cotterell, 1965). If advance cracks are located in these directions of maximum energy release, crack branching will be induced. Thus at high crack velocity, the crack would either branch or curve. The energy release ratio increases with crack speed and approaches a constant value as the velocity of crack propagation reaches a critical value (Sih, 1970).

The energy release rates just before branching must have the same value but after branching it is equal to 4γ rather than 2γ, where γ is the specific fracture energy (Eshelby, 1970).

For a running crack the stress intensity factor K is normally used to express the dynamic elastic fields in the vicinity of the crack tip. The dynamic intensity factor depends on the velocity of cracks but their relationship is not unique. In general, they behave in the same manner. Bradley and Kobayashi (1970, 1971) reported that the variation in crack velocities lag slightly behind the variations in the dynamic stress-intensity factor. Once the value of energy release rate (G) is obtained, the dynamic stress intensity factor K_I, under plane strain conditions, can be determined from the relationship

$$G = \frac{(1 - \nu^2)K_I^2}{E},$$

where E is the modulus of elasticity of the material and ν, the Poisson's ratio. According to Congleton and Petch (1967), the stress intensity factor is constant, as predicted by Sih (1970). High fracture stress and high crack velocity are essential for crack branching but it is not necessary that crack branching always occurs when these conditions are satisfied. Crack branching occurs in order to release the excess energy accumulating at the tip of a running crack. This large amount of strain energy is associated with a high stress intensity factor. Under dynamic tensile stress, fracture toughness increases with crack velocity but not proportionally (Cotterell, 1968). The increase in fracture toughness is a criteria for the increase in the roughness of the fracture surface, which, in turn, corresponds to the increasing stress intensity factor. Consequently, the surface roughness seem to have a closer relationship with the dynamic stress intensity factor than the crack velocity (Bradley and Kobayashi, 1971).

It was mentioned previously that the fracture surfaces of the rocks tested were very rough, indicating high fracture toughness of the rocks and high stress intensity factors in the specimens. Although we cannot compare the roughness of the fracture surfaces in the interior of the specimens to that close to the surface, perhaps due to the small sizes of the specimens, it can be expected that the stress intensity factor is higher in the interior than at the surface. If more secondary cracking, coalescence and chip forming are observed in the interior of the specimen than at the surface, this means that the stress intensity factor and crack velocity are higher in the interior. This is a reason why the cracks at the interior are larger than at the surface.

Considering together the large magnitude of the applied stress, the bluntness of the tip of the slot, the roughness of the fracture surface, and the availability of the 'advance cracks', crack branching in rock is not difficult to form, even at short distances of crack propagation.

Fracture Pattern Developed in Sample Geometry II.

In small specimens, the cracks start at the tip of the initial crack and propagate along a circular path. For larger specimens, other cracks in addition to the curved one may develop in a zig-zag pattern. A typical fracture pattern developed in Sample Geometry II is shown in Figure 18 above. The curved path of crack propagation is similar to the circular failure of slope normally found in soils and some rocks. By using the theory of circular failure, we developed a simple graphical procedure to fit the experimental results.

This was tested on the crack pattern in a quartz–diorite specimen (Figure 18). The curved path of the crack drawn using the graphical method fits very well with the actual crack path on the surface of the speciman. The method is therefore valid for this case, and we should be able to predict the directions of crack propagation by using the graphical method. However, it will not help us to understand rock fracture unless we consider the mechanisms involved. To use the proposed graphical method, we must accept certain rules and the phenomena found in fracture mechanics, especially at the critical point.

It is known that branching of a moving crack occurs at a point when the crack velocity approaches a critical value depending on the properties of the material (Yoffe, 1951; and many others), when the dynamic stress intensity factor is constant (Congleton and Petch, 1967), when the ratio of the stress intensity factor at branching to the fracture toughness is within a critical limit (Anthony and Congleton, 1968). To use the graphical method, the mechanisms at crack branching must be assigned positions, directions, and magnitudes (if known). Thus the maximum stresses and the maximum energy release at a crack tip were chosen as the basic criteria, because they fulfil the requirements for the graphical method. We assume that these criteria apply at the critical point where changing of the direction of crack propagation, or branching can be observed.

Consider the model in Figure 19A first. With the centre B, one can draw one circle passing through A. Further, there is only one tangent (AF) to the circle at A. Similarly, with the circle O, AE is the only tangent that is possible at point A. The lines AE and AF are unique at a critical moment. At high velocities of crack propagation and/or branching, the crack propagates in the direction of maximum energy release. There are two such directions for a crack propagating at high velocity (Cotterell, 1965). Since point A is the intersection of the circles whose centres are at O and B, the lines AE and AF are the only two possible tangents to the circles. At a critical moment, the crack will propagate along the arc AB. Thus, AE must represent the direction of maximum energy release. The line OA perpendicular to AE represents the direction of maximum stress on one side of the crack. Similarly, AD is the direction of maximum stress on the other side of the crack and AF the relative direction of maximum energy release. The direction AB cannot represent the direction of maximum stress on the other side of the crack, otherwise there will be two directions of maximum stress on the same side of the crack. For rapid recognition of the directions of maximum stress, ellipses representing isochromatic loops of shear stresses analogous to those seen in photoelastic studies of stress field at the crack tip are added to AO and AD. Thus, our model is in agreement with Cotterell (1966) who showed that the isochromatic loops can lean backwards only in the case of a crack which progressively deviates from a straight line. It can be seen from the graphical model that the influence of the centre O on the circular path of the crack is more than that of the centre B. This is possibly due to the fact that the centre B has some influence on the farther side of the crack. Beyond the critical point the crack propagates in a circular path about the centre O. Thus, we can state that a rapidly moving crack that tends to deviated from its previous path at a critical point, possesses a centre of maximum energy release located on the same side of the crack on which a change in direction or curvature of the path is observed. The crack

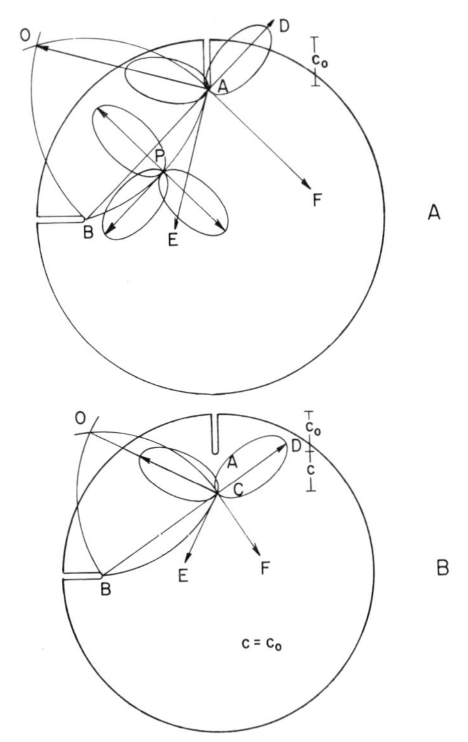

Figure 19. Model stress field and crack path for Sample Geometry II. (A) Critical point coincides with the tip of the charging slot. (B) Critical point is at an arbitrary point along the crack path. For nomenclature, see Table 2.

propagation is controlled by the origin of maximum energy release (OMER) which may become the origin of a failure circle if the fracture contributes to rupture of the specimen. Beyond the critical point, the curved path represents the line of local symmetry, which is the direction of maximum energy release. The circular crack will terminate at the tip of the virgin slot or not depending upon the released energy. If the released energy is not enough to create new crack surfaces, the crack will arrest or continue to propagate if reinitiation of the stress occurs. However, the process of releasing energy is continuous, not stepwise (Glucklich, 1971; and Cotterell, 1965). Figures 19B and 20 show the validity of the theoretical model to the case of a crack that propagates from the tip of the charging slot before a critical point can be observed. Once a critical point is about to develop, the origin of maximum energy release (OMER) forms and exists until the process of releasing energy is complete or another similar system is about to develop. If the energy release along AF is not enough to form new crack surfaces, the crack will not propagate in this direction. Consequently, the direction of crack propagation will deviate in the direction

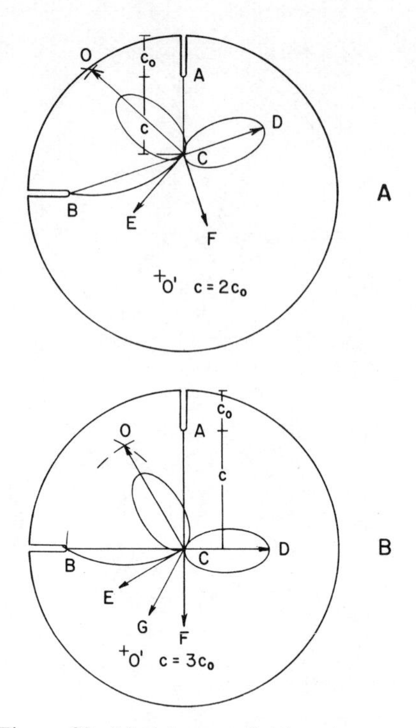

Figure 20. Model stress field and crack path for Sample Geometry II. (A) Critical point is at a greater distance from the charging slot than in Figure 19 B. (B) The critical point is opposite the fulcrum. Beyond this point the crack path is controlled by G. For nomenclature see Table 2.

controlled by the OMER only. If the main crack does not curve or branch, it will continue to propagate. According to the model, the stress history along the path of a moving crack before a critical point occurs, can be observed by assuming that a critical point can occur at any point along the path of crack propagation. Although, the directions of maximum stress and energy release at a point prior to a critical point may be different from those recorded by the photoelasticity method, they are comparable among themselves. Figure 20 above shows the pre-determined directions of the maximum stresses and their relative energy release, and paths of a propagating crack at particular points where the critical conditions are assumed. In Figure 20, it is shown by the graphical method that if a crack does not deviate from its previous path or branching does not occur, the direction of crack propagation ahead of the crack tip is controlled by the resultant of the directions of maximum energy release because the directions of maximum energy release lie on the same side of the main crack.

This interpretation is based on the graphical model. It should be noted that

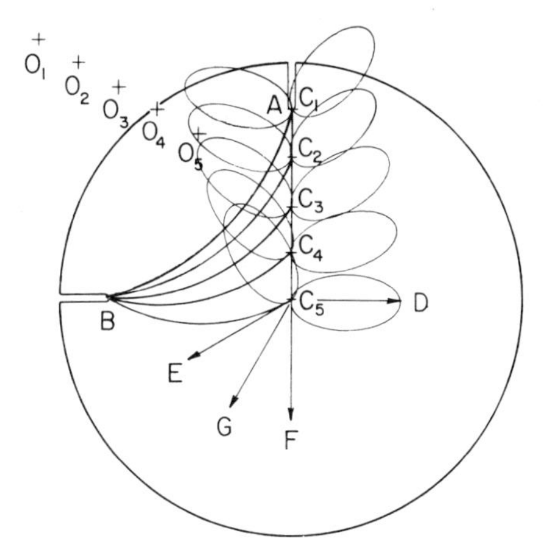

Figure 21. Changes in stress field in in Sample Geometry II with crack propagation. The OMER for each critical point changes as the crack progresses. For nomenclature, see Table 2.

the ratios of the increased crack length to the initial crack length throughout the analyses are shown for illustration only. The comparative curved crack paths and directions of maximum stress and relative energy release at particular points, predicted in the graphical method, are shown in Figure 21. The model also predicts that when a crack propagates to the centre of the specimen without branching or changing its path, the direction of crack propagation ahead of the crack tip is controlled by the direction of the resultant energy release ($\vec{G}$), which is the bisector of the angle between the directions of maximum energy release, i.e. $\vec{E}$ and $\vec{F}$. The resultant energy release is influenced by the origin of maximum energy release (OMER) and the fulcrum.

It should be noticed that the location of the OMER is unique for a particular critical point. It can be seen that if the crack does not deviate from its previous path or branching does not occur, the OMER moves while the crack propagates.

The origin becomes static when the critical conditions are satisfied. This means that for a particular system the amount of energy required to create a crack pattern is a definite maximum which depends on the strength, and the boundaries of thc bulk system, not only on the individual particles in the system. Thus, when the energy released during crack propagation reaches the critical value required for a system, according to our model, an OMER develops and governs the crack pattern. If excessive stored energy is available, another origin of the same kind will be arranged for the other systems. However, the formation of these origins of maximum energy release should be simultaneous. They cannot form in a step manner because the energy release is continuous. Since the circular path is controlled by definite OMER, the crack path along this line is stable.

It should be mentioned here that not only the OMER develops for a system

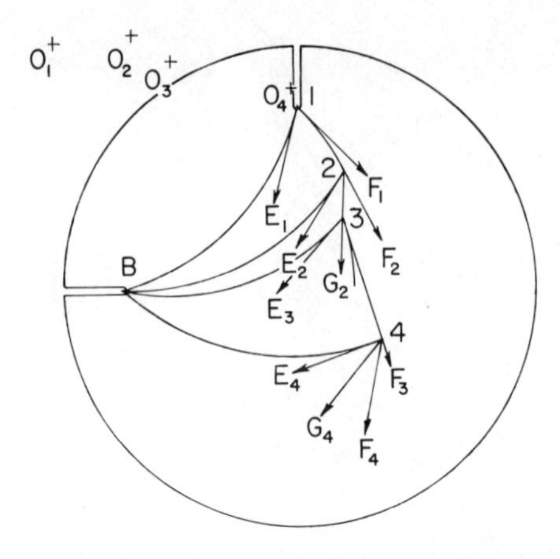

Figure 22. Application of stress model to a step branching crack in Sample Geometry II. For nomenclature, see Table 2.

but its opponent which is called the origin of maximum energy conservation, also develops. The origin of maximum energy conservation (OMEC) is included in this paper to explain the mechanisms that caused particular fracture patterns developed on the surface of the specimens of the other geometries.

A schematic diagram of a progressive crack pattern that may develop on the specimens of Geometry II is shown in Figure 22. At point 1, if branching does not occur, the direction of a circular crack will be directed by $\vec{E}_1$. If branching occurs the crack propagation will be controlled by $\vec{E}_1$ and $\vec{F}_1$. If the crack whose direction is controlled by $\vec{F}_1$, propagates to point 2 and neither deviation of its path nor branching occurs, the crack will propagate along $\vec{G}_2$ because $\vec{E}_2$ and $\vec{F}_2$ lie on the same side of the crack tip. At point 3, where branching occurs, the circular path will be controlled by $\vec{E}_3$ while the other branch is guided by $\vec{F}_3$. The conditions for crack propagation at point 4 will be the same as those at point 2. An actual progressive crack pattern is shown in Figure 18B. The point O_1, O_2 and O_3 represent the origins of maximum energy release of the first, second, and third systems respectively. According to the crack pattern shown on the photograph, it can be seen that crack branching occurs at the tip of the slot. One branch follows $\vec{E}_1$, the other does not follow $\vec{F}_1$ as predicted by the graphical method but $\vec{G}_1$ instead. It was found later that the crack pattern developed on the other slab face of the same specimen did show that the crack followed $\vec{F}_1$ instead of $\vec{G}_1$. This showed that the crack did not initiate and propagate at the same time. The crack pattern on the face of the slab shown in Figure 18A, indicates that the crack following $\vec{E}_1$ was induced but the branch along $\vec{F}_1$ did not develop. An independent crack initiated and propagated to point 2 where it changed its direction along $\vec{F}_2$. It is not certain whether the crack propagated along $\vec{E}_2$ from the photograph. However, the crack propagated to point 3 where branching developed again.

At this time, the crack propagated along $\vec{E}_3$ and $\vec{F}_3$ as predicted by the model. Small variations of the crack paths from the theoretical patterns are probably controlled by individual mineral grains. Thus, the theoretical model fits very well with the actual crack pattern and the mechanisms that control the crack propagation are now understandable. Even more, the crack paths controlled by the OMER are very stable.

Fracture Pattern Developed in Sample Geometry III.

After branching, cracks propagate either along straight lines, or circular paths, or somewhere within these two limits. The mechanisms that control crack propagation and cause the fracture pattern to develop, are shown by comparing the graphical model with the actual rock fracture patterns.

The specimen of Geometry III contains two collinear uncharged slots along the diametral plane in the direction perpendicular to the axis of the charging slot. The theoretical model for Geometry III is modified from that for Geometry II as shown in Figures 23 and 24. By using the same procedure we obtain two origins of maximum energy release for both halves of the specimen, i.e., O_L for the left half and O_R for the right half of the specimen. Based on the graphical method used in the previous sample geometry, it can be said that if the crack does not propagate along either one of the circular paths (arc CB_L or arc CB_R) or branching does not occur, the direction of the propagation of the crack is predicted by $\vec{G}$ due to the symmetry of the loading and the specimen. However, the crack patterns developed on the surface of the specimens do not show circular paths after branching. The cracks seemed to propagate along straight lines in definite directions. These lines can be drawn on the theoretical model by joining the tips of the moving cracks to those of the uncharged slots (lines CB_L and CB_R).

In Figure 23, it can be seen that these lines coincide with $\vec{F}_R$ and $\vec{F}_L$. When the crack propagates from the tip of the charging slot, $\vec{F}_R$ and $\vec{F}_L$ are separated from CB_R and CB_L respectively (see Figure 23B). However, the cracks still propagate along CB_L and CB_R after branching. This shows that there must be some factors controlling the directions of crack propagation besides the OMER. By including the mirror images of the OMER and their relative circular paths, it was found that the lines CB_L and CB_R are the lines bisecting the areas surrounded by the circular paths and their relative mirror images. After branching, the cracks propagate from the critical point along the mirror planes and terminate at the tips of the uncharged slots. The opponents of the origins of maximum energy release develop in order to resist rotation or any movement of the systems under consideration. If the centers of rotation are called the origins of maximum energy release (OMER), their opponents should be called the origins of maximum energy conservation (OMEC). Since the latter tends to preserve the stability of the systems, the cracks will not propagate along the mirror image of the circular paths. If the symmetry of loading and specimen exists, the cracks can form and propagate along the mirror planes if the stresses acting

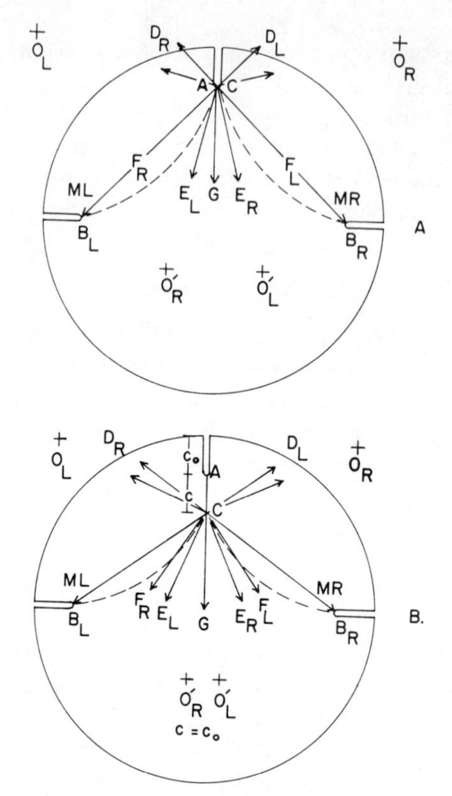

Figure 23. Model stress field and crack path for Sample Geometry III. (A) The critical point coincides with the tip of the charging slot. (B) The stress field is at an arbitrary point. For nomenclature, see Table 2.

perpendicular to the mirror planes exceed the tensile strength of the material along the mirror planes. The directions of crack propagation along these planes are designated hereafter as $\vec{M}$. Should the origin of maximum energy conservation (OMEC) lose its power to its opponent (OMER) completely, the crack will propagate along the circular path controlled by the OMER only. However, partial loss of power of the OMEC is possible. The location of each OMEC is defined by the location of its relative OMER. Thus, the OMEC can move. From the diagrams it can be seen that the stress fields in the specimen are complex. There are two overlapping maximum stress loops on each side of the crack. According to our model, as the crack propagates without curving or branching, the directions of maximum stress change and so do the stress fields. There will be a point along the crack path where the stress loops on each side of the crack coincide, and where crack branching is expected to occur. Beyond the critical point, the crack will propagate along either $\vec{M}$ or circular paths,

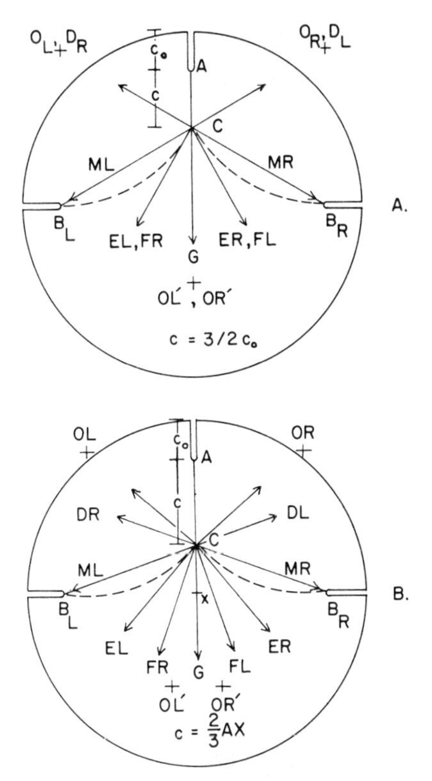

Figure 24. Model stress field and crack path for Sample Geometry III. (A) Stress situation where crack branching occurs. (B) Stress field as the crack approaches the centre of the specimen. For nomenclature see Table 2.

of somewhere between these two limits, depending on the interaction between the OMERs and OMECs. If the critical conditions are not satisfied, the direction of crack propagation is predicted by $\vec{G}$. Figure 25 shows an example of the crack pattern developed in a gabbro specimen of Geometry III. The graphical method indicates that our theoretical model is applicable. The partial loss of power of the OMEC can be seen, because after branching, the directions of crack propagation fall between $\vec{M}$ and the circular paths. However, the crack paths are still stable.

At this point, the real meanings of all the sides of the equiangular triangles BOA or BOC become apparent. The side OA or OC indicates the direction of maximum principal stress at the tip of a crack pointing toward the OMER. Consequently, the relative direction of maximum strain energy release can be obtained. Once the value of the maximum strain energy release on this side of the crack is known, one can compute the magnitude of the relative maximum

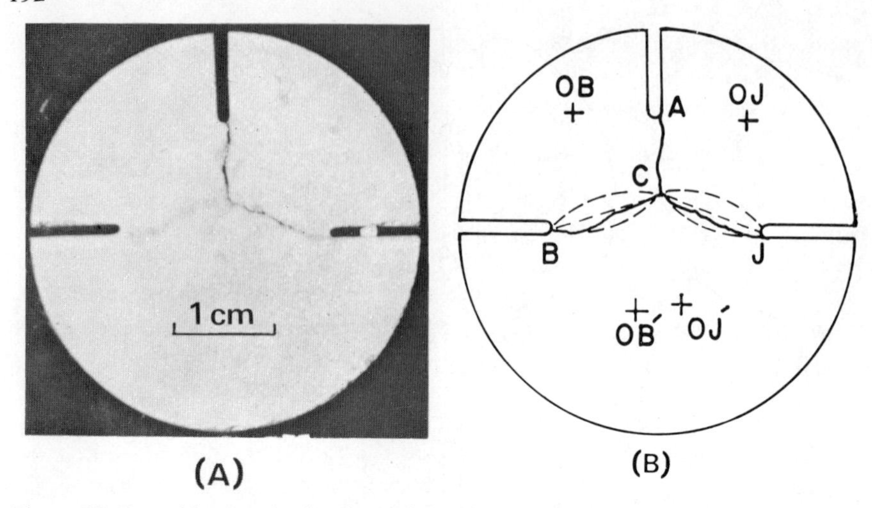

Figure 25. Branching cracks developed in gabbro of Sample Geometry III. The photo shows the stained sample. Note the major and minor cracks. The drawing shows the graphically determined limits of cracking. The crack stays within the predicted limits. For nomenclature, see Table 2.

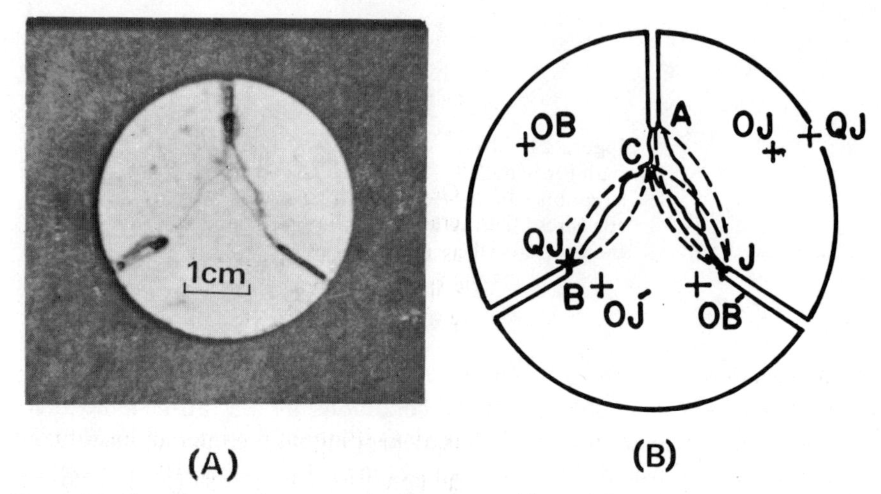

Figure 26. Complex crack pattern developed in gabbro of Sample Geometry VI. Two cracks initiate from the charging slot. The larger crack is along GL which coincides with MR. For nomenclature see Table 2.

stress. Thus the critical conditions for the system can be obtained. The sides OB serves the same purpose as OA or OC for the alternate direction of crack propagation. In this case, the magnitude of maximum stress in $\overrightarrow{OB}$ should be equal to that in $\overrightarrow{OA}$. If the critical conditions are kept constant, and the sample geometry and size of specimen are the same, the crack will theoretically propagate along the same circular path no matter whether the crack propagates from either A or B, because there is only one definite circular path controlled by an OMER.

The side AB or CB, represents the line of symmetry $\vec{M}$ between the OMER and its relative OMEC, and the crack will propagate along this line. If there is any lack of symmetry in either loading condition, sample geometry or material itself, the crack will propagate along the circulr path controlled by the OMER. However, we have not because of time limitations proved the complete validity of this interpretation from the theoretal model.

According to our model, we do not consider the interaction between the two collinear uncharged slots. Based on his experimental results, Theocaris (1972) has reported that the interaction between two collinear cracks or a crack and a straight boundary is not significant if the inside distance between the collinear cracks of the tip of the single crack and the boundary is longer than twice the length of each crack. Thus, the effect of interaction in both cases depends on the length of the cracks and the distance between inside tips of the cracks.

It can be seen that although there are some changes in sample geometry and stress field, the graphical method is still the same and applicable to the experimental results (as in Geometry VI applications, Figures 26 and 27). A theoretical model can be constructed for this sample geometry and the complex

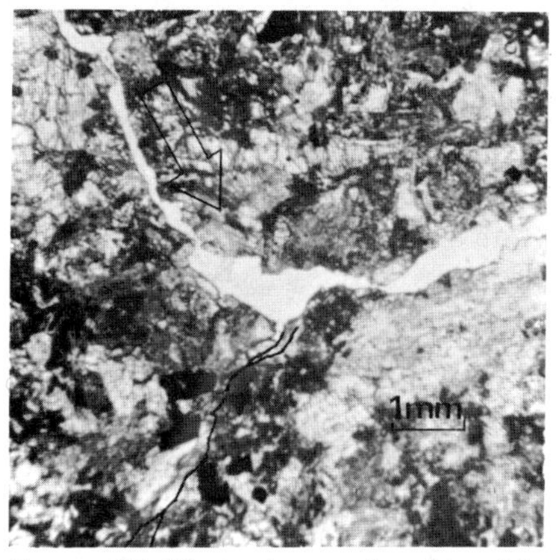

Figure 27. Photomicrograph of a thin section under crossed polars showing crack branching in gabbro of Sample Geometry VI. The arrow indicates the direction of cracking and the axis of the charging slot. Note the differences in crack size after branching. The crack branches at the junction of an integranular crack and a grain which fails by intragranular cracking. The main branch follows the intragranular crack. This is the common situation in branching cracks in rocks which fail by both mechanisms.

stress field at the tip of a running crack in the specimen can be visualized from the model. However, photoelastic studies should be done in future research to check the stress field predicted by the model. After branching, the cracks are controlled not only by the OMER, but also the OMEC depending on the symmetry of the sample geometry, loadings, and the material of the sample.

CONCLUSIONS

The exploding wire technique allows the study of fracture process from initiation through propagation, branching, and coalescence, and arrest of the cracks by controlling the sample geometry and detonation energy. With the aid of blunt initial cracks (charging slots) crack branching could be observed in small specimens.

A technique which will be very useful to study petrofabrics, porosity, and rock fractures in thin sections, has been developed and successfully used in this research. This technique combines staining the thin sections of rocks and the opened fractures (cracks, microcracks, and other planar defects). Staining the thin sections provides a means for rapid identification of some common minerals. Staining fracture improves the contrast between the rock particles and the opened cracks or microcracks. Discontinuous parts of a crack could be observed.

Examination of the fracture surfaces of the broken specimens indicated that although the bulk specimens of the rocks exhibited brittle failure, the microstructures developed on the fracture surfaces indicated both brittle and semi-brittle failure depending on the directions of crack propagation with respect to crystallographic orientations of mineral grains. The plastic deformations are normally associated with transgranular or cleavage crack propagation through the mineral grains. Rib marks and cleavage steps are microstructures commonly found on tensile fracture surfaces.

Based on the theory of circular failure used in soil mechanics, a graphical method was developed to fit the experimental results. It was assumed that under dynamic tensile stress, the deviation of a crack path or branching of a crack occurs at a critical point from which the crack propagates along the direction(s) of maximum energy release and perpendicular to the direction(s) of maximum stress. The following hypotheses were set up:

(a) A fast moving crack that tends to deviate from its previous path at a critical point, possesses a centre of maximum energy release located on the same side of the crack that a change in direction or curvature of the path is observed.

(b) The crack propagation beyond the critical point is controlled by the origin of maximum energy release, which may become the origin of a failure circle at rupture of the specimen.

(c) For symmetric conditions (symmetry in loading, sample geometry and material) the crack propagation beyond the critical point is controlled by the origin(s) of maximum energy release and its (their) opponent(s), the origin(s) of maximum energy conservation.

(d) For asymmetric conditions, the crack propagation is solely controlled by the origin(s) of maximum energy release.

By using the graphical method and theoretical models, almost all the fracture patterns developed in the sample geometries used in this research were explained. However, we have not proved the universal validity of the above hypotheses.

REFERENCES

Anthony, S. R. and Congleton, J. (1968). *Metal Sci. J.*, **2**, 158–160.
Anthony, S. R., Chubb, J. P. and Congleton, J. (1970). *Phil. Mag., Ser. 8*, **22**, 1201–1216.
Attewell, P. B. (1963). *Colliery Engineering*, July, 289–294.
Baëta, R. D. and Ashbee, K. H. G. (1967). *Phil. Mag., Ser. 8*, **15**, 931–938.
Berry, J. P. (1963). *J. Appl. Phys.*, **34**, no. 1, 62–68.
Bieniawski, Z. T. (1967). *Ph.D. Thesis*, University of Pretoria, Pretoria, South Africa, 246 pp.
Bieniawski, Z. T. (1968). *Felsmechanik und Ing.*, **6**, no. 3, 113–125.
Brace, W. F. and Walsh, J. B. (1962). *Amer. Min.*, **47**, 1111–1122.
Bradley, W. B. and Kobayashi, A. S. (1970). *Experimental Mech.*, **10**, **3**, 106–113.
Bradley, W. B. and Kobayashi, A. S. (1971). *Int. J. Eng. Frac. Mech.*, **3**, no. 3,317–332.
Brown, E. T. and Hudson, J. A. (1971). *Symp. Rock Fracture. Int. Soc. Rock Mech.*, Nancy, France, Paper 2–20, 11 pp.
Bullen, F. P., Henderson, F. and Wain, H. L. (1970). *Phil. Mag.*, **21**, 689–699.
Caslavsky, J. O. and Vedam, K. (1970). *Phil. Mag., Ser. 8*, **22**, 255–268.
Congleton, J. and Petch, N. J. (1967). *Phil. Mag., Ser. 8*, **16**, 749–760.
Cotterell, B. (1965). *Int. J. Fract. Mech.*, **1**, 96–103.
Cotterell, B. (1966). *Int. J. Fract. Mech.*, **2**, no. 3, 526–533.
Cotterell, B. (1968). *Int. J. Fract. Mech.*, **4**, no. 3, 209–217.
Cotterell, B. (1970). *Int. J. Fract. Mech.*, **6**, no. 2, 189–192.
Dyke, P. V., Conn, A. F. and Dagan, G. (1969). *National Technical Information Service*, Springfield, Virginia: AD-695 776, 132 pp.
Erdogan, F. (1971). *Int. J. Eng. Fract. Mech.*, **3**, 231–240.
Eshelby, J. D. (1970). *Inelastic Behavior of Solids*, (ed.) Kanninen *et al.*: McGraw-Hill Book Company, pp. 77–116.
Finkel', V. M., Guz', I. S., Kutkin, I. A., and Volodarskii, A. Ya (1970). English Trans., Society Phys.—*Solid State*, **12**, no. 8, 2300–2305.
Forwood, C. T. (1968). *Phil. Mag., Ser. 8*, **17**, 657–667.
Friedman, M., Perkins, R. D., and Green, S. J. (1968). *General Motors Technical Center*, Warren, Michigan, 23 pp.
Gardner, R. D. and Pincus, H. J. (1968). *Int. J. Rock Mech. Min. Sci.*, **5**, 155–158.
Gash, P. J. S. (1971). *Tectonophysics*, **12**, 349–391.
Glucklich, J. (1971). *Int. J. Eng. Fract. Mech.*, **3**, 333–344.
Hakalehto, K. O. (1970). *Int. J. Fract. Mech.*, **6**, no. 3, 249–256.
Handin, John, Friedman, M., and Logan, J. M. (1969). *Texas A & M University*, 40 pp.
Hanson, M. E. and Sanford, A. R. (1970). *National Technical Information Service*, Springfield, Virginia: AD—729 933, 36 pp.
Hasselman, D. P. H. and Fulrath, R. M. (1966). *J. Am. Ceramic Soc.*, **49**, no. 2, 68–72.
Hoagland, R. G. and Hahn, G. T. (1971). *Battelle Memorial Inst.*, Columbus, Ohio, 38 pp.
Jaeger, J. C. and Hoskins, E. R. (1966). *J. Geophys. Res.*, **71**, 2651–2659.
Kaechele, Lloyd E. (1967). *Ph.D. Thesis*, Stanford University, 214 pp.
Kelly, A., Tyson, W. R., and Cottrell, A. H. (1967). *Phil. Mag., Ser. 8*, **15**, 567–586.

Ko, Kyung C., Haas, C. J., and Clark, G. B. (1970). *Rock Mechanics and Explosives Research Center*, University of Missouri-Rolla, RMERC-TR-70-106 p.
Kutter, H. K. and Fairhurst, C. (1971). *Int. J. Rock Mech. Min. Sci.*, **8**, 181–202.
Lange, F. F. (1967). *Phil. Mag., Ser. 8*, **16**, 761–770.
Lange, F. F. (1970). *Phil. Mag., Ser. 8*, **22**, 983–992.
Lange, F. F. and Lambe, K. A. D. (1968). *Phil. Mag., Ser. 8*, **18**, 129–136.
Laniz, R. V., Stevens, R. E., and Norman, M. B. (1964). *U. S. Geol. Survey Prof. Paper 501-B*, pp. B152–B153.
Lanuron, D. (1966). *Extrait des Annales de la Societe Geologique de Belgique*, **89**, Bull. No. pp. 5–10.
Lewis, P. H. (1963). *Department of Electrical Engineering*, Michigan Technological University, 7 pp.
Lutton, R. J. (1970). *Proc. 12th Symp. Rock Mechanics*, A.I.M.E., pp. 561–572.
McWilliams, J. R. (1966). *A.S.T.M. STP 402*, Am. Soc. Testing Mats., pp. 175–189.
Moavenzadeh, F., *et al.* (1966). *Department of Civil Engineering*, Massachusetts Institute of Technology, 94 pp.
Murgatroyd, J. B. (1942). *J. Soc. Glass Tech.*, Sheffield (out of print), **26**, 155–171.
Oka, Y. and Majima, H. (1970). *Canadian Met. Quart.*, *9*, no. 2, 429–439.
Pawlowicz, R. (1970). *National Technical Information Service*, Springfield, Virginia: PNE–G–69, 76pp.
Petersen, C. F. and Shockey, D. A. (1971). *Stanford Research Institute*, Menlo Park, California, 28 pp.
Rad, P. F. and Moavenzadeh, F. (1968). *Department of Civil Engineering*, Massachusetts Institute of Technology, 127 pp.
Rad, P. F. (1971). *Tunnels and Tunnelling*, pp. 442–450.
Rosenfield, A. R. and Kanninen, M. F. (1970). *Phil. Mag., Ser. 8*, **22**, 143–154.
Sanders, W. T. (1972). *Int. J. Eng. Fract. Mech.*, **4**, 145–153.
Sih, G. C. (1970). In *Inelastic Behavior of Solids*, (ed.) Kanninen, *et al.*: McGraw-Hill Book Company, pp. 607–639.
Sriruang, Somsakdi (1972). *M.S. Thesis*, Michigan Technological University.
Steverding, B. and Lehnigk, S. H. (1970). *Int. J. Fract. Mech.*, **6**, no. 3, 223–232.
Tanaka, K., Mori, T., and Nakamura, T. (1970). *Phil. Mag., Ser. 8*, **21**, 267–279.
Tetelman, A. S. and McEvily, A. J., Jr. (1967). *Fracture of Structural Materials:* John Wiley & Sons, Inc., 697 pp.
Theocaris, P. S. (1972). *Int. J. Fract. Mech.*, **8**, no. 1, 37–47.
Tipnis, V. A. and Cook, N. H. (1966). *Proc. 5th U. S. Nat. Cong. Appl. Mech.*, Amer. Soc. Mech. Engrs., 595 pp.
Weber, P. and De Montille, G. (1971). *Rock Mechanics Abstract*, **2**, no. 2, 145.
Willard, R. J. and Hjelmstad, K. E. (1971). *Int. J. Rock Mech. Min. Sci.*, **8**, 529–539.
Willard, R. J. and McWilliams, J. R. (1969a). *Int. J. Rock Mech. Min. Sci.*, **6**, no. 1, 1–12.
Willard, R. J. and McWillams, J. R. (1969b). *Int. J. Rock Mech. Min. Sci.*, **6**, no. 4, 415–421.
Youash, Younathan (1969). *G.S.A. Bull.* **80**, 303–306.

Theoretical Calculation of Compliances of a Porous Medium

N. Warren and R. Nashner
Institute of Geophysics and Planetary Physics
University of California
Los Angeles, California 90024

INTRODUCTION

A number of important problems in geophysics require the ability to predict accurately the effects of cracks and textures on rock properties under effective pressures of up to a few kilobars. The input data would involve mineralogic and petrographic observables including mineralogy, mineral elastic moduli, and petrological data such as texture (fracture features) and grain and pore orientations and distributions. Considerable progress has been made toward developing such theories in recent years, and two new areas of research involve petrographic and rock physics studies of cracks and rock textures, and development of general methods of calculating the effects of structure on the elastic properties of geological materials. Although much has been written on the problem of calculating elastic properties for special cases of structured media, further development of generalized computational formats which allow calculation of elastic properties over a wide range of variables of geophysical interest is needed. Prediction of the elastic properties of cracked rock, soils, and foams as functions of structure and pressure should be formulated within a consistent theoretical framework.

Warren (1973) has discussed an approach for calculating the relative compressibility (β/β_0) of a porous medium over the whole range of porosities. In this paper, we present an extension of that method to the other elastic moduli. The calculation is explicitly designed to allow flexibility in modelling a wide range of structural effects over the entire range of porosities, and it allows various levels of approximation to be used. The effective relative elastic moduli are presented in the form of equations for the elements of a compliance tensor S_{ij}. A computer program has been developed for the explicit evaluation of the S_{ij} over a range of tensor loading conditions and pore geometries.

Solid isotropic rock models are first calculated from mineralogy and mineral physics data, and the additional effect of texture on these models is then computed. The solid-rock models are based on Hashin and Shtrikman upper- and lower-bound calculations. Texture effects, due to fractures and pores, are computed using a theory based on calculations of stress concentrations due to the pores.

The general method involves writing the expressions for the static or long-wavelength elastic moduli in terms of strain or stress concentrations, and then calculating the concentrations, so permitting the use of stress concentration data from the engineering literature.

The first section of this paper deals with the theoretical equation for S_{ij}; the second discusses pore strain computation; and the third presents results for rock-like materials with welded matrices. Upper bounds of elastic moduli, pore–pore interactions, and pore fluid effects are discussed briefly. The theory is used to predict velocity as a function of pressure for a lunar rock, compared with experiment.

THEORY

The compliance tensor elements S_{ij} are defined as the two subscript contractions of the compliance tensor **S** with elements satisfying.

$$\varepsilon_{kl} = S_{klmn}\sigma_{mn} \tag{1}$$

where ε_{kl} are strains and σ_{mn} are stresses. In contracted notation (Sokolnikoff, 1956)

$$S_{ij} \equiv S_{klmn} \quad \begin{cases} i \text{ or } j = 1, 2, 3, \text{ for } kl \text{ or } mn = 11, 22, 33 \\ i \text{ or } j = 4, 5, 6, \text{ for } kl \text{ or } mn = 23, 31, 12 \end{cases} \tag{2}$$

The relative compliances are given by S_{ij}/S_{0ij} where S_{ij} is a compliance of the bulk or porous material, and S_{0ij} is the equivalent compliance of a solid reference material. Throughout this paper, the subscript zero denotes solid material parameters.

The computational equations for S_{ij}/S_{0ij} are to be in a form compatible with three basic requirements. First, the equations must be general and must be restricted only by the definitions of the moduli (i.e., linear, first-order differentials). Second, the equations must be expressed directly in the modulus-porosity plane. Third, the equations must be compatible with infinitesimal strain theory over the entire porosity domain.

Although the equations are to be expressed as explicit functions of porosity, the nature of the porosity will not be restricted. That is, a pore may be a void, a solid, or a liquid inclusion, with or without pore pressure. A pore may also have zero volume, the equations being re-expressible for the limiting case of zero-volume cracks.

Additional constraints needed for modelling various types of materials are imposed as constitutive relations which must be satisfied simultaneously with the general equations. Examples of additional restrictions include requirements such as whether the matrix material is to be welded or composed of loose grains; whether the pore distribution in the material causes collapse of the material at some porosity less than 1.0; or whether the pore distribution is such that the material is foam-like and connected even for porosities approaching 1.0.

Additional restrictions are introduced by computational requirements. These restrictions are of two types: those which involve the order of approximation to which a calculation is to be made, and those which implicitly or explicitly restrict evaluation of the general equation. The second restriction includes the computational approximation of applying stress loads at infinity. In addition, in this paper, the general equations are not expressed in frequency and temperature as explicit variables. Computation is done with reference to isothermal and static or long-wavelength moduli. This restriction is not an inherent limitation in the general approach.

General Equations

The three requirements for the general equations for S_{ij}/S_{0ij} are satisfied by requiring the form of the equations to be compatible with an expression for a general modulus, Φ, which can be expressed as function of referred porosity, ω, (Warren, 1973)

$$\omega = 1 - V_0/V \tag{3}$$

where V is the bulk volume of the porous material under any loading condition, and V_0 is the volume of a reference sample with the same mass as the matrix of the bulk sample, under the same external loading conditions. The general compliance Φ is operationally defined by applying some arbitrary stress state, denoted by dR, to the sample and then measuring the volumetric strain $\delta V/V$

$$\Phi \equiv -\frac{1}{V}\frac{\delta V}{\mathrm{d}R} \tag{4}$$

Using referred porosity, Φ can be expressed directly in the porosity domain as a function of the compliance of a reference-solid, Φ_0, and of the porosity and its derivative.

$$\Phi(\Phi_0,\omega) \equiv \Phi_0 - \frac{1}{1-\omega}\frac{\delta\omega}{\mathrm{d}R} \tag{5}$$

By using the general volumetric strain compliance Φ, compatible expressions for linear and volume-independent compliances can be written as functions of porosity given the result of infinitesimal strain theory, that Φ can be expressed as the sum of linear strain quantities. In equation (5) Φ_0 and $\delta\omega$ are expanded in terms of linear strain quantities in three mutually perpendicular directions.

$$\Phi_0 \equiv {}_1\phi_0 + {}_2\phi_0 + {}_3\phi_0 \tag{6a}$$

where

$$ {}_i\phi_0 \equiv -\frac{\delta_i l_0}{{}_i l_0 \mathrm{d}R} \qquad i=1,2,3 \tag{6b} $$

and ${}_i l_0$ is the reference sample length in the i th direction.

$$ \frac{\delta\omega}{\mathrm{d}R} = (1-\omega)\left[\frac{1}{V}\frac{\delta V}{\mathrm{d}R} - \frac{1}{V_0}\frac{\delta V_0}{\mathrm{d}R}\right] \tag{7} $$

and

$$ \frac{\delta\omega}{\mathrm{d}R} = (1-\omega)\sum_{i=1}^{3}\left[\frac{{}_i S \delta_i l}{V \mathrm{d}R} - \frac{\delta_i l_0}{{}_i l_0 \mathrm{d}R}\right] \tag{8} $$

where ${}_i S$ is the cross-section area normal to the i th direction.

The volume strain or linear strain of the bulk material is written as the sum of terms over the matrix material and over the pore regions.

$$ \frac{1}{V}\frac{\delta V}{\mathrm{d}R} \rightarrow \frac{1}{V}\frac{\delta V_t}{\mathrm{d}R} + \frac{1}{V}\frac{\delta V_h}{\mathrm{d}R} \tag{9} $$

and

$$ \frac{{}_i S \delta_i l}{V} \rightarrow \sum_j \frac{{}^j\delta_i l_t \, {}^j \mathrm{d}_i S}{{}_i l_i S} + \sum_j \frac{{}^j\delta_i l_h \, {}^j \mathrm{d}_i S}{{}_i l_i S} \tag{10} $$

Subscript **t** denotes matrix quantities, and subscript **h** denotes pore quantities. The sums in equation (10) are over the cross-sectional elements ${}^j \mathrm{d}_i S$.

From equations (5) through (10), expressions for Φ/Φ_0 and ${}_i\phi/{}_i\phi_0$ are simply

$$ \Phi/\Phi_0 = \frac{\varepsilon_t}{\varepsilon_0}\left[\frac{V_t}{V}\right] + \frac{\varepsilon_h}{\varepsilon_0}\left[\frac{V_h}{V}\right] \tag{11} $$

$$ \frac{{}_i\phi}{{}_i\phi_0} = \frac{{}_i\varepsilon_t}{{}_i\varepsilon_0}\left[\frac{{}_i l_t \, {}_i S_t}{{}_i l_i S}\right] + \frac{{}_i\varepsilon_h}{{}_i\varepsilon_0}\left[\frac{{}_i l_h \, {}_i S_h}{{}_i l_i S}\right] \tag{12} $$

The quantities $\varepsilon_t/\varepsilon_0$ and $\varepsilon_h/\varepsilon_0$ are the ratios of the matrix and pore strains of the bulk sample to the strain of the reference sample under identical external loading. Subscript **i** denotes linear strains, for example,

$$ \varepsilon_h \equiv -\frac{1}{3}\delta V_h/V_h $$

and

$$ {}_i\varepsilon_h \equiv -\delta_i l_h/{}_i l_h $$

Average values of the quantities on the right side of equations (11) and (12) are implied. For a sufficiently large volume cut, ${}_i l_i S$, the coefficients in (12) are simply V_t/V and V_h/V. The expressions for the relative compliances S_{ij}/S_{0ij} are obtained by linear superposition of expressions for Φ and ${}_i\phi$, assuming appropriate stress loads.

In this paper we consider the following three cases:

(a) *Compressibility*

$$\beta = (S_{11} + S_{22} + S_{33}) + 2(S_{12} + S_{23} + S_{31}) = \frac{-\delta V}{V\mathrm{d}P} \tag{13}$$

P hydrostatic

(b) *Uniaxial Compliance*

$$S_{ii} = -\frac{1}{{}_il}\frac{\delta_i l}{\mathrm{d}_i T} \qquad \begin{cases} T \text{ uniaxial} \\ i = 1,2,3 \end{cases} \tag{14}$$

(c) *Shear Compliance*

$$S_{jj} = \frac{\delta_a l/_a l - \delta_b l/_b l}{\mathrm{d}Q} \qquad \begin{cases} Q \text{ shear} \\ j = 4, 5, 6 \\ \hat{a} \text{ and } \hat{b} \text{ at } 45^\circ \text{ rotation} \\ \text{about a reference axis in} \\ \text{the } \hat{1}, \hat{2}, \hat{3} \text{ coordinate frame} \end{cases} \tag{15}$$

which in terms of the generalized compliances yield

$$\beta/\beta_0 = \Phi/\Phi_0 \tag{16}$$

$$S_{ii}/S_{0ii} = {}_i\phi/{}_i\phi_0 \qquad \text{under } T;\ i = 1, 2, 3 \tag{17}$$

$$S_{jj}/S_{0jj} = \frac{1}{2}\left[{}_a\phi/{}_a\phi_0 + {}_b\phi/{}_b\phi_0\right] \qquad \text{under } Q;\ j = 4, 5, 6 \tag{18}$$

Welded Matrix

If the matrix of the porous sample is welded, the average matrix strains are obtained using Betti's reciprocal theorem (Love, 1944). If an external pressure perturbation, $\mathrm{d}P$, and a pore pressure perturbation, $\mathrm{d}P_h$, are applied

$$\frac{\varepsilon_t}{\varepsilon_0} = \frac{V}{V_t} - \frac{\mathrm{d}P_h}{\mathrm{d}P}\frac{V_h}{V_t} \tag{19}$$

Under an additional external uniaxial stress, $\mathrm{d}_i T$, applied along $i = 1$

$$\frac{{}_1\varepsilon_t}{{}_1\varepsilon_0} = \frac{V}{V_t} - \frac{\mathrm{d}P_h}{\mathrm{d}T + \mathrm{d}P(1 - 2\nu_0)} \cdot \left[\frac{{}_1l_h\,{}_1S_h}{V_t} - \nu_0\left(\frac{{}_2l_h\,{}_2S_h}{V_t} + \frac{{}_3l_h\,{}_3S_h}{V_t}\right)\right] \tag{20}$$

where Poisson's ratio is denoted by ν.
Under pressure plus an additional external shear couple, $\mathrm{d}Q$,

$$\frac{{}_a\varepsilon_t}{{}_a\varepsilon_0} = \frac{V}{V_t} - \frac{\mathrm{d}P_h}{\pm|\mathrm{d}Q|(1 + \nu_0) + \mathrm{d}P(1 - 2\nu_0)} \cdot \left[\frac{{}_al_h\,{}_aS_h}{V_t} - \nu_0\left(\frac{{}_bl_h\,{}_bS_h}{V_t} + \frac{{}_cl_h\,{}_cS_h}{V_t}\right)\right] \tag{21}$$

The choice of the sign of $\pm |dQ|$ depends on whether the direction $i = a$ has positive or negative strain under dQ.

For porous materials, with or without pore fluids such that the matrix strains are given by (19), (20), and (21), the compliance equations take the forms:

(a) *Compressibility*

$$\beta/\beta_0 + \frac{dP_h}{dP}\frac{V_h}{V} = 1 + \left\langle \frac{\varepsilon_h}{\varepsilon_0} \right\rangle \frac{V_h}{V} \tag{22}$$

(b) *Uniaxial*

$$\begin{aligned} \frac{S_{ii}}{S_{0ii}} = \frac{{}_il_t\,{}_iS_t}{{}_il_iS}\Bigg[V/V_t - \frac{P_h}{dT + dP(1 - 2v_0)} \Bigg\{ \frac{{}_il_h\,{}_iS_h}{V_t} \\ - v_0\left(\frac{{}_jj_h\,{}_jS_h}{V_t} + \frac{{}_kl_h\,{}_kS_h}{V_t} \right) \Bigg\} \Bigg] + \frac{{}_ie_h}{{}_ie_0}\left[\frac{{}_il_h\,{}_iS_h}{{}_il_iS} \right] \end{aligned} \tag{23}$$

(c) *Shear*

${}_a\phi/{}_a\phi_0$ and ${}_b\phi/{}_b\phi_0$ are of the form of equation (23, but with

$$dT \rightarrow \pm |dQ|(1 + v_0) \tag{24}$$

In equation (22) the coefficient V_h/V is the ratio of the pore volume to the bulk volume of the sample under all loading conditions. Therefore, it is the structural porosity η used by Walsh (1965a—see Warren, 1973). As first order quantities ω and η are equal. However, if the material is highly prestressed, the coefficients in (22) are correctly set equal to η. Similarly, the coefficient in (23) refers to the dimensions of the bulk sample under all loading conditions. This distinction is used in calculation of the pressure-dependence of the moduli.

An important consequence of equations (22), (23), and (24) is that for rock-like materials which can be characterized as having welded matrices, the calculations of the relative compliances, as functions of porosity, require only calculation of the pore strains.

PORE STRAIN COMPUTATION

Pore strains can be calculated to various orders of approximation. The pore strain terms in the equations for the compliances can be expanded as a sum over all pores or over all pore types with or without pore–pore interactions. Evaluation of the sum, however, may be very difficult if the computation is to be done for the pores in a finite volume V, with the external stress boundaries at finite distances from the pores. If, on the other hand, the average values of the pore strain terms are determined from calculations in which the stresses are applied at infinity, the resulting matrix and pore stresses and strains correspond to those of an infinitesimal pore in a volume V. This can be compensated for in an average sense by transforming the volumetric coefficients in the compliance equations. The matrix volume, V_t, *transforms as*

$$V_t \rightarrow V_t' = V$$

and

$$V_h \rightarrow V_h' = V_h \left[\frac{V_t'}{V_t} \right]$$

Under the transformation, the ratio of pore volume to matrix volume remains constant (Warren, 1973).

For stress application at infinity and assuming zero pore pressure, the equations for the compliance all take the simple form

$$_i\phi / _i\phi_0 = 1 + \sum_\alpha \frac{_\alpha\omega}{1 - {}_t\omega} \left[\frac{_{i\alpha}\varepsilon_h'}{_i\varepsilon_0} \right] \tag{25}$$

where the subscript **i** denotes the direction of the linear quantity, as before, $_\alpha\omega$ is the partial porosity of the α th type of pore, $_t\omega$ is the total porosity, and $_{i\alpha}\varepsilon_h'$ is the linear pore strain in the $\hat{i}$ direction, for the α th type of pore, calculated for stresses applied at infinity. The pore strains in (25) are to be evaluated. In the following we assume zero pore pressure.

No Pore–Pore Interactions

The basic assumption is made that if pore–pore interactions are neglected, the stress field acting on the matrix surrounding the α th type of pore, averaged over a sufficient number of pores, is equal to the applied stress field. In general, calculation of the $_{i\alpha}\varepsilon_h'$ then requires input parameters of the surrounding matrix moduli and the pore geometries.

The pore strain can be expressed in analytic form for a number of pore geometries. In this paper, pores are approximated as oblate spheroids with aspect ratios $\boldsymbol{\alpha}$ (major to minor axis) varying between unity (spheres) and infinity (flat, volumeless cracks). A computer program has been written, based on the solution of Edwards (1951) to evaluate the pore strains. Detailed discussion of the method is given in Nashner and Warren (1974). A summary is presented in the Appendix.

Single-class Pore Distributions

For oblate pores, the calculated pore strains are a strong function of pore aspect ratio and stress orientation. Figure 1 shows a plot of volumetric relative strain $\varepsilon_h'/\varepsilon_0$ for oblate spheroids as functions of aspect ratio and Poisson's ratio of the matrix material (ν). For aspect ratios over about 100, trapped pore fluid (with a bulk modulus of 25 kb) greatly reduces the compressive pore strain.

Figure 2 shows a similar plot for the compression of oblate spheroids under

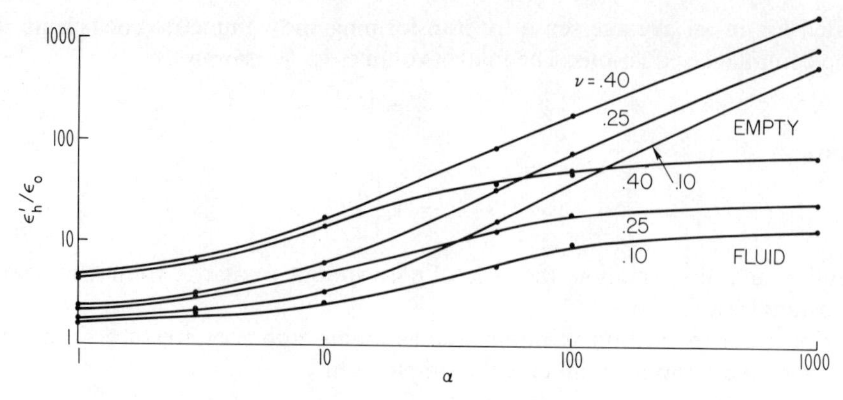

Figure 1. Normalized volumetric strain for empty and fluid-filled single pores in a matrix with Poisson's ratio of 0.10, 0.25, and 0.40, and a shear modulus of 374 kb.

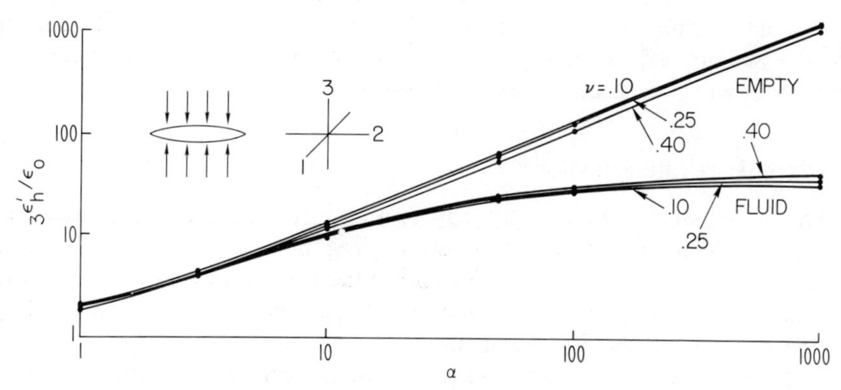

Figure 2. Linear strain curves for pores under uniaxial compression.

uniaxial loading normal to the equator. For large oblateness the relative uniaxial pore strain $_3\varepsilon_h'/_3\varepsilon_0$ for the empty pore is nearly proportional to α, in good agreement with the analytic solutions of Green and Sneddon (1948) and Sack (1946). The values of $\varepsilon_h'/\varepsilon_0$ under hydrostatic loading also have about the same slope. For highly oblate pores the pore strain $_3\varepsilon_h'$ becomes essentially independent of the type of stress field applied since the main component of stress that the pore is sensitive to is that which is normal to the plane of the pore. For stresses applied in the plane of the oblate pore, the value of the relative uniaxial strain in the plane becomes close to 1.

Averaged elastic moduli $\langle S_{ij}/S_{0ij}\rangle^{-1}$ can be evaluated from equations (16), (17), (18), and (25) and the calculated strains. The averaged moduli can be interpreted as upper bounds for isotropic, homogeneous media as functions of pore aspect ratio. The Hashin and Shtrikman upper bounds for the stiffness moduli (Hashin and Shtrikman, 1963) are the uppermost bounds for macro-

Table 1. Comparison of upper bound moduli for a porous medium from Hashin and Shtrikman (H & S) and from Warren (W) for Poisson's Ratio = 0.25

Porosity	Bulk modulus K/K_0		Young's modulus E/E_0		Shear modulus μ/μ_0		
		(W)		(W)		(W)	
ω	H & S	direct	H & S[a]	direct	H & S	from K, E	direct
0.10	0.800	0.800	0.817	0.818	0.821	0.822	0.828
0.20	0.640	0.640	0.666	0.667	0.671	0.673	0.681
0.30	0.509	0.509	0.537	0.539	0.543	0.543	0.555
0.40	0.400	0.400	0.427	0.429	0.433	0.435	0.444
0.50	0.308	0.308	0.333	0.333	0.338	0.338	0.345
0.60	0.229	0.229	0.249	0.250	0.254	0.255	0.262
0.70	0.160	0.160	0.176	0.177	0.179	0.180	0.186
0.80	0.100	0.100	0.111	0.111	0.113	0.113	0.118
0.90	0.047	0.047	0.052	0.053	0.053	0.054	0.056

[a] From H & S calculations for μ/μ_0 and K/K_0

scopically isotropic homogenous media. As a function of aspect ratio, the averaged upper-bound stiffness moduli will be equal to the Hashin and Shtrikman curves for spherical pores and will fall below them for increasing oblate or prolate pores since the pore strain term becomes increasingly dominant in the moduli calculations. This result is illustrated in Table 1 and in Figures 2 above and 3 below.

Table 1 shows a comparison of the Hashin and Shtrikman upper-bound shear modulus (μ/μ_0), bulk modulus (K/K_0), and Young's modulus (E/E_0) compared to the calculated values of these moduli using the method of this paper. Empty spherical pores are assumed, and the matrix material has a Poisson's ratio of $v_0 = 0.25$. Slight disagreements in the numerical values are due in part to round-off errors which have not been corrected for here, and to numerical difficulty in calculating the limiting core of $\alpha \rightarrow 1$ (see Appendix).

Since the net strain effect of an oblate pore is highly anisotropic, the upper-bound shear and Young's moduli calculated for oblate pores only have a meaning in an average isotropic-homogeneous sense. The required average pore strains are obtained from calculations of ${}_{i\alpha}\varepsilon_h'$ as functions of pore orientation with respect to the applied stress fields.

Although this procedure is important for aspect ratios $1 < \alpha \lesssim 10$, for highly oblate pores, the rotationally averaged pore strain is given to good approximation by

$$\frac{{}_{i\alpha}\varepsilon_h'}{{}_{i}\varepsilon_0} = \frac{1}{3}\left[\frac{{}_{3}\varepsilon_h'}{{}_{3}\varepsilon_0} + 2\right] \tag{26}$$

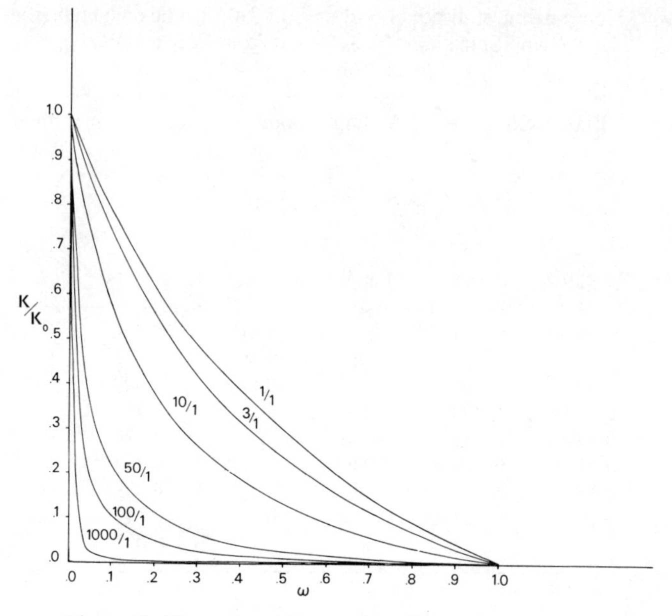

Figure 3. Upper-bound curves for bulk modulus as functions of porosity.

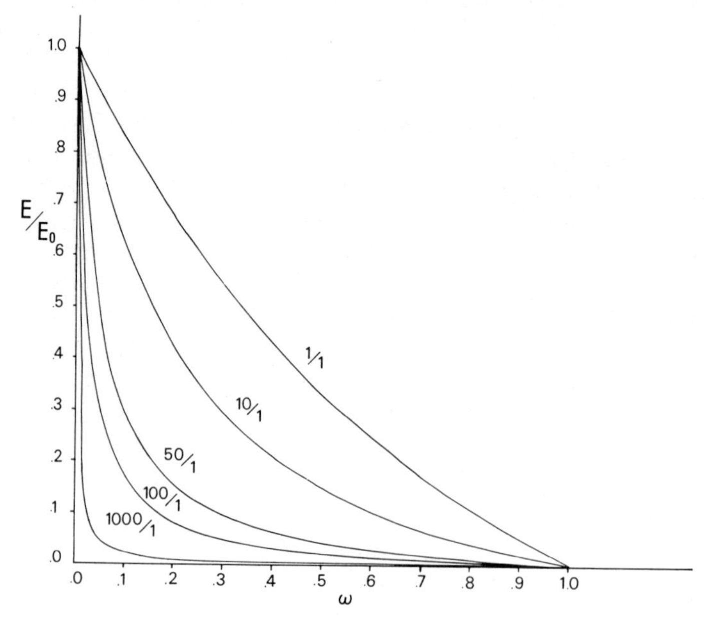

Figure 4. Upper-bound curves for isotropic Young's moduli as functions of porosity.

Figures 3 and 4 show plots of upper-bound K/K_0 and E/E_0 averaged as functions of porosity of a material with a matrix Poisson's ratio of $\nu_0 = 0.25$.

Poisson's Ratio Without and With Pore Fluids

Increasing the aspect ratio of the pores at constant porosity lowers the effective Poisson's ratio of the material if the pores are empty. If the pores are filled with fluid, Poisson's ratio, ν, recovers or increases. The change in and final value of ν is dependent on assumptions of transmissity of pore pressures between pores under the non-hydrostatic loading cases. This trend can be illustrated using upper-bound moduli. For a medium with matrix $\nu_0 = 0.25$ and a porosity $\omega = 0.01$, upper-bound calculations give $\nu = 0.248$ for empty pores with $\alpha = 1$, and $\nu = 0.192$ for empty pores with $\alpha = 100$. If the pores are filled with fluid, $\nu = 0.249$ to 0.250 for $\alpha = 1$ and $\nu = 0.231$ to 0.295 for $\alpha = 100$.

The depression of ν with dry cracking, and the recovery of ν to a normal value with fluid filling has, of course, been used in the explanation of earthquake precursor seismic anomalies (Nur, 1972). The results in the paper, however, emphasize that simple generalized statements of crack effects on average properties cannot be easily applied to problems of heterogeneity. Variables such as pore aspect ratio, preferred orientation, pore fluid boundary conditions, and fluid flow have marked effects on anisotropic macroscopic properties.

Closed Cracks

Although the calculations here involve open cracks, the basic strain terms in equation (25) for flat cracks can be cast in terms of crack length and crack number density, by a transformation of the quantity

$$\frac{{}_{\alpha}\omega\,{}_{i\alpha}\varepsilon'_h}{{}_{i}\varepsilon_0}$$

For flat pores, $\langle {}_{i\alpha}\varepsilon'_h/\varepsilon_0 \rangle$ is controlled by the stresses acting normal to the pore plane, and takes the form

$$\left\langle \frac{{}_{i\alpha}\varepsilon'_h}{{}_{i}\varepsilon_0} \right\rangle \approx f(\nu).\alpha$$

The partial porosity can also be expressed in terms of the pore aspect ratio $\boldsymbol{\alpha}$

$${}_{\alpha}\omega = \frac{{}_{\alpha}V_h}{V} = {}_{\alpha}N\,\frac{4}{3}\pi a^3 \frac{1}{\alpha}$$

where ${}_{\alpha}N$ is the number density of the α th class of pores, and $\mathbf{a}$ is a characteristic diameter of the oblate pore. Therefore, the strain term expression becomes

$${}_{\alpha}\omega\,\frac{{}_{i\alpha}\varepsilon'_h}{{}_{i}\varepsilon_0} \to f(\nu)a^3\,{}_{a}N$$

and the moduli may be calculated in terms of crack lengths and crack-length distributions.

Pore–Pore Interactions

The effect of pore–pore interactions on elastic moduli cannot be calculated uniquely. The interaction contribution to pore strain depends on explicit modelling variables. For example, packed spherical pores can interact at some porosity, so the material crumbles at $\omega < 1.0$; intersect selectively to generate a foam-like material coherent to $\omega \to 1.0$; or pack, assuming a particular pore-size distribution, to generate a less-foamlike but still coherent material.

If the direct effect on the average strain of a pore can be estimated, then the strain term ${}_{\alpha}\varepsilon_h'$ in the moduli equations can be replaced by

$$ {}_{\alpha}\varepsilon_h' \to {}_{\alpha}\varepsilon_h' + \kappa \tag{30} $$

where κ is the additional strain. This quantity κ can be estimated in an average sense by calculating stress concentrations around one pore surrounded by other pores within some mean interaction distance (e.g., one pore diameter). This approach can be applicable for high-porosity, coherent materials. Examples of such an interaction term are given in Warren (1973).

Pore–pore interactions may also be taken into account by approximating their effect through constraints on the packing of the pores. Such packings can be very crudely approximated through the weighing function in equations for S_{ij} by neglecting portions of the matrix material which, as a function of the packing of the pores, do not strongly affect bulk strains (i.e. act more like uncoupled grains (Warren and Anderson, 1973).

A self-consistency approach (Wu, 1966) can also be used to estimate the effective moduli. The calculations for two moduli are reiteratively calculated so that the matrix surrounding the pores in the pore strain calculations becomes that of the averaged porous material.

Figure 5a and 5b show the result of such a self-consistency calculation for K/K_0 and μ/μ_0 for a material with empty spherical pores. The matrix material parameters are $\mu_0 = 374$ and $v_0 = 0.25$. In this calculation, the material tends to become soil-like at about 40 per cent porosity. With increasing porosity, the shear modulus decreases more rapidly than the bulk modulus with the result that Poisson's ratio first decreases and then increases to high values more typical of compacted grains.

Table 2 shows self-consistent moduli for a material with oblate empty pores of aspect ratios = 500. The self-consistent Poisson's ratio goes to zero at a crack porosity slightly over 0.004.

Self-consistent moduli are only one of the possible sets of moduli for materials with pore–pore interactions. Given a crack population, the values of self-consistent moduli can be below those of observed moduli since the moduli of the average matrix material acting around any pore type can fall anywhere from those of the solid material to values below the average bulk material

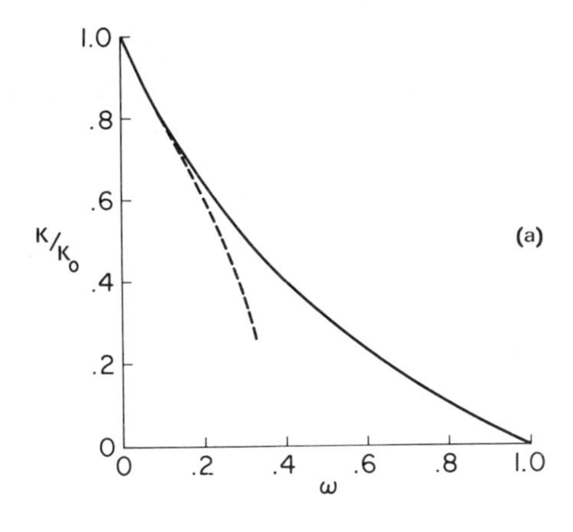

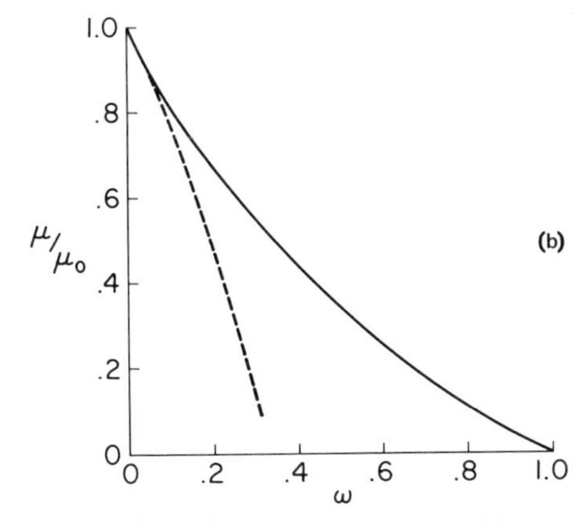

Figure 5. Self-consistent (a) K/K_0 and (b) μ/μ_0 as functions of porosities: solid curves are upper-bound moduli; dashed curves are self-consistent moduli.

Table 2. Self-consistent moduli

$\mu_0 = 374$, $v_0 = 0.25$, crack aspect ratio = 500

Porosity	K/K_0 Bulk modulus		E/E_0 Young's modulus		v Poisson's ratio	
	a	*b*	*a*	*b*	*a*	*b*
0.001	0.715	0.66	0.833	0.795	0.208	0.199
0.002	0.556	0.412	0.714	0.583	0.179	0.146
0.003	0.455	0.224	0.624	0.365	0.157	0.092

[a]Averaged upper bound—no pore–pore interaction
[b]Self consistent

moduli, depending on pore type, size, and local orientation distributions. Additional methods of calculating packing and interaction effects must be used.

MULTIPLE-CLASS PORE DISTRIBUTION: COMPARISON WITH EXPERIMENT

Calculations for materials with a distribution of pore types are made by summing pore strain times partial porosities over all pore types. Comparison of calculated elastic properties with experimentally measured properties can then be made given experimental input parameters of partial porosities, ${}_{\alpha}\omega$, aspect ratios or pore geometries, α, and the moduli of the solid pore-free material.

All of these parameters can be estimated. Good approximations of the solid material moduli can be obtained from mineralogy or from high-pressure direct elastic measurements. Given the mineralogy, theoretical moduli can be calculated as average moduli between the Hashin and Shtrikman upper and lower bounds. For silicates, the spread in the bounds is generally less than 10 per cent. Aspect ratios and partial porosities can, at present, be obtained from hydrostatic strain data. Progress is being made in obtaining them from petrographic data, in the form of either aspect-ratio-partial porosity or as crack-length-number density. Given mineralogy and strain data, one may predict the rest of the elastic moduli. Given mineralogy and petrographic structure data, one may predict all the elastic moduli.

No theoretical separation of static and dynamic moduli is made in this section. Although our program calculates static or zero-frequency theoretical elastic moduli, the moduli may be interpreted as dynamic moduli, as long as it is assumed that the wavelengths of interest are long compared to structural and grain dimensions (Mal and Knopoff, 1967). The basic solutions can then be interpreted as the time-independent solution to the nondispersive, infinitesimal, elastic equations. This point will be returned to later.

Zero-Pressure Moduli: Comparison with Measurement

In Table 3, comparison of theoretical and measured moduli are presented for a lunar sample and a granite. Pore–pore interactions have not been included in the calculations.

Lunar Sample 14310

Sample 14310 is a fairly homogeneous lunar anorthositic basalt. Elastic properties of samples of this rock have been measured as functions of pressure by two groups: 14310,72 by Todd *et al.* (1972) and 14310,82 by Trice *et al.* (1974). Mineralogies and theoretical solid-state moduli are available (Longhi *et al.*, 1972; Warren *et al.*, 1973). For the calculations here, the reference material parameters were obtained from the mineralogy. Total porosity was calculated from the theoretical and measured densities, partial porosities and pore aspect

Table 3. Zero-porosity elastic Moduli

Lunar sample 14310

$(P = O)$	α	P_c (kb)	Strain Ratios: Hydro	Strain Ratios: Uniaxial	Reference moduli[a]
0.051	1/1	—	2.65	2.012	$K_0 = 861$
0.005	600	1.7–2.0	500	250	$\mu_0 = 467$
0.0009	3000	0.5	2000	1000	$\nu_0 = 0.276$
					$E_0 = 1193$

Calculated	Observed (dynamic)[b]
$K/K_0 = 0.175$	$K/K_0 = 0.155$
$E/E_0 = 0.29$	$E/E_0 = 0.25$
$\mu/\mu_0 = 0.32$	$\mu/\mu_0 = 0.27$

om mineralogy
rice *et al.* (1974)

Westerly granite

$(P = O)$	α	P_c (kb)	Strain ratios: Hydro Dry	Hydro Stat.	Uniaxial Dry	Uniaxial Sat.[g]	Reference[c] moduli
0.007[d]	1/1	—	2.44	1.0	2.00	1.1	$K_0 = 564$
0.002[d]	500	2	400	20	200	55	$\mu_0 = 341$
0.0004[e]	5000	0.2	4000	27	2000	500	$\nu_0 = 0.248$
							$E_0 = 851$

	Calculated Dry	Calculated Sat	Nur[d] (Dynamic) Dry	Nur[d] (Dynamic) Sat	Brace[f] (Static) Dry	Simmons and Brace[c] (Dynamic) Dry
K/K_0	0.29	0.94	0.188	0.85	0.22	0.35
E/E_0	0.45	0.76	0.44	0.721	—	0.51
μ/μ_0	0.51	0.73	0.61	0.701	—	0.55

immons and Brace (1965). [d]Nur and Simmons (1969). [e]Warren (1973). [f]Brace (1965). [g]Assuming pores quilibrate with a pore pressure/applied stress ratio = 75 per cent.

ratios were calculated from hydrostatic, linear strain data (Trice *et al.*, 1974) using the method of Walsh (1965b). Strains in only a single direction are available. However, since the rock is fairly homogeneous, partial porosities were taken as three times the measured strains. Curvature of the strain curve at very low pressure was ignored. A three-aspect-pore population was assumed, spheres, and two classes of oblates with closures at pressures P_c of 500 bars and of 1.7 to 2.0 kb. The total porosity of the first class of oblates was slightly adjusted within the range of probability (${}_{\alpha}\omega = 0.005$ to 0.006) for agreement between calculated and observed K/K_0. The other moduli were calculated using the same partial porosities. The observed moduli were obtained from zero-pressure velocity measurements (Trice *et al.*, 1974).

Westerly Granite

For comparison of theoretical and experimental results, the solid material parameters were taken from the 10 kb velocity data of Simmons and Brace (1965). A three-aspect-ratio pore distribution was used, basically from Warren (1973), although the more recent porosity data of Nur and Simmons (1969) were used for the partial porosities. Although only single-aspect crack porosities have been published for westerly granite, Simmons has recently shown, by using differential stress analysis, that Westerly granite has a more continuous crack aspect ratio distribution (Simmons *et al.*, 1974).

Moduli as a Function of Pressure

Given the zero-pressure parameters for the elastic moduli, the effect of hydrostatic pressure on the moduli can be at least estimated. However, considerable approximations are involved. These arise due to the need to assume a closure law for the pores, and due to uncertainty in the values of real rock parameters. The moduli calculations are very sensitive to both aspect ratios and partial porosities of the oblate pores. Small variations in estimated crack closure pressures strongly affect the calculated values of the high-pressure moduli. Within the tolerances of these uncertainties, the zero-pressure results for 14310 are used to predict the velocity-pressure response of the sample.

A local relative modulus at some pressure P can be defined in terms of a stress perturbation dR applied at pressure P. A linear law for pore closure is assumed, so that the change in pore volume, $\Delta_{\alpha} V_h$, satisfies

$$\frac{\Delta_{\alpha} V_h}{V_{h(p=0)}} = P/P_c \qquad \text{for } P < P_c$$

and the partial porosity of the α th type pore decreases as

$${}_{\alpha}\eta(P) \approx {}_{\alpha}\omega(P=0)[1 - P/P_c]$$

Changes in $V(P)$ are neglected in the porosity term since $V >>> {}_{\alpha}V_h$ and $[V - \Delta V](P) >>> {}_{\alpha}V_h(P)$.

The relative pore strains are assumed to be essentially independent of pressure since all the stress-strain relations are assumed to be linear for $P < P_c$.

The predicted moduli versus pressure are therefore generated from equations of the form of equation (25), with the terms in the pore strain coefficient being given by

$$\frac{{}_{\alpha}\omega}{1 - {}_{t}\omega} \rightarrow \frac{{}_{\alpha}\eta(P)}{1 - {}_{t}\eta(P)}$$

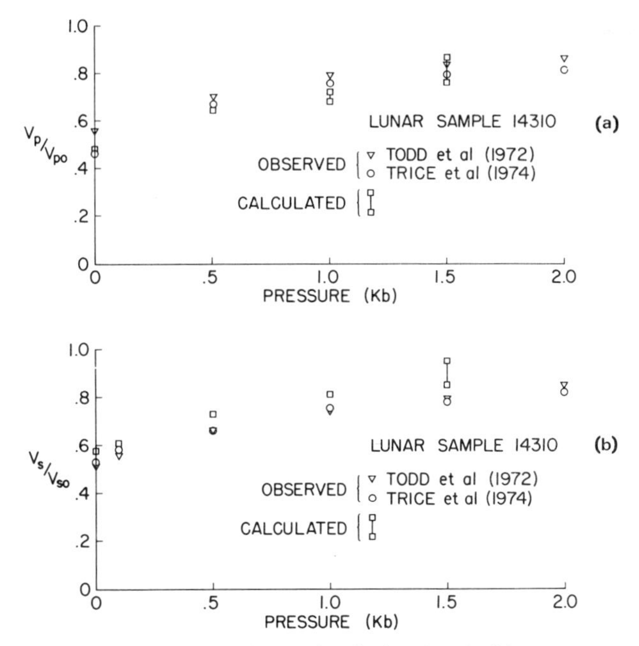

Figure 6. Comparison of calculated velocities as a function of pressure for lunar sample 14310. (a) Relative compressional velocity; V_{p0} is the theoretical compressional velocity of the solid rock; (b) V_{s0} is the theoretical shear velocity of the solid rock.

Figures 6a and 6b show predicted versus observed normalized velocity-pressure curves for 14310.

The effect of assuming closure of pores between 1.7 kb and 2 kb is illustrated by the increasing spread in the predicted velocity curves with pressure. A 'fit' could have been generated by assuming pore closure closer to 2.3 kb.

Although the calculation is very approximate, with no pore–pore interactions, the basic results indicate the strength of the method and the important implications of being able to solve the problem of petrographic crack study.

Dynamic Versus Static Moduli

The relation of static to dynamic moduli is still not well determined. Differences between dynamic and static results can involve relaxed and unrelaxed moduli and dispersion of velocities due to sample structure. In addition, if in the calculation of moduli versus pressure, a static modulus is obtained by a finite difference method around P, or by referring the calculation to the sample volume at $P = 0$, then the coefficients in the compliance equations are larger than those used here, and the predicted values of the stiffness modulus are decreased.

CONCLUSIONS

The problem of calculating the elastic moduli of rocks can be formulated in terms of mineralogical and petrographic observables. The formulation presented here requires input parameters of partial porosities, crack geometries, and rock mineralogy. Given these types of data, the long-wave-length or static isothermal elastic moduli of a rock can be predicted as a function of pressure. In general, both isotropic and nonisotropic properties can be modelled.

Present research trends, on the one hand, of developing methods of crack study using petrographic techniques and precise strain data, and on the other hand, of developing flexible computational methods for calculating elastic properties will lead to very strong predictive abilities in rock physics.

APPENDIX

Pore Strain Computation

A computer program based on Edwards' solutions has been written to calculate local stresses and displacements inside and outside of oblate and prolate spheroidal inclusions given stress loads applied at infinity. Five loads can be superimposed: plane hydrostatic, uniaxial, and pure shears in three mutually perpendicular planes. A spheroidal coordinate system is used for the calculations.

Pore strains are calculated from the displacements of any surface element of an inclusion. The displacements are in the form of u_{rp} representing the r th component of displacement resulting from the p th loading case. The u_{rp} are obtained from the three solutions of displacement equilibrium expressions first formulated by Boussinesq (1885), then illustrated and utilized by Sadowski and Sternberg (1947), and by Edwards (1951). The three general displacement solutions are

$$(u,v,\omega) = \frac{1}{2\mu} \text{grad T}$$

$$(u,v,\omega) = \frac{1}{\mu} \text{curl} [0,0,\xi]$$

$$(u,v,\omega) = \frac{Z}{2\mu} \text{grad } \lambda - \left[0,0,\frac{3-4v}{2\mu}\lambda\right]$$

where $\Delta T = \Delta\xi = \Delta\lambda = 0$
and (u,v,ω) are displacement vectors expressed in terms of the displacement potentials T, ξ, λ.

Averaged volume and linear strains are obtained by averaging elements over a quadrant of the inclusion. Normally a meridian from pole to spheroid equator is divided into 40 steps for these averages. Of course, if the strain state of the inclusion is uniform, this number of steps is not necessary.

The effect of pore orientation with respect to an applied load is obtained by imposing a succession of superpositions of the five loading cases.

The program input for each case study consists of the pore shape factor (aspect ratio), the five component load vectors, and the shear modulus and Poisson's ratio of the matrix, μ_t, v_t and of the inclusion μ_h, v_h. Pore fluids are approximated in the input by taking limiting values of the pore moduli $\mu_h = \mu_f$ and $v_h = v_f$, such that a finite fluid bulk modulus K_f exists

$$K = \lim_{\substack{\mu_f \to 0 \\ v_f \to 0.5}} \left[\frac{2\mu_f(1 + v_f)}{3(1 - 2v_f)}\right]$$

In this program, since the inclusion matrix boundary conditions are continuity of stress and displacement, the modelling of pore fluid effects on the moduli requires an additional modelling choice of the extent of pore interconnectiveness. Under nonhydrostatic loading, the calculated pore strains for fluid-filled pores depend on whether the fluids are trapped in the pores or can reach pore-fluid pressure equilibrium with other pores.

One computation problem in using Edwards is that the displacements and stresses are obtained as differences and ratios between very large numbers. Although good results are obtained for oblate and prolate inclusions, computation round-off can affect calculations for almost spherical pores under nonhydrostatic load. In our scheme, these effects are compensated for by heuristic adjustment of the magnitudes of some of the scalar coefficients in Edwards' equations by about one part in 2×10^4.

Computational accuracy is also limited for very oblate or prolate pores since the pore volumes and displacement cannot be integrated accurately using the current stepwise procedure. Highest accuracy is obtained for aspect ratios, α, on the order of 10^3 (flat cracks) to 10^{-3} (prolate needles), $\alpha \neq$ unity.

ACKNOWLEDGEMENTS

We wish to thank Drs. G. Sines and R. Westmann for important discussions. Work was partially supported under NASA grant NGL 05-007-330 and AEC, At (04-3)-34. Publication number, 1408. Institute of Geophysics and planetary Physics, University of California, Los Angeles, CA 90024.

REFERENCES

Boussinesq, J. (1885). *Application des Potentiels*. Gauthier Villars.
Brace, W. F. (1965). *Jour. Geophys. Res.*, **70**, (2), 391–398.
Edwards, R. H. (1951). *Jour. Appl. Mech.*, **18**.
Green, A. E. and I. N. Sneddon (1948). *Proc. Camb. Phil. Soc.*, **45**, (2) 252.
Hashin, Z. and S. Shtrikman (1963). *Jour. Mech. Phys. Solids*, **11**, 127.
Longhi, J., D. Walker and J. F. Hays (1972). *Proc. Third Lunar Science Conf., Geochim. Cosmochim. Acta, Suppl. 3*, Vol. 1, pp. 131–139. MIT Press.
Love, A. E. M. (1944). *A Treatise on the Mathematical Theory of Elasticity*. 4th ed. Dover pub.
Mal, A. K. and L. Knopoff (1967). *Jour. Inst. Maths. Applics.*, **3**, 376–387.

Nashner, R., and N. Warren, *Pore-stress, Strain Computer Computation*. Technical Report, available on request.
Nur, Amos and G. Simmons (1969). *Earth and Planetary Science Letters*, **7**, 183–193.
Nur, Amos (1972). *Bulletin of the Seismological Society of America* **62**, (5), 1217–1222.
Sack, R. A. (1946). *Proc. Phys. Soc. Lond.*, **58**, 729–736.
Sadowsky, M. A. and E. Sternberg (1947). *Jour. Appl. Mechanics*, **14**, 191–201.
Simmons, G., Geness, R. W. Siegfried III and M. Feves (1974). *Jour. Geophys. Res.*
Simmons, G. and Brace (1965). *Jour. Geophys. Res.*, **70**, (22) 5649–5656.
Sokolniknoff (1956). *Mathematical Theory of Elasticity*. McGraw-Hill Book Company.
Todd, T., H. Wang, W. S. Baldridge, and G. Simmons (1972). *Proc. Third Lunar Conf., Geochim Cosmochim. Acta, Suppl. 3*, Vol. 3, pp. 2577–2586. MIT Press.
Trice, R., N. Warren and O. L. Anderson (1974). *Proc. Fifth Lunar Science Conf. Geochim Cosmochim. Acta, Suppl. 5*, Vol. 3, pp. 2903–2911., Pergamon.
Walsh, J. B. (1965a). *Jour. Geophys. Res.*, **70**, (2), 381–389.
Walsh, J. B. (1965b). *Jour. Geophys. Res.*, **70**, (2), 399–411.
Warren, Nick (1973). *Jour. Geophys. Res.*, **78**, (2), 352–362.
Warren, N. R. Trice, N. Soga and O. L. Anderson (1973). *Proc. Fourth Lunar Sci. Conf., Geochim. Cosmochim. Acta, Suppl. 4*, Vol. 3, pp. 2611–2629., Pergamon.
Warren, Nick and O. L. Anderson (1973). *Jour. Geophys. Res.*, **78**, 6911–6925.
Wu, Tai Te (1966). *International Journal of Solids and Structures*, **2**, pp. 1–8.

Elastic Behaviour Near Phase Transitions

G. H. Shaw
Department of Geological Sciences
University of Washington
Seattle, Washington 98195

Recent theoretical work by Anderson (1970), Anderson and Liebermann (1970), Anderson and Demarest (1971), Demarest (1972), Thomsen (1971), Sammis (1970), and Weidner and Simmons (1972) has raised the possibility of theoretically examining elastic behaviour near phase transitions in some detail. These papers are based upon the general theory of crystal lattices, and various assumptions are made in the treatments, such as restriction to central forces and neglect of temperature effects (calculations apply strictly only to 0 K). For some of the treatments the repulsion potential is restricted to nearest neighbours, and a given type of repulsion law is assumed. These effects are discussed in detail by Anderson (1970), while Thomsen avoids these assumptions with his fourth order anharmonic theory by including a greater number of arbitrary parameters.

Probably the most interesting aspect of some of the lattice dynamic calculations is that they result in expressions for single crystal elastic constants and their pressure derivatives in terms of the bulk modulus and its first pressure derivative. This means that by making measurements of incompressibility and its pressure dependence (or more basically density as a function of pressure) one can predict the shear properties of a material. In practice K_0' is difficult to determine with precision by simple compression measurements. It is usually determined ultrasonically which means that the shear properties are determined directly as well.

Of great significance is the variation of the expressions for the constants with crystal structure, particularly for the shear constants and their pressure derivatives. According to the theoretical results it is possible to have negative pressure derivatives of the shear constants and the isotropic shear velocity. The implications of this theory to the earth's interior, and particularly the phase transition region, are very great. The detailed elastic behaviour near phase transitions in the earth may be predicted when the theory is extended to crystal classes of importance in the earth's interior.

Since Lattice theory is so important it is advisable to subject it to intensive testing. Anderson and Liebermann (1970) have checked the predictions of the law against experimental single crystal elasticity data, and Anderson (1970) has suggested possible corrections for noncentral forces based upon the experimental data for alkali halides. Reddy and Ruoff (1965) have performed similar calculations for RbBr, KI, and KBr using the Born (power law) repulsive potential and the exponential repulsive potential (for RbBr, KI, KBr, and CsBr). Their calculations were based upon the measured bulk moduli from single crystal measurements and the lattice parameters rather than the bulk moduli and their pressure derivatives. A negative pressure derivative of shear velocity has been found experimentally in a polycrystalline aggregate of ZnO (Soga and O. L. Anderson, 1967) which has a structure very similar to ZnS. There is good agreement between the experimental measurements and the theoretical calculations.

The lattice theory calculations may, in part, be tested by using elastic data on polycrystalline samples. In the present study polycrystalline samples of five halides (Agl, NH_4I, RbCl, RbBr, RbI) have been prepared and the elastic wave velocities measured as functions of pressure. The experimental procedure is given elsewhere (Shaw, 1974a,b and Birch, 1960). All five halides exhibit polymorphic phase transitions within the experimental pressure range, and the results provide a test of the lattice dynamic calculations as they apply to simple

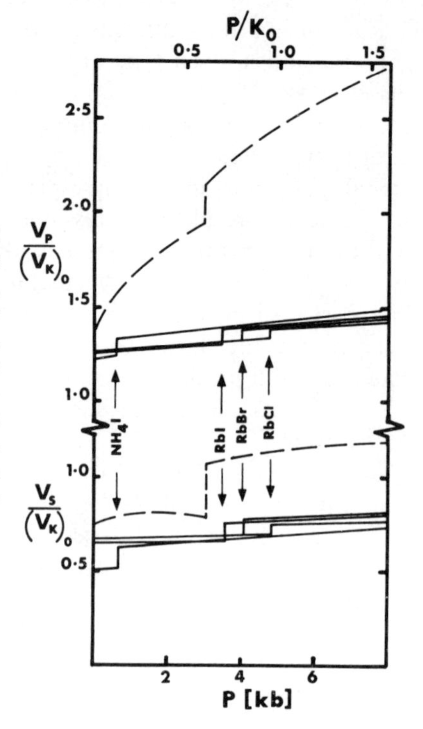

Figure 1. Reduced compressional and shear wave velocities versus pressure for the B1–B2 transition. The halides are indicated by reference to the transition pressure. The dashed lines are according to lattice theory with $K_0' = 5.5$ (Anderson and Demarest, 1971) and are plotted on the same reduced velocity scale using reduced pressure (P/K_0). Note that the two pressure scales are *not* equivalent.

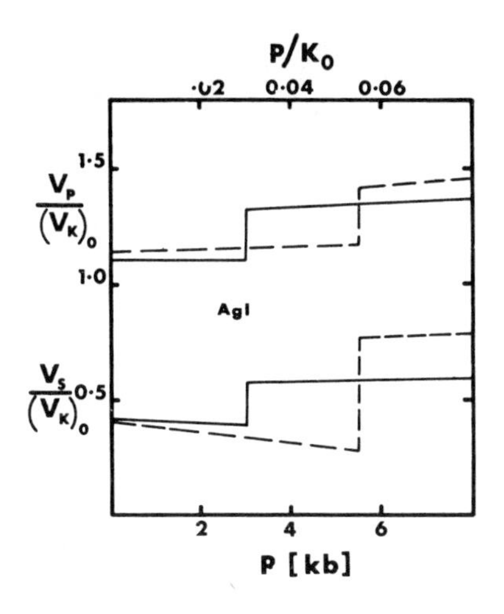

Figure 2. Reduced compressional and shear wave velocities versus pressure for the B3–B1 transition. The dashed lines are according to lattice theory with $K_0' = 5.5$ (Anderson and Demarest, 1971) and are plotted on the same reduced velocity scale using reduced pressure (P/K_0). Note that the two pressure scales are *not* equivalent.

polymorphs. The measurements on the high pressure phases in particular are very useful since single crystal elastic constants are not available for the high pressure forms. These data, therefore, provide a good test of the theoretical calculations made for polycrystalline aggregates of polymorphs (Anderson and Demarest, 1971).

Figures 1 and 2 show the experimental results for the samples: AgI, which illustrates the ZnS(B3)–NaCl(B1) transition and NH_4I, RbCl, RbBr, and RbI, which exhibit the NaCl(B1)–CsCl(B2) transition. The data have been corrected for compression and ΔV of transition. The procedure for doing this is given by Shaw (1974b). The range of pressure in the experiments is not sufficient to reveal non-linearity in the velocity-pressure curves. The data from these five halides have been compared with single crystal elastic constant data and isothermal compression measurements (Shaw, 1974b), and the agreement is quite good.

Comparison between the corrected velocity-pressure data in Figures 1 and 2 with the graphs of Anderson and Demarest (1971) (dashed lines) indicates that the lattice theory is qualitatively rather good at describing the velocity–pressure curves even to the extent of distinguishing the B3–B1 transition from the B1–B2 transition. The consistencies between the experimental data and the lattice dynamics results include the following:

For the B1–B2 transition:

(1) The shear velocity for the B1 phase has a lower pressure derivative than the compressional velocity.
(2) The derivatives of both velocities for the B2 phases are generally greater (positive) than the B1 phases.
(3) As K_0 decreases the pressure derivative of the Hill shear velocity decreases (in the lattice calculations negative dV_s/dP occurs in the vicinity of the phase transition as K_0 gets smaller and is positive for larger K_0' (Anderson and Demarest, 1971, Figure 11d).

(4) The per cent increase in shear velocity at the transition is about four times the per cent increase in compressional velocity according to the lattice theory and about three times according to the experimental data.

For the B3–B1 transition

(1) The pressure derivative of shear velocity for the B3 phase is negative, and the derivative for the compressional velocity of the B3 phase is very small.
(2) The pressure derivatives for both velocities in the high pressure phase are positive.
(3) The increase in shear velocity at the transition is about four to five times the per cent increase in compressional velocity according to the lattice theory and about three times according to the experimental data.

There are also inconsistencies between the theoretical and experimental results. According to the theory, a low value of K_0' should lead to higher transition pressures for both B3 and B1 structures, but this is in contradiction with the experimental data. The per cent increases in the velocities at the transition ($\Delta V_s/V_s$ and $\Delta V_p/V_p$) from the theory are two to three times the measured values (see Figures 1 and 2). Two of the experimentally determined values of K_0' (AgI LP phase and RbI HP phase) are inconsistent with the lattice formulae; they give negative values for either C_{44} or C_s in a region of known stability, an impossible situation (Born, 1940). In the case of RbI this may simply be due to experimental error in the determination of K_0' for the high pressure phase since the value determined would have to be raised only a few per cent to yield acceptable results in the lattice equations. This is not the case with AgI, and this has been discussed elsewhere (Shaw, 1974a).

These comparisons indicate that the lattice dynamics approach may be very useful in qualitatively and semi-quantitatively describing phase transitions. Improvement in prediction of transition pressures may be desirable. Whether precise quantitative results will be obtained with a small number of parameters

Table 1. Experimentally determined elastic parameters at the polymorphic transition pressures.

Phase	P	P/K_0	$V_p/(V_k)^\circ$	$V_s/(V_k)^\circ$	K_0'
AgI LP	3	0.0126	1.100	0.396	1.7
AgI HP	3	0.0087	1.183	0.519	6.8
NH_4I LP	0.6	0.0052	1.239	0.616	6.5
NH_4I HP	0.6	0.0044	1.314	0.718	7.0
RbCl LP	4.75	0.0285	1.333	0.675	6.6
RbCl HP	4.75	0.0233	1.349	0.729	6.0
RbBr LP	4	0.0292	1.329	0.688	5.6
RbBr HP	4	0.0252	1.389	0.772	6.0
RbI LP	3.5	0.0308	1.291	0.646	4.8
RbI HP	3.5	0.0252	1.345	0.731	4.7

is an open question, but, the success in yielding semiquantitative results for velocity changes at the phase transitions is striking. It should be pointed out that the use of reduced parameters (V_p/V_{K_0}, etc.) must be modified for specific cases where individual K_0, K'_0 (for both phases) and ΔV of transition data are available. For example, Table 1 gives the reduced parameters at the transitions for each phase of the halides in this study. It is clear that at the transition pressure the reduced pressure P/K_0 is not the same for both the low and high pressure phases. This is due to the difference in K_0 for each phase. This causes some problem since Anderson and Demarest joined their solutions for high and low pressure phases at the same reduced pressure for both phases. In practice this would have only a slight effect on their results. In addition, the reduced velocities may be referred to different $(V_k)_0$ values for each phase. This is not too serious a problem since the changes in ρ and K at the transition compensate to some extent (Anderson and Demarest, 1971).

In spite of the assumptions involved in the theoretical calculations, the calculated $\Delta V_s/V_s$ and $\Delta V_p/V_p$ have the correct relative values and very nearly the appropriate magnitudes when compared with experiment.

Although conclusions about silicates based on the behaviour of halides are open to question, recent earth models by Jordan (1972) and Dziewonski and and Gilbert (1972) indicate changes in compressional and shear velocity at the 400 km discontinuity which are consistent with both the theoretical and experimental results for halides. Liebermann's (1973) contradictory results from measurements on silicates indicate the need for caution in applying the present results to the earth's interior and, even more, the need for further experimental and theoretical studies.

ACKNOWLEDGEMENT

This research was carried out while the author was a National Science Foundation Graduate Fellow in the Department of Geological Sciences at the University of Washington. The author would like to thank N. I. Christensen for discussion and advice as well as the use of his laboratory. Support for this research came from NSF Grant GA-26513 and from a Sigma Xi Grant in aid of research.

REFERENCES

Anderson, O. L. (1970). *J. Geophys. Res.*, **75**, 2719–2740.
Anderson, O. L. (1971) and H. H. Demarest, *J. Geophys. Res.*, **76**, 1349–1369.
Anderson, O. L. and R. C. Liebermann (1970). *Phys. Earth and Planet. Interiors*, **3**, 61–85.
Birch, F. (1960). *J. Geophys. Res.*, **65**, 1083–1102.
Demarest, H. H. (1972). *J. Geophys. Res.*, **77**, 848–856.
Dziewonski, A. M. and F. Gilbert (1972). *Geophys. J. Roy. Astron. Soc.*, **27**, 393–446.
Jordan, T. H. (1972). *Ph.D. Thesis*, Calif. Inst. of Technol., Pasadena.
Liebermann, R. C. (1973). *J. Geophys. Res.*, **78**, 7015–7017.

Sammis, C. G. (1970). *Geophys. J. Roy. Astron. Soc.*, **19**, 258–297.
Shaw, G. H. (1974a). *J. Phys. Chem. Solids*, **35**, 911–916.
Shaw, G. H. (1974b). *J. Geophys. Res.*, **79**, 2635–2642.
Thomsen, L. (1971). *J. Geophys. Res.*, **76**, 1342–1348.
Weidner, D. J. and G. Simmons (1972). *J. Geophys. Res.*, **77**, 826–847.

Seismic Anisotropy in the Crust and Upper Mantle

D. Bamford
Department of Geological Sciences
University of Birmingham
Birmingham, U.K.

INTRODUCTION

In geophysical studies of the earth's interior, one normally assumes that the physical properties of rocks are isotropic. Thus elastic properties which are measurable *in situ* by seismological methods, are represented in terms of elastic isotropy; in particular, seismic wave velocities are assumed to be independent of measurement direction.

In reality, however, many rocks are not adequately described by isotropic assumptions. Stratified sedimentary rocks, for example, are often anisotropic (Uhrig and Van Melle, 1955) and single crystals of rock-forming minerals such as olivine show considerable velocity anisotropy (Verma, 1960); petrofabric studies show that many probable upper mantle source materials, many of them olivine-rich, show fabric and/or preferred orientations. Moores (1973) has pointed out that deformation fabrics are ubiquitous in Alpine ultramafics and in nodules, and (selecting two examples from the many available) Christensen (1971) has demonstrated the existence of anisotropy in the Twin Sisters Dunite, Washington, and Peselnick *et al.* (1974) have done the same for the Lanzo massif (lherzolite) in the Italian Alps. Furthermore, many laboratory investigations of the elastic properties of rocks up to pressures of 10 kb or more reveal a dependence of compressional and shear wave velocity upon direction (for example, Christensen, 1965, 1966a, 1966b, 1974; Babůska, 1972) and other anisotropy—related phenomena such as acoustic double refraction (see section 2 of this paper)—Christensen and Ramananantoandro (1971), Tilmann and Bennett (1973). These studies indicate that if elastic anisotropy exists at pressures of more than 1 or 2 kb and certainly if it is observed in the lower crust or upper mantle, then this anisotropy is controlled either by preferred mineral orientation or mineral layering (Todd *et al.*, 1973).

The measurement of elastic anisotropy *in situ* therefore offers as a new opportunity to study the earth's interior; in particular, although various mechanisms are proposed to account for anisotropy in the lithosphere (e.g. Francis, 1969; Ave'Lallemant and Carter, 1970), they all relate in some way to the movement of plates and so the study of anisotropy may be of special significance for geodynamics.

MEASURABLE EFFECTS OF ELASTIC ANISOTROPY

The theory of elastic wave propagation in anisotropic media is well known. Briefly, any generalized anisotropic material may be completely described by the specification of its density and the 21 independent elastic coefficients which form the cartesian components of a fourth-order elastic tensor. For an isotropic elastic material, the elastic tensor is isotropic and thus only two independent constants are needed to completely describe it. In comparison, therefore, the complete description of a generally anisotropic material is a demanding task and even the simplifying assumption of transverse isotropy (Stoneley, 1949) leaves 5 elastic coefficients and the density to be determined.

To the observational seismologist, the important point about elastic wave propagation in anisotropic media is that the distinction between compressional (P) and shear waves (S)—observed in isotropic media—no longer holds. Instead, for any propagation direction, there can exist a quasi-compressional wave and *two* quasi-shear waves; although the particle motions of these three waves will be mutually perpendicular, they will in general be neither parallel nor perpendicular to the propagation direction. The two shear waves travel with different velocities—this phenomenon is termed acoustic double refraction by Todd *et al.* (1973)—and all three waves show a dependence of velocity upon direction of propagation.

The distinction between Rayleigh and Love waves also breaks down in anisotropic media and a generalized surface is propagated with elliptical particle motion in a vertical plane inclined to the propagation direction—Crampin (1967) refers to these as coupled waves. Surface wave velocities will also depend on direction.

Three types of seismic measurement are likely to demonstrate the existence of anisotropy and provide estimates of the elastic constants required for complete specification of the anisotropic material: these are the measurement of the seismic velocities of body and surface waves in different propagation directions; the measurement of double shear wave travel—time differences in different propagation directions; and particle motion studies of body and surface waves. In principle, the first of these measurements involves little more than an extension of normal seismic methods. In practice, however, there are several difficulties.

First, each propagation path will be different, and in a laterally inhomogeneous earth this may make it very difficult to separate the effects of directional variations from those associated with lateral variations along and between propagation paths; this will be particularly so for surface waves which effectively sample the earth over a large range of depths. There exists, however, a sophisticated method for the interpretation of the travel-times of refracted body waves—the so called 'time-term' method (Morris *et al.* 1969; Bamford, 1973)—which can separate from within the travel-times the effects of different types of variation and, in particular, measure velocity anisotropy in the refracting layer. Various types of model including, for example, the possibility of lateral and/or anisotropic variations in refractor velocity are fitted by a least-squares method to the observed travel-times and the model which best explains the data can be selected on a statistical basis. This method is very powerful but requires the input of accurate travel-time data from very carefully planned observations. These requirements are likely to be met for compressional waves which are usually first-arrivals (with a good signal-to-noise ratio) on seismograms but are unlikely to be met for shear waves which are notoriously difficult to observe with good signal-to-noise ratios, masked as they so often are by compressional and surface wave energy. Thus, albeit with some difficulty, velocity–direction measurements can be obtained for compressional waves; similar shear wave information is much more difficult to obtain.

Second, velocities measured in seismic refraction studies are obtained from the travel-times of body waves that have travelled more or less horizontally through the top of a particular refractor; thus the top of the anisotropic layer will be well located but the information obtained will be incomplete, describing variations in one plane only and giving little or no indication of the vertical extent of the anisotropy. One can imagine that rather more complete information might be obtained using a combination of refraction and reflection techniques (vertically incident reflections sample vertical velocities) but this would require a considerable effort, and shear wave information would still be excluded.

Todd *et al.* (1973) have rightly pointed out that *in situ* observations of acoustic double refraction have several advantages. Problems regarding lateral variations are reduced because the pair of shear waves will travel by the same propagation path; lateral variations along the path will effect each shear velocity equally and so the observed time-difference will reflect only the gross anisotropy. Furthermore, this difference is explicitly related to the magnitude and symmetry of the mineral anisotropy (that is, the information is not restricted to a single plane) and thus a relatively simple series of observations of twin shear velocities are potentially powerful for the measurement of *in situ* anisotropy. The quality of shear wave information can be improved using data processing techniques based on particle motion analyses, for example by use of polarization filters (Montalbetti and Kanasewich, 1970). Particle motion studies themselves are reasonably straightforward to carry out and have the advantage that they offer the opportunity to recognize the presence of anisotropy without requiring extensive measurements but they are presumably very difficult to invert, that is, to turn into a quantitative description of a particular anisotropic layer. In detailed studies, therefore, measurements of seismic velocities in different propagation directions, and of double shear waves may be best performed in conjunction.

In situ measurement of velocity anisotropy

There are several reports of velocity anisotropy in the crust and upper mantle and it is useful to distinguish between reports for oceanic and continental regions.

Oceanic Regions

The first report of velocity anisotropy in an oceanic area was by Hess (1964) who compiled the results of a large number of linear seismic refraction profiles in the Pacific; qualitatively (Figure 1) one sees that the total anisotropic variation in upper mantle compressional velocity might be as much as 0.6 km/s (7–8 per cent)—the high velocities being sub-parallel to the oceanic fracture zones. This suggestion prompted a series of special refraction experiments to study upper

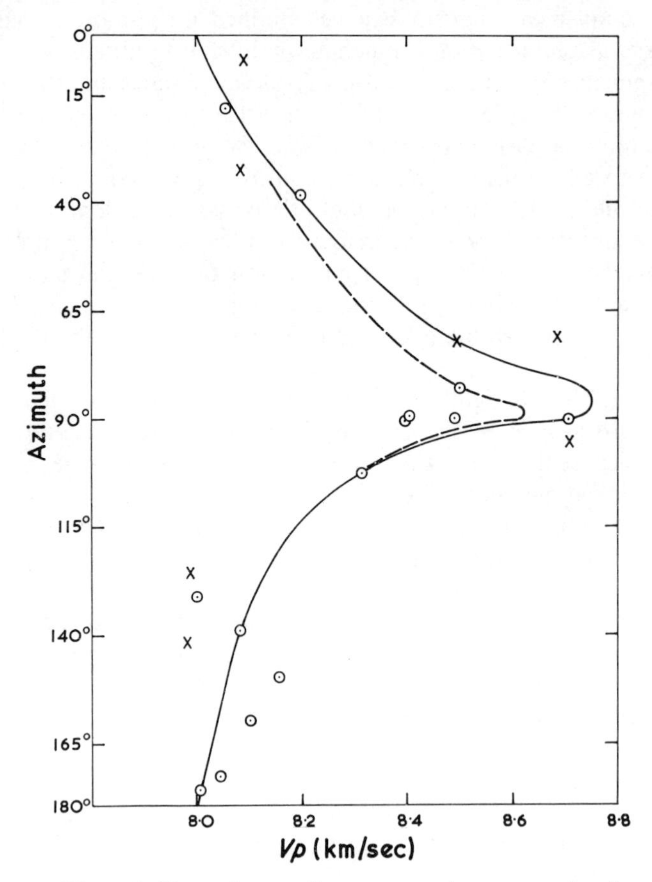

Figure 1. Dependence of upper mantle compressional wave velocity on direction for refraction profiles in the Pacific Ocean (Reproduced with permission from H. H. Hess, *Nature*, **203**, 629–631 (1964).)

mantle compressional wave anisotropy and its existence was demonstrated conclusively by Raitt *et al.* (1969) who measured an overall variation of 0.3 km/s off California, and by Morris *et al.* (1969) who measured an overall variation of 0.6 km/s near Hawaii. Further work quickly followed, with reports of upper mantle compressional wave anisotropy elsewhere in the Pacific (Keen and Barrett, 1971) and near the mid-Atlantic ridge (Keen and Tramontini, 1970). In each case, the highest velocities are found more or less parallel to the fracture zones.

The effort involved in these measurements was considerable. In order to prove the existence of anisotropy, the observations should represent different propagation directions through essentially the same material; at sea this can be achieved quite easily by the use of shot-station patterns such as that in Figure 2. In addition, the data must contain sufficient redundancy for the separation within the travel-times of variations due to structure, lateral and anisotropic

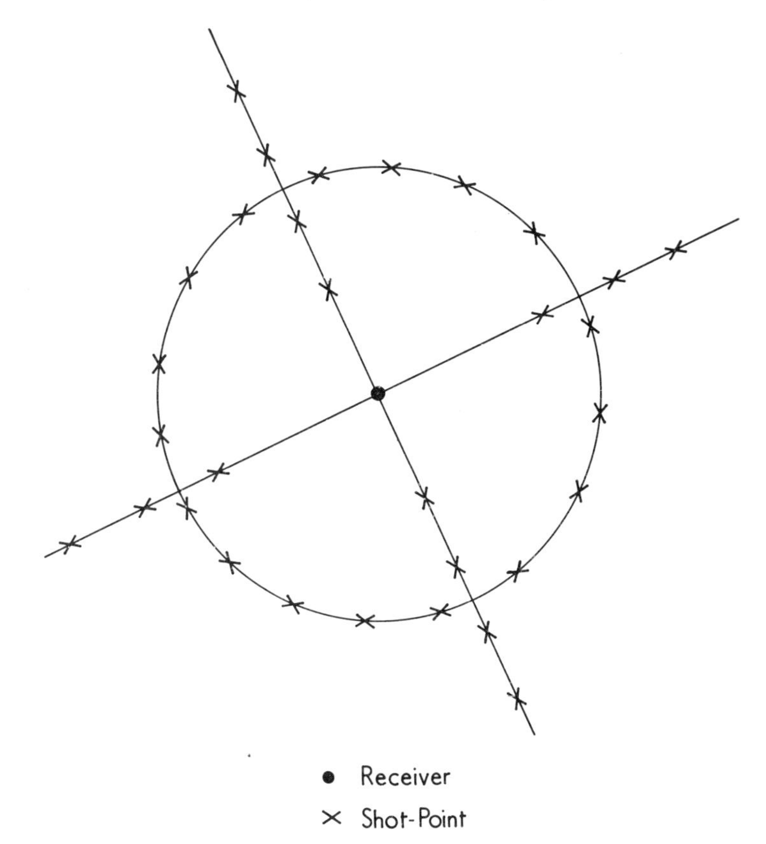

Figure 2. A shot-station pattern for measurement of velocity anisotropy (Reproduced from R. W. Raitt, G. G. Shor, T. J. G. Francis and G. B. Morris, *J. Geophys. Res.*, **74**, 3099 (1969) copyright by American Geophysical Union).

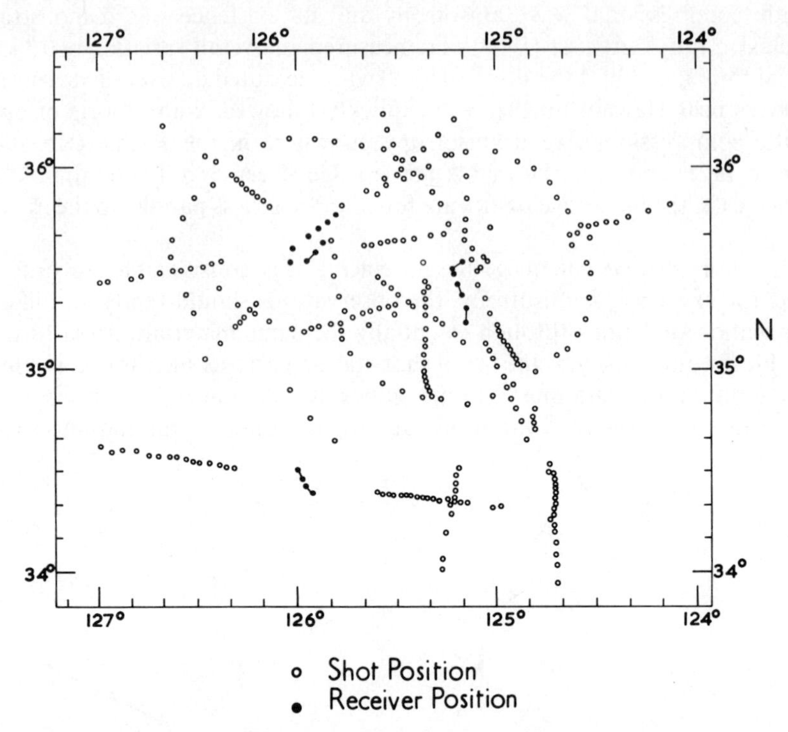

Figure 3. The shot-station pattern used in the 'Flora' area (Monterey Fan region of the Pacific). (Reproduced from R. W. Raitt, G. G. Shor, T. J. G. Francis and G. B. Morris, *J. Geophys. Res.*, **74**, 3098 (1969) copyright by American Geophysical Union).

velocity variations and so on. Thus, the directional coverage must be supplemented by an adequate areal coverage—Raitt *et al.* (1969) used the shot-station distribution shown in Figure 3 for part of their study, while in their time-term analysis, Morris *et al.* (1969) included the travel-times of 1147 compressional wave arrivals that had been refracted through the upper mantle. Studies of this size are very large in comparison with typical seismic programmes; perhaps the point is best illustrated by the work of Whitmarsh (1971) who, in an attempt to observe velocity anisotropy in the Atlantic upper mantle, used only 70 travel-times obtained with an observational scheme similar to that shown in Figure 2 and was unable to reach a positive conclusion.

In an analysis of oceanic lower crustal velocities, again based on a compilation of values obtained from linear refraction profiles, Christensen (1972) has suggested that the observed range of compressional velocities (Figure 4) may be due to an anisotropy of several per cent, the velocity maxima corresponding fairly sensibly with fracture zone directions; to date, however, this hypothesis has not been tested in special experiments.

Other types of measurement are rare in oceanic areas and certainly neither

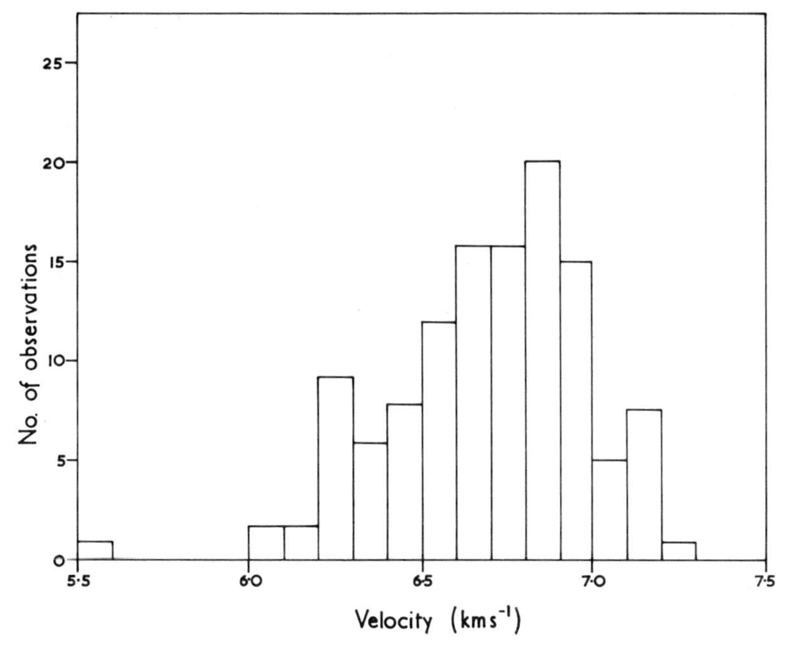

Figure 4. Histogram of observed velocities for oceanic lower crust (Reproduced with permission from N. I. Christensen, *Nature*, **237**, 450–451 (1972).)

shear wave nor particle motion studies have been reported. Forsyth (1973) has reported Rayleigh wave dispersion observations (period 15–170 seconds) as indicating an apparent 2 per cent velocity anisotropy in the eastern Pacific ocean, again with the maximum velocity parallel to the fracture zones, but he does no more than place the source of this anisotropy somewhere in the lithosphere.

There are virtually no indications of the total vertical extent of anisotropy in the oceanic lithosphere; by their nature, the seismic refraction experiments have sampled only the mantle that lies just beneath the crust. It has been suggested (e.g. Kovach and Press, 1961) that this uppermost layer may be only a few kilometres thick but this has not been confirmed so far.

Continental Regions

The various reports of velocity anisotropy in continental regions have, until recently, been somewhat less detailed than those for oceanic regions. Both Neumann (1930) and Byerly (1934, 1938), for example, reported the occurance of two distinct shear wave arrivals (SV and SH) on earthquake seismograms; in one case (Byerly, 1938) an SV arrival on a Massachussets recording of a California earthquake occurred approximately 14 seconds after the SH arrival. As Todd *et al.* (1973) point out, this example of *in situ* acoustic double refraction is difficult to interpret in terms of anisotropy because of the lateral variations evident across the United States although the hypothesis of mineral anisotropy in the upper mantle appears to be supported by the observation of anomalously low phase velocities for Rayleigh waves (SV) in comparison with those of Love waves (SH); perhaps a closer study of these surface waves would in fact reveal the coupling discussed in section 2 of this paper. Crampin (1967) has reported the observation of coupled Rayleigh–Love second modes (which cannot exist in isotropic media) for several paths through Eurasia and explains these observations by placing a comparatively thin anisotropic layer at the base of the crust; the layer must be thin as the coupling was not observed for the fundamental modes which sample a larger depth range.

Detailed reports of *in situ* acoustic double refraction are rare although interest appears to be growing rapidly because it is now recognized (Todd *et al.* 1973) that the application of non-hydrostatic stress to rocks containing cracks (i.e. at pressures of less than 1 or 2 kb) can induce velocity anisotropy; observations of double shear waves may therefore provide a simple earthquake prediction method. Gupta (1973) has demonstrated just how simple it is to make such measurements using only three-component seismometer stations. In a study of Nevada earthquakes and at a range of 100 to 130 kms from the source, he observed SV–SH travel-time differences of 0.3–0.8 seconds (i.e. a few per cent) for propagation paths through the upper crust; similar effects could be expected due to mineral anisotropy at greater depths.

Measurements of directionally dependent velocities in continents have only recently begun to appear in the press. The practical difficulties of carrying out deep seismic surveys on land are mainly responsible for this. Explosion seismic measurements in continental areas are of course very extensive; the difficulty is that nearly all of these measurements are carried out along linear profiles and only rarely are single experiments performed which even permit the study of possible effects due to anisotropy let alone the accurate measurement of it (e.g. Bamford, 1971; Berry and Fuchs, 1973). Certainly no specially designed experiments have been carried out to study velocity anisotropy at depth within continents.

On occasions, of course, rather inadequate data has been used to consider the possibility, usually at the cost of neglecting other sources of variation. Thus the results described by Dorman (1972) for the crust in Georgia (U.S.A.), which correspond to a velocity variation with azimuth of 0.75 km/s for compressional and 0.45 km/s for shear waves (i.e. about 13 per cent in each case), should be considered bearing in mind that each result is based on only nine observations. In contrast, and with more data—this time in the western Alps, Peterschmitt (1969) has reported upper mantle compressional wave velocities which vary with direction from 7.5 km/s to 8.6 km/s and has chosen to explain these variations in terms of a fairly intense topography on the crust—mantle boundary; although recent results (Peselnick *et al.*, 1974) indicate that this may have been an incorrect choice, it is clear that the results obtained from inadequate data may simply reflect the personal preferences of the interpreter.

In some areas of the world, however, and notably in Europe (Figure 5), the

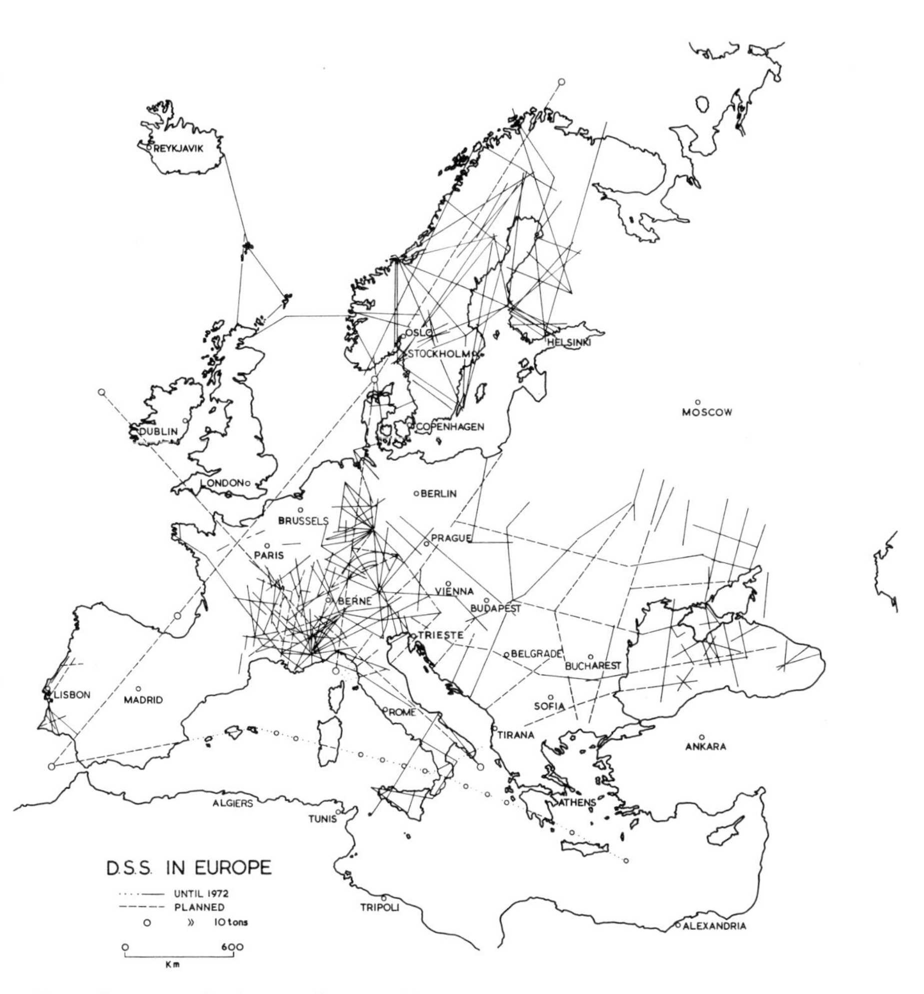

Figure 5. Deep-seismic-sounding profiles in Europe to 1972; compiled by Professor C. Morelli, Trieste.

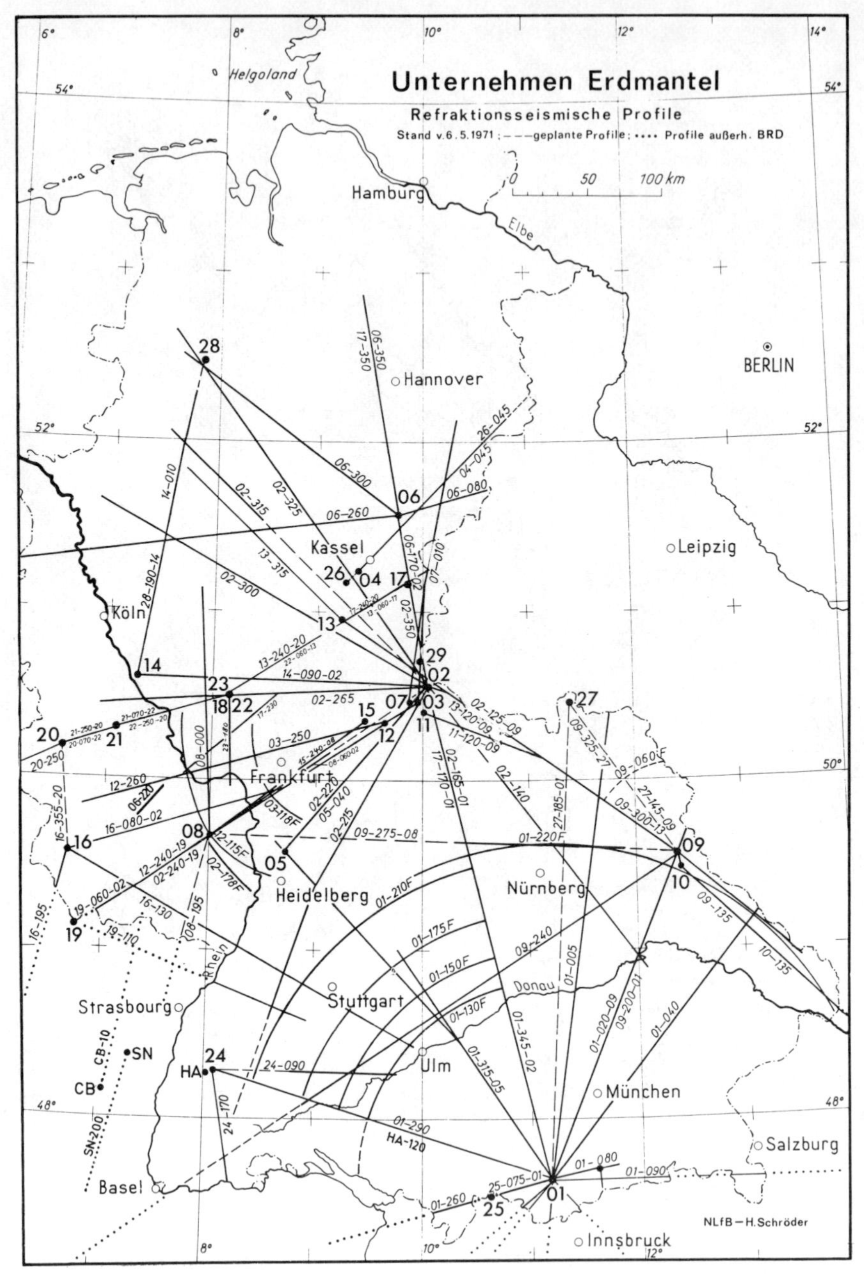

Figure 6. Seismic refraction profiles in western Germany; compiled by H. Schröder, Hannover. Coding scheme for profiles and fans: Profiles: Shot—approximate azimuth —(reversing shot). Fans: Shot—approximate distance—*F*.

number of linear seismic profiles that penetrate as deep as the upper mantle is now so large, and their directions so diverse, that more or less by accident the overall coverage meets the basic requirement for an objective study of velocity anisotropy which is that the observations should represent a wide variety of propagation directions through the same material. The best of these composite networks is perhaps that available is western Germany (Figure 6) where since 1958, over 80 seismic refraction profiles (and fans) have been recorded using mainly quarry blasts as sources. This data is of good quality and in particular contains many clear compressional wave arrivals that have been refracted through the upper mantle and which, nearly always being the first-arriving energy on the seismograms at appropriate distances (Figure 7), can be accurately timed. The apparent velocities of these arrivals vary from 7.4 to 8.6 km/s between profiles, and Giese and Stein (1971) have interpreted these variations only in terms of lateral variations in upper mantle velocity and structural variations at and above the crust–mentle boundary. If, on the other hand these velocities are plotted as a function of the azimuth from individual shot-points or groups of shot-points, a clear azimuthal dependence emerges (Figure 8).

Elsewhere (Bamford, 1973) I have described how the time-term method employed by Morris *et al.* (1969) and others can be used to choose between structure and lateral and/or anisotropic velocity variations as an explanation of these observations. In fact the 640 compressional wave travel-times available on this composite network are best explained by a model which includes both structure and a considerable upper mantle velocity anisotropy. The mean velocity is 8.1 km/s with a total anisotropic variation which may exceed 0.5 km/s —the maximum velocity direction is approximately 20° East of North. The structure obtained is quite sensible; the crust–mantle boundary deepens rapidly beneath the Alps and also has a minimum regional depth (beneath the Strasbourg–Nürnberg area) with a strike that corresponds to Hercynian directions in this area. Thus, while the composite data has various deficiencies in comparison with specially designed surveys (Bamford, 1973), there is considerable support for the hypothesis of upper mantle anisotropy in this area. This type of study has the great advantage that investigations can be pursued without additional cost; similar studies are in progress for other areas.

There are no other important demonstrations of velocity anisotropy in continental areas and, apart from the rather convincing work of Crampin (1967) previously mentioned, there is no direct evidence as to the depths to which the phenomena extends although there are reports (e.g. Hirn *et al.*, 1973) that the uppermost mantle layer might be less than 10–15 kms thick. Indeed there are several suggestions (Ansorge and Mueller, 1973; Kind, 1974) that the lower lithosphere, when studied with a long-range linear profile, appears to consist of alternating layers of high and low velocity; speculatively, this could be a reflection of vertical variations in petrofabric structure or in velocity anisotropy.

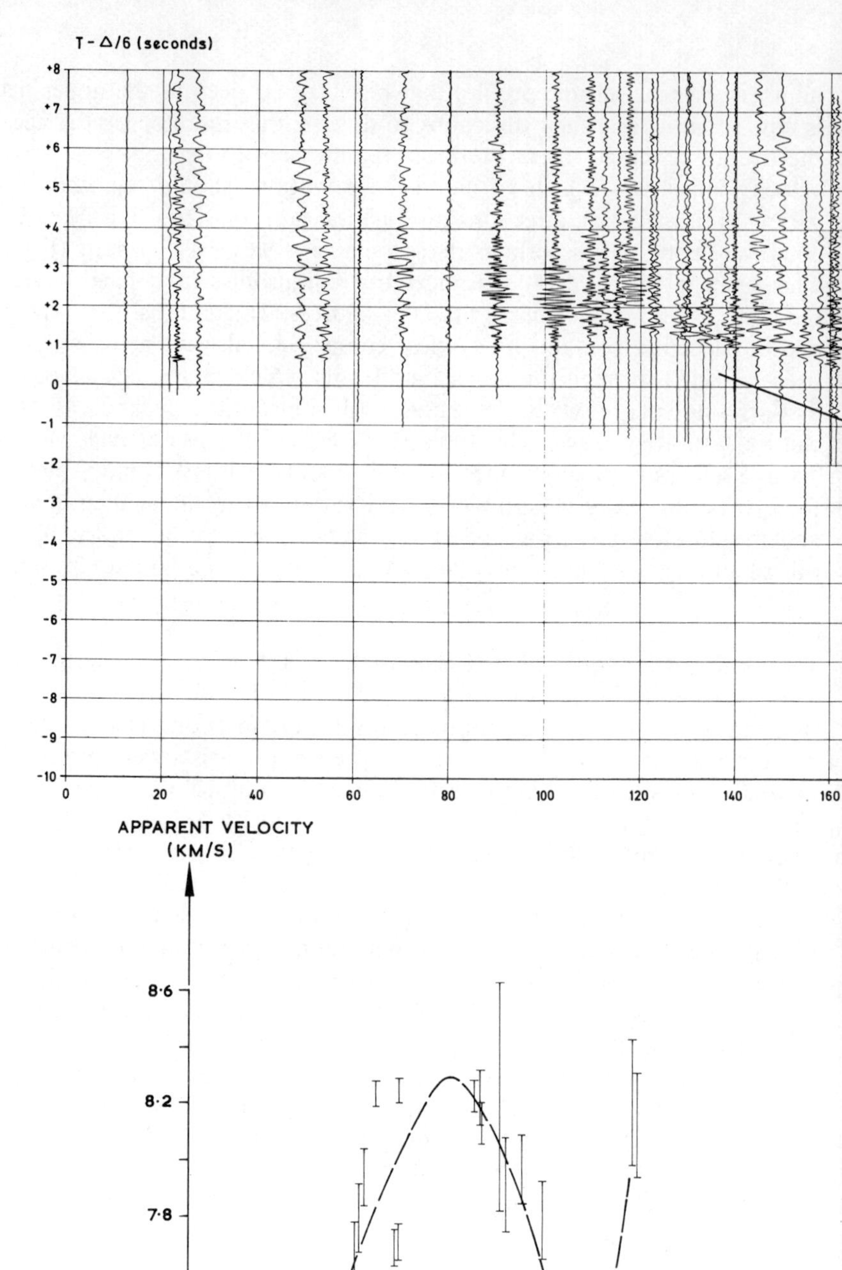
T - Δ/6 (seconds)
+8
+7
+6
+5
+4
+3
+2
+1
0
-1
-2
-3
-4
-5
-6
-7
-8
-9
-10
0
20
40
60
80
100
120
140
160
APPARENT VELOCITY
(KM/S)
8·6
8·2
7·8
7·4
7·0
0
60
120
180
240
300
360
AZIMUTH
(DEGREES)

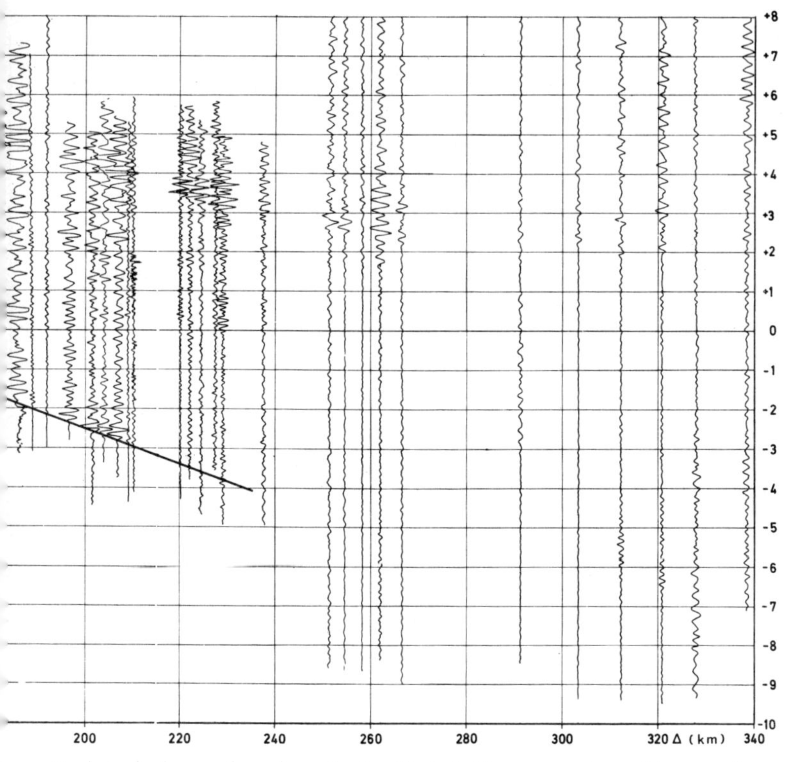

gure 7. A typical record section. Profile 02–165–01, now published by permis-
on of K. Fuchs and W. Kaminski. The phase P_n is identified by the line drawn on the section.

gure 8. Variation of apparent P_n velocity as a function of azimuth from shots 02, 03, 06, 11, 12, 15, 17.

DISCUSSION

The question of elastic anisotropy in the interior of the earth has clearly been only partially studied. Although its existence has been demonstrated in both oceanic and continental lithospheres, its true world-wide extent is unknown; it is possible that the suggested anisotropic variation of several per cent in upper mantle velocities could at least partially explain the observed spread of world-wide upper mantle velocities, a spread that has been previously attributed entirely to lateral variations in physical and/or chemical properties. Furthermore, really detailed measurements have been made only horizontally in the layer just beneath the crust; thus, this anisotropy has been only partly measured and its vertical extent is more or less unknown.

The successful recognition and description of *in situ* velocity anisotropy greatly complicates seismic measurements but the results are very rewarding. In the first place, the existence of mineral anisotropy restricts the range of possible upper mantle materials, enabling Forsyth and Press (1971), for example to choose peridotite (which contains olivine) instead of eclogite (which does not) as the major constituent of their model oceanic lithosphere; presumably a precise measurement of elastic parameters is even more constraining. Secondly, we have the opportunity to enhance our understanding of geodynamics.

The observation of upper mantle anisotropy in oceanic areas, with the maximum velocity (i.e. the preferred olivine orientation) parallel to the fracture zones (i.e. parallel to a direction of shear) could be readily explained by assuming that the preferred orientation was somehow 'frozen' into the lithosphere during its generation at mid-oceanic ridges. The recognition that the continental lithosphere may also be anisotropic greatly complicates this convenient picture; it may be that mechanisms such as that proposed by Ave'Lallemant and Carter (1970) might be more appropriate. They show that when a plate is dragged from beneath, the stress field near its base would promote recrystallization with a preferred orientation. If so, then one has to ask why the velocity anisotropy measured in western Germany is not related in any simple way to obvious stress directions such as the Alpine or Hercynian, and why it is so far removed from the supposed present-day base of the lithosphere. It is possible that the anisotropy was 'frozen' in under the action of an ancient stress field at a time when the base of the lithosphere was somewhat shallower than it is today. If so, then detailed anisotropy studies of a whole lithospheric plate should yield valuable information about the stress (and thermal) history of that plate.

The order of priorities seem, therefore, to be fairly clear. We must first establish whether or not elastic anisotropy is, world-wide, a typical property of the earth's interior; second, we must examine the vertical variations of the symmetry and magnitude of the anisotropy in several distinct areas (oceans, shields etc); third, we must attempt, in one or two places at least, to specify completely the anisotropic elastic constants of a particular layer. The last of these requires a massive effort: however, the first two objectives could be achieved by an extension of normal deep seismic sounding methods to obtain reliable shear wave information.

REFERENCES

Ansorge, J. and Mueller, S. (1973). *Z. Geophys.*, **39**, 385–394.
Ave'Lallemant, H. G. and Carter, N. L. (1970). *Geol. Soc. Am. Bull.*, **81**, 2203–3220.
Babuska, V. (1972). *Z. Geophys.*, **38**, 461–467.
Bamford, D. (1971). *Bull. Seis. Soc. Am.*, **61**, 1013–1032.
Bamford, D. (1973). *Z. Geophys.*, **39**, 907–927.
Berry, M. J. and Fuchs, K. (1973). *Bull. Seis. Soc. Am.*, **63**, 1393–1432.
Byerly, P. (1934). *Bull. Seis. Soc. Am.*, **24**, 81–99.
Byerly, P. (1938). *Bull. Seis. Soc. Am.*, **28**, 1–13.
Christensen, N. I. (1965). *J. Geophys. Res.*, **70**, 6147–6164.
Christensen, N. I. (1966a). *J. Geophys. Res.*, **71**, 3549–3556.
Christensen, N. I. (1966b). *J. Geophys. Res.*, **71**, 5921–5931.
Christensen, N. I. (1971). *Bull. Geol. Soc. Am.*, **82**, 1681–1694.
Christensen, N. I. (1972). *Nature*, **237**, 450–451.
Christensen, N. I. (1974). *J. Geophys. Res.*, **79**, 407–412.
Christensen, N. I. and Ramananantoandro, R. (1971). *J. Geophys. Res.*, **76**, 4003–4010.
Crampin, S. (1967). *Geophys. J. R. Astr. Soc.*, **12**, 229–335.
Dorman, L. M. (1972). *Bull. Seis. Soc. Am.*, **62**, 39–45.
Forsyth, D. W., and Press, F. (1971). *J. Geophys. Res.*, **76**, 7963–7979.
Forsyth, D. W. (1973). *Geophys. J. R. Astr. Soc.*, **35**, 376.
Francis, T. J. G. (1969). *Nature*, **221**, 162–165.
Giese, P. and Stein, A. (1971). *Z. Geophys.*, **37**, 237–272.
Gupta, I. N. (1973). *Science*, **182**, 1129–1132.
Hess, H. H. (1964). *Nature*, **203**, 629–631.
Hirn, A., Steinmetz, L., Kind, R. and Fuchs, K. (1973). *Z. Geophys.*, **39**, 363–384.
Keen, C. E. and Tramontini, C. (1970). *Geophys. J. R. Astr. Soc.*, **20**, 473–491.
Keen, C. E. and Barrett, D. L. (1971). *Can. J. Earth. Sci.*, **8**, 1056–1064.
Kind, R. (1974). *Z. Geophys.*, **40**, 189–202.
Kovach, R. L. and Press, F. (1961). *Geophys. J. R. Astr. Soc.*, **4**, 202–216.
Montalbetti, J. F. and Kanasewich, E. R. (1970). *Geophys. J. R. Astr. Soc.*, **21**, 119–130.
Moores, E. M. (1973). *Earth-Sci. Rev.*, **9**, 241–258.
Morris, G. B., Raitt, R. W. and Shor, G. G. (1969). *J. Geophys. Res.*, **74**, 4300–4316.
Neumann, F. (1930). *Bull. Seis. Soc. Am.*, **20**, 15–32.
Peselnick, L., Nicolas, A. and Stevenson, P. R. (1974). *J. Geophys. Res.*, **79**, 1175–1182.
Peterschmitt, E. (1969). *I ASPEI General Assembly*, Madrid.
Raitt, R. W., Shor, G. G., Francis, T. J. G. and Morris, G. B. (1969). *J. Geophys. Res.*, **74**, 3095–3109.
Stoneley, R. (1949). *Royal Astr. Soc. Mon. Notices, Geophys. Suppl.*, **5**, 343–353.
Tilmann, S. E. and Bennett, H. F. (1973). *J. Geophys. Res.*, **78**, 7623–7629.
Todd, T., Simmons, G. and Baldridge, W. S. (1973). *Bull. Seis. Soc. Am.*, **63**, 2007–2020.
Uhrig, L. F. and Van Melle, F. A. (1965). *Geophysics*, **20**, 774–779.
Verma, R. K. (1960). *J. Geophys. Res.*, **65**, 757–766.
Whitmarsh, R. B. (1971). *Bull. Seis. Soc. Am.*, **61**, 1351–1368.

Elastic Properties of the Lower Lithosphere, Obtained by Large-Scale Seismic Experiments in France

C. Prodehl[a], A. Hirn[b], R. Kind[a], L. Steinmetz[b] and K. Fuchs[a]

[a]*Geophysikalisches Institut, Universität, Hertzstrasse 16, D-75 Karslruhe 21, Germany*

and

[b]*Institut de Physique du Globe, Université 6, 11 quai Saint-Bernard, F-75230 Paris, France*

In September 1971 a seismic-refraction survey was carried out in order to obtain detailed knowledge of the fine structure of the lower lithosphere by a long-range profile of 900 km length. The experiment is described in detail by Groupe Grands Profils Sismiques and German Research Group for Explosion Seismology (1972). It consists of two parts: (a) Four profiles for crustal studies of 200–300 km along the line from the Bretagne to the Provence, (b) a long-range profile recorded along the same line from a shot-point in the Atlantic Ocean, 140 km west of Brest (see Figure 1 of Groupe Grands Profils Sismiques and German Research Group for Explosion Seismology, 1972; Hirn *et al.*, 1973; Sapin and Prodehl, 1973).

The interpretation of these measurements is described in detail by Hirn *et al.* (1973), Kind (1974), and Sapin and Prodehl (1973). Here only the main features of velocity–depth structure of the lower lithosphere will be described and compared with results from other long-range observations in Europe and North America.

The evaluation of the crustal profiles reveals a true velocity of the upper mantle immediately below the Mohorovičić discontinuity of 8.10–8.15 km/s and a more or less uniform crustal thickness of about 30 km. Though the fine structure of the crust shows some variations, the average crustal velocity is about constant along the whole line of the long-range profile. So the evaluation of phases which have travelled through the mantle can be made under the assumption that the crust is homogeneous (Sapin and Prodehl, 1973).

A first glance at the record section of the long-range profile (Hirn *et al.*, 1973, Figures 2 and 3) indicates that the first-arrival energy propagates with a mean

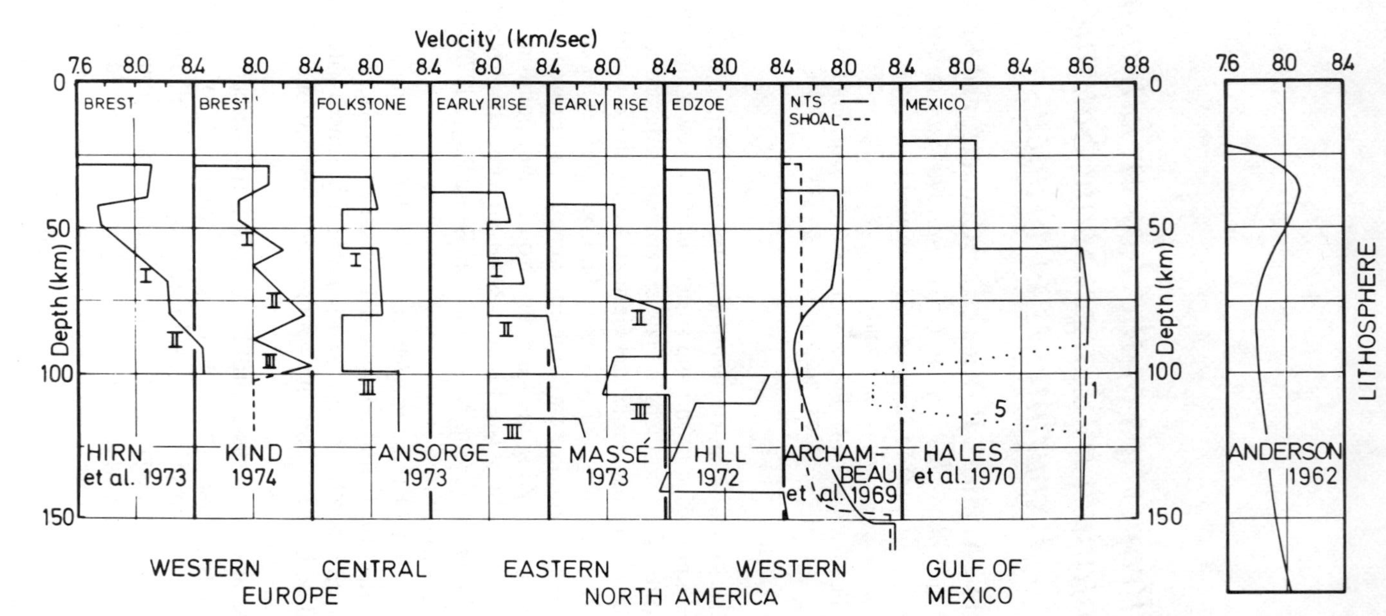

Figure 1. Models for *P*-wave velocities of the uppermost mantle in Europe and North America from long-range explosion observations.

velocity of about 8.1 km/s between 150 and more than 1000 km distance from the shot-point. However, the close station separation of only 5 km allows us to recognize that the first arrivals do not align on a single travel-time curve, but that have to be correlated by several travel-time curves which are clearly separated from each other: P_n (phase velocity 8.1 km/s, observed distance range 140–200 km), P_{I} (phase velocity 8.1–8.3 km/s, observed distance range 320–480 km), and P_{II} (phase velocity 8.3–8.6 km/s, observed distance range 450–620 km).

The inversion of the corresponding travel-time and amplitude data into velocity-depth values, using also synthetic seismograms (Hirn *et al.*, 1973, Figure 8) and neglecting changes of velocity in a horizontal direction, shows the following result which is presented in Figure 1 (first model from the left): 1. At a depth of 40–50 km there exists a low-velocity zone of 10–20 km thickness, the velocity decreasing from 8.1 to 7.7 km/s at about 35 km depth and increasing again up to 8.2 km/s between 50 and 70 km depth. 2. Between 80 and 90 km depth a second boundary zone is found, the velocity increasing from 8.2 to 8.45 km/s. The seismological significance of this new model of the lower lithosphere is the following: Thin layers of high velocity apparently form a trap for high-frequency seismic energy while low-frequency energy propagates through this fine structured part of the lower lithosphere into the earth's mantle below the asthenosphere.

The interpretation by Hirn *et al.* (1973) takes into account only the observations up to a distance of 600 km. Kind (1974), in an expanded reinterpretation, tries to explain in terms of fine structure of the lower lithosphere in addition a third high-energy phase (P_{III}) which is observed in the distance range from 580 to 740 km (Kind, 1974, Figure 1), assuming vertical variation of velocity only. He draws special attention to the fact, that each of the observed high-velocity phases can be correlated over a distance range of 100–200 km only and then rapidly vanishes because of lack of energy. The model which according to Kind (1974, Figure 12 and Table 1) fits best the observed data is shown in Figure 1 (second model from the left) and consists of several zones of low (7.9 and 8.0 km/s) and high velocity (8.2, 8.35, and 8.4 km/s). The numbers I, II, and III in Figure 1 correspond to the phases P_{I}, P_{II}, and P_{III}, respectively. By the introduction of a second velocity inversion below the boundary zone I, the depth range of boundary zone II as interpreted by Kind is decreased in comparison with that found by Hirn *et al.* (1973). Also, in order to explain the vanishing energy of phase P_{III} beyond 740 km distance, Kind assumes a half space with 8.0 km/s below 102 km depth.

Besides the two models obtained from the long-range observations in France in 1971, Figure 1 shows other models of the lower lithosphere also which were published by other authors for central Europe and North America. In some of these models the numbers I, II, and III are added in order to try a correlation of the phases identified by different authors in different areas for depth ranges which may correspond to each other. It should be pointed out that all models discussed here were obtained under the assumption of vertical variation of velocity only, i.e. neglecting changes of velocity–depth structure in horizontal direction along the lines of observation.

Ansorge has evaluated models of the structure of the lower lithosphere in central Europe using all long-range observations available in central Europe prior to the experiment in France in 1971 (see Ansorge and Mueller, 1971, 1973), one of which is presented in Figure 1—based on data from an explosion near Folkstone. Though the available data are much more sparse, which also explains the absence of a boundary zone II, the model by Ansorge (third model from the left in Figure 1) shows comparable features in terms of fine structure of the lower lithosphere in central Europe to those found for western Europe.

The models for eastern North America as shown in Figure 1 are based on long-range profiles radiating from Lake Superior and reaching distances up to 3000 km which were observed during the experiment 'Early Rise' in 1966 (e.g. Massé, 1973). The record sections of the profiles in North America (see, e.g. Massé, 1973, Figures 2 and 13) show clearly several phases which may correspond to the phases P_{I}, P_{II}, and/or P_{III} described above, but the distance ranges, at which these phases are recorded apparently are about 100 km greater than the distances at which P_{I}, P_{II}, and P_{III} are observed in France. The two models by Ansorge (Ansorge and Mueller, 1971, 1973) and Massé (1973) shown in Figure 1 are calculated for different profiles, but show common features of the fine structure of the lower lithosphere, the absence of zone I in Massé's model being caused by a gap of observations in the corresponding distance range. A comparison of these models with those for western and central Europe shows in general a similar fine structure of the lower lithosphere, though apparently the corresponding depth ranges beneath North America are partly found at greater depths than beneath Europe.

Hill (1972) proposes a model of the lower lithosphere beneath the Columbia Plateaus of the western United States, based on a profile recorded in N–S direction from the shot-point Edzoe in British Columbia, Canada. Here the velocity in the upper mantle does reach values equal to or greater than 8.0 km/s only at a depth of 100 km, a fact which is also observed by other authors (see, e.g. Archambeau *et al.*, 1969). The strong velocity inversion proposed by Hill (1972) between 120 and 180 km depth covers a depth range which is not reached by the models discussed above for eastern North America and Europe.

For completeness, Figure 1 contains also some velocity–depth distributions of the lower lithosphere which were deduced by Archambeau *et al.* (1969) for the western United States and by Hales *et al.* (1970) for the region of the Gulf of Mexico. However, it must be pointed out that these models can only be regarded as rough average models, because for the corresponding depth range there were not more than one or two seismograms available. Rather Archambeau *et al.* (1969) put their main effort into the over-all investigation of the depth range down to 1200 km depth. Finally, the right column of Figure 1 shows the general *P*-velocity distribution for the upper mantle down to 175 km depth which was proposed, e.g. by Anderson (1962) representing the state of knowledge before more detailed seismic investigations became available.

Summarizing it can be stated that, based on recent seismic-refraction data, the fine structure of the lower lithosphere apparently is much more complicated than was formerly assumed. This basic result may lead to a new understanding of the physics of the motion of lithospheric plates.

It must be mentioned, however, that differences in details of the models shown in Figure 1, for example, may not always represent real differences in structure, but may also be caused by different interpretation methods and, especially, by differences in density of data. Insufficient density of data can easily lead to incorrect correlation of phases. The models for North America, for example, are based on profiles with an average station interval of 20–50 km, while the distance interval between recording stations in France was only 5–10 km.

ACKNOWLEDGEMENTS

The experiment in France 1971 is a common project of French and German Geophysical Institutes with the collaboration of Swiss and Portuguese participants. The project was supported by the Institut National d'Astronomie et de Géophysique of France and the German Research Association.

REFERENCES

Anderson, D. L. (1971). In Wilson, T. (ed), *Continents Adrift; Readings from Scientific American*, Freeman & Co., San Francisco, 28–35.

Ansorge, J. Mueller, S. (1971). *Proc. 12th Gen. Ass. Europ. Seismol. Comm.* (Luxembourg 1970), Obs. Roy. de Belgique, Comm. A–13, Ser. Geophys. No. 101, 196–197.

Ansorge, J., Mueller, S. (1973). *Z. Geophys.*, **39**, 385–394.

Archambeau, C. B., Flinn, E. A., Lambert, D. G. (1969). *J. Geophys. Res.*, **74**, 5825–5865.

Group Grands Profils Sismiques; German Research Group for Explosion Seismology (1972). A long-range profile in France from the Bretagne to the Provence. *Ann. Geophys*, **28**, 247–256.

Hales, A. L., Helsley, D. E., Nation, J. B. (1970). *J. Geophys. Res.*, **75**, 7362–7381.

Hill, D. P. (1972). *Geol. Soc. Am. Bull.*, **83**, 1639–1648.

Hirn, A., Steinmetz, L., Kind, R., Fuchs, K. (1973). *Z. Geophys.*, **39**, 363–384.

Kind, R. (1974). *Z. Geophys.*, **40**, 189–202.

Masse, R. P. (1973). *Bull. Seismo. Soc. Am.*, **63**, 911–935.

Sapin, M., Prodehl, C. (1973). *Ann. Geophys.*, **29**, 127–145.

A Physical Interpretation of Bullen's Compressibility-Pressure Hypothesis

S. K. Runcorn
Department of Geophysics and Planetary Physics,
University of Newcastle upon Tyne,
Newcastle upon Tyne NE1 7RU, UK

INTRODUCTION

An interesting empirical relationship, derived from seismic data, on the variation of the incompressibility (k) with depth in the interior of the Earth was discovered by Bullen (1949, 1965). He pointed out that k is nearly the same above and below the core-mantle interface and also that dk/dp is nearly continuous between mantle and core. This compressibility-pressure hypothesis was supposed to result from a property of all materials at high pressure, but, while there is little support for such a general speculation, this relation has proved valuable in determining the density distribution in the Earth.

More recently Lyttleton (1963, 1965) has shown that the dependence of k upon p is linear for the mantle below about 300 km depth and for the core, and the coefficient is nearly 3.5. He has used this relation to obtain density distribution models for the terrestrial planets. As I have pointed out (Runcorn, 1973), the unknown temperature distribution and composition of other terrestrial planets will result in uncertainty from using this relation, the former being of more consequence in the smaller bodies where the pressures are less. Thus a basis in theoretical physics for this relation is required before its use in such modelling is placed on a secure basis.

CRYSTAL STRUCTURE OF LOWER MANTLE AND CORE MATERIAL

Rejecting the idea that the relation is a general one at high pressures, the question arises as to what the silicate of the lower mantle and the iron in the core have in common. It is clear that the incompressibility–pressure relation is linear below about 300 km depth and is curved in the upper mantle: it is reasonable to suppose that this is a consequence of the phase changes which occur at these pressures.

Since the pioneering work of Goldschmidt it has been clear that the greater size of the oxygen ion compared to the silicon, iron and magnesium ions plays a great role in determining mineral structures. Anderson (1965) has therefore suggested that the density of a wide group of minerals is controlled by the volumes and packing factors of the oxygen ions. In the case of spinel, the oxygen ions are arranged in a face centred cubic lattice with the metal ions in the interstices. Thus the oxygen ions are essentially in closed packed hexagonal arrays. Although the core is a fluid, it seems likely that the iron ions are also similarly in close packed hexagonal arrays. Thus it seems reasonable to postulate that the interesting incompressibility–pressure relation arises from this basic similarity of the spatial arrangement of the ions, assuming that the silicon, iron and magnesium ions, being small, play little part in determining the elastic behaviour.

THEORY

The complex interactions between the ions of the silicate, the Colomb and van der Waals forces should in this problem be neglected, as they respectively vary as the inverse square and sixth power of the inter-atomic distances. Thus for compressions of the crystal lattice resulting in relatively small changes in the crystal distance, it seems likely that we are dealing with the repulsive forces between ions, the overlapping potentials, equal to a constant c times the inverse n th power of the interionic distance, where n is known to be about 11 or 12. We may suppose for ions with complete outer electronic shells the nature of this repulsive potential is somewhat insensitive to atomic number.

Thus if the cubic structure is imagined to be compressed from ordinary pressure, where the lattice constant is a, to pressure p in the deep Earth, where the constant is $a_0 - \delta a$, then

$$p = \frac{nc\delta a}{a^{n+1}}$$

Then $\log p = \log nc + \log(\delta a) - (\mathrm{n} + 1)\log a$, and

$$\frac{\mathrm{d}p}{p} = \frac{\mathrm{d}a}{\delta a} - (n+1)\frac{\mathrm{d}a}{a}$$

where $\mathrm{d}a$ is the incremental changes in lattice constant due to a change in pressure $\mathrm{d}p$. If the density is ρ at pressure p then $3\,\mathrm{d}a/a = -\mathrm{d}\rho/\rho$, thus

$$k = \rho\frac{\mathrm{d}p}{\mathrm{d}\rho} = \frac{a}{\delta a}p + \frac{n+1}{3}p$$

$$= \frac{nc}{a_0^{n+1}} + \frac{n+1}{3}p$$

Both the form and the pressure coefficient approximate those found in the mantle and core. Cook (1972) reviews the experimental data on compressibilities at high pressure: the pressure coefficients vary considerably and no general explanation for the incompressibility–pressure relation was found.

It is concluded that the Bullen relation results from rather fundamental ideas concerning the close packing of atoms and the lack of strong dependence on chemistry of the repulsive forces between ions.

REFERENCES

Anderson, O. L. (1965). In *Physical Acoustics*, vol. 3B, Academic Press, New York.
Bullen, K. E. (1949). *MNRAS Geophys. Supp. 5*, 355–368.
Bullen, K. E. (1965). *An Introduction to the Theory of Seismology*, Cambridge University Press, p. 381.
Cook, A. H. (1972). *Proc. Roy. Soc. Lond. A.*, **328**, 301–336.
Lyttleton, R. A. (1963). *Proc. Roy. Soc. Lond. A.*, **273**, 1–22.
Lyttleton, R. A. (1965). *Mon. Not. Roy. Astr. Soc.*, **129**, 21–39.
Runcorn, S. K. (1973). *Nature*, **241**, 521–523.

The Electrical Conductivity of Forsterite, Enstatite, and Albite

Al Duba, A. J. Piwinskii, H. C. Heard and R. N. Schock
University of California, Lawrence Livermore Laboratory, Livermore, CA, USA

Electrical conduction in the highly insulating silicate minerals is poorly understood. The presence of chemical impurities; point, line and planar defects; order–disorder phenomena; and phase transitions can markedly influence the electrical conductivity (σ) of these minerals. This report will deal with the σ of three common silicate minerals: forsterite, enstatite, and albite. The σ data for silicate minerals fit the empirical equation (Hughes, 1953):

$$\sigma = \sum_x \sigma_x \exp(-A_x/kT),$$

where σ_x is a constant, A_x an activation energy, k Boltzmann's constant, and T temperature. The subscript signifies that more than one conduction mechanism (x) may be operative.

Our purpose here is to summarize those conductivity–temperature data for forsterite and enstatite crystals which have been measured at a known f_{O_2} corresponding to the range of inferred f_{O_2} believed to exist in the earth, moon, and other terrestrial planets. These laboratory σ–T data will then be correlated with field σ measurements in order to infer the probable thermal gradient. In addition, we evaluate nonequilibrium, time-dependent, effects on the σ of albite.

FORSTERITE σ AND THE GEOTHERM

Recent petrological and geophysical investigations indicate that $Fo_{90}Fa_{10}$ is the most probable olivine composition in the earth's upper mantle (Fujisawa, 1968). Figure 1 is a compilation of literature σ data for single-crystal and polycrystal forsterite, with a composition of about $Fo_{90}Fa_{10}$. These data may be divided into two groups. One group encompasses about an order of magnitude in σ and includes both polycrystal and single crystal results. These data yield an activation energy for conduction of less than 1.0 eV up to the

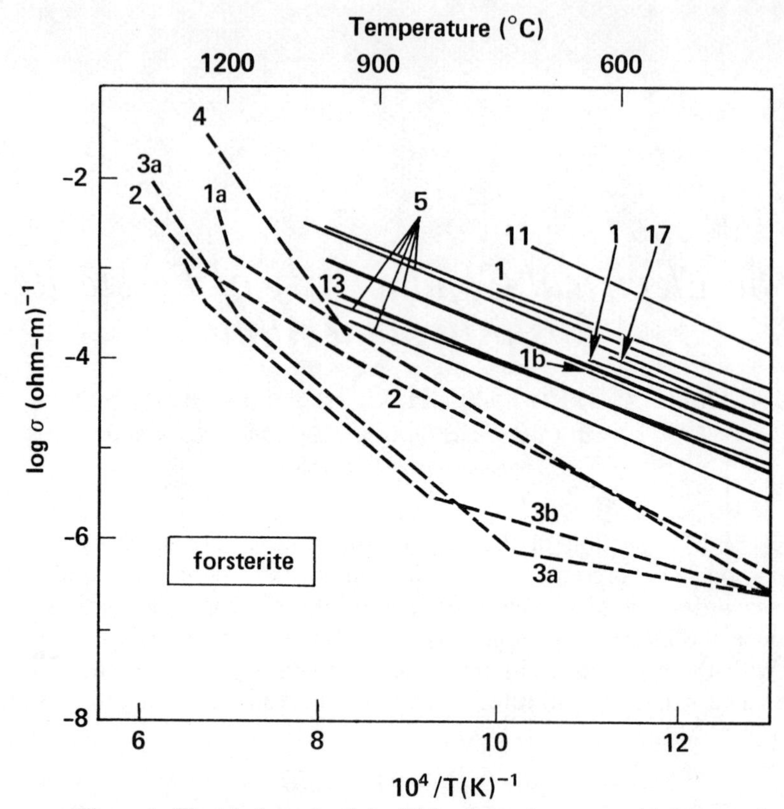

Figure 1. Electrical conductivity (σ) for forsterite, composition approximately $Fo_{90}Fa_{10}$. Numbers refer to the following studies. Single crystals: 1, 1a, 1b, Duba (1972); 2. Duba *et al.* (1972); 3a, 3b. Hughes (1953); 4. Hughes (1955); 5. Kobayashi and Maruyama (1971). Polycrystals: 11. Bradley *et al.* (1964); 13. Hamilton (1965); 17. Schult and Schober (1969).

highest temperature measured, 1000 °C. Another group, composed of olivines from the Red Sea area, has significantly lower σ and activation energy greater than 1.0 eV above 800 °C.

Using wet chemical analyses, electron microprobe, and Mössbauer techniques, Duba *et al.* (1973a) postulated that the major difference between σ of olivine from the Red Sea area (lines la, 2, 3a, 3b, and 4 in Figure 1) and that from San Carlos, Arizona (lines numbered 1 in Figure 1) was in Fe^{+3} content (See Table 1). This hypothesis was confirmed (Duba and Nicholls, 1973) in an apparatus in which f_{O_2} was controlled by a mixture of H_2/CO_2 flowing past the sample during σ measurement. A plot of f_{O_2} as a function of temperature for various H_2/CO_2 mixtures (dashed lines) is shown in Figure 2. Several oxidation–reduction reactions for pertinent minerals (solid lines) are also included to indicate equilibrium assemblages within this T–f_{O_2} range.

Figure 3 shows the σ of single crystal olivine from San Carlos, Arizona

Table 1. Olivine and pyroxene compositions

	San Carlos[a] Olivine	Red Sea[a] Olivine	Px 26[b]	(Gem) Px 173[b]	Px 1269[b]
SiO_2	40.7	40.5	57.96	57.56	57.23
TiO_2	tr.	tr.	0.03	0.16	< 0.05
Al_2O_3	0.2	0.3	0.19	0.58	0.18
Cr_2O_3	0.05	0.00	0.16	0.22	< 0.03
Fe_2O_3	0.16[c]	< 0.03[c]	—[d]	—[d]	—[d]
FeO	7.94	9.16	5.26	5.41	9.55
MnO	0.12	0.14	0.11	0.08	0.03
MgO	50.1	49.3	36.88	34.68	32.87
NiO	0.38	0.38	< 0.03	0.05	< 0.03
CaO	0.07	0.01	0.08	0.63	0.35
Na_2O	0.01	0.02	0.03	0.05	< 0.04
Total	99.73	99.84	100.70	99.42	100.18
	$Fo_{92}Fa_8$	$Fo_{90}Fa_{10}$	En_{94}	En_{93}	En_{85}

[a]Wet chemistry determination
[b]Electron microprobe determination
[c]Mössbauer determination
[d]Total iron reported as FeO

(line 1) as measured in argon to 7.5 kb (Duba, 1972) and also in air (line 7a) (Duba and Nicholls, 1973). In the latter experiment after equilibration at $f_{O_2} = 10^{-12}$b (1200 °C), well within the olivine stability field (Figure 2), σ (line 7b, Fig. 3) was nearly identical to that for the Red Sea forsterite (line 3a, Figure 3) (Hughes, 1953). The higher σ reported for the Red Sea olivine (line 2, Figure 1) (Duba *et al.*, 1972) is evidently due to oxidation because Duba *et al.* (1974) report a decrease in σ with time at about 1250 °C for the same specimen (line 6, Figure 3) at $f_{O_2} \sim 10^{-9}$ (1200 °C). The agreement amongst the three reported values of single crystal forsterite σ is remarkably good.

These results indicate that, at f_{O_2} within the olivine stability field, σ is similar for single crystal olivine of composition $Fo_{90}Fa_{10}$ from these two localities. This is in marked constrast to the observation (Duba, 1972) that σ of these two olivines differed by as much as three orders of magnitude when measured at unknown f_{O_2}. We therefore conclude that the intensive thermodynamic variable f_{O_2} can have a large effect on the σ of single crystal olivine. Up to at least 1500 °C, once f_{O_2} equilibrium is attained, σ is independent of time (Duba *et al.*, 1974; Duba and Nicholls, 1973). Furthermore, σ is independent of pressure to 8 kb (Duba *et al.*, 1974).

The temperature–depth profile for the earth's upper mantle can be derived (Figure 4) from the single crystal olivine σ measured along [010] to 1660 °C under controlled f_{O_2} (Duba *et al.*, 1974) and the σ-depth distributions of Banks (1969) and Parker (1970). This profile assumes forsteritic olivine is the dominant phase, that grain boundaries do not affect σ at the temperatures and pressures prevailing in the mantle, and σ along [010] is representative of σ in the poly-

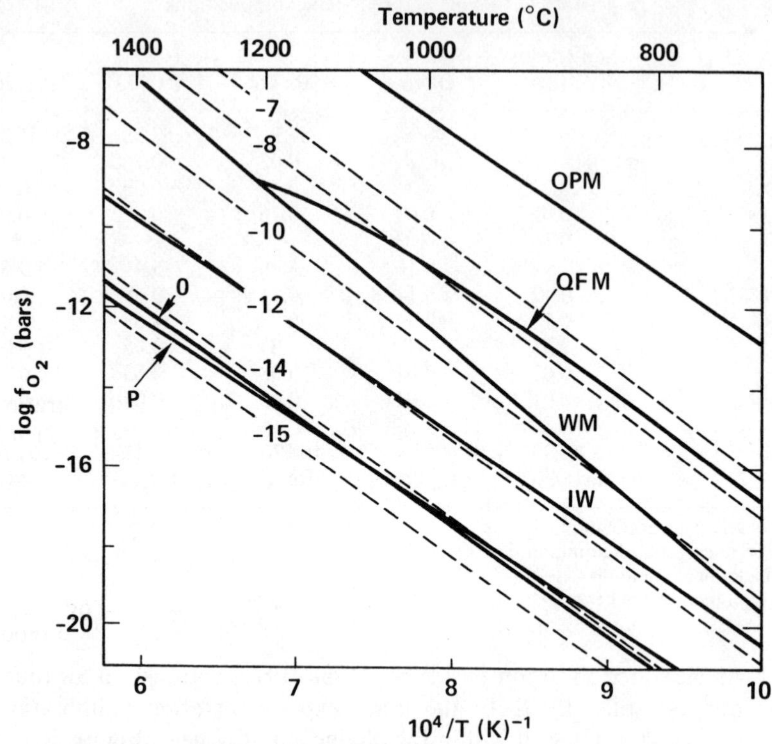

Figure 2. T–f_{O_2} paths following during σ measurements. Dashed lines are T–f_{O_2} paths calculated from thermodynamic data for constant H_2/CO_2 ratios. Solid lines are coded as follows:

IW : $FeO \rightleftharpoons Fe + \frac{1}{2}O_2$

O : $FeSi_{1/2}O_2 + MgSi_{1/2}O_2 \rightleftharpoons Fe + MgSiO_3 + \frac{1}{2}O_2$ (calculated for Fo_{90})

P : $FeSiO_3 + MgSiO_3 \rightleftharpoons Fe + SiO_2 + MgSiO_3 + \frac{1}{2}O_2$ (calculated for En_{90})

QFM: $2Fe_3O_4 + 3SiO_2 \rightleftharpoons 3Fe_2SiO_4 + O_2$

WM : $Fe_3O_4 \rightleftharpoons 3FeO + \frac{1}{2}O_2$

The above curves were based on Duba *et al.* (1973b)

OPM: $6FeSi_{1/2}O_2 + \frac{1}{2}O_2 \rightleftharpoons Fe_3O_4 + 3FeSiO_3$ (Nitsan, 1974) (calculated for Fo_{90})

(Reproduced from A. Duba, J. N. Boland and A. E. Ringwood, *J. Geol.*, **81**, 727–735 (1973) by permission of the University of Chicago Press).

crystalline aggregate. It further assumes there is no continuous framework formed by minor phases at these depths. The shaded regions in the two profiles represent the experimental error interpreted as the maximum possible pressure effect based on σ measured on the same specimen (8 kb, unknown f_{O_2}) (Duba *at al.*, 1974). The temperature spread associated with the uncertainties in the σ-depth distribution of Banks (1969) is about three times as large as the experimental error shown in Figure 4; the spread associated with the uncertainties

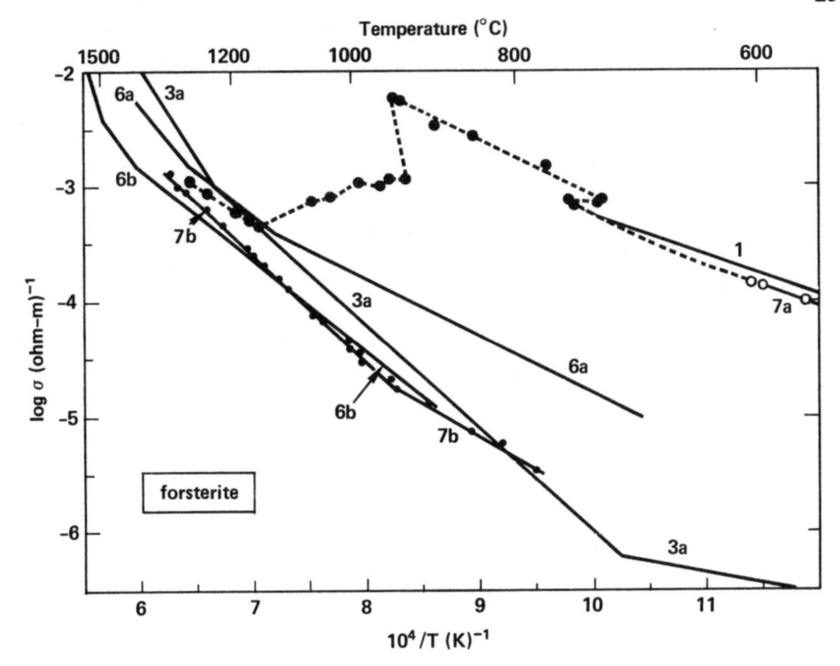

Figure 3. The electrical conductivity (σ) of single crystal San Carlos forsterite (lines 1, 7a, and 7b) compared to single crystal Red Sea forsterite (lines 3a, 6a, and 6b). Conditions of σ measurement are as follows:

Line 1 : [001] direction as measured at 1 kHz in argon of unknown f_{O_2} (Duba, 1972).

Line 7a: [001] direction, $f_{O_2} = 10^{-0.7}$ (air), dashed portion: $f_{O_2} = 10^{-12}$ (constant).

Line 7b: Same specimen as for 7a after equilibration in olivine stability field, $f_{O_2} = 10^{-12}$ (1200 °C).

Lines 7a, 7b, and dashed portion measured at 1.6 kHz (Duba and Nicholls, 1973).

Line 3a: [010] direction, measured (d.c) in air (Hughes, 1953).

Line 6a: [010] direction, measured at 1 kHz in argon of unknown f_{O_2} (Duba *et al.*, 1974)

Line 6b: Same specimen as for 6a after equilibration in olivine stability field, measured at 1.6 kHz $f_{O_2} = 10^{-8}$ (1200 °C) (Duba *et al.*, 1974).

reported by Parker (1970) is comparable to those in the figure. More reliable σ-depth data are necessary before a more precise geotherm can be derived.

ENSTATITE σ AND THE SELENOTHERM

Recent geochemical studies indicate that the dominant mineral phase in the lunar interior is orthoenstatite (Ringwood and Essene, 1970). Figure 5 shows the σ of several enstatite specimens measured (Duba *et al.*, 1973b) over a range of f_{O_2}. Pyroxene compositions, determined by electron microprobe, are indicated

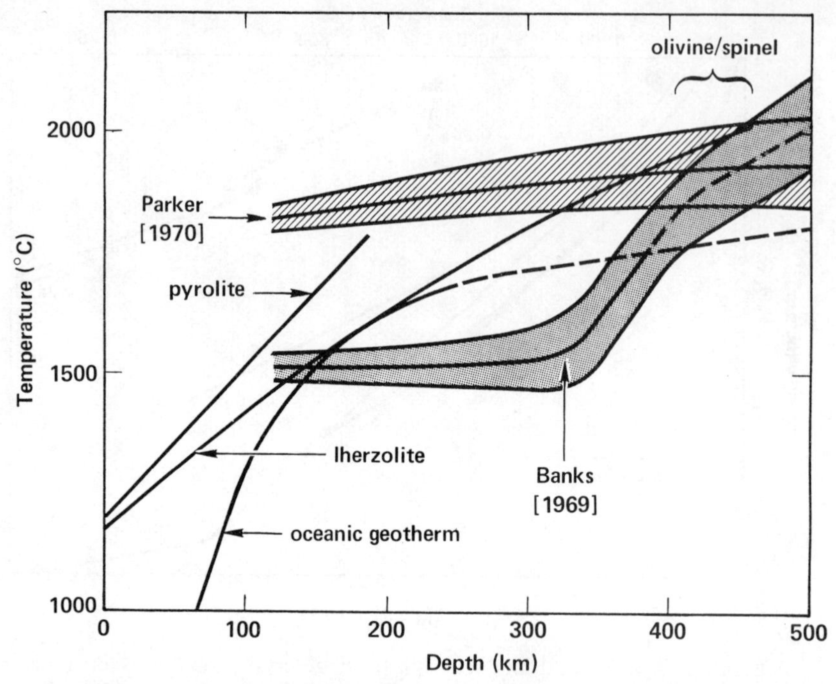

Figure 4. Temperature–depth profiles. Uncertainty, indicated by shaded areas, is based on measured pressure effect on σ (reproduced from A. Duba, H. C. Heard and R. N. Schock, *J. Geophys. Res.*, 72, 1672 (1974) copyright by American Geophysical Union). Also shown are the lherzolite solidus used by Ito and Kennedy (1967), as extended using the Simon equation (Griggs, 1972), the pyrolite solidus, and a proposed oceanic geotherm (Ringwood, 1966).

in Table 1. The most resistive pyroxene, Px 173 (line 3, Figure 5), is one of gem quality containing the least iron. These data are compared with the σ of single crystals of Bamle enstatite (line 5, Figure 5) measured by Hughes (1953) in air. The higher σ of the Bamle enstatite could result from the highly oxidizing atmosphere in which Hughes made his measurements.

Px 26 and Px 1269 (Table 1) possessed several cracks which had alteration products associated with them. In addition, these two pyroxenes showed a higher initial σ, which decreased with time at temperatures above 600 °C and after f_{O_2} equilibration became comparable to the σ of Px 173. Two major decreases in σ of the cracked pyroxenes were observed and are illustrated in Figure 6 for Px 1269. The first decrease in σ (line A, Fig. 6) was attributed to dehydration of talc in the cleavage cracks of Px 1269 and Px 26 (Duba *et al.*, 1973b). Electron microscope studies of recovered sample indicate that the second reduction in σ (line B, Figure 6) is related to re-solution of exsolution lamellae (Boland *et al.*, 1974). However, such changes could also be ascribed to other effects such as order–disorder phenomena or f_{O_2} equilibration. In contrast, the gem quality Px 173 had a small decrease in σ at about 840 °C

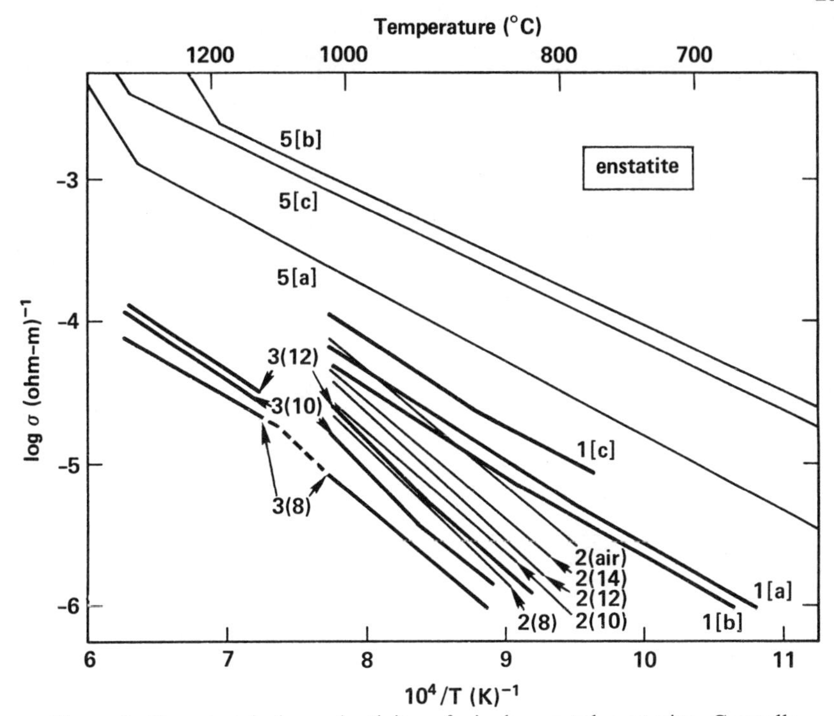

Figure 5. The electrical conductivity of single crystal enstatite. Crystallographic orientation, when known, is indicated in brackets; parentheses indicate the T–f_{O_2} path (Figure 2) followed during conductivity measurements. Lines 1, 2, 3 measured at 1.6 kHz (Duba *et al.*, 1973b). Lines are coded: 1. The conductivity of En_{85} as a function of crystallographic orientation at $f_{O_2} = 10^{-8}$ (at 1200 °C); 2. En_{94} [010]; 3. En_{93} random orientation; 5. Bamle enstatite (Hughes, 1953), measured (d.c.) in air. (Reproduced from A. Duba, J. N. Boland and A. E. Ringwood, *J. Geol.*, 81, 727–735 (1973) by permission of the University of Chicago Press).

which was attributed to re-solution of the minor amount of exsolution lamellae originally present.

The σ–T slope (Figure 5) for Px 173 *decreased* at temperatures above 1050 °C. This represents the ortho-clino phase transition (Duba *et al.*, 1973b). The limited temperature range of the orthoenstatite σ data makes calculation of the selenotherm subject to large uncertainty. Both σ_x and A_x for orthoenstatite above 1050 °C in its pressure stability field may be greater than that determined between 800 ° and 1050 °C. Furthermore, the σ for orthoenstatite may not be independent of pressure. Recognizing these restrictions, we illustrate (Figure 7) the range of selenotherms calculated from the σ-depth profile of Sonett *et al.*, (1971) and the σ data of Figure 5 for orthopyroxene at $f_{O_2} = 10^{-12}$ (1200 °C). This $T - f_{O_2}$ condition is deemed appropriate for the lunar interior (Sato *et al.*, 1973). It should be pointed out that Figure 7 is the result of considerable extrapolation of σ data with respect to temperature. The range of selenotherms results from uncertainties in the σ-depth profile and the variation of σ amongst the three

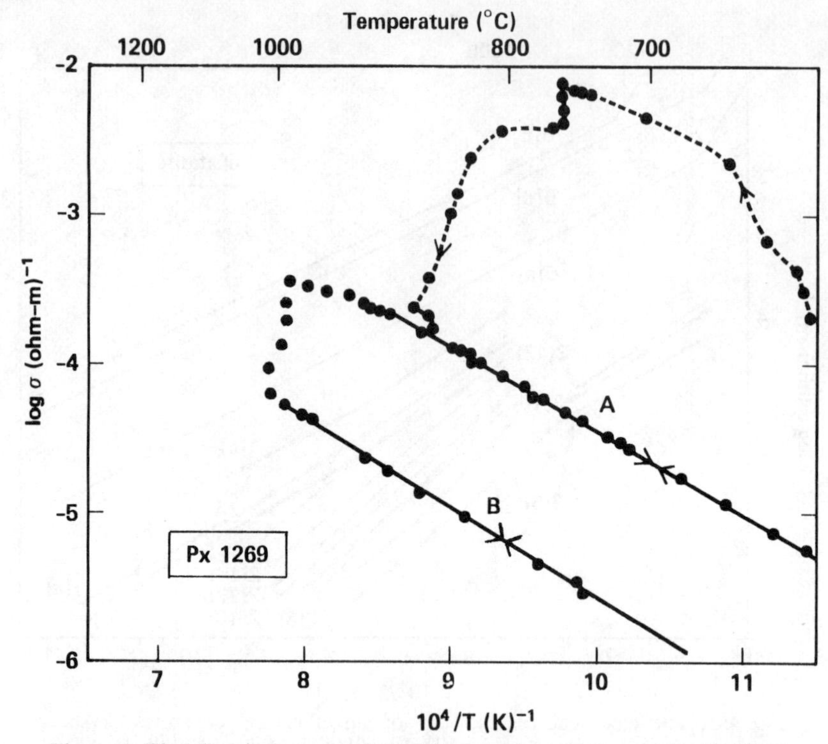

Figure 6. The electrical conductivity (σ) of single crystal enstatite, Px 1269, at $f_{O_2} = 10^{-8}$ (1200 °C). Lines A and B indicate regions where σ shows no time dependence. See text for discussion. (Reproduced from A. Duba, J. N. Boland and A. E. Ringwood, *J. Geol.*, 81, 727–735 (1973) by permission of the University of Chicago Press).

enstatite specimens investigated. The range of selenotherms calculated from single crystal forsterite σ is much smaller than that based on enstatite σ because of the lack of extrapolation of, and small variation in, single crystal forsterite σ data. A more recent σ-depth profile by Sonett *et al.*, (1972) yields a similar selenotherm whereas the higher values of σ at similar depths reported by Dyal and Parkin (1973) yields higher temperatures.

ALBITE σ AND TIME DEPENDENCY

A recent investigation (Piwinskii and Duba, 1974) of the σ of Amelia albite illustrates the influence of rate effects on the physical properties of this feldspar. At f_{O_2} close to the QFM buffer (Figure 2), a single crystal of Amelia albite was heated for 526 hours at 1080 °C followed by 1199 hours at 1100 °C and then 1486 hours at 1111 °C. Figure 8 shows the change of σ with time, at each of these temperatures.

Based on the results of X-ray studies (Prewitt, this volume) and solution

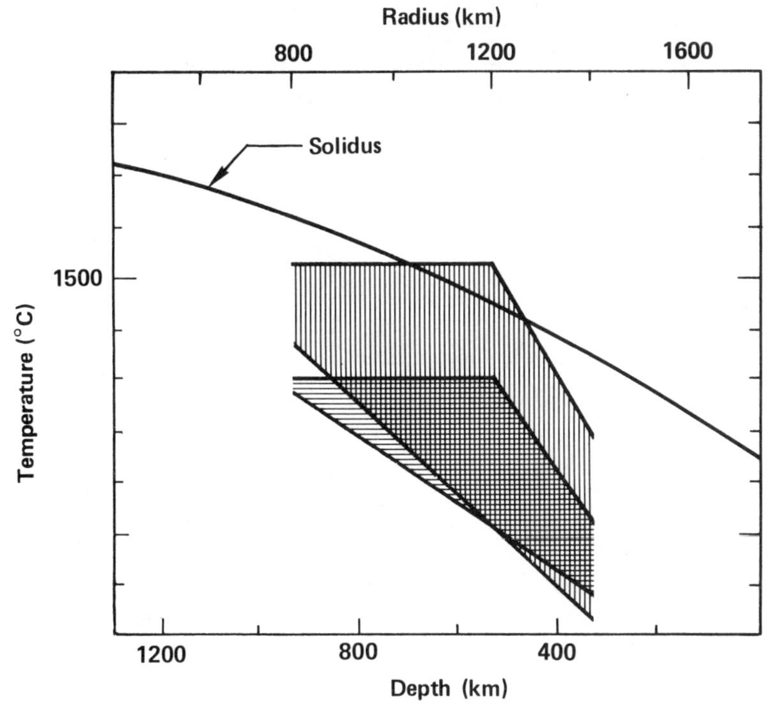

Figure 7. Range of possible selenotherms, based on pyroxene data (lines 1, 2 and 3) of Figure 5 (vertical lines) and olivine data (lines 6b and 7b) of Figure 3 (horizontal lines), calculated from σ-depth profile of Sonett *et al.* (1971). Shaded areas include uncertainty in σ-depth profile. The solidus of a lunar interior model is plotted for reference (Ringwood and Essene, 1970).

calorimetry (Kleppa, this volume), we believe that this increase in σ is due to Al/Si substitutional disorder in tetrahedral sites. If the σ is monitoring disorder, then the σ results indicate that even after more than 3200 hours at temperatures in excess of 1080 °C, the high albite is not completely disordered. Thus, Kleppa's ΔH between ordered and disordered albite probably would not represent the equilibrium value.

After 3200 hours at temperatures greater than 1080 °C, temperature cycling to 830 °C yielded σ–T plots with no hysteresis as shown in Figure 9. The remarkably low A_x (0.25 to 0.35 eV) has not been observed for other silicates at these temperatures. Figure 10 shows a further consequence of rate effects on σ. In this figure, the change in σ of polycrystalline albite upon melting (Khitarov and Slutskii, 1965) is compared with the σ for single crystal albite (Piwinskii and Duba, 1974) below the reported melting point at 1118 °C (Greig and Barth, 1938). It is evident that the sharp rise in σ (segment AB in Figure 10), which was attributed to the melting of albite by Khitarov and Slutskii, could be produced subsolidus if enough time were allowed. Certainly

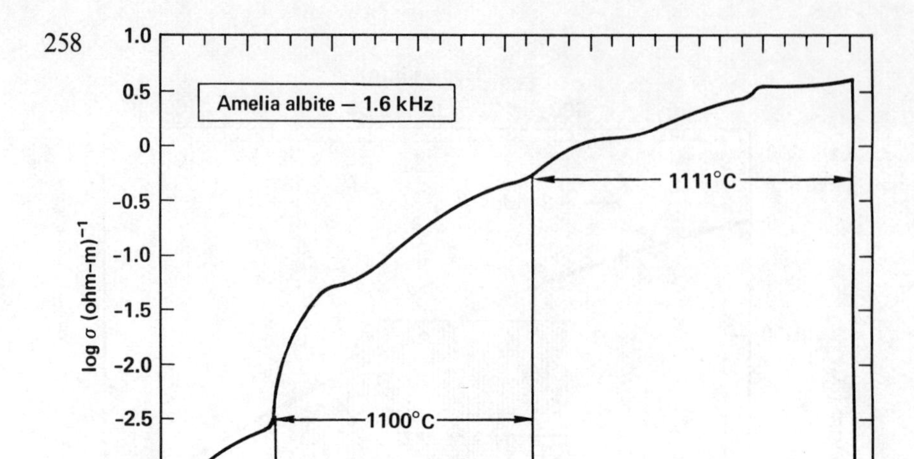

Figure 8. Electrical conductivity (σ) as a function of time for the Amelia albite at 1080 °C, 1100 °C, and 1111 °C with run durations of 526, 1199, and 1486 h, respectively.

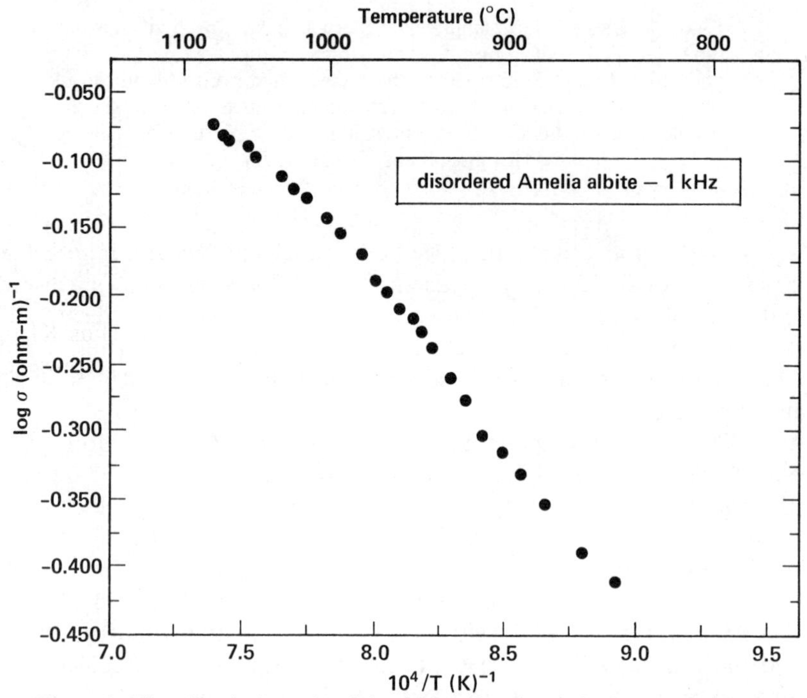

Figure 9. The electrical conductivity (σ) of disordered single crystal Amelia albite as a function of temperature after more than 3200 hours at temperatures greater than 1080 °C.

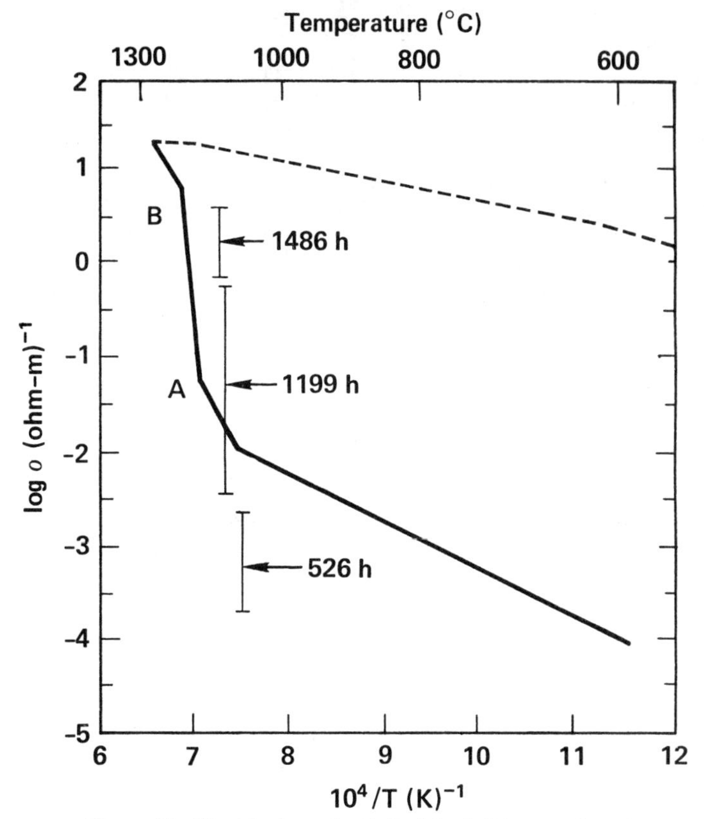

Figure 10. Electrical conductivity (σ) of albite as a function of temperature. The vertical lines indicate the change in σ for the Amelia albite at 1080 °C, 1100 °C, and 1111 °C for the times indicated (see Figure 8). The solid and dashed lines depict the change in σ for powdered albite (An_3Ab_{97}) on heating and cooling, respectively, at 2.8 kb (Khitarov and Slutskii, 1965). Khitarov and Slutskii attribute the sharp increase in σ, shown as line segment AB, to the melting of albite.

in some natural systems there is sufficient time for disordering to occur. It thus seems unnecessary to postulate any dramatic effects on σ due to melting of any silicates when rate effects are not considered.

ACKNOWLEDGEMENTS

This work was performed under the auspices of the U.S. Atomic Energy Commission. A portion of this work was supported by NASA Order Number W-13788. We thank B. Bonner and H. C. Weed for discussions.

REFERENCES

Banks, R. J. (1969). *Geophys. J. Roy. Astron. Soc.*, **17**, 457–487.
Boland, J. N., Al.Duba, and A. Eggleton (1974). *J. Geol.*, **82**, 507–514.
Bradley, R. S., A. K. Jamil, and D. C. Munro (1964). *Geochem. Cosmochim. Acta*, **28**, 1669–1678.
Duba, A. (1972). *J. Geophys. Res.*, **77**, 2483–2495.
Duba, A., and I. A. Nicholls (1973). *Earth Planet. Sci. Letters*, **18**, 59–64.
Duba, A., H. C. Heard, and R. N. Schock (1972). *Earth Planet. Sci. Letters*, **15**, 301–304.
Duba, A., H. C. Heard, and R. N. Schock (1974). *J. Geophys. Res.*, **79**, 1667–1674.
Duba, A., J. C. Jamieson, and J. Ito (1973a). *Earth Planet. Sci. Letters*, **18**, 279–284.
Duba, A., J. N. Boland, and A. E. Ringwood (1973b). *J. Geol.*, **81**, 727–735.
Dyal, P. and C. W. Parkin (1973). *Phys. Earth Planet. Interiors*, **7**, 251–265.
Fujisawa, H. (1968). *J. Geophys. Res.*, **73**, 3281–3294.
Greig, J. and T. F. W. Barth (1938). *Amer. Jour. Sci., 5th Series*, **35-A**, 93–112.
Griggs, D. T. (1972). In *The Nature of the Solid Earth*, ed. E. C. Robertson, McGraw-Hill.
Hamilton, R. M. (1965). *J. Geophys. Res.*, **70**, 5679–5692.
Hughes, H. (1953). *Ph.D. Thesis*, 131 pp., Univ. of Cambridge, Cambridge, England.
Hughes, H. (1955). *J. Geophys. Res.*, **60**, 187–191.
Ito, K., and G. C. Kennedy (1967). *Amer. Jour. Sci.*, **265**, 519–538.
Khitarov, N. and A. Slutskii (1965). *Geokhimiya, No.* 12, 1395–1403.
Kobayashi, Y., and H. Maruyama (1971). *Earth Planet. Sci. Letters*, **11**, 415–419.
Nitsan, U. (1974). *J. Geophys. Res.*, **79**, 706–711.
Parker, R. L. (1970). *Geophys. J. Roy. Astron. Soc.*, **22**, 121–138.
Piwinskii, A. J. and A. Duba (1974). *Geophys. Res. Letters*, **1**, 209–211.
Ringwood, A. E. (1966). Mineralogy of the mantle. In *Advances in Earth Sciences*, ed. P. M. Hurley, pp. 357–399, MIT Press, Cambridge, Mass.,
Ringwood, A. E. and E. Essene (1970). *Geochim. Cosmochim. Acta, Suppl. 1*, 769–799.
Schult, A., and M. Schober (1969). *Z. Geophys.*, **35**, 105–112.
Sonett, C. P., D. S. Colburn, P. Dyal, C. W. Parkin, B. F. Smith, G. Schubert, and K. Schwarz (1971). *Nature*, **230**, 359–362.
Sonett, C. P., B. F. Smith, D. S. Colburn, G. Schubert, and K. Schwarz (1972). *Proc. Third Lunar Science Conf. Geochim. Cosmochim. Acta Suppl. 3*, 2309–2336, MIT Press.

Electrical Properties of Rocks

G. R. Olhoeft
Department of Physics,
University of Toronto,
Toronto, Ontario

INTRODUCTION

Electrical properties of rocks are used in induced polarization, resistivity, and electromagnetic methods of mineral exploration (Keller and Frischknecht, 1966; Ward, 1967; Madden and Cantwell, 1967; Von Voorhis *et al.*, 1973), in crustal sounding (Hermance, 1973), in lunar and planetary sounding (Banks, 1969; Brown, 1972; Dyal and Parkin, 1973; Simmons *et al.*, 1972), in glacier sounding (Rossiter *et al.*, 1973), and in other applications. Studies of electrical properties in rocks have been performed as functions of frequency, temperature, applied field, pressure, oxygen fugacity, water content, and other variables (Keller, 1966; Ward and Fraser, 1967; Parkhomenko, 1967; Brace and Orange, 1968; Fuller and Ward, 1970; Alvarez, 1973b; Dvorak, 1973; Hansen *et al.*, 1973; Katsube *et al.*, 1973; Waff, 1973; Gold *et al.*, 1973; Schwerer *et al.*, 1973; Marshall *et al.*, 1973; Olhoeft *et al.*, 1974*b*; Duba *et al.*, 1974; Hoekstra and Delaney, 1974; and others).

This will be a brief review of the behaviour of electrical properties of rocks for several of the above parameters, concentrating mainly upon frequency, temperature, and water content in the context of likely mechanisms. There are many experimental techniques available and a discussion of them will not be attempted here (see von Hippel, 1954; Collett, 1959; ASTM, 1970; Hill *et al.*, 1969; Suggett, 1972; and Collett and Katsube, 1973). For geological materials, the best procedure is usually to follow a combination of techniques involving the observation of electrical properties as functions of applied field (to test for voltage–current non-linearity) and of frequency, all for variations of relevant environmental parameters (temperature, water content, etc.).

ELECTRICAL PROPERTIES

Electrical properties are derived from the solution of Maxwell's equations. Such solutions (Stratton, 1941, or other standard textbooks) yield an electromagnetic wave with a propagation constant

$$\gamma^2 = \omega\mu(\omega\varepsilon - j\sigma) \tag{1}$$

where

γ = propagation constant = reciprocal skin depth
ω = frequency (radians/second)
μ = magnetic permeability (henry/metre)
ε = dielectric permittivity (farad/metre)
σ = electrical conductivity (mho/metre)
$j = \sqrt{-1}$

All of the material properties (other than geometric boundary conditions) which are important to the description of the propagation of an electromagnetic wave (or lack of propagation) are described in the propagation constant by the quantities μ, ε, and σ.

The magnetic properties are completely described by the magnetic permeability (which may be complex, see Miles *et al.*, 1957; Olhoeft and Strangway, 1974b) which is dependent upon frequency, field strength, and temperature but is usually taken in geophysics to be a real constant and equal to the free space value ($\mu_0 = 4\pi \times 10^{-7}$ H/m). The electrical properties are completely described by the complex quantity $\omega\varepsilon - j\sigma$, in which ε may also be complex. The properties of this latter quantity will be the subject of discussion here.

Two general classes of solutions to Maxwell's equations thus result. At low frequencies, the dielectric permittivity is generally neglected in electromagnetic induction problems (magnetotellurics, planetary sounding, etc.). At high frequencies, the conductivity is generally neglected in electromagnetic propagation problems (radar sounding, microwave spectroscopy and radiometry, etc.). Very little work has been done in areas where both are significant (Olhoeft, 1975).

When permittivity is neglected, the induction skin depth is given by

$$d = (\mu\omega\sigma)^{-1/2}$$

which is the distance an electromagnetic wave penetrates into a material while being attenuated to 1/e of its surface magnitude. When the conductivity is neglected, the velocity of propagation is given by

$$v = (\mu\varepsilon)^{-1/2}$$

which is the speed of light in the material, and the characteristic impedance is

$$Z = (\mu/\varepsilon)^{-1/2}$$

which is approximately 377 ohms for free space. For further information regarding general electromagnetic problems and solutions, see Ward (1967) or Stratton (1941). Both the permittivity and the conductivity will be considered explicitly here, and examples will be given showing the regions where each dominates as well as the transition zone from induction to propagation where both are significant.

FREQUENCY DEPENDENCE

All frequency dependent electrical properties are here considered to be described in the complex permittivity or the complex resistivity (see discussions in Fuller and Ward, 1970, or Collett and Katsube, 1973). The conductivity is considered to be the value at zero frequency only and independent of frequency. Note that this means that the conductivity is not the reciprocal of the resistivity except in the limit of $\omega \rightarrow 0$ or when $\sigma \gg \omega\varepsilon$.

The complex dielectric constant is the ratio of the permittivity of the material to the permittivity of free space

$$K' - jK'' = \varepsilon/\varepsilon_0$$

where $\varepsilon_0 = 8.854 \times 10^{-12}$ F/m = free space permittivity

K' = real dielectric constant
K'' = imaginary dielectric constant

and the complex resistivity is

$$\rho' - j\rho'' = \frac{1}{\omega\varepsilon_0 K'} \frac{(D - j)}{(1 + D^2)} \tag{2}$$

where D is the loss tangent

$$D = \tan\delta = \frac{K''}{K'} + \frac{\sigma}{\omega\varepsilon_0 K'} = \frac{\rho'}{\rho''} \tag{3}$$

representing the measure of the phase shift between the electric and magnetic field vectors in an electromagnetic wave (Olhoeft and Strangway, 1974b) and a measure of the energy dissipation.

The general behaviour of electrical properties observed as a function of frequency is shown in Figure 1 (discussed below), and may be modelled by a circuit such as Figure 2. The conductance G represents the DC conductivity path, the capacitances C_L and C_H determine the low frequency limit of the dielectric constant, the resistor-capacitor pair RC_L determines the time constant of relaxation (rate of relaxation of an energy dissipative process), and the capacitance C_H determines the high frequency limit of the dielectric constant. This circuit has a single time constant, but, in general, a distribution of time constants is observed (see Ghausi and Kelly, 1968; Shuey and Johnston, 1973; and others). The general form of a complex dielectric constant with a distribution of time constants is (Gevers, 1945)

$$K' - jK'' = K_\infty + (K_0 - K_\infty) \int_0^\infty \frac{G(\tau)(1 - j\omega\tau)d\tau}{1 + \omega^2\tau^2} \tag{4}$$

where

$$K_0 = \lim_{\omega \rightarrow 0} K' = \text{low frequency dielectric constant}$$

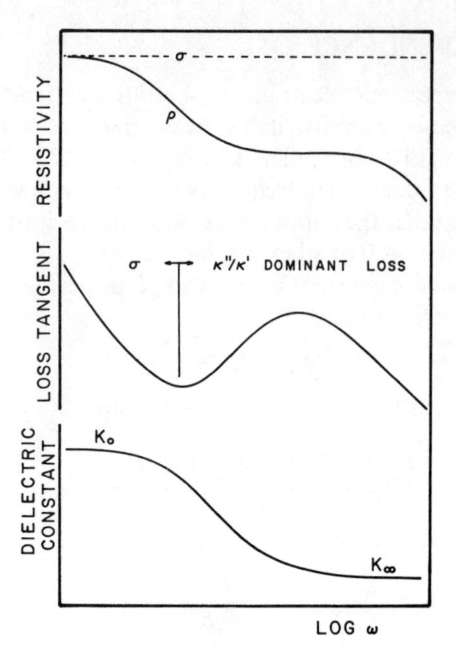

Figure 1. Schematic behaviour of the complex dielectric constant, loss tangent, and resistivity versus log frequency.

$$K_\infty = \lim_{\omega \to \infty} K' = \text{high frequency dielectric constant}$$

$$\tau = \text{time constant of relaxation}$$

and $G(\tau)$ is the time constant distribution function with the normalization

$$\int_0^\infty G(\tau)\mathrm{d}\tau = 1.$$

The high frequency dielectric constant is generally independent of temperature and water content, being mainly dependent upon density. In this context, the high frequency dielectric constant is valid for frequencies up to the optical where it becomes approximately equivalent to the square of the index of refraction (see von Hippel, 1954). The low frequency dielectric constant is primarily caused by the accumulation of charge at crystal boundaries and defects, and it will be further discussed below. The time constant distribution function may narrow with increasing temperature and generally tends to broaden with increasing water content, salinity, and defect structure.

Several specific distributions are found in the literature (Poole and Farach, 1971; de Batist, 1972; and others), one of the most useful being the Cole-Cole distribution (Cole and Cole, 1941)

$$K' - \mathrm{j}K'' = K_\infty + \frac{K_0 - K_\infty}{1 + (\mathrm{j}\omega\tau)^{1-\alpha}} \tag{5}$$

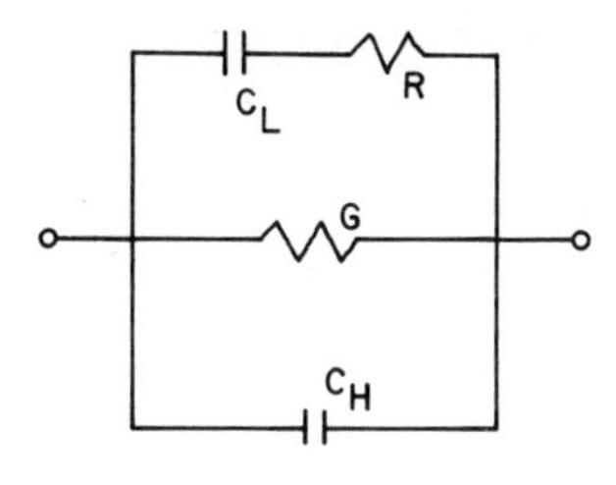

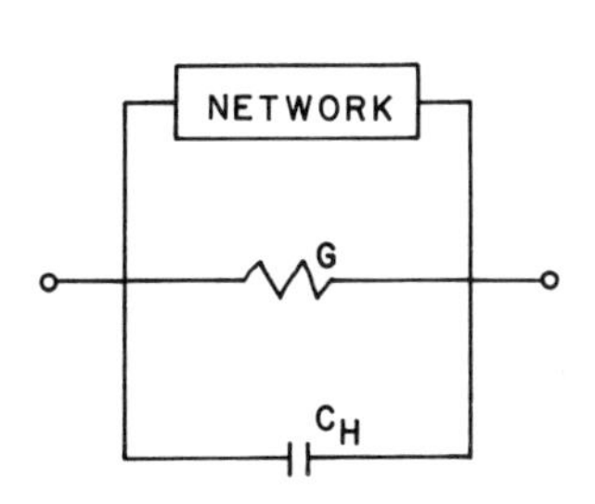

Figure 2. The equivalent circuit of a single relaxation with a DC conduction path (others are possible). A distribution of relaxation times substitutes a series of RC networks in place of RC_L.

where $1 - \alpha$ is the distribution breadth parameter with limits of $1 - \alpha = 1$ (a single relaxation) and $1 - \alpha = 0$ (an infinitely broad distribution). The slope of log loss tangent versus log frequency is equal to $1 - \alpha$ for $\omega\tau < 1$, and $-(1 - \alpha)$ for $\omega\tau > 1$. Examples of this distribution in rocks may be found in Saint-Amant and Strangway (1970), Alvarez (1973b), and Olhoeft *et al.* (1973, 1974a). The distribution parameter may be temperature dependent (Fuoss and Kirkwood, 1941; Olhoeft *et al.*, 1973, 1974a), tending to narrower distributions with increasing temperature. A common method of displaying data which fits the Cole–Cole type of distribution is the use of the Argand diagram which plots imaginary versus real dielectric constant. On this plot, the data form a semicircle with a depressed center. The amount of depression is a measure of $1 - \alpha$ (Poole and Farach, 1971; Alvarez, 1973b).

TEMPERATURE DEPENDENCE

As there are distributions of time constants describing the frequency dependence of electrical properties, so there are distributions of activation energies describing the temperature dependence. Both the time constants in the above distribution and the DC conductivity follow a generalized Boltzmann temperature dependence of the form

$$\tau = \sum_i \tau_i e^{E_i/kT} \tag{6}$$

where

τ_i = infinite temperature time constant (conductivity)
E_i = activation energy (positive for time constants, negative for conductivity, and possibly temperature dependent, see Olhoeft *et al.*, 1973)

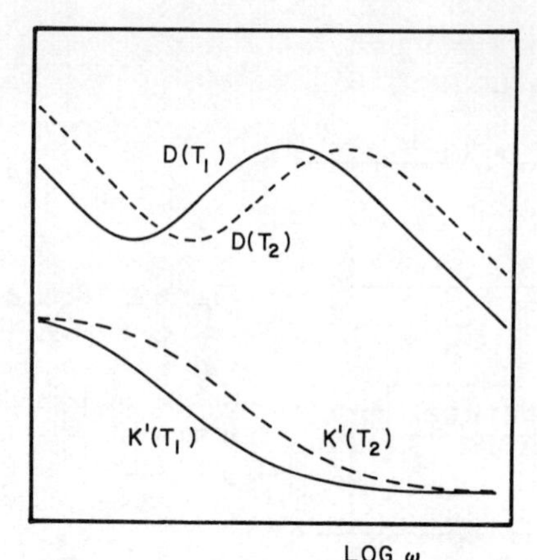

Figure 3. Schematic behaviour of the dielectric constant and loss tangent versus frequency for a distribution of time constants and a single thermal activation energy for two temperatures, T_2 greater than T_1.

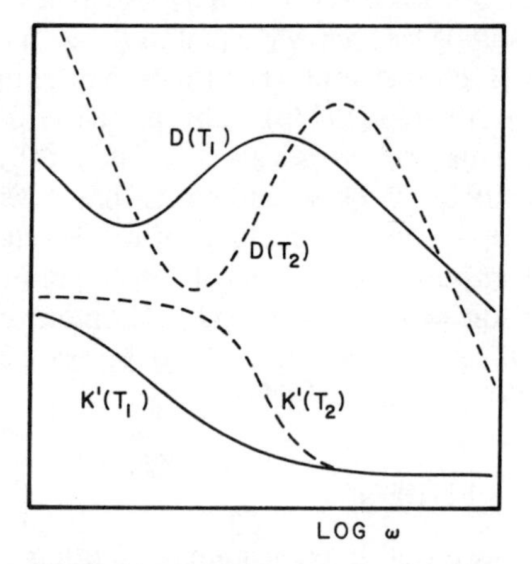

Figure 4. Schematic behaviour of the dielectric constant and loss tangent versus frequency for a distribution of time constants and a distribution of activation energies for two temperatures, T_2 greater than T_1.

k = Boltzmann's constant = 8.6176×10^{-5} eV/K
T = temperature (Kelvin).

The behaviour of a single term expression such as (6) with the Cole–Cole distribution (5) is shown in Figure 3. Increasing the temperature, increases the frequency of relaxation. More generally however, there may be a different expression (6) for each time constant in the distribution (4) thus generating behaviour as in Figure 4. Both the frequency of relaxation and the shape of the time constant distribution alter with increasing temperature. In equation (5) this would appear as a temperature variation in $1 - \alpha$.

VOLTAGE–CURRENT DEPENDENCE

The usual method of measuring electrical conductivity utilizes an application of voltage across a sample of known dimensions. The resultant current flow through the linear *VI* (voltage–current) relation of Ohm's law yields

$$\sigma = AI/V \tag{7}$$

where I is current, V is voltage, and A is a constant determined by geometry. Not all materials obey linear *VI* relations however. The best example of one which does not is a semiconductor diode with a *VI* relation of the form

$$I = I_s(e^{qV/kT} - 1) \tag{8}$$

where kT is as in (6), q is the electronic charge, and I_s is the saturation current (see Gray *et al.*, 1964, for details). This behaviour is not only nonlinear, but it is asymmetrical (positive and negative voltages produce different magnitudes of current). This latter type of nonlinearity is caused by accumulations of charge at regions where the material electrical properties change abruptly, such as from *p*-type to *n*-type semiconductor material in the diode above. Such regions (space charge layers, SCL) also occur at the edges of crystal lattices, at cracks and other defects, and at other abrupt changes in material energy.

As well as the cause of *VI* nonlinearity, these space charge layers are the cause of contact potentials and the Maxwell–Wagner frequency response of heterogeneous materials (Alvarez, 1973b), and of the complex interfacial effects between electrodes and samples (Covington, 1970, 1973; Baker and Buckle, 1968; Buck, 1969; Hampson, 1972, 1973; Hever, 1972; and others). The electrical properties within a space charge layer determine the rate of relaxation of the charge distribution. Thus, a heterogeneous material will have space charge layers in many different materials resulting in many time constants and a distribution of time constants as discussed above. Space charge layers are also the mechanism behind the low frequency dielectric constant, and the effects on low frequency properties are particularly pronounced in wet materials where the water forms the electrochemical double layer (space charge layer in a liquid; see Delahay, 1965; Schiffrin, 1970, 1972, 1973; Payne, 1973). In wet materials, the differentiation between the electrode-sample low frequency response and the response inherent in the interfacial properties of the sample is a particularly difficult problem in experiment design and data analysis.

Though *VI* nonlinearity in rocks is a well known and accepted phenomenon, it has not been subjected to any great scrutiny. Scott and West (1969) and Katsube *et al.* (1973) have observed *VI* nonlinearities in laboratory experiments using conductive minerals. Schwerer *et al.* (1973) and Olhoeft *et al.* (1974*b*) have reported nonlinear VI phenomena in very low conductivity lunar samples. Halverson *et al.* (1973) have reported nonlinear induced polarization responses in field surveys. These phenomenon have all been observed below 1000 Hz.

MOISTURE DEPENDENCE

Baldwin (1958), Von Ebert and Langhammer (1961), and McCafferty and Zettlemoyer (1971) have observed that a monolayer of water adsorbed onto a dielectric surfaces does not alter the dielectric constant of the bulk material from that in the dry state. Successively adsorbed layers drastically increase the low frequency dielectric constant, increase the frequency of relaxation, increase the time constant activation energy, and tend to broaden the width of the time constant distribution. Beyond about seven layers, free pore water begins to form and pore fluid conduction dominates the electrical properties. Adding alteration products and electrolytes to the water further increases the low frequency dielectric constant and the breadth of the time constant distribution. In addition, monolayers and successive layers increase the DC conductivity by several orders of magnitude. Examples of the above have been discussed by Saint–Amant and Strangway (1970), Strangway *et al.* (1972), Hoekstra and Doyle (1971), Alvarez (1973a), and others.

The process or mechanism of the modification of electrical properties by water is the formation of the electrochemical double layer. The monolayer is the strongly bonded Stern layer (on the order of 10 angstroms thick), and successively adsorbed layers become the Gouy–Chapman layer. The changes of electrical properties due to the addition of water are primarily within the space charge layer of the Gouy-Chapman layer (which may influence general physical properties to 1000 angstroms from the solid-liquid interface). These are further discussed and reviewed by Clark (1970), Schiffrin (1970, 1972, 1973), Sing (1973), and Payne (1973).

The additional problems caused by geometries in porous systems have been discussed by Davies and Rideal (1963), Rangarajan (1969), and Everett and Haynes (1973). Marshall *et al.* (1973) have remarked on anomalous dielectric properties attributed to fine parallel capillaries in rocks. The particular problem of connectivity with regard to electrical conductivity in porous media has been discussed by Shankland and Waff (1973) and Madden (1974) in the geological context.

A general discussion of electrolytes in aqueous solution may be found in Hasted (1972), Rao and Premaswarup (1969), and Pottel (1973). In such solutions, chemical reactions and Faradaic mass transport processes must also be considered in addition to the double layer (see Ott and Rys, 1973a, b; Schmidt, 1973a, b, c; DeVevlie and Pospisil, 1969; Reinmuth, 1972a, b, c; Armstrong and Firman, 1973a, b).

The effects of temperature and water content on electrical properties have been investigated by Baldwin (1958), Von Ebert and Langhammer (1961), Dransfeld *et al.* (1962), McCafferty and Zettlemoyer (1971). Hoekstra and

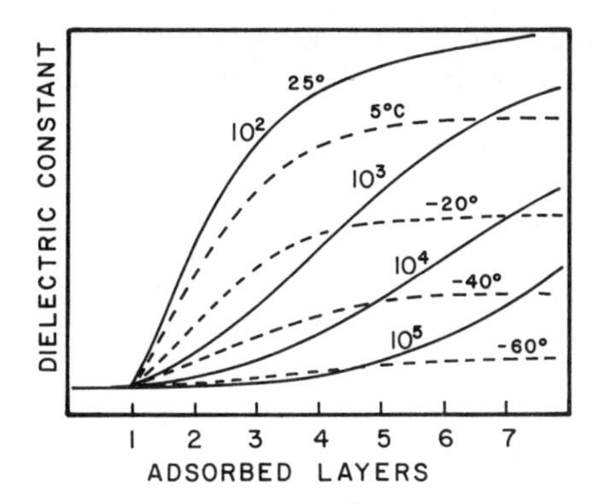

Figure 5. Schematic behaviour of the dielectric constant versus water content for various frequencies at room temperature (solid lines) and for various temperatures at 100 Hz (dashed lines).

Doyle (1971), and Hoekstra and Delaney (1974). The effects of temperature for a monolayer or less of water is the same as for a dry rock. For larger quantities of water, as the temperature is lowered to and through 0 °C, the relative alteration in electrical properties between wet and dry states becomes smaller. Below about − 60 °C, the effect of water in any quantity becomes extremely small and electrical properties are virtually indistinguishable from the dry state. The implications of these results are further discussed by McCafferty and Zettlemoyer (1971), Hoekstra and Doyle (1974), and Olhoeft and Strangway (1974a). Figure 5 quickly summarizes these findings.

The effects of temperature, water content, and pressure have been investigated by several groups with ambiguous results. Dvorak (1973), Dvorak and Schloessin (1973), Duba *et al.* (1974), Akimoto and Fujisawa (1965), and others have investigated the pressure effects in dry rocks, generally finding little or no change in electrical conductivity below 8 kilobars, but sometimes large and irreversible changes (possibly correlated with chemical changes) above 8 kb.

Brace *et al.* (1965), Brace and Orange (1968), and Stesky and Brace (1973) have observed large variation in electrical conductivity in wet samples with pressure to only 6 kbars. This alteration may be attributable to changes in pore volume and crack connectivity with pressure, but much more work has yet to be done (particularly with the additional complication of temperature as a variable). Holzapfel (1969) has observed that water under high pressure and high temperature becomes very conductive (more than 0.01 mho/m at 200 °C and 10 kbar), in some cases becoming more conductive than many rocks under similar conditions of temperature and pressure.

Frisillo *et al.* (1975) have observed the effects of temperature and pressure on the dielectric constant and loss tangent of soils in vacuum. After allowing for density changes, they found no change in loss tangent with pressure, though large changes were observed in the dielectric constant with only 30 bars pressure.

EXAMPLES

The variation of electrical properties with temperature has been discussed above. This is further illustrated in Figures 6 through 8 for a dry (outgassed in vacuum better than 10^{-7} torr), terrestrial pyroxene (salite) which has been partially discussed by Olhoeft *et al.* (1974b). The composition is shown in Table 1. The DC conductivity and the low frequency dielectric constant were tested and found to be linear in voltage–current from 123 to 5000 volt/cm of applied field.

The temperature and frequency response of the dielectric constant and loss tangent are shown in Figures 6 and 7 in vacuum as well as in air at room temperature. The complex resistivity and DC conductivity are shown in Figures 8 and 9. Also illustrated in Figure 9 are the equivalent AC conductivities calculated from Figures 6 and 7 by

$$\sigma_{AC} = \sigma_{DC} + \omega \varepsilon_0 K' \tan \delta$$

This data is parameterized by $K_\infty = 6.4$, a temperature dependent K_0 varying from 90 below 163 °C to 100 at 415 °C, and both a distribution of time constants and a distribution of activation energies. The distribution of time constants may be characterized by a temperature dependent Cole–Cole breadth para-

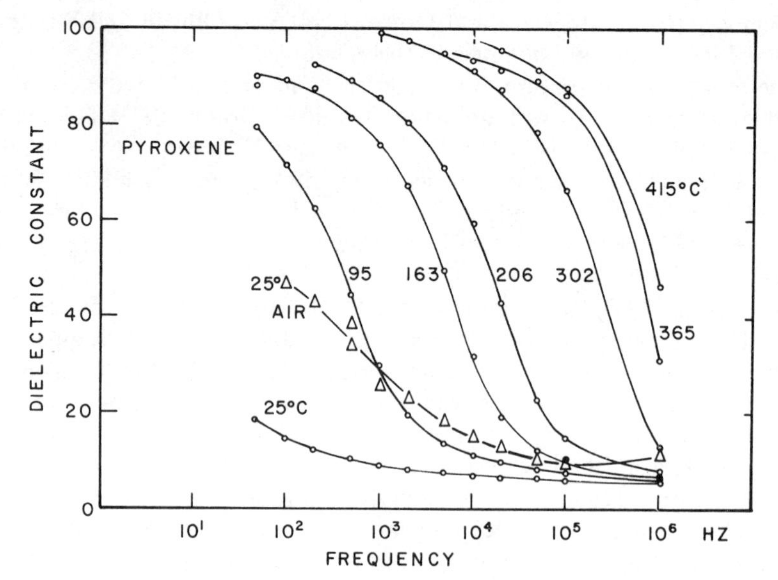

Figure 6. The dielectric constant versus frequency for a dry pyroxene in vacuum at various temperatures and in air at room temperature.

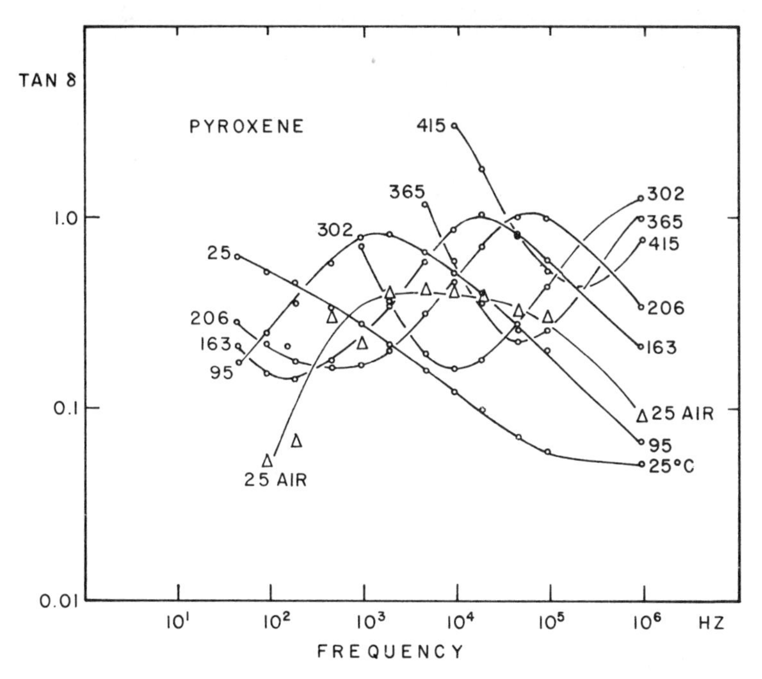

Figure 7. The loss tangent of a dry pyroxene in vacuum for various temperatures and in air at room temperature.

meter (indicative of the distribution of activation energies) $1-\alpha$ equal to 0.32 at 25 °C, 0.47 at 95 °C, and 0.63 above 302 °C. The activation energy of the apparent peak in the time constant distribution is constant at 0.60 eV, but the infinite temperature time constant at that point varies from 2.4×10^{-12} seconds at 95 °C to 4.1×10^{-12} seconds at 365 °C. This latter variation is indicative of a slight distribution of activation energies (as is also the variation in $1-\alpha$). The infinite temperature DC conductivity is 1.4 mho/m with an activation energy of -0.54 eV.

The electrical properties in Figure 8 clearly show the three regions where different terms dominate the solutions of Maxwell's equations. At 197 °C, the DC value of the resistivity (reciprocal of the DC conductivity) is asymptotically approached by the real part of the complex resistivity at frequencies below 100 Hz. Below 100 Hz in this sample is thus the region of electromagnetic induction where the permittivity may be safely neglected.

Continuing to higher frequencies, the loss tangent decreases in value and passes through a relative minimum near 4 kHz. Below 3 kHz, the DC conductivity is the dominant loss, and between 100 Hz and 3 kHz both the conductivity and permittivity are important to solutions of Maxwell's equations. Above 3 kHz, the DC conductivity may be safely neglected in the region of electromagnetic propagation (where dielectric mechanisms cause the dominant loss). The critical frequency separating the inductive region from the inter-

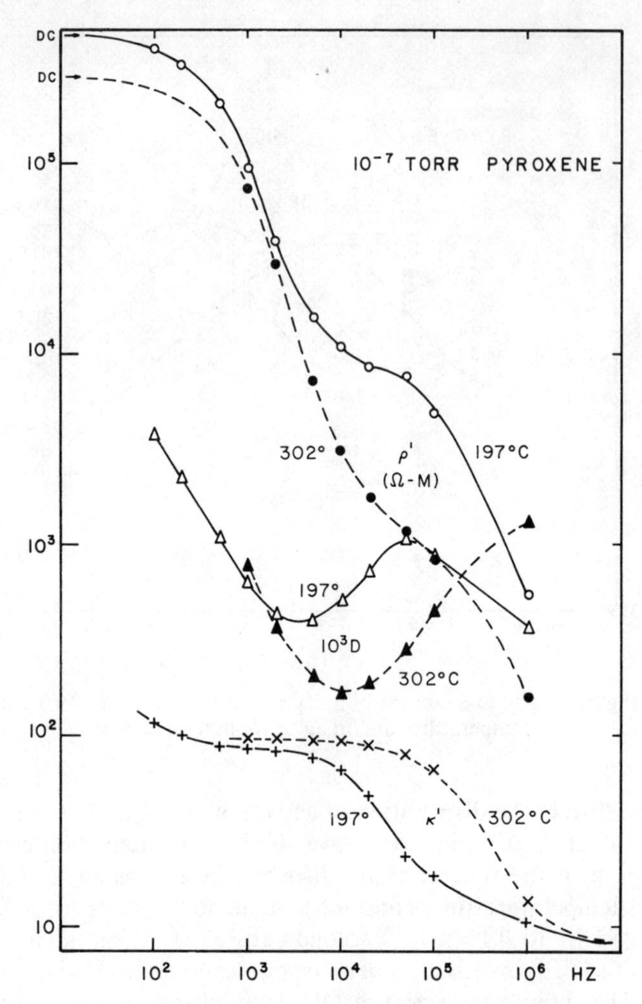

Figure 8. The dielectric constant, loss tangent, and real resistivity versus frequency at two temperatures to illustrate the critical frequencies (see text).

Table 1. Pyroxene Analysis (data from A. M. Reid, private communication)

SiO_2		52.67
TiO_2		0.05
Al_2O_3		0.94
FeO		9.62
MgO		12.04
CaO		23.94
	Total	99.26

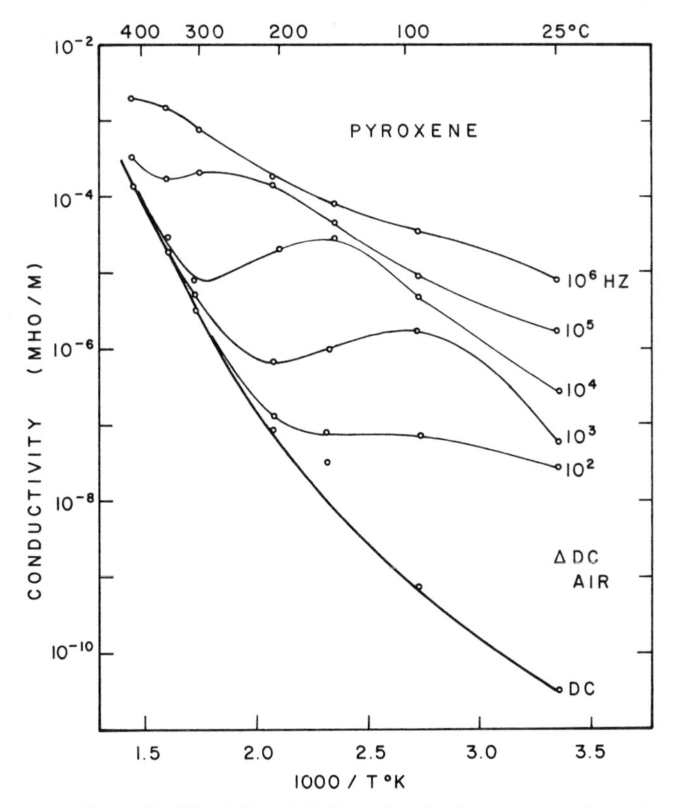

Figure 9. The AC and DC conductivities versus reciprocal temperature (see text).

mediate region occurs at $\omega = \sigma/\varepsilon_0 K'$, and the critical frequency separating the intermediate region from the propagation region occurs at $\omega = \sigma/\varepsilon_0 K''$.

Increasing temperature, increases the critical frequencies separating the regions, and the DC conductivity becomes relatively more important. This latter point may be seen in a slightly different manner in Figure 9 where the equivalent AC conductivities asymptotically approach the DC conductivity with increasing temperature or decreasing frequency.

Figures 10 and 11 show the dielectric constant and loss tangent of a granite at room temperature in solid and powder form (after Strangway *et al.*, 1972) in vacuum after outgassing at 200 °C and in 30 per cent atmospheric humidity. The difference between the solid and powder vacuum dielectric constants is directly attributable to the change in density. This may be calculated by using a formula of geometric mean between the dielectric constant of vacuum (equal to 1.0 at a density of 0.0) and the dielectric constant of the solid at the density of the solid by (Lichteneiker's formula, von Hippel, 1954)

$$K_p = K_s^{p/G} \tag{9}$$

where K_p and K_s are the dielectric constant of the powder and solid at densities

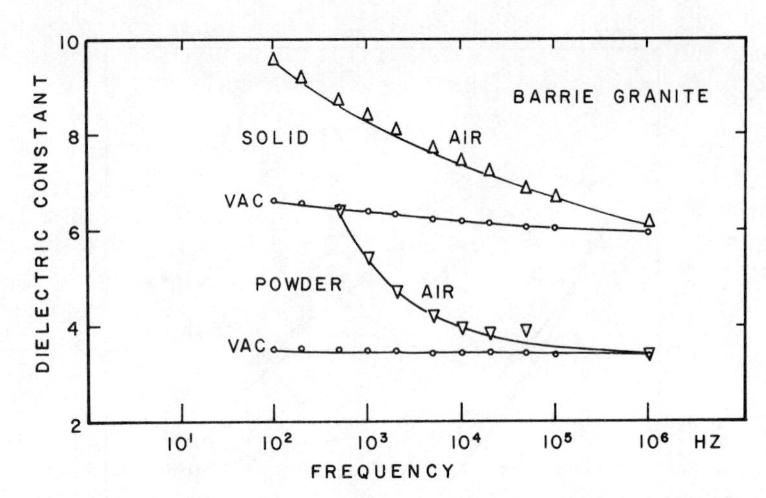

Figure 10. The dielectric constant of a solid and powder granite in vacuum and in air at room temperature.

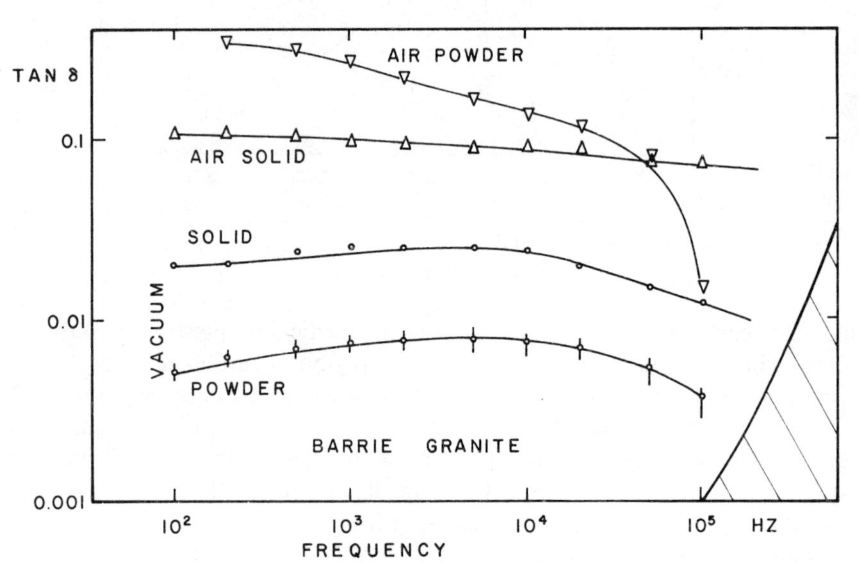

Figure 11. The loss tangent of a solid and powder granite in vacuum and in air at room temperature.

of p and G respectively. Allowing a complex dielectric constant, this formula yields a loss tangent of the form

$$\tan \delta_p = \tan (\delta_s p/G) \tag{10}$$

where $\tan\delta_p$ and $\tan \delta_s$ are the loss tangents of the powder and solid at densities of p and G respectively. The behaviour for the vacuum data in the dielectric constant of Figure 10 is consistent with formula (9), but the loss tangent then predicted through (10) for the transformation from solid to powder predicts a slightly higher powder loss tangent than is actually measured. As Olhoeft and Strangway (1975) have observed consistent fits by regression analysis to 91 lunar sample data points for both solids and soils in both formulas (9) and (10), this latter discrepancy in the granite powder loss tangent of Figure 11 may be attributable to the grinding process which produced the powder. This is under further investigation. The DC conductivity in vacuum reduced from 5.3×10^{-14} mho/m in solid form to 6.2×10^{-15} mho/m in powder form and may be attributable to both the change in density and the reduced grain contacts (connectivity) in the powder form.

As air of 30 per cent relative humidity enters into the sample chamber, all of the electrical properties change. The DC conductivity rises by several orders of magnitude, the dielectric constant increases at low frequencies, and the loss tangent increases. Of particular importance though is the change in shape of the frequency response and of the different changes observed in the solid and powder forms. Note that in both the dielectric constant and the loss tangent in vacuum that the solid and powder forms have similar frequency responses (remember that the slope of log frequency versus log loss tangent is proportional to the breadth of the time constant distribution). As water is added in the humid air, the frequency response of the dielectric constant and loss tangent in the solid is no longer similar to that of the powder. The powder response is now characteristic of a narrower distribution of time constants. In addition, the powder loss tangent increases in relative magnitude between vacuum and air (dry and wet) much more than the increase in the solid loss tangent. These latter effects may be attributable to the changed geometry of water adsorption (in pores for the solid and onto particulate grains in the powder) as well as the increased surface area in the powder over that available in the solid.

SUMMARY

This has been a brief introduction and review of some areas of current interest in the electrical properties of rocks. The different types of electrical properties and ways of presenting and parameterizing them have been discussed; the concepts of distributions of time constants and distributions of activation energies have been illustrated.

The transition from the inductive limit in the solution to Maxwell's equations where permittivity may be neglected to the region of propagation where the DC conductivity may be neglected has been shown by way of an experimental example, as have the specific effects of temperature on electrical properties. The effects of water on electrical properties have been reviewed with particular regard for the mechanism of the electrochemical double layer. The effects of pressure, electrolyte solutions, and voltage-current nonlinearity have been briefly mentioned, and it is hoped that a later paper may go into these and other similar phenomena in greater detail.

ACKNOWLEDGEMENTS

I thank Dr. A. L. Frisillo and Mr. H. N. Sharpe for assisting with the pyroxene measurements, which performed at the NASA Johnson Space Center while the author was with Lockheed Electronics Co., Houston, Texas. This work was greatly encouraged by Dr. D. W. Strangway and was partially supported by the Department of Energy, Mines and Resources.

REFERENCES

Akimoto, S. and H. Fujisawa (1965). *J. Geophys. Res.*, **70**, 443.
Alvarez, R. (1973a). *J. Geophys. Res.*, **78**, 1769.
Alvarez, R. (1973b). *Geophysics*. **38**, 920.
American Society for Testing and Materials (1970). *Annual Book of ASTM Standards*. D–150–70. pt. 29, 58–80, Philadelphia.
Armstrong, R. D. and R. E. Firman (1973a). *J. Electroanal. Chem.*, **45**, 3.
Armstrong, R. D. and R. E. Firman (1973b). *J. Electroanal. Chem.*, **45**, 257.
Baker, C. G. J. and E. R. Buckle (1968). *Trans. Faraday Soc.*, **64**, 469.
Baldwin, M. G. (1958). *Ph.D. Thesis*, Dept. of Chemistry, Univ. of North Carolina, Chapel Hill.
Banks, R. J. (1969). *Geophys. J.*, **17**, 457.
Brace, W. F. (1971). pp. 243–256 in *The Structure and Physical Properties of the Earth's Crust*. Geophys. Mono. Series, vol. 14, J. G. Hancock, ed., AGU, Washington, D.C.
Brace, W. F. and Orange, A. S. (1968). *J. Geophys. Res.*, **73**, 5407.
Brace, W. F., A. S. Orange, and T. R. Madden, (1965). *J. Geophys. Res.*, **70**, 5669.
Brown, W. E., Jr. (1972). pp. 243–257 in *Lunar Geophysics*, Z. Kopal and D. W. Strangway, eds., D. Rediel, Dordrecht-Holland.
Buck, R. P. (1969). *J. Electroanal. Chem.*, **23**, 219.
Clark, A. (1970). *The Theory of Adsorption and Catalysis*, Academic, New York.
Cole, K. S. and R. S., Cole (1941). *J. Chem. Phys.*, **9**, 341.
Collett, L. S. (1959). Ch. 5 in *Overvoltage Research and Geophysical Applications*, J. R. Wait, ed., Pergamon, New York.
Collett, L. S. and T. J. Katsube (1973). *Geophysics*. **38**, 76.
Covington, A. K. (1970). Ch. 2 in *Electrochemistry*, *Vol. 1*, G. J. Hills, ed., (Specialist Periodical Report) The Chemical Society, London.
Covington, A. K. (1973). Ch. 1 in *Electrochemistry*, *Vol. 3*, G. J. Hills, ed., The Chemical Society, London.
Davies, J. T. and E. K. Rideal (1963). *Interfacial Phenomena*. Academic Press. New York, Ch. 2.
de Batist, R. (1972). *Internal Friction of Structural Defects in Crystalline Solids*, North-Holland, Amsterdam, Ch. 2.
Delahay, P. (1965). *Double Layer and Electrode Kinetics*. Wiley–Interscience, New York.
DeLevie, P. and Pospisil, L. (1969). *J. Electroanal. Chem.*, **22**, 277.
Dransfeld, K., H. L. Frisch, and E. A. Wood (1962). *J. Phys. Chem.*, **36**, 1574.
Duba, A., H. C. Heard, and R. N. Schock (1974). *J. Geophys. Res.*, **79**, 1667.
Dvorak, Z. (1973). *Geophysics*, **38**, 14.
Dvorak, Z. and H. H. Schloessin (1973). *Geophysics*, **38**, 25.
Dyal, P. and C. W. Parkin (1972). pp. 97–121 in *Lunar Geophysics*, Z. Kopal and D. W. Strangway, eds., D. Reidel, Dordrecht-Holland.
Everett, D. H. and J. M. Haynes (1973). Ch. 4 in *Colloid Science*, vol. 1, D. H. Everett, ed., The Chemical Society, London.
Frisillo, A. L., G. R. Olhoeft, and D. W. Strangway (1975). *Earth Planet. Sci. Letters*, **24**, 345–356.

Fuller, B. D. and S. H. Ward (1970). *IEEE Trans. Geosci. Electronics*, GE-8, no. 1.
Fuoss, R. M. and J. G. Kirkwood (1941). *Am. Chem. Soc. J.*, **63**, 385.
Gevers, M. (1945). *Philips Res. Rep.*, **1**, 197–447.
Ghausi, M. S. and J. J. Kelly (1968). *Introduction to Distributed Parameter Networks.* Holt Rinehart and Winston. New York.
Gold, T., E. Bilson, and M. Yerbury (1973). p. 3093 in *Proc. of the Fourth Lunar Science Conf., Vol. 3*, Pergamon, New York.
Gray, P. E., D. DeWitt, A. R. Boothroyd, and J. F. Gibbons (1964). *Physical Electronics and Circuit Models of Transistors*, J. Wiley, New York.
Halverson, M. O., E. O. McAllister, and Y. C. Yates (1973). paper presented to the joint 43rd Annual International Meeting of the Soc. of Explor. Geophys. and the 5th Meeting of the Asoc. Mexicana Geof. de Explor., Mexico City, D. F. (abstract in *Geophysics.* **38**, 1203).
Hampson, N. A. (1972). Ch. 3 in *Electrochemistry, Vol. 2*, G. J. Hills, ed., The Chemical Society, London.
Hampson, N. A. (1973). Ch. 3 in *Electrochemistry, Vol. 3*, G. J. Hills, ed., The Chemical Society, London.
Hansen, W., W. R. Sill, and S. H. Ward (1973). *Geophysics*, **38**, 135.
Hasted, J. B. (1972). Ch. 5 in *Dielectric and Related Molecular Processes, Vol. I*, M. Davies, ed., The Chemical Society, London.
Hever, K. O. (1972). Ch. 19 in *Physics of Electrolytes*, J. Hladik, ed., Academic Press, New York.
Hill, N. E., W. E. Baughan, E. H. Price, and M. Davies (1969). *Dielectric Properties and Molecular Behaviour*. Van Nostrand, London.
Hoekstra, P. and A. Delaney (1974). *J. Geophys. Res.*, **79**, 1699.
Hoekstra, P. and W. T. Doyle (1971). *J. Coll. Interface Sci.*, **36**, 513
Holzapfel, W. (1969). *J. Chem. Phys.*, **50**, 4424.
Katsube, T. J., R. H. Ahrens, and L. S. Collett (1973). *Geophysics*, **38**, 106.
Keller, G. V. (1966). In *Handbook of Physical Constants*, pp. 553–577, Geol. Soc. Am. Memoir 97, Washington, D.C.
Keller, G. V. and F. C. Frischknecht (1966). *Electrical Methods in Geophysical Prospecting*, Pergamon Press, New York.
McCafferty, E. and A. C. Zettlemoyer (1971). *Disc. Faraday Soc.*, **52**, 239.
Madden, T. R. (1974). *EOS Trans. AGU*, **55**, 415.
Madden, T. R. and T. Cantwell (1967). pp. 373–400 in *Mining Geophysics, Vol. 2*, Society of Exploration Geophysicists, Tulsa, Oklahoma.
Marshall, L. C., R. E. Walker, and F. K. Clark, Jr. (1973). *EOS Trans. AGU*, **54**, 237.
Miles, P. A., W. B. Westphal, and A. R. Von Hippel (1957). *Rev. Mod. Phys.*, **29**, 279.
Olhoeft, G. R. (1975). *Ph.D. Thesis*, Dept. of Physics, Univ. of Toronto.
Olhoeft, G. R. and D. W. Strangway (1974a). *Geophys. Res. Letters, 1*, 141–143.
Olhoeft, G. R. and D. W. Strangway (1974b). *Geophysics*, **39**, 302.
Olhoeft, G. R. and D. W. Strangway (1975). *Earth Planet. Sci. Letters*, **24**, 394–404.
Olhoeft, G. R., A. L. Frisillo and D. W. Strangway (1974a). *J. Geophys. Res.*, **79**, 1599.
Olhoeft, G. R., A. L. Frisillo, D. W. Strangway and H. N. Sharpe (1974b). *The Moon*, **9**, 79.
Olhoeft, G. R., D. W. Strangway and A. L. Frisillo (1973). pp. 3133–3149. in *Proc. of the Fourth Lunar Sci. Conf., vol. 3*, Pergamon.
Ott, R. J. and P. Rys (1973a). *JCS Faraday Trans. 1*, **69**, 1694.
Ott, R. J. and P. Rys (1973b). *JCS Faraday Trans. I*, **69**, 1705.
Parkhomenko, E. I. (1967). *Electrical Properties of Rocks*, Plenum, New York.
Payne, R. (1973). *J. Electroanal. Chem.*, **41**, 277.
Poole, C. P. and H. A. Farach (1971). *Relaxation in Magnetic Resonance*, Academic Press, New York.

Pottel, R. (1973). pp. 401–432 in *Water: A Comprehensive Treatise, Vol. 3*, F. Franks, ed., Plenum Press, New York.

Rangarajan, S. K. (1969). *J. Electroanal. Chem.*, **22**, 89.

Rao, P. S. and D. Premaswarup (1970). Trans. Faraday Soc., **66**, 1974.

Reinmuth, W. H. (1972a). *J. Electroanal. Chem.*, **34**, 297.

Reinmuth, W. H. (1973b). *J. Electroanal. Chem.*, **34**, 313.

Reinmuth, W. H. (1973c). *J. Electroanal. Chem.*, **35**, 93

Rossiter, J. R., G. A. LaTorraca, A. P. Annan, D. W. Strangway and G. Simmons (1973). *Geophysics*, **38**, 581.

Saint-Amant, M. and D. W. Strangway (1970). *Geophysics*, **35**, 624.

Schiffrin, D. J. (1970). Ch. 6 in *Electrochemistry, Vol. 1*, G. J. Hills, ed., The Chemical Society, London.

Schiffrin, D. J. (1971). Ch. 4 in *Electrochemistry, Vol. 2*, G. J. Hills, ed., The Chemical Society, London.

Schiffrin, D. J. (1973). Ch. 4 in *Electrochemistry, Vol. 3*, G. J. Hills, ed., The Chemical Society, London.

Schmidt, P. P. (1973a). *JCS Faraday Trans. II*, **69**, 1104.

Schmidt, P. P. (1973b). *JCS Faraday Trans. II*, **69**, 1122.

Schmidt, P. P. (1973c). *JCS Faraday Trans. II*, **69**, 1132.

Schwerer, F. C., G. P. Huffman, R. M. Fischer, and T. Nagata (1973). pp. 3151–3166 in *Proc. of the Fourth Lunar Sci. Conf., vol. 3*, Pergamon, New York.

Scott, W. J. and G. F. West (1969). *Geophysics*, **34**, 87.

Shankland, T. and H. S. Waff (1973). *EOS Trans. AGU*, **54**, 1202.

Shuey, R. T. and M. Johnston (1973). *Geophysics*, **38**, 37.

Simmons, G., D. W. Strangway, L. Bannister, R. Baker, D. Cubley, G. LaTorraca, and R. Watts (1972). pp. 258–271 in *Lunar Geophysics*, Z. Kopal and D. W. Strangway, eds., D. Reidel, Dordrecht-Holland.

Sing, K. S. W. (1973). Ch. 1 in *Colloid Science, Vol. 1*, D. H. Everett, ed., The Chemical Society, London.

Stesky, R. M. and W. F. Brace (1973). *J. Geophys. Res.*, **78**, 7614.

Strangway, D. W., W. B. Chapman, G. R. Olhoeft, and J. Carnes (1972). *Earth Planet. Sci. Letters*, **16**, 275.

Stratton, J. A. (1941). *Electromagnetic Theory*, McGraw–Hill, New York.

Suggett, A. (1972). Ch. 4 in *Dielectric and Related Molecular Processes, Vol. 1*, M. Davies, ed., The Chemical Society, London.

Van Voorhis, G. D., P. H. Nelson, and T. L. Drake (1973). *Geophysics*, **38**, 49.

Von Ebert, G. and G. Langhammer (1961). *Kolloid Z.*, **174**, 5.

Von Hippel, A. R. (1954). *Dielectric Materials and Applications*, MIT Press, Cambridge.

Waff, H. S. (1973). *EOS Trans. AGU*, **54**, 1208.

Ward, S. H. (1967). pp. 10–196 in *Mining Geophysics, Vol. 2*, Society of Exploration Geophysicists, Tulsa, Oklahoma.

Ward, S. H. and D. C. Fraser (1967). pp. 197–223 in *Mining Geophysics, Vol. 2*, Society of Exploration Geophysicists, Tulsa, Oklahoma.

Electromagnetic Propagation Characteristics of Rocks

T. J. Katsube and L. S. Collett
Geological Survey of Canada,
601, *Booth Street,*
Ottawa, Ontario, Canada K1A OE8

1. INTRODUCTION

The electrical properties of rocks are being studied at the Geological Survey of Canada (G.S.C.) to assist the development of new and improvement of conventional electrical and EM sounding methods. Recent studies contributing to the understanding of the electrical and EM wave propagation mechanisms in the ground include those of Katsube and Collett, 1971, 1972, 1973, 1973a; Strangway *et al.*, 1972; Saint-Amant and Strangway, 1970; Fuller and Ward, 1970; Campbell and Ulrichs, 1969; Parkhomenho, 1967; Keller, 1966, 1967; Scott *et al.*, 1967; Howell and Licastro, 1961; Keller and Licastro, 1959; Marhsall and Madden, 1959; von Hippel, 1954; and Archie, 1952. The frequency range used at the G.S.C. is from 10^{-2} to 2×10^8 Hz at present, and preparations are being made to extend the frequency range up to 10^9 Hz. One of the most important applications of this work is to the development of radar sounding techniques.

Recent studies have indicated how and at what frequencies the electrical parameters (dielectric constant, resistivity and dissipation factor) of rocks are affected by factors such as: grain boundaries, moisture, porosity, mineral composition, conductive mineral content, and surface and electrode polarization effects (Katsube and Collett, 1971, 1972 and 1973). The theoretical relations between the electrical parameters and the EM wave propagation parameters are well known. This paper discusses recent work on the electrical properties and propagation characteristics of rocks.

2. ELECTRICAL AND EM PROPAGATION PARAMETERS

The theory of the electrical parameters of moist rocks is described in the paper by Katsube and Collett (1972), and ASTM (1968), Collett and Katsube

(1973) and von Hippel (1954) are also valuable references. The most important parameter is the absolute value of the complex resistivity (ρ^*).

$$\rho^* = \rho' + j\rho''$$

$$|\rho^*| = \sqrt{[(\rho')^2 + (\rho'')^2]}$$

where ρ' and ρ'' are the real and imaginary resistivity. The parallel resistivity (ρ_p) is the reciprocal of the real conductivity (σ'):

$$\rho_p = \frac{1}{\sigma'} \text{ and } \sigma^* = \sigma' + j\sigma'', \sigma^* = 1/\rho^*$$

where σ' and $\sigma =$ are the complex and imaginary conductivity. The dielectric constant (κ') is derived from the real relative permittivity in the following equations:

$$\kappa^* = \kappa' - j\kappa''$$

$$\sigma^* = j\omega\kappa^*\varepsilon_0$$

where κ^*, κ'', ε_0 and ω are the complex relative permittivity, relative imaginary permittivity, permittivity of air or vacuum and angular frequency. The dissipation factor (D) is determined by:

$$D = 1/\omega\kappa'\varepsilon_0\rho_p$$

The critical frequency (f_{cr}) is the frequency when $D = 1$.

The relevant theory can be found in Jordan and Bulman (1968), von Hippel (1954), or many other textbooks on EM theory.

The electric field strength at a distance of χ from the source is expressed by:

$$E(\chi) = E_0 \exp(-\gamma\chi)$$

where E_0 is the field strength at the source, and γ is the complex propagation coefficient:

$$\gamma = \sqrt{[(j\omega\mu)(1/\rho_p + j\omega\kappa'\varepsilon_0)]}$$

and is often expressed by:

$$\gamma = \alpha + j\beta$$

where α is the attenuation coefficient and β is the phase-shift factor. μ is the magnetic permeability. The propagation velocity (v) and wavelength (λ) are determined by:

$$v = \frac{\omega}{\beta} \text{ and } \lambda = \frac{v}{f}$$

where f is the frequency.

The intrinsic impedance (η) and depth of penetration (δ) are determined by:

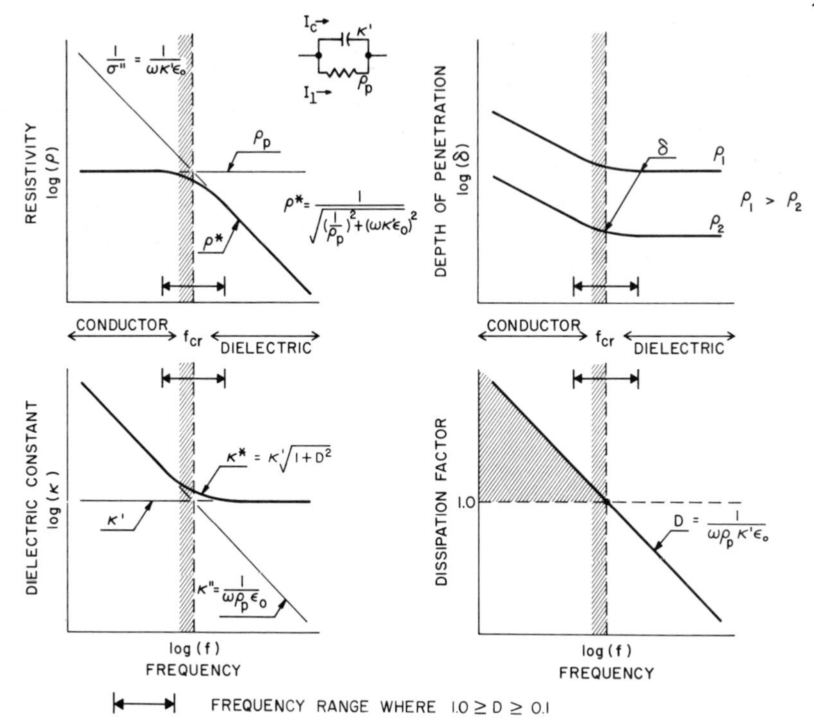

Figure 1. Frequency characteristics of electrical parameters for a parallel *RC* circuit, ρ^*: complex resistivity, ρ_p: parallel resistivity, σ'': imaginary conductivity, κ^*: complex relative permittivity, κ': dielectric constant, κ'': imaginary relative permittivity, ε_0: permittivity of air or vacuum, δ: depth of penetration, D: dissipation factor, f: frequency, ω: angular frequency.

$$\eta = \sqrt{\left(\frac{\mu}{\kappa'\varepsilon_0}\right)}$$

$$\delta = 1/\alpha$$

The reflection coefficient (R) is determined from

$$R = \frac{\eta_0 - \eta_G}{\eta_0 + \eta_G}$$

where η_0 and η_G are the intrinsic impedance of air and the ground.

Frequency characteristics of ρ^*, ρ_p, κ^*, κ', D and δ for a parallel *RC* equivalent circuit are shown in Figure 1.

3. INSTRUMENTATION AND SAMPLE PREPARATION

Rock samples are cut into discs of 1 inch in diameter and 5 millimetres in thickness. For moist measurements they are soaked in tap water, and for dry measurements they are dried in vacuum and measured in dry nitrogen atmosphere. Further details on sample preparation can be found in Katsube and Collett (1972 and 1973), and Collett and Katsube (1973).

An impedance measuring system is used for measurement of moist rocks in the frequency range from 10^{-2} to 10^5 Hz, and a capacitance bridge system for dry rocks in the frequency range from 10^2 to 10^5 Hz. For frequencies between 10^5 to 2×10^8 Hz, two *Q*- Meter systems are used for measurement of both dry and moist rocks. Further details of the measuring techniques are described in Katsube and Collett (1972), and the measuring equipment is described in Collett and Katsube (1973).

4. FREQUENCY SPECTRUM OF ELECTRICAL PARAMETERS

4.1 Dry Rocks

Dielectric constant (κ) dissipation factor (D) and parallel resistivity (ρ_p)

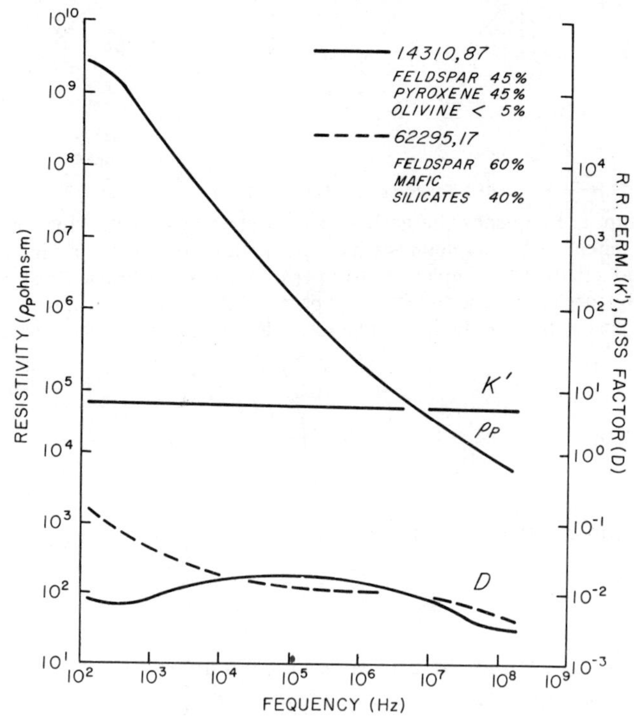

Figure 2. Typical frequency spectrum of electrical parameters: dielectric constant (κ'), parallel resistivity (ρ_p) and dissipation factor (D), for a dry rock (Lunar samples No. 14310, 87 and 62295, 17). (Reproduced with permission from T. J. Katsube and L. S. Collett, *Proc. Fourth Lunar Sci. Conference, Suppl.* 4, *Geochimica et Cosmochimica Acta, Vol. 3*, 3111–3131, 1973, Pergamon Press Inc.)

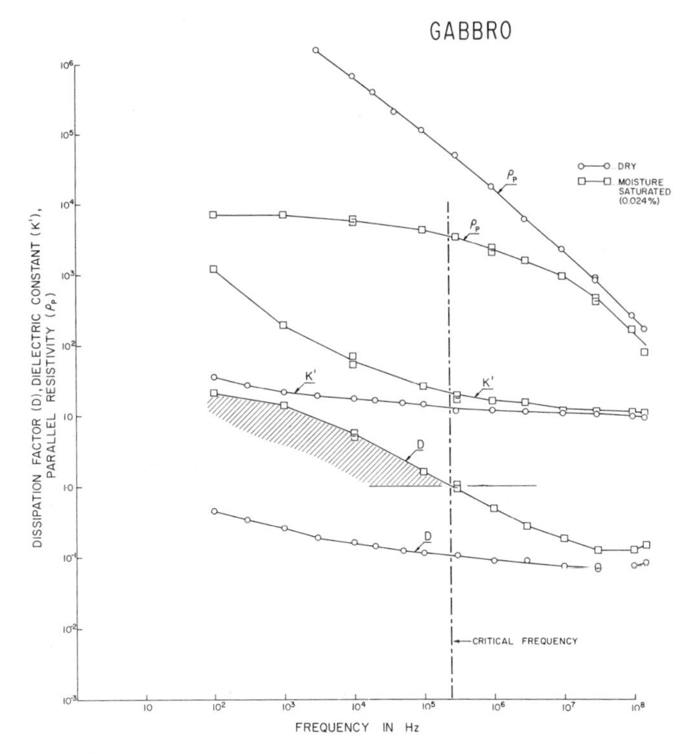

Figure 3. Typical frequency spectum for electrical parameters (κ', ρ_p and D) of a dry and moist rock (terrestrial gabbro).

are the parameters usually used to express the electrical characteristics of dry rocks. The frequency characteristics of these parameters have been generalized in the paper by Katsube and Collett (1973), and typical example is shown in Figure 2. It can be seen in this figure that as the frequency increases ρ_p decreases, κ' is generally constant, and D either decreases or shows a maximum in the 10^4 to 10^7 Hz range. Similar trends can be seen in Figure 3 for a dry terrestrial gabbro.

4.2 Moist Rocks

When the rocks are moist, with decrease in frequency, ρ_p increases but levels off from below the critical frequency (f_{cr}), κ' increases, and D rises well above 1, as can be seen in Figure 3. Complex resistivity (ρ^*) is the optimum parameter for expressing the electrical characteristics of a moist rock (Katsube and Collett, 1972). As shown in Figure 4, with increase in frequency, ρ^* is more or less constant until the frequency reaches f_{cr}, and above f_{cr} it decreases linearly. This is due to ρ^* being mainly dependent upon the ohmic characteristics ($|\rho^*| \simeq \rho_p$) of the rock below f_{cr}, and mainly dependent upon the dielectric

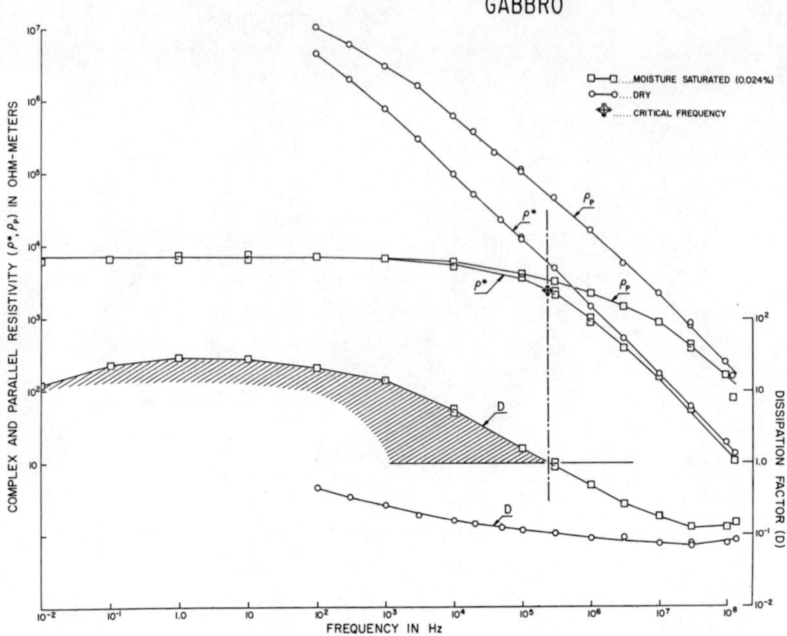

Figure 4. Typical frequency spectrum for electrical parameters: complex resistivity (ρ^*), parallel resistivity (ρ_p) and dissipation factor (D), of a dry and moist rock (terrestrial gabbro).

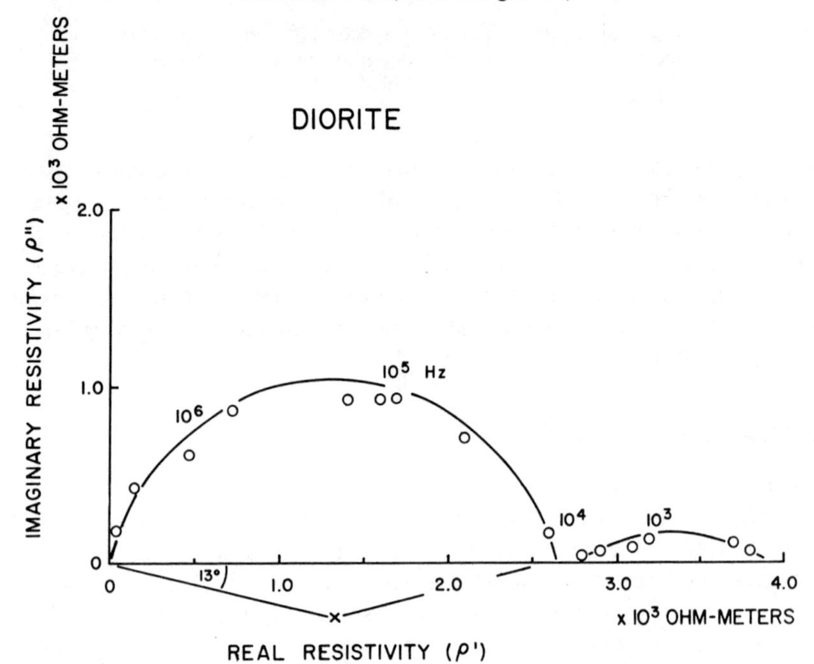

Figure 5. Cole–Cole diagram in the complex resistivity plane for a moist diorite sample.

characteristics of the rock ($\rho^* \simeq 1/\omega\kappa'\varepsilon_0$) above f_{cr}. D of the moist sample in Figure 4 shows a maximum in the frequency range of 0.1 to 10^2 Hz. The decrease of D with frequency below that at which D shows a maximum may be caused by polarization in the rock pores or at the electrodes. A Cole–Cole diagram in the complex resistivity plane for a moist diorite sample is shown in Figure 5. The frequency characteristics of ρ^*, ρ_p and D of this sample are similar to those of the gabbro sample in Figure 4, and have been published in Katsube and Collett (1973). The arc on the left hand side in Figure 5 indicates parallel RC circuit effect of ρ_p and κ' in the vicinity of f_{cr}. The arc to its right indicates the polarization effect due to either pore surfaces or the electrodes. No sulphides or other conductive minerals have been observed in these rocks.

5. ELECTRICAL MECHANISM OF THE CONDUCTIVE REGION

Models of the various rock pores are shown in Figure 6. Intergranular and vugular pores (sedimentary and volcanic rocks) partially consist of joint pores, as can be seen in this figure. It has been indicated by Katsube and Collett (1973) that the joint pores (crystalline rocks) have an ohmic or conductive

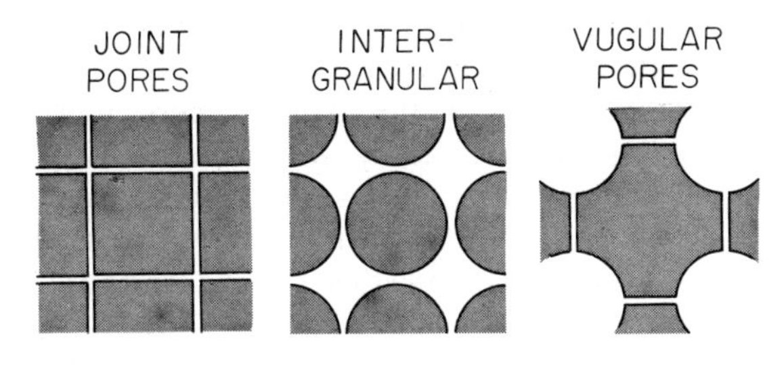

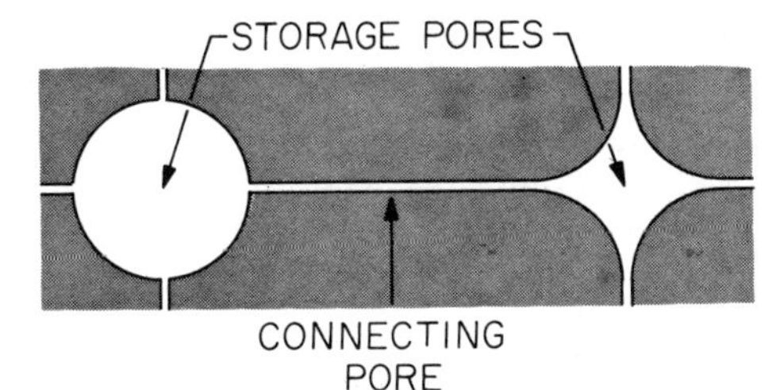

Figure 6. Model of various pores in rocks. Joint, intergranular and vugular pores are usually seen in crystalline, sedimentary and volcanic rocks, respectively. (Reproduced with permission from T. J. Katsube and L. S. Collett, *Proc. Fourth Lunar Sci. Conference, Suppl.* 4, *Geochimica et Cosmochimica Acta, Vol.* 3, 3111–3131, 1973, Pergamon Press Inc.)

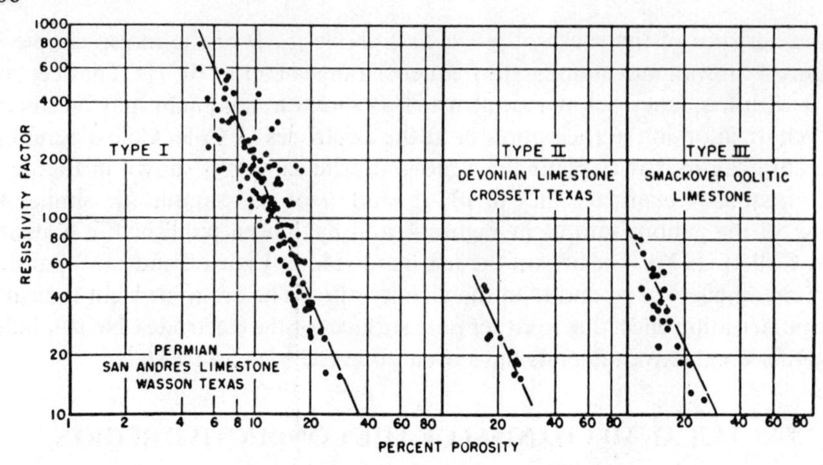

Figure 7. Resistivity *vs* porosity (Reproduced with permission from G. E. Archie, *Bull. Am. Assoc. Petrol. Geologists*, 36, 278–298 (1952).)

effect on the electrical characteristics of rocks which appears at frequencies below the critical frequency (f_{cr}). The factors included in the joint pore effect are thickness and porosity of joint pores, resistivity of liquid in the pores, configuration of pores, and the surface or membrane polarization (Marshall and Madden, 1959). Joint pore porosity and resistivity of the liquid in the pores are the main contributors to conductivity in moist rocks at these frequencies. The large difference between ρ_p of the dry and moist rocks at frequencies below f_{cr} (see Figure 3 or 4), is caused by these effects. The type of effect that the surface or membrane polarization has on the electrical characteristics of rocks is not very clear at present. However, from the mechanism of surface or membrane polarization, it is expected that this factor has a major effect on κ' and D at the frequencies below f_{cr}. An example of measurements on DC resistivity versus porosity by Archie (1952) is shown in Figure 7. This DC resistivity would be equivalent to ρ_p below f_{cr}, if it can be assumed that no surface or membrane polarization exists. The porosity in Figure 7 is not equivalent to the joint pore porosity. However, there perhaps exists a certain relationship between the joint pore porosity and the total porosity, and that relation can be expected to differ between different rock textures because of fluid resistivity. This is, perhaps, the reason that DC resistivity versus porosity relationship differs between different types of rocks, as shown in Figure 7.

6. ELECTRICAL MECHANISM OF THE DIELECTRIC REGION

At frequencies above f_{cr} the effect of the dielectric constant (κ') predominates and ρ_p has little effect on the general electrical characteristics of the rock. The porosity of the storage pores, the dielectric constant of the liquid in the pores, the existence of conductive minerals, chemical composition of the individual mineral grains are the factors which determine the electrical characteristics of the rock (Katsube and Collett, 1972, 1973). The joint pore porosity has little effect, unless the total porosity is very small (below 5 per cent).

In sedimentary and volcanic rocks, the porosity of the storage pores is the major contributor to the total porosity. Therefore, if the pores are saturated with water, the moisture content greatly affects the bulk dielectric constant (κ') since the dielectric constant of water is 80. The existence of conductive minerals within the mineral grains will cause large effects on the bulk dielectric constant of the rock. However, since the actual content of conductive minerals rarely exceed 10 per cent, their effect on the bulk dielectric constant is small. Small percentages of conductive minerals have a large effect on the dissipation factor (D) and parallel resistivity (ρ_p) in this region (Figure 8).

Mineral composition (Katsube and Collett, 1971, 1972) affects the dielectric characteristics of the mineral grains. However, this effect is small compared to that of water in the pores. The mineral composition has a large effect on the parallel resistivity of the mineral grains. At frequencies close to f_{cr}, water in the pores has a dominant effect on the bulk ρ_p of the rock. However, as

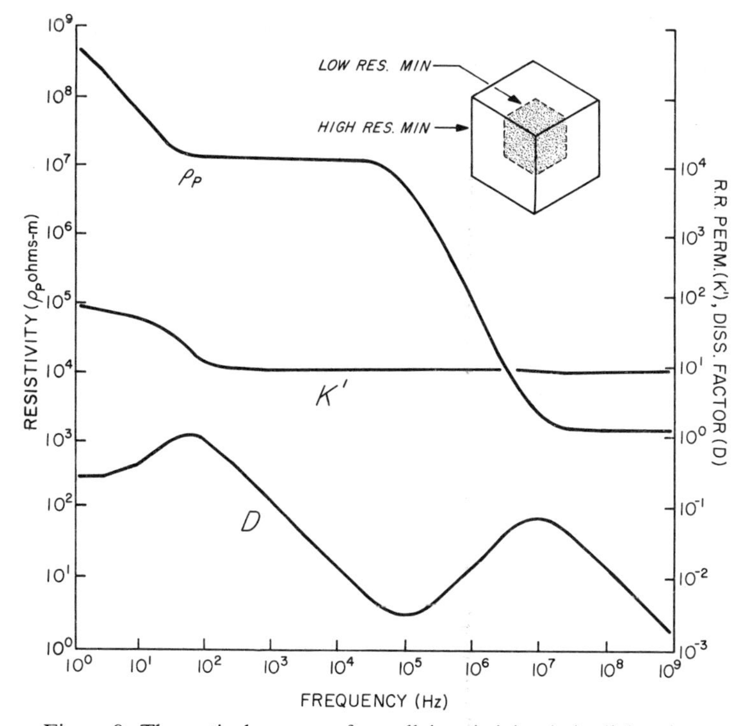

Figure 8. Theoretical curves of parallel resistivity (ρ_p), dielectric constant (κ') and dissipation factor (D) of a dry rock that contains 10 per cent of low resistance minerals (10^2 Ω-m) and 90 per cent of insulating minerals (10^7 Ω-m). (Reproduced with permission from T. J. Katsube and L. S. Collett, *Proc. Fourth Lunar Sci. Conference, Suppl.* 4, *Geochimica et Cosmochimica Acta, Vol. 3*, 3111–3131, 1973, Pergamon Press Inc.)

the frequency increases, the effect of ρ_p of the mineral grains increases (Katsube and Collett, 1972; Katsube, 1974).

Though such consideration has been given to the bulk ρ_p of the rock above f_{cr}, it should be stressed that its effect on the general characteristics are minor.

7. GENERAL ELECTRICAL MECHANISM

The critical frequency (f_{cr}) is, perhaps, the most important parameter when considering the electrical mechanism of moist rocks. At frequencies below f_{cr}, ρ_p is the most effective parameter, and at frequencies above f_{cr}, κ is the most effective parameter which determines the general electrical characteristics of rocks. ρ_p is determined mainly by joint pore porosity and the resistivity of the water in the pores. κ' is determined mainly by the storage pore porosity (assuming they are saturated with water), conductive mineral content and the dielectric properties of the mineral grains.

Low porosity rocks, such as crystalline rocks, contain mainly joint pores. High porosity rocks, such as sandstones or volcanic rocks with vugular pores, contain mainly storage pores. A good relationship exists between porosity (75 per cent) and dielectric constant when the rock is saturated with water (see Figure 9). ρ_p is dependent both upon the joint pore porosity and resistivity of the water in the pores, and not only on one of the two parameters. Therefore,

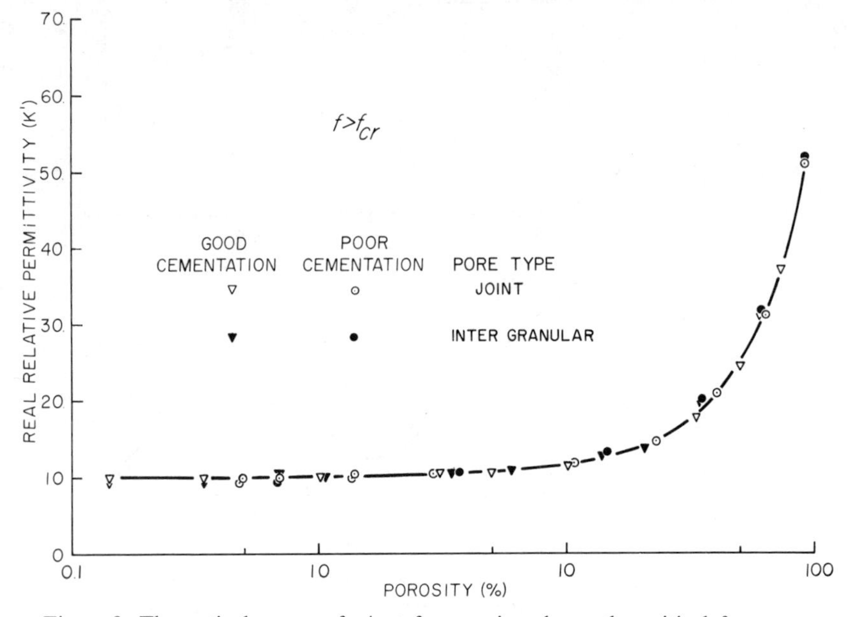

Figure 9. Theoretical curve of κ' at frequencies above the critical frequency for rocks with various types of pores and state of cementation. (Reproduced with permission from T. J. Katsube and L. S. Collett, *Proc. Fourth Lunar Sci. Conference, Suppl.* 4, *Geochimica et Cosmochimica Acta, Vol. 3*, 3111–3131, 1973, Pergamon Press Inc.)

the dielectric constant is perhaps the only parameter which can provide any unique information about porosity and water content.

The presence of conductive minerals and the chemical composition of the mineral grains affect the bulk dielectric constant of the rock. However, the effect of these two factors can be masked by the effect of water if the moisture content or porosity of the rock is large. In crystalline rocks which usually have low porosity, the dielectric constant may be a guide the chemical composition and conductive mineral content. Parallel resistivity of the mineral grains seem to have a large effect on the bulk parallel resistivity (ρ_p) or dissipation factor of moist rocks (Figure 4 and Figure 10, also Katsube, 1974). And ρ_p of the mineral grains have a close relation to their chemical composition and conductive mineral content (Katsube and Collett, 1972, 1973). Therefore, bulk dissipation factor or parallel resistivity measurements may provide additional information on these two factors.

Dissipation factor measurements at frequencies below f_{cr} may be useful. However, the problems in the conductive region are very complex, and much study is still required.

The wave length is usually much larger than the dimensions of the mineral

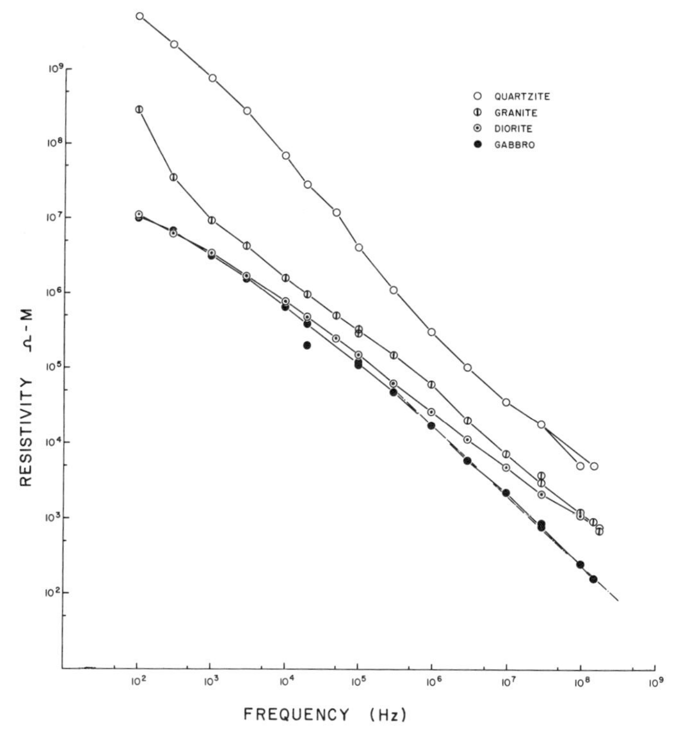

Figure 10. Parallel resistivity (ρ_p) of dry terrestrial rocks. (Reproduced with permission from T. J. Katsube, *Geol. Survey Canada*, Paper 74–1B, 1974).

grains, but when it reaches a value comparable with the diameter of the mineral grains, reflection and scattering must be taken into consideration. It is not yet known whether these effects can be used in geophysics, especially in the natural field and VLF cases.

Unterberger and Hluchanek (1973) has shown results of the window effect (low attenuation) in salt domes. This effect makes large depth of penetration possible, due to an anomalous decrease in dissipation factor values in the 10^8 to 10^9 Hz frequency range. Hoekstra and Delaney (1974) has indicated the existence of resonances due to water in soils, which considerably increases the dissipation factor values in the 10^8 to 10^{10} Hz frequency range. These window and resonance effects are due to a mechanism different from the mechanisms discussed above and are of much interest.

8. FREQUENCY SPECTRUM OF PROPAGATION PARAMETERS

The propagation characteristics of two moist crystalline rocks and one partially moist soil sample (mainly sand) has been investigated. Length of EM waves (λ) in these samples have been calculated from dielectric constant and dissipation factor measurements over the frequency range from 10^3 to 2×10^8 Hz (see Table 1). The results are shown in Figure 11. From this figure it can be seen that the wave length increases linearly with decrease in frequency until the critical frequency is reached (10^5 to 10^6 Hz in this case), and below f_{cr}

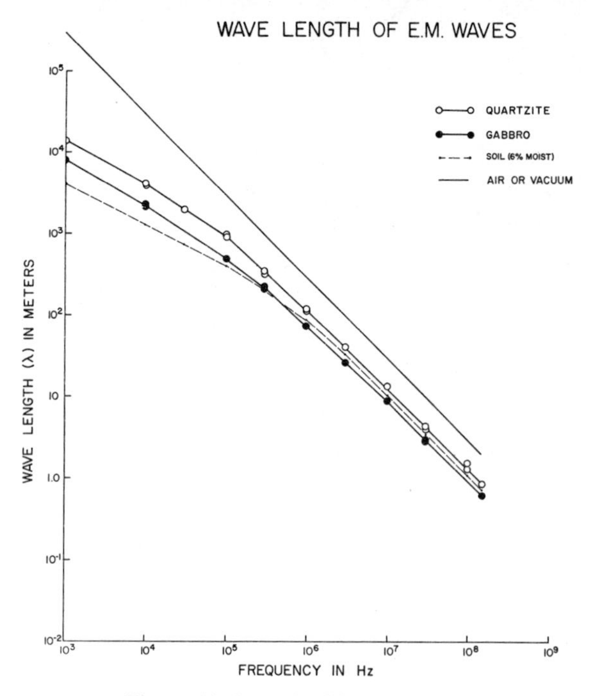

Figure 11. Length of EM waves.

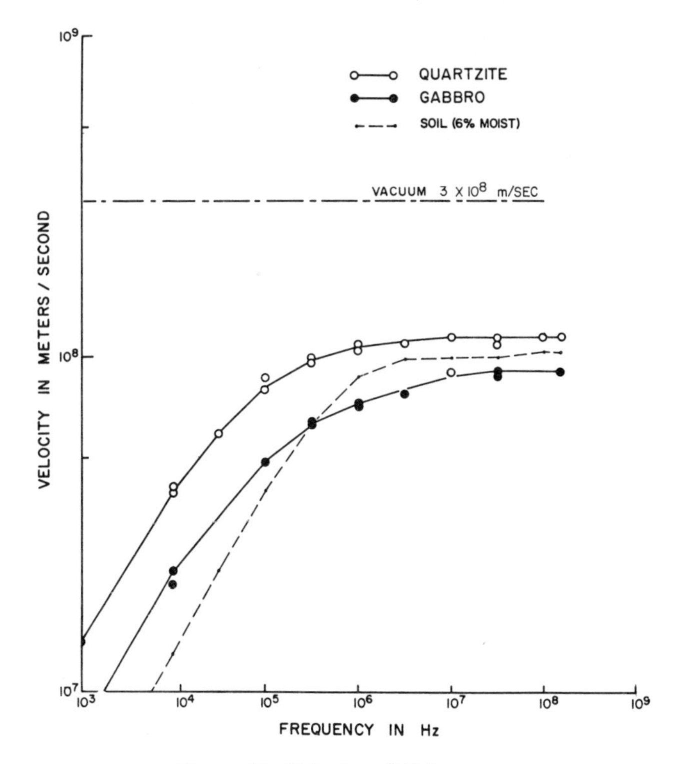

Figure 12. Velocity of EM waves.

the rate of increase declines. λ is mainly dependent upon the dielectric constant, but below f_{cr} the effect of the dissipation factor enters.

Propagation velocity (v) of EM waves for the same samples is shown in Figure 12. It can be seen that v is more or less independent of frequency above f_{cr}, but decreases rapidly with decrease in frequency below f_{cr}, v is mainly dependent on κ' above f_{cr}, but is affected by the dissipation factor to a great extent below f_{cr}.

Depth of penetration (δ) for EM waves with frequency for the same set of samples is shown in Figure 13. δ decreases with frequency and is mainly dependent upon the dissipation factor.

The intrinsic impedance versus frequency for the quartzite, gabbro and soil samples is shown in Figure 14. The general trend of these curves are very similar to those of the velocity curves (Figure 12). Curves for the reflection coefficient (R) versus frequency for these samples are shown in Figure 15. R is more or less independent of frequency at frequencies above 10 times f_{cr}. Below that frequency R increases with decrease in frequency and from below 10^4 Hz, R more or less saturates at values which are in the order of 0.85–0'98.

For gabbro the dielectric constant is larger and the parallel resistivity is

Table 1. Electrical parameters of moist quartzite, gabbro and soil samples. The measurements of κ' at frequencies below the critical frequency (in brackets) are less reliable than those above that frequency.

Parameter	Frequency (Hz)	Quartzite	Gabbro	Soil
MOISTURE CONTENT		0.10%[a]	0.024%[a]	6.0%[b]
Real Relative Permittivity (κ')	10^3	(18.0)	(190)	(550)
	10^4	9.0	(53)	(43)
	10^5	7.5	26	(14)
	10^6	6.5	16	8.0
	10^7	5.4	11	8.1
	10^8	5.6	4.2	7.8
Dissipation Factor (D)	10^3	52.0	14	18
	10^4	11.0	5.6	24
	10^5	1.4	1.6	22
	10^6	0.43	0.5	1.6
	10^7	0.096	0.18	0.34
	10^8	0.024	0.13	0.13
Parallel Resistivity (ρ_p)	10^3	1.9×10^4	6.8×10^3	1.8×10^3
	10^4	1.8×10^4	6.1×10^3	1.7×10^3
	10^5	1.5×10^4	4.2×10^3	1.7×10^3
	10^6	7.3×10^3	2.2×10^3	1.4×10^3
	10^7	3.5×10^3	9.0×10^2	6.2×10^2
	10^8	1.3×10^3	3.3×10^2	1.8×10^2

[a]Saturated
[b]Partially saturated

smaller than those of quartzite. Therefore, the propagation parameter curves for these two samples are similar for all of the five parameters over the entire frequency range, and no crossover of the curves is seen. However, the soil sample has a dielectric constant and parallel resistivity which are both smaller than those of the two crystalline samples. Therefore, at frequencies where the effect of the dissipation factor is dominant, the propagation parameters are in the same or in opposite order of their resistivity values. However, at frequencies where both the dielectric constant and dissipation factor affect the propagation parameters, the values of the propagation parameters are not in the same or opposite order or either the dielectric constant or dissipation factor. Crossovers for all the curves of the soil sample can be seen in Figure 11 to 15 above.

9. CONCLUSION

Recent studies on the electrical properties of rocks help to clarify a number of problems. Below the critical frequency the parallel resistivity is the most important parameter for characterizing the electrical properties. It is mainly

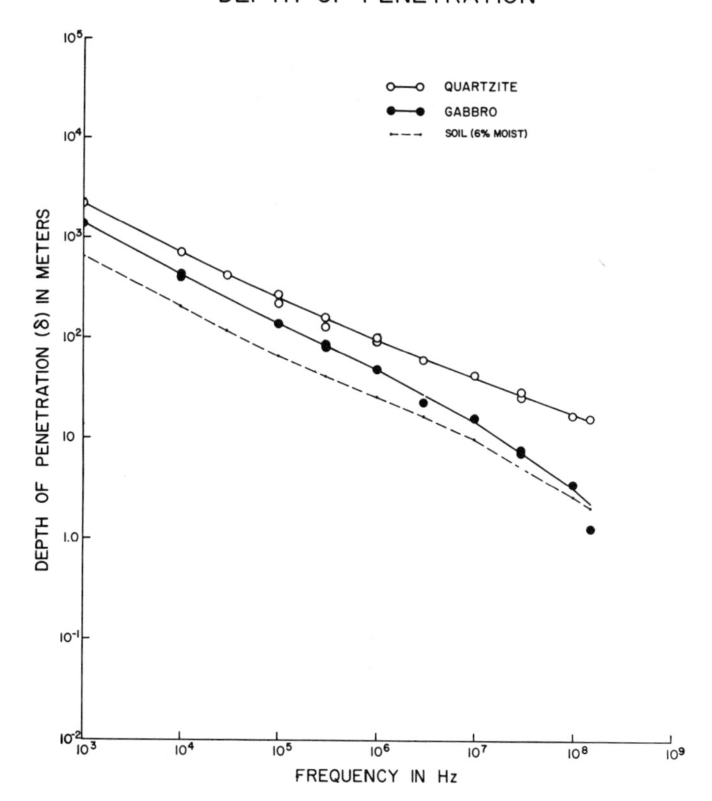

Figure 13. Depth of penetration for EM waves.

determined by the joint pore porosity and the resistivity of the water in the pores. Below the critical frequency, surface and electrode polarization effects do exist, but they are minor. Above the critical frequency, the dielectric constant is the most important parameter for characterizing the electrical properties. In moisture saturated rocks with porosities above 5 to 10 per cent, the dielectric constant is mainly dependent upon the porosity of the rock. In low porosity rocks, the dielectric constant is mainly dependent upon the conductive mineral content and mineral composition of the mineral grains. Dissipation factor and parallel resistivity measurements above the critical frequency may give additional information on the rock, however, the extent of this possibility is not clear at present. As a conclusion from this study, the dielectric constant is perhaps the parameter which can provide the most unique and reliable information on rocks. Studies on the EM wave propagation parameters have indicated that in most cases, the parallel resistivity or dissipation factor has a dominant effect at frequencies below the critical frequency. The same studies have also indicated that above the critical frequency, the dielectric constant has a major effect, and the dissipation factor has a minor effect on the propaga-

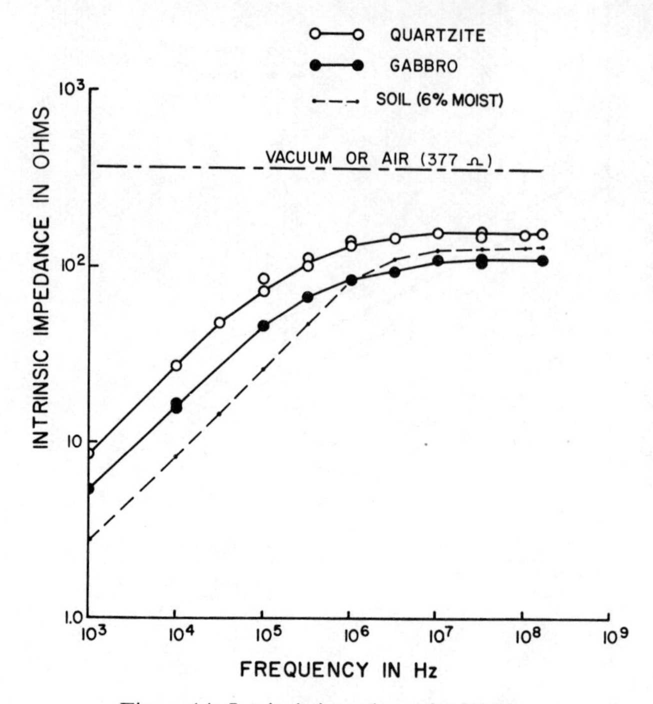

Figure 14. Intrinsic impedance for EM waves.

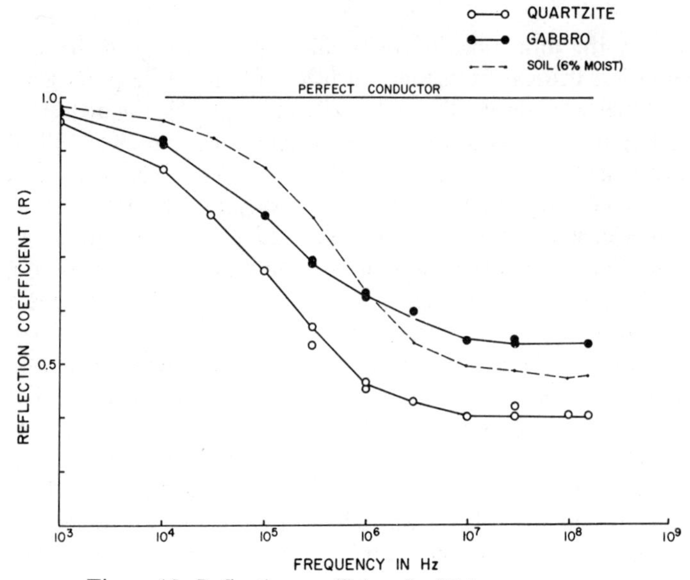

Figure 15. Reflection coefficient for EM waves.

tion characteristics of rocks and soils. Consideration of EM wavelength, window and resonance effects are important considerations for electrical and EM wave propagation studies in the future.

ACKNOWLEDGEMENTS

We are grateful to Dr. W. J. Chudobiac (Communications Research Centre, Ottawa) for his interest and discussions related to this paper. We also thank Mr. R. B. Gray (Communications Research Gentre) for his advice. We thank Mr. J. Frechette (Geological Survey of Canada) for making some of the measurements that were used in this paper. Most illustrations in this paper were drafted by Miss R. Dufour.

REFERENCES

Archie, G. E. (1952). *Bull. Am. Assoc. Petrol. Geologists*, **36**, 278–298.

ASTM (American Society for Testing Materials) (1968). *ASTM D 150–68*, 29–54.

Campbell, M. J. and Ulrichs, J. (1969). *J. Geophys. Res.*, **74**, 5867–5881.

Collett, L. S. and Katsube, T. J. (1973). *Geophysics*, **38**, 76–91.

Fuller, B. D. and Ward, S. H. (1970). *IEEE Transactions on Geoscience Electronics*, **GE-8**, 7–18.

Hoekstra, P. and Delaney, A. (1974). *J. Geophys. Res.*, **79**, 1699–1708.

Howell, B. F. and Licastro, P. H. (1961). *Am. Mineral.*, **46**, 269–288.

Jordan, E. C. and Balmain, K. (1968). *Electromagnetic Waves and Radiating Systems.* Prentice-Hall, 2nd edition.

Katsube, T. J. and Collett, L. S. (1971). *Proc. Second Lunar Sci. Conf., Geochim. Cosmochim. Acta, Suppl. 2*, MIT Press, Vol. 3, 2367–2379.

Katsube, T. J. and Collett, L. S. (1972). Presented at 42nd Annual International Meeting of S.E.G., Anaheim, California, Nov., 26–30.

Katsube, T. J. and Collett, L. S. (1973). *Proc. of Fourth Lunar Sci. Conference, Supplement 4, Geochimica et Cosmochimica Acta*, Pergamon Press, Vol. 3, 3111–3131.

Katsube, T. J. and Collett, L. S. (1973a). *Geophysics*, **38**, 92–104.

Katsube, T. J. (1974) *Geol. Surv. Can.*, Paper 74–1B.

Keller, G. V. and Licastro, P. H. (1959). *Bull. USGS*, 1052–H, 257–285.

Keller, G. V. (1966). In Handbook of Physical Constants, *Geol. Soc. Amer., Mem. 97*, (ed. S. P. Clark, Jr.) 533–571.

Keller, G. V. (1967). In *Electrical Properties of Rocks:* New York, Plenum Press, 263–308.

Marshall, D. J. and Madden, T. R. (1959). *Geophysics*, **24**, 790–816.

Parkhomenko, E. I. (1967). *Electrical Properties of Rocks.* Plenum Press.

Saint-Amant, M. and Strangway, D. W. (1970). *Geophysics*, **35**, 624–645.

Scott, J. H., Carroll, R. D. and Cunningham, D. R. (1967). *J. Geophys. Res.*, **72**, 5101–5115.

Strangway, D. W., Olhoeft, G. R., Chapman, W. B. and Carnes, J. (1972). *Earth Planet. Sci. Lett.*, **16**, 275–281.

Unterberger, R. R. and Hluchanek, J. A. (1973). Presented at the 43rd International Meeting of S.E.G., Mexico City, Mexico on Oct. 21–25.

von Hippel, A. R. (1954). *Dielectric Material and Applications.* Wiley.

Correlations Between Texture of Rock Samples and Induced Polarization Reaction

C. Gateau, J. M. Prevosteau, F. X. Vaillant

Bureau de Recherches Géologiques et Minières
B.P. 6009—45018 Orléans Cédex France

INTRODUCTION

The aim of our work is to obtain information concerning mineralization by means of electrical measurements and induced polarization. In particular we want to find out if it is possible to obtain information concerning the nature of conductive minerals, their relative compactness or dissemination, and their grain size. We used about fifty rock samples representing the different types of porphyry copper, and sandstones with galena. Specific methods were developed especially for the study, both for electrical measurements and for textural determination. In correlating the electrical and textural relations, a particular effort was made to achieve an initial approximation of the relationship between the conductive mineral content and phase variation or frequency effect. These results are still partial as the study is incomplete. This paper is therefore no more than a report of results obtained so far.

ELECTRICAL MEASUREMENTS

Preparation of Samples

The rock samples used are 2 cm^2 sections 2 cm long, cast in 'Araldite'. They are saturated, in a vacuum, with a solution of 0.003 N potassium chloride. The measuring cell is composed of four electrodes, two being copper current input electrodes, and two output impolarizable electrodes for measurement of potential.

Measurement Apparatus

Two techniques of pulse and frequency induced polarization measurements have been developed. These use a similar methodology based on a method of

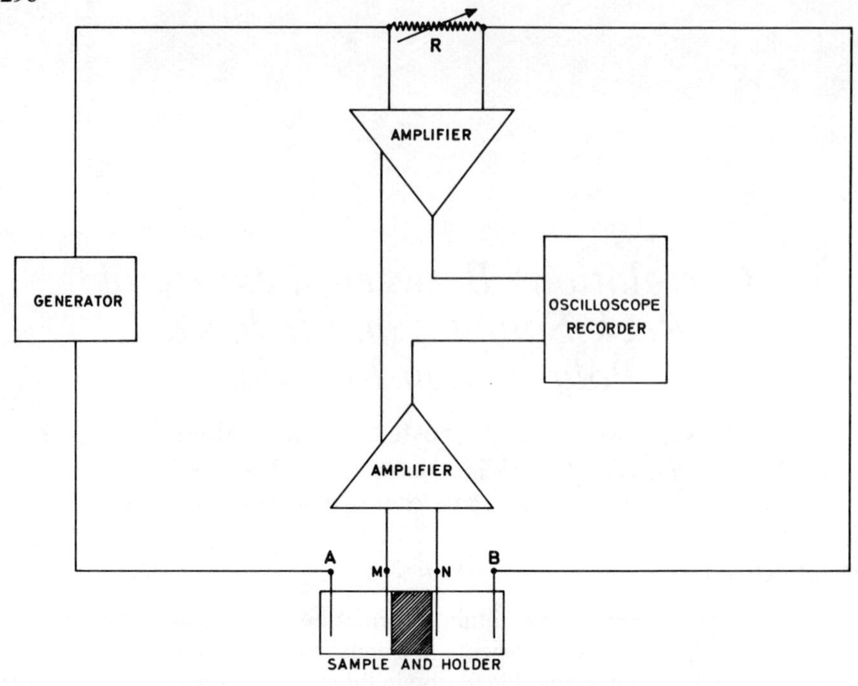

Figure 1. Induced polarization measurement system.

opposition. With the frequency technique (Figure 1) a I_{AB} sinusoidal current is sent through the sample. At the terminals of a resistance R a voltage is taken and opposed to the voltage V_{MN} taken at the terminals of the sample. After amplification, these signals are applied to the two inputs $X - Y$ of an oscilloscope. Strong variations of phase between the current I_{AB} and the voltage V_{MN} are measured by means of a Lissajous figure. For weak phase variations (< 200 mrd), the difference between the two signals is applied at the input Y, and one of the two signals is applied at the input X. By adjustment of the resistance R, when the signals have the same amplitude, an ellipse is obtained and the phase variation is then determined from the axial ratio of the ellipse.

The modulus and the phase of complex impedance for each rock sample is then measured and the variations between these two parameters (modulus and phase) is measured for different frequencies from 0.001 Hz to 100 Hz. With the pulse technique, the pulse voltage appearing at the ends of the sample, for both the input and cut off of the continuous polarization current, is recorded on an apparatus with high entry impedance. In general, taking into account the accuracy of the apparatus and the speed of measurement, we usually preferred the frequency technique.

Study of the Linearity of the Phenomena according to the Density of the Current. Passage from Pulse Method to Frequency Method

These studies were carried out on some natural samples of porphyry copper

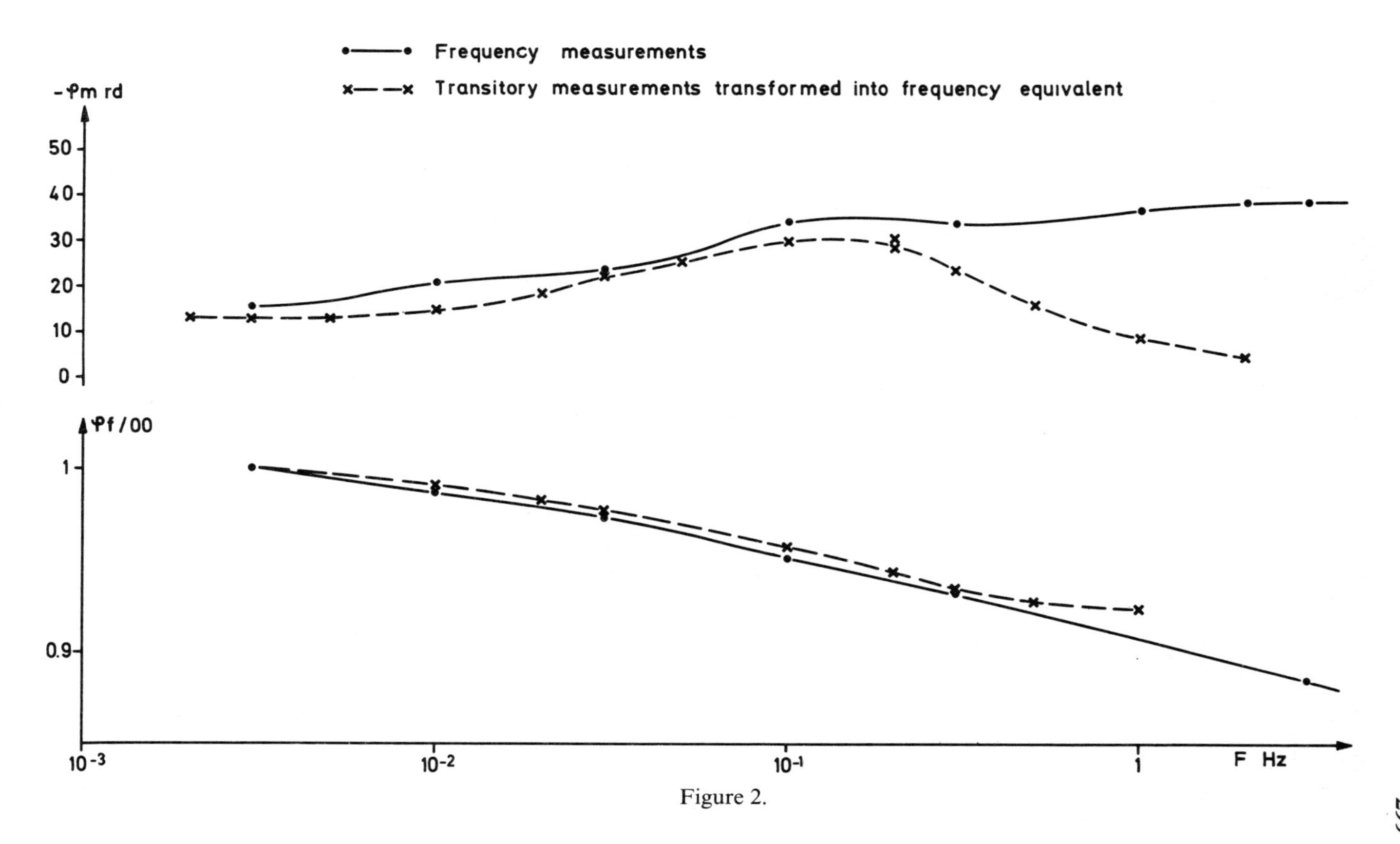

Frequency measurements
Transitory measurements transformed into frequency equivalent
-φm rd
50
40
30
20
10
0
φf/00
1
0.9
10-3
10-2
10-1
1
F Hz

Figure 2.

and sandstones with galena, having a disseminated mineralization (1 to 2 per cent). With current densities varying from 0.5 to 50 $\mu A/cm^2$ we observed only linear phenomena.

The equivalence of the two methods of study (pulse and frequency) could be shown approximately. Indeed, on Figure 2, it is possible to see that the modulus and complex impedance phase of the sample, calculated by the Fourier transformation of the pulse curve, are more or less equivalent to those observed directly by the frequency method for frequencies between 0.001 Hz and 0.3 Hz. The disparity of the two curves above 0.3 Hz is certainly the fault of the pulse measurement as our apparatus only records the pulse signal from 1 second after the cut off of the polarization current.

GEOMETRIC MEASUREMENTS

The measurement of the geometric parameters relative to the minerals and pores within the rock is made by analysis of images (QTM 720—IMANCO). After the determination of its electrical behaviour, each sample is cut into thin slices which are then mounted as polished sections. These sections, which are supposed representative of the initial rock, are used for textural analyses.

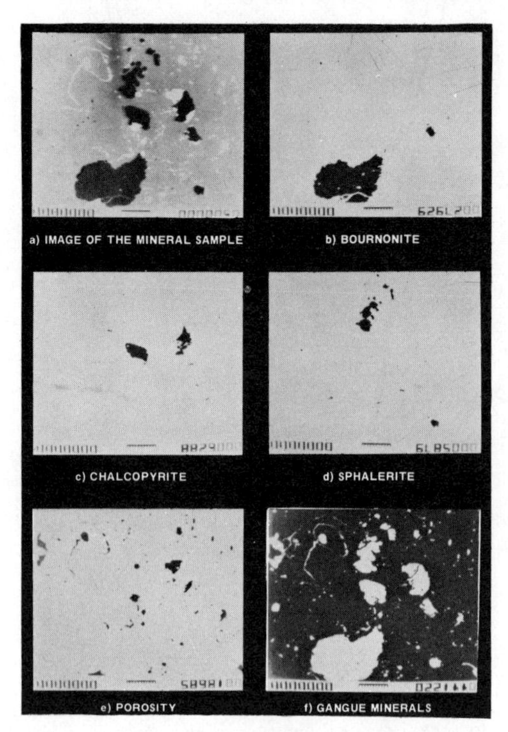

Figure 3. Detection and measurement of five different phases in a mineral sample (The full scale-bar represents 100 μm).

Figure 3 illustrates the detection and measurement of the rock: the polished section is placed under a microscope—the optical image thus obtained is filmed by a television camera and projected onto a control screen (normal image). At each point in this image, an analysis spot delivers an electric signal proportional to the luminous flux perceived. The comparison of this signal with the adjustable reference thresholds makes it possible to separate the field of analysis into zones of different reflecting power. The minerals present on the polished section can thus be isolated and differentiated.

The basic measurements concern the surfaces, perimeters, projections and counting of pre-selected figures. It is also possible to classify these figures according to the value of their chord, their area, their perimeter, a form factor, or even of the association of several of these parameters. For this study, we have measured the volume concentration of conductive minerals as well as the porosity by means of the surface function. The grain size distribution of the minerals present in the rock is measured by selection of the grains according to their specific area. Figure 9 below gives an example of granulometric curves.

CORRELATIONS BETWEEN ELECTRICAL AND TEXTURAL PARAMETERS

The electrical parameters recorded for each sample are the modulus and phase variations of complex impedance for a range of frequencies from 0.003 Hz to 100 Hz. The textural parameters that we try to correlate with the above data are initially the content and the grain size distribution of sulphide minerals, as well as the porosity.

The intensity of the induced polarization phenomena in the frequency technique is usually expressed either by a phase variation, or by a 'frequency effect' (F.E.), which is no more than a relative variation of resistivity between two

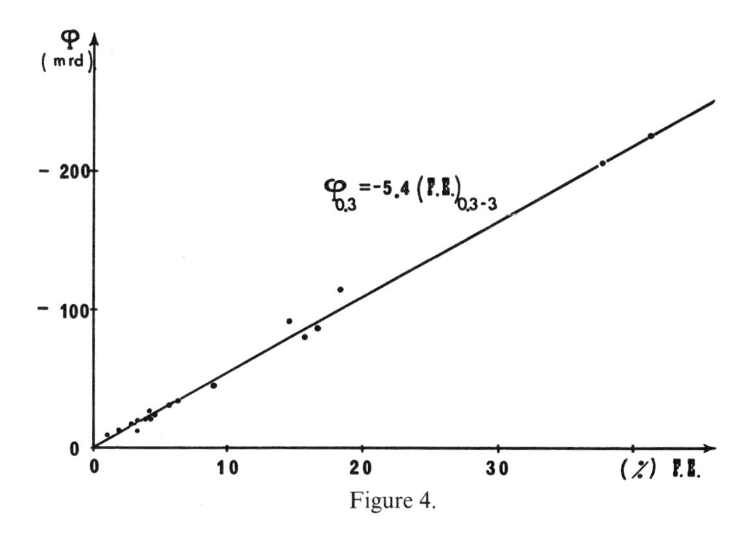

Figure 4.

frequencies. Indeed, it is possible to show experimentally the equivalence of these two parameters. Figure 4 represents the phase variation at 0.3 Hz in relation to the frequency effect measured between 0.3 and 3 Hz; the straight line obtained corresponds to a slope of 5.4 mrd/per cent for the analyses carried out.

It is interesting to note that the proportionality coefficient between phase variation and frequency effect is independent of the scale of the investigation (Figure 5). The phase spectrum (variation of phase differences with the frequency) usually presents a maximum, whose position, in the case of a massive mineralization, is directly connected with the nature of the mineral (Figure 6). However,

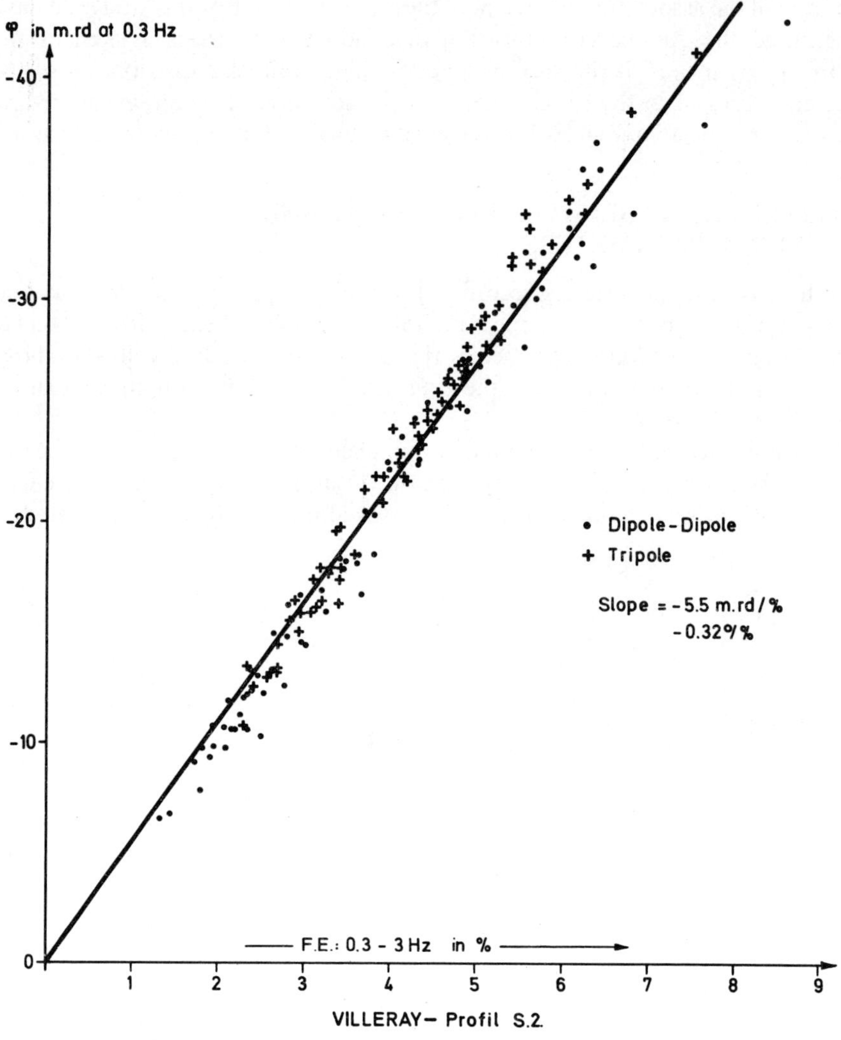

Figure 5. Field measurements (compare Villeray, Figure S.2)

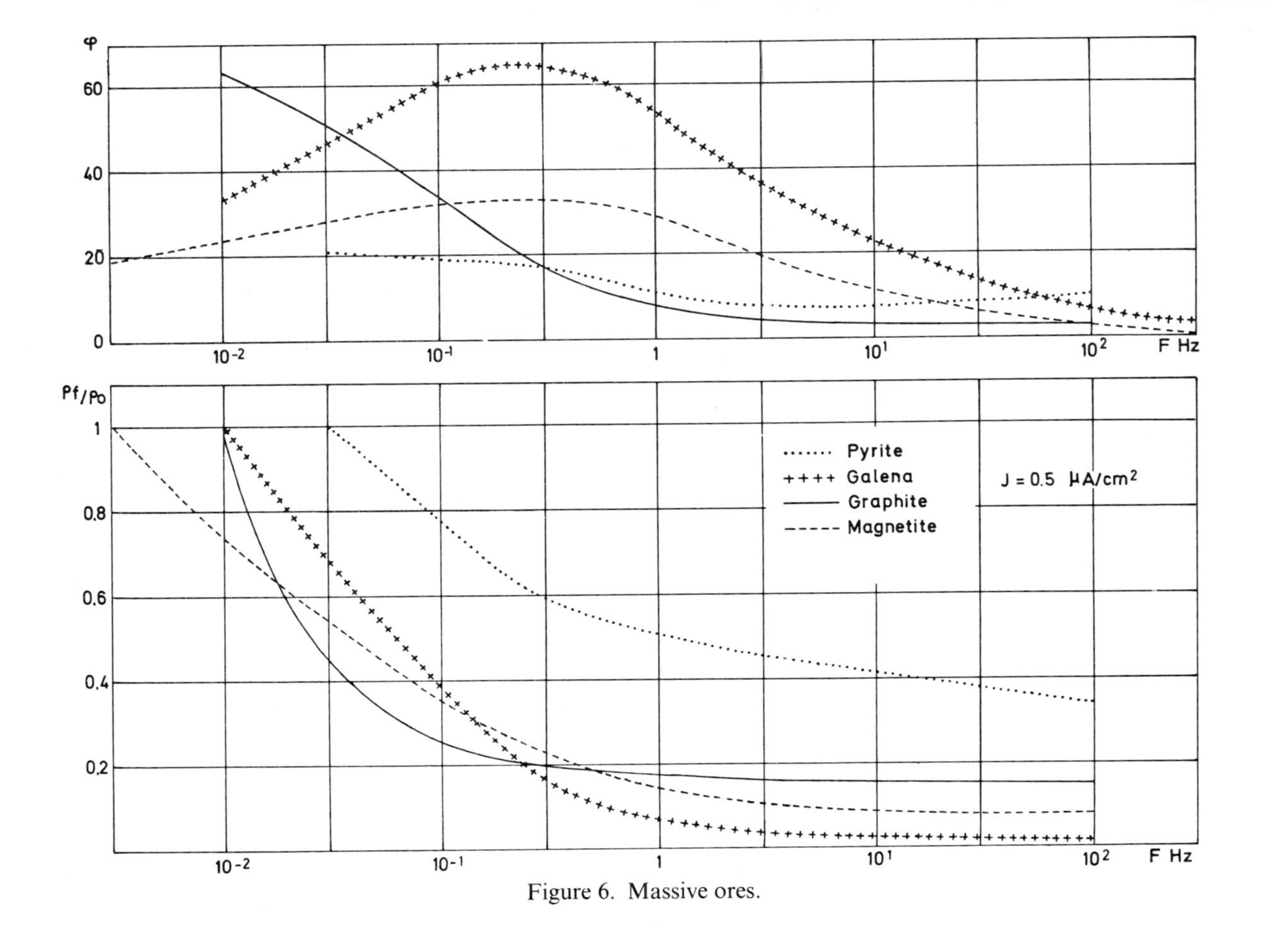

Figure 6. Massive ores.

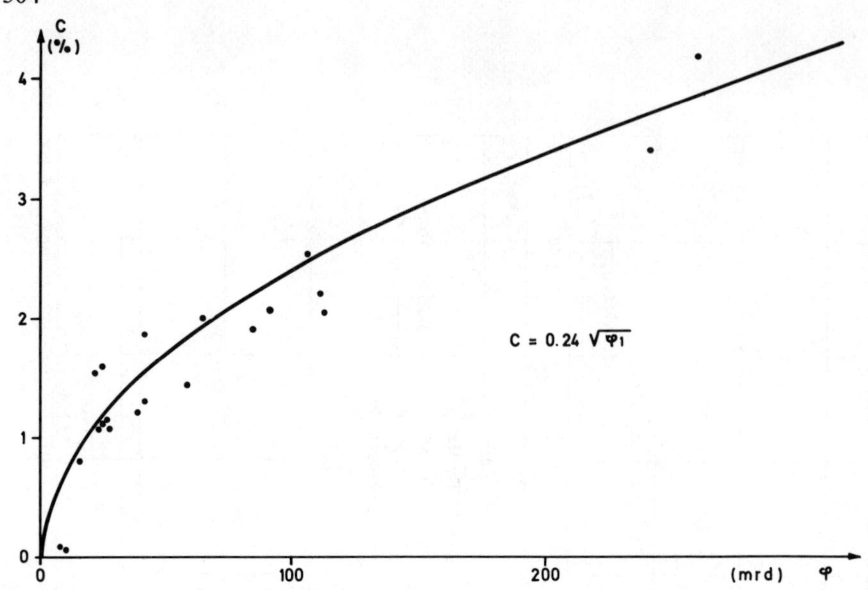

Figure 7. Correlation of phase shift with the content of the conductive minerals.

when the grade and the grain size distribution are identical, it is not possible to differentiate the low-grade disseminated sulphides from this maximum.

A preliminary relationship between electrical and textural parameters has been revealed by considering the phase variations ϕ in relation with the content C of the conductive minerals. Figure 7 shows, for example, that for a frequency of 1 Hz, we obtain the preliminary approximation:

$$C = 0.24\sqrt{\phi}$$

For the rocks studied up to now, (sulphides less than 5 per cent), the nature of the conductive mineral does not appear to affect the relationship.

The frequency at which the maximum phase variation takes place seems to be directly connected with the grade and the grain size distribution of the ore: Figure 8. It may be seen that this maximum decreases with the grade. It would seem that it is situated at a frequency which becomes lower as the size of the grains increases and the grade becomes higher. These observations, however, have not yet been translated into the form of a mathematical equation. The major obstacle to this is the difficulty in attributing an accurate granulometry to minerals which, in the natural samples, are usually very varied in grain size distribution (Figure 9).

In order to get a better knowledge of the textural parameters connected with the one (nature, grade, grain size) we prepared artificial samples. At the moment, we are using a cement containing calibrated grains of galena or pyrite, furthermore, we have limited ourselves to samples having as near as possible the same porosity spectrum as the porphyry coppers studied. In this way, by separating the contributions of the different textural parameters, we hope

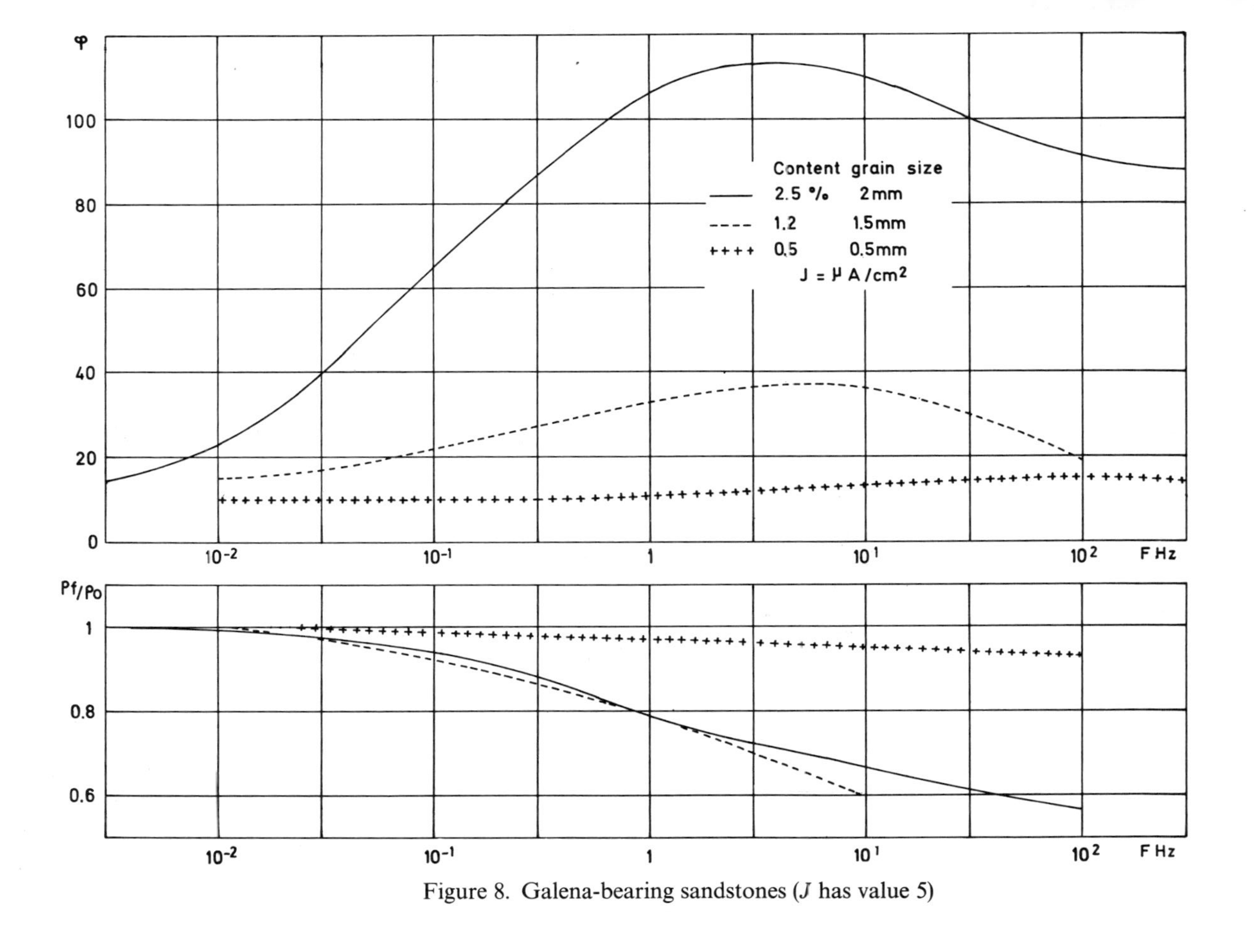

Figure 8. Galena-bearing sandstones (J has value 5)

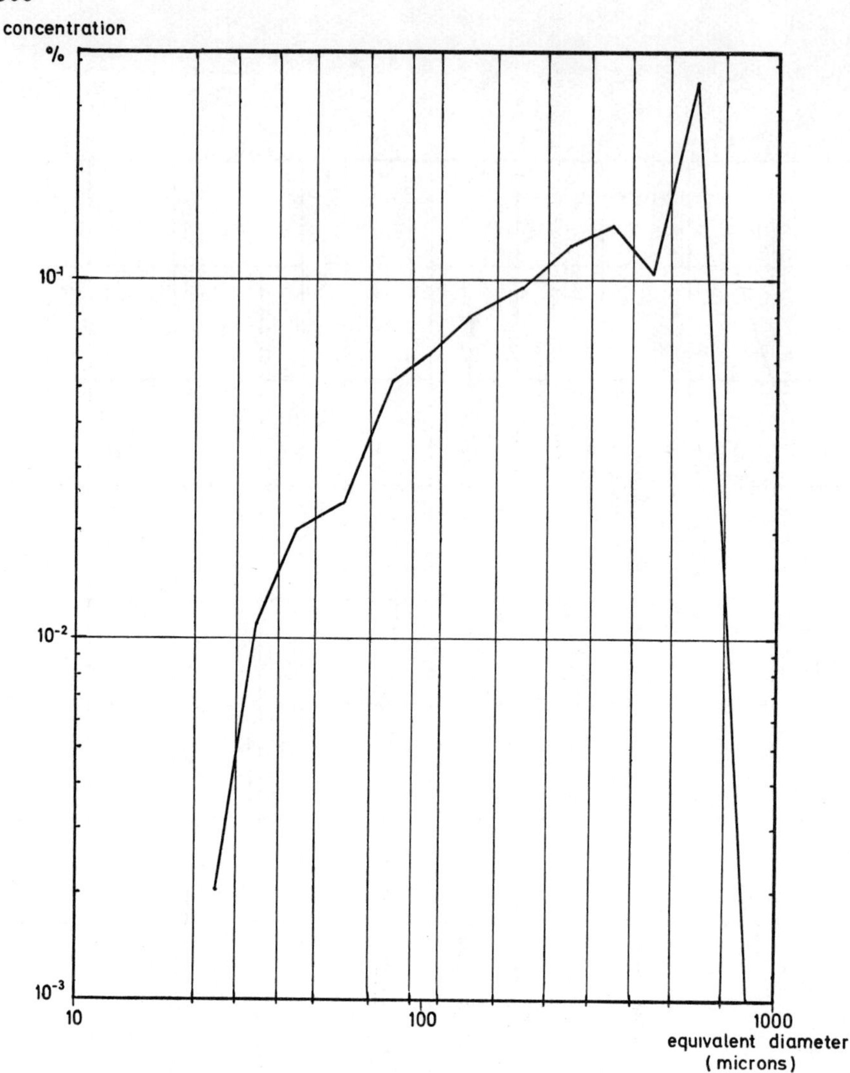

Figure 9. Size distribution by area of each particle in conductive minerals.

to be able to correlate them more easily with the electrical data. Unfortunately, the interpretation of these results can only be justified by analyses of a very large number of preparations.

To complete these studies of induced polarization phenomena in the laboratory, work programmes have been planned in order to plot the decrease curves in a pulse operation from results obtained by the frequency method (the reverse procedure is equally possible). In this way it will also be possible to analyse the results in terms of time constants: the coefficients and the time constants obtained by decomposition of the decrease curves with a sum of exponentials (3 to 5) are contrasted with the textural parameters. Bertin has obtained interesting results using an analogous, although simplified (2 exponentials), measuring technique.

CONCLUSION

The results acquired up to now have mostly been obtained by the frequency technique. They reveal, whatever the scale of the investigation, the existence of a linear relationship between phase variation and frequency effect. For the low grade disseminated sulphides a simple relationship is demonstrated between the grade and the phase variation. However, in these same conditions, it has not yet been possible to correlate the electrical behaviour of the rock with the nature of the sulphide.

It appears that a preliminary study conducted on artificial samples is necessary in order to determine the influence of the granulometry of conductive minerals on the electrical properties. The results of such a study should later be compared with those obtained from measurements carried out on natural samples.

ACKNOWLEDGEMENTS

This study has been carried out thanks to the financial aid of the DGRST within the sphere of agreement no. 72.7.0729 'induced Polarization'.

*Radioactive Heat Production; A Physical Property Determined by the Chemistry of Rocks**

L. Rybach

Institute of Geophysics, ETH Zurich, Switzerland

1. INTRODUCTION

The driving force behind all geodynamic processes is the internal heat of the earth. By determining the surface heat flow

$$q(0) = K\frac{\partial T}{\partial z} \quad (1)$$

where $q(0)$ is the heat flow, usually given in HFU ('Heat Flow') units (1 HFU = 10^{-6} cal/cm^2 sec), K the thermal conductivity (cal/cm sec °C) and $\partial T/\partial z$ the vertical temperature gradient (°C/cm), it is possible to gain insight into the earth's thermal regime. The correlation between heat flow and radioactivity of local rocks was mentioned as early as 1909 (Joly).

Radioactive heat production is a thermal property of rocks independent of in-situ temperature and pressure and is usually given in HGU ('Heat Generation') units (1 HGU = 10^{-13} cal/cm^3 sec). It contributes major amounts to the surface heat flow. On the average, surface heat flow is about 1.5 HFU and the heat flow from the mantle—in continental areas—is around 0.5 HFU; the difference is due to radioactive heat production in crustal rocks.

The key for interpreting continental heat flow patterns was given by Birch *et al.* (1968) and Roy *et al.* (1968) who demonstrated that in a given geologic province the *surface heat flow* $q(0)$ *varies linearly with surface heat production* $A(0)$:

$$q(0) = q_{red} + HA(0) \quad (2)$$

where q_{red} is the reduced heat flow characteristic of the province, and H is a characteristic depth which is of the order of 10 km. Due to the low heat production of deep crustal rocks q_{red} differs only slightly from the mantle heat flow.

*Contribution no. 93.

By applying (2) heat flow measurements can be reduced in order to relate individual $q(0)$ values to characteristic heat flow provinces.

2. RADIOACTIVE HEAT PRODUCTION

Radioactive decay converts mass into energy. The energy released appears at first as the kinetic energy of the particles and nuclei involved in the decay process (emitted α- and β-particles, recoil nuclei), further as the energy of the accompanying γ- and/or X-rays and of the neutrino. It has been demonstrated by Hurley and Fairbairn (1953) that—except the amount carried away by the neutrino—all energy is *converted to heat* in the immediate vicinity of the decaying nucleus.

All naturally radioactive isotopes generate heat to a certain extent. It can be shown, however, that the only significant contributions arise from the decay series of U^{238}, U^{235} and Th^{232}, and from the isotope K^{40}.

For heat flow interpretation the constants of Birch (1954) are used by most investigators. These constants give the amount of heat generated per gram U, Th, and K per unit time. These constants have been revised, based on the newest decay schemes, energies and mass differences (for details see Rybach, 1973). The new figures are listed in Table 1. The revised constants are significantly lower than the Birch values.

Table 1. Revised Heat Production Constants (cal/g.y)

Decay Series/Element	A (After Birch, 1954)	A (This work)
^{238}U	0.71	0.692
^{235}U	4.3	4.34
U (Natural)	0.73	0.718
^{232}Th	0.20	0.193
K (Natural)	27×10^{-6}	26.2×10^{-6}

The radioactive heat production of a given rock can be evaluated from the uranium, thorium and potassium concentrations c_U, c_{Th} and c_K, using the constants of Table 1:

$$A\,[HGU] = 0.317\,\rho\,(0.718c_U + 0.193c_{Th} + 0.262c_K) \tag{4}$$

where ρ is the density of the rock $[g/cm^3]$; c_U and c_{Th} are in weight ppm, c_K in weight per cent.

3. GEOCHEMICAL CONTROL OF HEAT PRODUCTION

It is evident from (4) that heat production in a given rock is governed by the amounts of uranium, thorium, and potassium present, which vary greatly with rock type but exhibit certain regularities due to the similar geochemical behaviour of U, Th, and K during the processes which determine the distribution of the natural radioelements (magmatic differentiation, sedimentation, metamorphism).

3.1 Igneous Rocks

During magmatic differentiation U and Th as well as K tend to successive enrichment in progressively more acidic phases. The similar behaviour of U and Th is due to their close correspondence in ionic size, high valence, electronegativity and coordination number with respect to oxygen.

Whereas abundances vary over several orders of magnitude according to the degree of differentiation in igneous rocks (cf. Table 2), the mean Th/U and K/U ratios remain fairly constant around 4 and 1×10^4, respectively. Due to this fact and also to the different heat production constants of U, Th and K (Table 1), U and Th contribute in most igneous rocks a comparable amount whereas K contributes an always smaller amount to total heat production, in proportions of about 40 : 45 : 15.

In rocks with nearly similar bulk chemistry, heat production is higher in the 'wet' facies (containing biotite and hornblende) than in the 'dry' phase (contain-

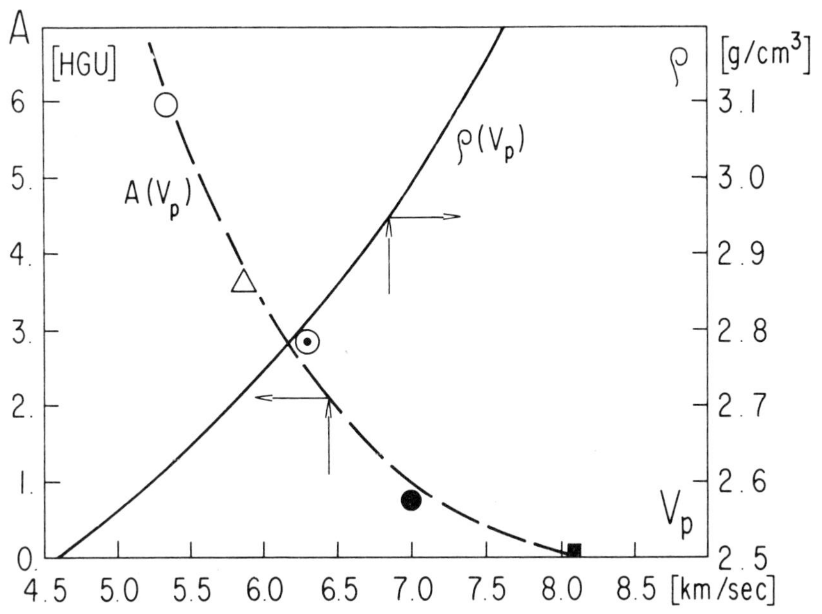

Figure 1. The relation between heat production A and seismic velocity v_p for the range of crustal rocks (dotted line) based on data from Table 2. ○: granite, △ : granodiorite, ⊙: diorite, ● : gabbro, ■ : peridotite. The solid line is based on the relation between v_p and ρ after Nafe and Drake (1959).

Table 2. Uranium, thorium and potassium abundances (from Clark *et al.*, 1966 and Wakita *et al.*, 1967) and radioactive heat production in igneous rocks

Rock	U ppm	Th ppm	K %	Th/U	K/U $\times 10^4$	ρ g/cm^3	A HGU	Δq[a] HFU
Granite/Rhyolite	3.9	16.0	3.6	4.1	0.9	2.67	5.85	5.9×10^{-2}
Granodiorite/Dacite	2.3	9.0	2.6	3.9	1.1	2.72	3.55	3.6×10^{-2}
Diorite, Quartzdiorite/Andesite	1.7	7.0	1.1	4.1	0.7	2.82	2.59	2.6×10^{-2}
Gabbro/Basalt	0.5	1.6	0.4	3.2	0.8	2.98	0.74	7.4×10^{-3}
Ultramafic Peridotite	0.02	0.06	0.006	3.0	0.3	3.23	0.028	2.8×10^{-4}
Ultramafic Dunite	0.003	0.01	0.0009	3.3	0.3	3.28	0.0045	4.5×10^{-5}

[a]Contribution to surface heat flow per 1 km thickness

Table 3. Uranium, thorium and potassium abundances (from Clark *et al.*, 1966; Rogers and Adams, 1969 and Heier and Billings, 1970) and radioactive heat production in sedimentary rocks

Rock	U ppm	Th ppm	K %	Th/U	K/U $\times 10^4$	ρ[a] g/cm^3	A HGU	Δq HFU
Carbonates						2.6		
Limestones	2.0	1.5	0.3	0.75	0.15		1.49	1.5×10^{-2}
Dolomites	1.0	0.8	0.7	0.80	0.7		0.87	8.7×10^{-3}
Sandstones						2.4		
Quartzites	0.6	1.8	0.9	3.0	1.5		0.78	7.8×10^{-3}
Arkoses	1.5	5.0	2.3	3.3	1.5		2.03	2.0×10^{-2}
Graywackes	2.0	7.0	1.4	3.5	0.7		2.43	2.4×10^{-2}
Shales	3.7	12.0	2.7	3.2	0.7	2.4	4.37	4.4×10^{-2}
Deep sea sediments	3.5	7.0	2.5	2.0	0.7	1.3	1.88	1.9×10^{-2}

[a]Broad average since ρ depends strongly on porosity

ing mainly pyroxene) as was clearly demonstrated by Smithson and Heier (1971) and Smithson and Decker (1973).

The densities in Table 2 have been converted to seismic velocities applying the Nafe and Drake (1959) curve. The plot of heat production A versus compressional wave velocity v_p (Figure 1) exhibits *strong correlation* in the range between 5 and 8 km/sec: A decreases with increasing v_p. Most crustal rocks fall within this range.

3.2 Sedimentary Rocks

Characteristic U, Th, and K abundances, as well as heat production values for the most common sedimentary rock types are given in Table 3. Uranium and thorium do not follow such parallel trends as in igneous rocks, mainly because U is more readily oxydized ($U^{4+} \rightarrow U^{6+}$) in aqueous solutions. Consequently, Th/U ratios show considerable variation which reflects pH–E_h conditions during sedimentation: one finds high ratios (Th/U > 6) in continental formations deposited in an oxydizing milieu but low ratios (Th/U < 2) in sediments of a reducing marine environment. There is no clear correlation between heat production and seismic velocity in sedimentary rocks.

3.3 Metamorphic Rocks

As metamorphic rocks are formed from igneous and/or sedimentary materials, the primary U, Th and K contents will be redistributed depending on the extent of the metamorphic transformation. As a general rule, the higher the metamorphic grade, the lower the heat production (Table 4).

The depletion of U and Th, caused by progressive metamorphism, is most markedly evident in rocks of the granulite facies. U^{4+} and Th^{4+} ions, which can only exist in 8-fold coordination with oxygen and, thus, can not be accommodated in the closely packed structures newly formed under the higher PT conditions, have a strong tendency towards upward migration and concentration (see for example Yermolaev and Zhidikova, 1966; Heier, 1973). Migration takes place at higher levels in the crust upon dehydration or near the base of the crust upon partial melting (migmatites). Since metamorphic mineral

Table 4. Uranium, thorium and potassium abundances (from Heier and Adams, 1965) and radioactive heat production in metamorphic rocks with granodioritic bulk chemistry

Metamorphic grade	U ppm	Th ppm	K %	Th/U	K/U × 10^4	ρ g/cm^3	A^a HGU	Δ Δq HFU
Greenschist/low Amphibolite facies	3.5	26.5	3.4	7.57	0.97	2.70	7.54	7.5×10^{-2}
High amphibolite facies	1.2	9.4	2.0	7.83	1.67	↓	2.83	2.8×10^{-2}
Low granulite facies	0.9	4.1	2.1	4.65	2.33	↓	1.75	1.8×10^{-2}
High granulite facies	0.4	0.9	2.9	2.25	7.25	2.90	1.07	1.1×10^{-2}

[a]Calculated for $\bar{\rho} = 2.75$ g/cm^3

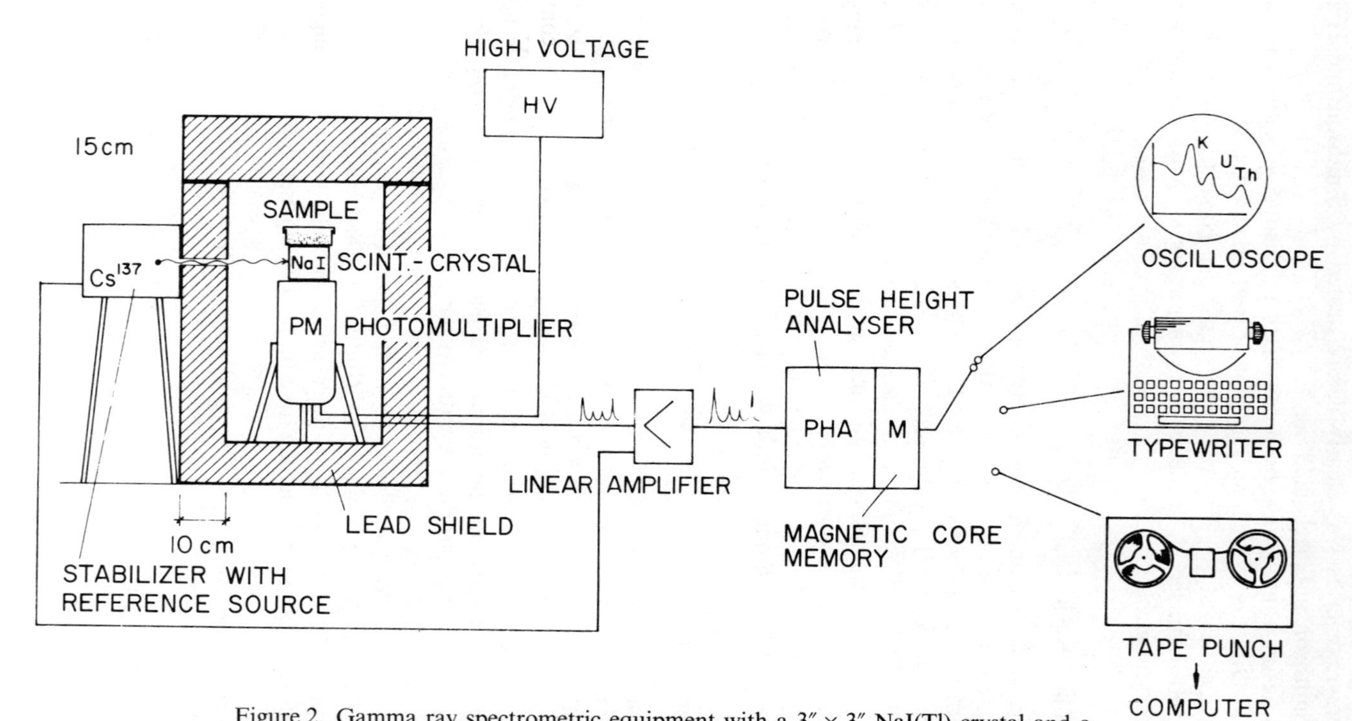

Figure 2. Gamma ray spectrometric equipment with a 3″ × 3″ NaI(Tl) crystal and a 1024-channel pulse height analyser. For details of the gain stabilizing unit, refer to Comunetti (1965).

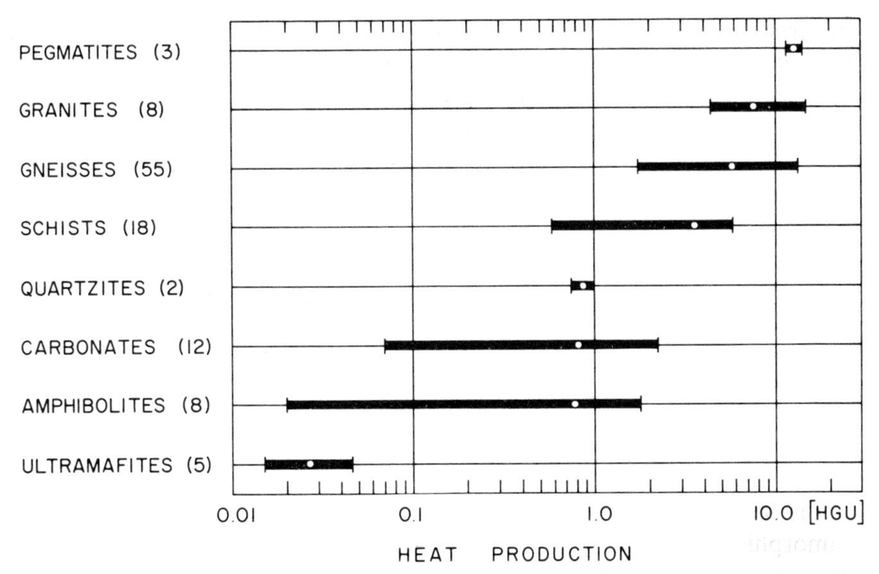

Figure 3. Variation of heat production and mean values (○) in rocks from the Swiss Alps. The number of samples per group (in brackets) corresponds roughly with the surface abundance of that rock type.

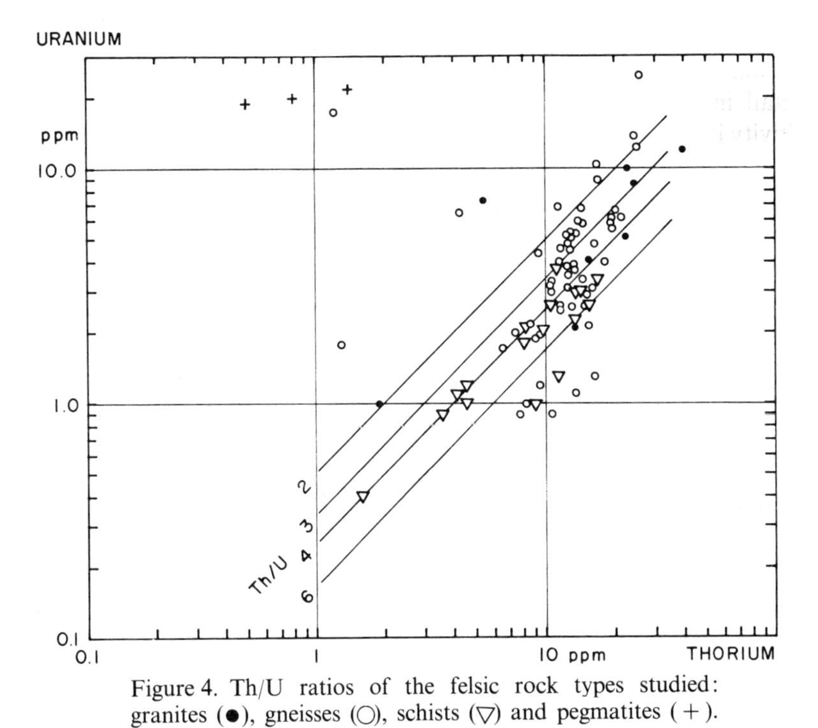

Figure 4. Th/U ratios of the felsic rock types studied: granites (●), gneisses (○), schists (▽) and pegmatites (+).

transformations (e.g. biotite → orthoclase + hypersthene + water) cause loss of volatiles, migration is accompanied by an H_2O and/or CO_2 phase.

Potassium seems to be more or less unaffected by these processes. The K/U ratio thus increases towards the granulite facies (Table 4). The density (and thus the seismic velocity) of metamorphic rocks increases with increasing metamorphic grade (Smithson, 1971); the heat production/seismic velocity correlation shows the same trend as in igneous rocks (Figure 1).

4. EXPERIMENTAL DETERMINATION OF HEAT PRODUCTION

In order to calculate heat production from (3) the uranium, thorium and potassium concentrations must be determined. Numerous analytical methods can be used for this purpose; *γ-ray spectrometry* has the advantage of *simultaneous* determination of all three radio-elements. Figure 2 shows the equipment which was used for the present study. The evaluation of the spectra is described in detail in Rybach (1971). Powered samples of about 300 g are required; sensitivity is 0.08 HGU.

In Figure 3 results of heat production determinations of 111 typical rocks from the Swiss Alps are given (for individual figures and sample descriptions see Rybach, 1973). The number of samples per group corresponds roughly to the relative abundance of that particular rock type in the Central Alps. Heat production in the ultramafic rocks was determined using atomic absorption and mass spectrometric isotope dilution as the analytical technique.

Heat production varies with rock type over several orders of magnitude. The highest values have been found in pegmatites according to the enrichment of U and Th in late, acidic derivates (see also Tilling *et al.*, 1970). Representative averages are: pegmatite, 12.7; granite, 7.6; gneiss, 5.8; metamorphic schist, 3.6; quartzite, 0.9; amphibolite, 0.8; carbonate, 0.8; ultrabasite, 0.03 HGU.

Thorium versus uranium in the more felsic rock types have been plotted in Figure 4. Most Th/U ratios lie between 2 and 6 for granites and gneisses; the Th/U ratio decreases with increasing Th content (see also Labhart and Rybach, 1971; 1974).

5. DISTRIBUTION OF RADIOACTIVE HEAT SOURCES IN THE CRUST

The vertical distribution of heat production $A(z)$ in the crust influences the geotherm $T(z)$ considerably. No constraint is posed on $A(z)$ by the linear heat flow–heat production relation $q(o) = q_{\text{red}} + HA(0)$ as long as

$$\int_0^\infty A(z)\mathrm{d}z = HA(0) \tag{5}$$

There is, however, geochemical (Lambert and Heier, 1967), thermodynamical (Turcott and Oxburgh, 1972) and direct empirical (Swanberg and Blackwell, 1973) evidence that *heat production decreases with depth.*

This is also evident from the function given in Figure 1 above, $A(v_p)$, since seismic velocity v_p increases with depth as a general rule. Lachenbruch (1968), while studying the granite batholiths of the Sierra Nevada in California, has proposed an *exponential model* for the decrease of heat production with depth:

$$A(z) = A(0) \exp(-z/H) \tag{6}$$

where H is the scale factor (logarithmic decrement) of the decrease. A systematic investigation of the radioactivity distribution in granitic units (Rybach and Labhart, 1973) confirmed this model; in most granitic bodies, radioactivity and thus heat production were found to decrease toward the center of the granite. Assuming a migration mechanism in a *temperature gradient* for distributing the radioactive elements (e.g. upward movement of U and Th with a vapour phase) we have an analogy with Fick's first law (see e.g. Girifalco, 1964). Also, by taking into account the additional driving force of the temperature gradient, we arrive at

$$j = -D\left(\frac{\partial c}{\partial z} + k\frac{\partial T}{\partial z}c\right) \tag{7}$$

where j is the diffusion current, D the diffusion constant and c the concentration (e.g. of uranium). The solution of (7) for a stationary state, $j \approx 0$ and an average, constant $\mathrm{d}T/\mathrm{d}z$ is

$$c(z) = c(0) \exp[-k(\mathrm{d}T/\mathrm{d}z)z] = c(0) \exp[-z/H] \tag{8}$$

which is formally identical with (6). The logarithmic decrement (scale factor) H of the decrease is, as it can be seen from (8), *inversely* proportional to the temperature gradient.

Deeply eroded granitic bodies exposed in the Swiss Alps, in which migration of the heat producing elements took place during cooling, represent suitable test cases: *exponential decrease* of A has been found in the Rotondo granite (Gotthard massif) and the Mont Blanc granite with a decrement of 4 km in both cases (Rybach und Labhart, 1973).

It should be noted that the mechanism described above applies to the 'mobile' uranium and thorium, i.e. to that portion of the bulk content which is not fixed in crystalline structures like zircon, allanite etc. Recent investigations on Swiss granites (Labhart und Rybach, 1974) show that up to 90 per cent of total uranium is leachable (i.e. mobile) in the rock units studied.

ACKNOWLEDGEMENT

The present study has been supported by the 'Schweizerischer Nationalfonds' (Grant No. 2.550.71).

REFERENCES

Birch, F. (1954). In· *Nuclear Geology*, Faul, H. ed., 148–174. Wiley & Sons, New York.

Birch, F., Roy, R. F. and Decker, E. R. (1968). In: *Studies of Appalachian Geology: Northern and Maritime*, Zen, E., White, W. S., Hadley, J. B. and Thompson, J. B., eds., 437–451. Interscience, New York.

Clark, S. P., Peterman, Z. E. and Heier, K. S. (1966). In: *Handbook of Physical Constants*, Clark, S. P. ed. Geological Society of America Memoir 97, Washington.

Comunetti, A. M. (1965). *Nucl. Instr. Meth.*, **37**, 125–134.

Girifalco, L. A. (1964). *Atomic Migration in Crystals*. Blaisdell Publ. Co., New York.

Heier, K. S. and Adams, J. A. S. (1965). *Geochim. Cosmochim. Acta*, **29**, 53–61.

Heier, K. S. and Billings, G. K. (1970). In: *Handbook of Geochemistry*. Wedepohl, K. H., ed. Vol. II/2, Springer–Verlag, Berlin.

Heier, K. S. (1973). *Fortschr. Miner.*, **50**, 174–187.

Hurley, P. M. and Fairbairn, H. (1953). *Bull. Geol. Soc. Amer.*, **64**, 659–673.

Joly, J. (1909). *Radioactivity and Geology*. Archibald Constable & Co., London.

Labhart, T. P. and Rybach, L. (1971). *Chem. Geol.*, **7**, 237–251.

Labhart, T. P. und Rybach, L. (1974). *Geol. Rdschau*, **63**, 135–147.

Lachenbruch, A. H. (1968). *J. Geophys. Res.*, **73**, 225–226.

Lambert, I. B. and Heier, K. S. (1967). *Geochim. Cosmochim. Acta*, **31**, 377–390.

Nafe, J. E. and Drake, C. L. (1969). In: Talwani, M., Sultton, G. A. and Worzel, J. L.: *J. Geophys. Res.*, **64**, 1545–1500.

Rogers, J. J. W. and Adams, J. A. S. (1969). In: *Handbook of Geochemistry*. Wedepohl, K. H., ed. Vol. II/1, Springer–Verlag, Berlin.

Roy, R. F., Blackwell, D. D. and Birch, F. (1968). *Earth Planet. Sci. Letters*, **5**, 1–12.

Rybach, L. (1971). In: *Modern Methods of Geochemical Analysis*. Wainerdi, R. E. and Uken, E. A., eds. 271–318. Plenum Press, New York.

Rybach, L. (1973). *Beitr. Geol. Schweiz*, Geotechn. Ser. Liefg. 51.

Rybach, L. und Labhart, T. P. (1973). *Schweiz. Min. Petr. Mittg.*, **53**, 379–384.

Smithson, S. B. (1971). *Geophysics*, **36**, 690–694.

Smithson, S. B. and Heier, K. S. (1971). *Earth Planet. Sci. Letters*, **12**, 325–326.

Smithson, S. B. and Decker, E. R. (1973). *Earth Planet. Sci. Letters*, **19**, 131–134.

Swanberg, C. A. and Blackwell, D. D. (1973). *Bull. Geol. Soc. Amer.*, **84**, 1261–1282.

Tilling, R., Gottfried, D. and Dodge, F. C. W. (1970). *Bull. Geol. Soc. Amer.*, **81**, 1447–1462.

Turcotte, D. L. and Oxburgh, E. R. (1972). *Science*, **176**, 1021–1022.

Wakita, H., Nagasawa, H., Uyeda, S. and Kuno, H. (1967). *Geochem. Journ.*, **1**, 183–198.

Yermolayev, N. P. and Zhidikova, A. P. (1966). *Geochemistry Internat.*, **3**. 716–731.

Convection in a Fluid of Variable Viscosity

F. W. Wray
Department of Geodesy and Geophysics,
University of Cambridge, Cambridge, UK

This paper deals with some preliminary experiments into two-dimensional thermally convective flow in a fluid (glycerine) with a viscosity which varies rapidly with temperature. The flow was observed by means of streak photography. While no attempt has been made to model the mantle, experiments of this type should give a better understanding of the processes which occur there.

The apparatus consisted of a nearly rectangular plastic tank (Figure 1) whose sides were 2 cm longer at the top than at the base. The tank was filled to a depth of 18.5 cm with glycerine, which was then covered by a thin layer of liquid paraffin to prevent surface absorption of water. A constant temperature metal plate was inserted into the top of the tank and place in contact with the top of the paraffin. This plate was maintained at some appropriate temperature by passing water through copper pipes, soldered to its upper surface. The temperature was controlled by a heating-proportional-to-error system. For temperatures below room temperature, the water was cooled by a refrigerating device before it was heated. The temperature variation was less than 0.4 °C during each experiment.

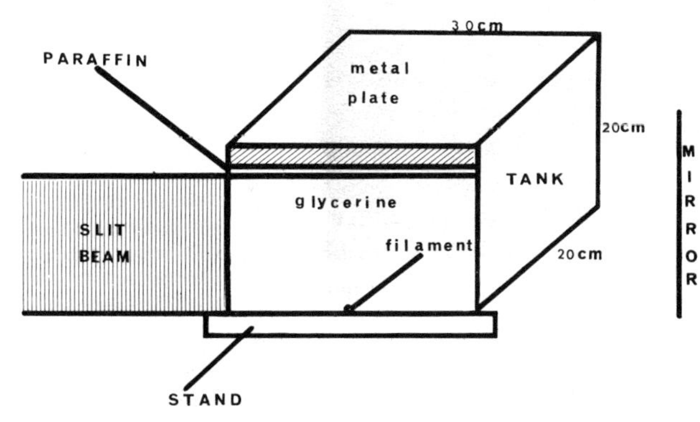

Figure 1.

The flow was driven by a filament (Figure 1) and, apart from a thin boundary layer at the front and back walls, was nearly two dimensional. The tank was lagged and levelled by placing it on a stand mounted on screws. During the course of an experiment the lagging was briefly removed and a slit beam, thin in the direction perpendicular to the front wall, passed through the tank, parallel to the front face and behind the boundary layer.

The beam was reflected back by a mirror to increase the illumination of particles in suspension in the glycerine. These scattered light out through the front face and the moving points of light were observed by streak photography. Various particles were tested and small glass beads (Ballotini, obtainable from Jencons (Scientific) Ltd., Mark Road, Hemel Hempstead, Herts.) proved to be the most satisfactory. The particles were photographed by exposing 400 ASA film for 20 seconds. The film was then specially developed to give an effective film speed of over 1000 ASA.

Two types of flow were observed. Figure 2 shows a steady state flow, which took about 16 hours to become established. The initial conditions were that the glycerine was at room temperature, the constant temperature top was set at 40 °C and the power into the filament was 40 watts.

Figure 2.

Figure 3.

Figure 3 shows a developing flow. Initially the glycerine was at room temperature, the constant temperature top was at 50 °C and the power into the filament was 75 watts. The photograph was taken after 10 hours.

From these preliminary experiments it seems that it takes about 16 hours for a steady flow to become established. A further set of experiments to observe steady flows will be undertaken. Also an attempt to determine temperatures within the fluid will be made to ascertain the effects of variable viscosity.

Core Formation in Mercury

H. G. Tolland

Department of Chemical Engineering
The University, Newcastle upon Tyne NE1 7RU, UK

The Mariner 10 results have shown that Mercury probably possesses a magnetic field of several hundred gammas at the surface, suggesting the presence of a liquid metallic core at some time in its history. The process of formation of this core may have differed considerably from that of the Earth's core, and possible differences are considered assuming that Mercury accreted as a uniform, or nearly uniform, mixture of core and mantle materials, and that one or both of these materials passed through a liquid or low viscosity phase.

Core formation in the Earth is thought to have followed accretion, and to have proceeded from an initially homogeneous mixture of 82 per cent by volume of silicates and 18 per cent of iron plus sulphides (Tolland, 1973). Infall of pieces of core material through the silicates of the proto-Earth would have occurred at a rate depending on their size and shape, and the silicate viscosity. From Tozer's (1967, 1970, 1971) rheological model the viscosity is virtually fixed at 10^{20} poise for any reasonable heat generation in a region several hundred kilometers thick. Differential rates of infall of larger and smaller pieces would have led to their meeting, and coagulating, with consequent acceleration of the fall.

Convection, which ensured that the Earth remained solid, would modify the core-forming process, but probably had little effect on its rate. According to Tozer (1967, 1970) the thermal response time for a planetary sized body is about 10^8 years, implying that the present thermal state is effectively decoupled from both the initial state, and from any early thermal event such as core formation.

In the Earth, agglomerates of core material are thought to have formed around infalling seeds (Tolland, 1973), leading to completion of core formation in about 10^8 years for a mantle viscosity of 10^{20} poise, and a seed diameter of 1 m. Core formation in all terrestrial planets except Mercury is expected to follow a similar process.

Astronomical data (Reynolds and Summers, 1969; Lyttleton, 1969; Anderson and Kovach, 1967) indicate that Mercury contains at least 30 volume per cent of core material, and the Mariner 10 results suggest a higher figure. Such a proportion would imply a very different process of core formation, because the critical concentration (percolation probability, p_c) for a binary mixture would have been approached initially, and may even have been exceeded. This critical concentration, 25 to 35 volume per cent (Tolland and Strens, 1972) is that above which each particle of core material has a finite probability of being part of an infinite cluster, and its value can be calculated using percolation theory (Broadbent and Hammersley, 1957; Dean, 1957). With much of the core material less interconnected, any liquid light material could have moved rapidly upward. More important there would have been a marked tendency for the solid light silicates to float upwards. This could have occurred only because the interconnection of the core material allowed flow. The ability of the core material to move downwards and the tendency of the light silicates to move upwards, would have contributed strongly to the core forming process. This would have become more marked with increase in the ratio of core material to silicates.

In the extreme case where p_c is greatly exceeded the situation might be envisaged as analagous to a number of cork particles initially uniformly distributed in water. The mantle material (cork) would rise through the low viscosity core material (water) with rapid release of gravitational energy.

If p_c was not exceeded initially, the some central concentration by a process similar to that envisaged for the Earth (Tolland, 1973) would be expected, until p_c was exceeded over a vast central region. For intermediate concentrations of core material both processes might have occurred simultaneously. With infall increasing the central concentration of core material, upward floating of mantle material would have become more important.

REFERENCES

Anderson, D. L. and Kovach, R. L. (1967). *Earth Planet. Sci. Lett.*, **3**, 19.
Broadbent S. R. and Hammersley J. M. (1957). *Proc. Cambridge Phil. Soc.*, **53**, 629.
Dean, P. (1963). *Proc. Cambridge Phil. Soc.*, **59**, 397.
Lyttleton, R. A. (1969). *Astrophys. Space Sci.*, **5**, 18.
Reynolds, R. T. and Summers, A. L. (1969). *Journ. Geophys. Res.*, **74**, 2494.
Tolland, H. G. (1973). Nature. Phys. Sci., **243**, 141.
Tolland, H. G. and Strens, R. G. J. (1972). *Phys. Earth Planet. Interiors*, **5**, 380.
Tozer, D. C. (1967) in T. F. Gaskell (Ed.) *The Earth's Mantle*. Academic Press, London.
Tozer, D. C. (1970) *Journ. Geomag. Geoelect.*, **22**, 35.
Tozer, D. C. (1971) *Proc. Conf. Lunar Geophys.* Lunar Science Institute, Houston.

PART II

High-Pressure Crystal Chemistry of Orthosilicates and the Formation of the Mantle Transition Zone

S. Akimoto
Institute for Solid State Physics, University of Tokyo
Roppongi, Minato-ku, Tokyo 106, Japan

Y. Matsui
Institute for Thermal Spring Research, Okayama University
Misasa, Tottori-ken 682–02, Japan

Y. Syono
Research Institute for Iron, Steel and Other Metals, Tohoku University
Katahiracho, Sendai 980, Japan

INTRODUCTION

It is widely accepted that olivine $(Mg,Fe)_2SiO_4$ is the most abundant mineral in the earth's upper mantle. Accordingly, an accurate knowledge of the phase relationships of olivine at high pressures and temperatures is required for understanding the structure and properties of the deeper regions of the mantle. Since the suggestion by Bernal (1936) that common olivine might transform to a new polymorph possessing the spinel structure in the deep mantle, high-pressure phase transformations of olivine have received much attention in the discussion of the transition zone of the mantle. The hypothesis of the olivine–spinel transformation was adopted by Jeffreys (1937) and Birch (1952) to explain a sudden increase in the gradient of the seismic wave velocity at a depth near 400 km. Birch (1952) predicted that the high-pressure breakdown of the silicate spinel to simple oxides is partly responsible for the formation of the zone of high wave velocity gradients.

The development of modern high-pressure high-temperature apparatus has made it possible to verify these predictions in the laboratory. Various silicates and germanates have been subjected to pressures above 200 kbar, and many new types of phase transformation have been found. In the present article, the pressure-dependent phase transformations of orthosilicates (most of which crystallize in the olivine structure at atmospheric pressure) are reviewed, with special emphasis on the experimental data obtained in our laboratory, and then the structure of the transition zone is discussed on the basis of the recent experimental investigation of the Mg_2SiO_4–Fe_2SiO_4 system.

HIGH PRESSURE CHEMISTRY OF ORTHOSILICATES

1. Crystal Chemistry of Olivine

Structure

A classification of the structure types based on ionic radii is shown in Figure 1 for compounds of the $A_2^{2+}B^{4+}O_4$ type.

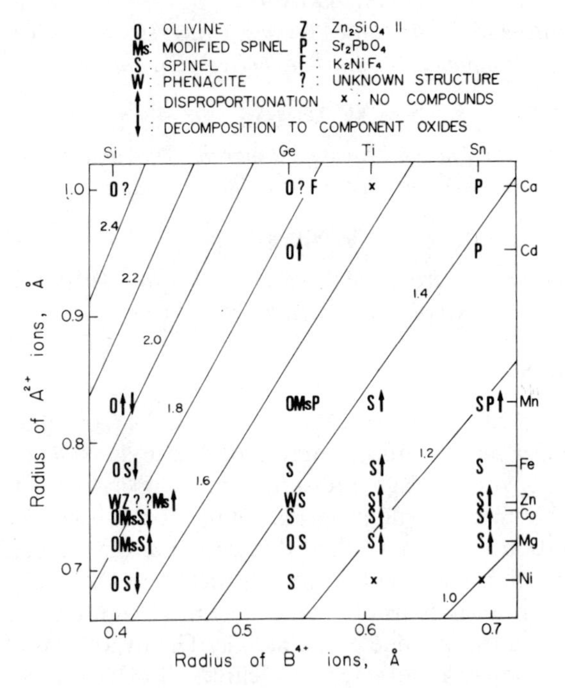

Figure 1. Crystal structure types in $A_2^{2+}B^{4+}O_4$ compounds. High pressure polymorphs are indicated in a row from left to right with increasing pressure. Numerals attached to the oblique lines indicate the ionic radius ratio r_A/r_B Data sources for orthosilicates are in the text. Those for germanates, titanates and stannates are: Wyckoff (1965), Akimoto and Syono (1967), Akimoto (1970), Ringwood (1970), Nishikawa and Akimoto (1971), Akimoto and Syono (unpublished data).

In the figure the horizontal and vertical axes are the ionic radii of smaller tetravalent B^{4+} ions and larger divalent A^{2+} ions, respectively. Ionic radii compiled by Shannon and Prewitt (1969, 1970) are used throughout the present article. As for the compounds with high-pressure polymorphs, several symbols are shown in a row from left to right with increasing pressure. It is generally seen that the upper left region with high ionic radius ratio r_A/r_B, i.e. orthosilicates and orthogermanates, is favourable for the olivine structure at atmospheric pressure.

The olivine structure is based on a nearly hexagonal-closest packed arrangement of oxygen ions, in which B^{4+} ions are placed in tetrahedral coordination, and A^{2+} ions in octahedral coordination. A stereographic projection on the (100) plane of synthetic forsterite (Mg_2SiO_4 olivine) based on the recent structure analysis by Smyth and Hazen (1973) is shown in Figure 2, where the relationship between MgO_6 octahedra and SiO_4 tetrahedra is seen. Two kinds of non-equivalent octahedra, i.e. $Mg(1)O_6$ and $Mg(2)O_6$, are also distinguished easily in the figure. Each $Mg(1)O_6$ octahedron shares four edges with adjacent Mg-octahedra and two edges with Si-tetrahedra, while each $Mg(2)O_6$ octahedron shares two edges with adjacent Mg-octahedra and one edge with a Si-tetrahedron. The $Mg(1)O_6$ octahedra are linked to each other by sharing their edges and make a chain along the c-axis. The $Mg(2)O_6$ octahedra are slightly larger and more distorted than the $Mg(1)O_6$ octahedra. The shortening of polyhedral edges shared between MgO_6 octahedra and SiO_4 tetrahedra is to be expected in ionic crystals (Pauling, 1960). This separates Mg^{2+} ions further from Si^{4+} ions and expands the structure, as the Si^{4+}–Mg^{2+} vector is oblique

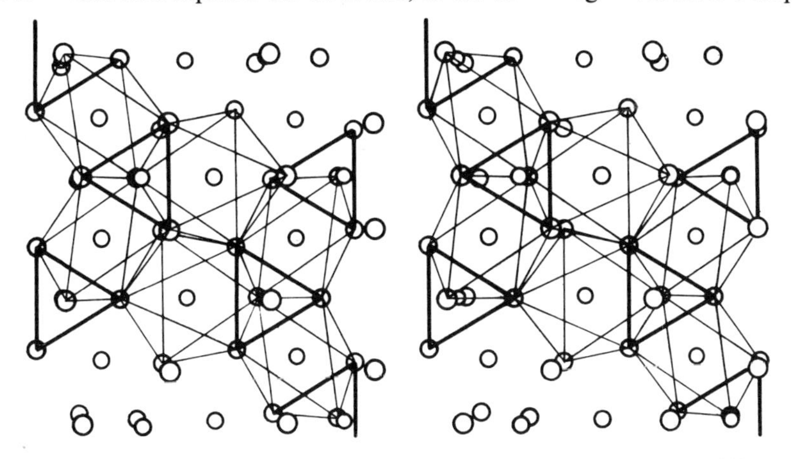

Figure 2. Stereographic projection of the structure of forsterite, αMg_2SiO_4 on (100).
Thick line: Edges of SiO_4 tetrahedron which are shared with MgO_6 octahedra.
Fine line: Edges of MgO_6 octahedra which are not shared with any SiO_4 tetrahedra.
Note the severe elongation of the $Mg(1)O_6$ octahedron induced by Si—Mg(1)—Si repulsion with directions diagonal to the orthorhombic unit cell.

Table 1. Parameters of olivine–spinel and olivine-modified spinel transformations

Compound	Transformation pressure (kbar)	Temperature (°C)	Cell parameters in olivine (Å)	Cell parameters in spinel (Å)	Cell parameters in modified spinel (Å)	Density increase (%)	dP/dT (bars/°C)	Entropy change (cal deg^{-1} mol^{-1})	Reference
Mn_2GeO_4	44	1000	$a=$ 5.061 (1) $b=$ 10.719 (1) $c=$ 6.295 (1)		$a=$ 6.025 (2) $b=$ 12.095 (4) $c=$ 8.752 (2)	7.1	31		1
Mg_2GeO_4	0	810	$a=$ 4.915 $b=$ 10.295 $c=$ 6.020	$a=8.255$		8.3	40		2
Fe_2SiO_4	51	1000	$a=$ 4.821 (1) $b=$ 10.477 (1) $c=$ 6.086 (1)	$a=8.234$ (1)		10.0	26	2.66	3
Zn_2SiO_4	130[a]	900	($a=$ 4.79[b] $b=$ 10.31 $c=$ 6.02)		$a=$ 5.740 (1) $b=$ 11.504 (1) $c=$ 8.395 (2)	(7.2)[c]			4
Co_2SiO_4	68	900	$a=$ 4.7823 (3) $b=$ 10.3044 (8) $c=$ 6.0036 (6)	$a=8.138$ (1)		9.8	32	2.19	5

Co_2SiO_4	74	1200	$a =$ 4.7823 (3) $b =$ 10.3044 (8) $c =$ 6.0036 (6)		$a =$ 5.753 (1) $b =$ 11.522 (4) $c =$ 8.337 (2)	7.1	17	1.20	5
Mg_2SiO_4	125	1000	$a =$ 4.7553 (6) $b =$ 10.1977 (14) $c =$ 5.9820 (7)		$a =$ 5.694 (1) $b =$ 11.467 (2) $c =$ 8.248 (2)	7.7	47	3.51	6, 7
Mg_2SiO_4	170^d	1000	$a =$ 4.7553 (6) $b =$ 10.1977 (14) $c =$ 5.9820 (7)	$a =$ 8.076 (1)		10.2		6.43	6, 8
Ni_2SiO_4	31	1000	$a =$ 4.7287 (3) $b =$ 10.1314 (6) $c =$ 5.9153 (3)	$a =$ 8.043 (1)		8.5	12	1.00	9

Notes:

[a]Synthesis pressure

[b]Cell parameters of virtual olivine extrapolated from the composition dependence of cell parameters in the Mg_2SiO_4–Zn_2SiO_4 system.

[c]Virtual density change.

[d]Extrapolated value from the phase diagram on the Mg_2SiO_4–Fe_2SiO_4 system.

References:

1. Akimoto (1970)
2. Dachille and Roy (1960)
3. Akimoto, Kodama and Kushiro (1967)
4. Syono, Akimoto and Matsui (1971)
5. Akimoto and Sato (1968)
6. Akimoto (1972)
7. Akimoto and Kawada, unpublished data
8. Ito, Matsui, Suito and Kawai (1974)
9. Akimoto, Fujisawa and Katsura (1965)

to all three crystallographic axes. Cell parameters of typical olivine-group compounds are listed in Table 1.

Olivine Group Solid Solutions

As noted above, the olivine structure provides two kinds of octahedral site for divalent cations. Cation distribution among non-equivalent positions is generally controlled by the sizes of the ions concerned, and further by the difference in the crystal field stabilization energies in different sites in the case of transition metal ions. Since thermodynamic considerations indicate that solid solutions with site-preference phenomena are necessarily nonideal (Banno and Matsui, 1967; Thompson, 1969), the properties of olivine group solid solutions, such as unit cell parameters, are affected by the degree of site-preference.

Changes in unit cell parameters with composition in the olivine group solid solutions are shown in Figure 3, for $(Mg,Mn)_2SiO_4$, $(Mg,Fe)_2SiO_4$, $(Mg,Zn)_2SiO_4$, $(Mg,Co)_2SiO_4$ and $(Mg,Ni)_2SiO_4$. Except for the Mg_2SiO_4–Zn_2SiO_4 system, these solid solutions were prepared at atmospheric pressure by the usual sintering technique using powder mixtures of the olivine type Mn_2SiO_4, Fe_2SiO_4, Co_2SiO_4 and Ni_2SiO_4 with Mg_2SiO_4 olivine as starting materials (Akimoto and Fujisawa, 1968; Matsui and Syono, 1968; Nishizawa and Matsui, 1972). Since Zn_2SiO_4 crystallizes in the phenacite structure at atmospheric pressure, the $(Mg,Zn)_2SiO_4$ solid solution with the olivine structure was synthesized at high pressures ranging from 70 to 90 kbar and at 1200 °C (Syono, Akimoto and Matsui, 1971). Although the Zn_2SiO_4 olivine could not be stabilized even at high pressures, the solubility limit at high pressures, which amounts to 75 per cent at 90 kbar, is much greater than that at atmospheric pressure which is reported to be only 24 per cent (Sarver and Hummel, 1962). An extrapolation yields unit cell parameters of 'virtual' Zn_2SiO_4 olivine as listed in Table 1 above.

As is seen in Figure 3b, the unit cell parameters a, b, c and V in $(Mg,Fe)_2SiO_4$ olivine obey Vegard's law almost completely, whereas in $(Mg,Mn)_2SiO_4$ and $(Mg,Ni)_2SiO_4$ cell edge lengths deviate remarkably from Vegard's law with a definite positive excess volume of mixing (Figures 3a and 3e). The behaviour of the unit cell parameters in $(Mg,Co)_2SiO_4$ (Figure 3d) is intermediate between the former two cases, and that in $(Mg,Zn)_2SiO_4$ (Figure 3c) appears to be similar to that in $(Mg,Co)_2SiO_4$. With a random distribution of cations, unit cell parameters should be nearly linear functions of composition, and the results shown in Figure 3 suggest that (i) in $(Mg,Fe)_2SiO_4$ olivine Mg^{2+} and Fe^{2+} ions occupy the two non-equivalent octahedral positions M(1) and M(2) nearly at random; (ii) cation distributions in $(Mg,Mn)_2SiO_4$ and $(Mg,Ni)_2SiO_4$ are highly ordered among M(1) and M(2); and (iii) the site-preference in $(Mg,Co)_2SiO_4$ and $(Mg,Zn)_2SiO_4$ is intermediate between the above two cases.

It is highly probable that the site-preference in $(Mg,Mn)_2SiO_4$ olivine is due to the significant difference in ionic radius between Mg^{2+} and Mn^{2+}, the

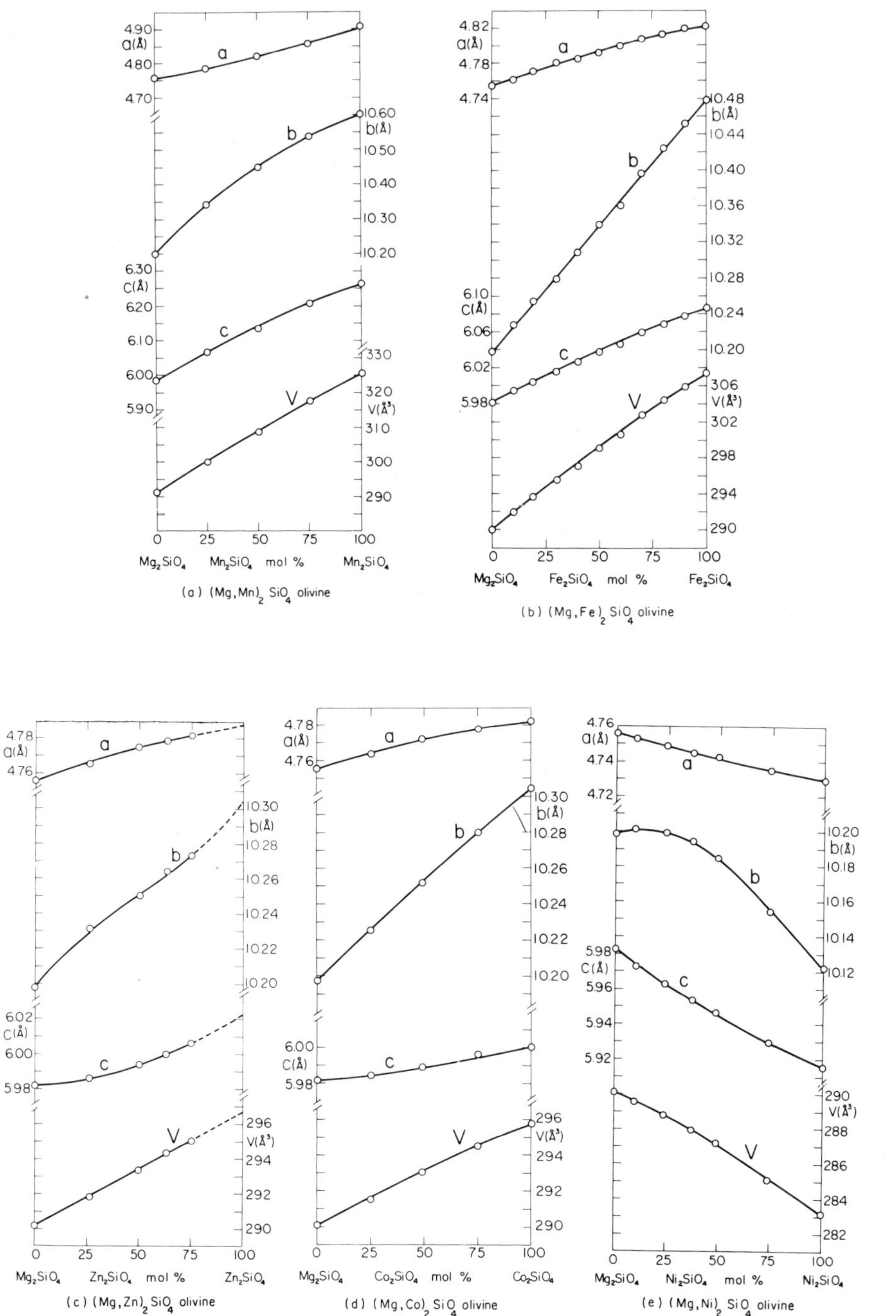

Figure 3. Change in unit cell parameters with composition in the olivine group solid solutions.

larger M(2) position being favoured by Mn^{2+} and the smaller M(1) position by Mg^{2+}. On the other hand, the ordered distribution of Ni^{2+} and Mg^{2+} in $(Mg,Ni)_2SiO_4$ olivine is presumably due to the combination of size effect (smaller Ni^{2+} prefers M(1)) and the crystal field effect of Ni^{2+} (crystal field stabilization of Ni^{2+} may be greater in M(1) than in M(2)). Summarizing the above discussion, it is suggested that the tendency of divalent cations to occupy the smaller M(1) position in the olivine structure becomes more pronounced in the order $Ni^{2+} > Co^{2+} \simeq Zn^{2+} > Fe^{2+} \simeq Mg^{2+} > Mn^{2+}$. This situation is schematically displayed in Figure 4, where the horizontal axis indicates size of ion (arbitrary scale) and vertical axis represents the M(1) preference energy (arbitrary scale) of a divalent cation (M) relative to Mg, which can be written in terms of free energies of hypothetical structures: $-[G_{Mg(1)M(2)SiO4} - G_{M(1)Mg(2)SiO4}]$, where, for example, $M(1)Mg(2)SiO_4$ is a hypothetical olivine in which M occupies the M(1) position only and Mg the M(2) position only (Banno and Matsui, 1967; Thompson, 1969). The effect of differences in size of the M(1) and M(2) positions may be taken to be proportional to the difference in ionic radii for M and Mg, so that smaller ions prefer the M(1) position, as shown in the diagram by a straight line passing through Mg (by definition) and Mn (free from crystal field effect). On the other hand, the effect of crystal field stabilization may be evaluated roughly to be proportional to the crystal field stabilization energy of the isolated cation in octahedral coordination, which is greatest in Ni^{2+} and zero in Mn^{2+}, the ratio being approximately 3 : 2 : 1 for Ni^{2+}, Co^{2+} and Fe^{2+} as is shown in the diagram. Thus the near-random distribution of Mg^{2+} and Fe^{2+} in $(Mg,Fe)_2SiO_4$ olivine is a result of cancelling out of two competing factors; the crystal field effect which makes M(1) prefer Fe^{2+}, and the size effect which renders M(2) favourable for Fe^{2+}. Direct evidence for the near-random distribution in $(Mg,Fe)_2SiO_4$ olivine has

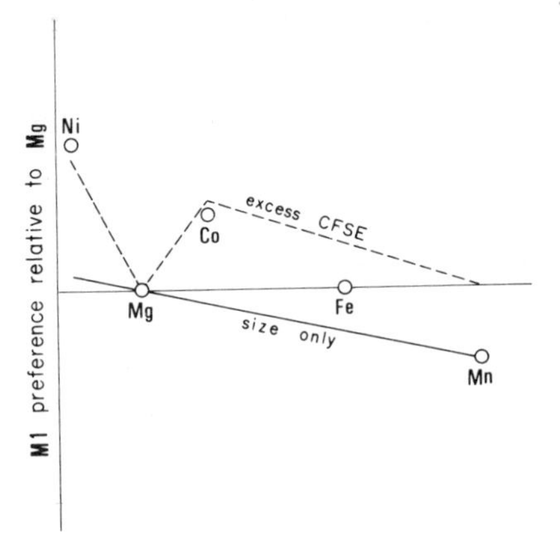

Figure 4. Preference energy of divalent cations relative to Mg^{2+} occupying the smaller M(1) position in the olivine structure.

been obtained from Mössbauer spectroscopy (Bush *et al.*, 1967; Virgo and Hafner, 1972) and X-ray diffraction measurements (Birle *et al.*, 1968; Finger, 1971; Finger and Virgo, 1971; Brown and Prewitt, 1973; Smyth and Hazen, 1973).

2. High Pressure Transformations in Olivine

The first experimental confirmation of the olivine–spinel transformation in orthosilicates was achieved by Ringwood (1958b) for Fe_2SiO_4. He published a series of papers in which the discovery of the spinel polymorph was described for Ni_2SiO_4 and Co_2SiO_4 (Ringwood, 1962; 1963). He also predicted the physical conditions of formation of the Mg_2SiO_4 spinel polymorph from a thermodynamic analysis of the olivine–spinel transformation in $(Mg,Ni)_2(Si,Ge)O_4$ solid solution (Ringwood, 1958a). These pioneering investigations have stimulated a more comprehensive investigation of the stability and phase relations of olivine type compounds and solid solutions. In the course of the systematic study of the olivine–spinel transformation, several new high-pressure polymorphs of olivine were found. The discovery of the modified spinel structure in some orthosilicates and orthogermanates was particularly important.

Ringwood and Major (1966a) first pointed out some complexity in the high-pressure transformations in the Mg-rich side of $(Mg,Fe)_2SiO_4$ olivine. They found that at pressures greater than 150 kbar and at approximately 1000 °C, pure Mg_2SiO_4 transformed completely to a birefringent phase (β phase) with a complex X-ray diffraction pattern, showing apparent resemblance to that of spinels but with many extra lines. (In the present article, the terms α, β and γ phase in A_2BO_4 type compounds are used for olivine, modified spinel and true spinel, respectively). This phase was also observed in $(Mg_{0.85}Fe_{0.15})_2SiO_4$ composition, and its occurrence seemed to depend upon the quenching procedure. From these features Ringwood and Major suggested that the β phase might be a 'distorted' or 'modified' spinel, and tentatively concluded that the original true spinel phase which was actually stable at high pressure transformed to a distorted modification when pressure was released.

Following the discovery of the β phase of Mg_2SiO_4 by Ringwood and Major, the isotype of the β phase was soon found in Co_2SiO_4 by Akimoto and Sato (1968) and in Mn_2GeO_4 by Akimoto (1970). Syono, Akimoto and Matsui (1971) also reported a stable field of the β phase in their comprehensive investigation of the high-pressure transformation of Zn_2SiO_4. As will be described later, successful synthesis of single crystals of the β phase of Co_2SiO_4 and Mn_2GeO_4 made it possible to assign the crystal symmetry as orthorhombic, and the structure determination justified the description of the β phase as a 'modified' spinel (Morimoto *et al.*, 1969 and 1970).

The phase diagram for the olivine-modified spinel and/or true spinel transformation of orthosilicates has been studied in greater detail at the Institute for Solid State Physics, University of Tokyo. The transformation diagrams

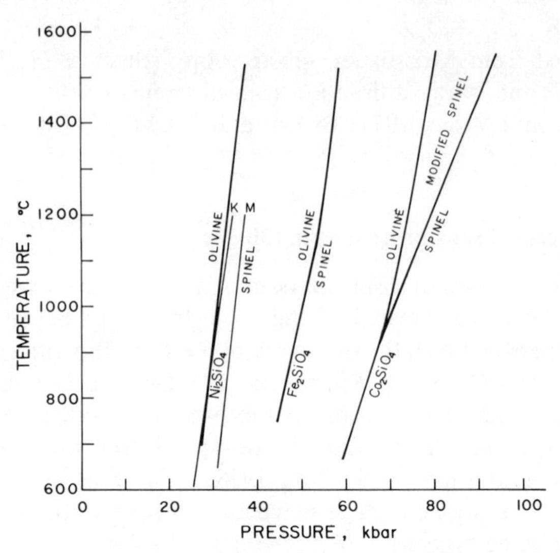

Figure 5. Pressure–temperature phase diagram for olivine–spinel (modified spinel) transformation of Fe_2SiO_4, Co_2SiO_4 and Ni_2SiO_4. Data sources are as follows: Fe_2SiO_4, Akimoto, Komada and Kushiro (1967); Co_2SiO_4, Akimoto and Sato (1968); Ni_2SiO_4, Akimoto, Fujisawa and Katsura (1965). The thin lines designated by K and M represent the equilibrium curves determined by Kirtel (1972) and Ma (1974). respectively.

of Ni_2SiO_4, Fe_2SiO_4 and Co_2SiO_4 were summarized in Figure 5 (Akimoto *et al.*, 1965 and 1967; Akimoto and Sato, 1968). Recently the stability diagram of Ni_2SiO_4 was reinvestigated by Kirtel (1972) and Ma (1974). Their data on the equilibrium curve for the olivine–spinel transformation in Ni_2SiO_4 are shown in the figure. It is seen that these three independent determinations using different types of high-pressure apparatus are in good agreement.

The occurrence of βCo_2SiO_4 complicates the phase diagram. At temperatures below approximately 930 °C, the olivine polymorph of Co_2SiO_4 transforms directly to the spinel polymorph, while at the higher temperatures, the modified spinel phase is observed in the pressure range between the olivine and the spinel fields.

The equilibrium curve for the olivine–spinel transformation of Fe_2SiO_4 shown in Figure 5 could not be approximated by a simple straight line, the deviation of the curve from the straight line being remarkably large for temperatures above 1150 °C. This marked change in slope of the transformation curve was interpreted qualitatively by Strens (1969) to be due to the decrease in entropy change of the reaction with increasing temperature. He emphasized the importance of entropy terms arising from the distribution of the *d* electrons over the t_{2g} levels of the octahedral sites. Since the splittings of the t_{2g} levels are larger in olivine than in spinel, the difference between the electronic entropies of αFe_2SiO_4 and γFe_2SiO_4 will decrease with increasing temperature, resulting in the decrease of the overall entropy change.

The appearance of an unquenchable new high-pressure phase of Fe_2SiO_4 at high temperatures offers another explanation for the observed behaviour of the olivine-spinel transformation. Although the usual quenching experiments give no indication of the existence of the modified spinel phase of Fe_2SiO_4, it is highly desirable to reinvestigate the stability of Fe_2SiO_4 *in situ* over a wide range of pressure and temperature.

The high-pressure transformation behaviour of Mg_2SiO_4 has long been controversial. Just after the discovery of βMg_2SiO_4 by Ringwood and Major (1966a), Akimoto and Ida (1966) also succeeded in the partial transformation of forsterite to a high pressure phase at pressures above 155 kbar at 800 °C. Since they found several X-ray diffraction lines attributable to the spinel structure and failed to identify the diffraction lines peculiar to the β phase, they claimed to have synthesized a true Mg_2SiO_4 spinel. All the subsequent experiments on the high-pressure transformations in the Mg_2SiO_4–Fe_2SiO_4 system (Ringwood and Major, 1970; Kawai *et al.*, 1970; Ito *et al.*, 1971; Akimoto, 1972; Suito, 1972), however, indicate that the β phase is the first high-pressure polymorph of forsterite to appear and further suggest that the stabilization of γMg_2SiO_4 (pure spinel polymorph of Mg_2SiO_4) should require a large additional pressure. Very recently, definite evidence of the existence of γMg_2SiO_4

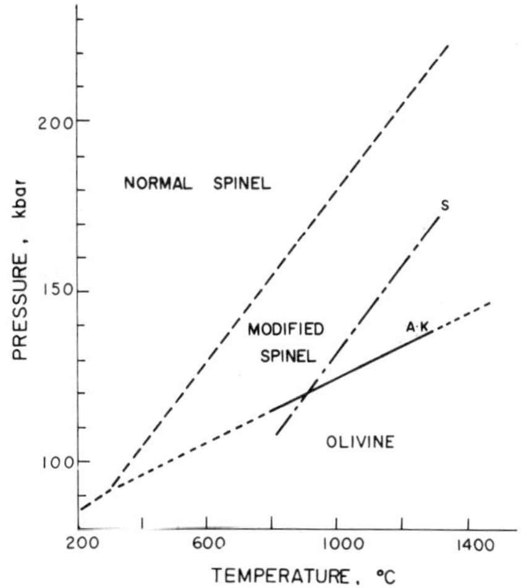

Figure 6. Pressure–temperature stability fields for the olivine–modified spinel–spinel equilibrium of Mg_2SiO_4. The lines designated by A·K and S represent the boundary curve for the olivine–modified spinel transformation of Mg_2SiO_4 determined by Akimoto and Kawada (unpublished data) and Suito (1972), respectively.

was presented by Ito, Matsui *et al.* (1974). They succeeded in the synthesis of γMg_2SiO_4 at approximately 250 kbar and at 1000 °C. The diffraction pattern of the product demonstrated in their paper is undoubtedly that of γMg_2SiO_4 with a minute amount of residual forsterite.

Stability relations among α, β, and γ phases of Mg_2SiO_4 have not yet been fully clarified owing to the great difficulty in the quantitative analysis of the experimental data at pressures above 150 kbar. Only a few preliminary phase diagrams have hitherto been reported on the olivine-modified spinel transformation in Mg_2SiO_4. In Figure 6 the phase boundary curve published by Suito (1972) is compared with our preliminary data (Akimoto and Kawada, unpublished data). Considerable discordance is seen between the two curves. Although it is very difficult to decide which curve represents the actual equilibrium curve between α and βMg_2SiO_4, the gradient (dP/dT) determined by Suito seems to be too high.

Parameters in olivine–spinel and olivine-modified spinel transformations in various orthosilicates and orthogermanates are summarized in Table 1 above. The modified spinel is approximately 7 to 8 per cent denser than the olivine structure, and the density increase associated with the olivine–spinel transformation ranges from 8.3 to 10.2 per cent among various compounds. The data on (dP/dT) in the table are those provided from the phase equilibrium curves such as shown in Figure 5. Information on (dP/dT) and the unit cell volume enables us to estimate the entropy change associated with the olivine–spinel and the olivine-modified spinel transformation using Clapeyron–Clausius relation $(dP/dT) = (\Delta S/\Delta V)$. The estimated values for typical compounds are also shown in the table.

In the case of Co_2SiO_4 and Mg_2SiO_4, the volume change of the olivine–spinel transformation is 1.3 times larger than that of the olivine-modified spinel transformation. In Co_2SiO_4 the entropy change of the olivine–spinel transformation is 1.8 times larger than that of the olivine-modified spinel transformation. If we assume that the same relation holds for Mg_2SiO_4, we can calculate the entropy change of the olivine–spinel transformation of Mg_2SiO_4 as 6.43 cal deg^{-1} mol^{-1}. Based on the thermodynamic data given in Table 1. Nishizawa and Akimoto (1973) predicted for Mg_2SiO_4 the phase boundary between modified spinel and spinel, and the triple point at which olivine, modified spinel and true spinel coexist. The tentative phase boundary is also indicated in Figure 6.

As illustrated in Figure 1, at high pressures some olivine–type compounds transform to structures different from both the spinel and the modified spinel structure. Typical examples are (i) olivine–K_2NiF_4 structure transformation found in Ca_2GeO_4 at high temperatures (Ringwood and Reid, 1968a; Nishikawa and Akimoto, 1971), (ii) olivine–unknown structure transformation found in Ca_2SiO_4 (Ringwood and Reid, 1968a), and in Ca_2GeO_4 at low temperatures (Nishikawa and Akimoto, 1971), and (iii) high-pressure breakdown of olivine found in Cd_2GeO_4 (Ringwood and Reid, 1968a) and in Mn_2SiO_4 (Ito, Matsumoto *et al.*, 1974). The Cd_2GeO_4 olivine disproportionates under high pressure into $CdGeO_3$ (perovskite) plus CdO (rocksalt). Mn_2SiO_4 olivine first disproportionates to $MnSiO_3$ (garnet) plus MnO (rocksalt) at approximately 140 kbar and at 1000 °C and finally decomposes to MnO (rocksalt) plus SiO_2 (stishovite) at pressures above 300 kbar and at 1000 °C.

3. Crystal Structure of Silicate Spinel

It is well known that the spinel structure is based on a nearly cubic-closest packed arrangement of oxygen ions with the cations in tetrahedral and octahedral coordination. Since the cations in the spinel structure occupy the special positions (8*a*) and (16*d*) the structure can be thoroughly described by only one parameter *u*, the *x*-coordinate for the oxygen atom, if the distribution of cations among (8*a*) and (16*d*) positions is known. Recent high-pressure syntheses of single crystals of the silicate spinel made it possible to solve these problems directly by X-ray structure analysis. Morimoto *et al.* (1974) first studied the crystal structure of the spinel polymorph of Co_2SiO_4, and Yagi *et al.* (1974) made a detailed study of the spinel polymorphs of Fe_2SiO_4 and Ni_2SiO_4.

The values of the oxygen parameter *u* for Fe_2SiO_4, Co_2SiO_4 and Ni_2SiO_4 spinel determined from the single crystal X-ray data are listed in Table 2. Ma (1974) also attempted independently to determine the oxygen parameter of Ni_2SiO_4 spinel using powder diffraction data. His value 0.367(3) is in rather good agreement with the value 0.3687(2) obtained at the Institute for Solid State Physics. These *u*-parameter values are decidedly smaller than 0.375 corresponding to the ideal closest-packed arrangement of oxygen atoms, the situation being reversed in normal spinels. In Figure 7 the *u*-parameter of three silicate spinels, γFe_2SiO_4, γCo_2SiO_4 and γNi_2SiO_4 is plotted against unit cell dimension. The *u*-parameter decreases with increasing cell dimension. The value of the *u*-parameter of γMg_2SiO_4, of which no single crystal has yet been obtained, can be estimated as approximately 0.368 by interpolation.

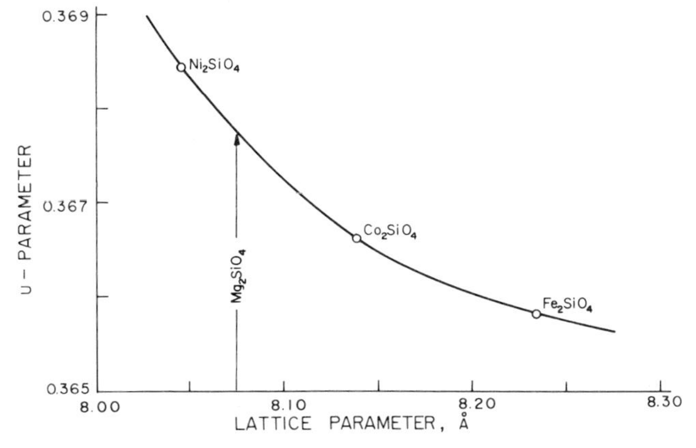

Figure 7. Change in *u*-parameter with lattice parameter for silicate spinels.

Table 2. Crystal structure parameters in silicate spinels

	Fe_2SiO_4 [1]	Co_2SiO_4 [2]	Ni_2SiO_4 [1]
Cell parameter	8.234 (1) Å	8.138 (1) Å	8.044 (1) Å
u-parameter	0.3658 (2)	0.3666 (3)	0.3687 (2)
Occupancy factor for Si in (16*d*) sites, α[a]	0.023 (10)	0.032 (8)	0.005 (12)
Interatomic distances and angles			
SiO_4 tetrahedron			
Si—O [× 4]	1.652 (2) Å	1.646 (3) Å	1.654 (2) Å
O—O [× 6]	2.697 (2) Å	2.688 (4) Å	2.701 (2) Å
O—Si—O [× 6]	109.5 (1) °	109.5 (1) °	109.5 (1) °
AO_6 octahedron			
M—O [× 6]	2.137 (2) Å	2.104 (3) Å	2.063 (2) Å
O—O (shared edge) [× 6]	3.125 (2) Å	3.066 (4) Å	2.987 (2) Å
O—O (unshared edge) [× 6]	2.915 (2) Å	2.880 (4) Å	2.846 (2) Å
O—O (mean)	3.020 Å	2.973 Å	2.917 Å
O—M—O [× 6]	94.0 (1) °	93.6 (1) °	92.8 (1) °
O—M—O [× 6]	86.0 (1) °	86.4 (1) °	87.2 (1) °
O—M—O (mean)	90.0 °	90.0 °	90.0 °

References: 1. Yagi, Marumo and Akimoto (1974)
2. Morimoto, Tokonami, Watanabe and Koto (1974)

[a]Parameter α in the expression $(M_{2-\alpha}\,Si_\alpha)\,(Si_{1-\alpha'}\,M_\alpha)O_4$

It has long been assumed that silicate spinel will crystallize in the normal spinel structure, where the A^{2+} cations are in octahedral sites and the Si^{4+} ions in the tetrahedral sites. Recent site-occupancy refinements indicate 2.3 ± 1.0 per cent of the total Si ions at the octahedral sites for γFe_2SiO_4, 3.2 ± 0.8 per cent for γCo_2SiO_4 and only 0.5 ± 1.2 per cent for γNi_2SiO_4. From these small values of the occupancy factors for Si in octahedral sites, it is concluded that the silicate spinel can practically be regarded as normal type.

A stereographic projection of γMg_2SiO_4 on the (111) plane is shown in Figure 8, where MgO_6 octahedra with surrounding SiO_4 tetrahedra are clearly seen. In the spinel structure, each MgO_6 octahedron shares six edges with adjacent MgO_6 octahedra. This results in a three dimensional framework of the chains of octahedra running parallel to the six face diagonals of the cubic cell. There is no edge-sharing between MgO_6 octahedra and SiO_4 tetrahedra. This is in remarkable contrast to the olivine structure.

Interatomic distances and angles in the silicate spinels are listed in Table 2. As a result of the small u-parameters ($u < 0.375$) the shared edges of AO_6 octahedra are very much longer than the unshared ones. As pointed out by Kamb (1968), this feature does not satisfy Pauling's third rule for edge-sharing by coordination polyhedra, and this would be the reason why silicate spinels are unstable at atmospheric pressure. Under high pressures, it is likely that the Si—O distance may remain virtually unchanged while A—O distance may decrease more easily. Consequently, it is possible that the actual u-parameter in silicate spinels is greater under very high pressures, where they are stable, than that under atmospheric conditions. However, we cannot account for the difference of the olivine–spinel transformation pressure of various orthosili-

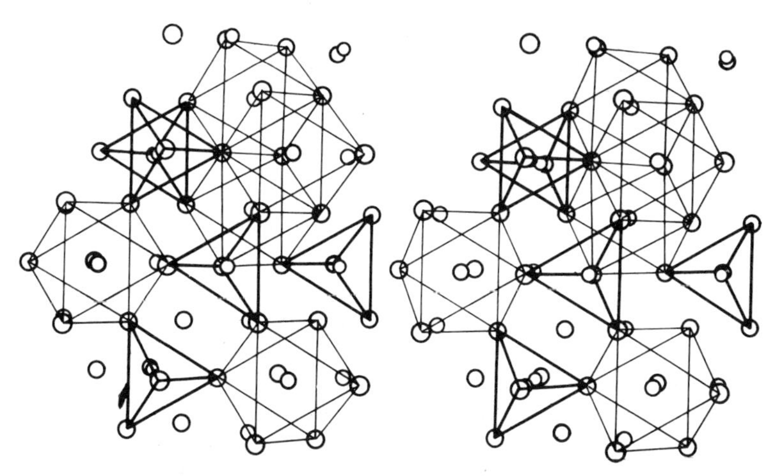

Figure 8. Stereographic projection of the structure of silicate spinel γMg_2SiO_4 on (111).
Thick line: Edges of SiO_4 tetrahedra.
Fine line : Edges of MgO_6 octahedra.

cates in this manner alone. The transformation pressure at 1000 °C from olivine to spinel increases in the sequence Ni_2SiO_4, Fe_2SiO_4, Co_2SiO_4 and Mg_2SiO_4 (Table 1 above and Figure 12 below), whereas the u-parameter decreases in the sequence Ni_2SiO_4, Mg_2SiO_4, Co_2SiO_4 and Fe_2SiO_4. As will be seen later, it is necessary to consider the crystal field effect in discussing the transformation pressure.

Since the olivine and spinel structures of orthosilicates are based on the hexagonal and cubic closest packing of oxygen atoms, respectively, and since both involve divalent cations in six-coordination and Si^{4+} in four-coordination, the coordination number of both cations and anions is conserved for the olivine–spinel transformation. This distinguishes the olivine–spinel transformation from the conventional pattern of the high-pressure phase transformations in ionic compounds, in which an increase in the coordination number of the cations takes place. Therefore, an explanation is needed for the fairly large density-increase associated with the olivine–spinel transformation, amounting to nearly 10 per cent. Kamb (1968) suggested that the greater density increase is chiefly attributable to the extraordinarily reduced density of the olivine structure resulting from the shared edge shortening. The density of forsterite ($\rho_{Mg2SiO4} = 3.21$ g/cm^3) is similar to that of enstatite ($\rho_{MgSiO3} = 3.20$ g/cm^3) despite the higher value of the metal oxygen ratio. This clearly indicates the poor packing of the olivine structure. On the basis of hexagonal closest packing of oxygen ions, we can readily derive a hypothetical 'ideal' olivine structure. Taking the O—O distance of 3.0 Å, we have the following unit cell parameters, $a = 4.77$ Å, $b = 10.26$ Å, $c = 6$ Å which compare quite satisfactorily with actual cell parameters in forsterite (see Table 1). Similarly, in the hypothetical 'ideal' spinel (i.e. $u = 0.375$), we find $a = 8.47$ Å assuming the O—O distance of 3 Å, whereas the actual unit cell length in γMg_2SiO_4 is only 8.076 Å, which corresponds to the mean O—O distance of 2.86 Å. The mean O—O distance in γMg_2SiO_4 is markedly smaller than in αMg_2SiO_4. This results in the high density of the silicate spinel.

4. Crystal Structure of Modified Spinel

As described before, β phase, a high-pressure polymorph of olivine, has hitherto been found in Mn_2GeO_4 (Akimoto, 1970), Co_2SiO_4 (Akimoto and Sato, 1968), Zn_2SiO_4 (Syono, Akimoto and Matsui, 1971), as well as in the magnesium rich side of the Mg_2SiO_4–Fe_2SiO_4 system (Ringwood and Major, 1966a, 1970; Kawai *et al.*, 1970; Ito *et al.*, 1971; Akimoto, 1972; Suito, 1972). Structure refinement was first carried out on βMn_2GeO_4 (Morimoto *et al.*, 1969, 1970, 1972) and then on βCo_2SiO_4 (Morimoto *et al.*, 1970, 1974) using the single crystals synthesized at the Institute for Solid State Physics. These analyses revealed that the β phase belongs to orthorhombic system with the space group *Imma*.

In Figure 9a the structure of βCo_2SiO_4 is projected on the (100) plane using the atomic coordinates reported by Morimoto *et al.* (1974). The oxygen atoms approximate cubic closest packing as in spinel. Although Si^{4+} ions occupy the tetrahedral sites and Co^{2+} ions the octahedral sites in both structures, the arrangements of the tetrahedral and octahedral sites in βCo_2SiO_4 are different from those in γCo_2SiO_4. For comparison, the structure of γCo_2SiO_4 projected on the $(1\bar{1}0)$ plane is shown in Figure 9b. Further, the structure of βCo_2SiO_4 viewed parallel to (101) is stereographically illustrated in Figure 9c.

In the modified spinel structure, two SiO_4 tetrahedra, which were isolated in the spinel structure, share one of their oxygen atoms (O(2) in Figure 9a) and form a Si_2O_7 group. The formation of the Si_2O_7 groups leaves an oxygen atom not bonded to any Si atom (O(1) in Figure 9a) and results in the lower orthorhombic symmetry of the β phase. The structural relationships between spinel and modified spinel are illustrated again in Figure 10 on the basis of the conventional unit cell of the cubic spinel structure. It is easily seen from these figures that, in order to obtain the structure of βCo_2SiO_4, we must modify the spinel structure by displacing four Si and four Co atoms out of the eight Si and sixteen Co atoms in the γCo_2SiO_4 cell. The term 'modified' spinel for the β phase was derived from these structural features. Moore and Smith (1969, 1970) independently reached the same conclusion on the crystal structure of the β phase by analysing the X-ray powder data of the β phase of $(Mg_{0.9}Ni_{0.1})_2SiO_4$ on the basis of the unit cell dimensions and space groups of βCo_2SiO_4 reported by Akimoto and Sato (1968).

In the modified spinel structure, edge-sharing occurs only between AO_6 octahedra as in the spinel structure. A difference is seen in the mode of connection of the AO_6 octahedra. In modified spinel the AO_6 chains run parallel to the *a* and *b* axes, corresponding to the diagonals of one of the three faces of the cubic cell of the spinel. We can therefore regard the modified spinel structure as an assemblage of two-dimensional networks of octahedral chains. This is a remarkable contrast to the three dimensional framework of chains in the spinel structure and to the one-dimensional *c*-axis chains in the olivine structure.

The average length of the shared and unshared edges of the CoO_6 octahedra in βCo_2SiO_4 is 2.99 A and 3.00 Å respectively (Morimoto *et al.*, 1974). This is not inconsistent with Pauling's third rule for edge-sharing by coordination polyhedra. From such a view-point the structure of the modified spinel is not so unusual compared with the silicate spinel mentioned before.

The modified spinel structure, however, partly violates Pauling's second rule or the electrostatic valence rule for some oxygen atoms. The charge balance for O(1) and O(2) in Figure 9a above is $-1/3$ and $+1/3$, respectively. Consequently, in comparison with the structures of olivine and spinel, both of which satisfy the rule, there remains doubt as to whether the modified spinel structure is really stable from the crystal-chemical viewpoint. Since all the modified spinel phases have been obtained as products of quenching experiments, the possibility that they were metastable and were produced during quenching of an original spinel has not yet been eliminated.

However, violation of the electrostatic valence rule is not unusual even in common silicates. The most obvious example is calcium-poor pyroxene such

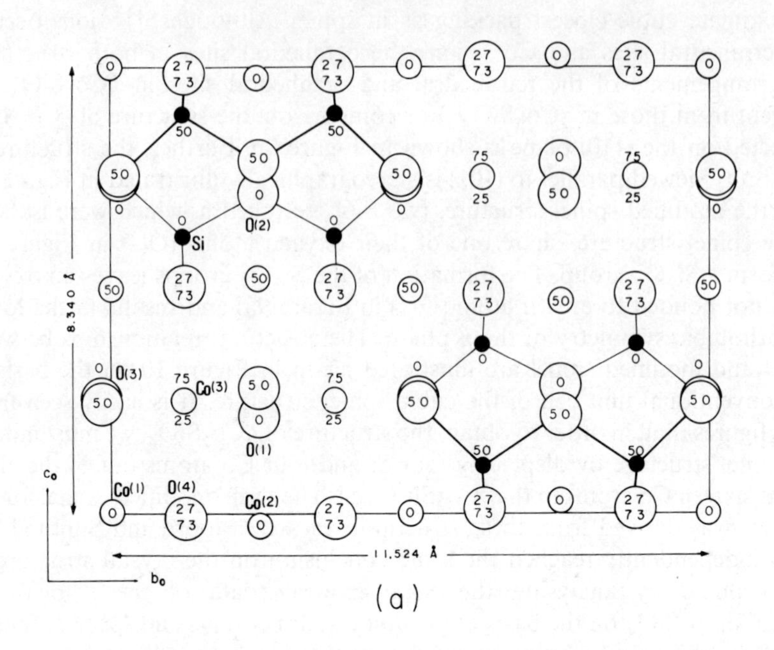
8.340 Å
11.524 Å
c_o
b_o
Si
O(2)
O(3)
Co(3)
O(1)
Co(1)
O(4)
Co(2)
(a)

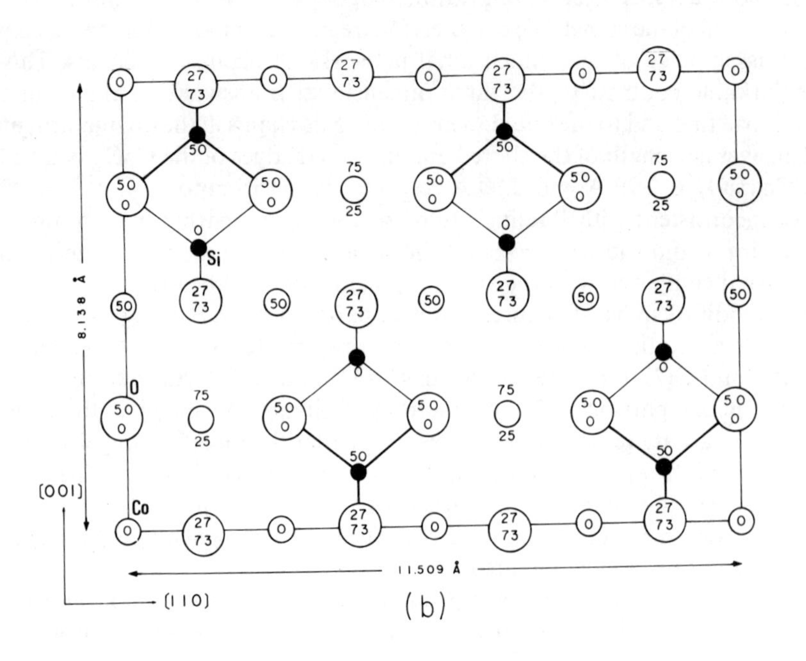
8.138 Å
11.509 Å
Si
O
Co
[001]
[110]
(b)

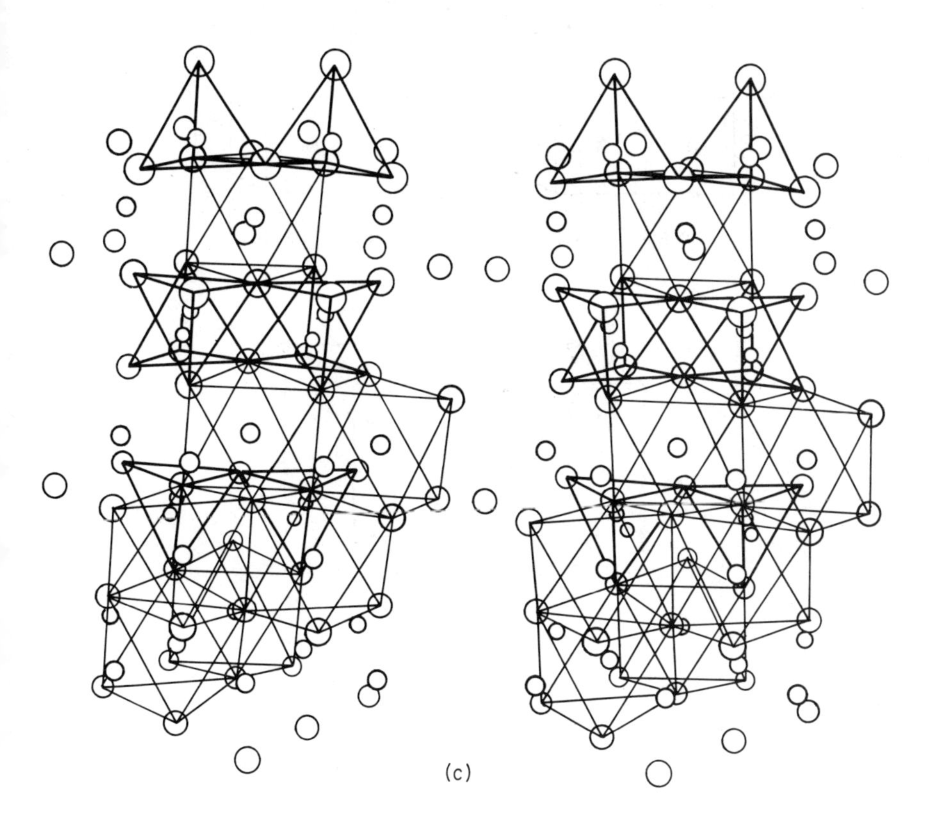

Figure 9. (a) Structure of βCo_2SiO_4 projected on (100).
(b) Structure of γCo_2SiO_4 projected on $(1\bar{1}0)$.
(c) Structure of βCo_2SiO_4 stereographically viewed parallel to (101).
Thick line: Edges of SiO_4 tetrahedra.
Fine line : Edges of CoO_6 octahedra.

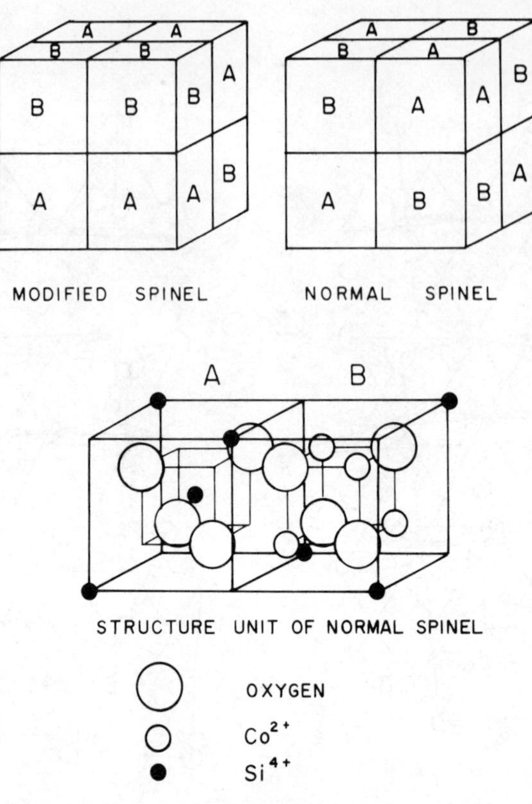

Figure 10. Structural relationship between normal and modified spinel. Note the systematic difference in the arrangement of A and B units between normal and modified spinel.

as enstatite in which the electrostatic imbalance upon oxygen atoms is of the same magnitude as in modified spinel.

To clarify these points, the stability of the modified spinel structure was examined *in situ* by a high-pressure and high-temperature X-ray diffraction technique. Very recently Akimoto and Sato (1975) successfully observed the transformation of Mn_2GeO_4 from olivine to modified spinel *in situ* using cubic anvil type of high-pressure apparatus installed with an X-ray analysis system. In Figure 11, the diffraction pattern at 45 kbar and at 780 °C (Figure 11b) is compared with that at 45 kbar and at room temperature (Figure 11a). In the figure, 2θ angles of the characteristic powder diffraction peak of αMn_2GeO_4 and βMn_2GeO_4 at atmospheric pressure and at room temperature are shown for reference by black and white arrows respectively. The transformation from olivine to modified spinel is obvious. The results shown in Figure 11a and b are also in good harmony with a previous study of the high-pressure and high-temperature stability diagram of Mn_2GeO_4 (Akimoto, 1970) which was determined by the usual quenching method.

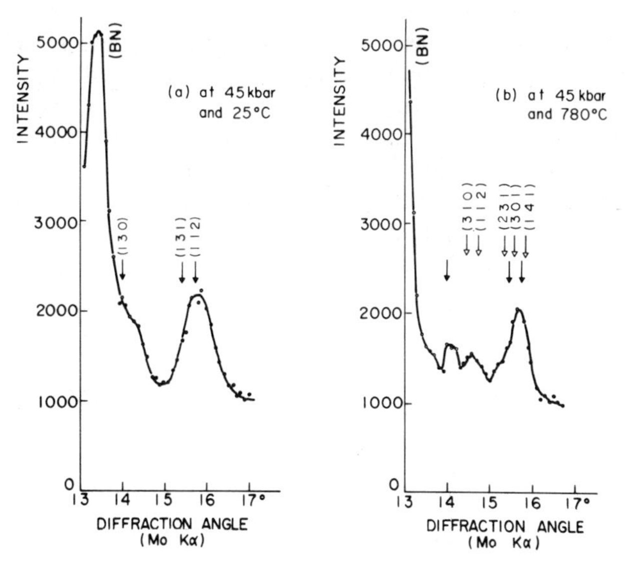

Figure 11. Left: (a) X-ray diffraction pattern of Mn_2GeO_4 at 45 kbar and at room temperature.
Right: (b) X-ray diffraction pattern of Mn_2GeO_4 at 45 kbar and at 780 °C.
The 2θ diffraction angles in Mo Kα radiation of αMn_2GeO_4 and βMn_2GeO_4 at atmospheric pressure and at room temperature are denoted by black and white arrows, respectively. The very strong peak around 2θ angles of 13 to 14° is due to the (002) reflection of hexagonal boron nitride which was used as the pressure transmitting material.

Recently, Tokonami *et al.* (1972) calculated the lattice energies (the Madelung energy corrected for the Born repulsive term) of the α, β, and γ polymorph of Co_2SiO_4 and concluded that the modified spinel structure is not necessarily unstable compared with the true spinel structure. All these recent studies are favourable to the view that the modified spinel represents a really stable high-pressure polymorph of olivine.

5. Crystal-field Effect on the Olivine–Spinel Transformation

It was shown in Figure 1 how the ionic radius ratio controls the crystal structures of $A_2^{2+}B^{4+}O_4$ type compounds. It is seen in the figure that the compounds with low ionic radius ratio, r_A/r_B around $1.0 \sim 1.2$, such as stannates and titanates, crystallize in the spinel structure at atmospheric pressure. At high pressures the spinel territory expands to the region with the higher ionic radius ratio. The modified spinel structure appears exclusively in orthosilicates and orthogermanates in a relatively narrow region of ionic radius ratios around 1.5 to 1.9 forming almost a boundary zone between olivine and spinel territories.

As described in the arguments on the *u*-parameter of the spinel structure, complicated experimental stability fields of olivine, modified spinel and spinel in orthosilicates and orthogermanates cannot be fully understood only by the ionic radius ratio. Syono, Tokonami and Matsui (1971) first pointed out explicitly that the crystal-field stabilization effect also plays an important role in the transformation of transition metal olivines to spinel. Particular features of the transition metal compounds are more clearly demonstrated in Figure 12, where the transformation pressures at 1000 °C from olivine to spinel or to

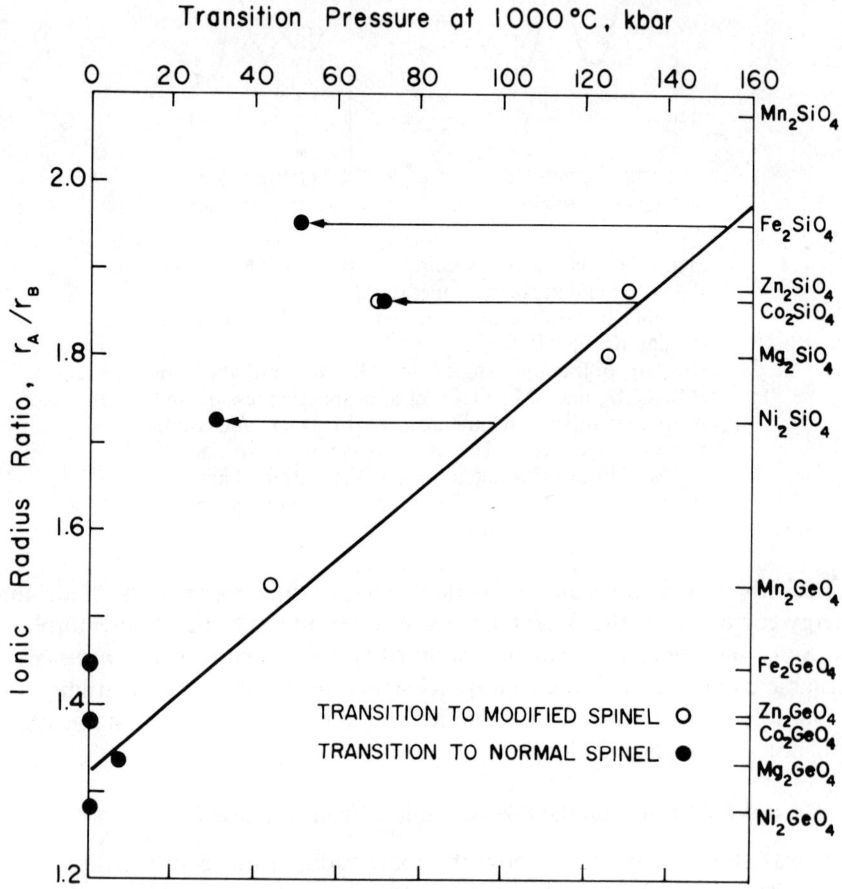

Figure 12. Transformation pressure at 1000 °C either to cubic normal spinel or to modified spinel in silicates and germanates as a function of the ionic radius ratio, r_A/r_B. The arrow indicates excess crystal field stabilization in spinel over olivine. It is to be noted that the transformation to the modified spinel form of Zn_2SiO_4 is not from the olivine polymorph. Data sources for silicates are seen in the text and the foot-note of Table 1. Those for germanates are as follows: Mn_2GeO_4, Akimoto (1970); Mg_2GeO_4, Dachille and Roy (1960); Fe_2GeO_4, Co_2GeO_4, Ni_2GeO_4, Durif–Varambon, Bertaut and Pauthenet (1956).

modified spinel are plotted against the ionic radius ratio. The cubic normal spinel structure for orthosilicates and orthogermanates is realized only for transition metal ions of Fe^{2+}, Co^{2+} and Ni^{2+} except for Mg_2GeO_4, and the modified spinel structure for Mg^{2+}, Mn^{2+} and Zn^{2+} besides Co_2SiO_4. It is also found in Figure 12 that the transformation pressure increases almost linearly with increasing ionic radius ratios, if ions which are sensitive to the crystal-field effect, such as Fe^{2+}, Co^{2+} and Ni^{2+} are excluded. In the case of Fe^{2+}, Co^{2+} and Ni^{2+} ions, the transformation takes place at pressures much below this general trend. In Figure 12 the degree of lowering is shown by the arrows. In the present discussion, we have implicitly assumed that there is no discrimination between the transitions to the cubic spinel and the modified spinel. This assumption may be warranted by the fact that both the modified spinel and the cubic spinel forms of Co_2SiO_4 become stable at much the same pressure.

Syono, Tokonami and Matsui (1971) interpreted this lowering in the formation energy of the cubic normal spinel to be due to an excess crystal field stabilization energy in the octahedral sites in spinel over olivine. It is also suggested that the excess stabilization energy in the cubic normal spinel is large for Fe^{2+} and Ni^{2+} and somewhat reduced for Co^{2+}, on the basis of the detailed comparison of the 3 *d*-orbital levels in the octahedral site of cubic normal spinel with those in the M(1) and M(2) sites of the olivine structure. This is the reason why the transformation pressure of Fe_2SiO_4 is somewhat lower than that of Co_2SiO_4. Further, appearance of the modified spinel in Co_2SiO_4 was successfully attributed to the weak Jahn–Teller character of Co^{2+} ions.

Recently, Mao and Bell (1972) measured the absorption spectrum of γFe_2SiO_4 using a newly developed high-resolution technique. Comparing the spectrum with that of fayalite, αFe_2SiO_4 (Burns, 1970), they determined the excess crystal-field stabilization energy to be 249.3 kbar cm^3/mole. Using this value they estimated the lowering of the transformation pressure for Fe_2SiO_4 to be 98 kbar. This calculation is in markedly good agreement with the value predicted in Figure 12, amounting to approximately 100 kbar.

6. Post-spinel Transformations

In both the spinel and the modified spinel structure, Si^{4+} ions are still tetrahedrally coordinated as in the olivine. Accordingly, it is very likely that at high pressures silicate spinels and modified spinels transform to denser structures characterized by octahedral coordination of Si^{4+} ions. High-pressure synthesis of stishovite, SiO_2 with the rutile structure (Stishov and Popova, 1961), in which Si^{4+} ions are octahedrally coordinated, is also favourable to the possible post-spinel transformation of orthosilicates.

Ringwood and Reid (1968b) have compressed olivines of Fe_2SiO_4, Co_2SiO_4 and Ni_2SiO_4 to pressures up to 150 kbar at about 1000 °C but no further transformation to phases other than spinels could be observed in these silicates. Until quite recently, it was difficult to subject samples to sustained pressures above 150 kbar simultaneously with temperatures above 1000 °C. Hence the first experimental approach to the post-spinel transformation of orthosilicates was made in an indirect way using analogous compounds as a model of silicates. Systematic study of the high-pressure transformation of the $A_2^{2+}B^{4+}O_4$ spinel was initiated by Akimoto and Syono (1967) on titanates. Ringwood and Reid (1968b) also investigated the transformation of a large number of spinels at high pressure in an attempt to clarify the systematic crystal-chemical and thermodynamic factors governing the spinel transformations. On the basis of the experimental data accumulated through these investigations, three major possibilities were suggested for the post-spinel transformation in orthosilicates: (i) Transformation into single dense phase such as Sr_2PbO_4-type structure found in Mn_2GeO_4 (Wadsley *et al.*, 1968; Akimoto, 1970), (ii) Disproportionation into ilmenite plus rocksalt phases, as seen in titanate spinels (Akimoto and Syono, 1967); (iii) Complete demomposition into mixed simple oxides. Some stannate spinels are classified into this category (Ringwood and Reid, 1968b). Ringwood (1970) suggested in the preliminary stage of the experimental investigation that the most probable mode of transformation of β or γMg_2SiO_4 will be to the Sr_2PbO_4-type structure.

Very recently direct evidence for the further phase transformations in silicate spinels was provided in a few super-high-pressure laboratories. Using diamond anvil apparatus, Bassett and Takahashi (1970) first announced the decomposition of γFe_2SiO_4 into wüstite and stishovite at pressures exceeding 250 kbar and at unknown temperatures. Their results were later confirmed by Mao and Bell (1971) and Bassett and Ming (1972) through further studies on Fe_2SiO_4. Ming and Bassett (1974) have extended these experimental studies to the Mg_2SiO_4-Fe_2SiO_4 system ranging up to 80 mole per cent of Mg_2SiO_4 and found the decomposition into (Mg, Fe)O plus stishovite at pressures above 250 kbar and at 1500 °C. Liu (1975) observed the decomposition of γNi_2SiO_4 into NiO plus stishovite at pressures of 140–190 kbar at temperatures of 1400–1800 °C. Using newly devised super-high-pressure apparatus with the multiple anvil sliding system (Kumazawa, 1971), Kumazawa *et al.* (1974) have claimed that a similar phase transformation occurs in pure Mg_2SiO_4 at 330 kbar and at 1000 °C.

Substantial progress has also been made by the research group of the Institute for Thermal Spring Research, Okayama University in the post-spinel transformations of orthosilicates, using the two-stage split-sphere type of super-high-pressure apparatus (Kawai and Endo, 1970). Ito has succeeded in observing the decomposition of γFe_2SiO_4, γCo_2SiO_4 and γNi_2SiO_4 into stishovite plus corresponding monoxide with the rocksalt structure (Ito, 1975). He has also reported the super-high-pressure synthesis of silicate ilmenites of $MgSiO_3$ (Kawai *et al.*, 1974) and $ZnSiO_3$ (Ito and Matsui, 1974). Such successful syntheses of silicate ilmenite harmonize well with their observation of the disproportionation of βZn_2SiO_4 into $ZnSiO_3$ ilmenite plus ZnO with the rocksalt structure (Ito and Matsui, 1974), and suggest the possibility of the disproportionation of γMg_2SiO_4 into $MgSiO_3$ ilmenite plus MgO (periclase) at moderately high pressures prior to the complete decomposition claimed by Kumazawa *et al.* (1974). As to Mn_2SiO_4 Ito, Matsumoto and Kawai (1974) have already established that the olivine polymorph directly disproportionates into $MnSiO_3$ garnet plus MnO and then decomposes into MnO plus stishovite without transforming to modified spinel or spinel polymorph. It is remarkable that contrary to Ringwood's prediction no evidence for the synthesis of Sr_2PbO_4-type silicates was obtained during these recent studies.

Based on the experimental results obtained in the Institute for Thermal Spring Research, the transformation scheme of orthosilicates at approximately 1000 °C up to nearly 350 kbar are schematically illustrated in Figure 13 together with the schemes for metasilicates, monoxides and SiO_2. Comparing these experimental data with those obtained with the diamond anvil apparatus, it is noticed that there still remains some ambiguity in the absolute values of the transformation pressures of silicate spinels. Further, it must be noted here that there is no consensus of opinion on the absolute pressure values above 150 kbar at high temperatures, and it is not yet possible to derive quantitative conclusions from results such as shown in Figure 13. The transformation sequence above 150 kbar should be regarded only as of reconnaissance nature and should not be understood as the established one. However, it is obvious from the diagrams that the transformation scheme of the transition metal silicate spinels differs distinctly from that of Mg_2SiO_4, Zn_2SiO_4 and Mn_2SiO_4. It is interesting that the synthesis of the silicate ilmenite was limited to magnesium and zinc, both of which are free from the crystal field effect.

Estimation of the equilibrium phase boundaries for these spinel-mixed oxides reactions has hitherto been made by a few investigators on the basis of the thermodynamic calculations (Ahrens and Syono, 1967; Anderson, 1967; Mao, 1967). The most remarkable feature of these calculations is the discovery of the negative slope of the phase boundary for Fe_2SiO_4. On the other hand Liu (1975) calculated the positive slope for the decomposition of γNi_2SiO_4 into NiO plus stishovite. As to the post-spinel transformation of Mg_2SiO_4, the uncertainties of the thermochemical data make it still difficult to determine whether the slope is positive or negative. Since it is well-known that the negative slope of the phase boundary results in the mechanical instability of the downgoing lithospherical plate, the accurate experimental determination of the phase boundary for the post-spinel transformation in the Mg_2SiO_4–Fe_2SiO_4 system is urgently needed for the further detailed understanding of the nature of the transition zone of the Earth's mantle.

It is very interesting to note that magnesium pyroxene, enstatite $MgSiO_3$, breaks down into β or γMg_2SiO_4 plus stishovite under moderately high pressure, but again constitutes a compound, $MgSiO_3$ ilmenite, at higher pressures (see Figure 13, Ito *et al.*, 1972; Kawai *et al.*, 1974). We should therefore be careful not to conclude that the reaction under high pressures occurs only in the direction of disproportionation or decomposition.

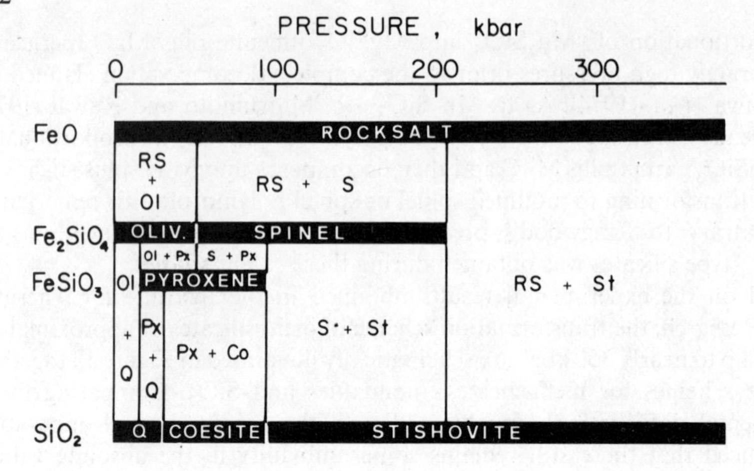

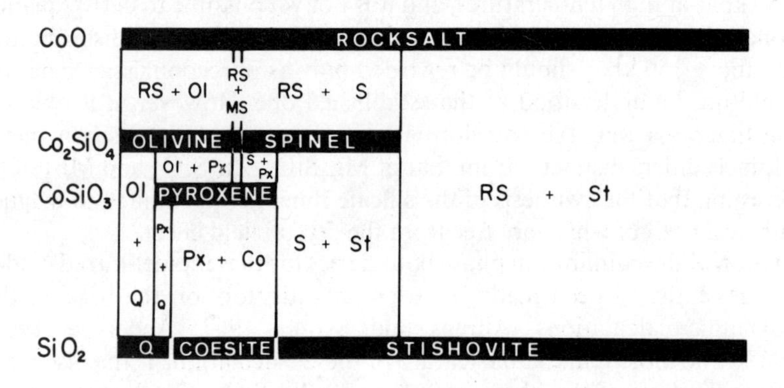

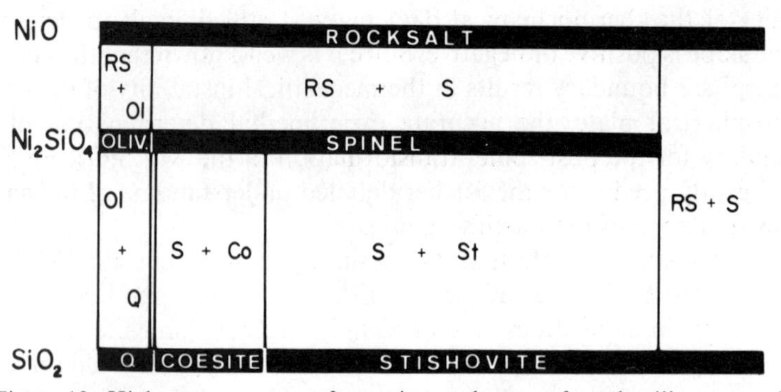

Figure 13. High-pressure transformation scheme of orthosilicates and metasilicates at approximately 1000 °C. The transformation pressures and sequence above 150 kbar are of reconnaissance nature. Data sources are seen in the text.

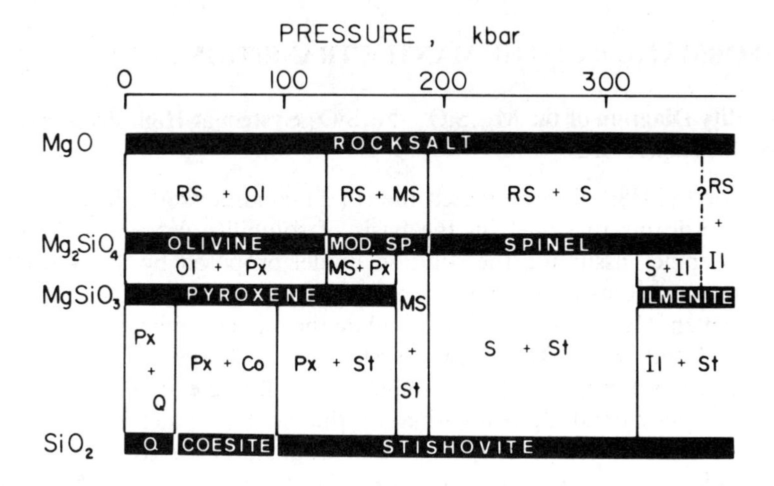

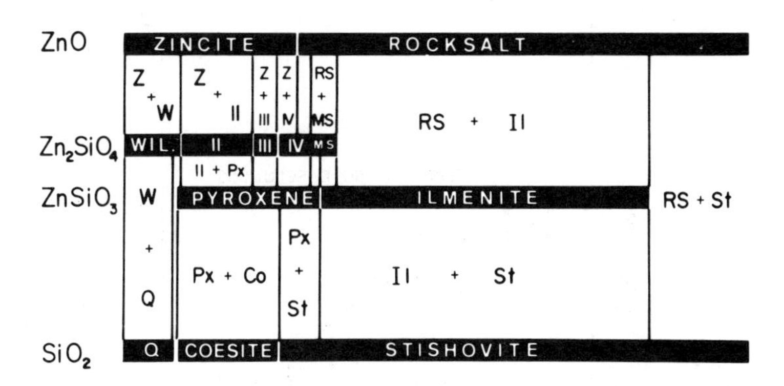

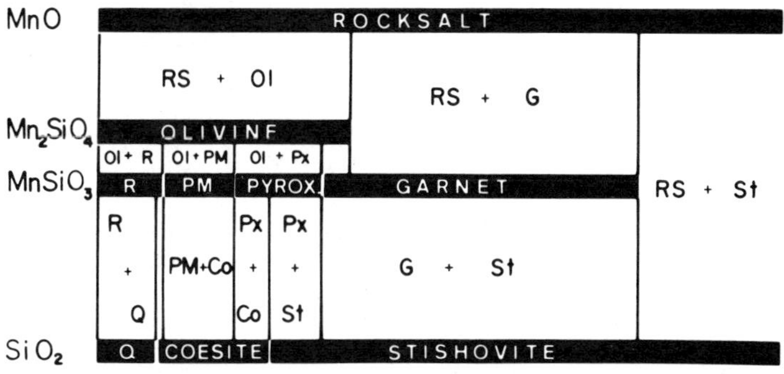

Note: Phase relations in $MgSiO_3$ above 300 kbar are only provisional. Disproportionation of spinel into rocksalt plus ilmenite has not yet been realized. Kumazawa *et al.* (1974) reported the decomposition of spinel into rocksalt plus stishovite at 330 kbar and 1000 °C, the claim being not supported by later studies (Kawai *et al.*, 1974; Ito, personal communication).

7. Stability Diagram of the Mg_2SiO_4–Fe_2SiO_4 System at High Pressures and High Temperatures

Ringwood (1958a) first estimated the (P, T) conditions at which Mg_2SiO_4 olivine transforms to spinel on the basis of solubility data of Mg_2SiO_4 in Ni_2GeO_4 spinel, assuming the solution model proposed by Temkin (1954). Soon after the successful synthesis of γFe_2SiO_4 (Ringwood, 1958b), Meijering and Rooymans (1958) attempted to calculate the olivine–spinel transformation diagram of the Mg_2SiO_4–Fe_2SiO_4 system using Ringwood's data. These works served as a guide to olivine–spinel solid solution equilibria.

Direct experimental determination of the olivine–spinel transformation diagram was first attempted by Boyd and England (1960) for the extremely Fe-rich side of the system, i.e. from Fe_2SiO_4 to $(Mg_{0\cdot 14}Fe_{0\cdot 86})_2SiO_4$ using

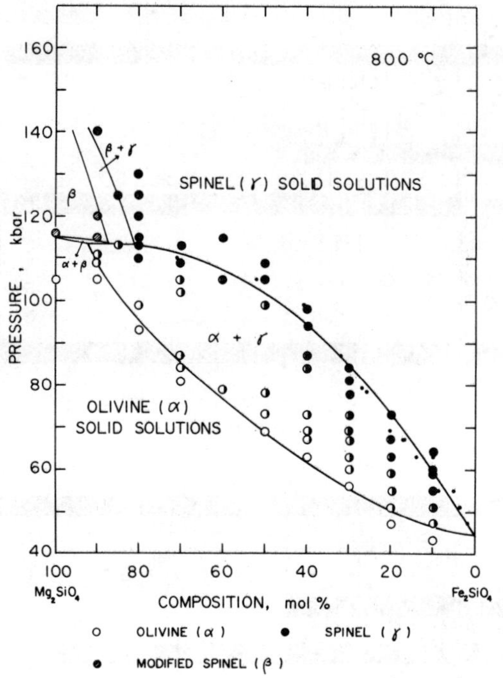

Figure 14. Phase relationships in the system Mg_2SiO_4–Fe_2SiO_4 at pressures up to 140 kbar and at 800 °C. Detailed tables showing results of individual runs will be published elsewhere (Akimoto and Kawada, in preparation). The nature of phases observed in individual runs is shown by symbols. Dots indicate the spinel composition determined from the lattice parameter.

the piston-cylinder type of high pressure apparatus. This work was followed by a series of investigations using apparatus of higher capabilities (Sclar and Carrison, 1966; Ringwood and Major, 1966a, 1970; Akimoto and Fujisawa, 1966, 1968; Akimoto and Ida, 1966; Kawai *et al.*, 1970; Ito *et al.*, 1971; Akimoto, 1972; Suito, 1972; Ito, Matsui *et al.*, 1974). Historical developments through these investigations are critically reviewed in recent publications by Ringwood and Major (1970) as well as by Akimoto (1972).

Among these investigations, the most comprehensive data have been provided from the Institute for Solid State Physics and the Australian National University. The latest versions of the phase relationships in the Mg_2SiO_4–Fe_2SiO_4 system obtained at the Institute for Solid State Physics are shown in Figures 14, 15 and 16, where the nature and proportions of phases observed in the high-pressure and high-temperature experiments are indicated in terms of isothermal sections at 800, 1000 and 1200 °C, respectively. Interrelationships among the stability field of olivine (α), modified spinel (β) and spinel (γ) solid solutions were first clarified over the whole range of composition and over wide ranges of pressure and temperature. These diagrams arc constructed by adding the new data obtained with the aid of the Bridgman-anvil type apparatus to previous data from the tetrahedral anvil press (Akimoto and Fujisawa, 1968).

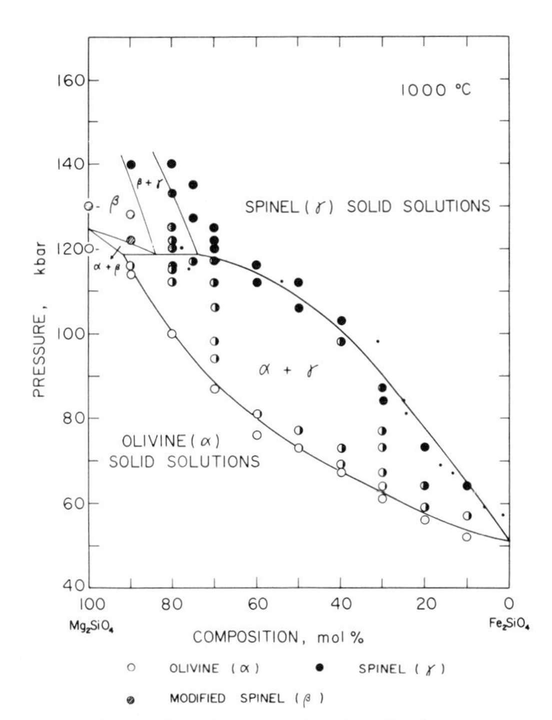

Figure 15. Phase relationships in the system Mg_2SiO_4–Fe_2SiO_4 at pressures up to 140 kbar and at 1000 °C.

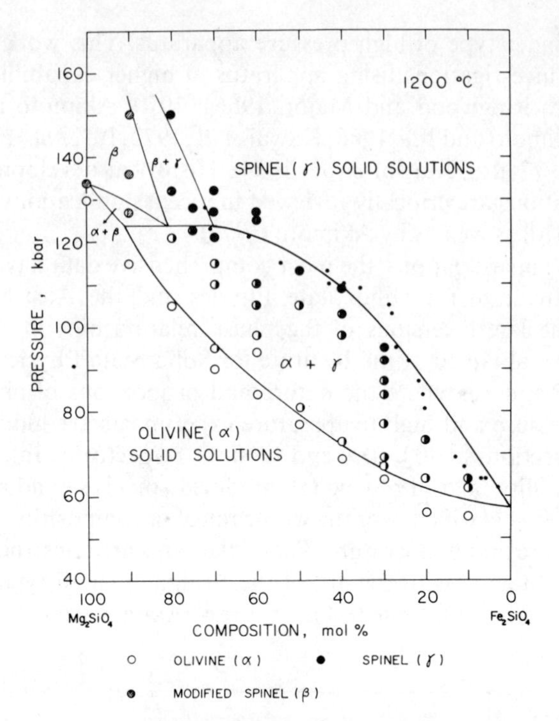

Figure 16. Phase relationships in the system Mg_2SiO_4–Fe_2SiO_4 at pressures up to 150 kbar and at 1200 °C.

The most remarkable feature of these phase diagrams is the appearance of the modified spinel phase exclusively in the Mg-rich side of the system, which was first discovered by Ringwood and Major (1966a, 1970). As described in the earlier section, the stability of the modified spinel structure is now established from a recent high-pressure and high temperature X-ray diffraction analysis. The stability field for $\beta(Mg, Fe)_2SiO_4$ is highly temperature-dependent. A remarkable expansion of the $\beta(Mg, Fe)_2SiO_4$ field is clearly seen as the temperature increases from 800 to 1200 °C. Although the investigation is progressively being extended to the higher pressure side, it is still difficult to determine the precise value of the transformation pressure from modified spinel to spinel in pure Mg_2SiO_4. Detailed experimental data used for preparing these diagrams will be published separately (Akimoto and Kawada, in preparation).

Using natural olivine with 93 mole per cent Mg_2SiO_4 and 7 mole per cent Fe_2SiO_4, Ito *et al.* (1971) determined the reaction boundary at which the natural olivine breaks down to spinel solid solution and olivine solid solution as P (kbar) $= 61 + 0.054\ T$ (°C). This determination is reasonably consistent with the phase relationships shown in Figures 14–16. The 1000 °C isothermal section of the Mg_2SiO_4–Fe_2SiO_4 system has also been reported by Ringwood and Major (1970) and Suito (1972). Phase boundaries shown in Figure 15 are in fairly good agreement with those obtained by Ringwood and Major (1970) in the pressure range below approximately 130 kbar. However, the phase boundaries between modified spinel solid solution and spinel solid solution in the figure are appreciably different from those reported by Ringwood and Major (1970) and Suito (1972). Ringwood and Major never succeeded in synthesizing the γ phase of $(Mg, Fe)_2SiO_4$ containing more than 90 per cent Mg_2SiO_4, and hence they show an almost pressure-independent phase boundary between the (β) and ($\beta + \gamma$) fields. Suito determined the β–γ transformation pressure of Mg_2SiO_4 to be approximately 220 kbar, taking the experiments with compositions ranging from pure Mg_2SiO_4 to $(Mg_{0.7}Fe_{0.3})_2SiO_4$ into consideration. The predicted transformation pressure from Figure 15 is about 50 kbar lower than his determination. As mentioned before, Suito further reported the unacceptably high dP/dT value of 0.13 kbar/°C for the α–β transformation of Mg_2SiO_4.

These serious disagreements between the experimental results above 130 kbar may be attributable to the difference in the sample assembly which is inherent in the different types of high-pressure apparatus. Pressure values at high temperatures are greatly affected by the temperature-dependent variation in the mechanical properties of the solid pressure-transmitting material and by the thermal expansion and the phase transformation of samples as well as pressure cell. Therefore, it is likely that the difference in the inner structure of the pressure cell leads to the discrepancy of the pressure values. Development of an accurate pressure calibration method at high temperature is an urgent necessity for the quantitative analysis of the phase diagram of the Mg_2SiO_4–Fe_2SiO_4 system.

8. Structure of the Upper Half of the Transition Zone in the Earth's Mantle

Since the phase diagram for the Mg_2SiO_4–Fe_2SiO_4 system has been approximately completed as shown in Figures 14–16, application of this diagram to geophysical problems will be discussed in this section. The main purpose of the analysis is to examine how satisfactorily the present phase diagram explains the observed seismic discontinuities.

The chemical and mineralogical constitution of the upper mantle has been discussed extensively by Clark and Ringwood (1964) and Ringwood (1966). They concluded that olivine is the major constituent of the upper mantle and its chemical composition is close to 89 mole per cent Mg_2SiO_4 and 11 mole per cent Fe_2SiO_4 down to depths around 200 km. In the following discussion, as a first approximation, the single-component olivine mantle was assumed and no further change in the Mg/Fe ratio in the olivine was considered to occur at greater depths in the mantle. The analysis of the mantle transition zone along this line has already been carried out by a number of investigators (Anderson, 1967; Akimoto and Fujisawa, 1968; Fujisawa, 1968; Ringwood and Major, 1970). However, the experimental uncertainty of the phase diagram used by them gives rise to considerable ambiguities in their conclusions. Particularly,

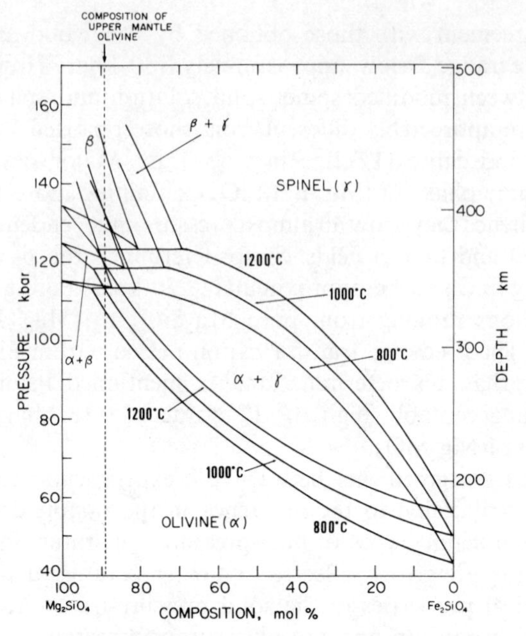

Figure 17. Olivine–spinel (modified spinel) transformation diagram for the system Mg_2SiO_4–Fe_2SiO_4.

the lack of the precise information on the gradient dP/dT for the α–β and the β–γ transformation was fatal to the detailed discussion of the transition zone.

The diagrams shown in Figure 14–16 are rearranged in Figure 17 in order to show that effect of temperature more explicitly. It is seen in the figure that the phase change of the mantle olivine with a composition of 89 mole per cent Mg_2SiO_4 occurs in the following sequence with increasing depth:

$$(\alpha) \to (\alpha + \gamma) \to (\alpha + \beta) \to (\beta) \to (\beta + \gamma) \to (\gamma).$$

In the first step of the phase change, olivine begins to equilibrate with the Fe-rich spinel solid solution with a composition near $(Mg_{0.5}Fe_{0.5})_2SiO_4$. However, the pressure range at which the olivine solid solution coexists with the spinel solid solution is extremely narrow, being estimated to be less than 5 kbar at temperatures around 1500 °C. As pressure increases, continuous phase change from $(\alpha + \gamma)$ via $(\alpha + \beta)$ to (β) takes place. The amount of the Fe-rich spinel solid solution is always far smaller compared with the coexisting olivine solid solution. Therefore, the density increase associated with the $(\alpha) \to (\alpha + \gamma)$ transformation is not so large compared with that during the second and third step of the phase change, $(\alpha + \gamma) \to (\alpha + \beta) \to (\beta)$. It should be remembered that the density increase in the α–β transformation in Mg_2SiO_4 is nearly 8 per cent. Pressure values at which the above sequence of phase change starts depend entirely upon temperature. Figure 17 indicates that at temperatures ranging from 1000 to 1500 °C, the transformation pressure varies from approximately 115 kbar to 132 kbar, corresponding to depths of 350 to 400 km. It also suggests that the total transformation width from α to β requires a pressure range of approximately 10 kbar around 1500 °C.

All these implications are consistent with recent seismic models for the transition zone of the Earth's mantle shown as Figure 18, which were constructed from the models proposed by Kanamori (1967), Archambeau *et al.* (1969), Whitcomb and Anderson (1970) and Helmberger and Wiggins (1971). Figure 18 indicates that the first step of the sharp velocity increase occurs at depths between 380 to 430 km and that a relative increase in *P*-wave velocity of 7 to 11 per cent is involved. Recent laboratory measurement of the ultrasonic wave velocities for the high-pressure polymorphs of Mn_2GeO_4 reveals that the *P*-wave velocity jump across the α–β transformation actually amounts to 5 to 18 per cent (Liebermann, 1973a, b; Mizutani, personal communication).

Based on these observational and experimental data, it is quite natural to

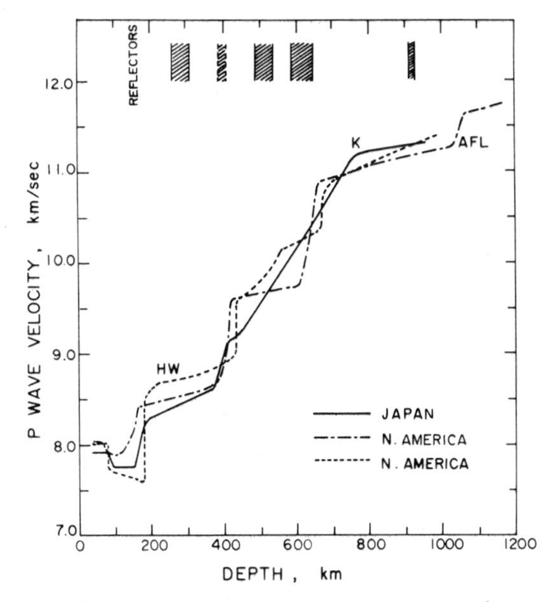

Figure 18. Distribution of seismic P wave velocities in the outer 1200 km of the mantle. Curves designated by K, AFL and HW represent the models proposed by Kanamori (1967) for the Japanese Islands, by Archambeau, Flinn and Lambert (1969) for the Western continental United States, and by Helmberger and Wiggins (1971) for the midwestern United States, respectively. Reflecting zones proposed by Whitcomb and Anderson (1970) are also indicated in the upper part of the figure.

conclude that the seismic discontinuity between 380 and 430 km is due to the olivine-modified spinel transformation. Accordingly, if the first step of the discontinuity at a depth around 380 km is taken as a faithful reflection of the beginning of the $(\alpha) \rightarrow (\alpha + \gamma) \rightarrow (\alpha + \beta) \rightarrow (\beta)$ transformation, the temperature at that depth can be determined as approximately 1400 °C. Figure 17 further indicates that the transformation from α to γ is likely to range over about 75–90 kbar (corresponding to 200–250 km width) under the geothermal gradient of 1.0–2.0 degree/km. There is little doubt that the modified spinel phase of $(Mg,Fe)_2SiO_4$ is the major constituent in the upper half of the transition zone from about 400 km to a depth just above the second distinctive seismic discontinuity occurring at about 650 km.

Helmberger and Wiggins (1971) suggested an inflection at 500 km in their seismic model of the upper mantle structure (Figure 17). Whitcomb and Anderson (1970) also recognized seismic reflectors at depths around 520 km. The phase diagram shown in Figure 17 strongly suggests that this discontinuity is due to the β–γ transformation, provided that a geothermal gradient of of about 1 °/km is adopted. This model also leads to the conclusion that the spinel solid solution with a composition of 89 mole per cent Mg_2SiO_4 exists as a single phase in a rather limited interval of the depths from about 580 km to 650 km, if the 650 km discontinuity is assumed to result from the post-spinel transformation. At the present stage of the super-high-pressure experiments, it is still difficult to compose the phase diagram for the post-spinel transformation of the Mg_2SiO_4–Fe_2SiO_4 system. Detailed analyses of the lower half of the transition zone is left for future investigation.

Although the present phase diagram of the Mg_2SiO_4–Fe_2SiO_4 system satisfactorily matches the seismic observations, it is obvious that the single component olivine model discussed above is too simple to be applied to the Earth's mantle. The presence of other components such as orthopyroxene, clinopyroxene and garnet must be considered. Recent laboratory experiments demonstrate that phase transformations take place in these minerals too, possibly in the transition zone. Ringwood and Major (1966b) and Ringwood (1967) claimed that the pyroxene-garnet transformation is partly responsible for the formation of the first step of the transition zone. Although the detailed phase diagram has not yet been completed, they suggest that it occurs at depths between 350 and 400 km. High-pressure disproportionation of pyroxene into spinel (or modified spinel) plus stishovite has been established for $MgSiO_3$ (Ito *et al.*, 1972) and for $FeSiO_3$ (Ringwood and Major, 1966c; Akimoto and Syono, 1970). Very recently Kawai *et al.*, 1974) succeeded in synthesizing the ilmenite polymorph of $MgSiO_3$. Some of these reactions and transformations may be responsible for the seismic wave-velocity increase in the lower-half of the transition zone. In the present state of our knowledge, it should be emphasized that we are actually ignorant of even the ratio (Mg + Fe)/Si in the lower transition zone. It follows that if this ratio is close to unity, the metasilicate stoichiometry should acquire greater importance. The possibility that the silicate ilmenite may be a major constituent in the lower mantle cannot be completely ruled out.

Further, it must be emphasized that reactions among the coexisting minerals have a considerable influence on the structure of the transition zone. Recently Ahrens (1973) attempted to calculate the distribution of Fe^{2+} and Mg^{2+} between coexisting garnet, pyroxene, olivine and spinel on the basis of the simple ideal solid solution model. He showed that the partition coefficient for Fe^{2+} and Mg^{2+} between garnet and olivine (spinel) varies systematically in the transition zone of the mantle. Experimental investigations along this line also started recently. Nishizawa and Akimoto (1973) concluded that the relative concentration of Fe^{2+} in spinel in the pyroxene–spinel system is likely to cause some change in the chemical composition of the modified spinel or spinel in the upper-half of the transition zone. This compositional change may result in a considerable decrease in the transformation pressure of (β) to ($\beta + \gamma$) and ($\beta + \gamma$) to (γ). The partition reaction of the mantle minerals seems to become one of the important clues for the investigation of the wave-velocity discontinuity.

ACKNOWLEDGEMENTS

The authors gratefully appreciate receiving private communications upon post-spinel transformations in orthosilicates from Eiji Ito of Institute for Thermal Spring Research in advance of publication.

REFERENCES

Ahrens, T. J. (1973). *Phys. Earth Planet. Inter.*, **7**, 167–186.
Ahrens, T. J. and Y. Syono (1967). *J. Geophys. Res.*, **72**, 4181–4188.
Akimoto, S. (1970). *Phys. Earth Planet. Inter.*, **3**, 189–195.
Akimoto, S. (1972). *Tectonophys.*, **13**(1–4), 161–187.
Akimoto, S. and H. Fujisawa (1966). *Earth Planet. Sci. Lett.*, **1**, 237–240.
Akimoto, S. and H. Fujisawa (1968). *J. Geophys. Res.*, **73**, 1467–1479.
Akimoto, S. and Y. Ida (1966). *Earth. Planet. Sci. Lett.*, **1**, 358–359.
Akimoto, S. and Yosiko Sato (1968) *Phys. Earth Planet. Inter.*, **1**, 498–504.
Akimoto, S. and Yosiko Sato (1975). Proc. of the First International Conference on Kimberlites, *Phys. and Chem. of the Earth vol. 9*, 837–843.
Akimoto, S. and Y. Syono (1967). *J. Chem. Phys.*, **47**, 1813–1817.
Akimoto, S. and Y. Syono (1970). *Phys. Earth Planet. Inter.*, **3**, 186–188.
Akimoto, S., H. Fujisawa and T. Katsura (1965). *J. Geophys. Res.*, **70**, 1969–1977.
Akimoto, S., E. Komada and I. Kushiro (1967). *J. Geophys. Res.*, **72**, 679–686.
Anderson, D. L. (1967). *Science*, **157**, 1165–1173.
Archambeau, C. B., E. A. Flinn and D. G. Lambert (1969). *J. Geophys. Res.*, **74**, 5825–5866.
Banno, S. and Y. Matsui (1967). *Proc. Japan Acad.*, **43**, 762–767.
Bassett, W. A. and T. Takahashi (1970). *EOS Trans. AGU*, **51**, 828.
Bassett, W. A. and L. Ming (1972). *Phys. Earth Planet. Inter.*, **6**, 154–160.
Bernal, J. D. (1936). Discussion, *Observatory*, **59**, 268.
Birch, F. (1952). *J. Geophys. Res.*, **57**, 227–286.
Birle, J. D., G. V. Gibbs, P. B. Moore and J. V. Smith (1968). *Amer. Mineral.*, **53**, 807–824.

Boyd, F. R. and J. L. England (1960). *Carnegie Inst. Year Book*, **59**, 48–49.
Brown, G. E. and C. T. Prewitt (1973). *Amer. Mineral.*, **58**, 577–587.
Burns, R. G. (1970). *Mineralogical Application of Crystal-Field Theory*, Cambridge Univ. Press, Cambridge.
Bush, W. R., S. S. Hafner and D. Virgo (1970). *Nature*, **227**, 1339–1341.
Clark, S. P. Jr. and A. E. Ringwood (1964). *Rev. Geophys.*, **2**, 35–88.
Dachille, F. and R. Roy (1960). *Am. J. Sci.*, **258**, 225–246.
Durif-Varambon, A., E. F. Bertaut and R. Pauthenet (1956). *Ann. Chim. (Paris)* [13], **1**, 525.
Finger, L. W. (1971). *Carnegie Inst. Year Book*, **69**, 302–305.
Finger, L. W. and D. Virgo (1971). *Carnegie Inst. Year Book*, **70**, 221–225.
Fujisawa, H. (1968), *J. Geophys. Res.*, **73**, 3281–3294.
Helmberger, D. and R. A. Wiggins (1971). *J. Geophys. Res.*, **76**, 3229–3245.
Ito, E. (1975). High-pressure decompositions in cobalt and nickel silicates, *Phys. Earth Planet. Inter.*, **10**, 88–93.
Ito, E. and Y. Matsui (1974). *Phys. Earth Planet.*, **9**, 344–352.
Ito, E., T. Matsumoto and N. Kawai (1974). *Phys. Earth Planet. Inter.*, **8**, 241–245.
Ito, E., T. Matsumoto, K. Suito and N. Kawai (1972). *Proc. Japan Acad.*, **48**, 412–415.
Ito, E., Y. Matsui, K. Suito and N. Kawai (1974). *Phys. Earth Planet. Inter.*, **8**, 342–344.
Ito, K., S. Endo and N. Kawai (1971). *Phys. Earth Planet. Inter.*, **4**, 425–428.
Jeffreys, H. (1937). *Mon. Not. Roy. Astron. Soc. Geophys. Suppl.*, **4**, 50–61.
Kamb, B. (1968). *Amer. Mineral.*, **53**, 1439–1455.
Kanamori, H. (1967). *Bull. Earthq. Res. Inst. Tokyo Univ.*, **45**, 657–678.
Kawai, N. and S., Endo (1970). *Rev. Sci. Instr.*, **41**, 1178–1181.
Kawai, N., S. Endo and K. Ito (1970). *Phys. Earth Planet. Inter.*, **3**, 182–185.
Kawai, N., M. Tachimori and E. Ito (1974). *Proc. Japan Acad.*, **50**, 378–380.
Kumazawa, M. (1971). *High Temp.–High Pressures*, **3**, 243–260.
Kumazawa, M., H. Sawamoto, E. Ohtani and K. Masaki (1974). *Nature*, **247**, 356–358.
Kirtel, A. (1972). *Ph.D. Thesis*, Universität Bonn.
Liebermann, R. C. (1973a). *J. Geophys. Res.*, **78**, 7015–7017.
Liebermann, R. C. (1973b). *Nature*, **244**, 105–107.
Liu, L.-G. (1974). Silicate Perovskite from phase transformations of pyrope–garnet at high pressure and temperature, *Geophys. Res. Lett.*, **1**, 277–280.
Liu, L. G. (1975). *Earth Planet. Sci. Lett.*, **24**, 357–362.
Liu, L. G. (1975). High-pressure disproportionation of Co_2SiO_4 spinel and implications for Mg_2SiO_4 spinel, *Earth Planet. Sci. Lett.*, **25**, 286–290.
Ma, Che-Bao (1974). *J. Geophys. Res.* **79**, 3321–3324.
Mao, H. K. (1967). *Ph.D. Thesis*, Univ. of Rochester, New York.
Mao, H. K. and P. M. Bell (1971). *Carnegie Inst. Year Book*, **70**, 176–177.
Mao, H. K. and P. M. Bell (1972). *Carnegie Inst. Year Book*, **71**, 527–528.
Matsui, Y. and Y. Syono (1968). *Geochem. J.*, **2**, 51–59.
Meijering, J. L. and C. J. M. Rooymans (1958). *Proc. K. Ned. Akad. Wet.*, **B61**, 333–344.
Ming, L. C. and W. A. Bassett (1974). *EOS Trans. AGU*, **55**, 416.
Ming, L. C. and W. A. Bassett (1975). The post-spinel phases in the Mg_2SiO_4–Fe_2SiO_4 system, *Science*, **187**, 66–68.
Ming, L. C. and W. A. Bassett (1975). High-pressure phase transformations in the system of $MgSiO_3$–$FeSiO_3$, *Earth Planet. Sci. Lett.*, **27**, 85–89.
Moore, P. B. and J. V. Smith (1969). *Nature*, **221**, 653–655.
Moore, P. B. and J. V. Smith (1970). *Phys. Earth Planet. Inter.*, **3**, 166–177.
Morimoto, N., S. Akimoto, K. Koto and M. Tokonami (1969). Science, **165**, 586–588.
Morimoto, N., S. Akimoto, K. Koto and M. Tokonami (1970). *Phys. Earth Planet. Inter.*, **3**, 161–165.

Morimoto, N., M. Tokonami, K. Koto and S. Nakajima (1972). *Amer. Mineral.*, **57**, 62–75.
Morimoto, N., M. Tokonami, M. Watanabe and K. Koto (1974). *Amer. Mineral.*, **59**, 475–485.
Nishikawa, M. and S. Akimoto (1971). *High Temp.–High Pressures*, **3**, 161–176.
Nishizawa, O. and Y. Matsui (1972). *Phys. Earth Planet. Inter.*, **6**, 377–384.
Nishizawa, O., and S. Akimoto (1973). *Contr. Mineral., Petrol.*, **41**, 217–230.
Pauling. L. (1960). *The Nature of the Chemical Bond*, 3rd edn., Cornell Univ. Press. Ithaca.
Ringwood, A. E. (1958a). *Geochim. Cosmochim Acta*, **13**, 303–321.
Ringwood, A. E. (1958b). *Geochim. Cosmochim. Acta*, **15**, 18–29.
Ringwood, A. E. (1962). *Geochim. Cosmochim. Acta*, **26**, 457–467.
Ringwood, A. E. (1963). *Nature*, **198**, 79–80.
Ringwood, A. E. (1966). In: P. Hurley (ed.), *Advances in Earth Science*, 357–398, MIT Press, Boston.
Ringwood, A. E. (1967). *Earth Planet. Sci. Lett.*, **2**, 255–263.
Ringwood, A. E. (1970). *Phys. Earth Planet. Inter.*, **3**, 109–155.
Ringwood, A. E. and A. Major (1966a). *Earth Planet. Sci. Lett.*, **1**, 241–245.
Ringwood, A. E. and A. Major (1966b). *Earth Planet. Sci. Lett.*, **1**, 351–357.
Ringwood, A. E. and A. Major (1966c). *Earth Planet. Sci. Lett.*, **1**, 135–136.
Ringwood, A. E. and A. Major (1970). *Phys. Earth Planet. Inter.*, **3**, 89–108.
Ringwood, A. E. and A. F. Reid (1968a). *Earth Planet. Sci. Lett.*, **5**, 67–70.
Ringwood, A. E. and A. F. Reid (1968b). *Earth Planet. Sci. Lett.*, **5**, 245–250.
Sarver, J. F. and F. A. Hummel (1962). *J. Am. Ceram. Soc.*, **45**, 304.
Sclar, C. B. and L. C. Carrison (1966). *Trans. Am. Geophys. Union*, **41**, 207.
Shannon, R. D. and C. T. Prewitt (1969). *Acta Cryst.*, **B25**, 925–946.
Shannon, R. D. and C. T. Prewitt (1970). *Acta Cryst.*, **B26**, 1046–1048.
Smyth, J. R. and R. Hazen (1973). *Amer. Mineral.*, **58**, 588–593.
Stishov. S. M. and S. V. Popova (1961). *Geokhimiya*, 1961 (10), 837–839.
Strens, R. G. J. (1969). In: S. K. Runcorn (ed.), *The Application of Modern Physics to the Earth and Planetary Interiors*, 213–220, Wiley-Interscience., London.
Suito, K. (1972). *J. Phys. Earth*, **20**, 225–243.
Syono, Y., S. Akimoto and Y. Matsui (1971). *J. Solid St. Chem.*, **3**, 369–380.
Syono, Y., M. Tokonami and Y. Matsui (1971). *Phys. Earth Planet. Inter.*, **4**, 347–352.
Temkin, M. (1945). *Acta Phys.-Chim. USSR*, **20**, 411.
Thompson, J. B. (1969). *Amer. Mineral.*, **54**, 341–375.
Tokonami, M., N. Morimoto, S. Akimoto, Y. Syono and H. Takeda (1972). *Earth Planet. Sci. Lett.*, **14**, 65–69.
Virgo, D. and S. Hafner (1972). *Earth Planet Sci. Lett.*, **14**, 305–312.
Wadsley, A. D., A. F. Reid and A. E. Ringwood (1968). *Acta Cryst.*, **B24**, 740–742.
Whitcomb, J. H. and D. L. Anderson (1970). *J. Geophys. Res.*, **75**, 5713–5728.
Wyckoff, R. W. G. (1965). *Crystal Structure* Vol. 3, 2nd edn., Wiley, New York.
Yagi, T., F. Marumo and S. Akimoto (1974). *Amer. Mineral.*, **59**, 486–490.

New Applications of the Diamond Anvil Pressure Cell: (II) Laser Heating at High Pressure

W. A. Bassett and L. C. Ming
Department of Geological Sciences
University of Rochester
Rochester, New York 14627

LASER HEATING IN THE DIAMOND CELL

Laser heating of samples under pressure has been carried out in a fairly standard style of diamond cell which we call the UP type (Figure 1). Diamond anvils (~ 25 mg) having faces 0.3 mm and 0.5 mm in diameter are mounted on rockers which permit the alignment of the diamond anvils so that their faces are parallel to each other. The rockers in turn are mounted in troughs,

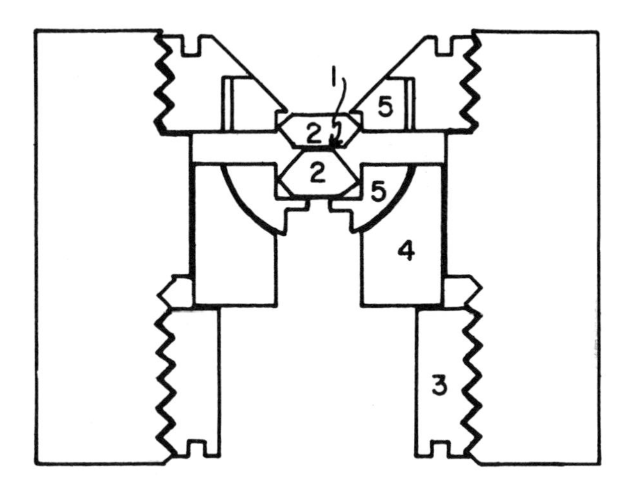

Figure 1. The UP type of diamond cell. A sample (1) placed between the diamond anvils (2) is subjected to pressure when driver screw (3) applies a force to sliding piston (4). The rockers (5) are used to align the anvil faces parallel to each other.

one fixed firmly in place, the other on the end of a sliding piston. A polycrystalline sample is subjected to pressure when a screw applies a force to the sliding piston which then drives one diamond anvil against the other trapping the sample between.

The sample is easily viewed through the upper anvil by means of a microscope when light is introduced through the lower anvil. An infrared laser beam ($\lambda = 1.06$ μm) from a YAG laser can be introduced into the microscope by reflecting it from a 45° dichroic mirror placed in the light path in the microscope as shown in Figure 2. The laser beam then travels down the microscope and is focused onto the sample through the upper anvil by the objective. When a protective filter is employed, the sample can be observed through the third ocular of a trinocular head with an optical pyrometer as the sample is being heated with the laser beam. Rough temperature measurements are made by comparing the intensity of the incandescent sample with that of a reference

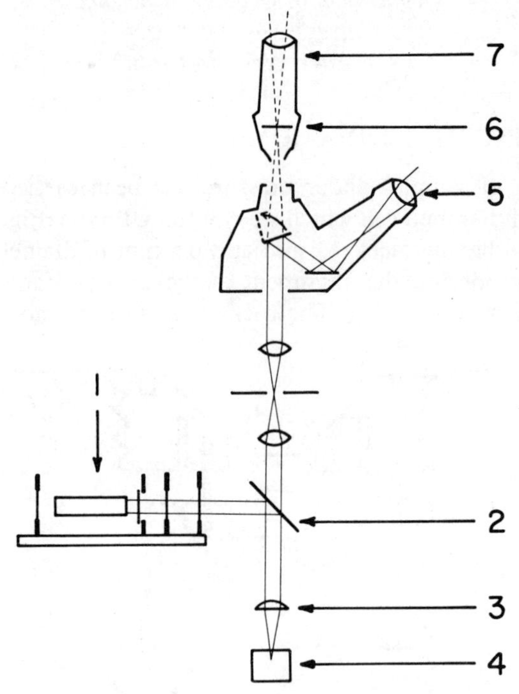

Figure 2. The optical system used to introduce an infrared laser beam into the diamond anvil cell. The laser beam produced by the YAG laser (1) is reflected by a dichroic mirror (2) through the microscope objective (3) onto the sample under pressure in the diamond cell (4). The sample may be observed during heating by the pair of oculars (5) and the temperature can be estimated by comparing the intensity of the image with that of the filament (6) in an optical pyrometer (7).

filament in the pyrometer. Although a correction is made for the effect of microscope optics on the intensity of the image, the uncertainty of true temperature is large due to the lack of information on emissivity of the sample.

The heated portion of the sample is about 10–50 μm across. The focused infrared beam can be moved about the sample simply by moving the diamond cell while the laser is on. The method of laser heating in the diamond cell is described in greater detail by Ming and Bassett (1974).

DISPROPORTIONATION OF OLIVINES TO CLOSE-PACKED OXIDES

The method of laser heating in the diamond cell has been applied to the problem of resolving the nature of the post-spinel phases of the olivine series. The approach was to place the sample under pressure in the diamond cell, sweep the laser beam back and forth across it while monitoring the temperature with the pyrometer and observing changes with the microscope, cool the sample, release the pressure, and remove the sample for X-ray analysis in a small Debye-Scherrer camera. The temperatures employed were typically 1200–1600 °C, low enough to avoid melting of the sample at the pressures used. Pressure was measured by calibration of the load against know phase transitions or calculated from lattice parameter measurements on NaCl mixed with the sample. The results show conclusively that fayalite disproportionates to 2FeO plus SiO_2 (stishovite) at 160 ± 10 kbars and 1600 ± 100 °C while forsterite disproportionates to 2MgO plus SiO_2 (stishovite) at a pressure estimated to be above 220 kbars at 1600 ± 100 °C. Intermediate compositions of $Fo_{80}Fa_{20}$, $Fo_{60}Fa_{40}$, and $Fo_{50}Fa_{50}$ disproportionate to magnesiowustite plus stishovite at intermediate pressures and 1600 ± 100 °C.

These results are consistent with those reported for pure fayalite by Bassett and Ming (1972) and for pure forsterite reported by Kumazawa *et al.* (1974).

ACKNOWLEDGEMENT

This research was supported by NSF grant GA–38056 X.

REFERENCES

Bassett, W. A. and L. C. Ming (1972). *Phys. Earth Planet. Interiors*, **6**, 154–160.
Kumazawa, M., H. Sawamoto, E. Ohtani, K. Masaki (1974). *Nature*, **247**, 356.
Ming, L. C. and W. A. Bassett (1974). *Rev. Sci. Instru.*, **45**, 1115.

Mineralogical Applications of High-Temperature Reaction Calorimetry

O. J. Kleppa
The James Franck Institute, and the Departments of Chemistry and Geophysical Sciences
The University of Chicago, Chicago, Illinois 60637, U.S.A.

INTRODUCTION

Mineralogical applications of reaction calorimetry may involve either (a) *direct combination and transformation studies*, or (b) *solution calorimetry*. Among these approaches the latter is the more generally useful. While solution calorimetry using acid solvents at or near room temperature has proved useful in some cases, it has very serious limitations when the minerals involved are dense and refractory. This is the principal domain of high-temperature solution calorimetry and of the modern heat flux calorimeter.

One of the crucial problems in high-temperature solution calorimetry is the choice of a suitable solvent, and some criteria used in the selection of oxide melt solvents for work on refractory minerals are outlined. Examples of applications of reaction calorimetry to mineral systems include studies of (a) enthalpies of formation from component oxides, (b) phase transformations, (c) disorder problems, and (d) mineralogical reactions and equilibria.

For reacting system at fixed composition or for a simple, first order phase transformation, we have the well-known thermodynamic relation

$$\Delta G_T = \Delta G_T^0 + \int_1^{P_{eq}} \Delta V_T \mathrm{d}P = \Delta H_T^0 - T\Delta S_T^0 + \int_1^{P_{eq}} \Delta V_T \mathrm{d}P \quad (1)$$

At equilibrium $\Delta G_T = 0$. Thus we have also

$$\int_1^{P_{eq}} \Delta V_T \mathrm{d}P = -\Delta G_T^0 \quad (2)$$

In these expressions, ΔG_T^0 is the standard Gibbs energy change at temperature T, ΔH_T^0 is the standard enthalpy and ΔS_T^0 is the standard entropy change;

ΔV_T is the volume change associated with the reaction. Due to compressibility ΔV_T will, in general, be different from ΔV_T^0 the volume change at one atmosphere pressure. However, as long as we are dealing with condensed phases and moderate pressures, we have

$$\Delta V_T \approx \Delta V_T^0$$

and

$$P_{eq} \simeq -\frac{\Delta G_T^0}{\Delta V_T^0} \tag{3}$$

Thus, if ΔG_T^0 and ΔV_T are known, P_{eq} may be calculated or estimated. The Clapeyron equation

$$\left(\frac{dP}{dT}\right)_{eq} = \frac{\Delta S}{\Delta V} \simeq \frac{\Delta S^0}{\Delta V^0} \tag{4}$$

similarly provides a link between the temperature derivative of the equilibrium pressure and the change in entropy.

The implication of these relationships is clear. If we have information on ΔH^0, ΔS^0, *and* ΔV^0 for a given reaction or phase transformation, i.e. if we have thermodynamic data valid at one atmosphere pressure, we can readily estimate the pressure–temperature equilibrium curve. If we also know ΔV as a function of temperature and pressure, we can obtain more precise values.

Our principal interest in the present paper will be focused on the purely thermal quantities ΔS^0 and ΔH^0. For completely ordered phases S^0 and ΔS^0 in principle may be derived from low temperature heat capacity data, via the third law

$$S^0 = \int_0^T C_P \mathrm{dln} T \tag{6}$$

Where the relevant heat capacity data are lacking, approximate values of S^0 for condensed phases at high temperatures often can be estimated. However, complications arise where the substânces involved are disordered. Under these circumstances configurational entropies sometimes can be estimated from detailed structural information.

It is more difficult to make realistic predictions or estimates of ΔH^0, which usually must be obtained from data based on reaction calorimetry.

SOLUTION CALORIMETRY OF MINERALS

In the years immediately following World War II, considerable progress was made in the general area of mineral thermochemistry. This progress was sparked by the development by Torgeson and Sahama (1948) of precision hydrofluoric acid solution calorimetry. The thermodynamic basis of this method is very straightforward:

$$\mathrm{A + B + Solvent = Solution};\ \Delta H_1 \tag{7a}$$
$$\mathrm{AB + Solvent = Solution};\ \Delta H_2 \tag{7b}$$

$$\mathrm{A + B = AB};\ \Delta H^0 = \Delta H_1 - \Delta H_2 \tag{7c}$$

Thus the standard enthalpy change for a given reaction (here A + B = AB) may be obtained from the difference between two measured enthalpies of solution. Following the work of Torgeson and Sahama, hydrofluoric acid solution calorimetry was adopted by a number of investigators in the silicate field, notably in studies of feldspars by Kelley *et al.* (1953), Kracek and Neuvonen (1952), and others. More recent work based on this method has been carried out, e.g. by Waldbaum (1966).

It is believed that the first purely thermochemical calculation of the *P*–T curve for an important silicate equilibrium was that of Kracek, Neuvonen and Burley (1951) on the stability of jadeite. This work preceded by several years the experimental equilibrium work of Robertson, Birch and MacDonald (1957), and of Birch and Le Conte (1960). While the agreement between the calculated and the experimental values was far from perfect, these studies served to illustrate both the potential of the thermochemical approach, and also some of its obvious limitations. Thus, Kracek, Neuvonen and Burley found that jadeite, as most other dense minerals, dissolves only with considerable difficulty in hydrofluoric acid, even at 75 °C. Their results also indicated that the observed enthalpies of solution to some extent reflected the past treatment of the jadeite samples, i.e. whether the jadeite powders had been ground in an agate or a mullite mortar. Another problem, which is common to all work on hydrofluoric acid solution calorimetry, is the fact that the observed enthalpies of solution are large numerical quantities compared to the enthalpy changes in typical mineral reactions and phase transformations. As a result of these various problems it soon became recognized that HF solution calorimetry, although clearly useful for certain types of mineral equilibrium studies, has a somewhat limited range of application. A new and more powerful approach was provided by oxide melt solution calorimetry.

In this method, which originated with the work of Yokokawa and Kleppa (1964, 1965), the aqueous acid solvent of conventional, room temperature solution calorimetry, is replaced by a high-temperature oxide melt solvent. The technique has certain obvious advantages compared to the room temperature methods:

(1) It allows thermochemical study of a wider range of oxidic substances, and, particularly of dense and refractory minerals.

(2) The observed enthalpies of solution are numerically much smaller than in acid solution work. Without loss in the precision of the *enthalpies of reaction* this permits the use of heat flux calorimetry.

(3) It is possible to work with much smaller samples than in conventional acid solution calorimetry. With samples of the order of 25 mg, it is feasible to combine solution calorimetry with high pressure-high temperature synthesis and equilibrium work.

However, there are also disadvantages. For example, it is not convenient to use oxide melt solution calorimetry for the study of the hydrous minerals which frequently participate in low-grade metamorphic reactions.

REACTION CALORIMETRY AT HIGH TEMPERATURE

Two basically different ideas have provided the basis for design of calorimeters operated at high temperature. The idea behind the *adiabatic* and '*conventional*' (or quasi-adiabatic) designs has been to reduce as far as possible (ideally to eliminate) the heat flux ('heat leak') between the calorimeter and the surroundings.

There are several different circumstances which tend to limit the effectiveness both of adiabatic and quasi-adiabatic calorimeters at high temperature. The most serious undoubtedly are related to the increasing importance of radiation in heat transfer as the temperature is increased.

In the design of *heat flux* calorimeters no attempt is made to reduce the heat leak. Instead one tries to fix the design parameters in such a way that the integrated heat flux can be measured reproducibly and conveniently. Figure 1 illustrates typical time temperature curves for adiabatic, conventional and heat flux calorimeters. In our own work we have found it convenient to design our heat flux calorimeters in such a way that the time constant of the calorimeter is of the order of 3 minutes. Under these conditions a temperature change which occurs in the calorimeter at time 0, will have decayed to less than 1/1000

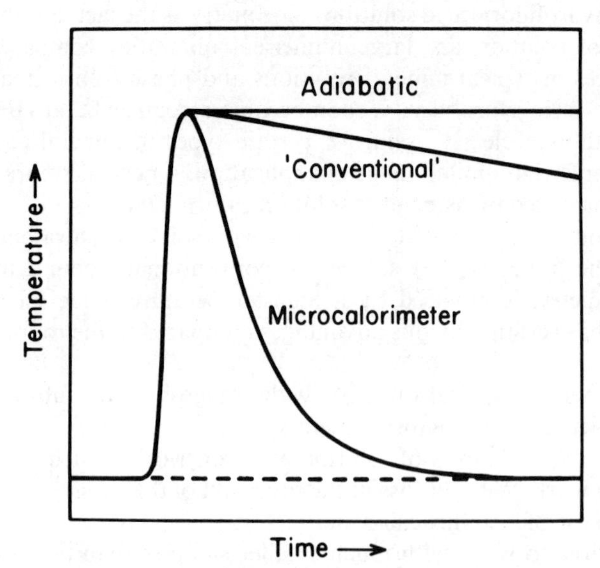

Figure 1. Schematic time–temperature curves for adiabatic, conventional (i.e. 'quasi-adiabatic') and micro (i.e. 'heat flux') calorimeters.

of its maximum value in about one-half hour. This usually is a convenient time scale for solution calorimetry.

Under favourable circumstances (for each) the heat flux calorimeter cannot compete with the adiabatic calorimeter with respect to experimental precision. In fact, we have found it very difficult to achieve a precision better than about ± 0.5 per cent in heat flux calorimetry carried out at any temperature. However, while the difficulties associated with adiabatic and conventional calorimetry increase dramatically with temperature, those of heat flux calorimetry increase much more slowly. Also, what heat flux calorimetry may lose in precision, it may regain in simplicity, convenience and reliability.

In solution (as distinct from bomb) calorimetry, the calorimetric process usually is not instantaneous, but may require a significant period of time. In order to monitor the progress of the dissolution of the solid sample, and the

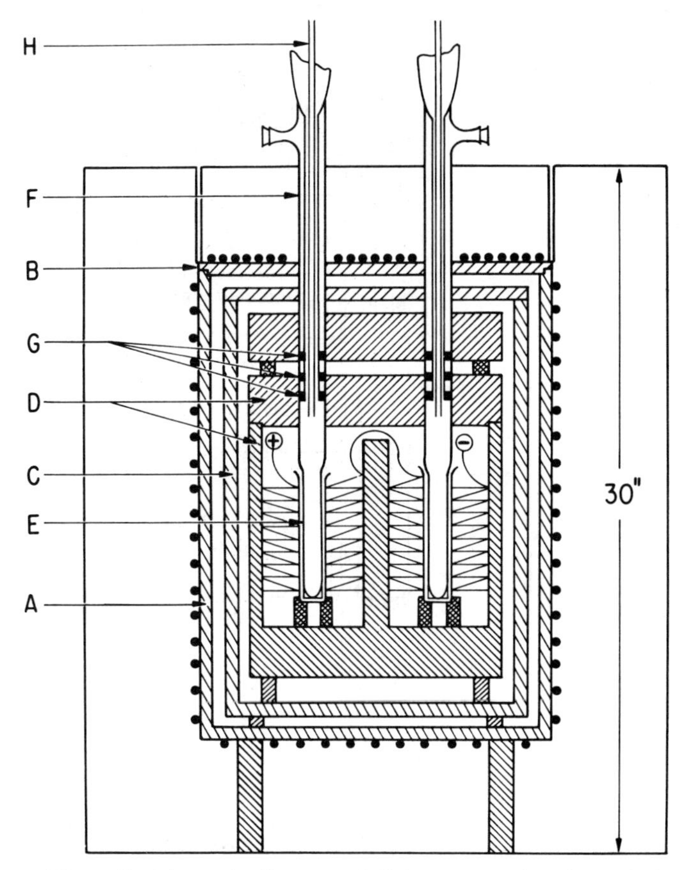

Figure 2. Schematic diagram of Calvet-type twin microcalorimeter for temperature up to about 800 °C. A: Main heater; B: Top heater; C: Heavy shield; D: Nickel block; E: Calorimeter; F: Protection tube; G: Radiation shields; H: Manipulation tube; + — –: Thermopile.

completion of the dissolution process, it is desirable to use a heat flux calorimeter with very good baseline stability. Such an apparatus is, e.g. the Calvet-type twin micro-calorimeter shown in Figure 2 (Kleppa, 1960). The twin construction, with two essentially identical calorimeters and thermopiles connected in series but backed against each other, is particularly useful in reducing the need for very close temperature control. This can be a very difficult problem at high temperatures. The apparatus illustrated in Figure 2 has a calorimeter block ('thermostat') constructed from pure nickel, which is a suitable material for operation in air at temperatures up to 7–800 °C. By choosing other construction materials such as inconel, stainless steel, nichrome, Kanthal, or aluminium oxide, the upper temperature limit of the apparatus may be increased to that indicated for the chosen material. As the temperature of operation is raised, problems frequently arise in connection with electrical noise in the thermopile system. Careful shielding of the thermopile system is indicated.

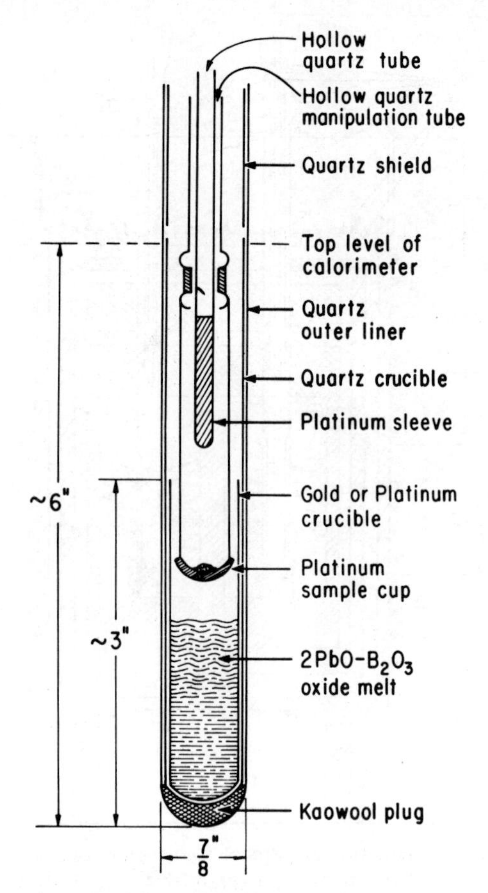

Figure 3. Schematic diagram of experimental arrangement used in oxide melt solution calorimetry.

Various experimental arrangements have been tried within the calorimeter proper. Our own preferred set-up is shown in Figure 3. Prior to the solution experiment the powdered sample (150–250 mesh; ~ 25 mg) is held in a shallow platinum cup above the surface of the solvent. Dissolution of the sample is initiated by dipping the cup into the melt. Stirring is manual by moving the platinum cup up and down several times.

Oxide Melt Solvents

In their initial work, Yokokawa and Kleppa (1964, 1965) explored the potential of liquid vanadium pentoxide at a calorimetric solvent at 700 °C. The results of this early exploration are summarized in Figure 4, which gives the

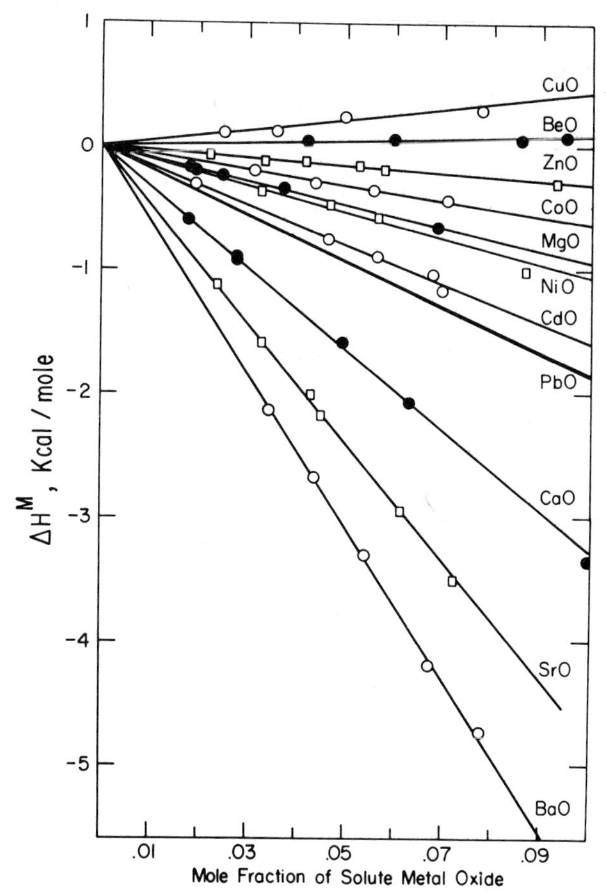

Figure 4. Molar integral enthalpies of mixing of liquid V_2O_5 + solid oxides at about 700 °C (Reprinted with permission from T. Yokkoawa and O. J. Kleppa, *Inorg. Chem.*, **4**, 1806 (1965). *Copyright by the American Chemical Society*).

enthalpy changes associated with the solution of some simple oxides MO in V_2O_5. Note in particular that for the oxides of the alkaline earth metals the enthalpies of solution change systematically in the sequence

$$BeO > MgO > CaO > SrO > BaO$$

with small positive values for BeO and quite large negative values for BaO. This is of course the order of increasing ionic radius of the cation and of increasing basicity of the solute oxide, which is consistent with the fact that the solvent V_2O_5 is a very strongly acidic oxide. These early results immediately suggest that an understanding of the elements of acid–base chemistry of melts is essential in any meaningful discussion of possible solvents for oxide melt solution calorimetry.

It is convenient to discuss the acid–base character of oxides and oxidic melts in terms of the Lux–Flood approach, which is analogous to the well-known Brönsted–Lowry scheme for protonic acids. According to Lux (1939) and Flood and Førland (1947), a suitable measure of the basicity of the melt is the oxygen ion activity, which is determined by the equilibrium

$$\text{Acid} + O^{-2} = \text{Base}.$$

The stronger the base, the higher the oxygen ion activity.

For the vanadium (V)-oxygen system we may in this scheme write a series of possible (but not necessarily realized) consecutive acid–base steps, which involve species such as V^{+5}, VO^{+3}, VO_2^+, V_2O_5, VO_3^-, $V_2O_7^{-4}$, and VO_4^{-3}. Similarly we may for each oxide MO consider the species M^{+2}, MO, MO_2^{-2}, etc. When a basic solute MO is added to liquid V_2O_5, this will give rise to the formation of vanadate species such as VO_3^-, $V_2O_7^{-4}$, and VO_4^{-3} (or polymeric forms of these). To a first approximation the magnitude and sign of the enthalpy change associated with the solution process

$$MO(s) + V_2O_5(l) = M^{+2} + \text{Vanadate (in } V_2O_5)$$

may be considered a measure of the tendency of this acid–base reaction to proceed from left to right, i.e. of the relative strength of V_2O_5 and M^{+2} as acids or of MO and vanadate as bases. The stronger the base MO (or the weaker the corresponding acid M^{+2}) the more negative the partial enthalpy of solution (ΔH_{MO}).

If the solute oxide M_nO_m is very acidic, e.g. has an acidity which approaches that of the solvent V_2O_5, the acid–base character of the dissolution process in large measure will be suppressed. Under these conditions the enthalpy of solution no longer will serve as a meaningful measure of the relative acid–base strengths of the solutes. For this purpose one should of course instead use an oxidic solvent of basic character, such as, for example, liquid PbO.

The vanadate–V_2O_5 mixtures which result from the dissolution of basic oxides in V_2O_5 are examples of buffered acid–base melts. Since most mixed oxide compounds contain two or more components of different acidity, it usually will be most convenient to use buffered melts as calorimetric solvent.

This also has the practical advantage that in the buffered regions the enthalpies of solution of acidic or basic oxides tend to vary less with composition.

We shall illustrate these problems further by considering the binary liquid system PbO–B_2O_3. At 700–800 °C, which is a convenient temperature for oxide melt solution calorimetry, the liquid range in this system extends from about 15 to 100 per cent B_2O_3 (Geller and Bunting, 1937). However, there is a miscibility gap in the high B_2O_3 region, which is also characterized by a very high viscosity. The system is glassforming, which has proved an advantage in checking that oxide samples do in fact go into solution.

Holm and Kleppa (1967) published the results of a calorimetric investigation of this system carried out at 800 °C. In this work the partial enthalpies of

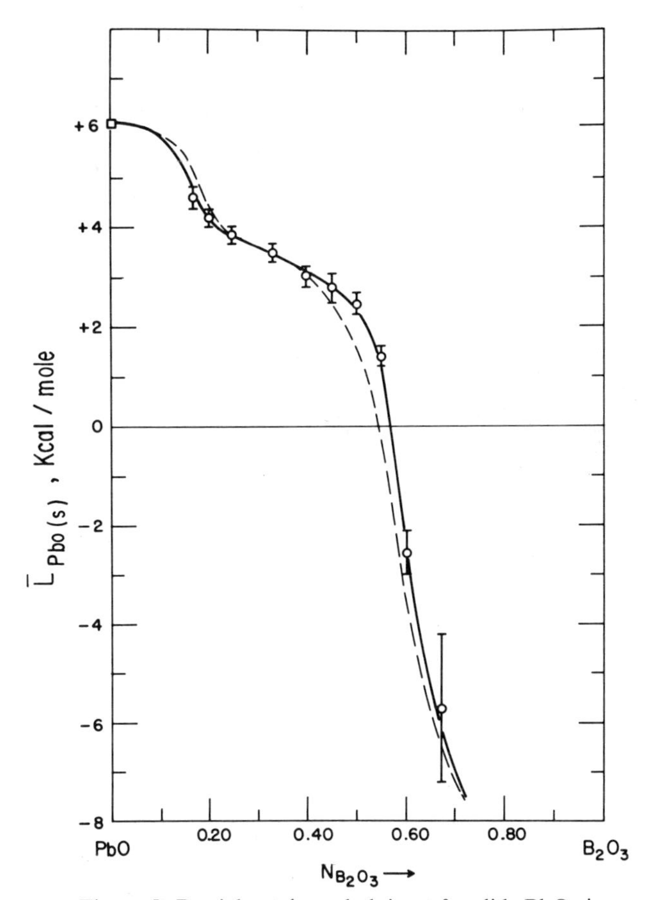

Figure 5. Partial mola-enthalpies of solid PbO in PbO–B_2O_3 melts at 800 °C □ Rodigina, Gomel'skii, and Luginina (1961); – – – By Gibbs–Duhem integration from $\bar{L}_{B_2O_3}$. (Reprinted with permission from J. L. Holm and O. J. Kleppa, *Inorg. Chem.*, 6, 645 (1967). Copyright by the American Chemical Society).

solution (actually $\Delta H/\Delta n$) of PbO(*s*) and $B_2O_3(l)$ were measured at compositions between about 17 mole per cent and about 67 mole per cent B_2O_3. At higher B_2O_3 contents the melts were too viscous to allow precise calorimetry by the direct mixing approach.

The results reported by Holm and Kleppa are shown in Figures 5 and 6. Since the partial enthalpies of both components were measured, it was possible to check the internal consistency of the result by means of the Gibbs–Duhem equation. The results of this check are indicated as broken lines in the figures. On the whole there is reasonable agreement between the calculated and measured data.

The data suggest interesting changes in the acidity of the melt as the composi-

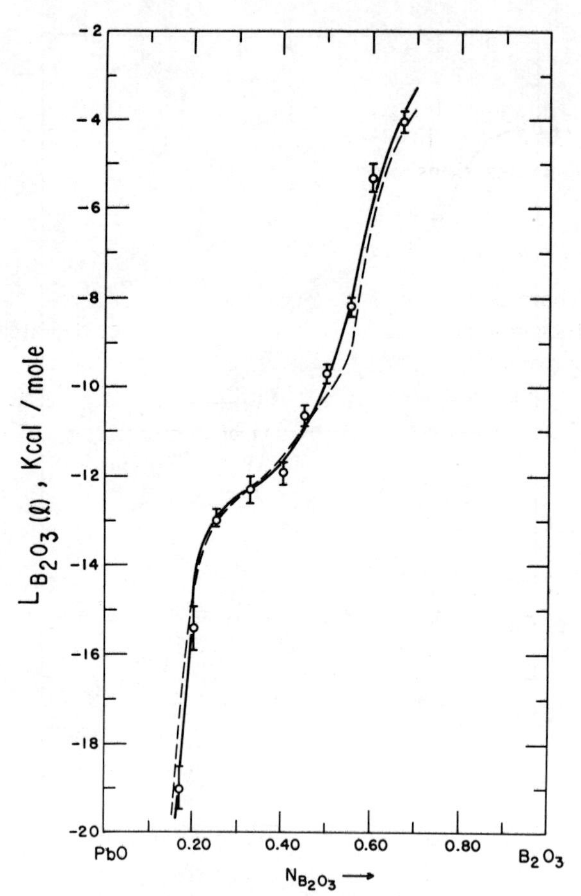

Figure 6. Partial molar enthalpies of liquid B_2O_3 in PbO–B_2O_3 melts at 800 °C– – –By Gibbs–Duhem integration from $\bar{L}_{PbO}$. (Reprinted with permission from J. L. Holm and O. J. Kleppa, *Inorg. Chem.*, **6**, 645 (1967). Copyright by the American Chemical Society).

tion varies from basic in the high PbO region to acidic in the high B_2O_3 range. We interpreted the results to indicate that the liquid lead borate system contains at least two complex borate anions, which have oxygen–boron ratios of about 3.5–4.0 and 1.84–1.91, respectively. These ratios were obtained by attempting to locate the two steeply rising parts of the partial enthalpy versus composition curves. Between these two compositions the curves are flatter, which presumably reflects the presence of both of the two borate anions in comparable amounts, i.e. a buffer region. Most of our early calorimetric work was based on the use of melts with a $MO:B_2O_3$ ratio of 3:1. However, as a result of the work by Holm and Kleppa we now prefer a somewhat more acid melt with a ratio of 2:1.

Mineralogical Applications

The mineralogical applications of high temperature reaction calorimetry fall into the following main but related categories: (a) enthalpies of formation of refractory mixed oxides, (b) phase transformations, (c) disorder problems, and (d) mineralogical reactions and equilibria.

(a) *Enthalpies of Formation*

The thermodynamics of formation of refractory mixed oxides from its component oxides is not easily studied experimentally since this area is almost inaccessible to conventional room temperature calorimetric methods. During

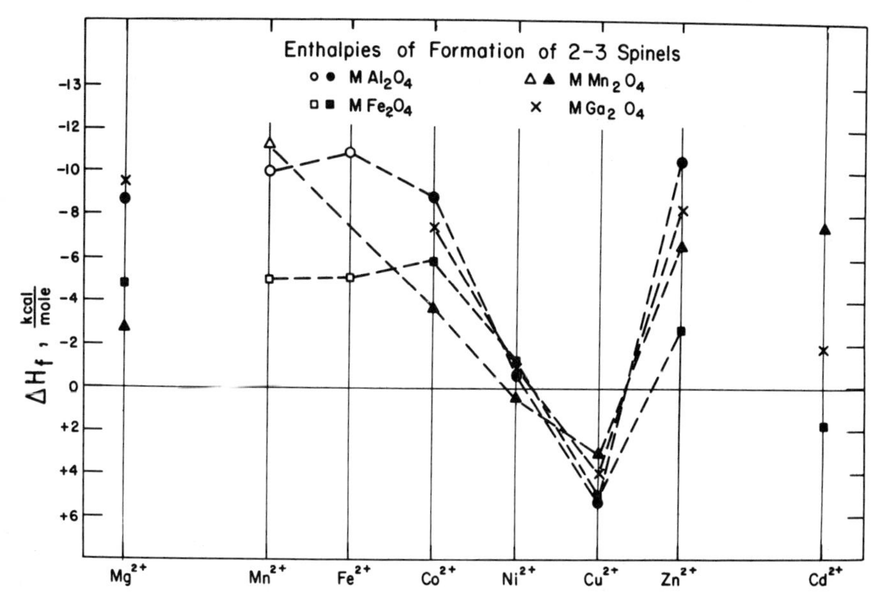

Figure 7. Enthalpies formation of 2–3 spinels from component oxides (Reprinted with permission from A. Navrotsky and O. J. Kleppa, *J. Inorg. Nucl. Chem.*, **30**, 479 (1968) Pergamon Press Ltd.)

the past 10–15 years some progress has been achieved through solid–gas equilibrium methods, and in particular by means of the solid oxide electrochemical cells introduced by Kiukkola and Wagner (1957). However, as is so commonly the case, while work of this type may provide reliable data on the Gibbs energy of formation, it generally does not allow an unambiguous separation of ΔG_f into its appropriate enthalpy and entropy components. Oxide melt solution calorimetry has represented a new 'handle' on this problem. A particularly significant example is the investigation by Navrotsky and Kleppa (1968) of the thermodynamics of spinel formation, from which we take Figure 7, which gives enthalpy of formation data for a series of refractory 2–3 spinels

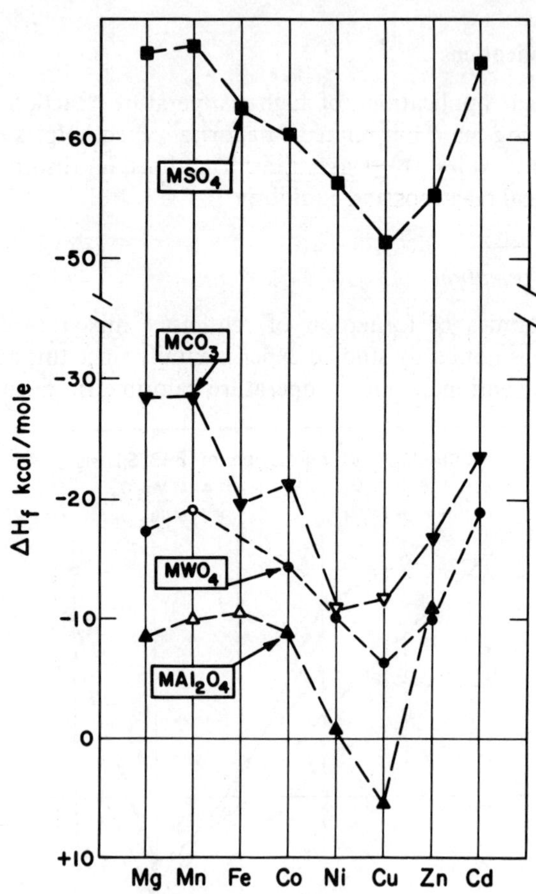

Figure 8. Enthalpies of formation of spinels (MAl_2O_4), tungstates (MWO_4), carbonates (MCO_3), sulphates (MSO_4) from component oxides (Reprinted with permission from A. Navrotsky and O. J. Kleppa, *Inorg. Chem.*, 8, 756 (1969). Copyright by the American Chemical Society).

taken from the families MAl_2O_4, MGa_2O_4, MMn_2O_4, and MFe_2O_4. More recently this work was extended by Müller and Kleppa (1973) to the corresponding compounds MCr_2O_4, which were studied by solution calorimetry at 900 °C. The systematic variation of ΔH_f with the cation is of considerable interest. However, Figure 8 which gives the enthalpies of formation of the compounds MAl_2O_4 compared with corresponding data for MWO_4, MCO_3, and MSO_4 shows that this variation with cation is nothing specific. The generally parallel character of the curves is apparent. The shift in ΔH_f to more negative values in the series

$$MAl_2O_4 < MWO_4 < MCO_3 < MSO_4,$$

simply reflects the increasing acidity of the second oxide in the sequence

$$Al_2O_3 < WO_3 < CO_2 < SO_3.$$

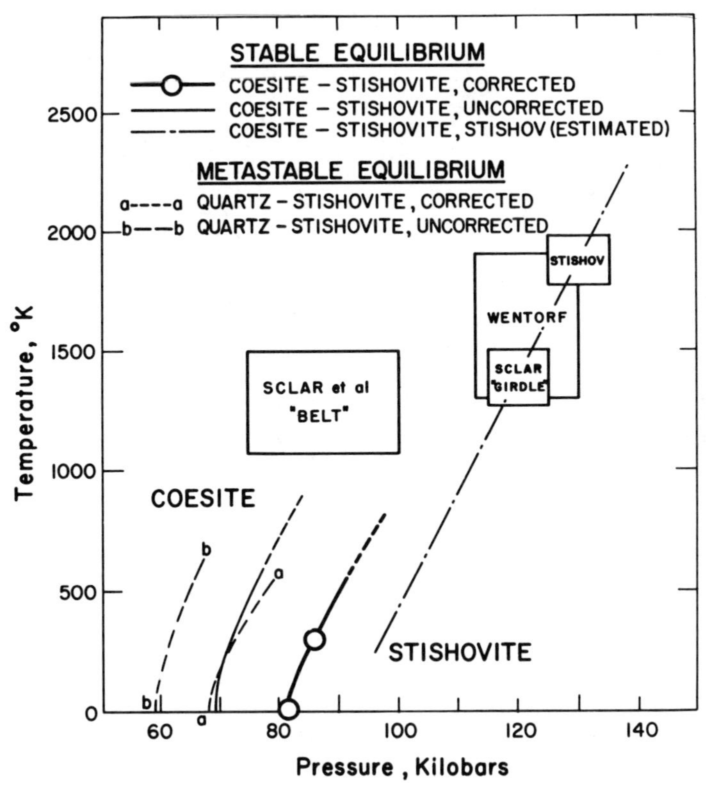

Figure 9. P–T stability fields for the coesite-stishovite equilibrium and for the metastable equilibrium quartz-stishovite calculated from thermodynamic data Sclar *et al.* refers to Sclar, Young, Carrison, and Schwartz (1962); Wentorf refers to Wentorf (1962); Stishov refers to Stishov (1963). (Reprinted with permission from J. L. Holm, O. J. Kleppa and E. F. Westrum, *Geochim. Cosmochim. Acta*, **31**, 2289 (1967) Pergamon Press Ltd.)

(*b*) *Phase Transformations*

It is frequently found that phase transformations in solid mineral systems are very sluggish, and that therefore equilibrium can be achieved in the laboratory only at quite high temperatures. For this reason quantitative equilibrium studies carried out *at temperature* may be very difficult and at times even impossible. However, there are no such restrictions on the thermodynamic calculations, which in principle will apply at any temperature.

Examples of important calculated phase equilibria in mineral systems are reported by Holm, Kleppa and Westrum (1967) in their work on the high pressure phases of silica. Figure 9 gives the thermodynamically calculated P–T curve for coesite–stishovite, and for the quartz–stishovite transformation, along with some static equilibrium data for the former equilibrium.

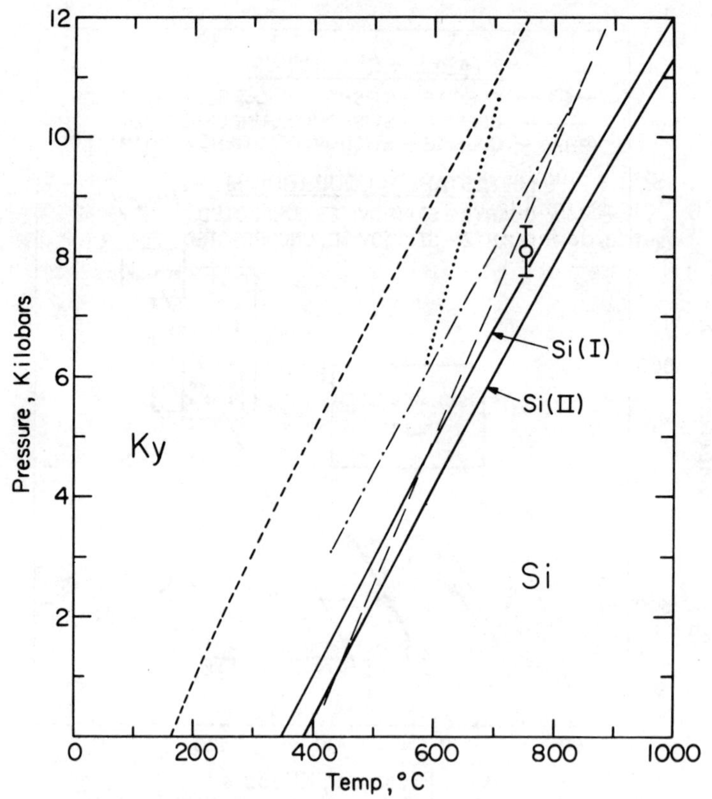

Figure 10. Comparison of calculated P–T curves for the kyanite–sillimanite equilibrium with some experimental curves. ——— Anderson and Kleppa (1969); – – – – Holm and Kleppa (1966); –·–·–·– Weill (1966); —— —— Richardson, Bell, and Gilbert (1968); ...... Althaus (1967); ϕ Newton (1966). (Reproduced with permission from P. A. M. Anderson and O. J. Kleppa, *Amer. J. Sci.*, 267, 285 (1969).)

The thermochemical results for the quartz–stishovite transformation are of particular interest, since this equilibrium occurs in the stability field of coesite and cannot be studied by the usual static equilibrium methods. However, this transformation takes place in shock wave experiments (e.g. see, Davis, 1972).

Another phase transformation, which has been rather extensively studied by high temperature-high pressure methods, and for which good thermochemical data are available both on ΔH^0 and on ΔS^0, is the kyanite–sillimanite equilibrium. Figure 10 gives a summary of some of the P–T data for this transformation taken from Anderson and Kleppa (1969). Note that while there is now reasonable agreement between calculated and experimental equilibrium pressures, this agreement does not extend to the dP/dT slopes which are somewhat higher in the equilibrium experiments than in the calculations. There is now good reason to believe that this discrepancy is due to some Al–Si disorder in the sillimanite structure, which makes the entropy of sillimanite somewhat larger than given by the third law value (e.g. see, Holdaway, 1971).

(c) *Disorder Problem*

It has been recognized for some time that structural disorder is an important phenomenon in a number of mixed phases of refractory character such as, e.g. in the spinels, mullite, pyroxenes, feldspars. As an example, let us first consider the disordering problem in spinel, $Mg[Al]_2O_4$. At low temperature the equilibrium state of this mineral has all the magnesium ions located on the tetrahedral interstitial sites in the essentially close packed array of oxygen ions; similarly, the aluminium ions are located on octahedral sites. However, when this mineral is heated above 800 °C, a disordering transformation occurs which moves a fraction (x) of the magnesium ion to the octahedral sites and a corresponding number of aluminum ions to the tetrahedral sites (Hafner and Laves, 1961). The state of this partially disordered spinel may be described as

$$Mg_{1-x}Al_x[Al_{2-x}Mg_x]O_4.$$

Figure 11 gives a plot of the configurational entropy associated with disorder in a spinel, as a function of the disorder parameter x. Note in particular the very sharp (infinite slope) rise of the configurational entropy for small values of x. It is this sharp initial rise of the entropy with x which makes structural disorder a very general problem in refractory mixed oxides at high temperature.

To a first approximation it may be assumed that the enthalpy change associated with the disordering process in spinels is proportional to x, i.e.

$$\Delta H = x\Delta H_{\text{int}}, \tag{8}$$

where the interchange enthalpy, ΔH_{int}, is a constant characteristic of the considered spinel. If ΔS is equal to the configurational entropy

$$\Delta S = -R\left[x\ln x + (1-x)\ln(1-x) + x\ln\frac{x}{2} + (2-x)\ln\left(1-\frac{x}{2}\right)\right] \tag{9}$$

and we may set

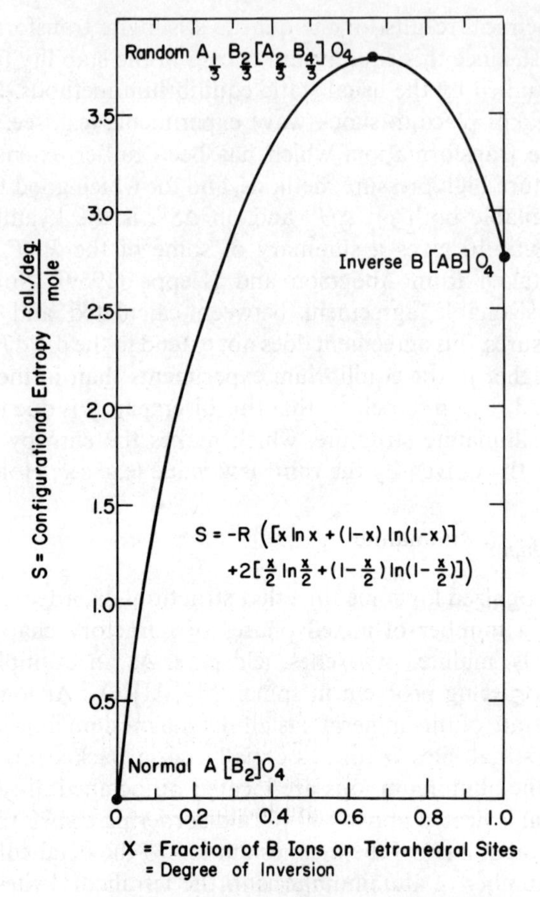

Figure 11. The dependence of the configurational entropy in spinels on the degree of inversion, x (Reprinted with permission from A. Navrotsky and O. J. Kleppa, *J. Inorg. Nucl. Chem.*, **29**, 2701 (1967) Pergamon Press Ltd.)

$$\Delta G = \Delta H - T\Delta S. \tag{10}$$

By imposing the equilibrium requirement $d\Delta G/dx = 0$, we obtain

$$-\frac{\Delta H_{int}}{RT} = \ln \frac{x^2}{(1-x)(2-x)}, \tag{11}$$

i.e. a relation between x, ΔH_{int}, and T. Navrotsky and Kleppa (1967) applied this model to $MgAl_2O_4$, and made a direct determination of ΔH for a natural spinel by means of oxide melt solution calorimetry. From these results and other information they concluded that for $MgAl_2O_4$ at 1000 °C, $\Delta H_{int} \approx 10$ kcal/mole and $x \approx 0.1$.

More recently a modified form of this simple disordering model was adopted by Holdaway (1971) in order to resolve some of the internal discrepancies between the experimental observations and the thermodynamic data for the kyanite–sillimanite–andalusite (Al_2SiO_5) system. Specifically, Holdaway assumes partial cation exchange between Al and Si atoms located on their respective tetrahedral sites in the sillimanite structure, according to the reaction

$$Al_{Al} + Si_{Si} = Si_{Al} + Al_{Si}. \tag{12}$$

His principal numerical input was his experimentally measured dP/dT slope of the sillimanite–andalusite phase boundary, which gave him an interchange enthalpy, ΔH_{int} of about 14.75 kcal/mole. With this interchange enthalpy, one would predict about 20 per cent disorder in sillimanite at 1400 °C.

Quite recently oxide melt heat of solution measurements on samples of natural sillimanite heat-treated at 1200–1700 °C at pressures of 16–23 kbar have been reported by Navrotsky, Newton, and Kleppa (1973). The results are shown in Figure 12 and indicate a distinct enthalpy of solution decrement relative to unheated sillimanite of about 1.3 kcal/mole by samples run at 1400–1550 °C. Analysis of these results in terms of the mentioned disordering model yields a value of $\Delta H_{int} = 16 \pm 1$ kcal/mole, in good agreement with Holdaway's value derived on entirely different grounds. At temperatures above 1550 °C, larger heat of solution decrements were observed. It is possible that these are due to some unknown and even more profound disordering process.

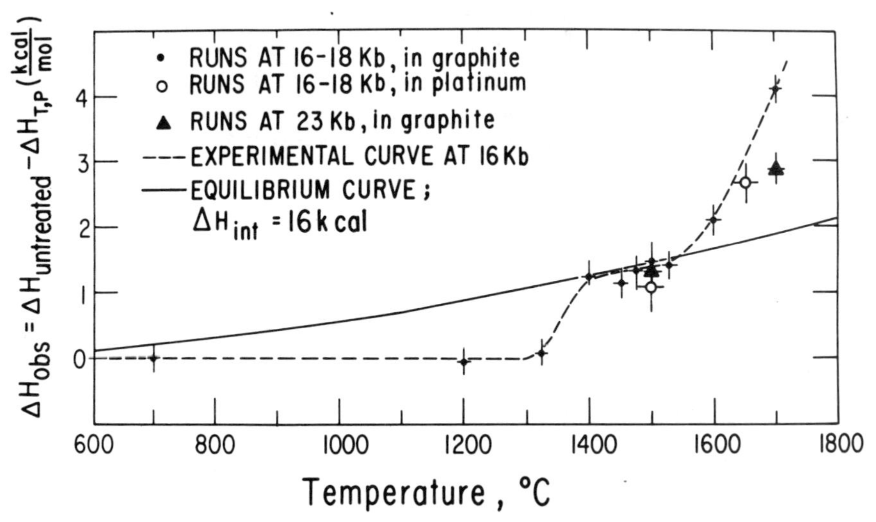

Figure 12. Heat of solution data on quenched samples of sillimanite heat-treated at high pressure. Solid line is calculated from simple Al–Si tetrahedral disorder (Eq. (12)) with $\Delta H_{int} = 16$ kcal/mole (Reprinted with permission from A. Navrotsky, R. C. Newton and O. J. Kleppa, *Geochim. Cosmochim. Acta*, 37, 2497 (1973) Pergamon Press Ltd.)

(*d*) *Mineralogical Reactions and Equilibria*

The discussion above has focused on only one of the serious problems associated with the application of thermochemical methods to mineral reactions and equilibria, i.e. the problem of structural disorder. Another problem, which we have not discussed, has to do with the effect of chemical impurities. For example, if we have thermodynamic data for pure, ordered phases only, how do the results of a thermodynamic calculation apply to a real system in which significant amounts of impurities as well as structural disorder may be present? Along the same line of reasoning, what will be the influence on the considered equilibrium of more extensive solid solution formation or deviation from stoichiometry, which at times may vary significantly with changes in temperature and pressure? While it is possible to give partial and qualitative answers to some of these questions on the basis of chemical knowledge and physical intuition, more quantitative answers generally will require detailed thermodynamic information on the phases which actually participate in the considered reaction. As an example let us consider the hypothetical reaction

$$\mathrm{A + B = AB.}$$

If carried out in the laboratory, this reaction may well be associated with solid solubility of A in B and of B in A, various impurities in both A, B and AB, and perhaps some structural disorder and solid solubility in the compound AB. Since the extent of these effects may be unknown or difficult to assess, it may be very difficult to make a realistic quantitative prediction about the equilibrium.

Let us assume that this reaction is very sluggish and can be studied only at very high temperatures and pressures. Through equations (2) and (3) this will in effect furnish a single value of ΔG^0. However, the equilibrium work does not cover a sufficient range in temperature to establish the slope of the P/T curve. Therefore, without information on ΔS or on ΔH, we are unable to trace the P/T curve to lower temperatures and pressures. At this point solution calorimetry, *if carried out on the phases which actually participate in the reaction* (i.e. on the reactants A + B and on the product AB), may provide information which will permit us to calculate ΔH, and hence also ΔS and the P/T slope.

The procedure which has been outlined here for the hypothetical reaction A + B = AB is, in fact, that adopted by Newton (1972) in his work on the breakdown of cordierite to sapphirine + quartz (+ enstatite?). More recent and more extensive work by Newton, Charlu, and Kleppa (1974) has led to a much more detailed, quantitative understanding of this reaction.

ACKNOWLEDGEMENTS

Our work on fused salt chemistry, which led to the development of oxide melt calorimetry has been supported by the National Science Foundation. The applications of this method to the study of refractory solids has been supported by the Army Research Office-Durham. We are also indebted to the NSF for the general support of Materials Science at the University of Chicago.

REFERENCES

Althaus, E. (1967), *Contr. Mineralogy and Petrology*, **16**, 29–44.
Anderson, P. A. M. and O. J. Kleppa (1969), *Am. J. Sci.*, **267**, 285–290.
Birch, F. and P. LeConte (1960), *Am. J. Sci.*, **258**, 209–217.
Davies, G. F. (1972), *J. Geophys. Res.*, **77**, 4920–4933.
Flood, H. and T. Førland (1947), *Acta Chem. Scand.*, **1**, 592–604.
Geller, R. F. and E. N. Bunting (1937), *J. Res. Natl. Bur. Stds.*, **18**, 585–593.
Hafner, S. and F. Laves (1961), *Z. Krist. Min.*, **115**, 321–330.
Holdaway, M. J. (1971), *Am. J. Sci.*, **271**, 97–131.
Holm, J. L. and O. J. Kleppa (1966), *Amer. Mineral.*, **51**, 1608–1616.
Holm, J. L. and O. J. Kleppa (1967), *Inorg. Chem.*, **6**, 645–648.
Holm, J. L., O. J. Kleppa and E. F. Westrum (1967), *Geochim. Cosmochim. Acta*, **31**, 2289–2307.
Kelley, K. K., S. S. Todd, R. L. Orr, E. G. King and K. R. Bonnickson (1953), *U. S. Bur. Mines Rept. Inv.*, 4955.
Kiukkola, K. and C. Wagner (1957), *J. Electrochem. Soc.*, **104**, 308–316.
Kleppa, O. J. (1960), *J. Phys. Chem.*, **64**, 1937–1940.
Kracek, F. C. and K. J. Neuvonen (1952), *Am. J. Sci.*, Bowen Vol., 293–318.
Kracek, F. C., K. J. Neuvonen and G. Burley (1951), *Washington Acad. Sci. J.*, 41, 373–383.
Lux, H. (1939), *Z. Elektrochem.* **45**, 303–309.
Muller, F. and O. J. Kleppa (1973), *J. Inorg. Nucl. Chem.*, **35**, 2673–2678.
Navrotsky, A. and O. J. Kleppa (1967), *J. Inorg. Nucl. Chem.*, **29**, 2701–2714.
Navrotsky, A. and O. J. Kleppa (1968), *J. Inorg. Nucl. Chem.*, **30**, 479–498.
Navrotsky, A. and O. J. Kleppa (1969), *Inorg. Chem.*, **8**, 756–758.
Navrotsky, A., R. C. Newton and O. J. Kleppa (1973), *Geochim. Cosmochim. Acta*, **37**, 2497–2508.
Newton, R. C. (1966), *Science*, **151**, 1222–1225.
Newton, R. C. (1972), *J. Geol.*, **80**, 398–420.
Newton, R. C., T. V. Charlu and O. J. Kleppa (1974), *Contr. Mineral. Petrol.*, **44**, 295–312.
Richardson, S. W., P. M. Bell, and M. C. Gilbert (1968), *Am. J. Sci.*, **266**, 513–541.
Robertson, E. C., F. Birch and G. J. F. Macdonald (1957), *Am. J. Sci.*, **255**, 115–137.
Rodigina, E. N., K. Z. Gomel'skii, and V. F. Luginina (1961), *Russ. J. Phys. Chem.*, **35**, 884.
Sclar, C. B., A. P. Young, L. C. Carrison, and C. M. Schwartz (1962), *J. Geophys. Res.*, **67**, 4049–4052.
Stishov, S. M. (1963), *Doklady Akad. Nauk S.S.S.R.*, **148**, 1186–1188.
Torgeson, D. R. and TH. G. Sahama (1948), *J. Am. Chem. Soc.*, **70**, 2156–2160.
Waldbaum, D. R. (1966), *Ph.D. Thesis*, Harvard University, 247 pp.
Weill, D. F. (1966), *Geochim. Cosmochim. Acta*, **30**, 223–237.
Wentorf, R. H., Jr. (1962), *J. Geophys. Res.*, **67**, 3648.
Yokokawa, T. and O. J. Kleppa (1964), *Inorg. Chem.*, **3**, 954–957.
Yokokawa, T. and O. J. Kleppa (1965), *Inorg. Chem.*, **4**, 1806–1808.

Synthetic $3d^{3+}$-Transition Metal Bearing Kyanites, $(Al_{2-x}M_x^{3+})SiO_5$.

K. Langer
Mineralogisch-Petrologisches Institut der Universität Bonn
D 5300 Bonn, Germany

1. INTRODUCTION

Kyanite is one of the three polymorphs of Al_2SiO_5, which differ in oxygen arrangement and the coordination number of aluminium and, therefore, also in their densities:

Andalusite	$Al^{[6]}Al^{[5]}Si^{[4]}O_5$	3.144 g cm^{-3}
Sillimanite	$Al^{[6]}Al^{[4]}Si^{[4]}O_5$	3.247 g cm^{-3}
Kyanite	$Al^{[6]}Al^{[6]}Si^{[4]}O_5$	3.674 g cm^{-3}

The structure of kyanite (Náray-Szabo *et al.* 1929, Burnham 1963) is a slightly distorted cubic close packed oxygen arrangement with aluminium solely in sixfold coordination (distorted octahedra) and is, therefore, the high pressure polymorph of Al_2SiO_5 as evident from the phase diagram of the pure system Al_2SiO_5 (Richardson *et al.*, 1969).

Simple transition metal substituted kyanites (and sillimanites or andalusites), $(Al_{2-x}M_x^{3+})SiO_5$, can be represented on the joins Al_2SiO_5–M_2SiO_5 of the ternary systems Al_2O_3–M_2O_3–SiO_2, where M stands for the substituting $3d^{3+}$-ions. The corresponding end members M_2SiO_5 are only hypothetical, because it has proved impossible to synthesize the pure M_2SiO_5 kyanites.

Though transition metal substitutions of octahedral aluminium in kyanite play a prominent role in the origin of colour and pleochroism of the mineral (White and White, 1967; Faye and Nickel, 1969; Rost and Simon, 1972), natural kyanites normally contain less than 0.5 mole per cent of the theoretical end members. One exception is an Fe^{3+}-bearing kyanite from Glen Clova, Scotland, coexisting with hematite and quartz, in which Chinner *et al.* (1969) found iron contents corresponding to about 2.5 mole per cent Fe_2SiO_5. In addition, Sobolev *et al.* (1968) found kyanites with up to 18 mole per cent Cr_2SiO_5 end member in a chromium rich grospydite nodule (Al-rich eclogite) in a kimberlite from Yakutia, Siberia. Furthermore, chromium was enriched in coexisting minerals of this rock in the order kyanite > garnet > clinopyroxene. This showed that contents of transition metal ions may be very high in kyanites at appropriate conditions of pressure, temperature and bulk composition and encouraged a systematic experimental study aimed at solving the following problems:

1. Synthesis of $3d^{3+}$-substituted kyanites.
2. Determination of solid solution limits on the joins Al_2SiO_5–M_2SiO_5.
3. Influence of transition metal substitutions on the kyanite-sillimanite and kyanite–andalusite transformations (extent of divariant reaction fields).
4. Influence of transition metal substitutions on kyanite lattice dimensions and colour. Structural characterization of substituted kyanites (octahedral site occupancy).

2. EXPERIMENTAL

It is well known from the numerous phase equilibrium studies in the pure system Al_2SiO_5 (see Holdway, 1971, for earlier references) that kyanite can only be synthesized at reasonable rates at high temperatures (some 900 °C or higher) and pressures some 3 kb higher than the equilibrium transformation pressures at the temperatures applied. Therefore, the synthesis and equilibrium runs were done in the temperature and pressure range 900–1400 °C and 15–30 kb in a piston-cylinder-apparatus of the Boyd and England type (1960) using talc and boron nitride as high pressure media and graphite tubes as heating elements (Johannes *et al.* 1971, Langer and Seifert, 1971).

Intimate mixtures of gem quality andalusite (0.40 per cent Fe_2O_3) the requisite oxide M_2O_3 and high purity SiO_2 made up following the general reaction scheme.

$$(1 - x/2)Al_2SiO_5 + x/2\, M_2O_3 + x/2\, SiO_2 = (Al_{2-x}M_x)SiO_5$$

served as starting materials in the synthesis runs. The charges were held in tight noble metal containers together with a little water to promote crystal growth. Run products were examined with powder X-ray, optical and electron microprobe methods.

3. ADJUSTMENT OF OXYGEN FUGACITY, f_{O_2}

In systems containing transition metal oxides such as V_2O_3, Cr_2O_3, Mn_2O_3, and Fe_2O_3, one has to be sure that the oxidation state of the transition metal is retained during the experiment, i.e. oxygen fugacities, f_{O_2}, have to be kept at values within the f_{O_2}–T-stability fields of these oxides at the pressures applied. In the high pressure cells used, with graphite as heating element and the water bearing phase talc as pressure medium, the oxygen fugacity is controlled by complex equilibria between $C_{Graphite}$, O_2, H_2, H_2O, CO, CO_2, CH_4....

In order to obtain some experimental information on oxygen fugacities in the high pressure cells used, solid state oxygen buffer mixtures Mn_2O_3/Mn_3O_4,

Fe_2O_3/Fe_3O_4, NiO/Ni, Fe_3O_4/FeO, and FeO/Fe (Huebner, 1971) with water were run in Pd- or Pt-capsules at 20 kb and different temperatures. Some results of this work, the details of which have still to be published (Frentrup and Langer, in preparation) are shown in Figure 1. Here, the equilibrium curves of oxidation–reduction reactions for the above systems are plotted in the $\log f_{O_2}$–T-field for a pressure of 20 kb. Constants A, B, and C of the equation

$$\log f_{O_2} = A - \frac{B}{T} - C\frac{P-1}{T}$$

representing these curves are taken from Huebner (1971). Triangles, the corners of which point towards the individual curves, represent runs performed with the corresponding buffer mixture. It is obvious from Figure 1 that the oxygen fugacity in the high pressure cells used lies approximately between 10^{-8} and 10^{-11} atm at 600 and 1200 °C, respectively, as indicated by the dashed line.

It is known from the work of Ulmer and White (1966) and confirmed by that of Haggerty *et al.* (1970) that Cr_2O_3 is table with f_{O_2} down to some $10^{-13.3}$–$10^{-10.5}$ atm, while the upper f_{O_2}-stability limit lies at about 10^3 atm at 600 °C (Shibasaki *et al.*, 1973). Thus, no precautions need to be applied, when working with Cr_2O_3 in our high pressure cells. Indeed, additional Cr^{2+}-bearing phases, such as Cr_2SiO_4, which was synthesized by Tsvetkov *et al.* (1964) and by Scheetz and White (1972), were never observed. On the other hand, it is evident from Figure 1 that the transition ions would become reduced, when

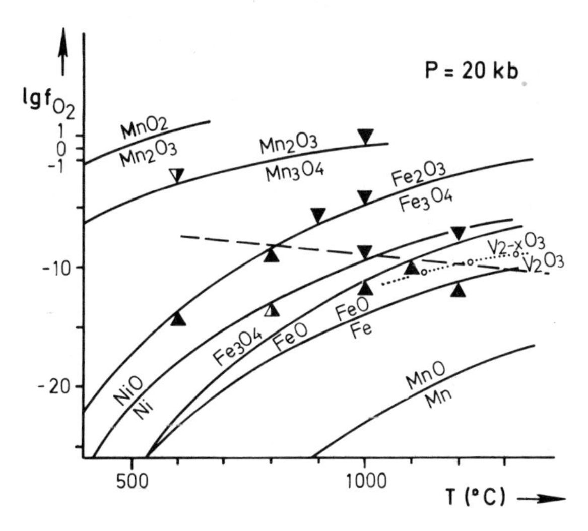

Figure 1. log f_{O_2}–T-relations of some solid state f_{O_2}-buffer systems at 20 kb and results of runs with these buffer mixtures in a graphite heated solid media high pressure cell (cf. text). Full triangles: complete reaction, half filled triangles: partial reaction. Triangles pointing downwards (upwards): growth of reduced (oxidized) phase.

working in the Mn_2O_3- and Fe_2O_3-systems, while V_2O_3 will be oxidized, because the upper f_{O_2}–T-stability limit of stoichiometric corundum type V_2O_3-phase (Wakihara and Katsura 1970) represented as a dotted curve in Figure 1 lies below the f_{O_2}-conditions in the pressure cells at synthesis temperatures of 900–1000 °C.

Therefore, runs in the V_2O_3-, Mn_2O_3-, and Fe_2O_3-bearing systems were performed with solid state oxygen buffers, namely 90/10-mixtures of Mn/MnO, MnO_2/Mn_2O_3, or Mn_2O_3/Mn_3O_4, respectively, plus water in a similar setup as was originally applied by Eugster (1957) in hydrothermal experiments with iron bearing systems. Details of the buffer technique will be published elsewhere (Frentrup and Langer, in preparation). Indeed, with these precautions syntheses of transition metal substituted Al_2SiO_5-polymorphs could be achieved without oxidation or reduction in the V_2O_3-, Mn_2O_3-, and Fe_2O_3-bearing systems.

4. SYNTHESIS AND MAXIMUM KYANITE SOLID SOLUBILITY

Using the above buffer techniques, syntheses of V^{3+}-, Cr^{3+}-, Mn^{3+}-, and Fe^{3+}-substituted kyanites and also Mn^{3+}-substituted andalusite (viridine) were achieved in runs with durations of up to about 70 hr at 20 kb and temperatures of up to 1100 °C. In longer runs, the buffer capacities were normally exhausted. If so, the run charges were completely oxidized in case of the vanadium bearing system or reduced in case of the mangenese or iron bearing systems. In the first one vanadium-free kyanite plus rutile type VO_2 plus quartz were found, in the two latter ones Mn-free kyanite plus spessartite plus quartz or Fe-free kyanite plus staurolite plus quartz, respectively. In the Cr-bearing system runs without buffer with durations of up to some 600 hr were performed.

Attempts to synthesize Ti^{3+}-substituted kyanites applying a TiO/Ti_2O_3-buffer have failed so far, because buffer and Ti_2O_3-bearing charges became oxidized even in 1 hr-runs. However, preliminary runs at 20 kb with waterfree charges and without buffer yielded blue run products consisting of kyanite plus rutile and quartz. Colour and slight shifts of the X-ray reflections of the kyanite could indicate that at least minor amounts of Ti^{3+} (less than 1 mole per cent Ti_2SiO_5) were incorporated. The system needs further study, however, it will be shown that pressures appreciably higher than 20 kb are required in the system Al_2SiO_5–Ti_2SiO_5 in order to obtain higher substitutional degrees than those corresponding to the normal 1 to 2 per cent miscibilities.

Concerning the maximum kyanite miscibilities, let us now start with the chromium bearing system Al_2SiO_5–Cr_2SiO_5. In this system, rate studies were done to determine the equilibrium solid solubility at different pressures and temperatures using mixtures with excess Cr_2SiO_5. Figure 2 shows the results. It can be seen that increasing reaction time increases the amounts of the theoretical Cr_2SiO_5-end member incorporated, their contents coming to a saturation value. It is also evident from the figure that pressure is by far the most important factor governing the maximum kyanite solid solubility:

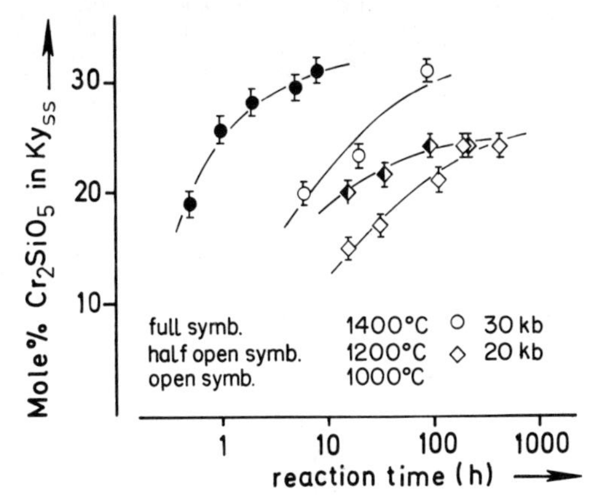

Figure 2. Rate studies in the system Al_2SiO_5–Cr_2SiO_5 at 20 and 30 kb aimed at determining the maximum kyanite solid solubility under different conditions.

Table 1. Miscibility limits of kyanite solid solution in the systems $Al_2SiO_5 - M_2SiO_5$

M	*P* (kb)	*T* (°C)	mol % M_2SiO_5
V^{3+}	20	900	14
Cr^{3+}	20	1000	24
		1200	24
	30	1000	31
		1400	31
Mn^{3+}	20	1000	6
Fe^{3+}	20	900	6.5
		1100	6.5

a pressure increase from 20 to 30 kb causes the maximum miscibility to increase by about 7 mole per cent Cr_2SiO_5, while a temperature increase from 1000 to 1200 °C at 20 kb or from 1000 to 1400 °C at 30 kb has no significant effect on the maximum solid solubility. The saturation values obtained in the Cr^{3+}-system at 20 and 30 kb are given in Table 1 together with maximum solid solubilities in the V^{3+}-, Mn^{3+}-, Fe^{3+}-systems at 20 kb, which were determined from similar experiments.

If one plots the solid solubility limits in the different systems Al_2SiO_5–M_2SiO_5 at 20 kb/900–1000 °C against the effective ionic radii (Shannon and Prewitt, 1969) of the trivalent transition metal atoms substituting for Al, a linear relationship between solubility (*s*) in mole per cent M_2SiO_5, and ionic radius (IR) in Å, is obtained (Figure 3):

$$s = 380 - 576 \cdot IR$$

The extrapolation of this relation to IR = 0.67 Å, the ionic radius of Ti^{3+}

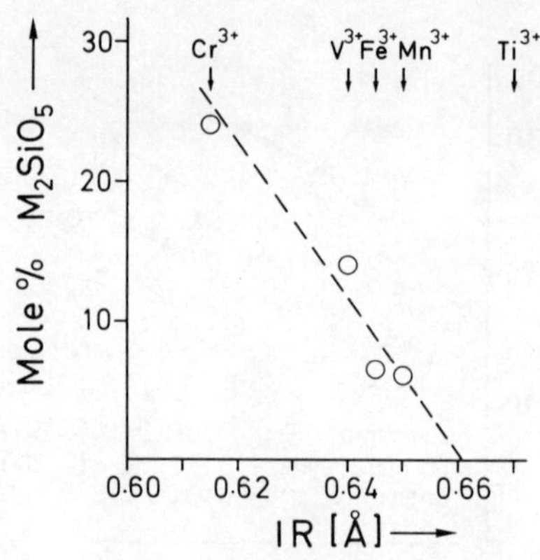

Figure 3. Maximum kyanite solid solubility at 20 kb/ca. 1000 °C as a function of the effective ionic radii, IR, of the substituting ions M^{3+} (IR from Shannon and Prewitt 1969, high-spin configuration for Mn^{3+} and Fe^{3+}).

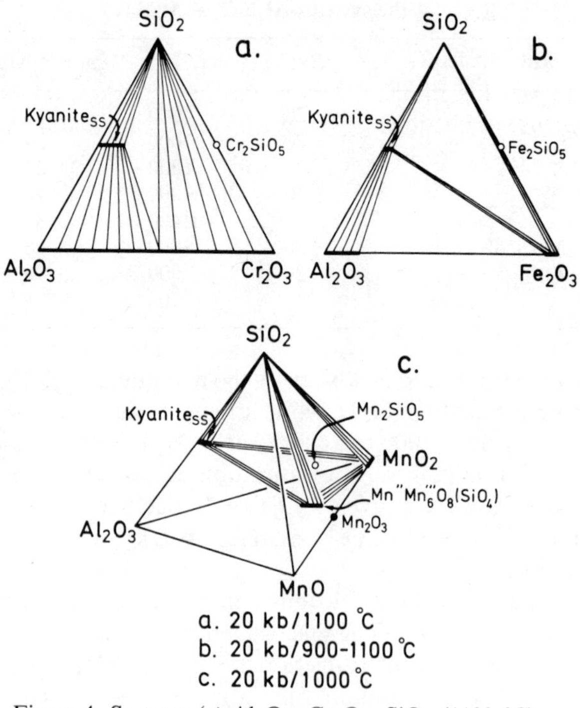

Figure 4. Systems (a) Al_2O_3–Cr_2O_3–SiO_2 (1100 °C), (b) Al_2O_3–Fe_2O_3–SiO_2 (900–1100 °C) and (c) Al_2O_3–MnO–MnO_2–SiO_2 (1000 °C) at 20 kb showing the phase relations outside the ranges of kyanite solid solubility (cf. text).

(Shannon and Prewitt 1969), shows that this ion can substitute for Al at 20 kb/1000 °C, if at all, only in trace amounts. Assuming a similar influence of pressure on the system Al_2SiO_5–Ti_2SiO_5 as found in the system Al_2SiO_5–Cr_2SiO_5, pressures higher than 30 kb would be required to obtain miscibilities of some 5 mole per cent in the Ti^{3+}-bearing system.

Outside the ranges of solid solubility on the joins Al_2SiO_5–M_2SiO_5, maximum kyanite mixed crystals were found to coexist with corundum type $(Al,M)_2O_3$-mixed crystals and quartz where $M = V^{3+}, Cr^{3+}$, and Fe^{3+} as shown for the last two cases in Figure 4(a) and (b) (Seifert and Langer, 1970; Langer and Frentrup, 1973). With $M = Mn^{3+}$, excess trivalent manganese disproportionated partially to produce Mn^{2+}-containing braunite plus Mn^{4+}-bearing pyrolusite besides SiO_2 as stable phases as shown in Figure 4(c) (Abs-Wurmbach and Langer, 1975). In braunites coexisting with Mn^{3+}-bearing kyanites approximately 1/6 of the Mn^{3+} was substituted by Al, causing the tetragonal lattice constants to decrease compared with pure braunite $Mn^{2+}Mn_6^{3+}O_8(SiO_4)$ (Abs-Wurmbach and Langer, 1975).

5. INFLUENCE OF TRANSITION METAL SUBSTITUTIONS ON POLYMORPHIC TRANSFORMATIONS OF KYANITE

Transition metal substitutions cause the univariant transformation curves to become divariant fields of ($ky_{ss} + sill_{ss}$)-, ($ky_{ss} + and_{ss}$)-, and ($and_{ss} + sill_{ss}$)-coexistence, because of unequal $3d^{3+}$-concentrations in the different structures. Chinner *et al.* (1969) have shown for example that Fe^{3+} is enriched in sillimanite in natural ky/sill-parageneses and in andalusite in ky/and-parageneses. For an experimental study of these divariant transformation fields, the equilibrium phase relations and phase compositions in Tx- or Px-sections of the aluminous parts of the joins Al_2SiO_5–M_2SiO_5 at constant pressures or constant temperatures, respectively, have to be established.

So far, a Tx-section with $M = Cr^{3+}$ has been worked out at 20 kb (Seifert

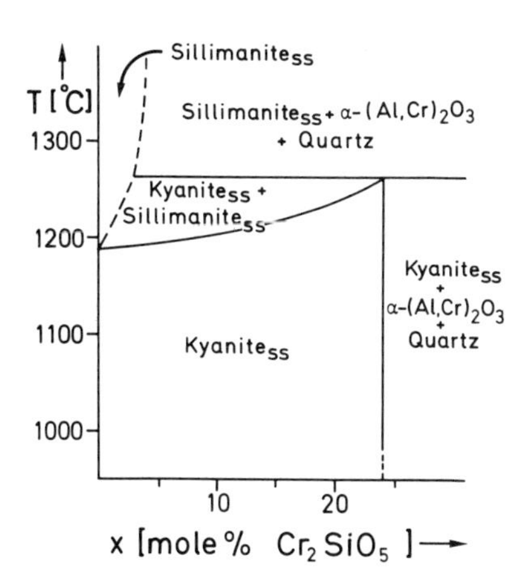

Figure 5. Tx-section of the aluminous part of the system Al_2SiO_5–Cr_2SiO_5 at 20 kb (Seifert and Langer 1970) showing the divariant field of (ky_{ss} + $sill_{ss}$)–coexistence and the stabilization of the kyanite structure at the expense of the sillimanite type due to $Al^{3+} \rightarrow Cr^{3+}$ substitution.

and Langer, 1970) proving the influence of Cr-substitutions on the kyanite–sillimanite transformation. This Tx-section is shown in Figure 5. Due to the smaller extent of substitution in sillimanite than in kyanite, we obtain the divariant (ky_{ss} + $sill_{ss}$)-field. Pure kyanite was found to transform into sillimanite at 1190 °C, which is in fair agreement with the data or Richardson *et al.* (1969) who found 1217 °C at 20 kb. The most Cr-rich kyanite transforms at 1270 °C into Cr-poor sillimanite, α-$(Al,Cr)_2O_3$-mixed crystals and quartz. Thus, at 20 kb we have a stabilization of the kyanite type at the expense of sillimanite by 80 °C due to 24 mole per cent Cr-substitution. This temperature difference between transformation of the Cr-free kyanite and the most Cr-rich kyanite solid solution was found to increase with pressure (Seifert and Langer, 1970).

Recently, Abs-Wurmbach and Langer (1975) found a similar stabilization of the andalusite type at the expense of both kyanite and sillimanite, caused by Mn^{3+}-substitutions forming the mineral viridine. Though equilibrium phase relations and phase compositions on the join Al_2SiO_5–Mn_2SiO_5 have still to be established in detail (work under way), the results obtained thus far may tentatively be represented in a preliminary Px-section at 900 °C shown in Figure 6. The exact positions of phase boundaries in this section are still to be worked out, however, the topology of the diagram is correct: from the experiments done so far, it can be concluded that maximum andalusite solid solubility

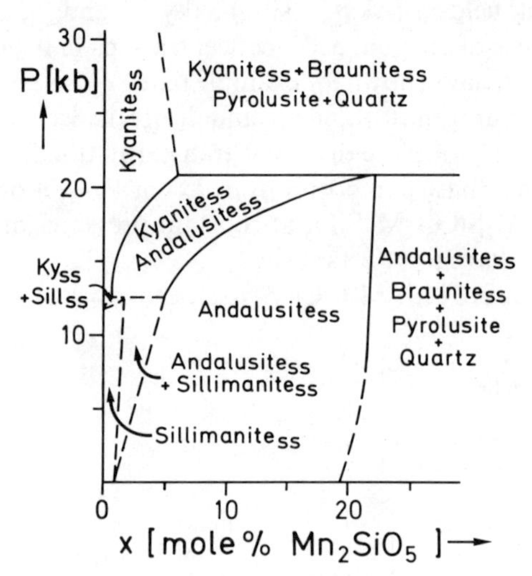

Figure 6. Tentative and preliminary Px-section of the aluminous part of the system Al_2SiO_5–Mn_2SiO_5 at 900 °C showing divariant transformation fields (and_{ss} + ky_{ss}), (and_{ss} + $sill_{ss}$), (ky_{ss} + $sill_{ss}$) and the stabilization of the andalusite structure type at the expense of kyanite and sillimanite due to $Al^{3+} \rightarrow Mn^{3+}$ substitution (cf. text).

lies near 22 mole per cent Mn_2SiO_5 and that this maximum andalusite transforms near 20 kb into Mn-poor kyanite with about 6 mole per cent Mn_2SiO_5, braunite, pyrolusite and quartz. The extension of the two phase field (and_{ss} + ky_{ss}) was determined by microprobe analyses of coexisting kyanite and andalusite obtained on a 2 mole per cent Mn_2SiO_5 bulk composition. Thus, the solid lines in Figure 6 are drawn on the basis of preliminary experimental evidence. The position of the broken lines can only be assumed thus far. The Px-section shows clearly one essential feature of the system, namely the stabilization of the low pressure-low temperature polymorph andalusite by Mn^{3+}-substitution. Strikingly enough, the andalusite type, which is unstable at 900 °C in the pure system Al_2SiO_5 becomes a stable phase even at pressures higher than 15 kb, i.e. far inside the stability field of pure kyanite if some 10–20 per cent of the Al-atoms are replaced by Mn^{3+}. This may probably be explained by the Jahn–Teller effect of this $3d^4$-ion in the andalusite matrix (Strens, 1966, 1968).

5. CRYSTALLOCHEMICAL PROPERTIES

The structure of kyanite (Náray–Szabó *et al.*, 1929; Burnham, 1963), is triclinic, space group $P\bar{1}$ with $Z = 4$ formula units per unit cell. There is a certain relationship between the $Al_2^{[6]}Si^{[4]}O_5$-kyanite structure type and the $M_2^{[6]}M^{[4]}O_4$-spinel type in that both structures contain oxygen in a cubic close packed arrangement, slightly distorted in kyanite, Al in octahedral and the other cation in tetrahedral interstices (Table 2). In spinel, we have a highly symmetrical ordering of occupied interstices, which is impossible in kyanite, because of the kyanite stoichiometry, $\Sigma\, M^{[4]}M^{[6]}/0 = 3/5$. The type of ordering in kyanite leads to a triclinic translation unit with symmetry $P\bar{1}$ within the principally cubic close packed oxygen arrangement. If an ideal cubic close packed oxygen arrangement in kyanite is assumed, triclinic axial angles as given in Table 3 can be calculated (Burnham, 1963). Comparison with those of synthetic Al_2SiO_5-kyanite shows that the angle distortion of the real kyanite cell is very small. Thus, the angles α, β, and γ result mainly from the type of ordering of occupied interstices, and they are not expected to change when Al is substituted by the larger $3d^{3+}$-ions. Indeed, we did not find deviations of α, β, and γ from values in Table 3b greater than $\pm 0.09°$, i.e. the subsititutions do not

Table 2. Comparison of occupied interstices in the spinel and kyanite structure types, both containing a cubic close packed oxygen arrangement.

Structure type	Unit cell content	Interstices per unit cell		Occupied interstices	
		Octahedra	Tetrahedra	Octahedra	Tetrahedra
Spinel	$M_{16}^{[6]}M_8^{[4]}O_{32}$	32	64	1/2	1/8
Kyanite	$Al_8^{[6]}Si_4^{[4]}O_{20}$	20	40	2/5	1/10

Table 3. Triclinic axial angles of kyanite unit cells, (a) calculated for an ideal cubic close-packed oxygen arrangement (Burham, 1963) and (b) found in synthetic Al_2SiO_5-kyanite (Langer and Frentrup, 1973)

Angle	(a) calculated for ideal fcc oxygen	(b) observed in synth. Al-kyanite
α	90.0°	89.97 ± 0.03°
β	100.9°	101.25 ± 0.03°
γ	105.5°	106.03 ± 0.03°

change the angles significantly. On the other hand, the cell volume V_0 and the cell dimensions a_0, b_0 and c_0 increase with increasing M_2SiO_5-contents (Table 4), and ΔV_0 is the same with V^{3+}, Mn^{3+}, and Fe^{3+} as substituting ions, the effective radii of which differ only by 0.005 Å, but is significantly smaller with the appreciably smaller Cr^{3+}-ion. As expected for substitutions in interstices of nearly cubic close-packed oxygen arrangement, increments Δa_0, Δb_0, and Δc_0 do not differ significantly, except Δa_0 in $(Al_{2-x}Cr_x^{3+})SiO_5$ and Δb_0 in $(Al_{2-x}Mn_x^{3+})SiO_5$.

Lattice constants of pure andalusite used as starting material and those of a synthetic andalusite mixed crystal with 17.3 mole Mn_2SiO_5 (viridine) are given in Table 5. If we assume a linear increase of cell volume with Mn_2SiO_5-contents of andalusite, a 10 mole per cent substitution would cause a relative volume increase of 1.59 per cent in andalusite compared with only 0.75 per cent in kyanite.

The kyanite structure contains four different octahedral sites M1, M2, M3, and M4, the first two forming chains parallel to c_0 by edge sharing. The geometry of the different octahedra is shown in Figure 7. The mean M–O distances are almost the same in all four types, namely about 1.91 A, i.e. the same as the mean Al–O in corundum (1.91 Å) and much less than Al–O in Al_2MgO_4-spinel (2.025 Å). Thus, the sizes of kyanite octahedra provide no reason for an Al/M^{3+}-ordering in the substituted mixed crystals $(Al_{2-x}M_x^{3+})SiO_5$. The Al/M^{3+}-distribution could be determined by powder X-ray methods only for $M = Cr^{3+}$ (25

Table 4. Increase of cell dimensions a_0, b_0, c_0, V_0 of kyanite solid solutions with 10 mole per cent M_2SiO_5; reference: pure Al_2SiO_5-kyanite. Triclinic axial angles remain unchanged. IR = effective ionic radii (Shannon and Prewitt, 1969).

M	Δa_0 (Å)	Δb_0 (Å)	Δc_0 (Å)	ΔV_0 (Å^3)	$IR^{[6]}$ (Å)
V^{3+}	0.024	0.028	0.026	2.2	0.640
Cr^{3+}	0.016	0.024	0.019	1.2	0.615
Mn^{3+} [a]	0.029	0 015	0.026	2.2	0.65[b]
Fe^{3+} [a]	0.029	0.029	0.022	2.2	0.645[b]

[a] Extrapolated values
[b] High-spin configuration

Table 5. Lattice constants of natural, Mn-free andalusite and synthetic Mn-bearing andalusite solid solution (viridine) with 17.3 mole per cent Mn_2SiO_5 (Abs–Wurmbach and Langer, 1975). Numbers in brackets: uncertainties in the last decimal point.

	Andalusites	
	Al_2SiO_5	$(Al_{1.65}Mn^{3+}_{0.35})SiO_5$
a_0 (Å)	7.796 (1)	7.914 (6)
b_0 (Å)	7.897 (1)	7.926 (6)
c_0 (Å)	5.557 (2)	5.604 (3)
V_0 (Å^3)	342.13 (8)	351.5 (3)

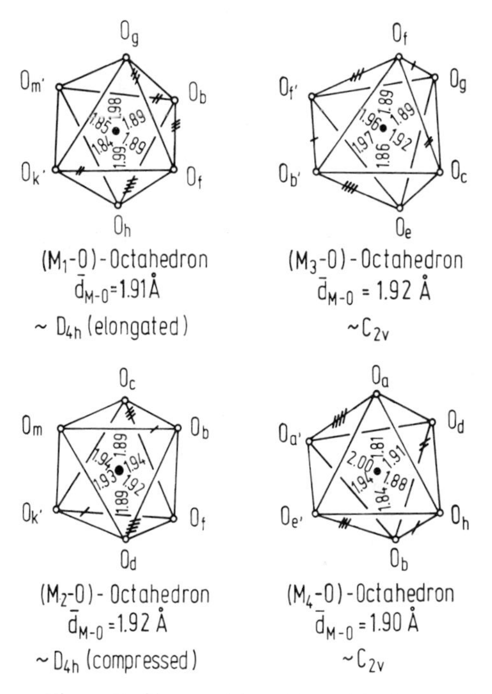

Figure 7. Shape of kyanite octahedra on the basis of Burnham's (1963) structure refinement. Simple, twofold, threefold, and fourfold signed edges indicate edge connection with neighbouring M1-, M2-, M3-, and M4-octahedra, respectively.

mole per cent Cr_2SiO_5) because of the lack of single crystals large enough for a structure refinement. From powder mounts of $(Al_{1.5}Cr_{0.5})SiO_5$, intensities of (300) and (200) reflections were measured and transformed to structure factors. These two reflections were selected because they did not overlap with

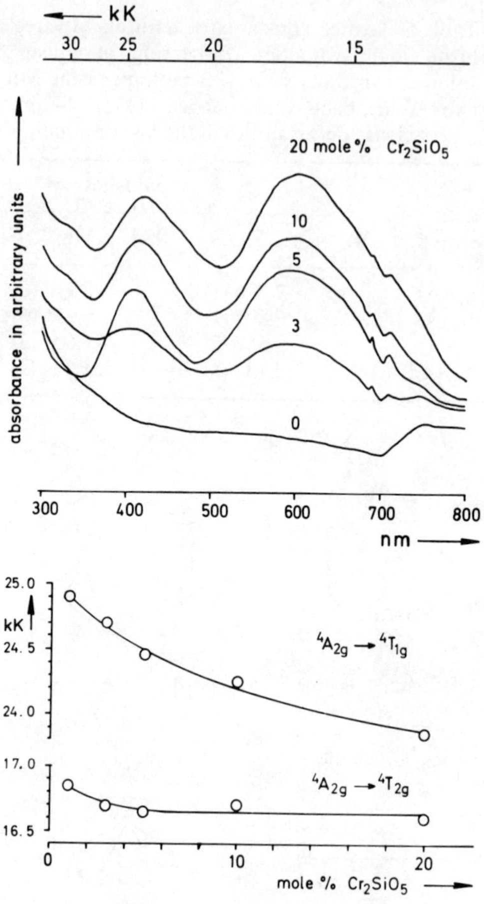

Figure 8. Diffuse reflectance spectra of Cr^{3+}-substituted kyanites differing in 'Cr_2SiO_5-contents (upper part) and dependence of the maxima of spin allowed transitions between the split terms of the ^4F-state in Cr^{3+} on Cr_2SiO_5 -contents (lower part).

other reflections in the complicated powder pattern of kyanite, and because intensities of both reflections are influenced to the same extent by preferred orientation (texture of powder mounts) and depend relatively strongly on the M1-, M2-, M3- and M4-occupancy. Values $|F_{300}/F_{200\text{ obs.}}|$ fitted the values $|F_{300}/F_{200\text{ cal.}}|$, calculated for different models of Al/Cr^{3+}-distribution the better the more Cr^{3+} was concentrated in the chain octahedra M1 and M2. The best numerical fit was obtained with (0.5Cr, 0.5A1) randomly distributed over M1 and M2, and M3 and M4 occupied only by Al (Langer and Seifert, 1971).

The colour of the transition metal bearing kyanites, $(Al_{2-x}M_x^{3+})SiO_5$, is light greyish-green with $M = V^{3+}$, deep emerald green with $M = Cr^{3+}$, light orange-yellow with $M = Mn^{3+}$ and slightly yellowish-green, almost colourless with

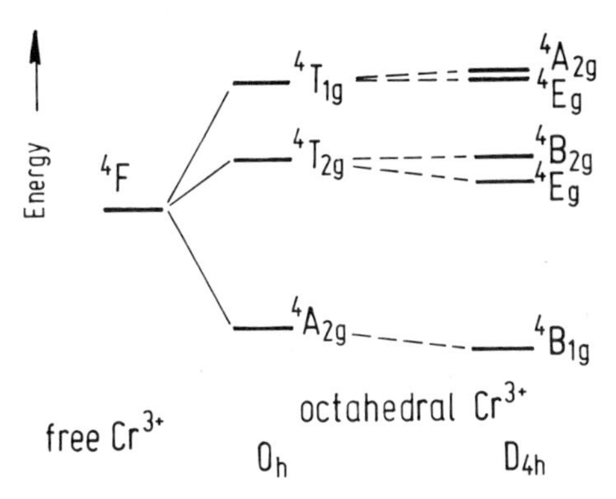

Figure 9. Split terms of the ^4F-state of free Cr^{3+} in cubic or tetragonally distorted octahedral fields.

$M = Fe^{3+}$. A detailed study of the single crystal absorption spectra of the substituted microcrystals is just beginning. So far, diffuse reflectance spectra have been measured of Cr^{3+}-substituted kyanites (Figure 8, upper part). These show the spin-allowed quartet-transitions in Cr^{3+}, designated $^4A_{2g} \rightarrow {}^4T_{2g}$ and $^4A_{2g} \rightarrow {}^4T_{1g}$ in case of true cubic symmetry of the Cr-bearing octahedra. In addition to these, Cr^{3+} quartet-doublet transitions $^4A_{2g} \rightarrow {}^2E_g$, $^2T_{1g}(G)$ are observed as weak, sharp bands at 14.10 and 14.45 kK. The most prominent feature of the spectra is the very high half band width ($\Delta E = 3.8$ kK) and low energy (16.7 kK) of the first, i.e. the 10 Dq-transition, compared with properties of the corresponding band in $(Al_{2-x}Cr_x)O_3$ mixed crystals of comparable substitutional degrees (Schmitz-DuMont and Reinen, 1959), while the second transition has almost the same energy in ky_{ss} and cor_{ss}. This could indicate that the 10 Dq-band of Cr^{3+} in kyanite consists of at least two unresolved, overlapping bands with maxima at about 17.6 (the energy of the corresponding Cr^{3+} -transition in cor_{ss}) and at about 15.6 kK. Such a splitting of the 10 Dq band of Cr^{3+} is to be expected in the case of a tetragonal distortion of the Cr^{3+}-bearing octahedra (Figure 9). Indeed, the chain octahedra M1 and M2, in which Cr^{3+} is concentrated (see above) approximate a symmetry D_{4h} (cf. Figure 7).

ACKNOWLEDGEMENTS

The experimental work referred to in the present paper was done, while the author was at the Institut für Mineralogie, Ruhr-Universität, Bochum. I thank Prof. W. Schreyer for making available the equipment of the petrology section of this institute, and Prof. H. Specker, Institut für Anorganische Chemie, Ruhr-Universität, Bochum for the use of a Cary 14 spectrophotometer and my colleagues and friends Dr. I. Abs-Wurmbach (Bochum), K. R. Frentrup (formerly at Bochum), and Prof. F. Seifert (Kiel) who cooperated in the project. Financial support of the project by Deutsche Forschungsgemeinschaft, Bonn-Bad Godesberg, is gratefully acknowledged.

REFERENCES

Abs-Wurmbach, I., and K. Langer (1975). *Contr. Mineral. and Petrol.*, **49**, 21–38.

Boyd, F. R., and J. L. England (1960). *J. Geophys. Res.*, **65**, 741–748

Burham, Ch. W. (1963). *Z. Kristallogr.*, **118**, 337–360.

Chinner, G. A., J. V. Smith, C. R. Knowles (1969). *Amer. J. Sci.*, **267-A**, Schairer Vol., 96–113.

Eugster, H. P. (1957). *J. Chem. Phys.*, **26**, 1760–1761

Faye, G. H., and E. H. Nickel (1969). *Canad. Mineralog.*, **10**, 35–46.

Haggerty, S. E., F. R. Boyd, P. M. Bell, L. W. Finger, W. B. Bryan (1970). *Proc. Apollo 11 Lunar Sci. Conf., Geochim. et Cosmochim. Acta, Suppl. 1*, **1**, 513–538

Holdaway, M. J. (1971). *Amer. J. Sci.*, **271**, 97–131.

Huebner, J. S. (1971). In *Research Techniques for High Pressure and High Temperature.* pp. 123–178, G. C. Ulmer ed., Springer Berlin 1971.

Johannes, W., P. M. Bell, H. K. Mao, A. L. Boettcher, D. W. Chipman, F. J. Hays, R. C. Newton, F. Seifert (1971). *Contr. Mineral. and Petrol.*, **32**, 24–38.

Langer, K., and K. R. Frentrup (1973). *Contr. Mineral. and Petrol.*, **47**, 37–46.

Langer, K., and F. Seifert (1971). *Z. anorg. allg. Chem.*, **383**, 29–39.

Náray-Szabó, St., W. H. Taylor and W. W. Jackson (1929). *Z. Kristallogr.*, **71**, 117–130.

Richardson, S. W., M. C. Gilbert, and P. M. Bell (1969). *Amer. J. Sci.*, **267**, 259–272.

Rost, F., and E. Simon (1972). *N. Jb. Miner. Mh.*, 383–395.

Scheetz, B. E., and W. B. White (1972). *Contr. Mineral. and Petrol.*, **37**, 221–227.

Schmitz-DuMont, O., and D. Reinen (1959). *Z. Elektrochem.*, **63**, 978–987.

Seifert, F., and K. Langer (1970). *Contr. Mineral and Petrol.*, **28**, 9–18.

Shannon, R. D., and C. T. Prewitt (1969). *Acta Cryst.*, **B25**, 925–946.

Shibasaki, Y., F. Kanamaru, M. Koizumi and S. Kume (1973). *J. Amer. Ceram. Soc.*, **56**, 248–250.

Sobolev Jr., N. V., I. K. Kuznetsova and N. L. Zyuzin (1968). *J. Petrol.*, **9**, 353–380.

Strens, R. G. J. (1966). *Mineral. Mag.*, **35**, 777.

Strens, R. G. J. (1968). *Mineral. Mag.*, **36**, 839.

Tsvetkov, A. J., Z. P. Ershova and N. A. Matveyeva (1964). *Izv. Akad. Nauk., SSSR, Ser. Geol.*, **29**., 3–14.

Ulmer, G. C., and W. B. White (1966). *J. Amer. Ceram. Soc.*, **49**, 50–51

Wakihara, M., and T. Katsura (1970). *Metallurg. Transact.*, **1**, 363–366.

White, E. W., and W. B. White (1967). *Science*, **158**, 915–917.

White, W. B. (1971). In *Research Techniques for High Pressure and High Temperature.* pp. 103–121, G. C. Ulmer ed., Springer Berlin 1971.

Systematic Studies of Interatomic Distances in Oxides

R. D. Shannon
Central Research Department
E. I. duPont de Nemours & Co.
Wilmington, Delaware

INTRODUCTION

Shannon and Prewitt (1969, 1970a) utilized Goldschmidt's methods in order to prepare an empirical set of effective ionic radii which reproduced moderately well ($\pm \sim 0.015$ Å) most accurate average interatomic distances found in modern crystal structure refinements of oxides and fluorides. These effective ionic radii showed certain regularities with valence (Figure 1), electronic spin (Shannon and Prewitt, 1969), and coordination (Figure 2) and provided in general a linear or regular curvilinear relationship between the cube of the ionic radius (r^3) and unit cell volumes (V) of isotypic series of compounds (Shannon and Prewitt, 1970b). However, there were certain systematic deviations from additivity of the Shannon–Prewitt radii which could not be explained at that time. These discrepancies involved: (1) compounds with high symmetry such as those having the rocksalt, fluorite, and perovskite structures, (2) certain compounds in isotypic series whose unit cell volumes did not lie on linear r^3–V plots, (3) compounds containing ions which generally formed highly distorted polyhedra such as Cu^{2+} and V^{5+} and (4) compounds containing tetrahedral oxyanions such as VO_4^{3-} and AsO_4^{3-}.

To find the source of these discrepancies, we look at several of the assumptions made in deriving the radii:

(1) The effect of covalency in shortening M–X distances is comparable for all M–F or M–O bonds.
(2) The average cation-anion distance over all similar polyhedra in a structure is constant.
(3) With a constant anion, the unit cell volumes of isotypic compounds are proportional to the volume of the cations.

It has become apparent that these assumptions are not always valid. In this paper, we will see how the breakdown of these assumptions manifests itself in deviations from additivity. In particular, we will see how covalency and the

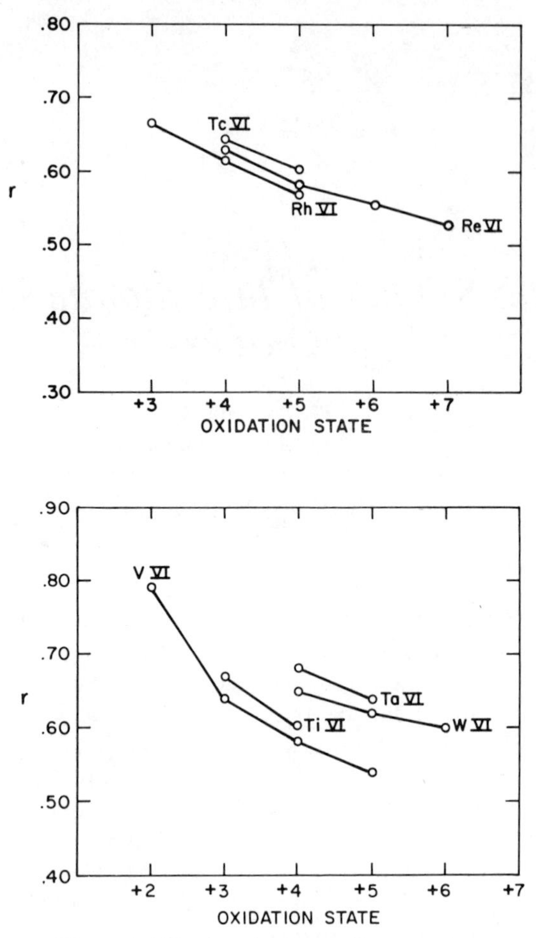

Figure 1. Effective ionic radii *vs.* oxidation state.

shape of the bond length-bond strength curves affects interatomic distances not only in oxides and fluorides, but in all halides and chalcogenides.

The majority of the radius values used in this paper are from Shannon and Prewitt (1969, 1970a); however, certain values are taken from a revised set of effective ionic radii (Shannon, to be published).

DEVIATIONS FROM ADDITIVITY OF EFFECTIVE IONIC RADII

A. Bond Distances in Highly Symmetric Structures

It was noted earlier (Shannon and Prewitt, 1969) that distances calculated for highly symmetric structures like the rocksalt, fluorite, and perovskite types were generally larger than those observed. Table 1 shows calculated and

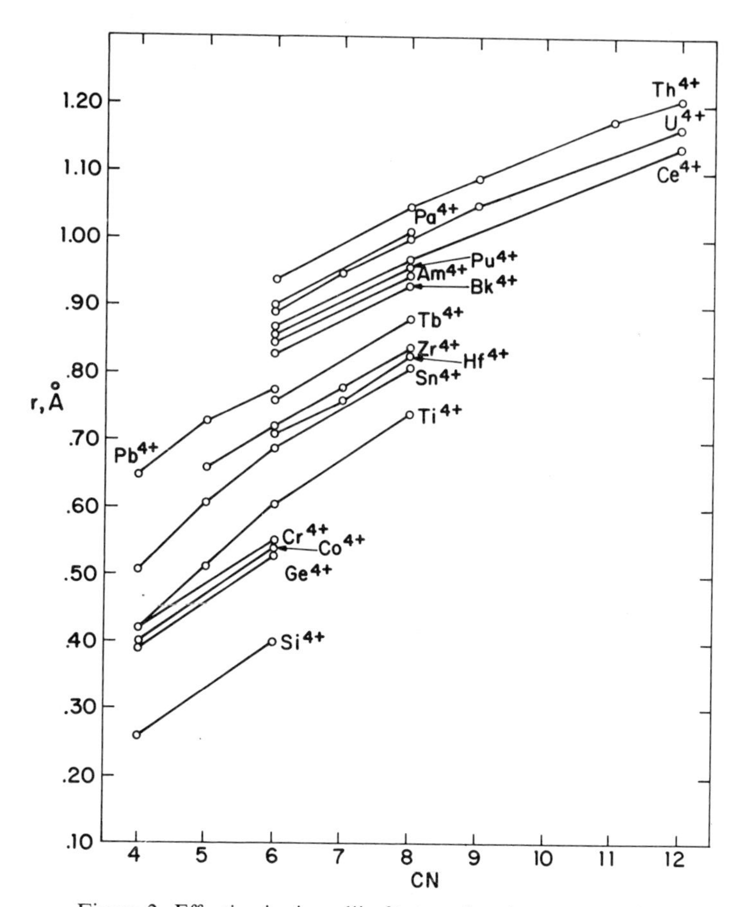

Figure 2. Effective ionic radii of tetravalent ions *vs.* coordination number.

observed distances in rocksalt and fluorite type oxides and fluorides. The discrepancies are not large for oxides but are significant for fluorides. The deviations diminish as the size of the cation increases and become negative for the oxides SrO, BaO, and EuO. These discrepancies are still not well understood and the only explanation offered is the decreased repulsion effects (Shannon and Prewitt, 1969).

Similar discrepancies between calculated and observed distances occur in ReO_3, $Cd_2Re_2O_7$ and $SrIrO_3$ (Table 2). It is possible that such differences are caused by the metallic nature of these oxides. The observed distances are close to those anticipated for Re^{7+}–O in ReO_3, Re^{6+}–O in $Cd_2Re_2O_7$, and Ir^{5+}–O in $SrIrO_3$ (See Table 2). Some or all of the electrons in the conduction band apparently do not contribute to the interatomic distance characteristic of the formal charge on the cation, i.e. Re^{5+} in $Cd_2Re_2O_7$. A similar effect has been observed in SmS when it undergoes a transition from the semiconducting to the metallic state, i.e. the observed distance in metallic SmS corresponds

Table 1. Calculated and observed interatomic distances in highly symmetric oxides and fluorides

Rocksalt

MO VI–VI	R_{Calc} M–O	R_{Obs} M–O	ΔR
NiO	2.09	2.084	0.006
MgO	2.12	2.105	0.015
CoO	2.145	2.133	0.012
FeO	2.18	2.155	0.025
MnO	2.23	2.222	+ 0.008
CdO	2.35	2.348	+ 0.002
CaO	2.40	2.405	− 0.005
SrO	2.54	2.58	− 0.04
BaO	2.72	2.762	− 0.042
EuO	2.52	2.572	− 0.052
MF VI–VI			
LiF	2.09	2.008	0.092
NaF	2.35	2.31	0.04
AgF	2.48	2.46	0.02
KF	2.71	2.673	0.037
CsF	3.00	3.00	0.00
	Fluorite		
MO_2 VIII–IV			
CeO_2	2.35	2.34	0.01
UO_2	2.38	2.37	0.01
ThO_2	2.43	2.42	0.01
MF_2 VIII–IV			
CdF_2	2.41	2.333	0.083
CaF_2	2.43	2.37	0.06
HgF_2	2.45	2.40	0.05
SrF_2	2.57	2.51	0.06
BaF_2	2.73	2.69	0.04

Table 2. Calculated and observed distances in some metallic oxides and sulphides

Compound	Observed distance, Å	Calculated distances, Å	
ReO_3	$\langle Re^{6+}-O\rangle = 1.87$	$\langle Re^{6+}-O\rangle = 1.90$	$\langle Re^{7+}-O\rangle = 1.88$
$Cd_2Re_2O_7$	$\langle Re^{5+}-O\rangle = 1.93$	$\langle Re^{5+}-O\rangle = 1.96$	$\langle Re^{6+}-O\rangle = 1.93$
$SrIrO_3$	$\langle Ir^{4+}-O\rangle = 1.975$	$\langle Ir^{4+}-O\rangle = 2.025$	$\langle Ir^{5+}-O\rangle = 1.97$
SmS(HP)	$\langle Sm^{2+}-S\rangle = 5.70$	$\langle Sm^{2+}-S\rangle = 5.94$	$\langle Sm^{3+}-S\rangle = 5.60$

more closely to what one expects for a $^{VI}Sm^{3+}$–S distance (Jayaraman *et al.*, 1970).

B. Non-linear $r^3 - V$ Plots

The majority of plots of unit cell volume against r^3 for isotypic structures are linear to a first approximation (Shannon and Prewitt, 1970b). This linearity generally allows both confirmation of unit cell volumes and interpolation of

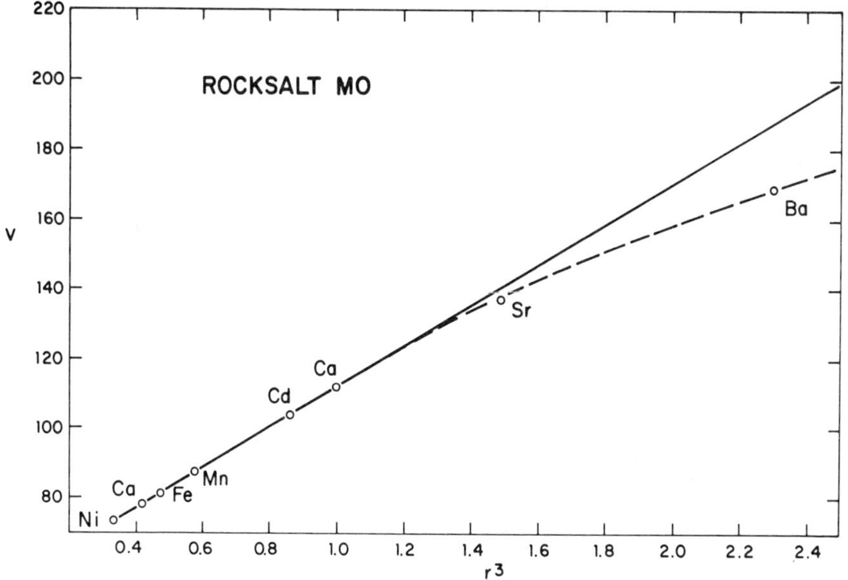

Figure 3. Effective ionic radius cubed (r_M^3) *vs.* unit cell volume (V) for the MO rocksalt structure.

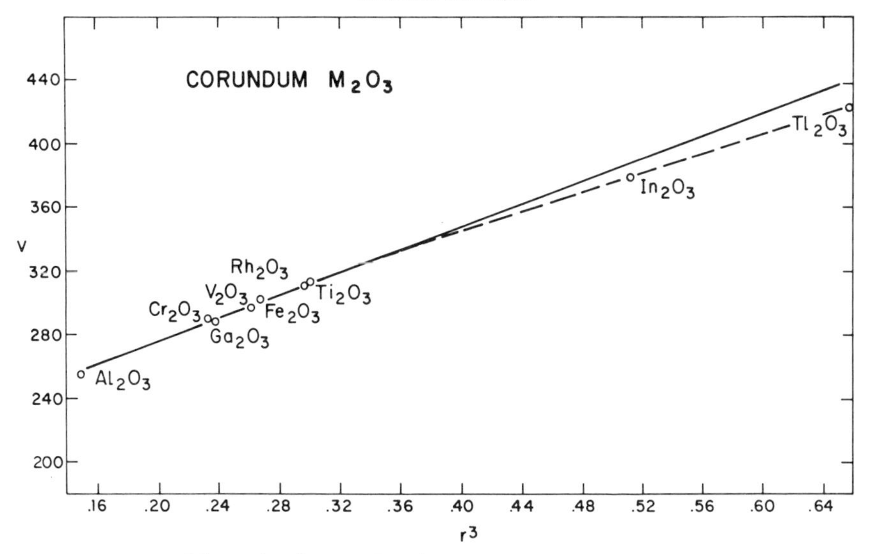

Figure 4. r_M^3 *vs.* V for the M_2O_3 corundum structure.

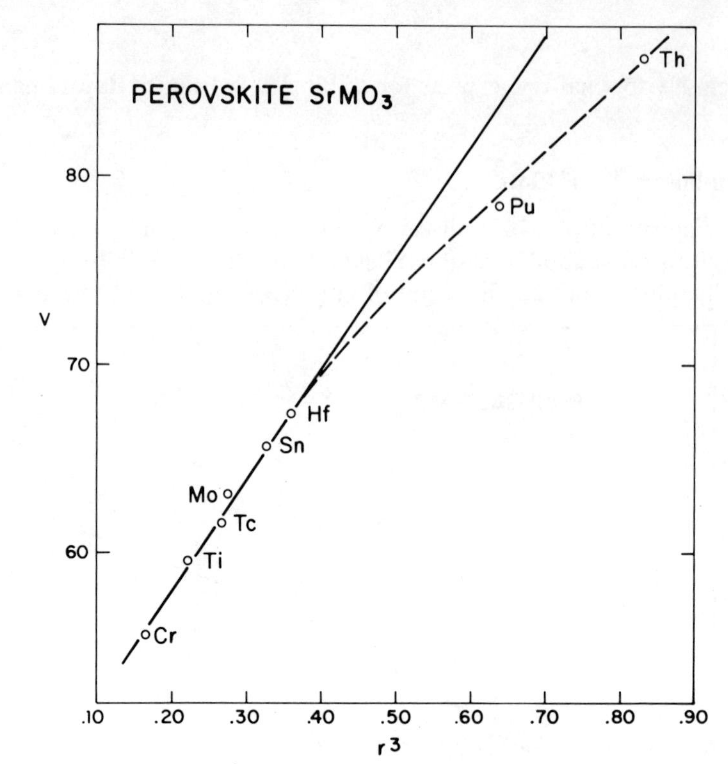

Figure 5. r_M^3 *vs.* V for the $SrMO_3$ perovskite structure.

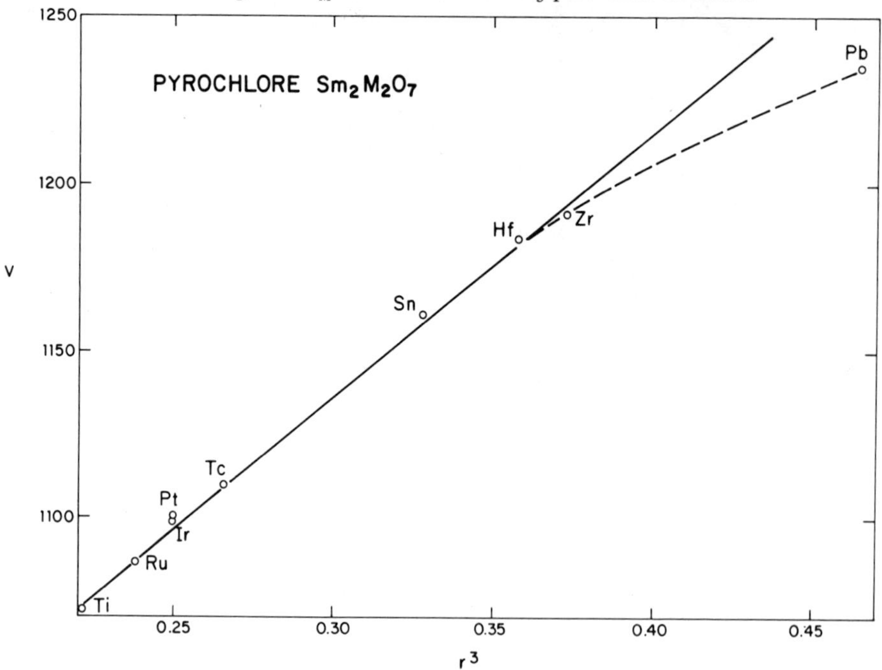

Figure 6. r_M^3 *vs.* V for the $Sm_2M_2O_7$ pyrochlore structure.

ionic radii when no experimental interatomic distances exist. However, it has recently been found by Fukunaga and Fujita (1973) for the $BaM^{4+}O_3$ perovskite series that this plot is distinctly nonlinear.

The same behaviour can be observed for MO rocksalt, M_2O_3 corundum, $SrMO_3$ perovskite, and $Sm_2M_2O_7$ pyrochlore structure types (Figures 3–6). Similarly, the unit cell of $NaInSi_2O_6$ was found to deviate from the otherwise linear plot for $NaMSi_2O_6$ pyroxenes (Prewitt *et al.*, 1972). It is possible that all such plots show this curvature and that it only becomes noticeable when the structure type is stable for cations having a wide range of radii. However, it should also be noted that the curvature becomes pronounced for compounds containing large, polarizable and relatively 'soft' cations such as Sr^{2+}, Ba^{2+}, In^{3+}, Tl^{3+}, Pb^{4+} and Th^{4+}. This curvature may be caused by a contraction accompanying the deformation of these cations.

C. Bond Distances in Distorted Polyhedra and Unusual Coordinations

It was noted (Shannon and Prewitt, 1969) that cations which are frequently found in distorted polyhedra such as V^{5+}, Cu^{2+}, and Mn^{3+} do not give consistent average interatomic distances. To understand this variation of mean distance in distorted polyhedra it is necessary to discuss the relationship between bond length and Pauling bond strength. We will also see how this relationship can be used to predict distances for cations in unusual coordinations.

The concept of mean bond strength ($\bar{s}$) was defined by Pauling (1929) as the valence (z) of a cation divided by its coordination (v). He formulated the electro-

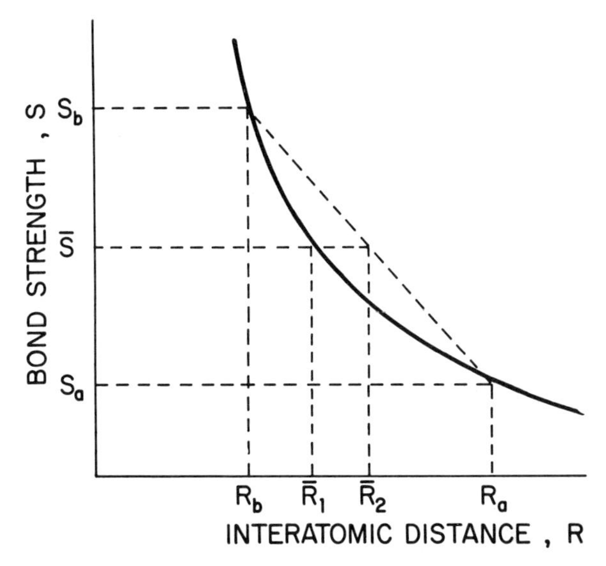

Figure 7. Typical bond length (R) *vs.* bond strength (s) curve.

static valence principle, by which the sums ($p = \Sigma\bar{s}$) of mean bond strengths around the cations and anions are approximately equal to their valence.

According to Pauling, the variation of bond strength with bond length is determined by the factor $1/v$. It is possible, however, to relate bond strength (s) to bond length (R) using analytical expressions (R–s curves). A review of these expressions is given by Brown and Shannon (1973). Using the expression $s = s_0(R/R_0)^{-N}$ where s = bond strength, R = bond length, R_0 and N are fitted parameters, and s_0 is the ideal bond strength associated with R_0, Brown and Shannon evaluated R–s relationships for the first 3 rows of the periodic table. Using these relationships, the sums of bond strengths about cations and anions were found to equal the valences with a mean deviation of about 5 per cent. Accepting the approximate validity of Pauling's second rule, it is possible to derive the effect of distortion of various polyhedra on their mean bond distances.

In Figure 7 a typical R–s curve is shown. An undistorted polyhedron results in an average bond strength $\bar{s}$ and a mean distance $\bar{R}_1$. A distorted octahedron with 3 bonds of length R_a and 3 of length R_b result in the same average bond strength $\bar{s}$ but a mean distance $\bar{R}_2 > \bar{R}_1$.

Strongly distorted octahedra like those containing V^{5+} and Cu^{2+} show a significant variation in mean distance with distortion (Brown and Shannon, 1973; Shannon and Calvo, 1973a). Figure 8 shows a plot of average bond length, $\bar{R}$ against Δ, the degree of distortion given by:

$$\Delta = \frac{1}{6}\sum\left(\frac{R_i - \bar{R}}{\bar{R}}\right)^2$$

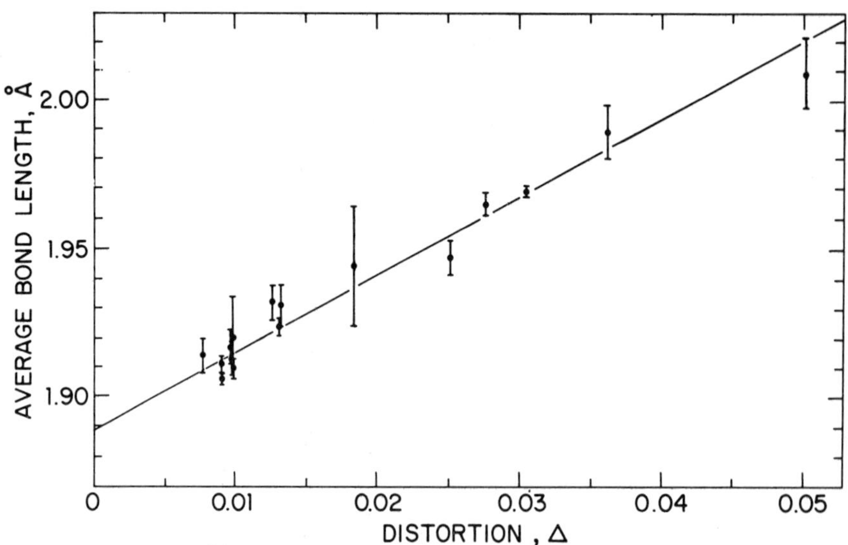

Figure 8. Mean V^{5+}–O distance *vs.* distortion. Distortion, Δ, is defined in text. Vertical bars represent average e.s.d.'s quoted by authors.

Table 3. Mean M1–O and M2–O distances in pyroxenes

Compound	M1–O	M2–O	Reference
$LiVO_3$	2.153	2.284	Shannon and Calvo (1973b)
$MgSiO_3$ clino	2.07	2.15	Morimoto *et al.* (1960)
$MgSiO_3$ ortho	2.070	2.158	Morimoto and Koto (1969)
$FeSiO_3$ clino	2.137	2.224	Burnham (1967)
$FeSiO_3$ ortho	2.145	2.240	Burnham (1967)

Octahedra containing Mg^{2+}, Zn^{2+}, Co^{2+} and Li^{+}, although generally less distorted than those of V^{5+} or Cu^{2+}, also show a similar dependence of mean bond length on Δ (Brown and Shannon, 1973). Using this concept, it has been possible to explain one of the unusual features of the pyroxene structure. In this structure there are large differences in mean distances in M_1 and M_2 sites. Table 3 summarizes the mean octahedral distances in $LiVO_3$, $MgSiO_3$ and $FeSiO_3$. The M_1 and M_2 sites are also characterized by different degrees of distortion. As with vanadates and cupric compounds, the average Li–O and Mg–O octahedral distances were found to be proportional to the degree of distortion (Shannon and Calvo, 1973b).

Distortions in tetrahedra are generally not so important as for octahedra but they do contribute to variations in mean tetrahedral distances. See for example the analysis of phosphates (Baur, 1974) and of amphiboles (Hawthorne, 1974). Similar behaviour will presumably be found for silicate tetrahedra. It is also apparent that the effects of distortion must be considered when using mean distances to estimate the site occupancy of polyhedra containing two atoms of different valence, but belonging to the same or neighbouring elements (e.g. Al^{3+} and Si^{4+} or Mn^{3+} and Mn^{4+}).

Bond length–bond strength curves of the form $s = s_0(R/R_0)^{-N}$ allow calculation of bond distances in coordination polyhedra generally found only at high pressures. Thus for R(Mg–O) in the hypothetical perovskite $^{XII}Mg^{VI}SiO_3$, we calculate:

$$\frac{2}{12} = 0.333\left(\frac{R}{2.098}\right)^{-5.0}:$$

$$R = 2.41\ \text{Å (Brown and Shannon, 1973).}$$

Miller (1972) has calculated M–O distances for unusual coordinations by a modification of Pauling's method. The values are in excellent agreement with those calculated by Brown and Shannon except at the very high and very low coordinations (Table 4).

D. Bond Distances in Highly Covalent Tetrahedral Oxyanions

It was originally believed that small tetrahedral ions were characterized by the most consistent radii. Indeed the radii of Si^{4+}, Ge^{4+}, Be^{2+} and Al^{3+}

Table 4. Calculated dependence of M–O distances on coordination number from data by Miller (1972)

	3	4	5	6	7	8	9	12
$Si–O^{II}$	1.532	1.609	1.686	1.763	—	—	—	—
$Si–O^{III}$	1.547 (1.52)	1.624 (1.625)	1.703 (1.708)	1.778 (1.778)	—	1.932 (1.896)	—	—
$Si–O^{IV}$	1.562	1.639	1.716	1.791	—	1.947	—	—
$Al–O^{IV}$	1.70 (1.66)	1.77 (1.76)	1.84 (1.84)	1.90 (1.91)	1.97 (1.97)	2.04 (2.02)	2.11 (2.07)	2.30 (2.20)
$Mg–O^{IV}$	1.87 (1.83)	1.95 (1.93)	2.03 (2.02)	2.11 (2.10)	2.19 (2.16)	2.27 (2.22)	2.35 (2.27)	2.59 (2.41)
$Fe^{2+}O^{IV}$	1.95 (1.90)	2.00 (2.00)	2.08 (2.08)	2.16 (2.15)	2.24 (2.22)	2.31 (2.27)	2.40 (2.32)	2.63 high-spin (2.44)

Distances in parentheses calculated $s = s_0(R/R_0)^{-N}$ from Brown and Shannon (1973).

were quite invariant from one structure to another. However it was pointed out by Banks *et al.* (1970) that R(V–O) (1.707 Å) in Ca_2VO_4Cl was considerably smaller than that predicted by the sum of the radii of $^{IV}V^{5+}$ and $^{IV}O^{2-}$ = 0.355 + 1.380 = 1.735 Å. The same is true for $Ba_3V_2O_8$ (Süsse and Buerger, 1970). If we compare the radii in several tetrahedral vanadates, $M_xV_yO_z$, we see an interesting correlation (Shannon, 1971; Shannon and Calvo, 1973c).

Ca_2VO_4Cl	0.327	0.325	$Cd_2V_2O_7$	0.358	0.355
$Ba_3V_2O_8$	0.322		$Zn_3V_2O_8$	0.350	
			$Zn_2V_2O_7$	0.359	

The difference between these two groups of compounds is the electronegativity of M. Since the V–O bond is very covalent, any cations which tend to remove electron density from the oxygen ions and consequently from the V–O bonds tend to lengthen the bond. In order to analyse the effect of the 'electron withdrawing' power of the other cations in vanadates and other tetrahedral oxyanions we have evaluated the average electronegativity

$$\bar{\chi} = \frac{x \cdot \chi_M + y\, \chi_T}{x + y}$$

for compounds $M_xT_yO_z$ and compared χ with average tetrahedral T–O distances (Shannon, 1971; Shannon and Calvo, 1973c). However, note that the

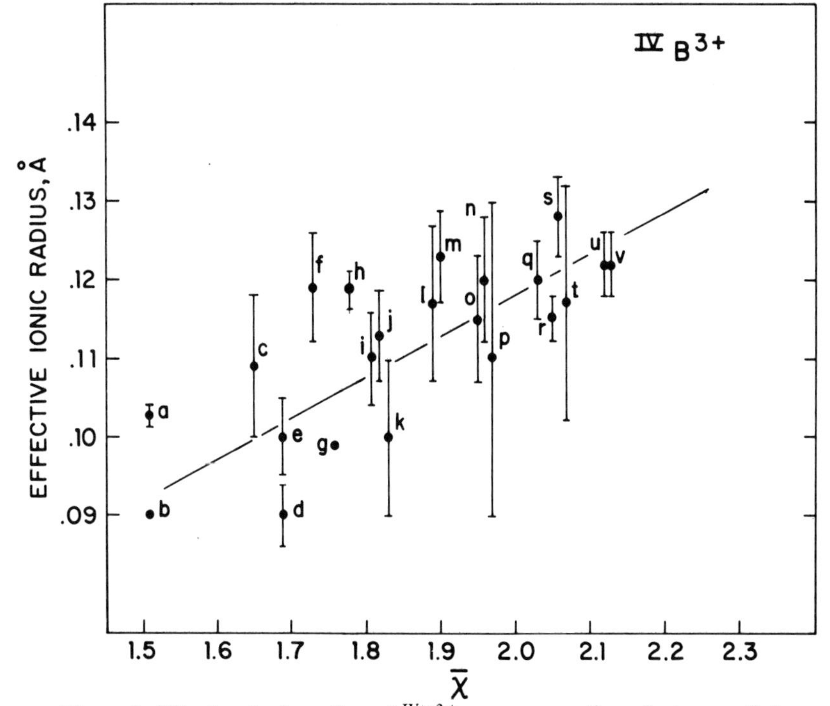

Figure 9. Effective ionic radius of $^{IV}B^{3+}$ *vs.* mean cation electronegativity. Vertical bars represent e.s.d.'s quoted by the authors.

Table 5. Interatomic distances and average electronegativities in borates

Compound	$\bar{\chi}$	$(M\text{–}O)_{av} - r(O^{2-})$	r_m	Reference[a]			
(a) γ-$LiBO_2$	1.51	$1.483 \pm 0.001 - 1.380$	0.103	66	JCPSA	44	3348
(b) $LiBO_2HP$	1.51	$1.47 \pm 0.001 - 1.380$	0.090	65	JPCSA	26	2083
(c) $MgAlBO_4$	1.65	$1.489 \pm 0.009 - 1.380$	0.109	65	MIASA	35	196
(d) CaB_2O_4 IV	1.69	$1.480 \pm 0.004 - 1.390$	0.090	69	ACBCA	25	965
(e) CaB_2O_4 III	1.69	$1.480 \pm 0.005 - 1.380$	0.100	69	ACBCA	25	955
(f) $NaBSi_3O_8$	1.73	$1.465 \pm 0.007 - 1.356$	0.119	65	AMMIA	50	1827
(g) $Ca_3Na_2Cl(SO_4)_2B_5O_8(OH)_2$	1.76	$1.487 \pm 0.007 - 1.388$	0.099	67	NJMMA	1967	157
(h) $CaBSiO_4(OH)$	1.78	$1.480 \pm 0.003 - 1.362$	0.118	73	AMMIA	58	909
(i) $Na_2B_8O_{13}$	1.82	$1.473 \pm 0.006 - 1.363$	0.110	67	ACCRA	22	815
(j) BaB_4O_7	1.82	$1.473 \pm 0.006 - 1.360$	0.113	65	ACCRA	19	297
(k) KB_5O_8	1.83	$1.46 \pm 0.01 - 1.360$	0.100	65	ACCRA	18	1088
(l) MgB_4O_7	1.89	$1.474 \pm 0.01 - 1.357$	0.117	66	NJMMA	1966	142
(m) $K_2B_5O_8(OH).2H_2O$	1.90	$1.483 \pm 0.006 - 1.360$	0.123	69	ACBCA	25	1787
(n) MnB_4O_7	1.91	$1.477 \pm 0.005 - 1.360$	0.117	74	JCPSA	60	1899
(o) $K_2B_4O_7.4H_2O$	1.96	$1.480 \pm 0.008 - 1.365$	0.115	63	ACCRA	16	975
(p) CdB_4O_7	1.97	$1.47 \pm 0.02 - 1.357$	0.113	66	ACCRA	20	132
(q) $KB_5O_8.4H_2O$	2.04	$1.477 \pm 0.005 - 1.357$	0.120	63	ACCRA	16	376
(r) B_2O_3 II	2.04	$1.475 \pm 0.003 - 1.356$	0.119	68	ACBCA	24	869
(s) $NaB(OH)_4.2H_2O$	2.06	$1.476 \pm 0.005 - 1.348$	0.128	63	ACCRA	16	1233
(t) $Ba[B(OH)_4]_2.H_2O$	2.07	$1.479 \pm 0.015 - 1.362$	0.117	69	ACBCA	25	1811
(u) γ-HBO_2	2.12	$1.472 \pm 0.004 - 1.355$	0.117	63	ACCRA	16	380
(v) β-HBO_2	2.12	$1.472 \pm 0.007 - 1.352$	0.120	63	ACCRA	16	385

[a]Codens for periodical titles, *Vol. II*, *ASTM Data Series DS 23B*, Phila., 1970.

Table 6. Interatomic distances and average electronegativities in silicates

Compound	$\bar{\chi}$	$(M{-}O)_{av} - r(O^{2-})$	r_m	Reference[a]			
$Li_6Si_2O_7$	1.21	1.639 ± 0.010 − 1.389	0.250	69	MOCHA	100	295
Na_2SiO_3	1.25	1.632 ± 0.003 − 1.386	0.246	67	ACCRA	22	37
γ-Ca_2SiO_4	1.30	1.644 ± 0.005? − 1.380	0.264	71	ACBCA	27	848
$CaMgSiO_4$	1.40	1.626 ± 0.005? − 1.380	0.246	65	TMPMA	10	34
β-$Na_2Si_2O_5$	1.42	1.624 ± 0.003 − 1.376	0.248	68	ACCRA	24	1077
$Na_2TiSi_4O_{11}$	1.43	1.620 ± 0.005? − 1.354	0.266	62	AMMIA	47	539
$CaSiO_3$	1.45	1.626 ± 0.00? − 1.374	0.252	63	MSAPA	1	293
Yb_2SiO_5	1.47	1.626 ± 0.009 − 1.368	0.258	70	SPHCA	14	854
$(Ca,Na)_2AlSi_2O_7$	1.48	1.622 ± 0.004 − 1.380	0.242	70	ZEKGA	131	314
$Ca_2ZnSi_2O_7$	1.49	1.618 ± 0.005 − 1.380	0.238	69	ZEKGA	130	427
$Ca_3Al_2Si_3O_{12}$	1.49	1.649 ± 0.001 − 1.380	0.269	66	ZEKGA	123	81
$Mg_2SiO_4.MgF_{1\cdot8}(OH)_{0\cdot2}$	1.49	1.630 ± 0.002 − 1.380	0.250	69	AMMIA	54	376
Mg_2SiO_4	1.50	1.634 ± 0.003 − 1.380	0.254	68	AMMIA	53	807
$NaScSi_2O_6$	1.50	1.632 ± 0.003 − 1.373	0.259	73	ACBCA	29	2615
$CaMgSi_2O_6$	1.50	1.635 ± 0.002 − 1.374	0.261	69	MSAPA	2	31
$Nd_2Si_2O_7$	1.52	1.626 ± 0.011 − 1.368	0.258	70	ACBCA	26	484
α-$Pr_2Si_2O_7$	1.52	1.616 ± 0.009 − 1.368	0.248	70	NATWA	57	452
$Ba_3Nb_6Si_4O_{26}$	1.53	1.621 ± 0.007 − 1.370	0.251	71	ACBCA	26	105
$Ca_5Si_2O_7(CO_3)_2$	1.54	1.624 ± 0.007 − 1.372	0.252	70	ZEKGA	132	288
$Ca_2NaHSi_3O_9$	1.55	1.630 ± 0.003 − 1.372	0.258	67	ZEKGA	125	298
$Gd_2Si_2O_7$	1.55	1.629 ± 0.011 − 1.367	0.262	70	ACBCA	26	484
$Er_2Si_2O_7$	1.57	1.621 ± 0.005 − 1.359	0.262	70	ACBCA	26	484
$BaBe_2Si_2O_7$	1.57	1.626 ± 0.015 − 1.380	0.246	71	AMMIA	56	1573
$Yb_2Si_2O_7$	1.58	1.626 ± 0.005 − 1.359	0.267	70	ACBCA	26	484
$CaMnSi_2O_6$	1.59	1.623 ± 0.005? − 1.366	0.257	62	ZEKGA	117	331
$NaAlSi_2O_6$	1.59	1.623 ± 0.002 − 1.374	0.249	66	AMMIA	51	956
$CaMnSi_2O_6$	1.59	1.644 ± 0.007 − 1.374	0.270	67	AMMIA	52	709

Table 6. Interatomic distances and average electronegativities in silicates (*Contd.*)

Compound	$\bar{\chi}$	$(M-O)_{av} - r(O^{2-})$	r_m	Reference[a]			
$NaCrSi_2O_6$	1.60	1.624 ± 0.004 − 1.374	0.250	69	MSAPA	2	31
$LiAlSi_2O_6$	1.60	1.618 ± 0.002 − 1.366	0.252	69	MSAPA	2	31
$Mg_3Al_2Si_3O_{12}$	1.61	1.635 ± 0.002 − 1.380	0.255	65	AMMIA	50	2023
$ZrSiO_4$	1.62	1.622 ± 0.005 − 1.360	0.262	71	AMMIA	56	782
$NaFeSi_2O_6$	1.62	1.628 ± 0.002 − 1.374	0.254	69	MSAPA	2	31
$MgSiO_3$	1.61	1.636 ± 0.009 − 1.366	0.270	69	ZEKGA	129	65
$Sc_2Si_2O_7$	1.63	1.621 ± 0.005 − 1.359	0.262	73	SPHCA	17	749
$NaInSi_2O_6$	1.63	1.633 ± 0.007 − 1.373	0.260	67	ACSAA	21	1425
$Ca_2Al_3(OH)Si_3O_{12}$	1.64	1.623 ± 0.012 − 1.369	0.254	68	AMMIA	53	1882
$Ca_2(Al,Mn,Fe)_3(OH)Si_3O_{12}$	1.64	1.624 ± 0.006 − 1.370	0.254	69	AMMIA	54	710
$Ca_2Al_3(OH)Si_3O_{12}$	1.64	1.620 ± 0.006 − 1.370	0.250	68	AMMIA	53	1882
$Ca_2(Al,Fe)_3(OH)Si_3O_{12}$	1.64	1.620 ± 0.007 − 1.370	0.250	71	AMMIA	56	447
$PbCa_2Si_3O_9$	1.67	1.641 ± 0.015 − 1.369	0.272	69	ZEKGA	128	213
$LiFeSi_2O_6$	1.67	1.620 ± 0.002 − 1.366	0.254	69	MSAPA	2	31
$(Mn,Ca)SiO_3$	1.68	1.628 ± 0.005 − 1.368	0.260	63	ZEKGA	119	98
Be_2SiO_4	1.70	1.631 ± 0.001 − 1.360	0.271	72	SPHCA	16	1021
$(Mg,Fe)Al_3SiBO_9$	1.71	1.619 ± 0.006 − 1.364	0.255	68	ACBCA	24	1518
Al_2SiO_5 (and)	1.71	1.627 ± 0.005 − 1.360	0.267	61	ZEKGA	115	269
Al_2SiO_5 (Ky)	1.71	1.628 ± 0.005 − 1.360	0.268	63	ZEKGA	118	337
Al_2SiO_5 (sill)	1.71	1.615 ± 0.006 − 1.358	0.257	63	ZEKGA	118	127
$(Mg_{0\cdot4}Fe_{0\cdot5}Ca_{0\cdot1})SiO_3$	1.72	1.632 ± 0.009 − 1.366	0.266	70	AMMIA	55	1195
$BaTiSi_3O_9$	1.73	1.622 ± 0.003 − 1.357	0.265	69	ZEKGA	129	222
$NaFe_3B_3Al_6Si_6O_{30}F$	1.77	1.618 ± 0.003 − 1.360	0.258	71	AMMIA	56	101
$Mn_5(OH)_2(SiO_4)_2$	1.77	1.627 ± 0.008 − 1.380	0.247	70	ZEKGA	132	1
$CaB_2Si_2O_8$	1.77	1.619 ± 0.010 − 1.359	0.260	74	AMMIA	59	79
$CaFe_3OHSi_2O_8$	1.78	1.636 ± 0.004 − 1.376	0.260	71	AMMIA	56	1573
$Al_2Be_3Si_6O_{18}$	1.77	1.606 ± 0.001 − 1.360	0.246	72	ACBCA	28	1899
$CaBSiO_4(OH)$	1.79	1.641 ± 0.011 − 1.364	0.277	73	AMMIA	58	909
$PbZnSiO_4$	1.80	1.630 ± 0.010 − 1.370	0.260	67	ZEKGA	124	115
$Fe_3Al_2Si_3O_{12}$	1.79	1.634 ± 0.002 − 1.380	0.254	71	ZEKGA	134	333

Table 6. Interatomic distances and average electronegativities in silicates (*Contd.*)

Compound	$\bar{\chi}$	$(M\text{–}O)_{av} - r(O^{2-})$	r_m	Reference[a]
$K_2Mg_5Si_{12}O_{30}$	1.82	$1.613 \pm 0.002 - 1.360$	0.253	72 ACBCA 28 267
Fe_2SiO_4	1.83	$1.638 \pm 0.003 - 1.380$	0.258	68 AMMIA 53 807
$Zn_2Mn(OH)_2SiO_4$	1.84	$1.626 \pm 0.007 - 1.365$	0.261	63 ZEKGA 119 117
$Cu_2Ca_2Si_3O_{10}.2H_2O$	1.84	$1.625 \pm 0.004 - 1.360$	0.265	71 AMMIA 56 193
$FeSiO_3$	1.85	$1.619 \pm 0.009 - 1.366$	0.253	67 CIWYA 65 285
$CaAl_2(OH)_2OH_2Si_2O_7$	1.87	$1.633 \pm 0.001 - 1.364$	0.269	71 AMMIA 56 1573
$PbSiO_3$	1.87	$1.628 \pm 0.003 - 1.36?$	0.268	68 ZEKGA 126 98
β-Ca_2SiO_4	1.89	$1.641 \pm 0.015 - 1.380$	0.261	70 PEPIA 3 161
Co_2SiO_4	1.89	$1.641 \pm 0.012 - 1.380$	0.261	70 PEPIA 3 161
SiO_2 (α-quartz)	1.90	$1.607 \pm 0.003 - 1.350$	0.257	62 ACCRA 15 337
SiO_2 (α-quartz)	1.90	$1.609 \pm 0.002 - 1.350$	0.259	65 ACCRA 18 710
SiO_2 (crist.)	1.90	$1.605 \pm 0.004 - 1.350$	0.255	65 ZEKGA 121 369
$Zn_4(OH)_2(OH_2)Si_2O_7$	1.92	$1.628 \pm 0.007 - 1.358$	0.270	67 ZEKGA 124 180
$Ca_2Al_4Si_{12}O_{32}.8H_2O$	1.95	$1.609 \pm 0.016 - 1.354$	0.255	69 ACBCA 25 1183
$Al_{13}(F,OH)_{18}Si_5O_{20}Cl$	1.95	$1.64 \pm 0.01 - 1.354$	0.286	60 ACCRA 13 15
$Al_{13}(F,OH)_{18}Si_5O_{10}Cl$	1.95	$1.635 \pm 0.015 - 1.354$	0.281	70 NJMMA 1970 552
$Na_2SiO_3.6H_2O$	2.01	$1.639 \pm 0.004 - 1.362$	0.277	71 ACBCA 27 2269
$Na_2SiO_3.9H_2O$	2.06	$1.631 \pm 0.010 - 1.358$	0.273	66 ACCRA 20 688

[a]Codens for periodical titles, *Vol. II, ASTM Data Series DS 23B*, Phila., 1970.

Table 7. Interatomic distances and average electronegativities in selenates

Compound	$\bar{\chi}$	$(M\text{–}O)_{av} - r(O^{2-})$	r_m	Reference[a]			
(a) K_2SeO_4	1.40	1.648 ± 0.010 – 1.395	0.249	70	ACBCA	26	1451
(b) Na_2SeO_4	1.50	1.654 ± 0.021 – 1.380	0.274	70	ACBCA	26	436
(c) $MnSeO_4$	2.05	1.652 ± 0.02 – 1.355	0.297	68	ZAACA	358	125
(d) $CoSeO_4$	2.21	1.642 ± 0.03 – 1.355	0.287	68	ZAACA	358	125
(e) $NiSeO_4$	2.23	1.670 ± 0.02 – 1.355	0.315	68	ZAACA	358	125
(f) $NiSeO_4.6H_2O$	2.21	1.660 ± 0.02 – 1.352	1.318	70	ZAACA	379	204

[a]Codens for periodical titles, *Vol. II, ASTM Data Series DS 23B*, Phila. 1970.

effects are small (~0.01–0.05 Å), and since the dependence of interatomic distance on oxygen coordination is of the same order of magnitude, it is necessary to make all correlations with the average interatomic distance corrected for oxygen coordination ($\bar{R}_{V-O}$–$r(O^-) = r(V^{5+})$. The systems P^{5+}–O, V^{5+}–O, and As^{5+}–O have been carefully analysed. For $\bar{\chi}$ *vs* $r(P^{5+})$, $r(As^{5+})$, and $r(V^{5+})$, there is a marked dependence of r on $\bar{\chi}$. Linear regression analyses resulted in correlation coefficients of 0.68, 0.85, and 0.80 respectively (Shannon and Calvo, 1973c). In Tables 5–7 similar data are shown for a number of accurately refined borate, silicate, and selenate structures. Figures 9–11 show the dependence of corrected mean interatomic distances on average cation electronegativity. Table 8 summarizes statistical data for many tetrahedral oxyanions. The slope (m)

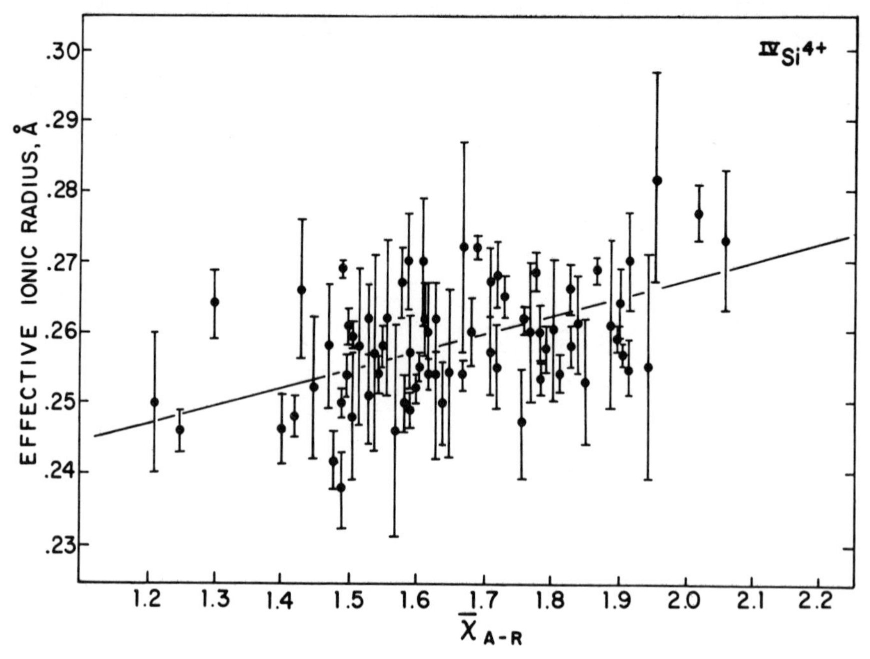

Figure 10. Effective ionic radius of Si^{4+} *vs.* mean cation electronegativity.

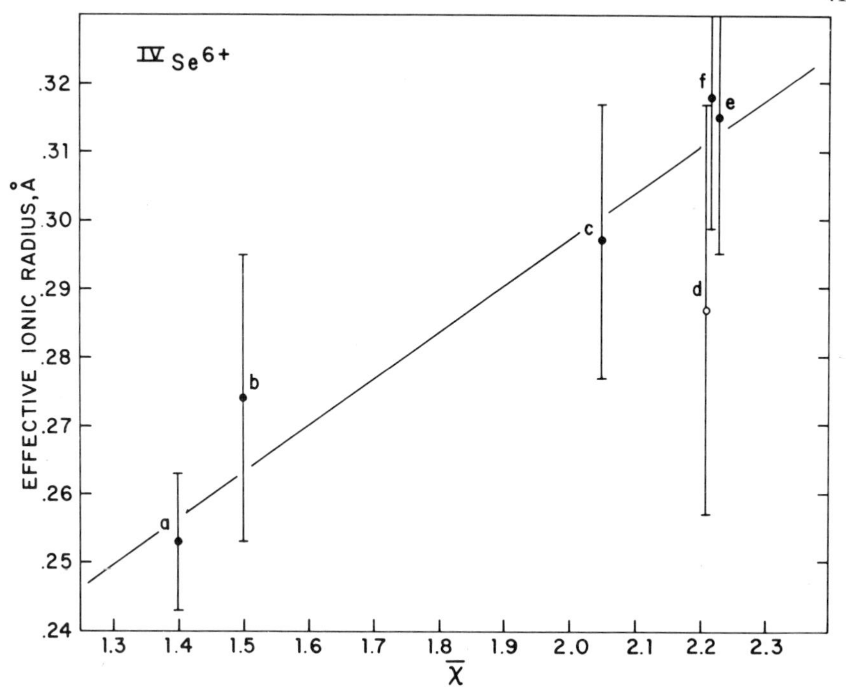

Figure 11. Effective ionic radius of $^{IV}Se^{6+}$ *vs.* mean cation electronegativity.

of r *vs* $\bar{\chi}$ for silicates is only 0.017, with a correlation coefficient of 0.40 (see Table 8). Thus, Si–O distances do not depend greatly on the nature of the other cations in the structure. However, distances in borates show a dependence on $\bar{\chi}$ similar to that of germanates, phosphates, arsenates, and sulphates

Table 8. Effect of mean cation electronegativity on corrected mean interatomic distance in tetrahedral oxyanions

Ion	N	Slope, m	Intercept, C_1	Correlation Coefficient, r_{cc}	Goodness of fit $\left[\frac{\Sigma(\Delta r)^2}{N-2}\right]^{1/2}$	Probability that $r > r_{cc}$
B^{3+}	27	0.0325	0.052	0.67	0.010	<0.001
Si^{4+}	74	0.017	0.231	0.40	0.009	<0.001
Ge^{4+}	16	0.030	0.333	0.62	0.010	~ 0.005
P^{5+}	62	0.029	0.120	0.68	0.010	<0.001
As^{5+}	19	0.032	0.250	0.85	0.009	$<<0.001$
V^{5+}	21	0.051	0.280	0.80	0.010	~ 0.001
S^{6+}	35	0.039	0.045	0.58	0.014	<0.001
Se^{6+}	7	0.053	0.185	0.92	0.013	~ 0.005
Cr^{6+}	14	0.037	0.230	0.56	0.020	~ 0.07
Mo^{6+}	15	0.037	0.340	0.57	0.016	~ 0.06
W^{6+}	7	0.029	0.373	0.60	0.009	~ 0.09
Cl^{7+}	6	0.073	0.077	0.87	0.017	~ 0.02

$r = m\bar{\chi} + c_1$

($m \simeq 0.3$). The correlation coefficients for these oxyanions are in the range 0.6–0.8. The vanadates, selenates, and chromates show the greatest dependence on $\bar{\chi}$ with $m = 0.5$–0.7 with correlation coefficients from 0.8–0.9.

Although tetrahedral distortion also contributes to the variation in mean interatomic distance in tetrahedral oxyanions, it is clear that the electronegativity of the second nearest neighbour is of importance. This dependence of r on χ is consistent with our ideas of covalence, i.e. increased covalence shortens bond distance. This effect is therefore described as 'covalent contraction'.

E. The Relative Size of Fe^{2+} and Mg^{2+} in Halides and Chalcogenides

The plot of r^3 *vs* V for the rutile fluorides is linear with the exception of MgF_2 and ZnF_2 (Shannon and Prewitt, 1970b), where $V(MgF_2)$ is less than $V(NiF_2)$ in contrast to the effective ionic radii (derived primarily from oxides) of Ni (0.69 Å) and Mg(0.72 Å). Similar behaviour was noted for the fluorides $CsNiF_3$ and $CsMgF_3$ (Longo and Kafalas, 1969). The explanation of this phenomenon became evident when the interatomic distances in Fe_2GeS_4 and Mg_2GeS_4 became available (Vincent and Perrault, 1971; Vincent and Bertaut, 1973) In these compounds the $\langle$Fe–S$\rangle$ distance was found to be shorter than the $\langle$Mg–S$\rangle$ distance in contrast to the radii of Fe^{2+} (0.78 Å) and Mg^{2+} (0.72 Å). Such deviations are again apparently caused by 'covalent shortening'.

In order to see the effect of covalency on M–X distances we derive a 'covalency contraction' parameter R_d which indicates the change of the distance of a covalent bond M–X relative to the distance of the less covalent bond Mg–X (Shannon and Vincent, 1975). We use two methods for comparing distances: (1) the use of ratios of the cube of the mean interatomic distances, R

$$R_d = \frac{\langle \text{Fe–X} \rangle^3}{\langle \text{Mg–X} \rangle^3}$$

in compounds having the same structure with differing anions, e.g. X = O, S and Se and (2) the use of ratios of unit cell volumes

$$R_v = \frac{V(\text{Fe}_m\text{X}_n)}{V(\text{Mg}_m\text{X}_n)}$$

for isotypic compounds with different anions. The use of (1) has the advantage of allowing comparison of actual distances but is not so useful as (2) because of the relative scarcity of structural data.

The unit cell data were obtained from Wyckoff (1960) and the ASTM card index unless specific references are given. Because the ratio of unit cell volumes is sensitive to the proportion of M to the other cations present, it is necessary to correct this ratio empirically (for the two hypothetical compounds Mg-$M_{99}O_{100}$ and Fe $M_{99}O_{100}$, $V(\text{Fe})/V(\text{Mg}) \simeq 1.00$). For example, in Fe- and Mg-sulphides we used $R_v = R_v' \pm 0.0002$ (P_M) where P_M = percentage of Fe or Mg relative to all other cations. The value of 0.0002 was derived from the series of

Table 9. Unit cell dimensions and R_v for isotypic Fe^{2+} and Mg^{2+} Compounds

Halides					Chalcogenides (continued)				
Compound	V	R'_v	R_v	$\bar{R}_v$	*Compound*	V	R'_v	R^a_v	$\bar{R}_v$
FeF_2	36.5	1.120	1.120	1.116	Fe_2GeS_4	530.7	0.930	0.924	0.964
MgF_2	32.6				Mg_2GeS_4	570.3			
$NaFeF_3$	60.7	1.084	1.13		$FeIn_2S_4$	1190.3	0.975	0.962	
$NaMgF_3$	56.0				$MgIn_2S_4$	1220.6			
$KFeF_3$	69.90	1.115	1.16		$FeYb_2S_4$	1273.0	0.968	0.955	
$KMgF_3$	62.71				$MgYb_2S_4$	1315.5			
$CsFeF_3$	81.3	1.068	1.118		$FeLu_2S_4$	1262.2	0.962	0.949	
$CsMgF_3$	76.1				$MgLu_2S_4$	1312.6			
K_2FeF_4	112.2	1.068	1.14		$FeSc_2S_4$	1165.9	0.971	0.958	
K_2MgF_4	104.1				$MgSc_2S_4$	1200.1			
Ba_2FeF_6	135.5	1.046	1.113		FeY_4S_7	522.9	0.985	0.969	
Ba_2MgF_6	129.6				MgY_4S_7	530.9			

Table 9. Unit cell dimensions and R_v for isotypic Fe^{2+} and Mg^{2+} Compounds (*Contd.*)

Compound	V	R'_v	R_v	$\bar{R}_v$
Halides				
$FeCl_2$	196.5	0.967		0.97
$MgCl_2$	203.2			
$FeBr_2$	75.22	0.954		0.95
$MgBr_2$	78.85			
FeI_2	96.02	0.940		0.94
MgI_2	102.12			
Chalcogenides				
FeO	80.79	1.082	1.08	
MgO	74.68			
Fe_2SiO_4	307.9	1.062		
Mg_2SiO_4	290.0			
$FeSiO_3$	438.9	1.056		
$MgSiO_3$	415.5			
$FeWO_4$	267.4	1.020		
$MgWO_4$	262.2			
$FeMoO_4$	264.2	1.010		
$MgMoO_4$	261.7			

Compound	V	R'_v	R^a_v	$\tilde{R}_v$
Chalcogenides (continued)				
$FeHo_4S_7$	519.2	0.990	0.974	
$MgHo_4S_7$	524.6			
$FeEr_4S_7$	513.7	0.989	0.973	
$MgEr_4S_7$	519.3			
$FeTm_4S_7$	509.5	0.993	0.977	
$MgTm_4S_7$	512.8			
Hydroxides				
$Fe(OH)_2$	42.33	1.035		1.02
$Mg(OH)_2$	40.90			
$FeSn(OH)_6$	466.7	0.999		
$MgSn(OH)_6$	467.1			

[a] $R_v = R'_v - 0.0002\ (P_m)$

sulphides listed in Table 9. When more than one value of R_v was available, we have taken the mean value, $\bar{R}_v$, as characteristic of a particular M–X bond.

To see the effect of covalence on the Fe–X distance relative to the Mg–X distance, we compare the ratio R_v or R_d to the electronegativity difference between Fe and X, $\Delta\chi_{Fe-X}$. Table 9 lists unit cell volumes and ratios of V_{Fe}/V_{Mg} in halides, chalcogenides, and hydroxides. Table 10 lists mean interatomic distances in structures containing Fe^{2+} and Mg^{2+} and the ratios R_d.

Table 9 shows that very different unit cell volume ratios, $\bar{R}_v$, result when the anion is changed. Both $\bar{R}_v$ and $\bar{R}_d$ show the same dependence on the anion electronegativity. In Figure 12 R_v and R_d are plotted against $\Delta\chi_{M-}$. It is evident that the degree of covalence of the Fe–X and Mg–X bonds is important in determining their relative bond lengths. The effect is particularly striking because the ratio $\bar{R}$ changes from a value > 1 to a value < 1 between O and Cl. Similar R_v and R_d data for halides and chalcogenides of Ni, Mn, Co, Zn and Cd have been analyzed (Shannon and Vincent, 1975). The slopes of the $\bar{R}$–$\Delta\chi$ curves are similar for Ni, Fe, Mn, Co and Cr. For Zn and Cd the slopes are significantly less. This difference in slope may be due to the difference in cova-

Table 10. Mean interatomic distances and R_d in compounds containing octahedral Fe^{2+} and Mg^{2+}

Compound	⟨M–X⟩	$\langle M\text{–}X\rangle^3$	$\bar{R}_d$	$\langle\bar{R}_d\rangle$
FeF_2	2.08	8.999	1.139	1.139
MgF_2	1.992	7.904		
$KFeF_3$	2.060	8.742	1.116	
$KMgF_3$	1.986	7.833		
$FeCl_2$	2.497	15.570	1.000	1.00
$KMgCl_2$	2.499	15.578		
$CsMgCl_3$	2.496			
FeO	2.1615	10.099	1.083	1.082
MgO	2.105	9.327		
Fe_2SiO_4	2.175	10.289	1.081	
Mg_2SiO_4	2.119	9.514		
FeS(20 °C)	2.505	15.719	0.893	
MgS(NaCl)	2.602	17.596		
FeS	2.419	14.155		
MgS(NaCl)	2.602	17.616	0.804	
Fe_2GeS_4	2.534	16.271	0.939	
Mg_2GeS_4	2.588	17.333		
Fe_2SiS_4	2.549	16.562	0.955	
Mg_2GeS_4	2.588	17.333		
FeSe(NiAs)	2.502	15.662	0.774	
MgSe(NaCl)	2.725	20.235		

(MiAS, 190 °C)

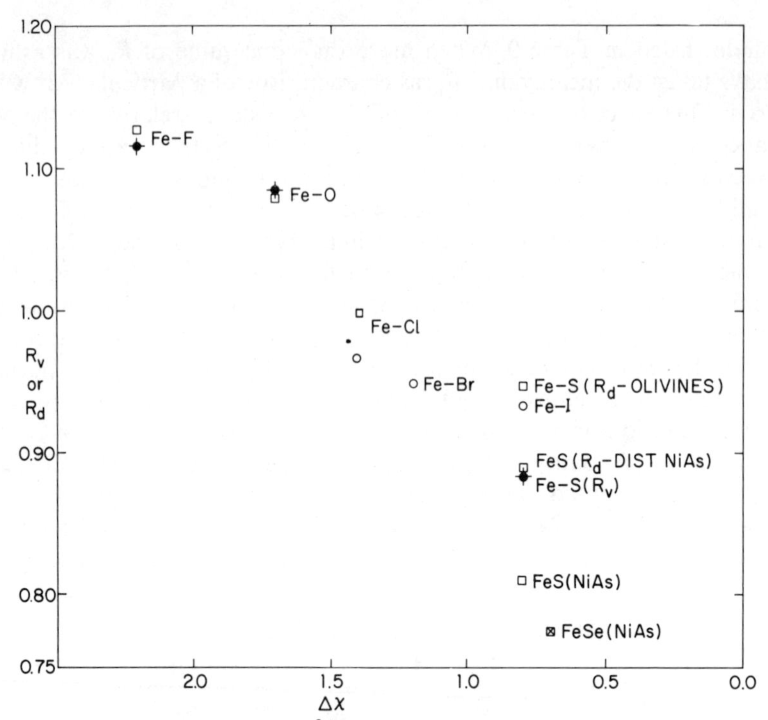

Figure 12. R_v *vs.* $\Delta\chi$ for Fe^{2+} compounds.
 ✦—R_v derived from many isotypic Fe and Mg compounds
 o—R_v derived from only a few isotypic Fe and Mg compounds
 □—R_d derived from several isotypic compounds
 ⊠—R_d derived from compounds with similar M coordination but different structures

lence of hybrid orbitals formed from metal *d* orbitals (Cr, Mn, Fe, Ni, Co) *vs.* metal *s-p* orbitals (Zn, Cd).

The relationship between $\bar{R}$ and $\Delta\chi$ is approximately linear except for NiAs phases containing S, Se and Te. Unfortunately, the sulfides, selenides and tellurides of Ni, Fe, Mn and Co are seldom isotypic with those of Mg, so that true values of R_v are not available for many of the highly covalent selenides and tellurides. It is apparent that R_v is structure dependent for highly covalent compounds, e.g. R_v for Fe^{2+}—S is 0.95 (M_2GeS_4 olivines), 0.89 (distorted NiAs structure) and 0.80 (NiAs structure).

Covalency affects plots of cell dimensions against ionic radius by lowering the more covalent ions relative to a line joining the Mg and Ca compounds. Thus, while Ni compounds have larger volumes than corresponding Mg compounds in fluorides (Figure 13), the Mg compounds have larger volumes in oxides and sulphides (Figures 14 and 15). Similarly, in fluorides and oxides Mn compounds have much greater volumes than the corresponding Mg compounds (Figures 13 and 14), whereas in sulphides the Mg and Mn compounds have approximately equal volumes (Figure 15), and in selenides, the volumes of the Mn compounds are smaller than those of Mg.

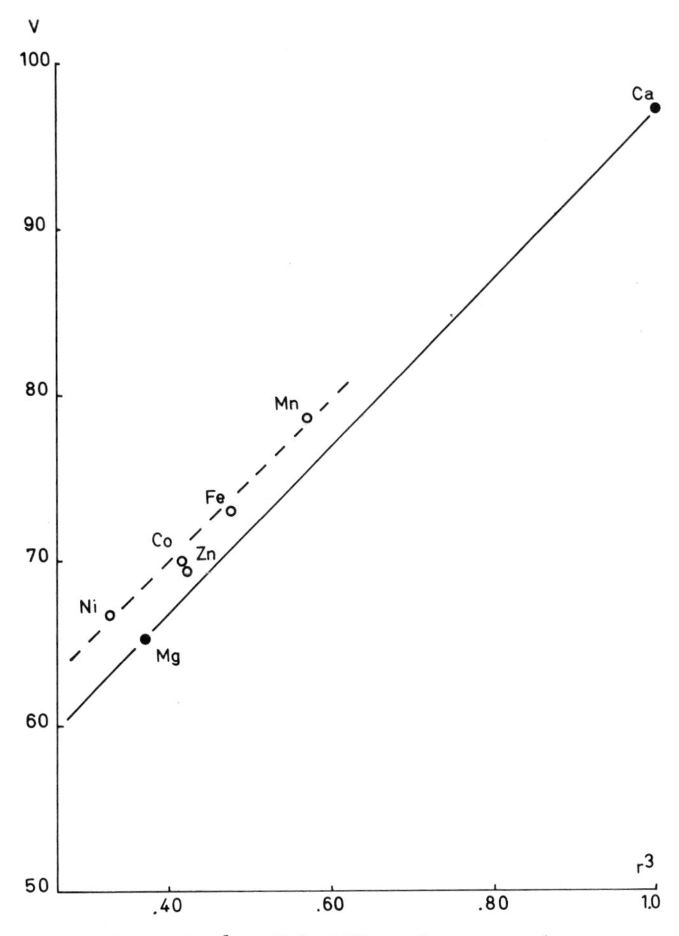

Figure 13. r_M^3 *vs.* V for MF_2 rutile compounds.

Biggar (1969) calculated unit cell volume ratios for isotypic Ni and Mg compounds in order to revise the Mg^{2+} and Ni^{2+} ionic radii of Ahrens (1952) and Pauling (1960). He concluded that $r(Ni^{2+}) = 0.97 \times r(Mg^{2+})$ in oxides and halides. It is interesting to note that Biggar found ratios for the halides F, Cl, Br, and I in agreement with the electronegativity dependence in this paper ($R_v = 0.060$, 0.971, 0.919, and 0.836 respectively) but he assumed that the deviations from 0.970 were the result of faulty data. We believe that these deviations are real, and are caused by different degrees of covalence.

The previous literature on the effects of partial covalence on interatomic distances in contradictory. Pauling (1960) cites the examples of CuF, BeO, AlN, and SiC where observed bond lengths are shorter than the sum of the covalent radii. He attributes these differences to partial ionic character and thus implies that partial ionic character shortens covalent bonds. This conclusion is in accord with the Schoemaker–Stevenson (1941) rule $D_{A-B} = r_A + r_B - C$

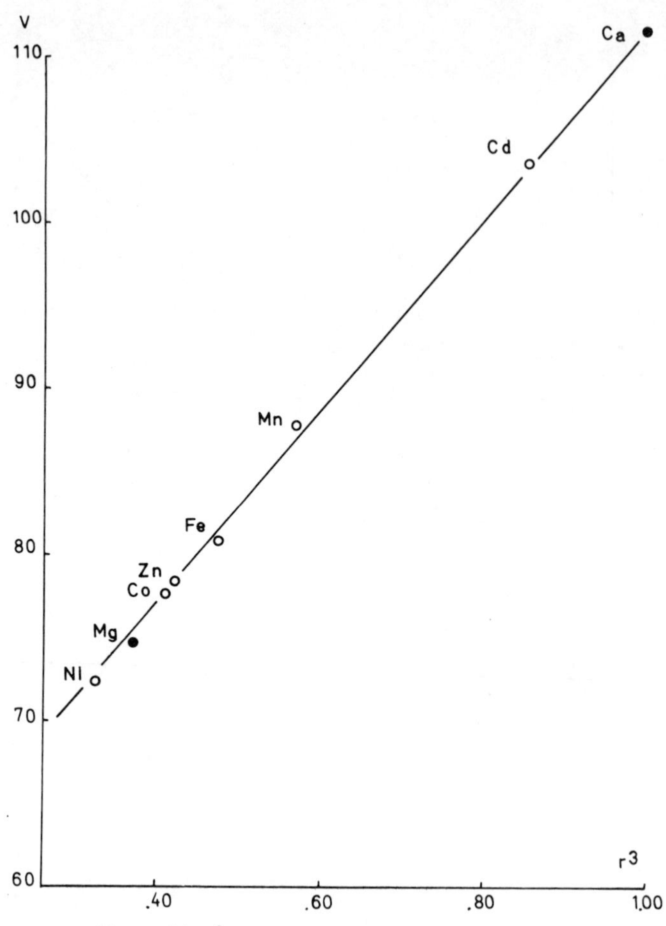

Figure 14. r_M^3 *vs.* V for MO rocksalt compounds.

$|\chi_A - \chi_B|$ where D = interatomic distance between A and B, r_A and r_B = covalent radii of A and B, χ_A and χ_B = electronegativity of A and B, and C = constant.

On the other hand Wells (1949) provided strong evidence against the validity of the Schoemaker–Stevenson rule. Furthermore, if the sums of Pauling's crystal and covalent radii are compared for $^{VI}Fe^{2+}$–X, the covalent distances are predicted to be shorter than the ionic distances. In a comparison of the sums of ionic radii with observed distances in the chalcogenides of Zn, Cd, and Hg, Roth (1967) also concluded that interatomic distances decrease as the amount of covalency increases. Similarly, Pauling also states that observed distances in FeI_2 (2.88 A) are greater than the sum of the covalent radii (2.58 A). The results for FeI_2 and those found by Roth correspond to our conclusions found by comparing ratios of distances and volumes as in Table 9 above.

Given the apparent relationship between covalence and contraction of the unit cell volume described here, it should be possible to relate R_v to the

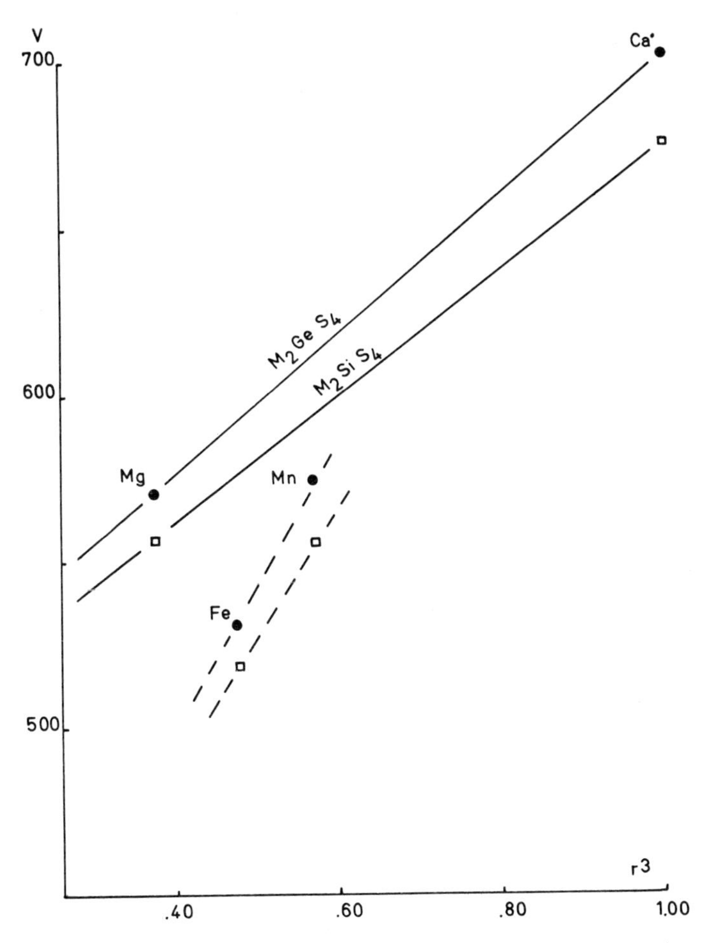

Figure 15. r_M^3 *vs.* V for M_2SiS_4 and M_2GeS_4 olivine compounds.

Mössbauer isomer shift for ferrous compounds. High-spin Fe^{2+} has the configuration $3d^6(t_{2g}^4e_g^2)$. We can look at R_v as a function of the occupation of $3d$ and $4s$ hybrid orbitals by ligand electrons as measured by the Mössbauer isomer shift.

The isomer shift δ is related to the density of s electrons at the nucleus of the source S and the absorber A by the expression:

$$\delta = \alpha \cdot (|\Phi(0)_A|^2 - |\Phi(0)_S|^2)$$

where α is a constant of proportionality equal to:

$$\alpha = \frac{4\pi}{5} 2e^2 R^2 S'(Z) \frac{\Delta R}{R}$$

with:

Z = number of electrons

R = radius of the nucleus

$\frac{\Delta R}{R}$ = relative variation of R between the excited and ground state

$S'(Z)$ = relativistic correction factor

and $|\Phi(0)_A|^2$ and $|\Phi(0)_S|^2$ are the two s electron densities at the nucleus of the absorber and source, respectively. If covalence results in electron transfer from ligands to Fe^{2+} $4s$ orbitals, the electron configuration can be written $3d^6 4s^x$, and there is a change of electron density s at the nucleus. If this compound is the absorber:

$$|\Phi(0)_A|^2 = \sum_{n=1}^{3} |\Phi_{ns}(0)|^2 + x|\Phi_{4s}(0)|^2$$

and δ varies linearly with x. Walker *et al.* (1961) have proposed a scale of x as a function of isomer shift δ for Fe^{2+}. Table 11 and Figure 16 show the relationship between $\bar{R}_v$ and δ where we have used the x proposed by Walker as abscissa.

These covalence effects involving octahedral Fe^{2+} may be important in determining the stability of certain structures at pressures present in the earth's interior. It is generally assumed that the covalence of a bond in a particular compound increases with the applied pressure. However, it should be noted that Huggins (1975) has found a slight increase in Mössbauer isomer shift for *tetrahedral* Fe^{3+} in several compounds. If we compare for example, Fe_2SiO_4 and Mg_2SiO_4, it may be said from our previous consideration that at high pressures the relative increased covalence of the Fe–O bonds will result in a greater shrinkage of these bonds and consequently a higher density than that anticipated under standard pressure and temperature. This apparently contradicts the results of Perez–Albuerne and Drickamer (1965) and Lewis *et al.* (1966) who find that increasing degrees of covalence lead to *lower* compressibilities.

The decrease in Fe–O distances relative to those of Mg–O at high pressures

Table 11. Mössbauer isomer shifts in Fe^{2+} compounds

Compound	R_v	R_d	δ mm/s[a]	$x\%$ ($4s^x$)	References
FeF_2	1.12	1.14	1.47	0	Wertheim *et al.*, 1967
FeO	1.08	1.08	1.24	7.7	Romanov, 1972
Fe_2GeO_4	1.06		1.20	9.0	Imbert, 1966
$FeCl_2$	0.99		1.19	9.3	Ono *et al.*, 1964
FeI_2	0.94		1.07	13.3	Hazony *et al.*, 1968
Fe_2GeS_4	0.92	0.93	0.98	16.3	Meyer, 1974
FeS		0.89	0.90	19.0	Ono *et al.*, 1962
FeSe		0.77	0.56	30.3	Ono *et al.*, 1962

[a]Isomer shifts δ are relative to stainless steel. The shifts contain contributions from the second order Doppler shift and the zero point energy, in addition to the isomer shift discussed here, as in the Walker publication.

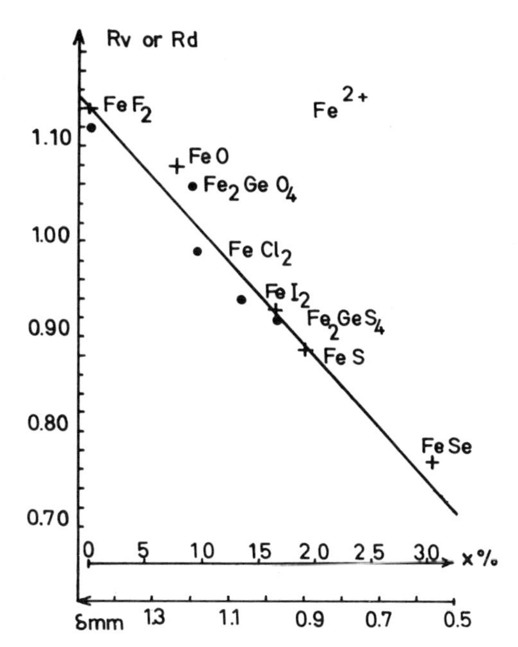

Figure 16. R_v *vs.* Mössbauer isomer shift in Fe^{2+} compounds.

also implies that transitions to more dense spinel phases will occur at lower pressures in the case of the Fe compound. This has indeed been observed by Syono *et al.* (1971) in his study of the olivine–spinel transitions of the M_2SiO_4 phases. They have attributed the lower pressures necessary for the olivine–spinel transition in the cases of Ni_2SiO_4, Co_2SiO_4 and Fe_2SiO_4 to crystal field stabilization effects. An even simpler explanation is that the transition pressure is lowered because of 'covalent shortening' of Ni–O, Co–O and Fe–O bonds relative to the Mg–O bonds.

When applied to the olivine $Mg_{0.9}Fe_{0.1}SiO_4$, the effects of covalent shortening at very high pressures will not be great but should perhaps be considered when calculating densities of any of the Fe^{2+} containing phases anticipated at very high pressures.

CONCLUSIONS

Ionic radii generally show systematic trends with valence, coordination number, and electronic configuration. Regularities of radii with coordination number led to empirical bond length–bond strength curves which (1) help to explain the increase in mean interatomic distances with degree of distortion in different structures and (2) can be used to calculate interatomic distances in unusual cation coordinations which may be found deep in the earth's interior.

Discrepancies which sometimes occur between calculated and observed distances in highly symmetric structures are not yet understood. Curvature observed in certain plots of r^3 against V for isotypic series of compounds may be caused by the polarizable nature of large cations such as Ba^{2+}, In^{3+}, Tl^{3+}, Pb^{4+} and Th^{4+}.

Mean tetrahedral T–O distances (corrected for oxygen coordination) in compounds $M_xT_yO_z$ depend on the covalency of the T–O bond. This covalency is affected by the average cation electronegativity, $\bar{\chi}$, which is a measure of the 'electron withdrawing' power of the M ions. This dependence of the distance on covalence is strong for selenates and vanadates, moderate for borates, germanates, phosphates, arsenates and sulfates, and weak for silicates.

Discrepancies in relative sizes of divalent transition metal ions and Mg^{2+} in fluorides, oxides and sulphides are explained by the covalent contraction parameter R_v. R_v for Fe^{2+}, derived only from relative cell volumes of isotypic Fe^{2+} and Mg^{2+} compounds, decreases as covalence increases, i.e. transfer of electrons from ligands to the cation reduces the cation–anion distance and contracts the cell. R_v is also proportional to the Mössbauer isomer shift in compounds containing octahedral Fe^{2+}. This concept explains the relative differences in cell volume for certain isotypic pairs such as NiF_2–MgF_2 and NiO–MgO, where the volume of the Ni compound is larger than that of Mg for fluorides and smaller for oxides. Similarly, for the pairs FeF_2–MgF_2 and Fe_2GeS_4–Mg_2GeS_4, it explains the fact that FeF_2 has a larger cell than MgF_2, whereas Fe_2GeS_4 has a *smaller* cell than Mg_2GeS_4.

REFERENCES

Ahrens, L. H. (1952), *Geochim. Cosmochim. Acta*, **2**, 155.
Banks, E., Greenblatt, M., and Post, B. (1970), *Inorg. Chem.*, **9**, 2259.
Batsonov, S. (1968), *Russ. Chem. Rev.*, **37**, 332.
Baur, W. H. (1974), *Acta Cryst.*, **B30**, 1195.
Biggar, G. M. (1969), *Min. Mag.*, **37**, 299.
Brown, I. D. and Shannon, R. D. (1973), *Acta Cryst.*, **A29**, 266.
Burnham, C. W. (1967), *Carnegie Inst. Wash. Yearbook*, **65**, 285.
Fukunaga, O. and Fujita, T. (1973), *J. Solid St. Chem.*, **8**, 331.
Hazony, Y., Axtmann, R. C., and Hurley, J. W. (1968), *Chem. Phys. Lett.*, **2**, 673.
Hawthorne, F., *Ph.D. Thesis*, McMaster Univ., Hamilton, Ontario, 1974.
Huggins, F. E. (1976), this volume.
Imbert, P. (1966), *C. R. Acad. Sci. Paris*, **263B**, 184.
Jayaraman, A., Narayanamurti, V., Bucher, E. and Maines, R. G. (1970), *Phys. Rev. Letters*, **25**, 1430.
Lewis, G. K., Perez-Albuerne, E. A. and Drickamer, H. G. (1966), *J. Chem. Phys.*, **45**, 598.
Longo, J. M., and Kafalas, J. A. (1969), *J. Solid St. Chem.*, **1**, 103.
Meyer, C. (1974), *Thèse de 3e Cycle*, Grenoble.
Miller, G. H. (1972). *M.Sc. Thesis*, University of Newcastle upon Tyne.
Morimoto, N., Appleman, D. E., and Evans, H. T. (1960), *Z. Krist.*, **114**, 120.
Morimoto, N., and Koto, K. (1969), *Z. Krist.*, **129**, 65.
Ono, K., Ito, A., and Hirahara, E. (1962), *J. Phys. Soc. Japan*, **17**, 1615.
Ono, K., Ito, A., and Fujita, T. (1964), *J. Phys. Soc. Japan*, **19**, 2119.
Pauling, L. (1929), *J. Amer. Chem. Soc.*, **51**, 1010.

Pauling, L. (1960), *The Nature of the Chemical Bond*, Cornell Univ. Press, Ithaca, New York.
Perez-Albuerne, E. A., and Drickamer (1965), *J. Chem. Phys.*, **43**, 1381.
Prewitt, C. T., Shannon, R. D., and White, W. B. (1972), *Contr. Mineral and Petrol.*, **35**, 77.
Romanov, V. P., and Checherskaya, L. F. (1972), *Phys. Stat. Sol.*, **496**, K183.
Roth, W. L. (1967), *Physics and Chemistry of II–VI Compounds*, M. Aven and J. S. Prener, Ed., North Holland Publ. Co.
Schoemaker, V. and Stevenson, D. P. (1941), *J. Amer. Chem. Soc.*, **63**, 37.
Shannon, R. D. and Prewitt, C. T. (1969), *Acta Cryst.*, **B25**, 925.
Shannon, R. D. and Prewitt, C. T. (1970a), *Acta Cryst.*, **B26**, 1046.
Shannon, R. D. and Prewitt, C. T. (1970b), *J. Inorg. Nucl. Chem.*, **32**, 1427.
Shannon, R. D. (1971), *Chem. Comm.*, 981.
Shannon, R. D. and Calvo, C. (1973a), *Acta Cryst.*, **B29,** 1338.
Shannon, R. D. and Calvo, C. (1973b), *Can. J. Chem.*, **51**, 266.
Shannon, R. D. and Calvo, C. (1973c), *J. Solid St. Chem.*, **6**, 538.
Shannon, R. D. and Vincent, H. (1974). *Structure and Bonding*, Vol. 19, 1.
Süsse, P. and Buerger, M. (1970), *Z. Krist.*, **131**, 161.
Syono, Y., Tokonomi, M. and Matsui, Y. (1971) *Phys. Earth Planet. Interiors*, **4**, 347.
Vincent, H. and Perrault, G. (1971), *Bull. Soc. France. Mineral. Crist.*, **94**, 551.
Vincent, H. and Bertaut, E. F. (1973), *J. Phys. Chem. Solids*, **34**, 151.
Wells, A. F. (1949), *J. Chem. Soc.*, 55.
Wertheim, G. K. and Buchanan, D. N. (1967), *Phys. Rev.*, **161**, 478.
Wyckoff, R. W. G. (1960), *Crystal Structures*, Vols. I–IV, Wiley, New York.

Crystal Structures of Pyroxenes at High Temperature

C. T. Prewitt
Department of Earth and Space Sciences
State University of New York
Stony Brook, New York 11794, USA

INTRODUCTION

Although for many years mineralogists and other earth scientists have been interested in the behaviour of mineral crystal structures at high temperatures, it is only recently that single-crystal diffractometers, computer control, suitable crystal heaters, and individual initiative have been combined to produce reliable high-temperature X-ray diffraction intensities. Now, cell expansion and atom parameter data are available for a wide range of silicate crystal structures including olivines, pyroxenes, amphiboles, feldspars, garnets, nepheline, natrolite, β eucryptite, quartz, tridymite, and cristobalite. Because there have been more high-temperature studies on pyroxenes than on any other mineral group, this paper discusses these studies and attempts to summarize the important conclusions which can be drawn concerning the nature of the pyroxene structures at high temperature.

EQUIPMENT FOR HIGH-TEMPERATURE DIFFRACTION EXPERIMENTS

The high-temperature diffraction experiments which have produced the pyroxene data discussed here have involved the use of computer-controlled X-ray diffractometers and single-crystal furnaces of various designs. In order for an experiment to produce useful data, a furnace must be constructed so that temperature can be maintained with reasonable certainty while at the same time allowing enough access for the incident and diffracted X-ray beams. One of the first successful diffractometer heaters was that of Foit and Peacor (1967) which used an internally-heated chamber for the crystal and which was mounted on a diffractometer with Weissenberg geometry. This design was good for temperature control, but restrictions imposed by the Weissenberg

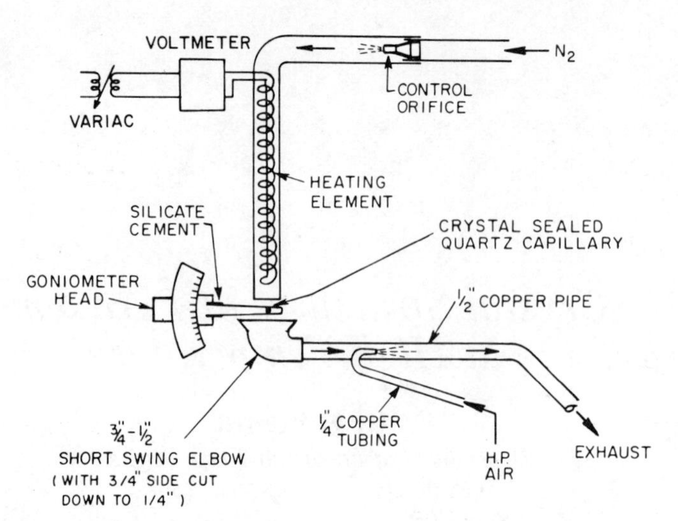

Figure 1. Schematic diagram of a device which uses hot nitrogen gas to heat a single crystal.

geometry make computer control difficult, thus reducing its attractiveness. In the first single-crystal work on the high-low phase transformation in clinopyroxene, Smyth (1969) used an oxy-hydrogen flame to heat the crystal. This has the advantage that very high temperatures can be obtained, but temperature control is difficult. Smyth could estimate his temperature to only $\pm$ 100 °C. Another design developed by the author and used for examining the high-low transition in cummingtonite (Prewitt, Papike, and Ross, 1970) is shown in Figure 1. Here, nitrogen gas is heated by a Sylvania Serpentine Gas Heater and then passed over the crystal and exhaused through a copper tube and away from the diffractometer. The advantage of this technique is that it is inexpensive and easy to construct. Its disadvantage is that it is clumsy and can attain maximum temperature of only about 900 °C. An improved heater based on this approach has been described by Smyth (1972).

A single-crystal heater designed at Stony Brook and described by Brown, Sueno, and Prewitt (1973) is shown in Figure 2. This heater consists of a coil of Pt–Rh (13 per cent) wire shaped in the form of a horsehoe and coated with polycrystalline ZrO_2. It allows considerable freedom for crystal orientation and a 2θ range on the diffractometer of 0°–90°. Although the crystal temperature is very stable if the heater is protected from air currents, the steep temperature gradients from the hot resistance wire to the crystal make exact temperature measurement difficult. Brown *et al.* (1973) reported an uncertainty of $\pm$ 20 °C in temperature measurement. However, this uncertainty is not too great for the reconnaissance-type structure work discussed here. Better accuracy is, of course, required for detailed phase-transition studies.

Most of these high-temperature experiments require that the crystal be mounted in a silica glass capillary in order to protect it from the atmosphere inside the furnace. For iron-bearing minerals such protection is essential and

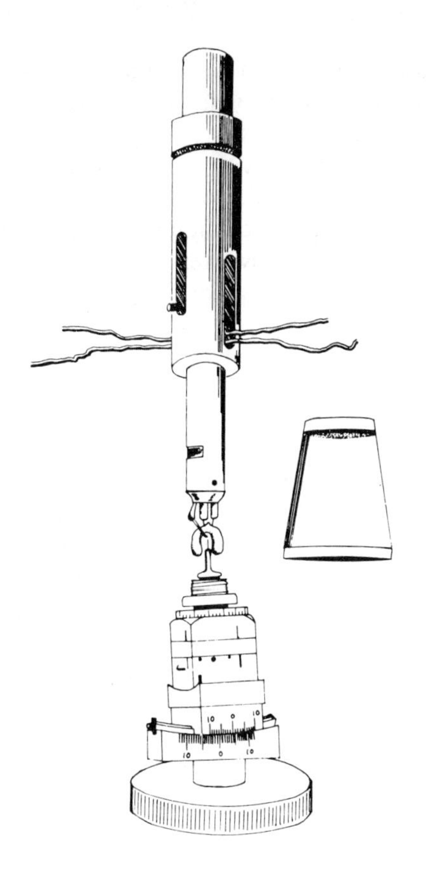

Figure 2. Resistance heater which can be used on a diffractometer or precession camera. This heater design allows great flexibility in selection of incident and diffracted X-ray beam angles (Reproduced with permission from G. E. Brown, S. Sueno and C. T. Prewitt, *Amer. Mineral.*, **58**, 698–704 (1973) copyright by the Mineralogical Society of America).

most investigators have apparently been successful in preventing oxidation of iron from affecting the experimental results. Evidence used to support this includes (1) the absence of colour changes in the crystal (2) no significant change in cell parameters, and (3) no change in composition (using the electron microprobe) between initial and final stages of the experiment. A few runs have been made in the author's laboratory in which powdered carbon or iron were included in the capillary with the crystal in order to keep iron in the ferrous state but equivalent results were obtained when the capillary was merely evacuated and sealed. More sophisticated techniques for controlling the partial pressure of oxygen could be devised and probably will be used in the future.

PYROXENE STRUCTURE STUDIES

Pyroxenes constitute one of the major rock-forming mineral groups and, because of their variable chemistry and structure, present a variety of crystal-chemical problems which are as yet poorly understood. Over the years a number of investigators have attempted to discover just what factors control pyroxene chemistry and structure and certain progress has been made. However, most of these investigators used structural data obtained at room temperature whereas pyroxenes grow at temperatures above 1000 °C and it seems that room-temperature structural data might be insufficient or misleading when used to illustrate what might be happening at these temperatures. Most high-temperature pyroxene studies have been initiated with this in mind and the discussion that follows attempts to provide a guide to the most significant results which these studies have provided. In this discussion it will be assumed that the reader is familiar with current literature on pyroxene crystal structures. For a summary of the various structure types, see Papike, Prewitt, Sueno and Cameron (1973).

Pyroxenes of the Clinoenstatite Type

Table 1 lists the pyroxene structures which have been studied using high-temperature X-ray diffraction data. Although most of these have been published since 1970, the first important pyroxene work was described in a series of three papers published in the 1960's. The first of these papers which was based on a high-temperature powder-diffractometer pattern, described a high-low inversion in clinoenstatite at approximately 995 °C (Perrotta and Stephenson, 1965). Although these authors indexed the high phase on a triclinic cell, Smith (1969) reindexed the pattern using the monoclinic space group *C2/c*, and Smyth (1969) confirmed the *C2/c* space group using single-crystal techniques (although with a crystal of different composition, $Mg_{0.30}Fe_{0.70}SiO_3$). Later, Prewitt *et al.* (1970) showed that a similar transition exists for pigeonite and also for

Table 1. Pyroxene crystal structures determined above room temperature

Crystal	Temp. Range	Reference
Clinopyroxene ($P2_1/c$)		
pigeonite, $Ca_{0.09}Mg_{0.39}Fe_{0.52}SiO_3$	RT–960	Brown *et al.* (1972)
clinohypersthene, $Ca_{0.015}Mg_{0.31}Fe_{0.67}SiO_3$	RT–825	Smyth (1974)
Clinopyroxene (*C2/c*)		
acmite, $NaFeSi_2O_6$	RT–800	Cameron *et al.* (1973)
diopside, $CaMgSi_2O_6$	RT–1000	Cameron *et al.* (1973)
hedenbergite, $CaFeSi_2O_6$	RT–1000	Cameron *et al.* (1973)
jadeite, $NaAlSi_2O_6$	RT–800	Cameron *et al.* (1973)
spodumene, $LiAlSi_2O_6$	RT–760	Cameron *et al.* (1973)
ureyite, $NaCrSi_2O_6$	RT–600	Cameron *et al.* (1973)
diopside	RT–800	Finger and Ohashi (1974)
Orthopyroxene (*Pbca*)		
orthohypersthene, $Ca_{0.015}Mg_{0.31}Fe_{0.67}SiO_3$	RT–850	Smyth (1973)
ferrosilite, $FeSiO_3$	RT–980	Sueno *et al.* (1975)
Protopyroxene (*Pbcm*)		
protoenstatite, $MgSiO_3$	1100	Sadanaga *et al.* (1969)
protoenstatite	1100	Smyth (1971)

In °C, RT = room temperature.

the amphibole cummingtonite. High-temperature structure refinements are now available for these clinopyroxenes which have a $P2_1/c \rightarrow C2/c$ transition: pigeonite $Ca_{0.09}Mg_{0.39}SiO_3$ (Brown, Prewitt, Papike, and Sueno, 1972) $Ca_{0.015}Mg_{0.31}Fe_{0.67}SiO_3$ (Smyth, 1974b), and a series on the ferrosilite-hedenbergite join (Ohashi, Burnham, and Finger, 1975). Surprisingly, a modern structure refinement of clinoenstatite at room temperature has not yet been reported and the only single-crystal high-temperature work on clinoenstatite is that of Sadanaga, Okamura, and Takeda (1969) and of Smyth (1974a). Although both of these papers discuss polymorphism in enstatite, neither gives new structure parameters.

Probably the most interesting aspect of the primitive ($P2_1/c$) clinoenstatites at high temperatures is the transition to a crystal structure similar to that of diopside with $C2/c$ symmetry. Prewitt, Brown and Papike (1971) showed that the transition in pigeonite is a function of composition with the transition temperature being lowered by the presence of larger cations Fe and Ca, relative to each other and to Mg. The transition temperature appears to drop from about 1000 °C for the most Mg-rich to about 500 °C for the most Fe-rich natural pigeonites. Ohashi *et al.* (1975) later showed that the transition is near room temperature for synthetic $Ca_{0.20}Fe_{0.80}SiO_3$. Brown *et al.* (1972) described the transition as a result of a 'straightening' of the *B* silicate chains as a function of temperature until they become crystallographically equivalent to the *A* chains at the transition temperature. The point of transition is determined by monitoring a class of X-ray reflections for which $h+k$ is an odd number (class *b* reflections). These reflections are inconsistent with space group $C2/c$ and become weaker than class *a* reflections ($h+k=$ even) as a function of temperature and finally disappear at the transition. In later work on $Ca_{0.015}Mg_{0.31}Fe_{0.67}SiO_3$ Smyth (1974b) reported discontinuities in the cell parameters and a discontinuous drop in the intensity of the 102 reflection at the transition temperature. He concluded from this that the transition is first-order. To the author's knowledge, there has been no independent confirmation of these discontinuities; high-temperature Guinier powder X-ray diagrams or DTA measurements should be made to provide additional evidence.

Pyroxenes of the Diopside Type

In contrast to the $P2_1/c$ pyroxenes, the $C2/c$ structures have only one silicate chain and either melt, decompose, or transform reconstructively to another phase at the upper limits of their high-temperature ranges. This limitation and the relatively large number of available end-member compositions permitted Cameron, Sueno, Prewitt and Papike (1973) to compare the structural thermal expansions of six different $C2/c$ pyroxenes, as listed in Table 1 above, over a wide range of temperatures. This work clearly illustrates the concept of differential polyhedral expansion whereby the large-cation octahedra expand significantly with temperature but the silicate tetrahedra are essentially constant in volume. This results in a straightening of the silicate chains to accommodate

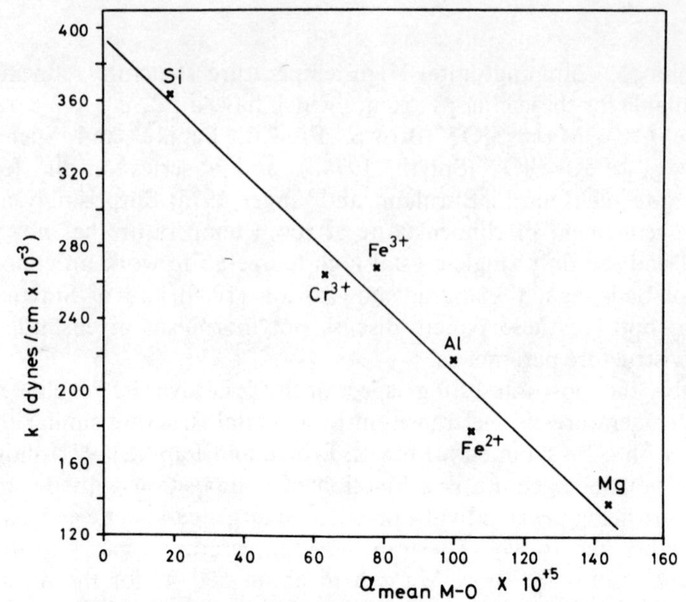

Figure 3. Plot of bond force constant *vs.* mean thermal expansion coefficient for Si–O and M–O bonds in *C2/c* pyroxenes (Reproduced with permission from M. Cameron, S. Sueno, C. T. Prewitt, and J. J. Papike, *Amer. Mineral.*, **58**, 594–618 (1973) copyright by the Mineralogical Society of America).

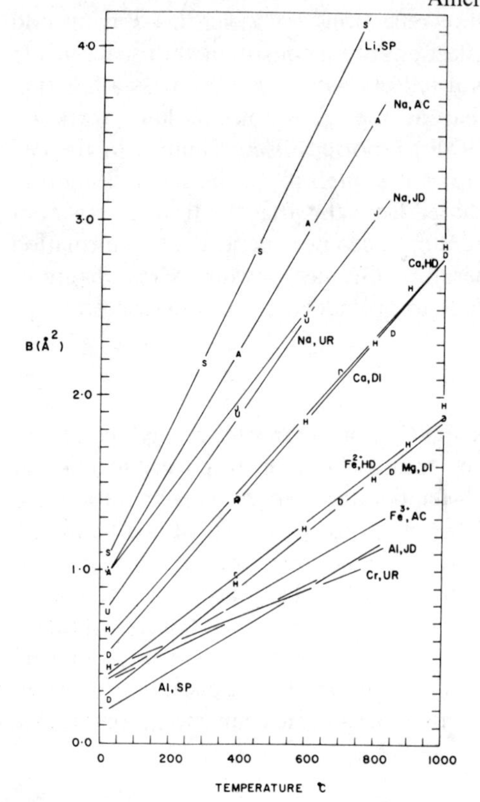

Figure 4. Variation of the isotropic temperature factors for cations in *C2/c* pyroxene structure *vs.* temperature. In general, the temperature factors increase most for singly-charged cations in the M2 sites. Abbreviations: HD = hedenbergite, DI = diopside, AC = acmite, UR = ureyite, JD = jadeite, SP = spodumene (Reproduced with permission from M. Cameron, S. Sueno, C. T. Prewitt, and J. J. Papike, *Amer. Mineral.*, **58**, 594–618 (1973) copyright by the Mineralogical Society of America).

the expanding octahedra. The relative rates of thermal expansion, not only of the octahedra *vs.* the tetrahedra, but also between different types of octahedra are probably the most interesting results with regard to the physical properties of these and other minerals. Cameron *et al.* (1973) found that mean thermal expansion coefficients could be assigned to cation-oxygen bonds which increase $Si^{4+}-O < Cr^{3+}-O < Fe^{3+}-O < Al^{3+}-O < Fe^{2+}-O < Na^{+}-O < Mg^{2+}-O < Ca^{2+}-O < Li^{+}-O$. Except for $Mg^{2+}-O$ and possibly $Li^{+}-O$, this series is about what one would expect, i.e. 'strong bonds' involving small cations with large formal valence expand at a lower rate than weak bonds of large cations with small formal valence. However, a comparison of independently determined force constants with mean thermal expansion coefficients (Figure 3) shows good relative agreement, even for Mg. Cation coordination also affects the thermal expansion coefficients, with higher coordination resulting in larger coefficients. Figure 4 shows that the isotropic temperature factor reflects all of these factors, cation size, charge, and coordination number.

Orthopyroxenes

High-temperature refinements have been reported only for orthopyroxenes with compositions $Ca_{0.015}Mg_{0.30}Fe_{0.68}$ (Smyth, 1973), and $FeSiO_3$ (Sueno, Cameron, Prewitt, and Papike, 1975). Because the orthopyroxene structure is similar to that of $P2_1/c$ clinopyroxene, the high-temperature results for orthopyroxene are comparable to those for Ca-poor clinopyroxenes, except there is no low-high transition for the orthopyroxene. Ohashi and Finger (1973) speculated on a possible transition which would not involve a change in space group. However, Sueno *et al.* (1975) show that it is geometrically impossible for the two silicate chains to become crystallographically equivalent in the orthopyroxene structure. What actually happens is as follows: The silicate tetrahedra rotate as the temperature is raised so that each chain becomes straighter and the O3–O3–O3 angles in the chain approach 180°. For ferrosilite the angles never do reach 180° but, instead, an oriented phase transformation takes place to high clinoferrosilite where the O3–O3–O3 angle is closer to 180°. This transformation is discussed more fully in another section of this paper.

The discussion above illustrates one of the problems one encounters in dealing with pyroxene structures. Even though the pyroxene structures have been examined and discussed in dozens of papers, no one has, as yet, published a paper which fully explains the geometrical relationships between the different pyroxene structures. Papike *et al.* (1973) did show that pyroxene structures are based on approximate close-packing of oxygen atoms, but the distortions present in real structures are not yet fully understood. Thus, the study of pyroxene structures and their fundamental crystal chemistry will continue in the future.

Protoenstatite

Protoenstatite is the $MgSiO_3$ phase that is apparently stable above 1000 °C.

It is orthorhombic and differs from other pyroxenes in the way in which octahedral and tetrahedral units are linked together (Papike *et al.*, 1973). The compositions for which this phase occurs are limited to those near $MgSiO_3$; in single-crystal studies, Smyth (1971) was unable to maintain the protoenstatite structure below 1000 °C (it inverts to mixtures of ortho- and clinoenstatite), but Boyd and Schairer (1957) reported that protoenstatite could be quenched to room temperature in hydrothermal experiments.

Two refinements of protoenstatite using data obtained above 1000 °C have been reported, one by Sadanaga, Okamura, and Takeda (1969) and one by Smyth (1971). Although the essential details of the structure as given by these papers are undoubtedly correct, there is uncertainty about their space groups and the R factors from the refinements are large (0.128 and 0.19). Previous to Smyth's work the accepted space group was *Pbcn*, but Smyth noticed reflections that violated the b glide and assigned the acentric space group $P2_1cn$ as most probable. Unless better refinements are obtained, it may not be possible to resolve two of the most interesting aspects of the protoenstatite structure, i.e. the mechanism of the transformation between it and the lower temperature forms, and the reason for the limited range in composition for protoenstatite.

RECONSTRUCTIVE PHASE TRANSFORMATIONS

It is important to note that there are two different reconstructive phase transformations which take place between ortho- and clinopyroxenes. One of these is found only in Mg-Fe pyroxenes and at relatively low temperatures ($\sim$ 600 °C). The transition from orthoenstatite to clinoenstatite is assisted enabled by the presence of a shearing stress (Coe, 1970) and requires the breaking of Mg–O bonds. Brown, Morimoto and Smith (1961), Coe and Muller (1973), and Smyth (1974a) have proposed models to explain this transformation. The other ortho-to-clino transformation takes place at high temperatures, but apparently only for compositions with a substantial amount of Fe substituting for Mg. Sueno, Cameron, and Prewitt (1973) found that an orthoferrosilite crystal transforms *rapidly and reversibly* to twinned high clinoferrosilite at 1025 °C without destroying its shape or losing its orientation. This is clearly a different situation from the transformation of orthoenstatite to low clinoenstatite and may take place by a different mechanism. Certainly, applied shear stress is not needed in this high-temperature transformation. Because there is confusion about these transformations and because they are of interest to tectonophysicists as well as mineralogists and crystallographers, it is important that the proper experiments and interpretations be performed which clearly distinguish between transformations and which give definitive information about how they take place.

DISCUSSION

Now that data are available for a number of pyroxene structures at high temperatures, it is worthwhile to assess what we have learned and to suggest further studies that might be made to improve our knowledge of the high-temperature behaviour of these minerals. First of all, these experiments have thus far shown that differential polyhedral expansion results from different thermal expansion rates of various cation-oxygen bonds and, in turn, greatly affects the bulk thermal expansion and polymorphism of pyroxenes. We have learned that thermal expansion rates are functions not only of atom type, but also of cation charge and coordination number. There are inverse correlations between bond strength and the rates of change of cation interatomic separations and isotropic temperature factors, thus giving essential information needed for the prediction of thermal expansion coefficients.

We now know that silicate chains do not behave in the same way in all pyroxene structures. For example, the angle representing the amount of kink in silicate chains (the O3–O3–O3 angle) changes very little in the $C2/c$ structures, but varies over a much wider range and reflects the amount of polymorphism in $P2_1/c$ and orthopyroxene structures. In fact, we should be very close to the point of being able to state in a very fundamental way why certain polymorphs are found for specific pyroxene compositions under given temperature conditions.

For the future, we expect to see better control and measurement of temperature for phase transition experiments, improved design of equipment so that equally good data can be obtained at temperatures high than 1200 °C, and, most important, application of these data to geophysical and geochemical problems. Also, needed are more fundamental explanations of thermal expansion and correlation of expansion and temperature factors with elastic constants. A further step will be to measure useful diffraction intensity data at both high pressure and high temperature.

ACKNOWLEDGEMENT

This work was supported by National Science Foundation Grant No. AO 41137.

REFERENCES

Boyd, F. R., Jr. and J. F. Schairer (1957). *Carnegie Inst. Washington Year Book*, **56**, 223–225.
Brown, G. E., C. T. Prewitt, J. J. Papike and S. Sueno (1972). *J. Geophys. Res.*, **77**, 5778–5789.
Brown, G. E., S. Sueno and C. T. Prewitt (1973). *Amer. Mineral.*, **58**, 698–704.
Brown, W. L., N. Morimoto and J. V. Smith (1961). *J. Geol.*, **69**, 609–616.
Cameron, M., S. Sueno, C. T. Prewitt and J. J. Papike (1973). *Amer. Mineral.*, **58**, 594–618.
Coe, R. S. (1970). *Contrib. Mineral. Petrol.*, **26**, 247–264.
Coe, R. S. and W. F. Müller (1973). *Science*, **180**, 64–66.
Finger, Larry W. and Yoshikazu Ohashi (1974). *EOS*, **55**, 463.
Foit, F. F., Jr. and D. R. Peacor (1967). *J. Sci. Instruments* **44**, 183–185.
Ohashi, Y., C. W. Burnham and L. W. Finger (1975). *Amer. Mineral.*, **60**. 423–434.

Ohashi, Y. and L. W. Finger (1973). *Carnegie Inst. Washington Year Book*, **72**, 544–547.
Papike, J. J., C. T. Prewitt, S. Sueno and M. Cameron (1973). *Z. Kristallogr.*, **138**, 254–273.
Perrotta, A. J. and D. A. Stephenson (1965). *Science*, **148**, 1090–1091.
Prewitt, C. T., G. E. Brown and J. J. Papike (1971). *Proc. Second Lunar Science Conf., Geochim. Cosmochin. Acta, Suppl. 2*, vol. 1, pp. 59–68.
Prewitt, C. T., J. J. Papike and M. Ross (1970). *Earth Planet. Sci. Lett.*, **8**, 448–450.
Sadanaga, R., F. P. Okamura and H. Takeda (1969). *Mineral. J. Japan*, **6**, 110–130.
Smith, J. V. (1969). *Nature*, **222**, 256–257.
Smyth, J. R. (1969). *Earth Planet. Sci. Lett.*, **6**, 406–407.
Smyth, J. R. (1971). *Z. Kristallogr.*, **134**, 262–274.
Smyth, J. R. (1972). *Amer. Mineral.*, **57**, 1305–1309.
Smyth, J. R. (1973). *Amer. Mineral.*, **58**, 636–648.
Smyth, J. R. (1974a). *Amer. Mineral.*, **59**, 345–352.
Smyth, J. R. (1974b). *Amer. Mineral.*, **59**, 1069–1082.
Sueno, S., M. Cameron and C. T. Prewitt (1973). *Geol. Soc. Amer., Abstr. Programs*, **4**, 829.
Sueno, S. M. Cameron, C. T. Prewitt and J. J. Papike (1975). *Amer. Mineral.* (in the press).

Modelling Crystal Structures

M. J. Dempsey and R. G. J. Strens
School of Physics, The University,
Newcastle upon Tyne NE1 7RU, England.

The usefulness of modelling methods lies in their speed and their ability to extract the maximum of useful information from the available experimental data. When the problem is overdetermined, i.e. when the number of independent input data exceeds the number of output variables, there is often a bonus in the form of improved accuracy, which may exceed that of any one experimental determination of the property in question. A recent example from this laboratory is the calculation by Wood and Strens (1972) of the (P,T,X) stability field of orthopyroxene from very limited data. The results were indistinguishable within experimental error from the 900° (P,X) data of Smith (1972) published later that year.

We now report some applications of a modified version of the distance least squares (DLS) program developed by Meier and Villiger at ETH Zurich. The original program is described fully in the manual available with the FORTRAN deck from Professor Meier, whose generosity in providing the program is gratefully acknowledged. Previous applications of the DLS program have been described by Meier and Villiger (1969), Barrer and Villiger (1969) and Baur (1972).

The modified program is written in ALGOL W for the IBM 360/67 computer at Newcastle, and takes about 20 seconds to produce coordinates (11) and unit cell parameters (3) for the olivine structure. Preparation of input data for each new structure takes from an hour for a very simple case, e.g. cubic perovskite, to a few days for a complex structure with 40 parameters. Each variation then requires from a few minutes to a few hours of preparation. These times are of the order of 10^{-1} to 10^{-2} of those required for a structure determination, and could be reduced by further programming changes.

Despite the great advances which have followed the introduction of fast digital computers and automated diffractometers there is still great scope for the use of modelling methods. The determination of the crystal structure of a single complex silicate mineral at room temperature and pressure is no trivial task, and the time required is increased when working at other temperatures, and especially when working at high pressure. There are thus useful applications of modelling methods to the study of the effects of changes in pressure, temperature and composition on crystal structure, always assuming the structure model to be realistic. There are also potential applications to the study of thermal vibrations, and to the determination of the probable properties of hypothetical phases, e.g. high-pressure phases such as silicate perovskites, which cannot at present be synthesised, but which are thought to be present in the Earth's lower mantle.

Progress has been made in all these fields in recent years. Detailed studies of the effects of changing composition on crystal structure have become commoner since the technique of site population refinement (Finger, 1969) was introduced, and data now exist for several important solid solution series. Temperature effects have received considerable attention recently (*Amer. Mineral.*, 58, 577–704, and structure determinations at high pressure have now started (Hazen and Burnham, 1974). Direct calculation of thermal vibration figures has been attempted with considerable success by Ohashi and Finger (1973) using an electrostatic model of ionic interactions. Current methods of estimating the volumes of high-pressure phases are rather primitive, and usually rely on a graphical or least-squares representation of the the dependence of volume on some function of the interatomic distances, although Baur (1972) has applied the DLS method to possible Mg_2SiO_4 structures.

THE DLS METHOD

The DLS method uses the fact that the number of crystallographically independent interatomic distances in most structures exceeds the number of variables (atomic coordinates and unit cell parameters), so that the latter may be determined by prescribing a sufficient number of the former. For the phases considered below, these 'prescribed interatomic distances' are of three types: metal–oxygen (M–O); oxygen–oxygen (O–O); and metal–metal (M–M), but there is no reason why others should not be added, for example hydroxyl and hydrogen bonds.

Average metal–oxygen bond lengths may be estimated for common cations and coordinations with an accuracy of about 1 per cent, using the data of Shannon and Prewitt (1969, 1970) and other sources although individual metal–oxygen distances in distorted polyhedra often vary by more than 10 per cent from bond to bond. Oxygen–oxygen and metal–metal distances are less easily estimated, but may usually be found to within 5 per cent from quite crude models of oxygen packing. Unit cell parameters, and distances between atoms in special positions, are frequently known to better than 0.1 per cent, and may usually be estimated to better than 2 per cent in even the least favourable cases.

Statisfactory estimates of interatomic distances and unit cell parameters having been made, weighting factors are introduced to allow for the very different strengths of MO, OO and MM interactions. The original program used empirically determined weights which did not vary with interatomic distance, and were typically in the ratio 10:5:2 for (Si,Al)–O, O–O and Si–Si interactions in framework silicates. In principle, the introduction of each new ion or coordination number required the determination of a new empirical weighting factor, restricting the usefulness of the program.

We have sought to generalize the weighting scheme for M–O and O–O interactions, and to allow the weights to vary with bond length, so that e.g. a long Si–O bond has less weight than a shorter one, and a short O–O distance has a heavier weight than a long one.

THE NEW WEIGHTING SCHEME

Metal–Oxygen Interactions: The obvious weighting scheme is one which assigns weights proportional to bond strengths (expressed in valence units) according to one of several equations relating bond length to bond strength. Perhaps the best known of these is that of Brown and Shannon (1973):

$$s = s_0(r_0/r)^n \tag{1}$$

where s is the bond strength in valence units (v.u.), r is the M–O distance, s_0 is the strength assigned for $r = r_0$, and usually $n \simeq 5$.

Another weighting scheme uses an equation derived by Strens (1967, in Miller, 1972) from Pauling's (1960) treatment of the carbon–carbon bond:

$$r = r_6 + (r_6 - r_4)f(k,x) \tag{2a}$$

where r_6 and r_4 are metal–oxygen distances in octahedral and tetrahedral coordination, k is the ratio of force constants for these two coordinations, and x is the degree of 6/4 bond character. Thus $x = 1$ for tetrahedral coordination, $x = 0$ for octahedral coordination, and $x = -0.5$ for eightfold coordination. Trials suggest that k is close to 6/4, i.e. 1.5, increasing to nearer 1.8 for bond strengths near 1 valence unit. The equation:

$$f(k,x) = (-kx/(k-1)x + 1) \tag{2b}$$

can be used to relate bond strength to bond length. Unit strength is that of an octahedral M–O bond of normal length, and is easily converted to valence unit by multiplying by a factor $z/6$, where z is the formal charge of the cation.

Both equations give very similar bond strengths in the region of practical interest ($4 \leqslant$ coordination number $\leqslant 8$), but they diverge at low and high coordinations. It is not yet known which better represents the real variation of bond strength with length under these conditions, although (2) yields reasonably good estimates of bond length for such very strongly bonded species as MgO^0, AlO^+ and SiO^{2+} formed in spark discharges. Both (1) and (2) gave good DLS results.

Because of the very wide range of (s_0,r_0,n) values now available, and the greater familiarity of other workers with the Brown and Shannon (1973) method, we now use (1) to provide weighting factors.

Oxygen–Interactions: Two problems which arise are the form of the variation of O–O repulsion with distance, and the magnitude of this repulsion relative to M–O bond strengths. At distances greater than two oxygen radii (2.80 Å), a small constant weight of 0.167 v.u. has been chosen by trial and error. We have modelled O–O interactions at shorter distances by analogy with the repulsion to be expected between two balloons of 'electron gas' which goes as:

$$f = (\pi/8)(4r_0^2 - r_{00}^2) \qquad (r_{00} < 2r_0)$$

and found that a weighting factor given by:

$$w = 0.167 + f$$

gave satisfactory results, i.e. prevented O–O distances from being reduced to unrealistic levels by the influence of the heavily-weighted MO interactions.

Metal–Metal Interactions: No satisfactory general weighting scheme has been found, and MM interactions were given a nominal value of 0.167 v.u. when used.

FUDGE Factor: The remaining modification to the weighting scheme involves the 'FUDGE' term, which multiples all changes in variables by a factor $0 <$ FUDGE $\leqslant 1$ to assist convergence. In the original program, FUDGE remained constant throughout the refinement. We found that this often led to a violent oscillation of the sum of squares if FUDGE was too high, or to slow convergence if it was too low, the effects usually showing after the first two or three cycles. After some trials a variable weighting scheme was used, given by:

$$\text{FUDGE} = 1/(c-2)^{1/2}$$

where c is the cycle number. Thus, FUDGE $= 1$ for the first three cycles, decreasing smoothly to 1/3 in the eleventh.

PROGRAM TESTING

Following these modifications, the program was tested extensively on the natrolite, olivine and β-phase structures. The coordinates alone were varied during the first three cycles, and both coordinates and cell edges were varied during the remaining eight cycles. A satisfactory and stable sum of squares was usually reached after the fifth cycle of joint refinement. Slower convergence usually implies an error in the data.

The results may be tested in three ways: direct comparison of observed and calculated coordinates and cell edges confirms that the program and weighting scheme used are working satisfactorily and that the model is realistic, and consideration of the magnitude of the sum of squares and the summation of bond strengths about each ion in the structure reveals defects in the input data. In our experience, the DLS method works surprisingly well, and appears capable of further development.

In what follows, we find it convenient to distinguish three types of structure which occur in DLS work: that determined by X-ray diffraction (the X-structure), that calculated using the DLS program (the D-structure), and the structure used as a base for calculation (the base structure). For example, in attempting to determine the likely structure of the hypothetical phase $NaAlSiO_4$ in the $CaFe_2O_4$ structure type, we calculated D-structures of $NaAlSiO_4$ using the X-structures of $CaFe_2O_4$ and $NaAlGeO_4$ as base structure. Much better results were obtained (judged by sum-of-squares and bond strengths) using $NaAlGeO_4$ as a base structure. This we attribute to the similarity of ionic size and especially of charge balance between $NaAlGeO_4$ and $NaAlSiO_4$.

Choice of base structure is important in determining the quality of the D-structure. In most cases, the base- and X-structures are identical, but this is not always so, as it is sometimes necessary to use coordinates from one source, and cell edges from another to provide the initial estimates. We normally obtained prescribed distances for hypothetical phases by replacing bonds of given (s,r,n) in the base structure with bonds of the same strength s, and appropriate (r,n). In dealing with the effect of changes in (P,T,X) on bond lengths, the interatomic distances in the base structure were simply scaled up or down according to the compressibility, thermal expansion, or change in radius on substitution. For shared polyhedral edges, the mean of the two estimates was used.

APPLICATIONS OF THE DLS METHOD

Given a D-structure for some known (P,T,X), it is in principle very easy to generate a D-structure for some different (P,T,X) by changing the prescribed

Table 1. Constants used in DLS calculations

	CN	r_0 (Å)	n	10^5 (deg^{-1})	K_0 (Mb)
Na^+	6	2.449	5.6	—	0.5^a
Mg^{2+}	6	2.098	5.0	1.442^c	1.54^b
Ca^{2+}	8	2.468	6.0	4.03^s	1.14^b
Fe^{2+}	6	2.155	5.5	1.048^c	1.54^a
Co^{2+}	6	2.118	5.0	—	—
Cd^{2+}	7	2.348	5.0	—	—
Al^{3+}	6	1.909	5.0	1.002^c	2.99^a
Fe^{3+}	6	2.012	5.3	—	—
Si^{4+}	4	1.625	4.5	0.192^c	4.00^m
Ti^{4+}	6	1.952	4.0	2.71	2.07^b
Ge^{4+}	4	1.750	5.4	—	—
Sn^{4+}	6	2.048	6.0	—	—

Values of coordination number (CN), r_0 and n are from Brown and Shannon (1973), and $S_0 = Z/CN$ where Z is the formal charge on the cation.

Thermal expansions are from [c]Cameron *et al.* (1973); [s]Skinner (in Clark, 1966).

Bulk Moduli (K_0) are from [b]Birch (in Clark, 1966); [a]Anderson and Anderson equations, [m]Miller (1972).

distances to values appropriate to the new conditions. The differences between the two D-structures then provide an estimate of the response of the real structure to the change in (P,T,X). The quality of this estimate will obviously depend on that of the original D-structure, and especially on the realism of the assumptions used to determine the changes in prescribed distances. Values of (s_0,r_0,n) and of thermal expansion and bulk modulus used in this work are given in Table 1.

Substitution: There is little scope for error in dealing with a simple homovalent substitution, e.g. Fe^{2+} for Mg^{2+}, in which the (r_0,n) values appropriate to both ions are known. The changes in the heavily-weighted M–O distances, and those in the O–O distances of the $[MO_n]$ polyhedron are quite accurately known, although some difficulties may arise in estimating relative changes in distorted polyhedra, and in dealing with O–O distances on shared edges. Heterovalent substitutions, e.g. 2Al for (Mg,Si) still present serious problems, and substitutions involving d^4 and d^9 ions have effects which are not at present

Table 2. Comparison of observed and calculated effects of changes in (P, T, X) on cell dimensions of olivines

(a) Effect of (Fe–Mg) substitution in α-Mg_2SiO_4 (%)

Forsterite ⟶ Fayalite		*a*	*b*	*c*	*V*
	obs.	1.13	2.39	1.75	5.34
	calc.	0.95	2.47	1.80	5.31

observed data from Birle *et al.* (1968)

(b) Effect of temperature change (20 to 900 °C) on α-Mg_2SiO_4 and α-Fe_2SiO_4 (%)

		a	*b*	*c*	*V*
Forsterite	obs.	0.67	1.20	1.06	2.93
	calc.	0.73	1.15	1.04	2.97
Fayalite	obs.	—	—	—	2.49
	calc.	0.54	0.84	0.87	2.27

Forsterite data from Smyth and Hazen (1973)
Fayalite data from Skinner (in Clark, 1966).

(c) Effect of 100 kb Pressure on α-Mg_2SiO_4 (%)

	a	*b*	*c*	*V*
obs.	1.87	3.58	2.29	7.76
calc.	1.52	2.06	1.91	5.41

'Observed' compressions calculated from data of Birch (in Clark, 1966) for a sample of composition $(Mg_{91.7}Fe_{9.3})_2SiO_4$.

modelled by the program, there being a strong Jahn–Teller effect in these cases.

Table 2a shows the observed and calculated changes in cell edges along the forsterite-fayalite series, using the (s_0,r_0,n) values in Table 1. The change in cell edge is quite accurately predicted, and the volume change is very close to the observed value.

Thermal expansion: It is more difficult to deal with thermal expansion, because accurate thermal expansion data are not yet available for many bonds, and little is yet known of the way in which thermal expansion coefficients depend on ionic charge, interatomic distance, and coordination number. We have used bond expansion coefficients determined by Cameron *et al.* (1973) wherever possible, and thermal expansion data for the simple oxides in other cases.

Table 2b shows observed and calculated thermal expansion between 300 and 1123 K for forsterite and fayalite. Again, agreement is quite good for expansion parallel to each cell edge, and very good for the volume expansions.

Thermal vibrations: A further modification was made to enable thermal vibration figures to be calculated on the assumption that the energy required to displace an ion to a point on a contour of constant sum-of-squares remains constant around that controur. The sum-of-squares was first calculated at each point on a grid, the original of which was the calculated equilibrium position of the ion considered. All other ions were kept in their calculated equilibrium positions. Contours of constant sum-of-squares were then drawn, and compared with thermal vibration data from the literature.

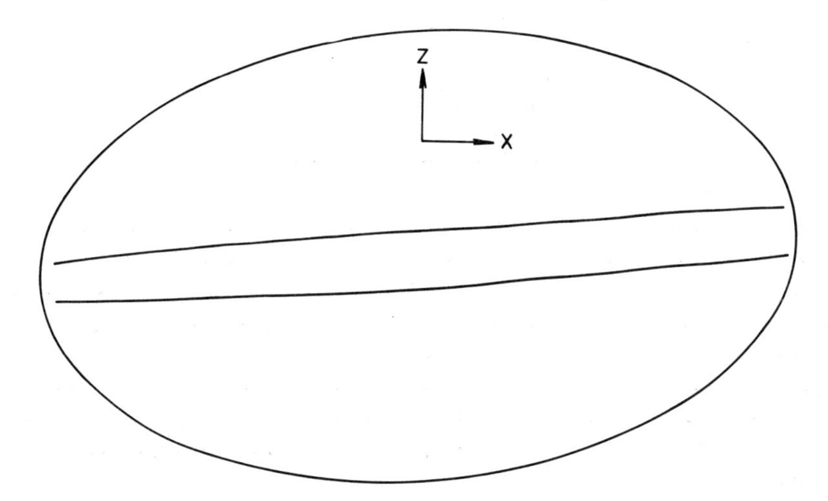

Figure 1. The axis of a deep valley in the DLS sum-of-squares plot for the O(1) atom in natrolite coincides with the major axis of the thermal vibration figure determined by Peacor (1973). The pair of lines is an arbitrary contour which closes outside the limits of the diagram.

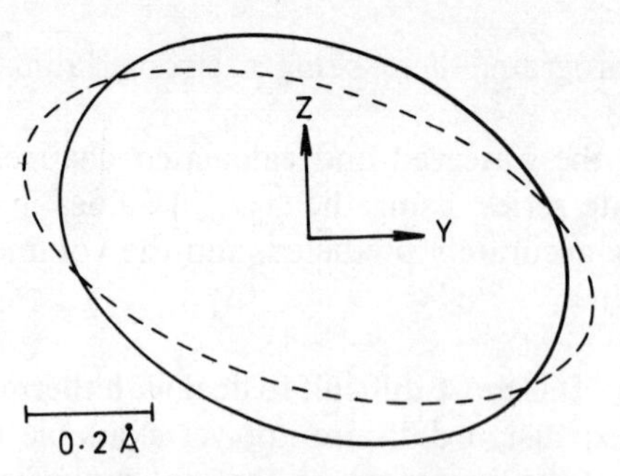

Figure 2. The DLS (solid) and observed (dashed) vibration figures for the Mg(1) ion in olivine (Smyth and Hazen, 1973), showing the approximate coincidence of the easy vibration directions.

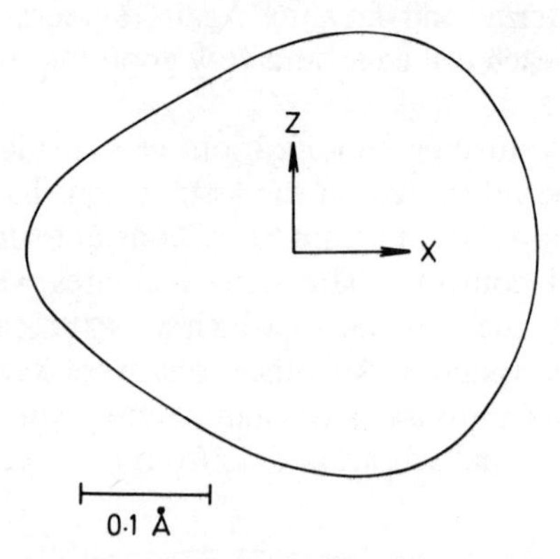

Figure 3. The DLS vibration figure for the Mg(2) ion in olivine is not elliptical, suggesting that a more complex description of thermal vibration figures may now be desirable.

Figure 1 shows the very clear correspondence between the axis of a deep valley in the sum-of-squares plot and the easy vibration direction of the O(1) ion in natrolite (Peacor, 1973). Figure 2 shows fair agreement between the shape and orientation of the thermal vibration ellipsoid of the Mg(1) ion in olivine and the corresponding contours of constant sum-of-squares. Finally, Figure 3 makes the point that thermal vibration figures are not necessarily ellipsoidal, and that with the increasing interest in high-temperature crystallography the time may have come to adopt a more complex description of thermal motions.

Compressibility: Direct determinations of bond compressibilities, comparable with the bond thermal expansions now becoming available (Cameron *et al.*, 1973) are not yet available, although volume compressibilities are known for most simple oxides, and can also be calculated using the Anderson–Anderson (1970) relation, written below in modified form:

$$K_0\,(\mathrm{Mb}) = (3.62 \pm 0.12) z^+ z^- N/V \tag{3}$$

where K_0 is the bulk modulus at zero pressure, z^+, z^- are the formal charges on cation and anion, N is the number of atoms in the formula unit, and V is the formula volume in cubic Ångstroms. Note that if V is taken to be proportional to r^3, then $K(r) \propto r^{-3}$, implying large differences between the compressibilities of long and short bonds within individual coordination polyhedra. The dependence of K on r may also be explored by starting with the simplified Birch–Murnaghan equation (Birch, 1952):

$$P = \frac{3K_0}{2}\left[\left(\frac{\rho}{\rho_0}\right)^{7/3} - \left(\frac{\rho}{\rho_0}\right)^{5/3}\right] \tag{4}$$

letting $\rho \propto 1/r^3$, the dependence of bulk modulus upon r becomes:

$$K(r) = K_0\left[7\left(\frac{r_0}{r}\right)^7 - 5\left(\frac{r_0}{r}\right)^5\right] \tag{5}$$

Evaluation of (3) and (4) for MgO at 250 kb shows that r/r_0 differs by only a fraction of one per cent, the difference increasing rapidly as $P/K_0 \rightarrow 1$. With the exception of the Na–O bonds in $NaAlSiO_4$, P/K_0 in our calculations is small, and we have adopted the simple assumption that:

$$K(r) = K_0(r_0/r)^3 \tag{6}$$

Table 2c compares calculated values of the compressibility of forsterite and fayalite with those deduced from Verma's (1960) ultrasonic data. The relative values of the principal compressibilities are in the right order, although absolute agreement is poor, the calculated volume compression being only 0.70 of the 'observed' value. Whether an actual measurement of principal compressibilities over the range 0–100 kb would give better agreement with the DLS figures remains to be seen. Mao *et al.* (1974) have noted that ultrasonic methods give bulk moduli of only 1.39 to 1.61 Mb for magnetite, compared with 1.83 Mb by static compression, and a similar discrepancy could well occur in olivine.

Response of Structure to Pressure: The inverse third power dependence of K on r might be expected to cause the long bonds in distorted coordination polyhedra to compress more than the short bonds, reducing the distortion at high pressures. This contrasts with the behaviour as temperature and composition change, when no systematic effect on the degree of distortion is to be expected.

Comparison of zero- and high-pressure D-structures of sphene ($CaTiSiO_5$) containing $[CaO_8]$, $[TiO_6]$ and $[SiO_4]$ polyhedra of very different compressibility and degree of distortion shows no systematic reduction in distortion

Table 3. Sphene $CaTiSiO_5$

(a) Coordinates and cell parameters

		$X^{a,b}$ $P=0$ $T=300$ K	D $P=0$ $T=300$ K	D $P=25$ kb $T=300$ K	D $P=0$ $T=1200$ K
x:	O2	9097	9097	9102	9087
	O3	3737	3737	3737	3734
y:	Ca	1677	1677	1678	1674
	Si	1818	1818	1818	1818
	O1	0668	0668	0674	0661
	O2	0689	0689	0681	0704
	O3	2111	2111	2110	2114
z:	O2	1838	1838	1835	1840
	O3	3982	3982	3982	3995
a		7.073	7.073	7.038	7.138
b		8.718	8.718	8.680	8.781
c		6.555	6.555	6.525	6.606
β		113° 58′	113° 58′	113° 58′	113° 53′
V		369.4	369.4	364.2	378.6

(b) Changes in interatomic distances from room *T* and *P* values

	RT	25 kb	1200 k
2 × Si–O2	1.6095	− 29[c]	+ 40[c]
2 × Si–O3	1.5990	− 34	+ 39
mean	1.6043	− 32	+ 39
2 × Ti–O1	1.8617	− 75	+ 148
2 × Ti–O2	1.9973	− 91	+ 184
2 × Ti–O3	2.0362	− 85	+ 175
mean	1.9651	− 84	+ 169
Ca–O1	2.4146	− 157	+ 256
2 × Ca–O2	2.4242	− 155	+ 261
2 × Ca–O3	2.6838	− 140	+ 278
2 × Ca–O3	2.4221	− 164	+ 288
mean	2.4821	− 153	+ 273

[a]Coordinates × 10^4
[b]*X* data from Mongiorgi and Sanseverino (1968)
[c]Changes in last digits quoted.

with pressure (Table 3). Similar results were obtained for $NaAlSiO_4$ in the $CaFe_2O_4$ structure (Table 4), which is built of $[NaO_8]$, $[AlO_6]$ and $[SiO_6]$ polyhedra of different properties.

It seems that changes in polyhedral distortions at the pressures involved (25 to 250 kb) may depend on factors other than bond compressibilities, e.g.

Table 4. $NaAlGeO_4$ X- and D-structures; $NaAlSiO_4$ D-structures

	$NaAlGeO_4$				$NaAlSiO_4$			
	X[a]		D		D ($P=0$)		D ($P=250$ kb)	
	x	y	x	y	x	y	x	y
Na	756	652	756	652	755	657	760	643
Al	933	889	932	888	927	886	931	888
Ge(Si)	580	894	580	894	576	898	579	895
O1	790	849	790	849	777	856	785	848
O2	886	516	886	516	891	514	886	515
O3	521	784	521	784	518	794	519	789
O4	574	573	574	573	576	577	579	574

Bond lengths in $NaAlSiO_4$

	$P=0$	$P=250$ kb
2 × Na–O1	2.4475	2.3849
2 × Na–O2	2.3162	2.0770
2 × Na–O4	2.2170	2.0807
Na–O3	2.4726	2.4070
Na–O3	2.3319	2.2075
	2.3457	2.2125
2 × Al–O1	1.9132	1.8123
2 × Al–O4	1.9151	1.8204
Al–O3	1.9888	1.8509
Al–O4	1.9376	1.8065
	1.9305	1.8205
2 × Si–O2	1.8301	1.7769
2 × Si–O3	1.7968	1.7386
Si–O1	1.7820	1.7363
Si–O2	1.8330	1.7931
	1.8115	1.7601

[a]From Reid *et al.* (1967)

the 'soft' $[NaO_8]$ and $[CaO_8]$ polyhedra deform to fill the spaces between the 'hard' $[SiO_4]$, $[SiO_6]$, $[TiO_6]$ and $[AlO_6]$ polyhedra, rather than deforming in the manner expected of isolated polyhedra not subjected to external constraints.

Hypothetical High-pressure Phases: The best DLS results have been obtained with structures based on close-packing of oxygen ions, including the β and γ-forms of Mg_2SiO_4 (Table 5b below). The method should therefore be suitable for estimating the volumes of hypothetical high-pressure phases such as

$MgSiO_3$ (perovskite form), $NaAlSiO_4$ (calcium ferrite), and Mg_2SiO_4 (strontium plumbate), which are of possible geophysical interest, although they have not yet been synthesised. Estimates of volumes have usually relied on graphical or least-squares treatments of the dependence of some function of the volume on some function of the metal–oxygen distances in a series of stable compounds of each structure type. Extrapolation to the $f(r)$ appropriate to the hypothetical phase then yields $f(V)$ and V.

The accuracy of these extrapolated volumes may be far worse than the standard deviations (typically 1 per cent) suggest. The very fact that a structure becomes unstable at some $f(r)$ suggests that some limiting factor, e.g. O–O repulsion, operates at that point, possibly changing the dependence of $f(V)$ on $f(r)$, and making the extrapolated volume a minimum estimate.

The DLS method is less subject to this problem, but the very good agreement between observed and calculated volumes of close-packed phases (Tables 5a, b, c) does not imply that the volumes of hypothetical phases are equally accurate. The estimates of prescribed interatomic distances are inevitably better for phases of known structure, and the deviation between the real and the D-structures can be expected to increase with the differences in volume, metal-oxygen distance and charge-balance between the base- and D-structures.

For these reasons the quality of the D-structures of Mg_2SiO_4 polymorphs in Table 5 probably decreases in the order $Sr_2PbO_4 > K_2NiF_4 > CaFe_2O_4$, the first involving only 6–8 per cent changes in metal–oxygen distances with no significant change in charge balance, and the last changes in r of 5, 13 and 17 per cent combined with major changes in charge-balance. The volume of the $NaAlSiO_4$ D-structure, with only an 8 per cent change in r for one ion in passing from the base- to the D-structure is probably accurate to 1 per cent.

The volumes of cubic silicate perovskites in Table 5c are probably high for two reasons: there is often some bond-shortening in symmetrical structures, and the high-pressure forms of perovskite usually have (smaller) orthorhombic cells in which the large ions have 8 short and 4 long bonds to oxygen rather than 12 in the cubic form. The orthorhombic perovskite D-structures (Table 5b) are based on $CaTiO_3$, which has a similar tolerance ratio to $MgSiO_3$.

Baur's (1972) calculations of D-structures of Mg_2SiO_4 in the K_2NiF_4 and Sr_2PbO_4 forms are not directly comparable with ours, as he assumed a 3 per cent shortening of the Si–O bond in the former (equivalent to a pressure of 360 kb), and constrained the $a:b:c$ ratio to be equal to that of Mn_2GeO_4 in the latter.

Miller (1972) has shown that lower mantle densities match those of assemblages such as $MgSiO_3$ (perovskite) + MgO (NaCl or CsCl) or $MgO + SiO_2$ (PbO_2-II or CaF_2), and there is increasing evidence that the spinel form of Mg_2SiO_4 breaks down first to ilmenite + MgO, and then to stishovite + MgO (see Akimoto *et al.*, this volume). The calcium ferrite, strontium plumbate and K_2NiF_4 forms of Mg_2SiO_4 may therefore be unimportant. The calcium ferrite form of $NaAlSiO_4$ almost certainly exists, and has a volume below that of such alternative assemblages as $Na_2O + Al_2O_3$ + stishovite or $NaAlO_2$ + stishovite. Orthorhombic $MgSiO_3$ perovskite is also likely to be an important phase.

Table 5 (a). Observed and calculated unit cell dimensions and volumes of olivines and (Mg, Fe, Co) β-phases

Compound	Structure type	Struct.	P (kb)	T (K)	a	b	c	V
Mg_2SiO_4	Olivine	X	0	300	4.762	10.225	5.994	291.9
		D	0	300	4.796	10.371	5.760	286.5
		X	0	1173	4.795	10.355	6.060	300.8
		D	0	1173	4.831	10.490	5.820	294.9
		D	100	300	4.723	10.157	5.650	271.0
		D	100	1173	4.757	10.275	5.715	279.4
Fe_2SiO_4	Olivine	X	0	300	4.816	10.469	6.099	307.5
		D	0	300	4.842	10.628	5.864	301.8
		X[a]	0	1173	4.798	10.387	6.055	301.8
		D	0	1173	4.868	10.717	5.915	308.6
		D	100	300	4.768	10.407	5.752	285.4
		D	100	1173	4.794	10.495	5.803	291.9
Mg_2SiO_4	β-phase	X	0	300	8.248	11.450	5.710	539.3
		D	0	300	8.242	11.471	5.680	537.0
Fe_2SiO_4	β-phase	D	0	300	8.405	11.673	5.801	569.1
Co_2SiO_4	β-phase	X	0	300	8.337	11.522	5.753	552.6
		D	0	300	8.299	11.627	5.684	548.5

[a]For a hortonolite $Mg_{0.75}Fe_{1.1}Mn_{0.15}SiO_4$

In olivines 300 *K* X-data from Birle *et al.* (1968), 1173 K X-data from Smyth and Hazen (1973). *β*-phase X-data from Moore and Smith (1970) and Morimoto *et al.* (1970) respectively.

Table 5 (b). Observed and calculated unit cell dimensions of other non-cubic A_2BO_4 and ABO_3 structures

Compound	Structure type	Struct.	P (kb)	a	b	c	V
Ca_2GeO_4	K_2NiF_4	X[a]	0	3.70	3.70	11.88	162.6
Ca_2GeO_4	K_2NiF_4	D	0	3.700	3.700	11.881	162.6
Mg_2SiO_4	K_2NiF_4	D	0	3.489	3.489	10.901	132.7
Mg_2SiO_4	K_2NiF_4	D	250	3.417	3.417	10.390	121.3
Cd_2SnO_4	Sr_2PbO_4	X[b]	0	5.564	9.888	3.193	175.7
Cd_2SnO_4	Sr_2PbO_4	D	0	5.641	9.922	3.122	174.7
Mg_2SiO_4	Sr_2PbO_4	D	0	5.154	8.906	2.754	126.4
Mg_2SiO_4	Sr_2PbO_4	D	250	4.874	8.545	2.700	112.5
$CaFe_2O_4$	$CaFe_2O_4$	X[c]	0	9.230	10.705	3.024	298.8
$CaFe_2O_4$	$CaFe_2O_4$	D	0	9.238	10.729	3.021	299.4
Mg_2SiO_4	$CaFe_2O_4$	D	0	8.918	10.362	2.789	257.7
Mg_2SiO_4	$CaFe_2O_4$	D	250	8.510	9.974	2.734	232.0
$NaAlGeO_4$	$CaFe_2O_4$	X[d]	0	8.871	10.402	2.840	262.1
$NaAlGeO_4$	$CaFe_2O_4$	D	0	8.870	10.398	2.840	261.9
$NaAlSiO_4$	$CaFe_2O_4$	D	0	8.641	10.161	2.734	240.1
$NaAlSiO_4$	$CaFe_2O_4$	D	250	8.117	9.693	2.627	206.7
$CaTiO_3$	$CaTiO_3$	X[e]	0	7.638	7.630	7.638	445.1
$CaTiO_3$	$CaTiO_3$	D	0	7.631	7.620	7.631	443.7
$CaSiO_3$	$CaTiO_3$	D	0	7.152	7.170	7.152	366.7
$FeSiO_3$	$CaTiO_3$	D	0	7.081	7.101	7.081	356.0
$MgSiO_3$	$CaTiO_3$	D	0	7.059	7.061	7.059	351.8

[a]Reid and Ringwood (1970); [b]Trömel (1967); [c]Decker and Kasper (1957); [d]Reid *et al.* (1967); [e]Monoclinic pseudo-cell $P2_1/m$, $\beta = 90° 40' \pm 1'$ (Naray Szabo, 1943)

Table 5 (c). Observed and calculated structure data for various cubic A_2BO_4 and ABO_3 compounds

Compound	Struct.	*P* (kb)	T (K)	*a*	*u*	*V*
γ-Mg_2SiO_2	X[a]	0	300	8.076	0.368	526.73
	D	0	300	8.0846	0.3667	528.42
	D	0	1000	8.1686	0.3658	545.06
γ-Fe_2SiO_4	X[b]	0	300	8.23	0.3658	557.44
	D	0	300	8.2344	0.3647	558.34
	D	0	1000	8.3002	0.3640	571.83
γ-Co_2SiO_4	X[c]	0	300	8.138	0.366	538.96
	D	0	300	8.1374	0.3660	538.84
$MgSiO_3$	D	0	300	3.5058	—	43.09
	D	250	300	3.4412	—	40.75
$FeSiO_3$	D	0	300	3.5270	—	43.87
$CaSiO_3$	D	0	300	3.6050	—	46.85
	D	250	300	3.5188	—	43.57
$SrTiO_3$	X	0	300	3.904	—	59.50
	D	0	300	3.9043	—	59.52

[a]Ito *et al.* (1974); [b]Akimoto (this volume); [c]Morimoto *et al.* (1970)

DISCUSSION

The DLS program described differs from that of Meier and Villiger (1969) mainly in the use of bond strengths which vary with bond length in place of constant empirical weights. It differs from that of Baur (1972) in the use of a different relationship between bond length and bond strength for the M–O bonds, and in the use of variable weights for O–O interactions. Our O–O weights are about half those used by Meier and Villiger, and twice those used by Baur, for long (> 2.8 Å) O–O distances.

The results obtained for the effects of Fe–Mg replacement, thermal expansion and 100 kb compression in olivine are in sufficiently good agreement with experiment to encourage use of the DLS method in similar cases. The worst agreement is for the 0–100 kb compression, but this may well be attributable to underestimation of compressibility by the ultrasonic method, rather than any error in the DLS results.

The calculated contours of constant sum-of-squares correlate well with the thermal vibration figures, and seem to provide an alternative approach to that of Ohashi and Finger (1973), although some problems of scaling remain. Some of the results suggest that thermal vibration figures are not adequately represented by triaxial ellipsoids, and that better representations may be desirable in modern high-precision structure determinations, and especially those at high temperatures.

The DLS method clearly has useful applications to the calculation of molar volumes and structures of hypothetical high-pressure phases, and provided a good base structure is available, cell edges accurate to ± 1 per cent and volumes to better than ± 3 per cent should be routinely attainable for close-packed structures. Results so far obtained suggest that expected systematic reduction of distortion with pressure may not occur, but more high-pressure structure determinations will be required to establish this.

There are potential applications of DLS structures to the calculation of crystal-field splittings in mantle phases (Burns, this volume), and it may prove possible to obtain better estimates of bulk moduli and their variation with (P,T,X). However, this will require a better understanding of the factors controlling thermal expansion in oxides and silicates, and the wider availability of high-pressure and high-temperature structure data.

REFERENCES

Anderson, D. L. and Anderson, O. L. (1970). *Journ. Geophys. Res.*, **75**, 3494.
Barrer, R. M. and Villiger, H. (1969). *Z. Krist.*, **128**, 352.
Baur, W. H. (1971). *Nature (Physical Science)*, **233**, No. 42, 135.
Baur, W. H. (1972). *Amer. Mineral.*, **57**, 709.
Birch, F. (1952). *Journ. Geophys. Res.*, **57**, 227.
Birle, J. D., Gibbs, G. V., Moore, P. B. and Smith, J. V. (1968). *Amer. Mineral.*, **53**, 807.
Brown, I. D. and Shannon, R. D. (1973). *Acta Cryst.*, **A29**, 266.
Cameron, M., Sueno, S., Prewitt, C. T. and Papike, J. J. (1973). *Amer. Mineral.*, **58**, 594.
Clark, S. P. (1966). *Handbook of Physical Constants*, Geol. Soc. Amer.
Decker, B. F. and Kasper, J. S. (1957). *Acta. Cryst.*, **10**, 332.
Finger, L. W. (1969). *Mineral. Soc. Amer.*, Special Paper 2, 95.
Hazen, R. M. and Burnham, C. W. (1974). *Amer. Mineral.*, **59**, 1166.
Ito, E., Matsui, Y., Suito, K. and Kawai, N. (1974). *Phys. Earth Planet. Interiors*, **8**, 342.
Mao, H-K., Takahashi, T., Bassett, W. A., Kingland, G. L. and Merrill, L. (1974). *Journ. Geophys. Res.*, **79**, 1165.
Meier, W. M. and Villiger, H. (1969). *Z. Krist.*, **129**, 411.
Miller, G. H. (1972). *M.Sc. Dissertation*, University of Newcastle upon Tyne.
Mongiorgi, R. and di Sanseverino, L. R. (1968). *Miner. Petrogr. Acta*, **14**, 123.
Moore, P. B. and Smith, J. V. (1970). *Phys. Earth Planet. Interiors*, **3**, 166.
Morimoto, N., Akimoto, S., Koto, K. and Tokonami, M. (1970). *Phys. Earth Planet. Interiors*, **3**, 161.
Naray-Szabo, St. V. (1943). *Naturwissenschaften*, **31**, 203.
Ohashi, V. and Finger, L. W. (1973). *Annual Report of the Director*, Geophysical Laboratory, 1972–73, 547.
Pauling, L. (1960). *Nature of the Chemical Bond*, 3rd Edn., Cornell, New York.
Peacor, D. R. (1973). *Amer. Mineral.*, **58**, 676.
Reid, A. F. and Ringwood, A. E. (1970). *Journ. Solid State Chem.*, **1**, 557.
Reid, A. F. Wadsley, A. D. and Ringwood, A. E. (1967). *Acta. Cryst.*, **23**, 736.
Shannon, R. D. and Prewitt, C. T. (1969). *Acta. Cryst.*, **B25**, 925.
Shannon, R. D. and Prewitt, C. T. (1970). *Acta. Cryst.*, **B26**, 1046.
Smith, D. (1972). *Amer. Mineral.*, **57**, 1413.
Trömel, M. (1967). *Naturwissenschaften*, **54**, 17.
Verma, R. K. (1960). *Journ. Geophys. Res.*, **65**, 757.
Wood, B. J. and Strens, R. G. J. (1972). *Mineral. Mag.*, **38**, 909.

New Applications of the Diamond Anvil Pressure Cell: (III) Single Crystal Diffractometer Analysis at High Pressure

W. A. Bassett
Department of Geological Sciences
University of Rochester
Rochester, New York 14627 USA

R. M. Hazen
Hoffman Laboratory
Harvard University
Cambridge, Massachusetts 02138
and

L. Merrill
High Pressure Data Center
Brigham Young University
Provo, Utah 84602

INTRODUCTION

Geophysicists are concerned with the effects of pressure and temperature on the physical properties of crystalline solids. Similarly, mineralogists and crystal chemists are interested in the variation of atomic arrangements with these intensive variables. Single-crystal X-ray diffraction has provided the best, and in some cases the only, technique for measuring certain physical properties including symmetry, lattice parameters, and structure, as well as cation order–disorder, site occupancy, thermal vibrations, and bond character. In addition, precise atomic positional parameters are essential for calculations of Madelung constants and site energies, which in turn are employed in theoretical calculations of elastic properties, transport properties, and lattice dynamical properties.

Recent technical advances and experimental results in high-temperature single-crystal X-ray diffraction have been summarized by Prewitt (1976) in these proceedings. While structure refinements of minerals at temperatures in excess of 1100°C are now in the literature, no analogous high pressure studies of minerals have been published. Recent development of a high-pressure single-crystal diamond cell (Weir *et al.*, 1969; Merrill and Bassett, 1974) now permits high-pressure crystallographic study using precession cameras and single-crystal diffractometers. This paper reviews recent progress in high-pressure single crystal X-ray diffraction experiments, and proposes several possible applications of the diamond anvil cell to problems in geophysics, mineralogy, and crystal chemistry.

PRINCIPLE OF THE DIAMOND ANVIL CELL

In the diamond anvil high pressure cell a sample is placed between two flat parallel faces of gem quality brilliant cut diamonds ($\sim$25 mg) which in turn are mounted on metal pistons or platens which are pushed or pulled together by screws. Holes in the metal parts allow visual observation of the sample through the diamond anvils.

Since pressure may be defined as force divided by area, a high pressure in an opposed anvil device may be achieved by employing a large force or a small area. The latter is the basis for the success of the diamond anvil cell. Clearly the principle is useful only if the methods for making observations on the samples are suitable for use with very small samples. Both the visible and X-ray portions of the spectrum have provided abundant information on microscopic samples while they are under pressure in the diamond anvil cell.

THE SINGLE CRYSTAL DIAMOND CELL

In the cell used for single crystal X-ray diffractometer studies, diamond anvils with 0.8 mm and 1.0 mm diameter faces are mounted on beryllium plates supported by stainless steel platens in such a way as to provide a conical dispersion angle of $\sim 52^\circ$ in both sides of the cell (Figure 1). The platens which are triangular are pulled together by three screws at the apices. The single crystal sample is suspended in glycerine or oil in a hole $\sim 300\ \mu$m in diameter in an inconel foil 250 μm thick placed between the anvils. The anvils then effectively trap the liquid in the hole and as the platens are pulled together, the sample is subjected to hydrostatic pressure.

The diamond cell may be mounted on a modified eucentric goniometer head (Figure 2). The crystal under investigation is easily seen through the diamonds, and may be centred in the X-ray beam in the standard way. However, several slight modifications to standard X-ray equipment may be required for most efficient use of the diamond cell. For example, on the precession camera a beam-stop with large clearance must be used to allow for adequate precession motion of the diamond cell. A standard 'C'-type beam-stop was found to satisfy this requirement. Approximately 50 per cent of reciprocal space between 5° and 50° 2θ is accessible. For crystals with small unit-cell dimensions (i.e. large reciprocal spacings) the use of wavelengths shorter than the standard Mo K_α ($\lambda = 0.7107$ Å) may be warranted. Silver K_α radiation

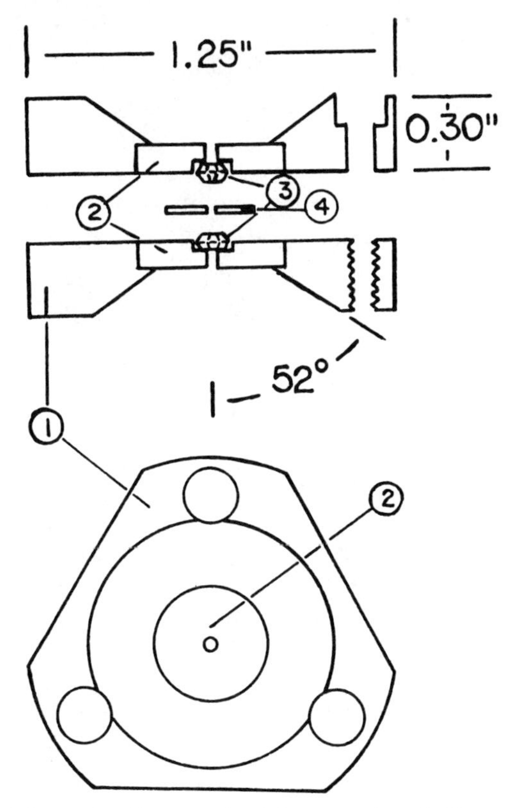

Figure 1. Diagram of miniature gasketed diamond anvil pressure cell. (1) stainless steel platens, (2) beryllium discs, (3) diamond anvils, and (4) inconel gasket.

($\lambda = 0.5609$ Å) would approximately double the accessible volume of reciprocal space.

Laue reflection of the continuum X-ray spectrum by the diamond anvils introduces an interference which can be minimized by either a pre-sample monochromator or by pulse-height discrimination. Automated data collection of diffracted X-ray intensities is straightforward using the miniature pressure cell. The only important mechanical modification required for the Picker four-circle automated diffractometer is replacement of the standard beam collimators by new ones approximately 18 mm shorter. This allows complete rotation of the pressure cell about the phi axis. X-ray intensity data may be reduced and analysed in the standard way. Corrections for absorption by the diamond and beryllium parts of the cell appears to be necessary for a structure analysis with a least squares residual less than 10 per cent, and a computer program for making this calculation is in preparation. A more detailed description of the single crystal pressure cell is given by Merrill and Bassett (1974).

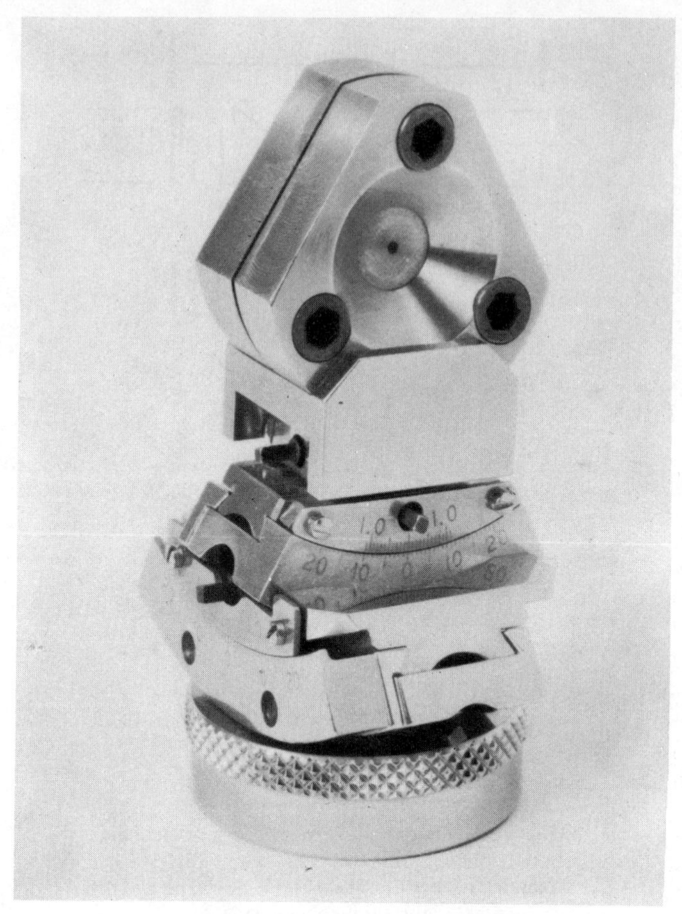

Figure 2. The diamond anvil cell mounted on a standard eucentric goniometer head.

APPLICATIONS

The most obvious application of the single crystal pressure cell to problems of interest to geophysicists, mineralogists, and crystal chemists is the determination of unquenchable high pressure crystal structures. Studies by Merrill and Bassett (1974) on the high pressure polymorphs of $CaCO_3$ and by Hazen and Burnham (1974) on the high pressure phase of gillespite ($BaFeSi_4O_{10}$) at 26 kb have demonstrated the effectiveness of this technique.

Structure of $CaCO_3$ II

Bridgman (1939) found that calcite compressed at room temperature undergoes a discontinuous volume change at 14 kb and another at 17 kb. These values have subsequently been revised to 15 kb and 22 kb. These represent

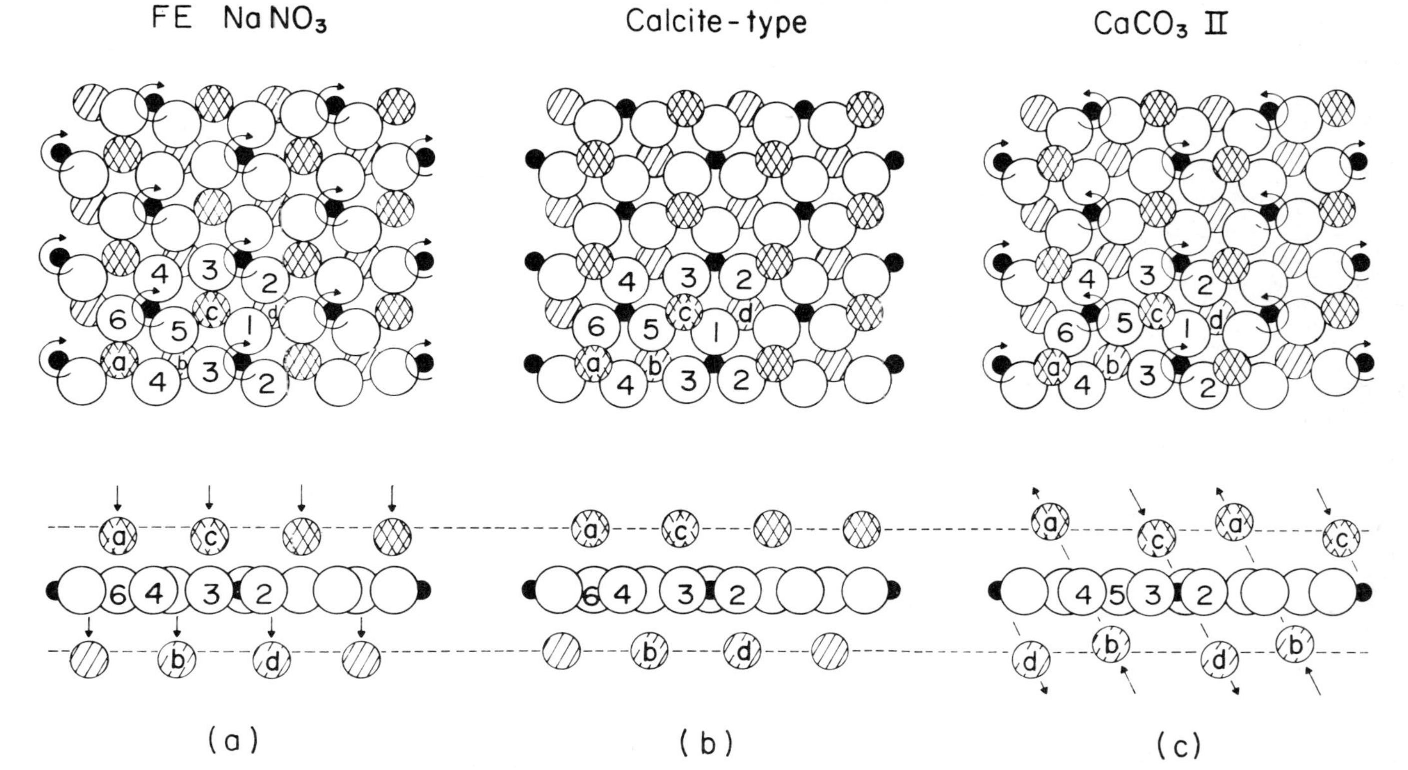

Figure 3. (a) Rotational and translational displacements in the ferro-electric (FE) $NaNO_3$ phase transformation. (b) a single anion layer of the calcite-type structure with cations above and below, (c) rotational and translational displacements in the $CaCO_3$ I—$CaCO_3$ II phase transformation. The open circles with numbers represent oxygen atoms, the shaded circles with letters represent Na and Ca atoms, the solid circles represent N and C atoms. Note that in $CaCO_3$ the carbonates in adjacent rows rotate in opposite directions while calcium atoms are displaced parallel to the $(10\bar{1}4)$ crystallographic plane as shown.

first order metastable phase transitions since they occur within the aragonite stability field. Van Valkenburg (1965) discovered that if he immersed a single crystal of calcite in glycerine in a hole in inconel foil between the faces of diamond anvils, as described above, he could observe under the microscope the first phase transition in calcite as a line travelling across the crystal as pressure was applied. The most remarkable aspect of his observation was the fact that the sample remained a perfect single crystal after the transition. It was this observation that encouraged us to attempt a single crystal analysis of $CaCO_3$ II.

One hundred and fifteen unique reflections were obtained on $CaCO_3$ II with a GE diffractometer using a pre-sample monochromator. Patterson and Fourier maps were used to analyze the data. The final refinement of the structure had a least squares residual of 10 per cent. In addition to the single crystal measurements, polycrystalline samples were suspended in glycerine gasketed in a diamond cell just as was done for the single crystal work. This was done in order to obtain more accurate lattice parameter data (Table 1).

Table 1. Lattice data for $CaCO_3$ II

$a = 6.334 \pm 0.020$ Å
$b = 4.948 \pm 0.015$ Å
$c = 8.033 \pm 0.025$ Å
$\beta = 107.9 \pm 10$ deg.
Space group $P2_1/c$

Both $CaCO_3$ and $NaNO_3$ have the calcite structure at 1 bar pressure and room temperature (Figure 3b). Barnett *et al.* (1969) reported that $NaNO_3$ goes through a second order transition at 40 kb and room temperature to a new phase in which all of the NO_3 groups are rotated in the same direction and all of the Na ions are displaced in the same direction parallel to the 3-fold axis (Figure 3a). This displacement of the cations with respect to the anions causes the high pressure phase of $NaNO_3$ to be ferroelectric.

The analysis of $CaCO_3$ II reveals that in this structure also, anion groups are rotated and the cations are displaced. In this structure, however, rows of anion groups rotated clockwise alternate with rows of anion groups rotated counterclockwise (Figure 3c). Coupled with this are cations alternately displaced in directions parallel to one of the $(10\bar{1}4)$ planes in the original hexagonal coordinate system. Since cations are displaced alternately in opposite directions the $CaCO_3$ II phase is not ferroelectric.

The pressure induced phase transformations in $NaNO_3$ and $CaCO_3$ are believed to take place as a result of an instability of a soft vibrational mode at the zone centre and zone boundary respectively.

Gillespite I and II

Gillespite is a rare barium iron silicate ($BaFeSi_4O_{10}$), which has an unusual

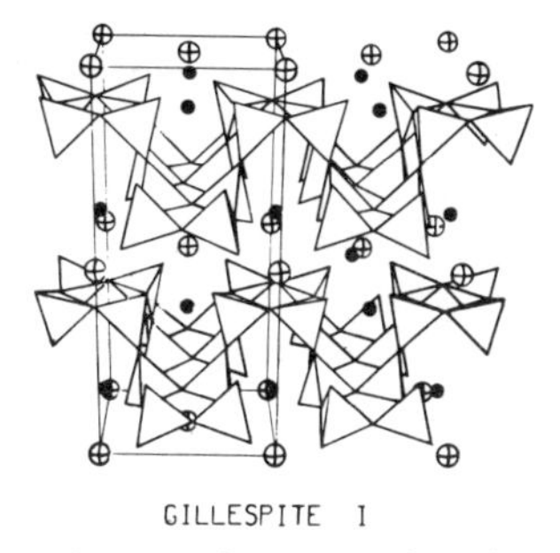

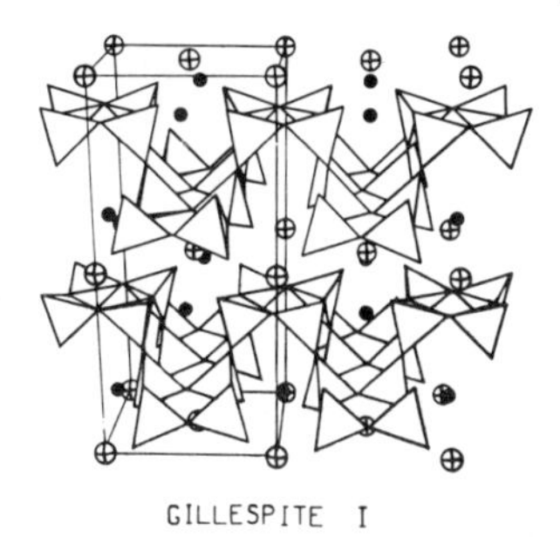

Figure 4. Stereoscopic pair representing the structure of gillespite I. The tetrahedra represent silica tetrahedra, the circles with pluses represent Ba atoms, and the solid circles represent Fe atoms.

tetragonal (space group $P\ 4/n\ 2_1/c\ 2_1/c$) layer structure at room temperature and atmospheric pressure (Figure 4). The atomic arrangement was first determined by Pabst (1943) who found that four-membered rings of silica tetrahedra form the basic building blocks. All apical oxygens of a given ring point in the same direction, and each ring is linked to four others with apices pointing in the opposite direction. In this way an infinite tetrahedral layer is constructed. Ferrous iron atoms lie on fourfold axes, and are bonded to four oxygens in square-planer coordination. Interlayer barium cations are in distorted cubic coordination. The fundamental layer repeat parallel to *c* is 8 Å. However, four-member rings of silica tetrahedra in adjacent layers are rotated approximately 15° with respect to each other, thus the true identity period parallel to *c* is 16 Å.

The unusual square-planar coordination of high-spin ferrous iron made gillespite a logical mineral for optical absorption and Mössbauer spectroscopic studies. Strens (1966) recorded the optical absorption spectra of gillespite at elevated pressures, and was the first to note a reversible transition from the red low pressure phase to what he described as a colourless high pressure phase. Abu-Eid, Mao, and Burns (1973) confirmed the reversible phase transformation, noting a colour change from red to blue at 26 kb. Changes in both optical absorption and Mössbauer spectra were observed, but the nature of the structural transformation remained unknown.

Single-crystal cone-axis precession photographs reveal that the *c*-axis of high-pressure gillespite (gillespite II) is approximately half that of gillespite at low pressures (gillespite I). Thus the high-pressure phase has a one-layer repeat parallel to *c* as opposed to the two-layer repeat of gillespite I. From *c*-axis precession photos there is seen to be little change in the *a* dimension; however systematic space group absence criteria corresponding to the symmetry of the low pressure phase are violated in gillespite II, indicating a change in space group. Careful analysis of the distribution of diffraction intensities revealed violations of fourfold symmetry, and the gillespite II symmetry is orthorhombic, $P2_12_12$.

Gillespite II intensity measurements were made on a computer controlled

4-circle single-crystal diffractometer. Due to geometrical constraints of the pressure cell, only *hk*0, *hk*1 and *hk*2 data were available—thus errors for z-atom coordinates were significantly greater than those for x and y coordinates. Furthermore, absorption corrections for diamond and beryllium were not made, and such omission leads to errors of up to 10 per cent in total absorption. Still, an R of 10.5 per cent was obtained for a gillespite II structure model.

While the change from gillespite I to gillespite II involves only slight changes in atomic coordinates, all cation coordination polyhedra are affected by the displacive transformation. Of considerâble interest to spectroscopists is the change in coordination and bond distances of ferrous iron. There is very little shortening of iron-to-oxygen bonds, and thus no strong evidence for a high-spin to low-spin transition as has been suggested. However, the square-planar coordination of the low-pressure phase is definitely distorted into what might be called an extremely flattened tetrahedral environment in gillespite II, with opposite oxygen–iron–oxygen angles changing from 178° to 167°.

An unusual aspect of the room-pressure gillespite structure is the nearly straight 178° silicon–oxygen–silicon bond angles between adjacent four-member rings. This implies maximum separation of these rings at low pressure. In gillespite II at 26 kb, however, these angles have been reduced to 157°, and the silicon–silicon distances between rings is considerably reduced. The resulting c-axis shortening appears sufficient to account for almost all of the 4 per cent volume reduction in the reversible phase transition.

The barium coordination in gillespite I is a distorted cube. The change from a two-layer structure (with 15° relative rotation of silicate-rings in adjacent layers) to a one-layer structure (with no relative rotation of these layers) at high pressure is associated with distortion of the barium coordination polyhedron to that approaching a square anti-prism. Assuming that the square anti-prism represents a more closely-packed oxygen group than a cube, this change in barium coordination may explain the change from a two-layer to a one-layer polytype at high pressure.

OTHER APPLICATIONS OF HIGH-PRESSURE X-RAY CRYSTALLOGRAPHY

Several possible applications of high-pressure single-crystal X-ray diffraction may provide important data for earth scientists. Of considerable interest is the continuous variation of atomic coordinates in known structures as a function of pressure. For example, while the bulk compressibilities of olivine and pyroxenes are well known, the variation of individual bond angles and distances as a function of pressure are not known. Accurate structure refinements at high pressures could yield several important relationships such as variation of effective ionic radius with pressure, polyhedral distortion *vs.* pressure, and thermal vibrations *vs.* pressure. Accurate atomic coordinates of high-pressure phases could also lead to precise calculations of Madelung constants and site energies, which in turn could give improved theoretical values for elastic, thermal, and transport properties of minerals at elevated pressures.

Observations on the orientations of crystal structures before and after phase transformations might yield information on the mechanisms of these phase transformations. Similarly, the variation of individual X-ray diffraction intensities as a function of time could reveal the rates of transformations of some minerals.

Shannon (1976) has noted that bond covalency should increase continuously as a function of pressure. It may be possible to obtain a quantitative plot of bond covalency *vs.* pressure, at least for some light atoms. Cooper *et al.* (1973) recently demonstrated the covalent character of boron-oxygen bonds, by first calculating an electron density map using high angle diffraction data only. These data presumably are the result of diffraction from electrons close to the nucleus, in essentially spherical symmetry about the atom centre. This map is then subtracted from a second map for all data; the difference represents the density of electrons occupying space between atom centres. Integration of inter-atomic electron density becomes a simple measure of bond covalency. A series of such plots for a structure as a function of pressure might reveal systematic variations of covalency with pressure.

An interesting application of the diamond cell might be found in diffusion or exchange experiments at high pressures. It is known that cell dimensions may be a sensitive function of certain cation ratios such as potassium to sodium. If a pressure fluid rich in one cation were used to hydrostatically compress a mineral rich in a second cation, then changes in cell dimensions *vs.* time would provide a measure of diffusion or exchange rates.

Perhaps the most exciting applications of high-pressure crystallography must await development of means to heat single crystals at high pressures. With such an experimental capability crystal structures and physical properties of minerals could be studied at conditions approximating those of the earth's interior. In fact, it may become possible to study phase equilibria as a function of temperature and pressure directly on single crystals.

ACKNOWLEDGEMENT

The high-pressure crystallographic study of gillespite and $CaCO_3$ were supported by NSF grants GA-12852, GA-41415, and GA-38056X.

REFERENCES

Abu-Eid, R. M., H. K. Mao, and R. G. Burns (1973). *Carnegie Inst. Wash. Year Book*, **72**, 564–567.

Barnett, J. D., J. Pack, and H. T. Hall (1969). *Trans. Amer. Cryst. Assoc., Proc. of the Symposium on 'Crystal Structure at High Pressure'*, vol. 5.

Bassett, W. A. and L. Merrill (1974). *Rev. Sci. Instr.*, **45**, 290–294.

Bridgman, P. W. (1939). *Am. J. Sci.*, **237**, 7–18.

Cooper, W. F., F. K. Larsen, and P. Coppens (1973). *Amer. Mineral.*, **58**, 21–31.

Hazen, R. M. and C. W. Burnham (1974). *Amer. Mineral.*, **59**, 1166–1176.
Merrill, L. and W. A. Bassett (1974). *Acta Cryst.*, **B31**, 343–349.
Pabst, A. (1943). *Amer. Mineral.*, **28**, 372–390.
Prewitt, C. T. (1976). This volume.
Shannon, R. D. (1976). This volume.
Strens, R. G. J. (1966). *Chem. Commun.*, **21**, 777–778.
Van Valkenburg, A. (1965). *Conference Internationale Sur-les-hautes Pressions*, Le Creusot, Saone-et-Loire, France au 6 Aout.
Weir, C. E., G. J. Piermarini and S. Block (1969). *Rev. Sci. Instr.* **40**, 1133.

An Empirical Method for Correlating and Predicting Melting and Solid-State Phase Transformations

R. R. Reeber
Department of Mineralogy and Petrology
Cambridge CB2 3EW, England

Thermodynamically first-order solid-state phase transformations are often difficult to characterize because of metastability or slow reaction kinetics. When these problems are compounded by effects of stoichiometry, relative differences in defect concentrations and sample corrosiveness, experimental determinations of the underlying equilibrium relations become very difficult. It is the aim here to illustrate an empirical method for correlating and predicting melting and solid-state phase transformations, which may serve as a guide for the experimental crystal chemist in isolating specific variables limiting the utility of such scaling methods for specific structure types. The origin of scaling laws lies in the early work of Lindemann (1910) and Einstein (1911). Lindemann derived a relationship between melting and atomic vibration from the basic assumption that at a critical atomic separation melting occurred. Einstein points out in his comments that Lindemann's work seems to indicate that the law of corresponding states for simple solids and liquids holds true within a remarkably good approximation. In later work Van Vechten (1972) has used scaling functions with two adjustable parameters to derive melting points from bond energies of A^nB^{8-n} crystals. Although his results are reasonable for some III–V compounds, they exceed the experimental values of Narita *et al.* (1970) for II–VI compounds by as much as 20 per cent. Here we are not attempting to calculate transformation temperatures directly, but instead to correlate them with a dimensionally correct scaling parameter, a characteristic temperature. This parameter is derived with simplifying assumptions from a simplified lattice model of the solid. The derivation illustrated below follows Reeber and McLachlan (1971). The temperature has a simple theoretical relationship to the lattice-vibration spectrum and relies on elastic constant and lattice parameter data, both of which are experimentally measurable with a fair degree of accuracy. An important assumption is that crystal anisotropy and structural arrangement have large influences on solid-state transformations. From this it follows that a limiting frequency of the lattice in a specific significant crystallographic direction is a good indicator of such transformations. This treatment is not limited to so-called soft-mode transitions but also relates 'hard' mode transformations. In such a transformation the modal population of an entire branch of the vibration distribution may be significantly different for the different possible arrangements. The possibility of hard-mode transformation exists when the structure does not have the best arrangement to minimize its changes in vibrational entropy with changes in the temperature, pressure and composition of the enclosing system. As McConnell (1972) points out, before experimentally observed behaviour can be interpreted in terms of the permissible behaviour of a system it is important to try to determine the underlying behaviour of the system when kinetic and mechanistic constraints are either eliminated or minimized.

In the wurtzite, sphalerite, and CsCl structures (000*l*) and (*lll*)-type planes are important features. Their longitudinal-acoustical (LA) planar waves are considered and their characteristic cutoff frequencies are calculated. A characteristic temperature is then determined from the relation $\theta = h\nu/k$, where h and k are Planck and Boltzmann's constants respectively and ν is the characteristic frequency. It is especially important to realize that the atomic arrangement is an important determinant of the lattice characteristic temperature. Similar lattices with similar electronic configurations should have to a first approximation similar vibrational frequency distributions. The work of Dowling and Cowley (1966) in general confirms this supposition. Thus, by restricting attention to what is an apparently conspicuous feature of a particular structure it is expected that some correspondence with transformation behaviour will be revealed.

Previous work has shown the Debye characteristic temperature to be a useful means of correlating and interpreting experimentally measured physical and thermal properties of solids (Blackman, 1955). If, however, certain crystallographic directions are particularly important, then the Debye frequency, which is an average over all directions many of which have little or no influence on the process being investigated, would be a less sensitive scaling parameter. Hewat (1972) illustrates by comparison with lattice-dynamical calculations that mean-square vibrational frequencies calculated with the Debye model are subject to two errors each of about 25 per cent. He claims that these are present even at low temperatures. Since his calculations utilized ZnS and BeO, substances considered here, it is additional justification for a different approach to the derivation.

CALCULATION OF LATTICE CHARACTERISTIC TEMPERATURES

We consider planar waves of atoms propagating along a specific direction of the crystal in this case the $\langle 111 \rangle$ direction of sphalerite. The (*lll*) planes spaced at $a\sqrt{3}/6$ have two atoms in each increment of area $a^2\sqrt{3}/2$. Since we are interested in the LA frequency, the larger mass (M_1) of the compound appears in the limiting frequency solution of the diatomic linear lattice, see equation 1. This solution is a standard textbook one and is illustrated by Brillouin (1953).

$$\theta_{111}^{\mathrm{LA}} \equiv \frac{h\nu_{\mathrm{B}}^{\mathrm{LA}}}{k} = \frac{h}{k\pi}\left(\frac{2KN}{2M_1}\right)^{1/2} \tag{1}$$

where $2N$ is the number of atoms in one mole of the crystal, K is the force constant along the $\langle 111 \rangle$ direction, M_1/N is the mass of one of the heavier atoms in the structure, and h and k are Planck and Boltzmann's Constants respectively. We determine K in the $\langle 111 \rangle$ direction from the definition of the elastic constant as follows:

$$C_{ij} \equiv \frac{\mathrm{d}F}{\mathrm{dA}}\left(\frac{\mathrm{d}l}{l}\right)^{-1} \approx \frac{\Delta F}{\Delta A}\left(\frac{\Delta l}{l}\right)^{-1} \tag{2}$$

Table 1. Characteristic and melting temperatures

Compound	T (°K)	θ^{LA} (°K)	θ_{D} (°K)	T_{fusion} (°K)
BeOw	298	1115	1280	2803
ZnOw	298	399	416	2060 predicted
ZnSw	298	349	351	1973
ZnS	77	352	352	
ZnS[a]	0	341	341	
ZnSe	0	301	278	1788
CdSw	0	230	219[b]	1643
ZnTe	0	219[b]	226	1563
CdSew	298	218	218	1513
CdTe	0	197	163	1365
CdTe[a]	0	201	164	
HgSe	0	168	153	1073
HgTe	0	161	144	943
HgTe[a]	0	162	142	
GaP	300	388	445	1740
GaAs	298	351	346	1511
InP	300	268	304	1331
AlSb	300	247	295	1338
GaSb	298	245	267	979
InAs	300	254	249	1215
InSb	300	220	204	803
CsCl	298	194	176	918
CsBr	298	186	151	909
CsI	298	174	127	899
TlBr	298	163	126	733
TlCl	293	163	139	702

[a]Vekilov and Rusakov (1971).
[b]Gerlich (1967).

where we assume that at the unit-cell level we can replace the differential quantities by incremental ones. Rearranging Equation (2) and defining $K = \Delta F/\Delta l$ as the force constant in a specific direction we obtain K in terms of known quantities, a_0 and C_{ij} the elastic constant along $\langle 111 \rangle$. The elastic constant along $\langle 111 \rangle$ in the cubic system is equal to $(1/3)(C_{11} + 2C_{12} + 4C_{44})$, ΔA is the incremental area of a (lll) plane considered and Δl is the distance between planes in the $\langle 111 \rangle$ direction. A similar calculation along $\langle 000l \rangle$ for the wurtzite structure is made with the simplifying assumption that the distance between (0001) planes is equal.

Lattice characteristics for sphalerite and wurtzite-structures listed in Table 1 and illustrated in Figure 1, 2, and 3 below have been systematically reduced by $1/\sqrt{2}$ for comparison with Debye temperatures.

RESULTS

Figure 1 illustrates the excellent correspondence of melting temperatures

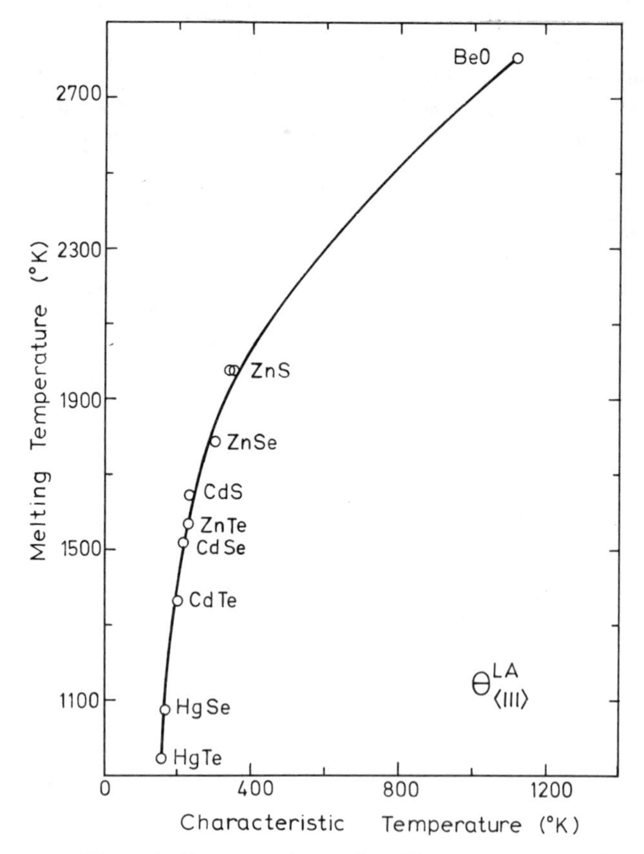

Figure 1. Correspondence of melting temperatures of II–VI compounds.

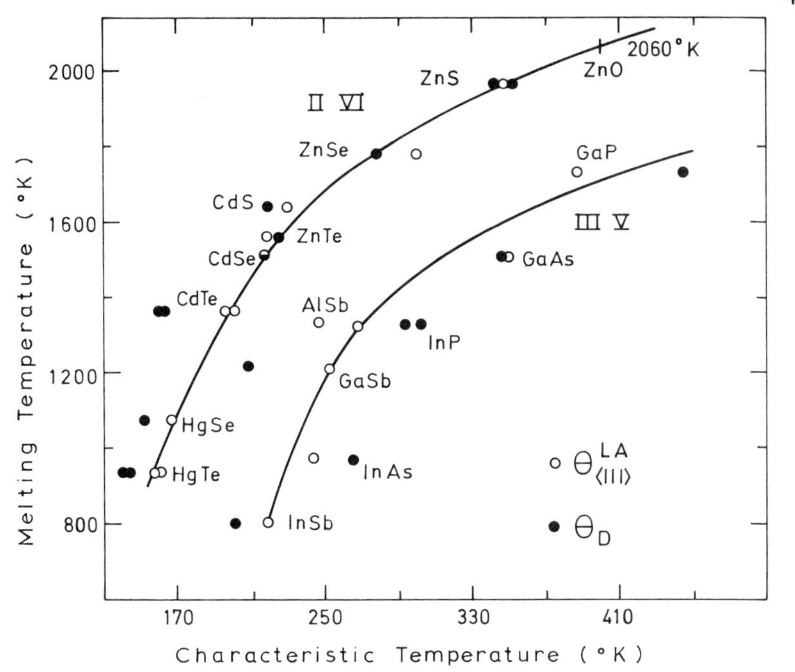

Figure 2. Comparison of scaling with lattice characteristic temperatures *vs.* Debye temperatures.

of II–VI compounds. For wurtzite-structure compounds the ⟨0001⟩ direction was taken as equivalent to the ⟨111⟩ of sphalerite. Figure 2, a similar plot on an extended scale, compares the correspondence obtained with the Debye characteristic. The Debye temperatures, also listed in Table 1, were calculated from the same elastic constant data with the VRH approximation (e.g. Chung (1963); Chung and Buessem (1967); Anderson (1963). References to II–VI elastic constants data are listed in Reeber and McLachlan (1971). Others are from the compilation of Simmons and Wang (1971). Melting points for III–V compounds are from the compilation of Madelung (1964). The melting point of ZnO is predicted to be at 2060 ± 80 °K with a pressure greater than atmospheric. Figure 3 illustrates a similar plot for caesium halides, data for these is from Kubaschewski and Evans (1965) and Simmons and Wang (1971). Data for low temperature thin film transformations are the work of Blackman and Khan (1961). Although the caesium halide thin film results are most probably not representative of equilibrium behaviour they illustrate that the internal energy of the two phases is not greatly different. The CsCl transformation is in general supposed to be relatively rapid, but investigations to be reported elsewhere show that metastability of more than 100 °C can occur in single crystal investigations. The range is indicated on Figure 3. Figure 4 illustrates the complexities of the sphalerite-wurtzite transformation. The lower limit could possibly be interpreted as the approximate 'equilibrium' trans-

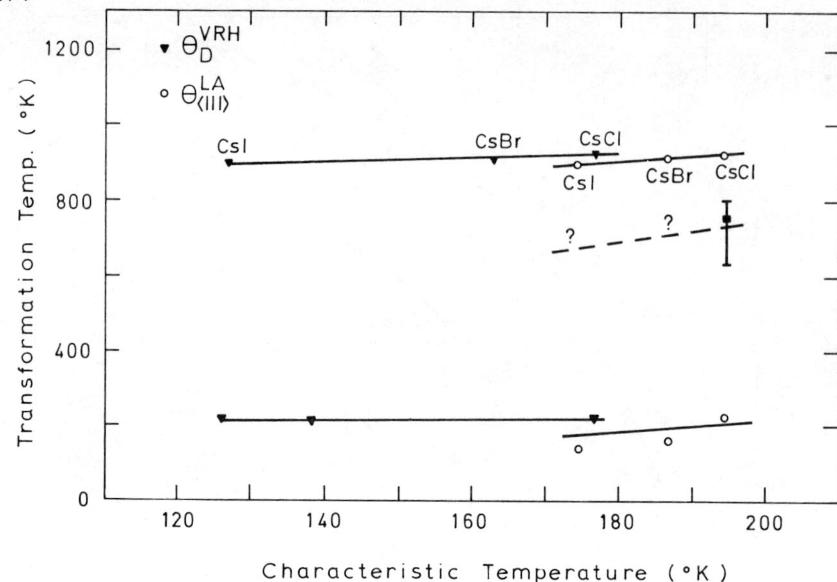

Figure 3. Correspondence of transformation temperatures of caesium halides.

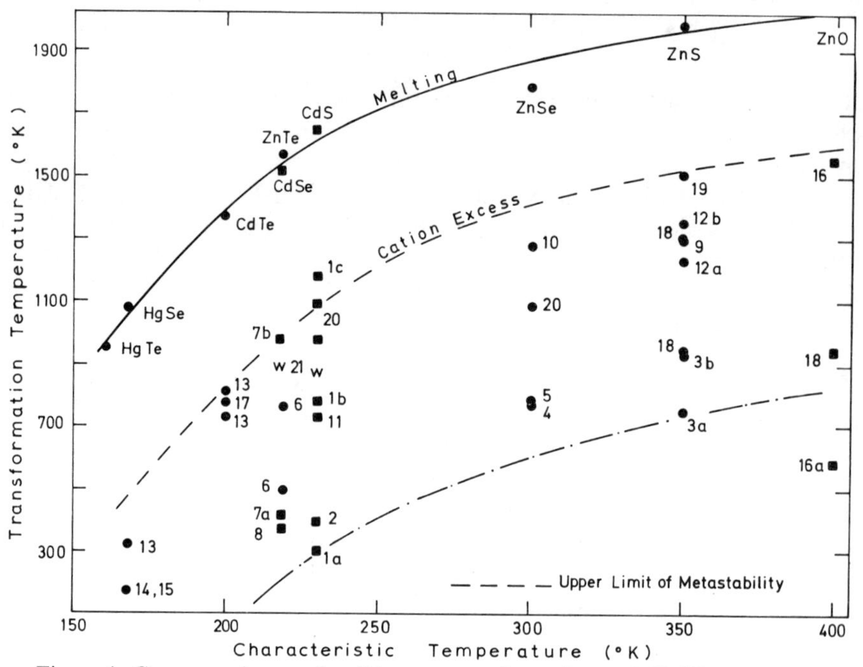

Figure 4. Correspondence of solid-state transformations of II–VI compounds. □ Hexagonal at room temperature ○ cubic at room temperature. – – – – – upper limit of metastability.

References and Notes

1. Rittner, E. S. and Schulman, J. H. (1943). *J. Phys. Chem.*, **47**, 537.
   (a) At 25 °C cubic CdS covered with concentrated solution of ammonium sulphide transformed to hexagonal.
   (b) Same transformation occurred dry after 1 week at 500 °C.
   (c) Same transformation occurred after several minutes at 700–900 °C.
2. Ahlburg, H. and Caines, R. (1962). *J. Phys. Chem.*, **66**, 185.
   Hydrothermal growth of hexagonal CdS at 120 °C.
3. Scott, S. D. and Barnes, H. L. (1972). *Geochim et Cosmochim. Acta*, **36**, 1275.
   (a) Hydrothermal reaction, conversion of ZnS cubic to hexagonal.
   (b) With excess sulphur conversion of ZnS cubic to hexagonal in dry reaction.
4. Korneeva, I. V. (1961). *Sov. Phys. Cryst.*, **6**, 505.
   Decomposition of $ZnSeN_2H_4$ complex at 480 ± 20 °C forming hexagonal ZnSe.
5. Williams, P. M. and Yoffe, A. D. (1972). *Phil. Mag.*, **25**, 247.
   Transformation of prior ion-bombarded hexagonal ZnSe crystal to cubic phase upon anneal in vicinity of 500 °C.
6. Shalimova, K. V., Andrushko, A. F., Spinulescu–Carnaru, I., Seredinskii, B. P. (1965). *Sov. Phys. Cryst.* **9**, 623.
   see also Spinulescu–Carnaru, I. 1966. *Phys. Stat. Sol.*, **18**, 769. Hexagonal ZnTe films evaporated with excess Zn.
7. Pashinkin, A. S. and Sapozhnikov, R. A. (1963). *Sov. Phys. Cryst.*, **7**, 501.
   (a) Noticeable conversion cubic CdSe to hexagonal after 10 hours at 130 °C.
   (b) Complete conversion after 18 hours at 700 °C.
8. Nagata, S. and Agata, K. (1951). *J. Phys. Soc. Japan*, **6**, 523.
   Thin cubic CdSe films become hexagonal after 1 hour heat treatment at 100 °C.
9. Allen, E. T., Crenshaw, J. L. and Merwin, H. E. (1912). *Am. Jour. Science*, **34**, 341.
   Cubic-hexagonal transformation at 1020 °C for ZnS heated in vacuum.
10. Fitzgerald, A. G., Mannami, M., Pogson, E. H. and Yoffe, A. D. (1966). *Phil. Mag.*, **14**, 197.
    Growth of ZnSe hexagonal crystals at 1000 °C.
11. Charbonnier, M. and Murat, M. (1972). *Thermal Analysis* Vol. 2 Ed. Oswald, H. R. and Dubler, E. Birkhaüser Verlag, Basel and Stuttgart, pp 547–557.
    DTA of cubic CdS shows an irreversible and slow transformation between 450 °C and 560 °C with maximum at 536 °C.
12. Lendvay, E. (1971). *J. Cryst. Growth*, **10**, 77.
    (a) Perfect cubic ZnS crystals, with NH_4Cl transport, formed on 950 °C substrate.
    (b) Perfect hexagonal ZnS crystals, with NH_4Cl transport, formed on 1070 °C substrate.
13. Reeber, R. R. DTA and X-ray indication of structural instability in cubic CdTe and HgSe, to be published.
14. Zhdanova, V. V., Lukina, V. I. and Novikova, S. I. (1966). *Phys. Stat. Sol.*, **13**, K19.
    Anomalously large negative thermal expansion in HgSe.
15. Reeber, R. R. small DTA effect.
16. Dereń, J., Nedoma, J. and Nowok, J. (1972). *Zeit. für Krist.*, **136**, 315.
    Cubic crystal of ZnO crystallizing at 573 °K transforms to zincite at 1273 °C.
17. Weinstein, M., Wolff, G. A. and Das. B. N. (1965). *Appl. Phys. Letters*, **6**, 73.
    Hexagonal CdTe film deposited epitaxially on CdS crystal at 500 °C.
18. Secco, E. A. (1958). *J. Chem. Phys.* **29**, 406.
    Anomalous changes in diffusion coefficient of Zn in ZnO and ZnS. (e.g. Azároff, L. V. 1961. *J. App. Phys.*, **32**, 1658; *ibid.* 1663. Structural interpretation of differences in diffusion coefficients).
19. Samelson, H. and Brophy, V. A. (1961). *J. Electrochem. Soc.*, **108**, 150.
    Transition of wurzite to pseudo-cubic phase in vacuum at 1240 °C.
20. Yim, W. M. and Stofko, E. J. (1972). *J. Electrochem. Soc.*, **119**, 381.
    (a) Reaction of Zn vapour and H_2Se gas deposited some hexagonal ZnSe wurtzite on a cubic GaAs substrate at 800 °C.
    (b) Reaction of Cd vapour and H_2S gas deposited CdS, grenockite, on cubic GaAs substrate at 690 and 810 °C.
21. Reeber, R. R. and Kulp, B. A. (1965). *Trans. AIME*, **233**, 698.
    Wurtzite phase of CdS after 2 week heat treatment at 600 °C.
    Reeber, R. R., *PhD dissertation*, The Ohio State University, 1968.
    Wurtzite phase of CdSe after 500-hour heat treatment at 622 °C.

formation temperature in the absence of mechanistic, kinetic and stoichiometric constraints. The diagram does not include all the transformations known in these compounds and several features, particularly in the Hg and Cd compounds, indicate structural instabilities of the cubic structure that may or may not be indicative of a transformation to wurtzite. At present experiments are being done to ascertain the nature of these and several other transformations. The diagram however illustrates in a general way that a correspondence of states principle does qualitatively correlate the available information.

DISCUSSION

An empirical method of correlating solid-state transformations has been illustrated that systematizes to some extent the transformation behaviour of isostructural solids. Complicating factors such as stoichiometry, metastability, changes in the type and number of defects, and the assumptions of the model prevent it from being a quantitative predictor of transformations *per se.* It does seem to be a useful qualitative guide for the experimentalist in that it isolates some of the most important parameters and difficulties attached to the study of a particular transformation. It also illustrates to some extent the interplay between structural arrangement and the dynamics of the lattice. Implications of the relationships with regard to the relative energies of corresponding groups of isostructural compounds will not be given here. Calculations have been made for other sphalerite structure compounds such as SiC. It would appear that with suitable scaling factors these can be plotted on a diagram similar to Figure 4 above. Suitable adaptations of the theory to consider pressure derivatives of elastic constants can possibly provide similar corresponding state relationships at high pressures. Deviations from regularity can possibly give some insight concerning changes in the nature of the interionic bonding forces. At present experiments are under way to clarify further the phase relations illustrated in this paper and also the calculations are being extended to other structures. It is felt that a bit more justification can be given to Einstein's comments concerning Lindemann's results, in quote: 'Es scheint daraus hervorzugehen, daß das Gesetz der übereinstimmenden Züstande für einfache Körper in festen und flüssigen Züstande mit bemerkenswerter Annäherung gilt.'

ACKNOWLEDGEMENT

The author would like to acknowledge the support of the U.S.–German Fulbright Commission and the Institut für Kristallographie, T. H. Aachen for the initial phases of this work and to Dr. J. D. C. McConnell and the Natural Environmental Research Council for the opportunity to continue it. Helpful discussions with Professor M. Blackman and Dr. McConnell are gratefully acknowledged.

REFERENCES

Anderson, O. L. (1963). *J. Phys. Chem. Solids*, **24**, 909.
Blackman, M. (1955). *Handbuch Der Physik VII (1)*, Springer–Verlag, Berlin, p. 374.
Blackman, M. and Khan, I. H. (1961). *Proc. Phys. Soc.*, **77**, 471.
Chung, D. H. (1963). *Phil. Mag.*, **8**, 833.
Chung, D. H. and Buessem, W. R. (1967). *J. Appl. Phys.*, **38**, 2535.
Dolling, G. and Cowley, R. A. (1966). *Proc. Phys. Soc.*, **88**, 463.
Einstein, A. (1911). *Annal. der Phys.*, **35**, 679.
Gerlich, D. (1967). *J. Phys. Chem. Solids*, **28**, 2575
Hewat, A. W. (1972). *J. Phys. C.*, **5**, 1309.
Kubaschewski, O. and Evans, E. Ll. (1965). *Metallurgical Thermochemistry*, 4th Edition, Pergamon Press, London and New York.
Lindemann, F. A. (1910). *Physik. Zeit.*, **11**, 609.
McConnell, J. D. C. (1972). *The Feldspars*, Proceedings of a NATO Advanced Study Institute, Manchester 1972, p. 460. Manchester University Press, Manchester 1974.
Madelung, O. (1964). *Physics of III–V Compounds*, John Wiley and Sons, New York.
Narita, K. Watanabe, H. and Wada, M. (1970). *Japan J. Appl. Phys.*, **9**, 1278.
Reeber, R. R. and McLachlan, D. (1971). *Canad. J. Phys.*, **49**, 2287.
Simmons, G. and Wang, H. (1971). *Single Crystal Aggregate Properties:* A Handbook MIT Press, Cambridge, Massachusetts and London, England.
Van Vechten, J. A. (1972). *Phys. Rev. Let.*, **29**, 769.
Vekilov, Yu. Kh. and Rusakov, A. P. (1971). *Fiz. Tver. Tela*, **13**, 1157; *Sov. Phys. Solid State*, **13**, 956, (1972).

Potassium in the Earth's Core: Evidence and Implications

K. A. Goettel
Department of Earth and Planetary Sciences
Massachusetts Institute of Technology
Cambridge, Massachusetts 02139

1. INTRODUCTION

Many previous authors, including Ringwood (1966), Gast (1960, 1972), Wasserburg *et al.* (1964), and Hurley (1968), have concluded that the Earth is markedly depleted in potassium relative to ordinary chondrites. However, recent progress (Larimer, 1967; Larimer and Anders, 1967; Lewis, 1972, 1973; Grossman, 1972) in understanding the condensation of planetary constituents from the primitive solar nebula has necessitated a re-examination of models for the chemical composition of the Earth. Lewis (1972, 1973) presented data in support of a model based on chemical equilibrium during the condensation process; this model predicts that the Earth has an Fe–FeS core and the full solar or chondritic K/Si ratio.

The purpose of the present paper is to review critically evidence bearing on the Earth's total potassium content and evidence bearing directly on the question of potassium in the Earth's core.

2. CHEMICAL MODELS FOR THE EARTH

The chemical equilibrium model (Lewis, 1972, 1973) appears to be the most realistic model for the formation of the terrestrial planets. The quantitative predictions of this model are consistent with the observed mean densities of Mercury, Venus, Earth, and Mars, and with existing data on the chemical composition of the terrestrial planets. In this model, differences in the major and trace element content, and in the content of volatiles among the terrestrial planet are primary features resulting directly from differences in condensation temperature. This model predicts that the Earth has the solar proportions of nearly all of the rock-forming elements, including refractories (Ca, Al, Ti, U, Th, etc.), Fe, Ni, Si, Mg, S, Na, and K. The Earth is thus predicted to have an

Fe–FeS core and the full solar K/Si ratio. There is not sufficient potassium in the crust and upper mantle for the entire Earth to have the solar K/Si ratio; therefore the chemical equilibrium model implies that a substantial fraction of the Earth's potassium must be in the lower mantle or in the core. The presence of sulphur in the Earth's core is particularly significant, since the arguments for potassium in the core are based on the chalcophile behaviour of potassium.

Strong, independent evidence for sulphur in the Earth's core was presented by Murthy and Hall (1970, 1972). These authors noted that an Fe–FeS melt will be the first melt formed in the early Earth; because the Fe–FeS eutectic temperature is so low (988 °C) and is virtually independent of pressure (Brett and Bell, 1969), core formation can occur very early in the Earth's history without necessitating a very hot thermal history for the Earth. Murthy and Hall compared the abundances of a number of volatile elements in the Earth's crust and upper mantle to the abundances of these volatiles in several meteorite classes. Sulphur was found to be depleted in the Earth's crust and mantle by a larger factor than H_2O, the halogens, or the rare gases. Segregation of sulphur into the core, as predicted by the chemical equilibrium model for the Earth, readily explains this anomalous apparent depletion of sulphur relative to other, more volatile elements.

Inhomogeneous accretion models (Turekian and Clark, 1969; Clark *et al.*, 1972) assume that planets accrete during the condensation process in the solar nebula. Planets accrete with initially layered structures: refractory condensates form a proto-core, followed by an Fe–Ni layer, a magnesium silicate layer, and finally a veneer of low temperature, volatile-rich condensates. However, this type of model cannot match the observed mean densities of all of the terrestrial planets without resorting to secondary processes such as *ad hoc* major element fractionations among the planets. The four orders of magnitude difference in the H_2O/CO_2 ratio of Earth and Venus is not compatible with the retention of volatiles as a late-stage veneer as postulated by the inhomogeneous accretion models; none of the carbonaceous chondrites exhibit such a drastic variation in H/C ratio. This type of model also fails to produce the required light element in the Earth's core, since neither S nor Si is predicted to be in the core, and these are the only two elements which have the right properties and are sufficiently abundant (Ringwood, 1966). Several other important aspects of the Earth's composition, including the mantle's FeO content and the patterns in the relative abundances of volatiles discussed by Murthy and Hall are difficult to reconcile with the predictions of inhomogeneous accretion models. The present author concludes that these models are not a realistic representation of the formation of the terrestrial planets.

Ganapathy and Anders (1974) have developed a chemical model for the Earth, based on the assumption that the Earth formed by the same processes as chondrites. Condensation and remelting in the solar nebular results in seven components: early condensate; remelted and unremelted Fe–Ni–Co alloy and (Fe,Mg)-silicates; FeS; and a carbonaceous, volatile-rich silicate. The proportions of these components in the Earth were estimated by applying assumed geochemical constraints such as K/U ratio, bulk U and Fe content, etc. Several other planetary models have been constructed from a mixture of meteorite types with the proportions adjusted to match the mean density and chemical composition of a planet. However, Ganapathy and Anders commented that the arbitrary, *ad hoc* nature of this approach is aesthetically objectionable. This same criticism may be applied to the model proposed by Ganapathy and Anders. Their model is completely flexible: the relative proportions of their seven components can be arbitarily varied to match the observed chemical composition of a planet. Another problem for this model is the great difference in H_2O/CO_2 ratio observed for Earth and Venus. In this model, volatiles are contained in a low temperature, carbonaceous silicate component; however, none of the carbonaceous chondrites displays such a large variation in H/C ratio. By a suitable choice of the proportions of the seven components, the model of Ganapathy and Anders could be adjusted to predict a composition for the Earth which is very similar to the composition predicted by the chemical equilibrium model. Other models such as the simple chondritic model, or models based on mixtures of meteorites also predict quite similar compositions. However, because the chemical equilibrium model predicts the correct densities and composition for each of the terrestrial planets without any *ad hoc* assumptions, the present author prefers the chemical equilibrium model.

The carbonaceous chondrite model proposed by Ringwood (1966) is the only model which unambiguously predicts that the Earth is markedly depleted in potassium relative to the potassium content of chondrites. In this model, each of the terrestrial planets has an initial composition similar to the composition of Type 1 carbonaceous chondrites. Density and composition differences among the terrestrial planets are attributed to varying degrees of reduction and volatile element loss. However, this model is incompatible with many observational data. The postulate that each of the terrestrial planets had initial compositions similar to carbonaceous chondrites is not compatible with current theories requiring an adiabatic temperature gradient in the primitive solar nebula (Cameron and Pine, 1973; Lewis, 1974); the steep temperature gradient in the nebula must produce distinct composition differences among the terrestrial planets. In Ringwood's model, *in situ* reduction of iron oxides to form planetary cores results in the generation of a massive, primitive atmosphere. The escape of these primitive atmospheres, a basic requirement of the carbonaceous chondrite model, is a very implausible event for which no quantitative theory has ever been produced. This model even fails to explain the observed densities of the terrestrial planets. The model attempts to explain the high density of Mercury by invoking a Hayashi phase of extreme solar super-luminosity to evaporate silicates from Mercury; however, current theories of stellar evolution (Larson, 1974) conclude that the Hayashi phase does not occur. The low density of Mars is explained by assuming that the primordial material for Mars contained little carbonaceous material. However, for this assumption to be valid, the entire solar system, from Mercury to the asteroid belt, except for Mars, would have to be rich in carbonaceous material; the implausibility of this postulate seems evident. The volatile content of the Earth, including the abundance of H_2O, the presence of radiogenic ^{129}Xe (Boulos and Manuel, 1971) and the patterns in volatile abundances discussed by Murthy and Hall (1970, 1972) are all very difficult to reconcile with the high temperature *in situ* reduction and degassing required by the carbonaceous chondrite model for the Earth. The basic postulates of this model appear to be highly implausible, and the model cannot explain the observed mean densities and volatile contents of the terrestrial planets. The present author concludes that this is not a viable model for the Earth. and it will not be considered further.

Notwithstanding the strong evidence favouring the chemical equilibrium model for the Earth, the conclusion that the Earth has a solar K/Si ratio is *not* entirely dependent on the assumptions of the chemical equilibrium model; in fact, departures from the chemical equilibrium model do not reduce the Earth's predicted potassium content. In the chemical equilibrium model, potassium condenses as alkali feldspar near 1000 K. In the inhomogeneous accretion model, potassium condenses at a lower temperature as the oxide or perhaps as the sulphide because of the stability of K_2S. However, the requirement that the formation temperature of the Earth be low enough for retention of volatiles means that the temperature must be low enough for complete condensation of potassium. Therefore, even if the assumption of chemical equilibrium during the condensation process is not entirely valid, the Earth is still predicted to have the full solar K/Si ratio. Other models, including the model of Ganapathy and Anders, the simple chondritic model, or models based on mixtures of meteorites, are also compatible with the Earth having the solar K/Si ratio. However, the chemical equilibrium model predicts nearly the same composition for the Earth without any *ad hoc* assumptions, and is thus the preferred model. The conclusion that the Earth has the solar K/Si ratio thus appears to be a very firm conclusion.

An extensive discussion of the evidence supporting the chemical equilibrium model for the Earth, and the evidence against inhomogeneous accretion models, the model of Ganapathy and Anders, and the carbonaceous chondrite model has been presented elsewhere (Goettel, 1976a).

3. POTASSIUM CONTENT OF THE EARTH

One of the classic arguments supporting the chondritic model for the Earth is the 'chondritic coincidence': the present heat production in an Earth with the chondritic or solar proportions of K, U, and Th closely matches the observed mean heat flow of the Earth (Urey, 1956; Hurley, 1957; Birch, 1958). The concentrations of K, U, and Th in the crust and upper mantle in continental regions are sufficient to produce the observed heat flow; however, in oceanic regions the concentrations of these elements are far too low to explain the observed heat flow. The recent revision of the mean heat flow in oceanic areas by Williams and Von Herzen (1974) has included the important contribution from the cooling lithosphere, and thereby greatly increased the estimated heat flow in oceanic areas, and the total heat flow of the Earth. This higher heat flow estimate compounds the difficulty which models predicting less than the solar proportion of potassium in the Earth have in explaining terrestrial heat flow data. Terrestrial heat flow data support the predictions of the chemical equilibrium model: (1) the Earth has the solar K/Si ratio, and (2) a substantial fraction of the Earth's potassium must be in the deep interior.

The K/U ratio in terrestrial rocks (Wasserburg *et al.*, 1964), Sr isotope evidence (Gast, 1960), and Ar degassing studies (Hurley, 1968) have all been interpreted as confirmation that the Earth, relative to chondrites, is strongly depleted in K and Rb. These data demonstrate convincingly that sufficient K and Rb are not contained in the crust and upper mantle for the entire Earth to have solar K/Si and Rb/Si ratios. However, the interpretation that these data may be applied to the entire Earth is based on overly restrictive assumptions about the geochemical behaviour of K and Rb. If a substantial fraction of the Earth's total K and Rb are partitioned into the core at the time of core formation very early in the Earth's history, as suggested by Lewis (1971) and Hall and Murthy (1971), then the Earth may still have the full solar proportions of K and Rb. In this case, a substantial fraction of the Earth's K, Rb, and radiogenic Ar would be effectively decoupled from the crust and mantle. Therefore, the K/U data, Sr isotope evidence, and Ar degassing studies do not contradict the basic prediction of the chemical equilibrium model: The Earth has the full solar K/Si ratio.

4. EVIDENCE FOR POTASSIUM IN THE CORE

(a) Stability of K_2S

Lewis (1971) emphasized that several elements, including potassium, which are normally considered lithophile have appreciable chalcophilic properties. Reactions of the form

$$K_2O + H_2S = K_2S + H_2O \quad (1)$$

were considered for the purpose of portraying the relative stabilities of pure sulphides and oxides. At 1000 °K, the equilibrium constant for reaction (1) is $10^{14 \cdot 8}$, this high value demonstrates the stability of K_2S and confirms that potassium may indeed have appreciable chalcophilic properties. If sodium is substituted for potassium in reaction (1), the equilibrium constant at 1000 °K is $10^{9 \cdot 57}$; thus Na appears to be significantly less chalcophile than K. K is depleted in the Earth's crust and upper mantle relative to chondrites, but Na is not depleted; the explanation of this difference in the geochemical behaviour of K and Na may be that K, but not Na, is partitioned into Fe–FeS melts at the time of core formation.

Hall and Murthy (1971) considered a variety of chemical reactions representative of both the present oxidized crust and the more reducing conditions existing during core formation. Data for reactions of the form

$$MeSO_4 + FeS = MeS + FeSO_4 \quad (2)$$

$$MeCO_3 + FeS = MeS + FeCO_3 \quad (3)$$

which reflect the highly oxidized present crust, suggest that all of the alkali metals will be lithophile under crustal conditions. However, data for reactions of the form

$$MeSiO_3 + FeS = MeS + FeSiO_3 \quad (4)$$

$$\tfrac{1}{2}Me_2SiO_4 + FeS = MeS + \tfrac{1}{2}Fe_2SiO_4 \quad (5)$$

which reflect the more reducing conditions existing during core formation, suggest that Li and Na remain lithophile, but that K, Rb, and Cs may be chalcophile and thus partitioned into the core. Uncertainty about the effects of Al and high pressure, and the paucity of relevant thermodynamic data preclude direct calculation of the amount of K expected in the core.

(b) Metallurgic Evidence

Chalcophile behaviour of potassium has been documented in metallurgic systems. Cissarz and Moritz (1933) analysed co-existing metal, sulphide, and silicate phases from the smelting of the Mansfeld copper ores. The sulphide matte contains 0.8 per cent K_2S; extremely low abundances of Si (0.05 per cent) and Al (0.01 per cent) in the sulphide matte indicate that the high K_2S value cannot be due to contamination by K-bearing silicates. The co-existing silicates contain 17 per cent Al_2O_3 which demonstrates that the presence of alumino-silicates does not preclude the formation of K_2S. The silicates also contain 4 per cent FeO; the oxygen fugacity is probably somewhat below the Fe–FeO buffer and is thus compatible with the oxygen fugacity expected in a differentiating, approximately chondritic Earth.

A recent microprobe study of blast furnace byproducts (vom Ende *et al.*, 1966) provided additional data on potassium in sulphide phases. A K–Fe–S compound and an alkali–Mn–S compound were identified as constituents of the lowest temperature melt. Textural evidence indicated that the K-bearing sulphide and the Mn sulphide crystallized from a common FeS-rich melt. These data varify the plausibility of partitioning potassium into an Fe–FeS melt. In both metallurgic systems discussed, K is several times more abundant than Na in the sulphide phases.

(c) Meteoritic Evidence

Djerfisherite, a potassium–iron–sulphide, has been discovered in several enstatite chondrites and achondrites, and in a sulphide nodule within an iron meteorite (Fuchs, 1966; El Goresy *et al.*, 1971). In enstatite chondrites, djerfisherite occurs in the silicate matrix and is always associated with troilite, and usually with both troilite and kamacite; troilite inclusions were observed in some grains. These observations suggest that djerfisherite may crystallize from an FeS-rich melt. In the Bishopville achondrite, a djerfisherite-troilite-kamacite inclusion was found in sodic plagioclase; this striking observation demonstrates the chalcophile behaviour of potassium even in the presence of feldspar. Although oxygen fugacities in enstatite meteorites are somewhat lower than expected for a differentiating, approximately chondritic Earth, the occurrence of the potassium sulphide djerfisherite is significant because it documents the chalcophile behaviour of potassium in meteoritic systems which have overall chemical compositions similar to the composition of the Earth. Djerfisherite contains some sodium substituting for potassium; however, the maximum Na content measured by El Goresy *et al.* was less than 10 per cent of the K content, and the Na content of many grains was below detection limits.

Shima and Honda (1967) measured the potassium content of component minerals of the Abee enstatite chondrite. They found 3.7 per cent of the total K in oldhamite (CaS) and 3.1 per cent in troilite. Oversby and Ringwood (1972) argued that the amount of K observed in FeS in Abee places an upper limit of about 2.5 per cent of the Earth's K in the core, since Abee has a greater proportion of FeS than the Earth and the Earth is more oxidized than Abee. However, this argument is not valid because the Abee data represent low temperature partitioning of potassium between solid silicates and solid FeS. The relevant partitioning for the Earth involves an Fe–FeS melt and solid silicates at temperatures hundreds of degrees above the Abee equilibration temperature (Murthy and Hall, 1970, 1972; Lewis, 1971). Thus, the Abee data do not place an upper limit on the amount of potassium in the Earth's core.

(d) Experimental Results

Goettel (1972) investigated the partitioning of potassium between solid roedderite ($K_2Mg_5Si_{12}O_{30}$) and an Fe–FeS melt. Roedderite and merrihueite (the Fe analogue) are the only two silicates containing essential potassium which have been identified in chondrites; therefore, roedderite is a reasonable choice for one of the potassium bearing phases in the early Earth. At 1030 °C, a K_2S/FeS weight ratio of $(3.340 \pm 0.015) \times 10^{-3}$ was measured; if the Earth has a solar K/Si ratio, a K_2S/FeS weight ratio of about 9×10^{-3} would suffice to extract all of the Earth's potassium into the core. The potassium content of the quenched Fe–FeS melt was found to increase significantly with increasing temperature. The oxygen fugacity in these experiments was somewhat below the Fe–FeO buffer and thus compatible with a differentiating, approximately chondritic Earth. Uncertainty about the identity of the K-bearing phases in the early Earth, and uncertainty about the effect of pressure on potassium partitioning prevent calculation of the K content of the Earth's core from these data. However, these data certainly demonstrate the plausibility of partitioning significant amounts of K into Fe–FeS melts at the time of core formation.

Partial melting experiments are presently being conducted on samples of the Forest City H5 chondrite to examine the partitioning of potassium between the Forest City silicates and Fe–FeS melts. The composition predicted for the Earth by the chemical equilibrium model is similar to the composition of ordinary chondrites. Therefore, these Forest City experiments were undertaken as a first-order approximation to the primary differentiation of the Earth into an Fe–FeS core and a silicate mantle. Preliminary results have confirmed the presence of potassium in the quenched Fe–FeS melts, but reliable, quantitative results are not yet available. The results of these experiments and results of an investigation of potassium partitioning between alkali feldspar and Fe–FeS melts have been presented elsewhere (Goettel, 1975, 1976b).

(e) Evidence Against Potassium in the Earth's Core

Oversby and Ringwood (1972, 1973) presented data on the partitioning of potassium between basaltic liquids and Fe–FeS melts, and argued against potassium in the Earth's core. However, experiments involving differentiated basaltic material are not particularly relevant to the partitioning of potassium at the time of core formation. The basaltic composition used by Oversby and Ringwood contained 13 per cent Al_2O_3; this proportion of Al_2O_3 greatly exceeds the proportion of Al_2O_3 in an Earth with a solar or chondritic Al/Si ratio, and strongly biases the experiment in favour of alkali alumino-silicate formation. The chemical and physical conditions which are relevant to the partitioning of potassium into the Fe–FeS core are the conditions which existed during the primary differentiation of the Earth into metal-sulphide core and silicate mantle, and not the conditions presently existing in the crust or upper mantle.

Oversby and Ringwood examined the partitioning of potassium between an Fe–FeS melt and a silicate liquid. However, at a pressure of even a few tens of kilobars, an Fe–FeS melt will form at temperatures several hundred degrees below the solidus temperature for mantle silicates. Therefore, experiments relevant to the partitioning of potassium into the core almost certainly involve solid silicates rather than silicate liquids. Hall and Murthy (1971) noted that K, Rb, and Cs are selectively enriched in the liquid phase during partial melting of silicates because their large ionic size makes them incompatible in many mineral structures; this ionic size incompatibility may be a major factor in the partitioning of these elements into Fe–FeS melts at the time of core formation.

Seitz and Kushiro (1974) conducted melting experiments on partially reduced samples of the Allende Type 3 carbonaceous chondrite. At 1350 °C and 20 kilobars, the silicate melt contained 0.1 per cent K_2O, while K in the sulphide melt was below the detection limit of 0.01 per cent. At 20 kilobars, the silicate solidus of the partially reduced Allende material is more than 300 °C above the Fe–FeS eutectic temperature at this pressure. As discussed in the preceding paragraph, experiments relevant to potassium in the Earth's core almost certainly involve partitioning of potassium between an Fe–FeS and solid silicates. Therefore, these experimental data may be of only marginal significance with respect to the potassium content of the Earth's core.

5. IMPLICATIONS OF POTASSIUM IN THE CORE

Two distinct periods in the geochemical differentiation of the Earth must be considered: (1) the primary differentiation into metal–sulphide core and silicate mantle, and (2) the subsequent differentiation of the silicate mantle. The geochemical behaviour of many elements, most notably potassium, may be markedly different in these two differentiation processes. Elements with significantly chalcophilic or siderophilic properties will be preferentially partitioned into the Fe–FeS core at the time of core formation very early in the Earth's history. The elements partitioned into the core may include K, Rb, and Cs, as well as many other elements such as Ni, Co, Pt-group metals, Pb, Cu, Ag, Au, Cd, Bi, etc. which are normally considered siderophile or chalcophile.

A major consequence of the partitioning of these elements into the core is that the composition of the Earth may be very similar to the composition of ordinary chondrites; or in the context of the chemical equilibrium model, that the Earth has the solar proportions of nearly all of the rock forming elements including refractories (Al, Ca, Ti, Ba, Sr, U, Th, rare earths, etc.), Fe, Ni, Mg, Si, S, Na, and K. Geochemical data such as K/U ratios, Sr isotope evidence, K–Rb–Ar data, and U/Pb isotope evidence must be interpreted in the context of the two-stage differentiation history of the Earth. Inferences drawn from the chemistry of the crust and upper mantle must be extrapolated very cautiously if they are to be extended to the entire mantle or the entire Earth.

The Fe–FeS eutectic temperature is so low (988 °C) and increases so slowly with pressure that core formation would occur very early in the Earth's history even if the Earth retained virtually none of its accretional gravitational potential energy; adiabatic compression alone is probably sufficient to initiate core formation. At high pressures, the solidus temperature for mantle silicates is so much higher than the Fe–FeS eutectic temperature that core formation may not necessitate large scale mantle melting.

Most previous thermal history calculations have assumed that K, U, and Th are geochemically coherent during the Earth's differentiation; partitioning of K into the core results in a major redistribution of the radiogenic heat sources in the Earth. One consequence of the postulated solar K/Si ratio for the Earth and the short half life of ^{40}K is that a large fraction of the total radiogenic heat production of the Earth is released early in the Earth's history. Low-temperature core formation and the possibility of potassium in the core necessitate a complete revision of existing thermal history models for the Earth.

Decay of ^{40}K in the core will provide a major internal heat source to drive core convection. For a K content of 0.1 per cent, the rate of heat production is about 2×10^{12} cal/sec (Verhoogen, 1973); this K content is about 40 per cent of the total K in an Earth with the solar K/Si ratio. The rate of heat production by ^{40}K is likely to be several orders of magnitude greater than the rate of dissipation of magnetic energy in the core; even if the efficiency of conversion of thermal of magnetic energy is very low, it appears that there will be sufficient energy from ^{40}K to maintain the geomagnetic field. Heat from ^{40}K in the core will produce a large heat flux through the mantle; this heat is an energy source to drive mantle convection and continental drift.

Despite the high heat flux through the lower mantle from ^{40}K in the core, the lower mantle may not convect. The recent data of Mao (1973) and Mao and Bell (1972) indicate that the thermal conductivity of silicates may increase by several orders of magnitude at high pressures. From simple geometric considerations, the heat flux per unit surface area in the mantle will decrease as the square of the distance from the core/mantle interface. The data of Mao and Bell suggest that the thermal conductivity of the mantle may decrease more rapidly than the heat flux. The transition between the lower, possibly nonconvecting mantle, and the upper, convecting mantle may be at the depth where the thermal conductivity has become low enough for the adiabatic gradient to be exceeded. Because of the short half life of ^{40}K, the heat flux through the lower mantle decreases with time. Therefore, the depth in the upper mantle to which convection extends may be decreasing with time. If this speculative hypothesis is valid, it may have important implications for paleo-continental drift, since the depth and rate of upper mantle convection will be strongly time dependent.

ACKNOWLEDGEMENTS

I am grateful to Professor John S. Lewis for many helpful discussions. Thus work was supported in part by NASA Grant NGL-22-009-521.

REFERENCES

Birch, F., (1958), *Bull. Geol. Soc. Amer.*, **69**, 483.
Boulos, M. S. and Manuel, O. K., (1971), *Science*, **174**, 1334.
Brett, P. R. and Bell, P. M., (1969), *Ann. Rep. Geophys. Lab Yearbook*, **67**, 198.
Cameron, A. G. W. and Pine, M. R., (1973), *Icarus*, **18**, 377.
Cissarz, A. and Moritz, H., (1933), *Metallwirtschaft*, **12**, 131.
Clark, S. P., Turekian, K. K., and Grossman, L. (1973), in: *The Nature of the Solid Earth*, ed. E. C. Robertson, McGraw-Hill, pp. 3–18.
El Goresy, A., Grogler, N., and Otterman, J., (1971), *Chem. Erde*, **30**, 77.
Fuchs, L. H., (1966), *Science*, **153**, 166.
Ganapathy, R and Anders, E., (1974), in: *Proceedings of the Fifth Lunar Science Conference*, **2**, 1181.
Gast, P. W., (1960), *J. Geophys. Res.*, **65**, 1287.
Gast, P. W., (1972), in: *The Nature of the Solid Earth*, ed. E. C. Robertson, McGraw-Hill, pp. 19–40.
Goettel, K. A., (1972), *Phys. Earth Planet. Int.*, **6**, 161.
Grossman, L., (1972), *Geochim. Cosmochim. Acta*, **36**, 597.
Goettel, K. A., (1975), *Eos, Trans. Amer. Geophys. Un.*, **56**, 387.
Goettel, K. A., (1976a), *Geophys. Surveys*, **2**, in the press.

Goettel, K. A., (1976b), *Geochim. Cosmochim. Acta*, submitted.
Hall, H. T. and Murthy, V. R. (1971), *Earth Planet. Sci. Lett.*, **11**, 239.
Hurley, P. M., (1957), *Bull. Geol. Soc. Amer.*, **68**, 379.
Hurley, P. M., (1968), *Geochim. Cosmochim. Acta*, **32**, 1025.
Larimer, J. W., (1967), *Geochim. Cosmochim. Acta*, **31**, 1215.
Larimer, J. W. and Anders, E., (1967), *Geochim. Cosmochim. Acta*, **31**, 1239.
Larson, R. B., (1974), *Fund. Cosmic Phys.*, **1**, 1.
Lewis, J. S., (1971), *Earth Planet. Sci. Lett.*, **11**, 130.
Lewis, J. S., (1972), *Earth Planet. Sci. Lett.*, **15**, 286.
Lewis, J. S., (1973), *Ann. Rev. Phys. Chem.*, **24**, 339.
Lewis, J. S., (1974), *Science*, **186**, 440.
Mao, H. K., (1973), *Carnegie Institution Yearbook*, **72**, 557.
Mao, H. K. and Bell, P. M., (1972), *Science*, **176**, 403.
Murthy, V. R. and Hall, H. T., (1970), *Phys. Earth Planet. Int.*, **2**, 276.
Murthy, V. R. and Hall, H. T., (1972), *Phys. Earth Planet. Int.*, **6**, 123.
Oversby, V. M. and Ringwood, A. E., (1972), *Earth Planet. Sci. Lett.*, **14**, 345.
Oversby, V. M. and Ringwood, A. E., (1973), *Earth Planet. Sci. Lett.*, **18**, 151.
Ringwood, A. E., (1966), *Geochim. Cosmochim. Acta*, **30**, 41.
Seitz, M. G. and Kushiro, I., (1974), *Science*, **183**, 954.
Shima, M. and Honda, M., (1967), *Geochim. Cosmochim. Acta*, **31**, 1995.
Turekian, K. K. and Clark, S. P. (1969), *Earth Planet. Sci. Lett.*, **6**, 346.
Urey, H. C., (1956), *Proc. Natl. Acad. Sci. U.S.*, **42**, 889.
vom Ende, H., Grebe, K., and Schmidt, B., (1966), *Arch. Eisenhüttenw.*, **37**, 433.
Wasserburg, G. J., MacDonald, G. J. F., Hoyle, F., and Fowler, W. A., *Science*, **143**, 465.
Williams, D. L. and Von Herzen, R. P., (1974), *Geology*, **2**, 327.

*Electronic Transition in Iron and the Properties of the Core**

M. Bukowinski and L. Knopoff
Institute of Geophysics and Planetary Physics
University of California, Los Angeles

I. INTRODUCTION

The postulate that the inner core of the earth is solid has remained virtually unchallenged since the discovery of the inner core by Lehmann (1936), Gutenberg and Richter (1938) and Jeffreys (1939a, b). Loosely, the argument for the solidity of the inner core rested on the fact that nothing could be thought of that was more dense than the liquid iron of the outer core, and at the same time as abundant cosmically as the solid phase of iron. Hence, the temperature and pressure at the inner core-outer core boundary were presumed to coincide with the melting curve of iron (Birch, 1940; Jacobs, 1953, 1954).

Until recently, some aspects of the model of solidity of the inner core presented some points of uncertainty. Among these early problems were the facts that (1) searches for the phase PKJKP were fruitless (Burke-Gaffney, 1953; Hutchinson, 1955; Bullen, 1956); (2) precise estimates of the melting temperature of iron at inner-core pressures and independent estimates of the temperatures in the core were unknown. Hence, a check on the hypothesis of the coincidence of these two temperatures was not possible. Indeed, the hypothesis has been used to estimate temperatures in the core. Estimates of the melting temperature of iron at these pressures varied widely from 2,000 to 10,000 K; (3) there were some models of the core which proposed that the outer core be molten silicate (Ramsey, 1949, 1950; Bullen, 1952). In this case, the chemical equilibrium between the outer and inner cores disappears and the temperature of the boundary, if the inner core is indeed iron, can be lower than the melting point. In a variation of this theme, MacDonald and Knopoff (1958) and others proposed, from interpretations of shock-wave data, that the outer core had a density greater than that of silicates and less than that of iron. If the inner core is solid iron, the chemical disequilibrium between the two regions exists again, and once again the temperature can be lower than the melting point.

* Publication No. 1362

Some of these doubts were subsequently removed, as we shall demonstrate below.

Perhaps the only counter-proposal to the concept of solidity was given by Cole (1957), who suggested that the inner core instead might be a liquid of the same composition as the outer core. The inner core was presumed to be the seat of the earth's magnetic field, and was also presumed to be a very good conductor, stiffened by the self-same magneto–hydrodynamic effects that produce the earth's magnetic field. Shortly thereafter, Knopoff and MacDonald (1958) showed, among other arguments, that if the outer core is also a good conductor, it would act as a shield for fluctuations in the magnetic field produced in the inner core. Hence, the fluctuations in the field observed at the surface must have their origin in a field in the inner core which is very large indeed; so large as to preclude the likelihood that the inner core was the seat of the earth's magnetic field.

As we have indicated, some of the uncertainties in the properties of both the inner and outer cores have been allayed by recent experimental developments. First, and most important, is the discovery that the inner core can support shear motions. This has been demonstrated by two methods: the elusive phase PKJKP has been identified (Julian *et al.*, 1972), and models of the earth which fit the spectrum of free oscillations, and especially radial overtones of higher order (Dziewonski and Gilbert, 1971, 1972), require that the inner core have a non-zero rigidity.

With regard to the temperature at the inner-outer core boundary, Higgins and Kennedy (1971) have investigated the systematics of melting conditions as a function of strain for a wide variety of materials in the laboratory and conclude that the melting point of iron at the inner core boundary is near 4000 K. The justification for making the seemingly large extrapolatory jump from the limits of laboratory pressures to inner core pressures is that the change in strain between the two pressures is not really very large.

Recent shock wave experiments on silicates as well as on iron, in conjunction with improved estimates of the density of the outer core from inversion of free oscillation data, have confirmed the earlier proposal: that the density of the outer core is too great for silicates and too small for pure iron or iron-nickel.

However, the resolution of some of these problems has introduced a new suite of problems that require understanding and possible explanation. The first and most prominent of the new dilemmas is that the shear properties of the inner core are rather unusual in comparison with ordinary solids at surface conditions. Specifically, the value of Poisson's ratio σ for the inner core is around 0.45, an extraordinarily high value for solids at the surface of the earth, except for some organic polymers and for glasses close to their melting point.

Second, the determination of the temperature at the top of the inner core has led to an implied conclusion that an extremely low temperature gradient exists in the outer core, one sufficiently low to be considerably less than the adiabatic temperature gradient. If the earth's magnetic field originates due to magnetohydrodynamic processes in the outer core, and if these in turn are due to thermally driven convection in the core, then the temperature gradient in the outer core must be equal to, or slightly greater than, the adiabatic gradient. Thus, the Kennedy–Higgins observation (1973) is sufficient to argue against an origin of the magnetic field in magnetohydrodynamic processes in the outer core. Birch (1972) has questioned whether the extrapolations of laboratory melting data are for the melting of the same phase of iron as in the inner core. However, we do not believe that the temperature at the inner-outer core boundary will be seriously affected by this argument, and the question of the origin of the magnetic field remains a serious one. Bullard and Gubbins (1971) have proposed that internal magnetohydrodynamic waves in a stably stratified outer core might be the cause of the magnetic field.

A third uneasiness concerns the density jump across the inner–outer core boundary. From the applications of inversion techniques for the periods of free oscillation to the determination of the interior structure of the earth, it has been found that the density contrast at this boundary is especially sensitive to the values of the periods of the higher radial overtones of the spherically symmetric modes of oscillation. The most common value quoted today for the density jump across this boundary is about 1 gm/cm^3 (Dziewonski and Gilbert, 1971; Bolt, 1973), although some models with smaller density jump have been obtained (Dziewonski, personal communication). The density change upon melting of iron at one atmosphere is only about 2.6 per cent, or 0.6 gm/cm^3. One would expect the density jump on melting to decrease, instead of increase, with rising pressure.

Finally, for some models of the inner core derived from inversion of the free modes of oscillation, the *S*-wave velocity has been computed to decrease with increasing depth in the inner core (Dziewonski, personal communication). Although the shear modulus has been found to decrease with increasing pressure on some single crystals of the zinc oxide structure (Soga and Anderson, 1967), we do not expect a decrease in *S*-wave velocity with increasing pressure in polycrystalline solid iron. This property of the inner core, as noted, is not a well-established result, and hence we cannot give this feature great weight in arguing for 'strangeness' in the inner core.

Some of the early catalogues of unusual properties for the inner-outer core interface remain. Most serious and unusual among these is the presence of several branches to the travel–time curves for *P*-waves in or near this depth. Gutenberg reported (1957) the observation of two phases, with different frequency content at some epicentral distances, which would indicate the presence of a relatively more complex boundary at that interface than is present at the outer core–mantle interface. Recent work on the structure of the inner–outer core boundary has shown that the travel–time curves have several branches which also implies a rather complex boundary (Bolt, 1964; Adams and Randall, 1964; Hannon and Kovach, 1966). More than one interfacial surface would seem to be indicated.

One way to avoid some of these new difficulties is to propose that the inner core is a dense liquid or, rather, a glassy phase of iron. By this assumption, the temperature of the inner-outer core interface would be raised well above the melting point of iron. Thus, a temperature gradient in the outer core could be accommodated which was at least at the adiabatic. Poisson's ratio for the inner core could then be close to 0.5, since a value of 0.45 is not unusual for glasses at temperatures above their softening points. A decrease of shear velocity with increasing pressure in the inner core could occur if the temperature in the inner core continues to increase with depth and causes the temperature to be even more remote from the melting curve than at the surface of the inner core.

In order to account for the sharpness of the inner–outer core boundary, we assume that a phase transformation takes place in the liquid iron. This differs only slightly in concept from the usual model, that the inner core boundary is at the melting transition. In our case, we assume that a structural transition takes place in the liquid. The only visible candidate for such a transition is a change in the electronic configuration of iron under pressure. The measured density change in the cases of the electronic transition in Ce and Cs are of the order of 11 per cent and hence suggestive that even the anomalous density change at the inner core surface may be consistent with the existence of an electronic transition.

Below, we discuss the nature of the proposed transition in liquid iron and its effect on some of the mechanical properties to be expected. The main part of this paper is concerned with the calculation of the mechanical properties of the core under the assumption that it is pure liquid iron, with different electronic configurations in the inner core and outer core regions. We are not able to take into account the effects of alloying elements, whether they be closely related elements such as nickel, or elements more remote from iron in the periodic table.

II. DYNAMIC PROPERTIES OF A VISCOUS LIQUID

When subjected to a low-frequency shear stress, linear solids and liquids behave differently: the former respond elastically while the latter flow. However, at high frequencies, a liquid may also appear to be rigid and to transmit shear waves with little absorption. These seemingly exclusive types of behaviour may be understood in terms of the geometric arrangement and type of motion of the constituent molecules. The rigidity of the solid results from the mutually fixed and well-ordered pattern of the equilibrium positions of the molecules. Once the solid melts, long-range order disappears and the molecules are free to move relative to each other. This motion is characterized by a relaxation time τ: if the applied stress has a period shorter than τ, the liquid will respond elastically but there will be no time for any significant rearrangement of the molecules. If, however, the period is much longer than τ, the molecules rearrange themselves to relieve the stress; any elastic response will be masked by the resultant flow.

Maxwell's theory of viscoelasticity combines theories of elasticity and viscosity to provide a unified description of a dense fluid, and allows us to study a liquid over its full range of dynamic properties. Although it is inappropriate at very high frequencies ($\omega > 10^{11}\ s^{-1}$), Maxwell's model should be quite sufficient for a qualitative, and often quantitative, discussion of liquids at lower frequencies.

If e_{ik} is the strain tensor and P_{ik} the stress tensor, the equations of motion may be written in the following covariant form:

$$-p = L\theta; \qquad P'_{ik} = 2Me_{ik} \tag{1}$$

where $p = -\frac{1}{3}P_{ii}$ is the hydrostatic stress, θ is the change of volume $\theta = e_{ii}$, and $P'_{ik} = P_{ik} + p\delta_{ik}$ is the shear stress deviator; this has been related to the shear strain deviator $e'_{ik} = e_{ik} - \frac{1}{3}\theta\delta_{ik}$ through the viscoelastic modulus M. The viscoelastic moduli L and M are

$$L = K_1 + K_2(1 - A_2^{-1}) \tag{2}$$

$$M = \mu(1 - A_1^{-1}) \tag{3}$$

where K_1 is the isothermal bulk modulus, K_2 is a 'relaxation' bulk modulus defined such that $K_1 + K_2$ is the adiabatic bulk modulus, i.e. it describes the immediate reaction of the liquid to a suddenly applied compression; μ is the instantaneous shear modulus and

$$A_i = 1 + \tau_i \frac{\mathrm{d}}{\mathrm{d}t} \tag{4}$$

where

$$\tau_1 = \eta_1/\mu \tag{5}$$

$$\tau_2 = \eta_2/K_2 \tag{6}$$

are relaxation times and η_1 and η_2 are the shear and bulk viscosities respectively.

From the equations of motion (1), we may immediately obtain expressions for the velocities and absorption of harmonic wave motions transmitted along one of the coordinate axes. In this case, differentiation with respect to time $\mathrm{d}/\mathrm{d}t$ is replaced by $\mathrm{i}\omega$, and the operators A, L and M become

$$A_i = 1 + \mathrm{i}\omega\tau_i \tag{7}$$

$$L = K_1 + \frac{K_2}{1 + (\mathrm{i}\omega\tau_2)^{-1}} \tag{8}$$

$$M = \frac{\mu}{1 + (\mathrm{i}\omega\tau_1)^{-1}} \tag{9}$$

The velocity of wave propagation is

$$v = (N/\rho)^{1/2} \tag{10}$$

where N is a viscoelastic modulus and is equal to M in the case of transverse (or shear) waves, and to $M + 2L$ in the case of longitudinal (or compressional) waves. Since N is a complex quantity, we may write it as

$$N = ne^{i\phi}, \qquad n = |N| \tag{11}$$

The velocity is now given by

$$v = (n/\rho)^{1/2}\{\cos(\phi/2) + i\sin(\phi/2)\} \tag{12}$$

It is evident from this expression that the wavelength is

$$\lambda = \frac{2\pi}{\omega}(\rho/n)^{-1/2}\cos(\phi/2) \tag{13}$$

and that the absorption coefficient per unit wavelength is

$$\alpha = 2\pi\tan(\phi/2) \tag{14}$$

In the case of shear waves

$$\tan\phi = \frac{1}{\omega\tau} \tag{15}$$

where $\tau = \tau_1$ (henceforth we drop the subscript 1 on τ) and

$$N = M$$

Therefore

$$n = \frac{\mu}{\sqrt{\{1 + (\omega\tau)^{-2}\}}} \tag{16}$$

For low frequencies, i.e. $\omega\tau << 1$, the magnitude of the complex modulus n goes to zero linearly and the attenuation factor α is so large that propagation becomes impossible; in other words, a liquid will not transmit low-frequency shear waves.

For high frequencies $\omega\tau >> 1$, $n \simeq \mu$ *and*

$$v = (\mu/\rho)^{1/2} \tag{17}$$

$$\alpha = 2\pi\tan(\phi/2) \simeq \pi\tan\phi = \frac{\mu}{\eta_1}\frac{\pi}{\omega} \tag{18}$$

Thus, at high frequencies, the absorption coefficient is inversely proportional to the viscosity! At high frequencies, a viscous liquid is stiffened and becomes more and more transparent to shear-wave motions.

An electronic phase transition occurring in liquid iron must have an effect upon the interatomic potential and, consequently, upon the value of the relaxation time τ of liquid iron. Thus, in our model, we assume that an electronic transition takes place in liquid iron at the pressure at the surface of the inner core. This changes the relaxation time of the iron in the denser state so that it may become transparent to shear-wave motions.

III. ELECTRONIC TRANSITIONS

Electronic transitions have been proposed for a number of phase transformations that have been observed in the laboratory. An electronic transition was first proposed (Lawson and Tang, 1949; Lawson, 1956; Likhter *et al.*, 1958; Herman and Swenson, 1958) to explain a discontinuous drop in resistance in cerium at 5 kb, accompanied by a discontinuous change in volume, but without change in crystal structure. It was postulated that, under pressure, a $4f$ electron is promoted to the $5d$ band.

A large increase in the resistance of cesium at 41 kb is also accompanied by an isostructural change in volume. Sternheimer (1950) proposed that the transition is due to the promotion, under pressure, of a $6s$ electron to the empty $5d$ band. Although Sternheimer's analysis is at best qualitative, his main conclusion was confirmed by Yamashita and Asano (1970) with an APW calculation.

Since the pioneering work on Ce and Cs, several other electronic transitions have been observed. A review of the state of the experimental evidence on electronic transitions is given by Drickamer (1965). A careful study of these observations, plus some simple theoretical arguments, indicate that all elements with unfilled inner electron bands must undergo pressure-induced electronic transitions. In the limit of infinite pressures, any atomic system must break down into a Fermi sea of electrons and nuclei.

A free iron atom has the ground state configuration $3d^6 4s^2$. Because of the incompletely filled d-shell, it is possible to alter the electronic structure to one of the form $3d^{8-n}4s^n$, where the value of n ranges from 0 to 2. The alteration of n is made by changing the environment around the iron atom. As an example, we note that crystalline iron at room conditions has the approximate structure $3d^7 4s^1$ (Stern 1959; Duff and Das, 1970).

The change in electronic structure from that of free iron may be understood by a simple application of perturbation theory. When N atoms are brought together from infinity into a compact solid, each of the degenerate ground state levels is split into N sublevels.Since N is of the order of 10^{23}, give or take several orders of magnitude, the number of sublevels is enormous; because of thermal effects, we can imagine that each level has been broadened into a continuous band. The net result is that both the $3d$ and $4s$ bands are incompletely filled; there has been a shift in the population toward a greater filling of the $3d$ levels.

Theoretical calculations (Wood, 1960) show that the electrons at the top of the d-band are more tightly bound than the d-electrons in the free atom, whereas those at the bottom of the band are more diffuse. There is evidence (Friedel, 1969) that the latter electrons contribute to the conduction band. This split character of the d electrons is expected to become more pronounced with further decrease of the interatomic distance. Eventually, iron will go into a $3d^8$ state; in the liquid phase, it may be argued on the grounds of the stability of half-filled shells that the valency of the new state is 3 (i.e. 5 of the d-electrons

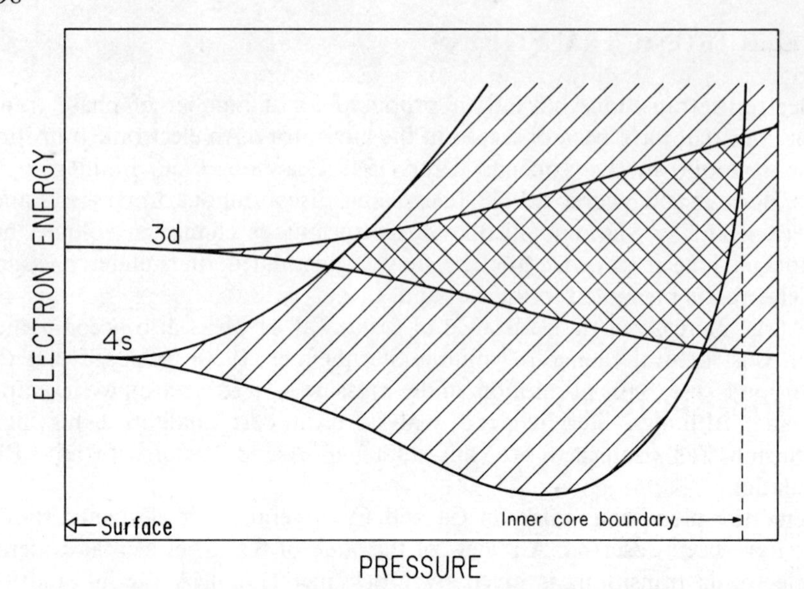

Figure 1. Schematic band structure of iron: electron energy as a function of reciprocal lattice spacing (pressure). The electronic transition occurs when the energy of 4*s* electrons becomes greater than that of 3*d* electrons. The shaded areas show the allowed energies.

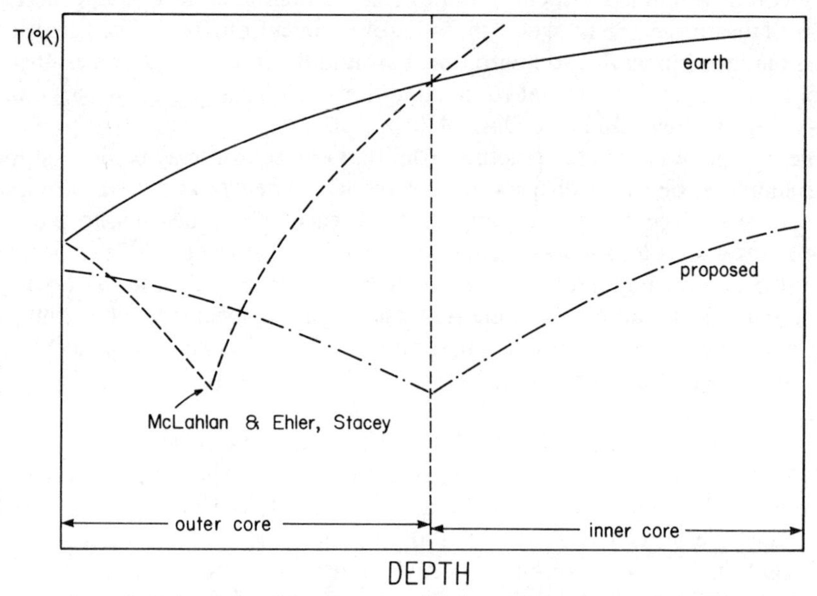

Figure 2. Schematic of temperatures in the core for the earth (solid curve) and for the melting relations proposed by McLachlan and Ehlers (1971) and in this paper.

are below the Fermi level). A sketch of the hypothetical band structure is shown in Figure 1 with electron energy plotted as a function of pressure, or what is schematically the same, as a function of reciprocal lattice spacing.

Figure 2 shows a hypothetical curve of the melting temperature of iron, assuming that the transition occurs at the inner-outer core boundary. For comparison, we also show the transition discussed by McLachlan and Ehlers (1971) and criticized by Stacey (1972). The former authors considered the melting transition to be responsible for the *solid* inner core. In our model, we assume that the melting temperature is everywhere below the actual temperature of the core.

IV. MODEL POTENTIAL

The most convenient way of computing interatomic potentials in metals makes use of 'pseudopotential' theory. It is based on the fact that electrons in a metal do not respond to the strong Coulomb potential of the ions, but instead 'see' a weak screened potential. A full quantum mechanical treatment starts with an exact transformation of the Schrödinger equation into one where the true potential $V(r)$ is replaced by a weak operator pseudopotential (Harrison, 1969). The ensuing calculations are involved and the results are rather complicated. For our purposes, it is sufficient to use a model that is based on the pseudopotential idea. We only present an outline of the model; interested readers are referred to Harrison (1969) for details (see also Faber 1972; March, 1968).

The interatomic potential $w(q)$ is composed of a direct Coulomb interaction and an indirect interaction via the conduction electrons,

$$w(q) = w_d(q) + w_i(q) \tag{19}$$

where q is the wave vector. The direct interaction, assuming the core overlap is negligible, is simply

$$w_d(q) = \frac{4\pi(Ze)^2}{q^2} \tag{20}$$

and is the Fourier transform of the point Coulomb repulsion.

The indirect interaction is the change in the structure-dependent energy which arises due to immersing the ions in a sea of conduction electrons. The latter interact with the ions and with each other, giving rise to a net attraction,

$$w_i(q) = \frac{-q_s^2}{4\pi e^2 \varepsilon} u_b(q) u_b(-q) \tag{21}$$

Here q_s is a screening parameter defined by

$$q_s^2 = 4\pi e^2(-\Delta\rho_e/\Delta U) \tag{22}$$

where $\Delta\rho_e$ is a deviation from the mean density of electrons ρ_e and ΔU is the screened potential 'seen' by the electrons. ε is the dielectric constant and $u_b(q)$

is a 'bare', i.e. unscreened, pseudopotential 'seen' by the electrons. Different models may now be constructed by postulating various forms of $u_b(q)$. One of the simplest is that of Ashcroft (1966) in which, well outside the atomic core, the pseudopotential is that of a point change. Inside the ion core of effective radius R_m, the pseudopotential is taken to be zero; this reflects the almost complete cancellation of the large field inside the ion core.

In wave-vector space, the Fourier transform of the pseudopotential becomes

$$u_b(q) = \frac{-4\pi(Ze)^2}{q^2}\cos(q_s R_m) \tag{23}$$

Substituting into (21) and (19), we obtain for the total interatomic potential

$$w(q) = \frac{4\pi(Ze)^2}{q^2}\left(1 - \frac{q_s^2}{q^2\varepsilon}\cos(q_s R_m)\right) \tag{24}$$

To obtain the potential in real space, we must make some assumptions about q_s and ε. q_s may be estimated by postulating that (a) it is possible to define a local value of the Fermi kinetic energy K_F, and (b) that the total Fermi energy E_F is uniform in space. Then, ignoring exchange and correlation, it follows that

$$q_s^2 = \frac{6\pi\rho_e e^2}{K_F}$$

The dielectric constant may be evaluated from the relation

$$\varepsilon = 1 + \frac{q_s^2}{q^2} \tag{26}$$

Substituting (25) and (26) into (24), and inverting to real space, we obtain the approximate interatomic potential

$$w(r) = \frac{(Ze)^2}{r}\cosh^2(q_s R_m)e^{-q_s r} \tag{27}$$

We note at once that this is simply a screened Coulomb potential enhanced by the factor $\cosh^2(q_s R_m)$. The latter term reflects the orthogonality of the conduction electrons to the core electrons: the larger the ion core, the less efficient the screening.

The potential of equation (27) will be used in the following section to estimate the dynamic properties of the earth's core. It must be emphasized that equation (27) is at best a qualitative estimate of the potential. Although it is rather successful in applications to simple metals, it is not expected to represent transition metals with the same degree of accuracy. However, for our purposes, we need only estimate the order of magnitude of the relaxation time τ; for this purpose, equation (27) is probably quite satisfactory.

V. CELL MODEL

To determine the effect of the transition on liquid iron, we apply the theory

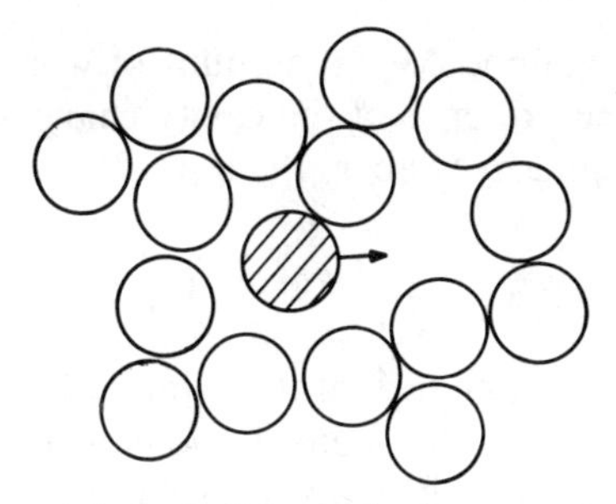

Figure 3. Schematic diagram of the mechanism of diffusion in a liquid structure. The energies in the diffusion process include the energy required to form a vacancy, the barrier potential and the entropic work $T\Delta S$.

of the preceding section to estimate the time τ that an atom spends at onc equilibrium position.

Measurements of diffusion coefficients of liquid metals indicate that, in many cases, they are well represented by an expression of the form

$$D = D_0 e^{-E/kT} \tag{28}$$

where E is some activation energy. Deviations from this formula are attributed to cooperative modes of atomic motion. We assume that the particle motion in the liquid follows an activation process similar to (28).

As may be estimated from equation (28), thc potential energy of liquid iron at core pressures is of the same (or higher) order of magnitude as the kinetic energy. Thus, a cell model should provide an adequate description of the liquid. We imagine each atom to reside in a cell defined by its nearest neighbours. The atom vibrates and exchanges energy with the cell 'walls'. Occasionally, a hole is created in the 'wall' into which an atom may move (Figure 3). If the energy necessary to open the hole is E_h, the potential barrier separating the atom from the hole is E_p, and the change in entropy of the liquid is ΔS, then the average number of diffusive 'jumps' per atom per sec is (neglecting a normalization factor of the order of unity)

$$P = \nu e^{(-E_p + E_h - TS)/kT} \equiv \nu e^{-Q/kT} \tag{29}$$

where ν is the average frequency of vibration of the atom, k is the Boltzmann constant and T is the temperature. E_p may be estimated by smearing the charge in the remaining part of the 'wall' into a uniform distribution σ and integrating the potential $w(R)$. If n_0 is the number of nearest neighbours and a and r_0 are the radii of the cell and hole respectively, then for $(r_0/a)^2 < < 1$,

$$E_p = \frac{n_0}{4\pi a^2} \int_{\substack{\text{wall}\\\text{minus}\\\text{hole}}} w(R)\,dS = \frac{n_0 (Ze)^2}{a} e^{-q_s a} \left[\frac{e^{-q_s(r_0 - a)} - e^{-q_s a}}{2 q_s a} \right] \cosh^2(q_s R_m) \tag{30}$$

To expand a cavity of volume ΔV, an amount of work $p\Delta V$ must be done against the pressure p. The presence of the cavity changes the configurational potential energy of the liquid by (Faber, 1972, p. 73)

$$\Delta W = \frac{1}{12\pi^3}\int_0^\infty [a(q) - 1]\frac{\mathrm{d}w}{\mathrm{d}q}\,\mathrm{d}q \tag{31}$$

where $a(q)$ is the interference function of the liquid. (Since ΔW is of an order of magnitude smaller than the other contributions to the exponent Q, it will be neglected in the numerical calculations.) Thus, the change in energy E_h is given by

$$E_h = p\Delta V + \Delta W \tag{32}$$

To compute the change in entropy ΔS, we note that

$$\mathrm{d}S \simeq \left.\frac{\partial S}{\partial V}\right|_T \mathrm{d}V \tag{33}$$

It follows from thermodynamic identities that

$$\left.\frac{\partial S}{\partial V}\right|_T = \left.\frac{\partial p}{\partial T}\right|_V = \alpha K_T \tag{34}$$

where α is the volume coefficient of thermal expansion and K_T is the isothermal bulk modulus. Therefore

$$\Delta S \simeq \alpha K_T \Delta V \tag{35}$$

where ΔV is the volume of the hole.

VI. NUMERICAL ESTIMATES OF τ_M

The average time an atom stays at one equilibrium positions is

$$\tau = \frac{1}{p} = \nu^{-1}\mathrm{e}^{Q/kT} \tag{36}$$

The quantity ν may be estimated from the equipartition theorem

$$m\omega^2 A^2 = kT$$

where A is the amplitude of the atomic vibration and m is the mass of the atom. A may be obtained by equating the kinetic and potential energies of the atom within the cell. A little reflection shows that τ must be closely related to the Maxwell relaxation time τ_M. Since we may at best only estimate the order of magnitude of τ within the core, we assume $\tau = \tau_M$. This is also suggested by the discussion in section II.

To proceed with the calculation, we must make some assumptions regarding the properties of the liquid core. For the density distribution, any of the current models will suffice; we have used the Bullen–Haddon (1967, 1968) model for the inner core. Since $kT \sim 10^{-2}K_F$ in the core, we assume that the Fermi surface is sharp and that the Fermi energy may be estimated by the usual free electron model

$$K_F = \frac{\hbar^2}{2m}(3\pi^2 \rho_e)^{2/3} \tag{37}$$

where $\rho_e = \rho_i Z$ and ρ_i is the number of atoms per unit volume. In our model, the valency $Z = 3$ in the inner core; in the outer core, it is probably close to zero. In view of the fact that Z is roughly 0.6 in iron at STP conditions (Stacey, 1972), the value $Z = 1$ used in the calculations provides an upper bound for τ_M in the outer core.

As pointed out above, the factor $\cosh^2(q_s R_m)$ in the potential reflects the effect of the Pauli exclusion principle upon the screening electrons. A typical value of R_m is 0.5 Å. This gives $\cosh^2(q_s R_m) \simeq 3.6$ in the outer core. Since the conduction electrons in the inner core are in a d-band, we expect their screening to be more efficient than that of plane-wave electrons. We may allow for the improved screening by decreasing the value of $\cosh^2(q_s R_m)$. This amounts to decreasing the effective size of the core, thus allowing the electrons to concentrate closer to the ions. Calculations were made with $\cosh^2(q_s R_m) = 3.6$, 2 and 1. The last value corresponds to a point ion and should result in a lower bound for the potential.

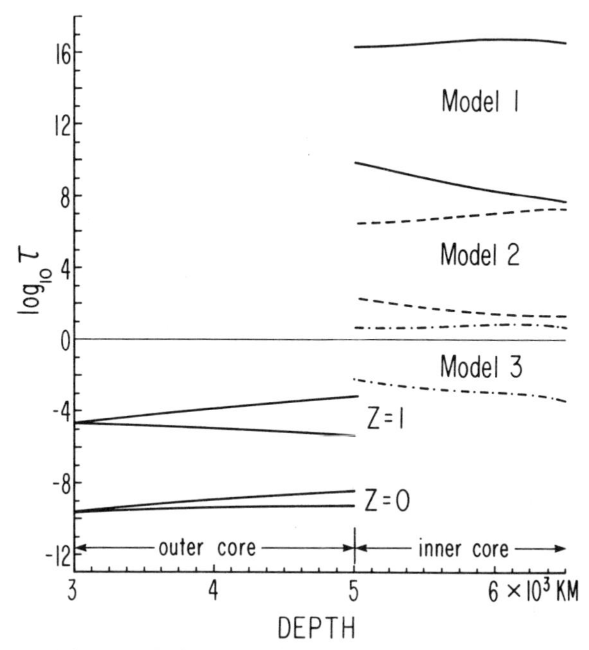

Figure 4. Relaxation times for different models of temperatures in the core. The core transmits shear waves for frequencies much larger than τ^{-1}.

Noting that the transition rate P is a sharply peaked function of r_0, we choose for r_0 the value which maximizes P. In the outer core, we find $r_0 \simeq 0.31a$, whereas in the inner core r_0 ranges from $0.34a$ to $0.47a$. In all cases, r_0 is larger than the ionic core sizes. (The radius of the iron ionic core is ~ 0.76 Å at STP conditions.)

Figure 4 shows values of τ computed by taking $\cosh^2(q_s R_m) = 3.6$, 2 and 1; these are denoted by model numbers 1, 2 and 3 respectively. Each model covers a band of values: the top of each band corresponds to the Kennedy–Higgins temperature distribution. The rest of the band was obtained by raising the temperature at the centre of the core while keeping it fixed at the outer core–mantle boundary. The lower edge of each band corresponds to a central core temperature of 6500 °K. For the outer core, we computed τ by using model number 1 for two values of the valency: $Z = 0, 1$.

The most striking feature of the curves is the large discontinuity across the inner-outer core boundary. In the outer core, the threshold period for propagation of shear waves is of the order of 10^{-8} sec, whereas in the inner core, the threshold may be as large as 10^{15} sec. In fact, the inner core should be transparent at all seismic frequencies for all but the point-ion model ($\cosh^2(q_s R_m) = 1$).

We consider next the amount of absorption of shear waves. We find, with the help of equations (5), (17) and (18), that the fraction of energy transmitted along an inner-core diameter is given by

$$\frac{I}{I_0} = \exp\left[-2\int_0^{r_{\text{i.c.}}} \frac{dx}{\tau\sqrt{\left\{\frac{\mu}{\rho}\left(1+\frac{1}{(\omega\tau)^2}\right)^{-1}\right\}}}\right] \tag{38}$$

For seismic periods of one second, and for all but the point–ion model, $\omega\tau >> 1$; the above expression simplifies to

$$\frac{I}{I_0} \simeq \exp\left[-2\int_0^{r_{\text{i.c.}}} \frac{dx}{\tau\sqrt{(\mu/\rho)}}\right] \tag{39}$$

Taking $(\mu/\rho)^{1/2}$ to be a constant with the value ~ 3.6 km/sec, we may estimate the transmitted energy as a function of the central temperature. The results are shown in Figure 5. Note that the intermediate potential ($\cosh^2(q_s R_m) = 2$) predicts a very strong dependence on temperature. This behaviour is characteristic of hot glasses and is expected of liquids under extreme pressure. The upper bound potential ($\cosh^2(q_s R_m) = 3.6$) predicts that I/I_0 is essentially independent of the temperature distribution within the range considered. This suggests that the iron in the inner core is no longer liquid. Although we feel that this potential function is unrealistic, when faced with the uncertainties involved, we cannot rule out the possibility that the change in electronic configuration triggers a freezing transition in the core.

To check whether the model is compatible with present knowledge about the core, we compute the shear viscosity of the liquid outer core and Poisson's ratio for the inner core. The latter is particularly interesting since it may be

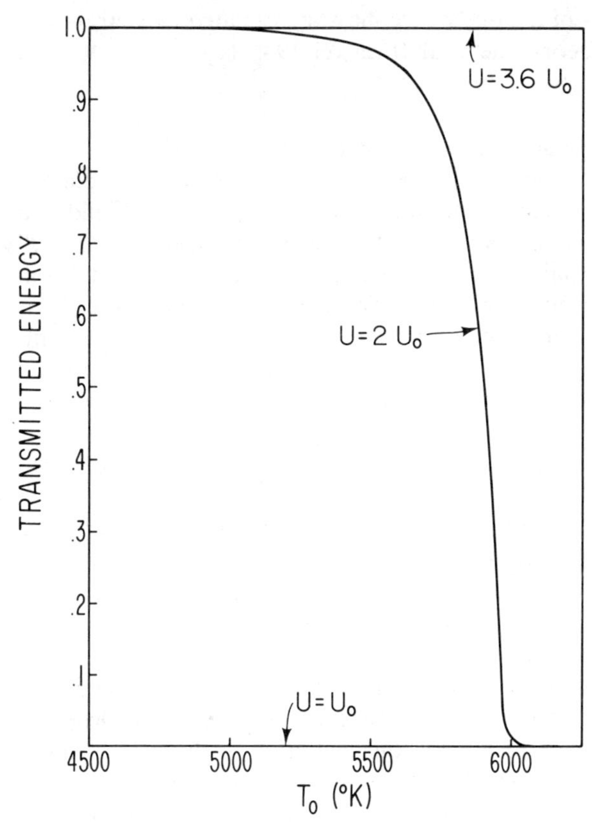

Figure 5. Fraction of seismic energy transmitted by inner core as a function of central temperature for three different models of the pseudopotential.

compared with the value deducted from seismic data. Using the expressions for the acoustic velocities in a viscoelastic liquid, we easily obtain two limiting expressions for Poisson's ratio σ_l:

$$\sigma_l = \frac{1}{2}\left(1 - \frac{1}{2}\frac{\mu}{K_1}\omega\tau\right) \qquad \omega\tau \ll 1 \tag{40}$$

$$\sigma_l = \frac{3K - 2\mu}{6K + 2\mu}\left(1 + \frac{\mu}{2(\omega\tau)^2}\frac{3K + 4\mu}{(3K + \mu)(3K - 2\mu)}\right) \qquad \omega\tau \gg 1 \tag{41}$$

As long as $\omega\tau \gtrsim 1$, σ_l is virtually independent of ω and has the value $(3K - 2\mu)/(6K + 2\mu)$, which is the ordinary value in the theory of elasticity. The ratio μ/K may be estimated by using standard results of liquid transport theory (Egelstaff, 1967). Using core data and the model potential, we obtain $10\mu \simeq K$. This results in a Poisson's ratio of ~ 0.45 for $\omega\tau >> 1$, in good agreement with the seismic observations.

The viscosity of the outer core may be estimated by a straightforward application of rate theory; the result (Frenkel, 1946) is

$$\eta = 6\rho_i kT\tau \tag{42}$$

where τ is the relaxation time. Using the computed value of τ in equation (42), we obtain a lower limit for the viscosity of the order of 10 poise. We feel that this value is more realistic than the much smaller values found in the literature, which do not take proper account of the effect of pressure (Gans, 1972).

To test the reliability of our model, we computed viscosities for some liquid metals at atmospheric pressure. The computed values agreed within a factor of two with experimental viscosities. Considering that the latter frequently show an even greater scatter, it is reasonable to assume that equation (42) gives at least the right order of magnitude of the core viscosity.

VII. DISCUSSION

In the discussion above, we have assumed that iron undergoes an electronic transition in the liquid state at the pressures at the top of the inner core. Thus, the validity of these arguments concerning the liquidity of the inner core depends crucially on the reality of the electronic transition occurring in nature at precisely the conditions at the inner core surface. We have studied the literature on shock-wave experiments in iron and find no evidence for a transition at these pressures; however, we also find no evidence for electronic transitions in shock-wave experiments in other materials, including cesium.

The only calculation to date of the conditions for the transition to the $3d^8$ state of iron has been made by Gandel'man (1963). He finds the transition to occur at a fourfold compression. If this result is correct, then the pressures at which the transition occurs are far greater than inner-core pressures and our conjecture must be rejected. However, we believe that there is considerable doubt concerning the reliability of his calculation, since (1) we have not been able to duplicate his result using computers which are more advanced than those available to him and, (2) the calculation is extremely sensitive to assumptions regarding the nature of the interatomic potential and truncation effects in the expansions of the wave functions. We consider the problem open at this time, and are carrying out detailed numerical calculations on this point ourselves.

Even if we take the most pessimistic view possible, namely that a liquid inner core does not occur in nature, then some of our results for the outer core are still of interest. In particular, we believe that our estimate of the lower bound of the viscosity in the outer core is probably more realistic than other theoretical estimates found in the literature.

Finally, we wish to summarize the geophysical consequences of our model. We have shown above that, if the inner core is a liquid in the $3d^8$ state of iron, there will be a significant change in the relaxation time, principally due to the change in the valency. As a consequence of the increase in valency, there is a change in the viscosity and hence in the transparency to shear-wave motion. We are in agreement with part of Stacey's assessment (1972) of the influence of such a transition on the properties of the core. Stacey criticized McLachlan and Ehlers' (1971) postulate that a $3d^8$ transition occurs in the outer core, on the grounds that the resistivity would become too high and thus would inhibit magnetohydrodynamic effects in the outer core. However, by placing the transition at the surface of the inner core, we have not restricted the nature of magnetohydrodynamic effects in the outer core. Some of the consequences of our model are that we can accommodate a transparency to shear waves, with a 'strange' high value of Poisson's ratio and with a possible decrease in S-wave velocity toward the center of the inner core. Because of uncertainties in the temperatures of the inner core, we cannot predict whether the shear-wave velocity in the inner core will rise or fall with depth. Finally, the large density change at the inner core transition is not inconsistent with the model of an electronic transition; we are confident that the calculation now in progress regarding the pressure at which the transition takes place will also yield information regarding the density change.

In our model, we have removed the constraint that the temperature at the inner core surface be at the melting point. If the temperature distribution in the outer core is adiabatic (or slightly greater than adiabatic), then the problem of getting rid of the heat from the outer core becomes vital (Kennedy and Higgins, 1972). Mantle-wide convection seems to be the only solution to this problem. If this is the case, then the inner core and plate-tectonic motions at the surface are more intimately connected than was supposed. The upper bound on the temperature at the inner core is thus defined: either it is given by the requirement that the temperature be low enough to guarantee transparency to shear waves, or it is limited by the heat flux from the surface of the earth.

REFERENCES

Adams, R. D., and Randall, M. J., *Bull. Seismol. Soc. Amer.*, **54**, 1299 (1964).
Ashcroft, N. W., *Phys. Lett.*, **23**, 48 (1966).
Birch, F., *Amer. J. Sci.*, **238**, 192 (1940).
Birch, F., *Geophys. J. Roy. Astron. Soc.*, **29**, 373 (1972).
Bolt, B. A., *Bull. Seismol. Soc. Amer.*, **54**, 191 (1964).
Bolt, B. A., *Scientific Amer.*, **228**, 24 (Mar. 1973).
Bullard, E. C., and Gubbins, D., *Nature*, **232**, 548 (1971).
Bullen, K. E., *Mon. Not. Roy. Astron. Soc., Geophys. Suppl.*, **6**, 383 (1952).
Bullen, K. E., *Bull. Seismol. Soc. Amer.*, **46**, 333 (1956).
Bullen, K. E., and Haddon, R. A. W., *Nature*, **213**, 574 (1967).
Bullen, K. E., and Haddon, R. A. W., *Proc. Nat. Acad. Sciences*, **58**, 846 (1966).
Burke-Gaffney, T. N., *Bull. Seismol. Soc. Amer.*, **43**, 331 (1953).
Cole, G. H. A., Observatory, **77**, 17 (1957).
Drickamer, H. G., *Solid State Phys.*, **17**, 1 (1965).
Duff, K. J., and Das, T. P., *Phys. Rev.*, **B3**, 192 (1970).
Dziewonski, A. M., and Gilbert, F., *Nature*, **234**, 465 (1971).
Dziewonski, A. M., and Gilbert, F., *Geophys. J. Roy. Astron. Soc.*, **27**, 293 (1972).

Egelstaff, P. A., *An Introduction to the Liquid State*, Academic Press (1967).
Faber, T. E., *Theory of Liquid Metals*, Cambridge Monographs on Physics (1972).
Frenkel, J., *Kinetic Theory of Liquids*, Oxford (1946).
Friedel, J., 'Transition Metals. Electronic Structure of the *d*-Band. Its Role in the Crystalline and Magnetic Structure', in *The Physics of Metals: 1. Electrons*, Ziman, J. M. (ed), Cambridge Press (1969) p. 340.
Gandel'man, G. M., *Sov. Phys. JETP (Engl. Transl.)*, **16, No. 1**, 94 (1963).
Gans, R. F., *J. Geophys. Res.*, **77**, 360 (1972).
Gutenberg, B., *Trans. Amer. Geophys. Union*, **38**, 750 (1957).
Gutenberg, B., and Richter, C. F., *Mon. Not. Roy. Astron. Soc., Geophys. Suppl.*, **4**, 363 (1938).
Hannon, W. J., and Kovach, R. L., *Bull. Seismol. Soc. Amer.*, **56**, 441 (1966).
Harrison, W. A., *Phys. Rev.*, **181**, 1036 (1969).
Herman, R., and Swenson, C. A., *J. Chem. Phys.*, **29**, 398 (1958).
Higgins, G., and Kennedy, G. C., *J. Geophys. Res.*, **76**, 1870 (1971).
Hutchinson, R. O., *Eq. Notes, Eastern Sec. Seismol. Soc. Amer.*, **26**, 45 (1955).
Jacobs, J. A., *Nature*, **172**, 297 (1953).
Jacobs, J. A., *Nature*, **173**, 258 (1954).
Jeffreys, H., *Mon. Not. Roy. Astron. Soc., Geoph. Suppl.*, **4**, 548 (1939a).
Jeffreys, H., *Mon. Not. Roy. Astron. Soc., Geoph. Suppl.*, **4**, 594 (1939b).
Julian, B. R., Davies, D., and Sheppard, R. M., *Nature*, **235**, 317 (1972).
Kennedy, G. C., and Higgins, G. H., *Tectonophysics*, **13**, 221 (1972).
Kennedy, G. C., and Higgins, G. H., *J. Geophys. Res.*, **78**, 900 (1973).
Knopoff, L., and MacDonald, G. J. F., *Geophys. J. Roy. Astron. Soc.*, **1**, 216 (1958).
Lawson, A. W., 'The Effect of Pressure on Electrical Resistivity of Metals', in *Progress in Metal Physics*, vol. 6, Chalmers, B. and King, R., (eds), Pergamon Press, New York (1966) p. 1062.
Lawson, A. W., and Tang, T. Y., *Phys. Rev.*, **76**, 301 (1949).
Lehmann, I., *Un. Geod. Geophys. Internat., Assoc. Seismol. Ser. A. Travaux Scientifiques, No. 14*, 87 (1936).
Likhter, A. W., Riabinin, N., and Vereschaguin, L. F., *Sov. Phys., JETP (Engl. Transl.)*, **6**, 469 (1958).
MacDonald, G. J. F., and Knopoff, L., *Geophys. J. Roy. Astron. Soc.*, **1**, 284 (1958).
McLachlan, D. Jr., and Ehlers, E., *J. Geophys. Res.*, **76**, 2780 (1971).
March, N. H., *Liquid Metals*, Pergamon Press, New York (1968).
Ramsay, W. H., *Mon. Not. Roy. Astron. Soc.*, **108**, 406 (1948).
Ramsay, W. H., *Mon. Not. Roy. Astron. Soc., Geophys. Suppl.*, **5**, 409 (1949).
Ramsay, W. H., *Mon. Not. Roy. Astron. Soc., Geophys. Suppl.*, **6**, 42 (1950).
Soga, N., and O. L. Anderson, *J. Appl. Phys.*, **38**, 2985 (1967).
Stacey, F. D., *Geophys. Survey*, **1**, 99 (1972).
Stern, F., *Phys. Rev.*, **116**, 1399 (1959).
Sternheimer, R., *Phys. Rev.*, **78**, 238 (1950).
Wood, J. H., *Phys. Rev.*, **117**, 714 (1960).
Yamashita, J., and Asano, S., *J. Phys. Soc. Japan*, **29**, 264 (1970).

*High Pressure Chemistry and Physics of Iron Compounds**

C. W. Frank
Sandia Laboratories, Albuquerque, New Mexico 87115, USA
and
H. G. Drickamer
School of Chemical Sciences and Materials Research Laboratory
University of Illinois, Urbana, Illinois 61801, USA

INTRODUCTION

Since iron compounds play a significant role in the mineral composition of the mantle, their physico-chemical behaviour under high pressure is of considerable interest. In this presentation we hope to illustrate the richness and variety of the high pressure phenomena observed in iron compounds. The material to be discussed represents recent advances in the high pressure research on iron compounds conducted at the University of Illinois since it is these results with which we are most familiar. A general review of the high pressure solid state chemistry of iron compounds may be found in *Electronic Transitions and the High Pressure Chemistry and Physics of Solids*, by H. G. Drickamer and C. W. Frank, Chapman and Hall, London (1973).

We will concentrate on phenomena which are fairly localized and will not discuss such purely cooperative behavior as transitions between paramagnetic and ferromagnetic or antiferromagnetic states. We will first consider changes of spin-multiplicity of Fe(II) which include both decreases from high- to low-spin and increases from low- to high-spin. This will be followed by an examination of the reduction of high-spin Fe(III) to Fe(II). Of these transitions, the high- to low-spin transformation and the reduction process would appear to have the most geological significance. The low- to high-spin transition is probably of lesser importance but is included to illustrate the variety of solid state behaviour under pressure. The problem of describing the high pressure behaviour will be approached in the context of a theory of pressure-induced electronic transitions. Such a treatment serves to unify the wide ranging phenomena into a coherent picture.

*This work supported by the U.S. Atomic Energy Commission.

The presentation will be organized as follows: First, our concept of an electronic transition will be defined and essential background information on the effect of pressure on electronic energy levels will be presented. Next, a phenomenological thermodynamic description of continuous electronic transitions will be outlined. This will be followed by consideration of the relationship between pressure-induced thermal processes and optical transitions. This material will serve as a foundation on which to build our analysis of the high pressure iron chemistry. Recent representative results have been selected to illustrate the electronic transitions of iron. The high- to low-spin transition will be examined in a substitutional impurity system of Fe(II) in MnS_2; the low- to high-spin transformation in a series of phenanthroline derivatives; and the reduction in a series of acetylacetonate and hydroxamate compounds. Finally, we briefly extend our analysis of the reduction to a non-iron compound to illustrate the generality of the phenomenon.

THE NATURE OF ELECTRONIC TRANSITIONS

Energy Level Displacements

The fundamental effect of pressure on electronic structure is to increase the overlap between adjacent orbitals. As a result of orbital interactions, this increased overlap can lead to the shift of the energy of one orbital with respect to another. In fact, since the radial extent, orbital shape and compressibility vary between atomic orbitals of different quantum numbers, we may expect the electronic response to pressure to vary and relative energy displacements to be common.

There is an important consequence of this energy level displacement which is of considerable generality. If the separation between the ground and first excited state is not too large, relative energy level shifts may be sufficient to establish a new ground state for the system or to modify greatly the properties of the original ground state by configuration interaction. We will define such an event as an *electronic transition.* Such transitions may take place discontinuously at a definite pressure or continuously over a pressure range. They are in some cases easily reversible; in others, sluggishly so with considerable hysteresis; and in still others, they may lead to the formation of a new stable compound. The transitions may involve cooperative phenomena to a greater or lesser extent. For example, when a particular site transforms to a new ground state, there may be an electrical polarization and/or mechanical strain introduced. This induced strain may act to accelerate or inhibit the transformation of neighbouring sites. We shall consider later a phenomenological treatment which qualitatively accounts for the site interaction and predicts the general features of the observed iron results.

Since the occurrence of energy level displacements is crucial to the concept of pressure-induced electronic transitions, we will now consider experimental evidence for such shifts. We will concentrate on those excitations of particular interest in the coordination chemistry of transition metal compounds. These include d–d transitions on the metal ion, π–π^* transitions on the surrounding ligands and charge-transfer between the ligands and central metal.

The interesting features of transition metals arise from the configuration of the partially filled d-shell. In the free ion the five orbitals are degenerate. However, this degeneracy is partially removed when the metal ion is placed in a crystal lattice as shown in Figure 1. Only the octahedral case will concern us here. Here the $d_{x^2-y^2}$ and d_{z^2} orbitals form a doublet of $e_g(\sigma)$ symmetry which is higher in energy than the triplet d_{xy}, d_{xz} and d_{yz} of $t_{2g}(\pi)$ symmetry. The orbital energy separation, represented by Δ, is a measure of the ligand field strength and is one factor which determines the electronic configuration of the metal ion. The other is the mean spin-pairing energy which accounts for

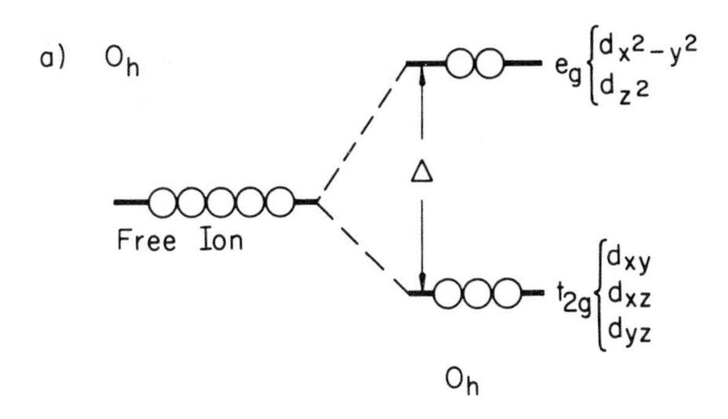

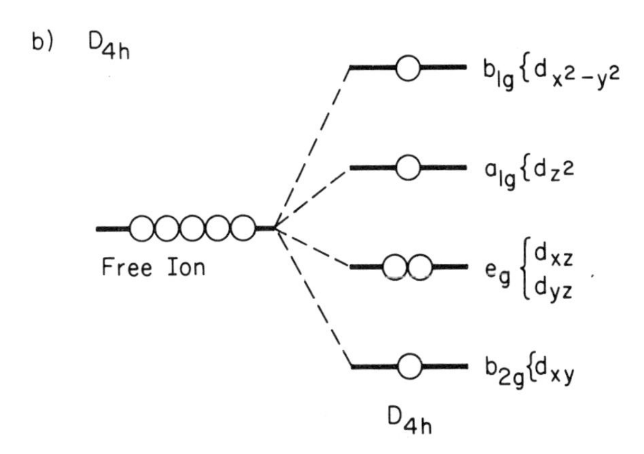

Figure 1. Ligand field splittings in O_h and D_{4h} symmetry. (Reproduced with permission from H. G. Drickamer and C. W. Frank, *Electronic Transitions and the High Pressure Chemistry and Physics of Solids*, Chapman and Hall, London, 1973, p. 15).

classical interelectronic repulsion and quantum-mechanical exchange. It may be formulated in terms of the experimentally determined Racah interelectronic repulsion parameters B and C. For octahedral symmetry only high- and low-spin states are allowed. Thus, if the ligand field splitting is less than the spin-pairing energy, a high-spin state will result; if greater, a low-spin state. The high-spin state for Fe(II) with six d-electrons corresponds to having four of the five orbitals contain one electron each with the remaining orbital at the lower energy containing two electrons with opposed spins. The low-spin state for Fe(II) will have all six electrons paired in the lower energy t_{2g} level. The calculated crossover region between high- and low-spin, based on the size of the ligand field, is actually fairly broad, up to 2000 cm^{-1}, due to the neglect of spin–orbit coupling and configuration interaction.

Several comments may be made concerning d–d transitions under pressure. For high-spin complexes the ligand field increases with increasing pressure up to 10 per cent in 100 kilobars, as a result of increased electrostatic repulsion and stronger covalent bonding. In fact, a simple point charge model would predict that Δ would increase as R^{-5}, where R is the metal-ligand separation.

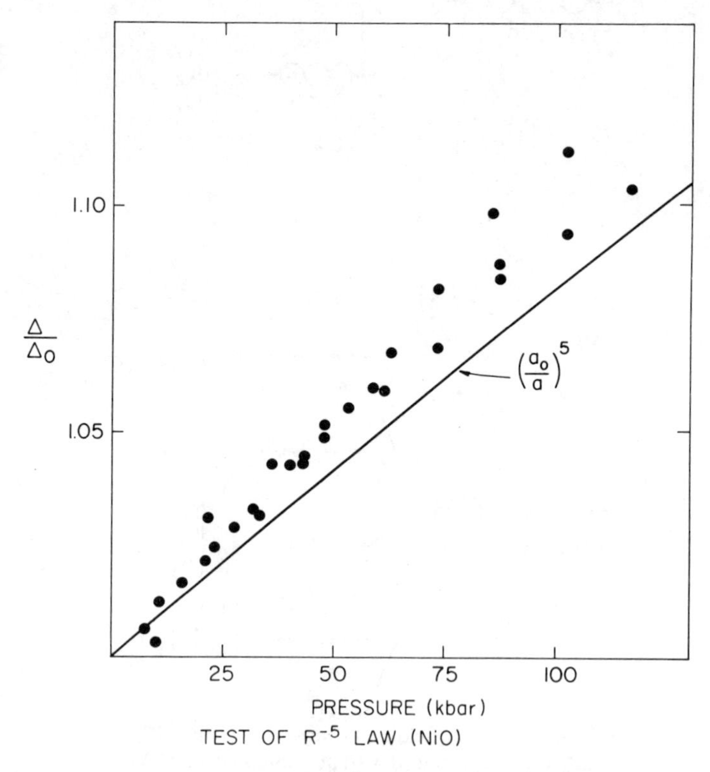

Figure 2. Δ/Δ_0 and $(a_0/a)^5$ versus pressure for NiO. (Reproduced from H. G. Drickamer, *J. Chem. Phys.*, **47**, 1880 (1967) by permission of The American Institute of Physics).

Figure 2 compares measured changes in Δ for Ni^{+2} in NiO with a prediction based on the assumption that the local compressibility is the same as that of the bulk crystal. Here, as in other systems, it is found that Δ generally increases faster than predicted. However, even this modest agreement must be considered fortuitous to some degree, since the point charge model is simply inadequate for an accurate description.

The Racah parameters *B* and *C*, which are essentially empirical measures of interelectronic repulsion, tend to decrease with increasing pressure by as much as 3 to 10 per cent in 100 kilobars. This is shown for $MnCl_2$ and $MnBr_2$ in Figure 3. This is generally associated with increased overlap of metal and ligand electrons (i.e. increased covalency) and consequent delocalization of the 3*d* electrons.

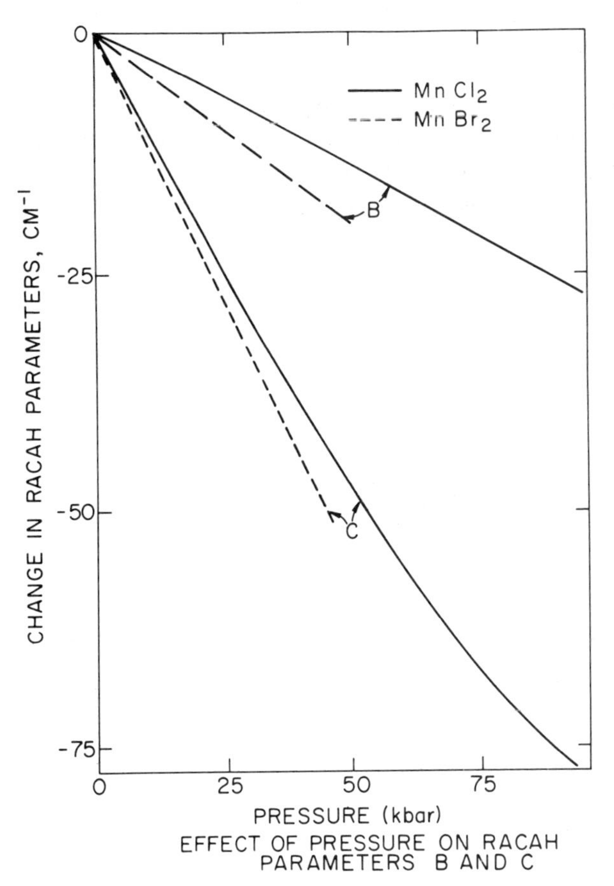

Figure 3. Racah parameters *B* and *C* versus pressure for $MnCl_2$ and $MnBr_2$. (Reproduced from J. C. Zahner and H. G. Drickamer, *J. Chem. Phys.*, **35,** 1489 (1961) by permission of The American Institute of Physics).

An additional experimental parameter which is rather sensitive to such metal–ligand covalency interactions is the Mössbauer isomer shift. The isomer shift is a measure of the *s*-electron density at the iron nucleus; by convention, a decrease in isomer shift corresponds to an increase in electron density. A delocalization of the 3*d*-orbitals as inferred from the pressure behaviour of the Racah parameters would be reflected in the isomer shift through a decrease in shielding of the 3*s*-electrons and a subsequent increase in electron density at the nucleus (or decreased isomer shift). This is demonstrated for several relatively ionic Fe(II) and Fe(III) compounds in Figure 4. Here the change in isomer shift in 200 kilobars is of the order of 10 per cent of the total difference between typical isomer shifts of ionic ferric and ferrous compounds, a fairly significant change. We note that changes in 4*s*-orbital occupation with pressure would also contribute to this behaviour; however, this is probably not a major factor for these materials. Thus, from the data on the Racah parameters and isomer shifts, it is clear that a primary effect of pressure is to promote a spreading or delocalization of metal electron density into the region of the surrounding ligands. This fact has considerable significance in the later discussion.

Since the pressure-induced electronic transitions depend intimately on interaction of the metal ion with the surrounding ligands, we must also consider the ligand behaviour under pressure. Several different ligand systems will be

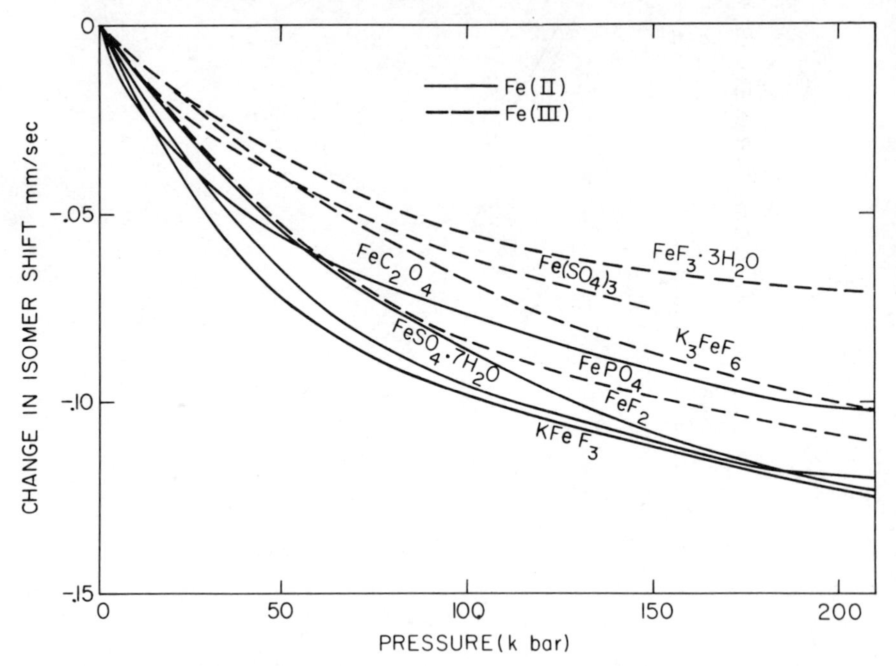

Figure 4. Isomer shift versus pressure for ionic compounds of Fe(II) and Fe(III). (Reprinted with permission from H. G. Drickamer, R. W. Vaughan and A. R. Champion, *Accounts Chem. Res.*, **2**, 43 (1969). Copyright by the American Chemical Society).

considered in the later discussion. However, they have the common characteristic that most are aromatic or quasiaromatic. Such aromatic hydrocarbons are characterized by a conjugated orbital system which is quite sensitive to compression. The electronic transition of interest is from a bonding π molecular orbital to an antibonding π^* orbital. Frequently these π–π^* transitions decrease in energy by 0.5 to 1.0 eV (4000 to 8000 cm^{-1}) in 100 kilobars. One ligand of particular interest for the study of spin transitions is the 1, 10 phenanthroline chromophore shown in Figure 5. This is a three-membered heterocyclic ring

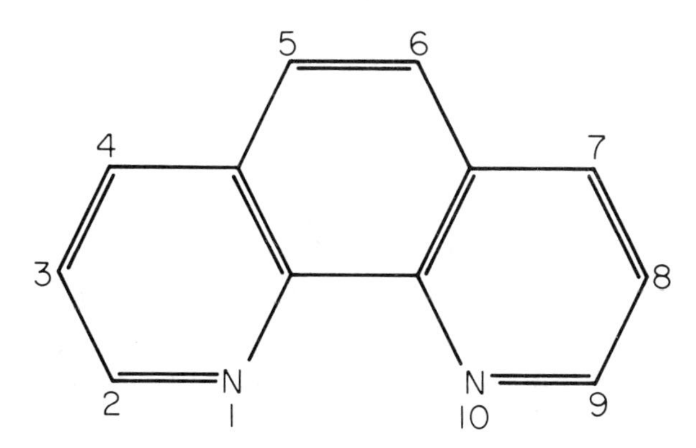

Figure 5. Structure of phenanthroline chromophore. (Reproduced from C. B. Bargeron and H. G. Drickamer, *J. Chem. Phys.*, **55**, 3472 (1971) by permission of the American Institute of Physics).

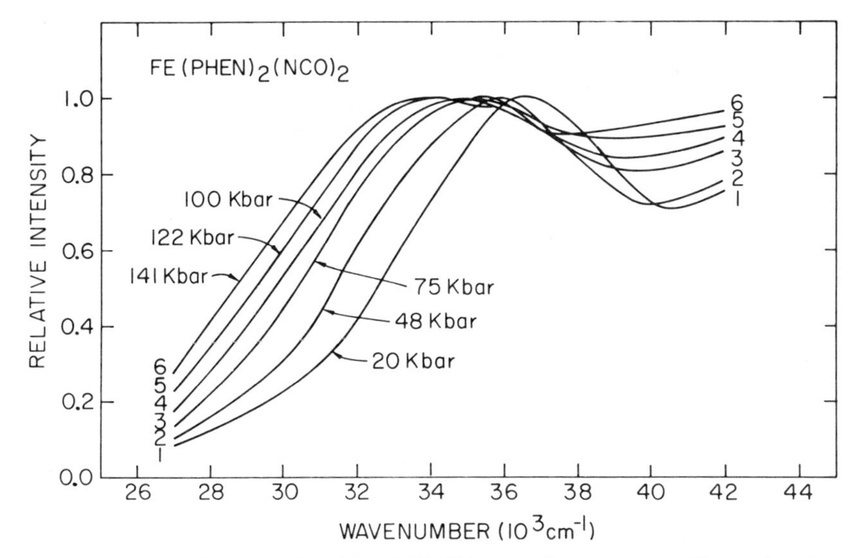

Figure 6. UV spectra of $Fe(phen)_2(NCO)_2$ at various pressures. (Reproduced from D. C. Fisher and H. G. Drickamer, *J. Chem. Phys.*, **54**, 4833 (1971) by permission of the American Institute of Physics).

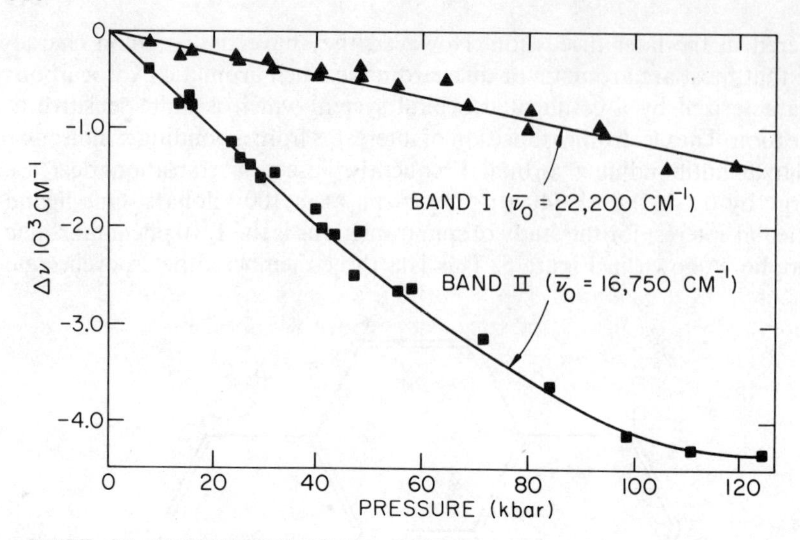

Figure 7. Shift of metal-to-ligand charge transfer peak versus pressure for K_2OsBr_6. (Reproduced from A. S. Balchan and H. G. Drickamer, *J. Chem. Phys.*, **35**, 357 (1961) by permission of the American Institute of Physics).

which coordinates octahedrally to the iron through the nitrogens at the 1 and 10 positions. Optical spectra for the π–π^* transition of this ligand in an iron complex are shown in Figure 6. There is a fairly large red shift and significant peak broadening. As will be discussed later, this red shift implies a higher probability of mixing between the ground and excited states and, hence, increased occupation of the π^* orbitals at high pressure.

Finally, we wish to consider ligand-to-metal or metal-to-ligand charge-transfer. For example, the halide complexes of the heavy metals osmium and rubidium exhibit a pair of peaks at 16–17,000 cm^{-1} and 21–22,000 cm^{-1} which Jorgenson has assigned to ligand-to-metal ($\pi \rightarrow t_{2g}$) electron transfer, split by spin-orbit coupling [Jorgenson 1959]. Figure 7 shows data for K_2OsBr_6 in which the centre of the peak system shifts to lower energy by about 2500 cm^{-1} in 120 kilobars while the spin-orbit splitting increases by 3000 cm^{-1}. Such red shifts are typical for this type of transition. Thus, the effect of pressure on electronic energy levels is to promote first, a delocalization of metal electron density; second, a reduction in energy separation between the highest bonding and lowest antibonding ligand π orbitals; and finally, a reduction in energy separation for ligand π and metal $d\,\pi$ orbitals.

Phenomenological Description of Continuous Electronic Transitions

With this background on energy level shifts established, we may now consider the nature of the electronic transitions in more detail. It is fairly common for discontinuous electronic transitions which are accompanied by a volume discontinuity to occur in metals. However, iron compounds which undergo changes of spin or oxidation state do so continuously over a range of pressure. In fact, the reduction of Fe(III) to Fe(II) has been frequently described in terms of an equilibrium constant defined as the ratio of the ferrous to ferric concentrations which depends on pressure in a power law type formulation. A phenomenological description of continuous electronic transitions has been developed which predicts all of the general features observed in high pressure iron chemistry [Slichter and Drickamer, 1972]. Although quantitative calculations have not been possible, it is instructive to consider the essential aspects of the theory.

We consider a two component system composed of unconverted and converted sites with Gibbs free energies G_0 and G_1, respectively. The free energy of the mixture is given by

$$G = N_0[(1-C)G_0(P,T) + CG_1(P,T) + \Gamma(P,T)C(1-C)] - T\sigma_{mix} \tag{1}$$

where N_0 is the initial concentration of iron sites, C is the site fraction converted, Γ accounts for interaction between sites in what is, in essence, a nonideal solution, and

$$\sigma_{mix} = k_B[N_0 \ln N_0 - N_0 C \ln N_0 C - N_0(1-C)\ln N_0(1-C)] \tag{2}$$

is the entropy contribution due to the variety of ways of selecting the converted sites. Manipulation of equation (1) yields the expression for the equilibrium constant shown in equation (3).

$$\ln K = -\frac{1}{k_B T}[\Delta G(P,T) + \Gamma(P,T)(1-2C)] \tag{3}$$

The problem then involves selecting models for ΔG and Γ. In the simplest case, no interaction between sites is allowed and the elastic response of the lattice is linear. This model predicts that at sufficiently high pressure the more compressible system will have a smaller volume, regardless of which system was smaller at zero pressure. Thus, the possibility of a maximum in the equilibrium constant versus pressure curve is demonstrated if the converted material is stiffer than the unconverted. In fact, the K plots for some systems do tend to deviate from linear behaviour in this manner. However, the results are generally quite qualitative due to the neglect of Γ and the well-known fact that the bulk modulus is definitely pressure dependent in the region under consideration. In an improved version, site interaction is still ignored but the system is treated as having a non-linear elastic response (i.e. a pressure dependent bulk modulus). Coulomb attraction, crystal field and covalent effects as well as closed shell repulsion factors are included in this model. Although the experimental data are approximated more accurately, a close cancellation of different contributions is required.

The most satisfactory description is one which takes into account interaction between adjacent sites. Thus, we expand Γ in a power series in pressure

$$\Gamma(P,T) = \Gamma_0(T) + P\Gamma_1(T) + P^2\Gamma_2(T) \tag{4}$$

where Γ_0 accounts for the fact that at zero pressure the conversion of a given ion or molecule requires a different free energy at low concentration from that at high concentration. Γ_0 represents a balance between the Madelung energy associated with charge-transfer and the elastic strain associated with misfit of converted molecules in the lattice. A positive Γ_0 corresponds to an attractive coupling between sites which leads to an increase in conversion; a negative Γ_0 to a repulsive coupling and a decrease in conversion. Γ_1 is associated with the fact that at zero pressure the volume change on conversion depends on the state of the neighbouring atoms. Γ_1 is proportional to the product of the fractional changes in volume and bulk modulus. Γ_2 corresponds to the fact that the bulk modulus change on conversion depends on what neighbours are present. It is proportional to the square of the fractional change in bulk modulus.

The effect of Γ_0 on the pressure and temperature dependence of $\ln K$ is shown in equations (5)–(7).

$$\frac{d\ln K}{d\ln P} = \frac{1}{\left[1 - \dfrac{\Gamma_0}{k_B T}\dfrac{2K}{(1+K)^2}\right]}\left(\frac{d\ln K}{d\ln P}\right)_{\Gamma_0 = 0} \tag{5}$$

For the pressure at which $K = 1$ this reduces to

$$\frac{d\ln K}{d\ln P} = \frac{1}{\left[1 - \dfrac{\Gamma_0}{2k_B T}\right]}\left(\frac{d\ln K}{d\ln P}\right)_{\Gamma_0 = 0} \tag{6}$$

Relatively small repulsive interactions (Γ_0 negative) may have a large effect on the slope (e.g. the slope is cut in half for $\Gamma_0 = -0.05$ eV). The temperature dependence is shown in equation 7.

$$\frac{d\ln K}{d(1/T)} = -\frac{1}{k_B}\frac{\Delta H + \left(\dfrac{1-K}{1+K}\right)\left(\Gamma - T\dfrac{d\Gamma}{dT}\right)}{\left[1 - \dfrac{2\Gamma}{k_B T}\dfrac{K}{(1+K)^2}\right]} \tag{7}$$

From this expression it is clear that attractive interaction (Γ positive) may cause the denominator to go to zero which would lead to a discontinuity at some temperature, as has been observed by Konig and Madeja for low to high spin changes in ferrous phenanthroline and ferrous bipyridyl compounds [König and Madeja, 1967A, 1967B, 1968]. The original work presents graphical solutions for the fraction converted at each level of approximation. With these the possibilities of cooperative phenomena, of the occurrence of discontinuous jumps in conversion at a specific temperature, and of hysteresis may be demonstrated.

Comparison of Thermal and Optical Transitions

The treatment just described provides justification for the simultaneous coexistence at equilibrium of two distinct sites in the solid state matrix and for a continuous transition between the two. This assumes, of course, that no problems of metastability are encountered. We now examine the question of the electronic driving force behind the transition more closely. We have already noted that it is the relative displacement of orbital energy levels which leads to the observed electronic transition. However, the energy level shifts of the relevant electronic orbitals are only modest fractions of the transition energies, as measured by optical absorption techniques. Resolution of the apparent dilemma rests on the recognition of the fact that the pressure-induced transitions are thermal processes and as such may require considerably less energy than the corresponding optical transitions. One such example of this energy difference is given by alkali halide colour centres which have ν_{max} in

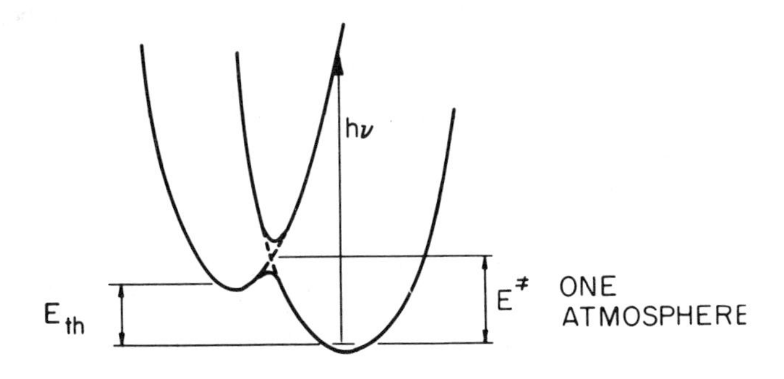

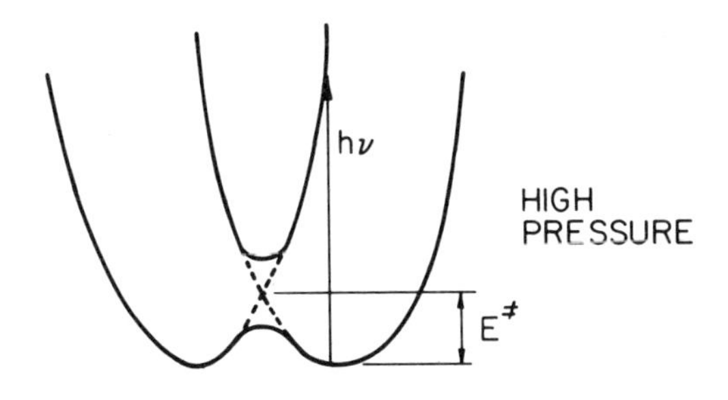

SCHEMATIC CONFIGURATION
COORDINATE DIAGRAM

Figure 8. Schematic configuration coordinate diagram. (Courtesy H. G. Drickamer, C. W. Frank and C. P. Slichter, *Proc. Nat. Acad. Sci.*, **69**, 933 (1972).)

the range of 2 to 4 eV, yet can be bleached thermally even below room temperature. In addition, redox reactions in solution between two oxidation states of the same ion exhibit optical absorption of several eV; yet E_{th} is zero. There are several reasons for this, two of which are illustrated in the schematic configuration coordinate diagram of Figure 8. Here energy is plotted vertically and the horizontal scale corresponds to a single configuration coordinate along some nuclear displacement.

The potential wells are shown near their minima where anharmonicity effects may be neglected. The two wells shown could represent, for example, the ligand π and π^* states or alternatively, the ligand π and metal *d*-orbital states. Relative energy displacements in the first case are important in the spin transition whereas the second example applies to the oxidation state change. Optical transitions between the ground and excited states must occur vertically on such a diagram by the Franck–Condon principle since electronic transitions are much more rapid than nuclear motions. Thermal processes, however, are not subject to this restriction so that E_{th} is simply the difference in energy between the respective minima. Thus, it should be possible that, as a result of pressure induced energy level shifts, the excited state potential will shift to lower energy, reducing the thermal energy to zero and establishing a new ground state for the system at high pressure.

Although this is the primary difference between thermal and optical transitions, there are others which may in some cases be of comparable importance. For example, configuration interaction between the ground and excited states may arise due to spin-orbit or electron-lattice coupling. This may significantly affect the energy difference. In addition, the parity selection rules of optical transitions are relaxed over the time scale characteristic of thermal processes. Finally, the diagram is oversimplified in that only one configuration coordinate is considered while the actual number of such coordinates is equal to the number of normal modes of the system. For a thermal process, the volume is the appropriate coordinate whereas an optical transition may involve a different set. The difference between the optical ($h\nu_{max}$) and thermal (E_{th}) transition energies depends on the square of the displacement of the excited state equilibrium from the ground-state equilibrium. An approximate relationship expressing this difference for a given transition is given in equation (8) [Drickamer *et al.*, 1972].

$$h\nu_{max} = E_{th} + \frac{1}{16 \ln 2} \frac{(\delta E_{1/2})^2}{k_B T} \left(\frac{\omega}{\omega'}\right)^2 \tag{8}$$

where $\delta E_{1/2}$ is the optical peak half width and ω and ω' correspond to the force constants for the ground and excited states. We will find this expression quite useful later in determining the likelihood of occurrence of an electronic transition. With sufficiently accurate data on peak positions and widths, it is possible to use other relationships from this analysis to predict the ratio of force constants ω/ω' and the separation between the ground and excited state potential wells along the configuration coordinate. The calculations indicate that one may generally assume $\omega/\omega' = 1$ even when the half width changes by 50 per cent in 100 kilobars. A large degree of internal consistency is demonstrated in that the same results are obtained using absorption or emission data or some combination thereof. Thus, rather detailed information about the potential energy curves may be obtained.

IRON BEHAVIOUR UNDER PRESSURE

Experimental Considerations

With this as a background, we are now in a position to examine the specific behaviour of iron compounds. Many of these results have been obtained using Mossbauer spectroscopy which has been widely used in examining mixed valencies and site occupancies of iron in mineral structures [Burns, 1972]. We begin by briefly reviewing the characteristics of the Mössbauer parameters of interest. The isomer shift has already been described as a measure of the *s*

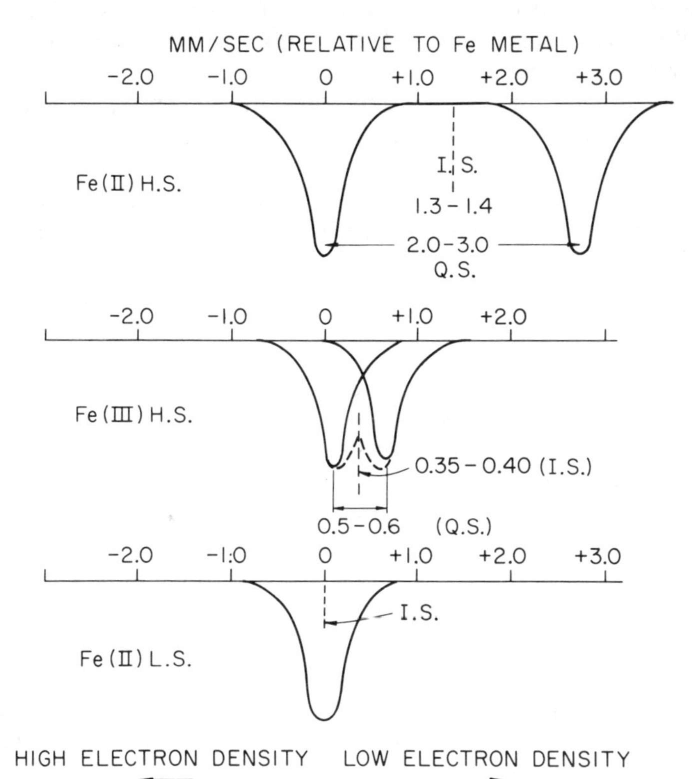

Figure 9. Typical Mössbauer spectra. (Reproduced with permission from H. G. Drickamer and C. W. Frank, *Electronic Transitions and the High Pressure Chemistry and Physics of Solids*, Chapman and Hall, London, 1973, p. 69).

electron density at the iron nucleus. The second parameter is the quadrupole splitting which measures the extent of interaction between an electric field gradient at the nucleus and the nuclear quadrupole moment. The electric field gradient has two components: an external lattice contribution from the surrounding ligand field and an internal valence contribution related to the symmetry of the d-electron shell. The latter effect is the more significant due to the short range nature of quadrupolar forces. Typical Mossbauer spectra are shown in Figure 9 for the three states of interest–high-spin Fe(II), high-spin Fe(III) and low-spin Fe(II). Representative ranges of isomer shifts and quadrupole splittings have been indicated. The $3d^6$ configuration of high spin Fe(II) leads to a $^5T_{2g}$ state which is not spherically symmetric; hence, a large quadrupole splitting is obtained. The large positive isomer shift indicates a relatively low electron density at the nucleus because of the six $3d$-electrons. The high-spin Fe(III) ion has a spherically symmetric $^6A_{1g}$ state so the modest quadrupole splitting is due solely to the ligand field contribution. The isomer shift is smaller than for high-spin Fe(II) due to the decreased shielding. Finally, we consider the low-spin Fe(II) ion which has a symmetrical $^1A_{1g}$ state and, therefore, very little quadrupole splitting. The low isomer shift arises from the extensive metal electron delocalization to excited ligand levels. We will consider this 'back donation' in more detail later. Since the isomer shifts and quadrupole splittings of these states are quite distinct, it is an easy task to follow the conversion for high- to low-spin, or the reverse, for Fe(II) and the reduction of high-spin Fe(III) to Fe(II). Relative concentrations of each state may be readily determined from areas under the Lorentzian curves resulting from least squares computer fits to the data.

There are some possible experimental difficulties, however. In the first case, one must assume that the f numbers at the unconverted and converted sites are equal. Also, self-absorption at high sample concentrations may affect the apparent amounts of ferric and ferrous states present. This may pose a serious restriction on the early data published before 1969. Finally, since the pressure is only quasi-hydrostatic in the Bridgman anvil device used, shear effects are possible. Although such effects may influence the sample during the loading process and be observed at low pressure, there are several compelling reasons to conclude that the primary effect is that of pressure. First, isobaric runs show a strong effect of temperature on conversion; shear effects should be minimized at constant pressure. Second, conversions in homologous series of materials with presumably similar shear characteristics may be correlated well with their electronic properties. Third, shear sensitive materials show little of the shear product under pressure. Finally, the pressure-induced transitions are reversible if efforts are made to remove residual mechanical strain in the sample pellet.

Decrease of Fe(II) Spin-State

In general, the iron results we discuss in the following will be obtained from several series of compounds in which the structural and bonding characteristics are similar for each member. The effect of electronic structure will be examined through appropriate chemical substitution on the basic structure. The first transition of interest—high-spin Fe(II) to low-spin Fe(II)—represents a transition from a paramagnetic to a diamagnetic ground state and is readily understood. We have already noted that the ligand field increases and the Racah parameters decrease with pressure for high-spin compounds. Thus, the potential energy required to occupy the higher lying $e_g(\sigma)$ orbitals is increasing while the spin pairing energy is decreasing. Both factors increase the probability of a high- to low-spin transition, although the increase in ligand field is probably more important. An example is given by Fe(II) as a dilute substitutional impurity in MnS_2 [Bargeron *et al.*, 1972]. In FeS_2 the iron is low-spin at all pressures.

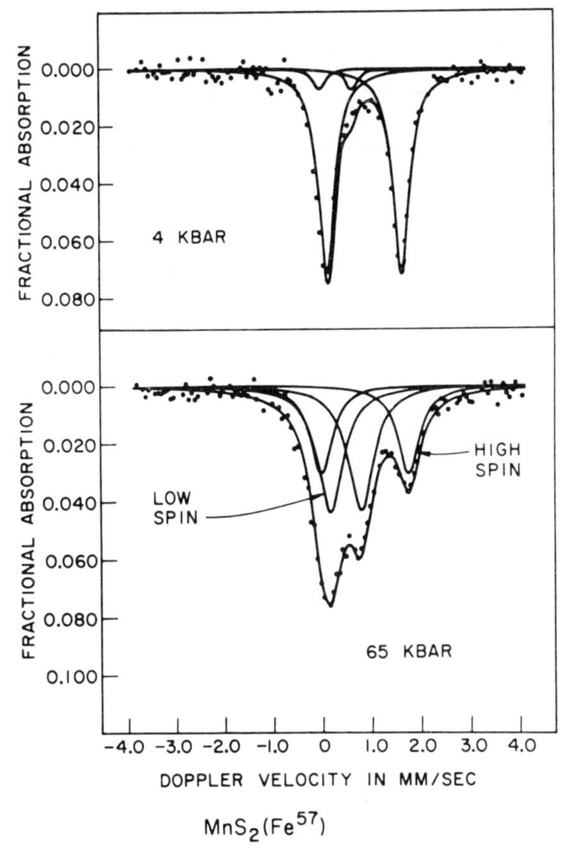

Figure 10. Mossbauer spectra of $MnS_2(^{57}Fe)$ at 4 kbar and 65 kbar. (Reprinted with permission from C. B. Bargeron, M. Avinor and H. G. Drickamer, *Inorg. Chem.*, **10**, 1339 (1971). Copyright by the American Chemical Society).

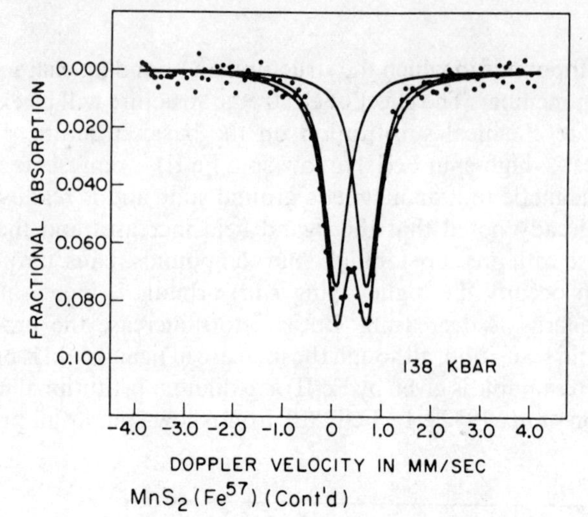

Figure 11. Mössbauer spectrum of MnS_2 (^{57}Fe) at 138 kbar. (Reprinted with permission from C. B. Bargeron, M. Avinor and H. G. Drickamer, *Inorg. Chem.*, **10**, 1339 (1971). Copyright by the American Chemical Society).

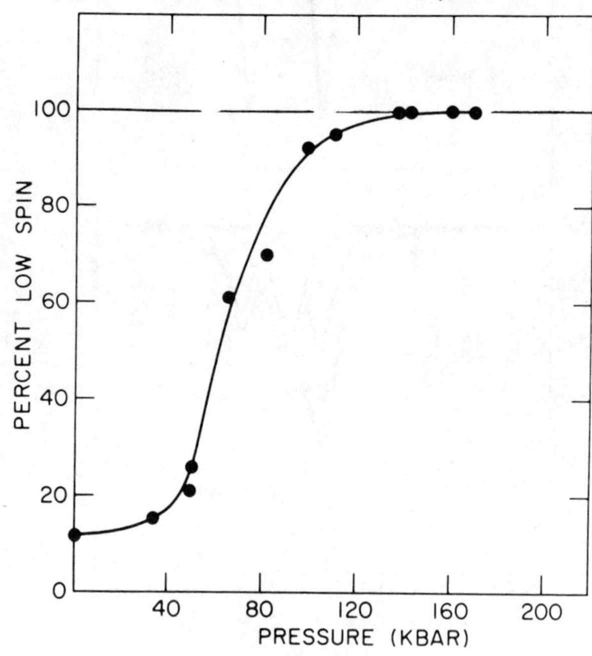

Figure 12. Conversion to low-spin form versus pressure for MnS_2(^{57}Fe). (Reprinted with permission from C. B. Bargeron, M. Avinor and H. G. Drickamer, *Inorg. Chem.*, **10**, 1339 (1971). Copyright by the American Chemical Society).

However, inclusion of the iron impurity in the MnS_2 host results in a high-spin state at atmospheric pressure because of the significantly larger lattice parameter of the host. Certainly the local lattice parameter near the substitutional impurity will be different from the bulk but it seems reasonable to visulize the iron as being in an expanded lattice relative to FeS_2. Mössbauer spectra for the impurity system are shown in Figure 10. The spectrum illustrates a typical high-spin Fe(II) system with a very small contribution due to the low-spin species. From atmospheric pressure up to about 40 kilobars there is no significant change. However, above that point the relative amount of low-spin Fe(II) begins to increase. At 65 kilobars significant conversion has taken place. Figure 11 shows that at still higher pressures the conversion is complete; only a low-spin Fe(II) spectrum is observed. The conversion to low-spin is plotted as a function of pressure in Figure 12. The attainment of 100 per cent conversion is unique to this system; the other spin and oxidation state changes do not go to completion. High- to low-spin transitions have also been observed in other systems. However, this one example illustrates the point quite well that the increase of ligand field splitting with increasing pressure can lead to a spin-pairing at the crossover region.

Increase of Fe(II) Spin-state

The other spin transition of interest is the inverse of the one just considered (i.e. an increase of spin-state is observed). Such increases in spin-multiplicity have been observed in phenanthroline derivatives, heavy metal ferrocyanides, phthallocyanine complexes and in certain biological derivatives, all of which are initially low-spin. That such an event should occur is, at first glance, surprising from both thermodynamic and electronic considerations. First, the low-spin state would normally be considered to have a smaller volume than the corresponding high-spin ion. In addition, one would expect the high-spin state to be less stable at high pressures due to the normally observed increase of ligand field.

All low-spin compounds which exhibit this phenomenon have a particular kind of metal-ligand bonding at one atmosphere. This is illustrated in Figure 13. The ligands have empty π^* states which lie fairly low in energy and are of correct symmetry to bond with the metal $3d_\pi$ orbitals. The donation of metal d_π electrons into these π^* levels increases the covalent bonding and tends to stabilize the d_π level. Thus, the ligand field increases from Δ to Δ' and a low spin state results. As a result of the electron delocalization, the shielding at the iron nucleus is considerably reduced relative to the high-spin case and a small isomer shift is obtained, as previously noted. In fact, there is a good correlation between the ligand field and Fe(II) isomer shift; a large field is associated with extensive delocalization and a small isomer shift.

Before considering the explanation of the transition, we present evidence that it does indeed occur. As an example, we examine tris complexes with 1, 10 phenanthroline [Fisher and Drickamer, 1971]. In these materials three

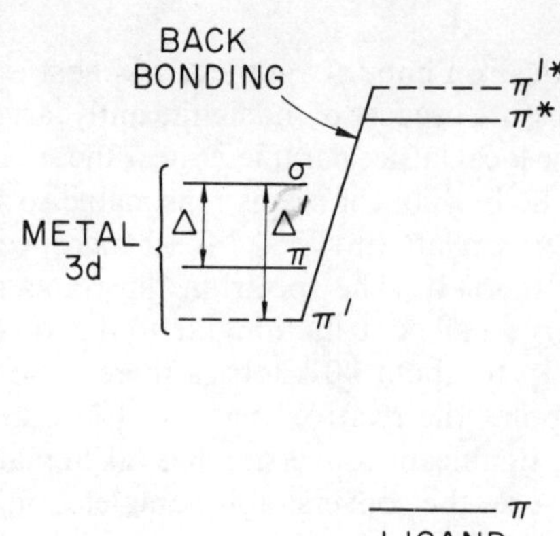

Figure 13. Effect of back donation on molecular orbitals. (Reproduced with permission from H. G. Drickamer and C. W. Frank, *Electronic Transitions and the High Pressure Chemistry and Physics of Solids*, Chapman and Hall, London, 1973, p. 130).

phenanthroline chromophores occupy all coordination sites on the iron. The anions are outside the coordination sphere. The appreciable degree of back donation to the phenanthroline π^* system leads to ligand fields in the range 16 to 19,000 cm^{-1} and a low-spin state. Mössbauer spectra of the tris chloride complex are shown in Figure 14. At low pressure only typical low-spin peaks with a small isomer shift and quadrupole splitting are observed. Near 40 kilobars contributions from the high-spin state are first measurable. The high pressure spectra shown in Figure 15 indicate approximately 20 per cent conversion at 95 kilobars which increases to 25 per cent at 153 kilobars. As shown in Figure 16, conversions were similar for other complexes in which the anion was an isocyanate, azide or thiocyanate. This indicates that the electronic distribution on groups outside the coordination sphere does not significantly affect the metal. In these cases inductive effects could conceivably be transmitted between molecules; however, they decrease rapidly with distance. Of course, resonance interaction requires the entry of the group into the π electron system. This would not be the case for the external anions.

On the other hand, if substituent groups are placed on the phenanthroline molecule such that electron conjugation with the extended ring system is possible, the bonding characteristics of the nitrogens are considerably altered [Bargeron and Drickamer, 1971]. This is illustrated in Figure 17 where groups with varying inductive and resonance properties are substituted at the 5 position, which is strongly conjugated to the coordinating nitrogen. Here the extents of

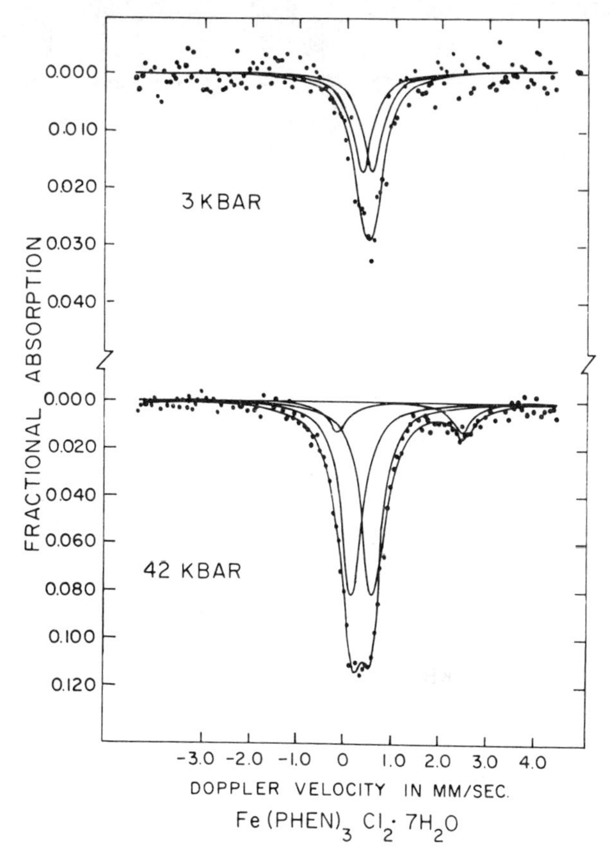

Figure 14. Mössbauer spectra of $Fe(phen)_3Cl_2 \cdot 7H_2O$ at 3 kbar and 42 kbar. (Reproduced from D. C. Fisher and H. G. Drickamer, *J. Chem. Phys.*, **54**, 4826 (1971) by permission of the American Institute of Physics).

conversion are in the same order as the electron withdrawing strength for the methyl, chloro and nitro derivatives. These data are for the tris complex with thiocyanate anions. Similar results were obtained with chloride anions.

Having thus demonstrated that the low- to high-spin transition does occur, we must consider the explanation. First, we examine the thermodynamic aspects with regard to the volume change. What must be borne in mind is that the criterion for the conversion of a process to increase with pressure is that the volume of the system *as a whole* decrease. It is not necessary that every bond shorten; some may shorten while others lengthen. Alternatively, it is possible that the requisite system volume reduction may be achieved by intermolecular rather than intramolecular means. Thus, the spacing between molecules or complexes may decrease due to increased attractive forces or

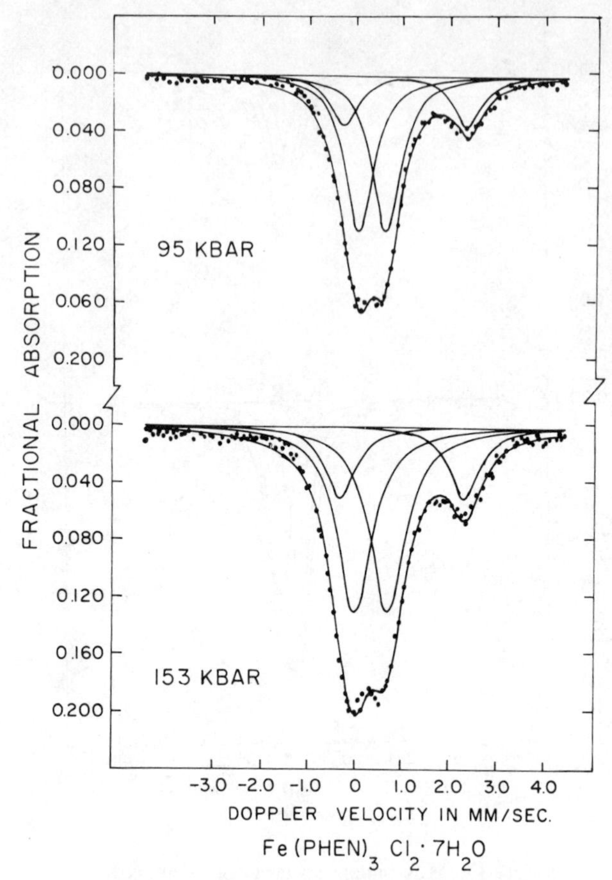

Figure 15. Mössbauer spectra of Fe(phen)$_3$Cl$_2$·7H$_2$O at 95 kbar and 153 kbar. (Reproduced from D. C. Fisher and H. G. Drickamer, *J. Chem.* Phys., **54**, 4826 (1971) by permission of the American Institute of Physics).

greater packing efficiency. In addition, it is possible that, while the low-spin state occupies a smaller volume at atmospheric pressure, its compressibility may be smaller so that the situation may be reversed at high pressure.

Justification for the transition on an electronic basis requires consideration of the effect of pressure on the back-donation process. Optical data for the phenanthrolines show large red shifts of 2100–2700 cm^{-1} for the π–π^* transition in unsubstituted tris compounds. In addition, the metal to ligand charge-transfer bands shift to lower energy by 1400–1850 cm^{-1}. Although the red shift is only a modest fraction of the transition energy, it is the change in thermal energy which is significant. Thus, we postulate a thermal occupation of the π^* level by π electrons. The assumption of such a thermal process is partially substantiated by isobaric runs at 85 kilobars for the thiocyanate complex with unsubsti-

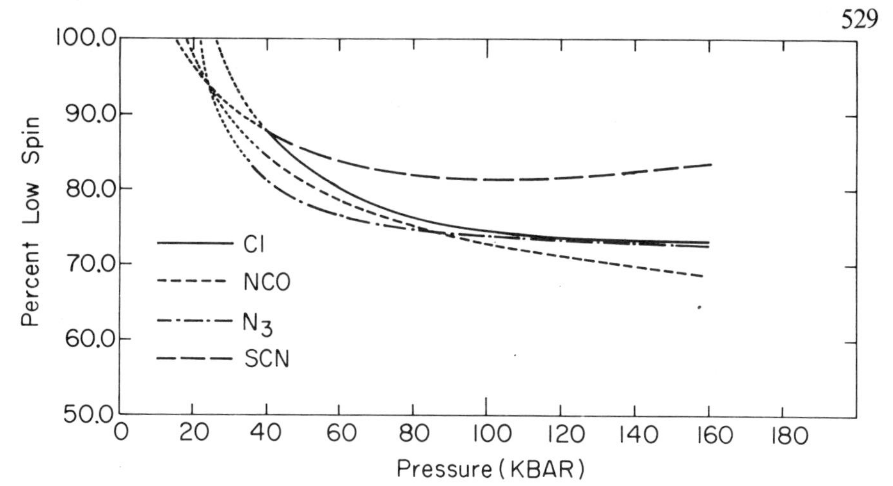

Figure 16. Conversion to high-spin form versus pressure for ferrous tris phenanthroline complexes. (Reproduced from D. C. Fisher and H. G. Drickamer, *J. Chem. Phys.*, **54**, 4832 (1971) by permission of the American Institute of Physics).

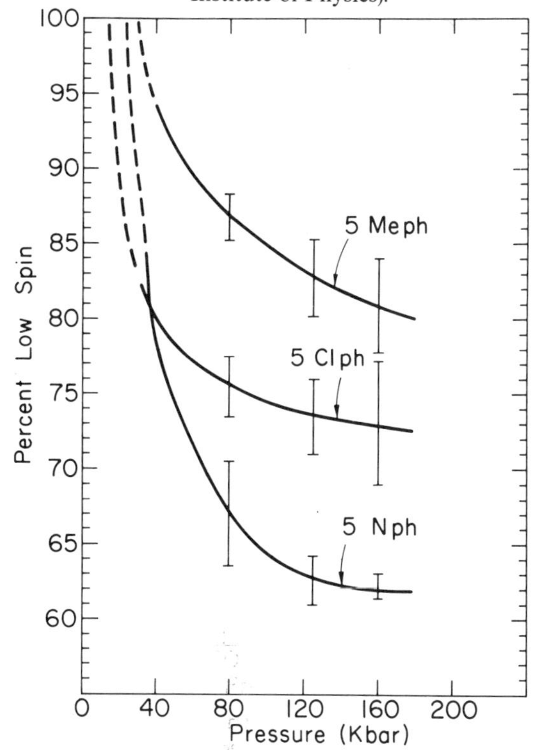

Figure 17. Conversion to high-spin form versus pressure for substituted tris phenanthroline complexes. (Reproduced from C. B. Bargeron and H. G. Drickamer, *J. Chem. Phys.*, **55**, 3473 (1971) by permission of the American Institute of Physics).

Table 1. Comparison of the energy associated with thermal versus optical excitation—π–π^* transition in phenanthroline (Courtesy H. G. Drickamer, C. W. Frank and C. P. Slichter, Proc. Nat. Acad. Sci., **69**, 937 (1972).)

P (*kbar*)	$h\nu_{max}$	$\delta E_{1/2}$	E_{th}
0	4.60	0.95	+ 1.35
50	4.45	1.05	+ 0.45
100	4.30	1.14	− 0.40
150	4.20	1.20	− 0.98

tuted phenanthroline. The data indicate a decrease in the low-spin component with increasing temperature. The material showed 87 per cent low-spin at 298 °K which decreased to 69 per cent at 420 °K.

If we assume that the force constants of the ground and excited states are approximately equal, the thermal energy corresponding to the population of the π^* level may be estimated, as shown in Table 1. At low pressure the π orbital is stable by over an electron volt but above 50 kilobars the sign changes, indicating a stabilization of the π^* orbital. Certainly the data are crude and the

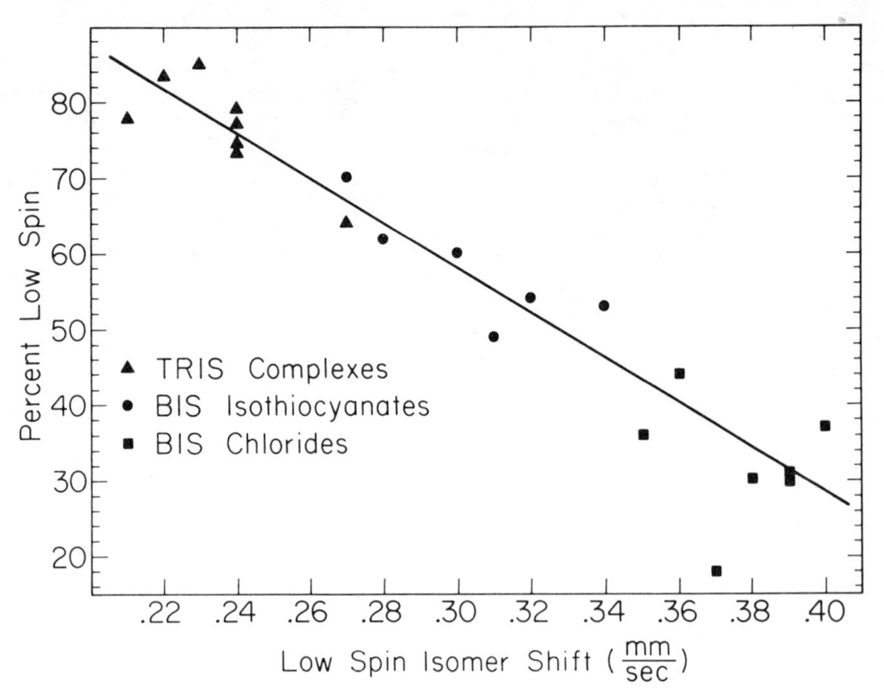

Figure 18. Per cent low-spin versus low-spin Fe(II) isomer shift for substituted phenanthroline complexes. (Reproduced from C. B. Bargeron and H. G. Drickamer, *J. Chem. Phys.*, **55**, 3479 (1971) by permission of the American Institute of Physics).

analysis is approximate. Nevertheless, it is clear that above 50 kilobars significant mixing of the π and π^* orbitals is not improbable. This increased occupation of the π^* orbitals by ligand π electrons reduces their availability for back donation by metal $3d$-electrons. This tends to destabilize the d_π orbitals and, hence, decrease the ligand field, resulting in the transition to the high-spin state. Unfortunately, we cannot verify this by direct measurements of the ligand field because of the presence of intense charge-transfer peaks in the same region. However, we noted previously that there is a good correlation between the low-spin isomer shift and the size of the ligand field. Low isomer shifts correspond to large ligand fields and thus should indicate relatively large amounts of low-spin. Such a correlation at 100 kilobars is presented in Figure 18 for the substituted phenanthrolines. The results are quite satisfactory. Thus, the low- to high-spin transition illustrates how an electronic transition in the ligand system can alter the metal-ligand bonding characteristics and thereby change the field at the iron site enough to allow for a spin rearrangement.

Reduction of Fe(III)

The third electronic transition of interest, and the most general one, is the reduction from the ferric to the ferrous state. Reduction has been observed in approximately fifty compounds since it was first discovered by means of Mössbauer resonance in 1967. Materials exhibiting the effect include the halide, thiocyanate and phosphate systems which have significant ionic character; hydrates in which the neighbours are H_2O molecules; ferricyanides; metallorganic compounds like the acetates and acetylacetonates; and biological prototypes such as the hydroxamates and porphyrin derivatives. Furthermore, the ferrates with higher oxidation states also have been shown to reduce with pressure. The reduction process involves the transfer of a ligand electron, probably from a nonbonding orbital, to the metal $3d_\pi$ orbital, leading a hole on the ligand. The electron and hole probably remain closely associated at the reduced iron site. Thus, the new ground state of the system consists of a ferrous ion and a collectively oxidized set of ligands. It is the presence of this hole in the ligand system which reduces the likelihood that the Fe(II) site will display typical ferrous properties such as infrared stretching frequencies, etc.

An understanding of the electronic basis for the reduction is best obtained from studies on homologous series in which the derivatives have similar structural and bonding characteristics. One such series consists of the acetylacetonate derivatives which form a group of rather covalent complexes, as shown in Figure 19 [Frank and Drickamer, 1972]. The nature and position of the side groups R_1, R_2 and R_3 on the quasi-aromatic ring have been changed, thus systematically varying the electronic environment at the oxygens and, hence, the nature of the metal-ligand interaction. The parent compound ACA is shown as the first ligand. As will be demonstrated later, the dipivaloyl (DPM), methyl (MACA), phenyl (PACA) and ethyl (EACA) derivatives lead to a higher

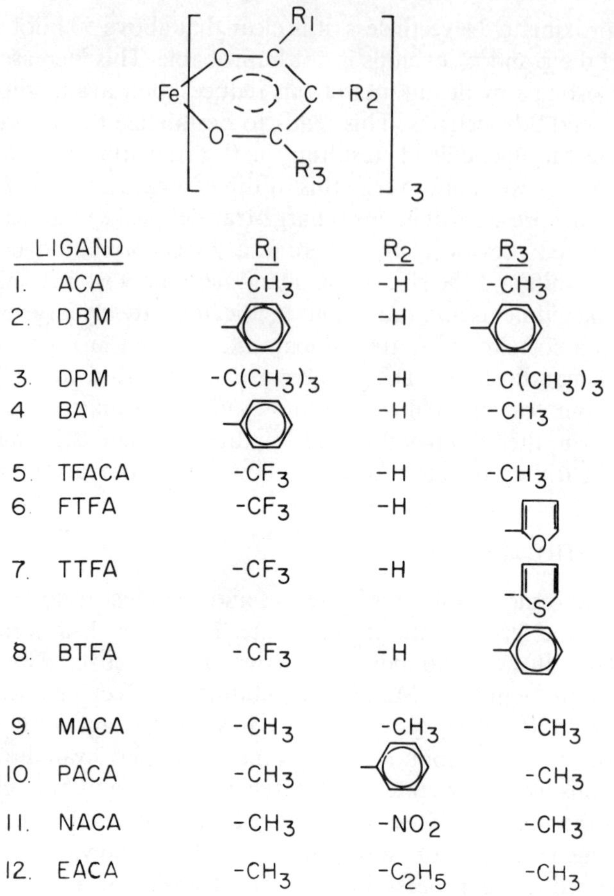

LIGAND		R_1	R_2	R_3
1.	ACA	$-CH_3$	-H	$-CH_3$
2.	DBM		-H	
3.	DPM	$-C(CH_3)_3$	-H	$-C(CH_3)_3$
4	BA		-H	$-CH_3$
5.	TFACA	$-CF_3$	-H	$-CH_3$
6.	FTFA	$-CF_3$	-H	
7.	TTFA	$-CF_3$	-H	
8.	BTFA	$-CF_3$	-H	
9.	MACA	$-CH_3$	$-CH_3$	$-CH_3$
10.	PACA	$-CH_3$		$-CH_3$
11.	NACA	$-CH_3$	$-NO_2$	$-CH_3$
12.	EACA	$-CH_3$	$-C_2H_5$	$-CH_3$

Figure 19. Structure of substituted β-diketones. (Reproduced from C. W. Frank and H. G. Drickamer, *J. Chem. Phys.*, **56**, 3552 (1972) by permission of the American Institute of Physics).

electron density on the oxygens than for the case of ACA. The dibenzoyl (DBM) and benzoyl (BA) derivatives have electron donor tendencies comparable to ACA and the fluorinated (TFACA, FTFA, TTFA and BTFA) and nitro (NACA) derivatives are poorer electron donors than ACA. Typical Mössbauer spectra for the parent compound of the series $Fe(ACA)_3$ are shown in Figure 20. At atmospheric pressure the spectrum shows only high-spin Fe(III). By 41 kilobars, measurable high-spin Fe(II) has appeared. At the highest pressures, as shown in Figure 21, approximately 70 per cent conversion to the ferrous state has occurred. Upon release of pressure and attempts at removal of residual mechanical strain by powdering the sample, the spectrum returns substantially to the original ferric state. The process used to remove the strain is rather inefficient; it is felt that a complete release of strain would result in an entirely

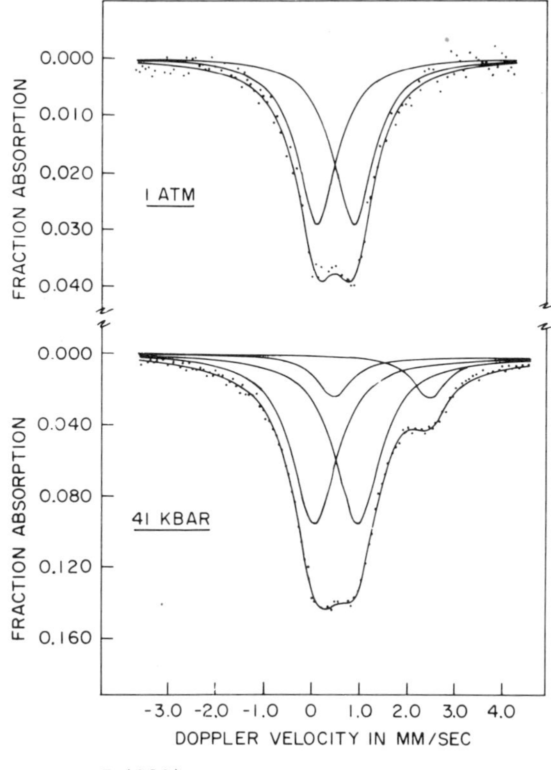

Figure 20. Mössbauer spectra of ferric acetylacetonate at 1 atm and 41 kbar. (Reproduced from C. W. Frank and H. G. Drickamer, *J. Chem. Phys.*, **56**, 3553 (1972) by permission of the American Institute of Physics).

ferric state. As noted earlier, interaction between sites is expected to play an important role in the conversion process. That this reduction is indeed cooperative to some degree is shown by the fact that the conversion occurs over a wide pressure range and that there is some degree of hysteresis upon release of pressure.

A convenient means of expressing the conversion is in terms of the equilibrium constant $K = C_{II}/C_{III} = AP^M$, as shown in Figure 22 for $Fe(ACA)_3$. It is difficult to measure accurately conversions less than 10 per cent. Thus, the low pressure data show some slight deviation from linearity. However, a linear log–log relationship is obtained over a wide pressure range. Smoothed K plot data for several of the acetylacetonate derivatives are shown in Figure 23. For these compounds, which are poor π acceptors but exhibit a range of σ donor tendencies, the data may generally be fitted by a straight line over much of the pressure range. There is a general tendency for the conversions to converge at

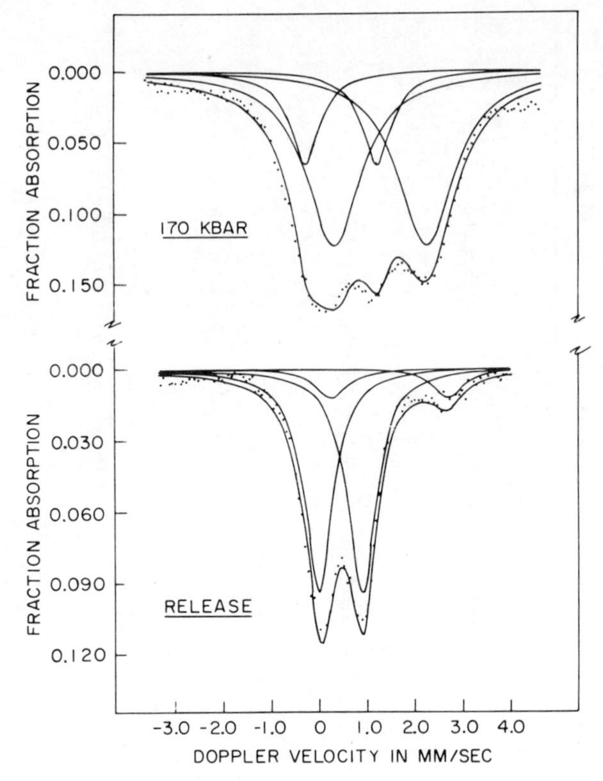

Figure 21. Mössbauer spectra of ferric acetylacetonate at 170 kbar and after release of pressure. (Reproduced from C. W. Frank and H. G. Drickamer, *J. Chem. Phys.*, **56**, 3553 (1972) by permission of the American Institute of Physics).

high pressure. Thus, derivatives with large initial conversions have small slopes, and vice versa. This may be demonstrated quite well with a plot of log A versus M in Figure 24. All twelve compounds are represented. Good electron donors tend to have large A and small M values whereas poor electron donors have small A and large M values.

Complementary evidence for the reduction process may be obtained from optical measurements of the areas under the charge-transfer peaks. Two effects are competing in the measurement. In the first place, a general increase in integrated charge-transfer band intensity reflects an increase in transition moment with decreasing interatomic distance, as shown by Mulliken [Mulliken, 1952A, 1952B]. On the other hand, a decrease in intensity would accompany the loss of ferric sites as they are converted to ferrous ones. Typical data are shown in Figure 25 for one of the derivatives—$Fe(PACA)_3$. The Mössbauer results are indicated by the dashed line; the optical data by the points and solid line. The agreement is quite satisfactory.

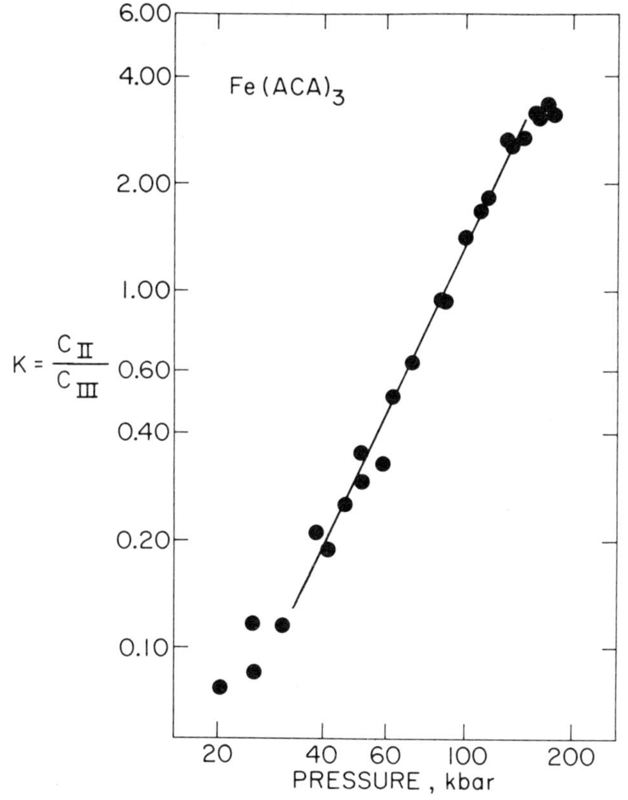

Figure 22. Log K versus log P for ferric acetylacetonate. (Reproduced from C. W. Frank and H. G. Drickamer, *J. Chem. Phys.*, **56**, 3558 (1972) by permission of the American Institute of Physics).

In analysing the factors influencing the degree of conversion of Fe(III) to Fe(II) with pressure, it is desirable to have a measure of the electron donor tendencies of the ligand system at high pressure. There are several such parameters at atmospheric pressure such as the Hammett σ electrophillic substitution constants, the acid dissociation constant, the half wave potential from polarography and the appearance potential from electron impact mass spectrometry. Unfortunately, these cannot be measured in the solid state at high pressure. However, good correlations exist between these parameters and the ferric isomer shift. Two of these are shown in Figure 26 where the numbers refer to the different diketone derivatives and the chemical parameters have been plotted such that movement to the right corresponds to an increase in electron donor tendency. In all cases an increase in electron donor ability corresponds to a decrease in isomer shift. Thus, it is possible to use the change

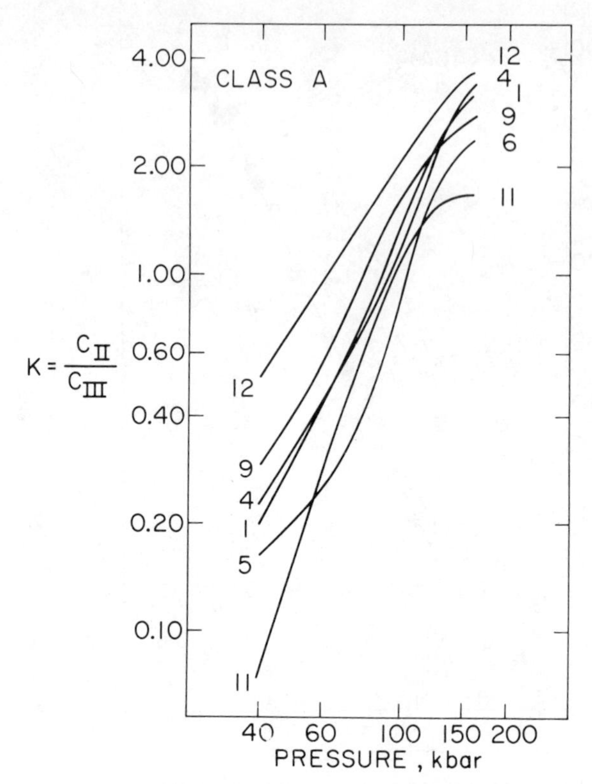

Figure 23. Log K versus log P for class A compounds. (Reproduced from C. W. Frank and H. G. Drickamer, *J. Chem. Phys.*, **56**, 3558 (1972) by permission of the American Institute of Physics).

in isomer shift to examine the influence of the changing donor ability on the conversion process at high pressure. We would expect an increase in conversion with pressure for all compounds due to the reduced separation between the ligand non-bonding and metal $3d_\pi$ levels. However, this should be modulated by the electron donor tendency of the ligand system. Such an effect is illustrated in Figure 27 where the change in conversion is plotted against the change in isomer shift. Those compounds which show a relatively large decrease in isomer shift (increase in electron donor ability) show a large increase in conversion while those which exhibit an increase in isomer shift show very little change in conversion.

The second series in which systematic variation of electronic properties has been made is that of the hydroxamate derivatives with general formula [Grenoble and Drickamer, 1971].

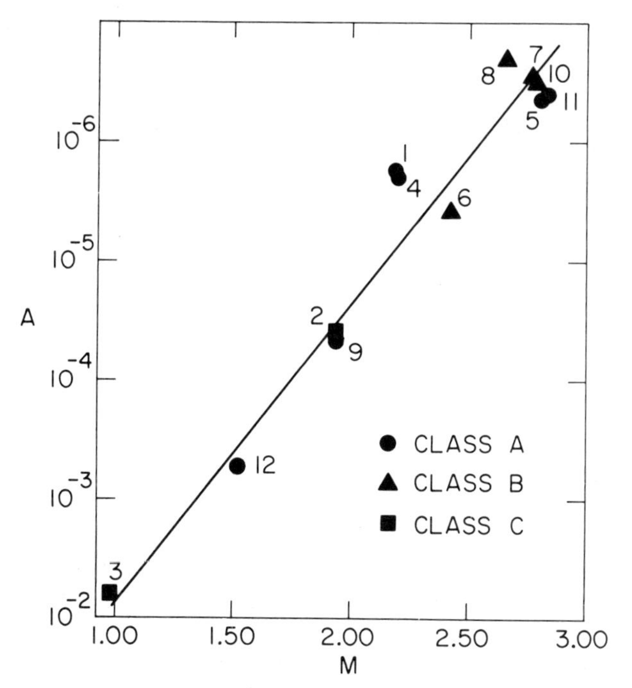

Figure 24. Log A versus M for all diketone derivatives. (Reproduced from C. W. Frank and H. G. Drickamer, *J. Chem. Phys.*, **56**, 3560 (1972) by permission of the American Institute of Physics).

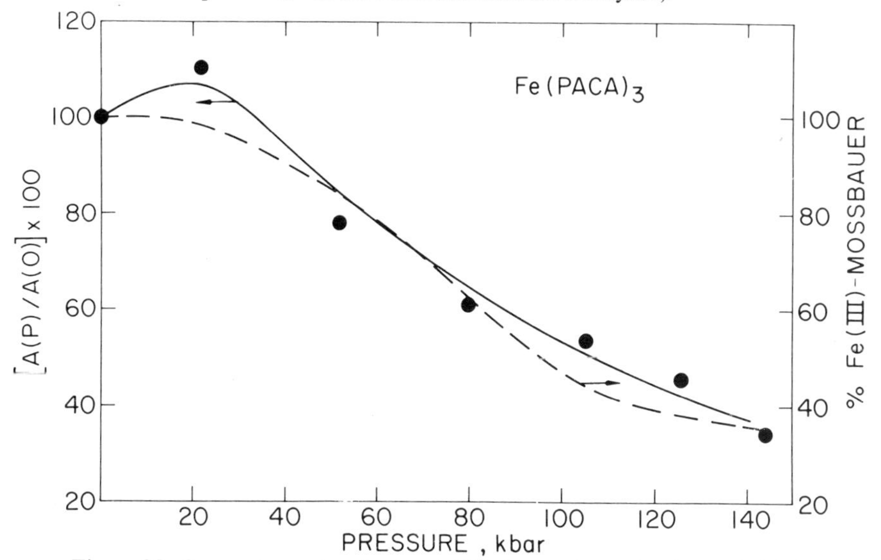

Figure 25. Comparison of optical and Mössbauer determinations of Fe(III) to Fe(II) conversion for $Fe(PACA)_3$. (Reproduced from C. W. Frank and H. G. Drickamer, *J. Chem. Phys.*, **56**, 3562 (1972) by permission of the American Institute of Physics).

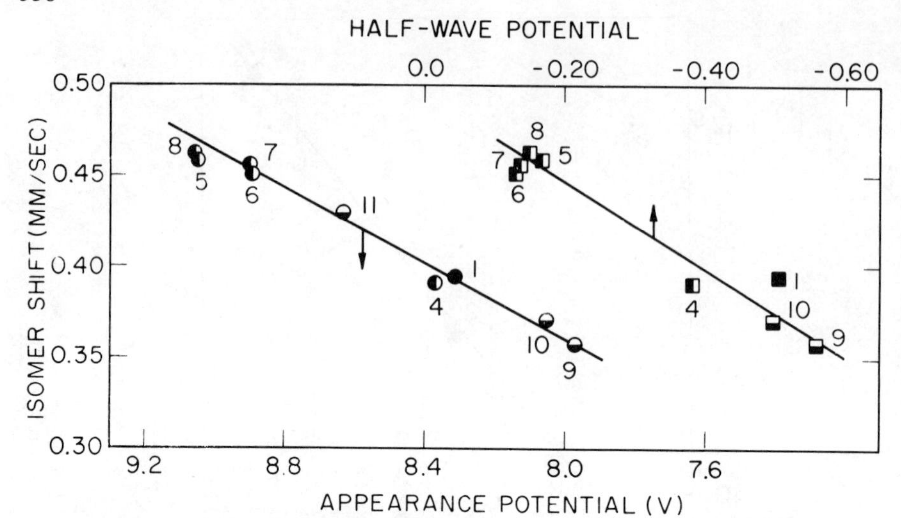

Figure 26. Fe(III) isomer shift versus half-wave potential and appearance potential. (Reproduced from C. W. Frank and H. G. Drickamer, *J. Chem. Phys.*, **56**, 3556 (1972) by permission of the American Institute of Physics).

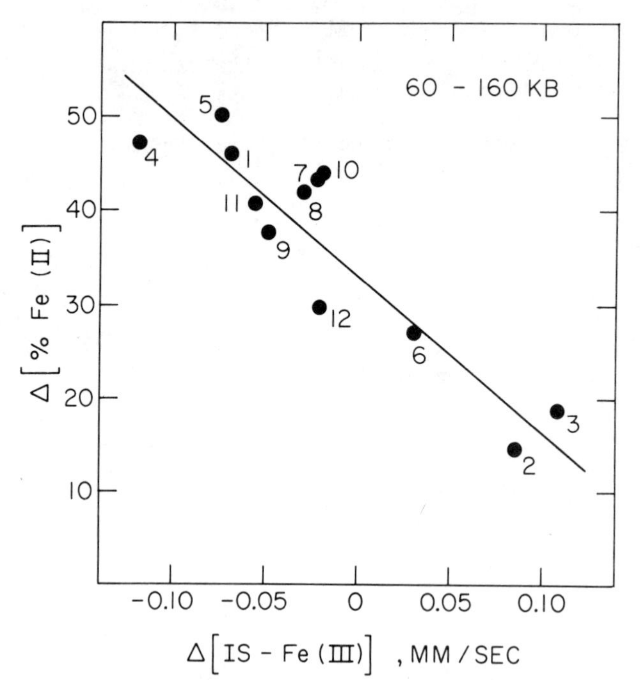

Figure 27. Change in conversion versus change in isomer shift (60 to 160 kbar). (Reproduced from C. W. Frank and H. G. Drickamer, *J. Chem. Phys.*, **56**, 3561 (1972) by permission of the American Institute of Physics).

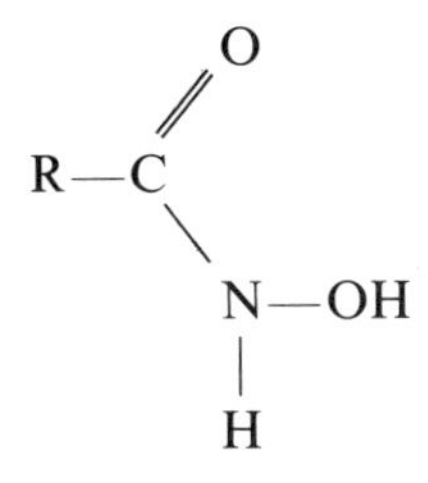

As in the acetylacetonates, the coordination to the iron is through the oxygens after removal of the acidic proton. We include these data in the discussion for two reasons: First, they provide a good illustration of the effect of temperature on the reduction process. Second, the optical charge-transfer bands may be assigned unambiguously to ligand-to-metal charge-transfer so that analysis of the thermal energy is possible. Figure 28 shows the room temperature Mössbauer spectra of the salicyl derivative at 4 and 138 kilobars. The asymmetry

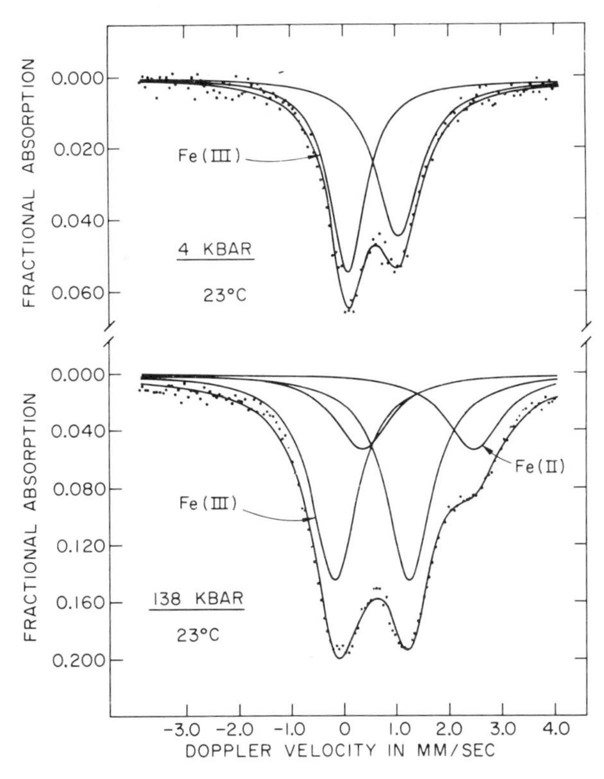

Figure 28. Mössbauer spectra of $Fe(SHA)_3$ at 4 kbar (23 °C) and 138 kbar (23 °C). (Courtesy D. G. Grenoble and H. G. Drickamer, *Proc. Nat. Acad. Sci.*, **68**, 550 (1971).)

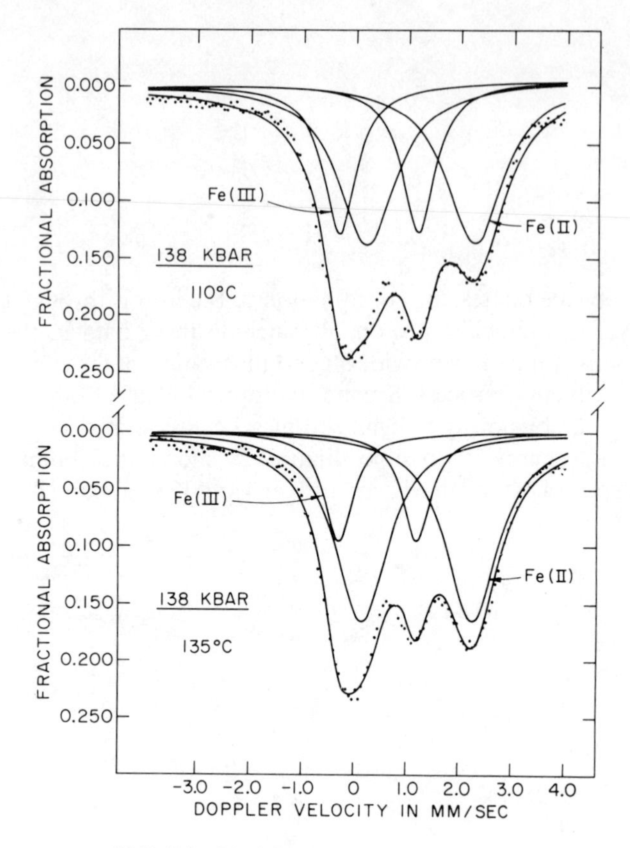

Figure 29. Mössbauer spectra of $Fe(SHA)_3$ at 138 kbar (110 °C) and 138 kbar (135 °C). (Courtesy D. G. Grenoble and H. G. Drickamer, *Proc. Nat. Acad. Sci.*, **68**, 550 (1971).)

of the ferric doublet at low pressure is attributed to spin-spin coupling between iron sites. As site-to-site distance decreases with increasing pressure, the spin-spin relaxation time is shortened and the asymmetry disappears. The effect of temperature on the conversion is seen by comparing the 23 °C high pressure

Table 2. Optical versus thermal excitation: ligand-metal charge-transfer in ferric hydroxamates and ferrichrome A. For 10 per cent reduction of Fe(III). (Courtesy H. G. Drickamer, C. W. Frank and C. P. Slichter, Proc. Nat. Acad. Sci., 69, 936 (1972).)

Compound	Pressure	$h\nu_{max}$	$\delta E_{1/2}$	E_{th}
AHA	125	2.80	0.900	−0.11
BHA	105	2.70	0.875	−0.06
SHA	70	2.54	0.840	−0.02
FA	37	2.65	0.835	+0.11

spectra with those at the same pressure and temperatures of 110 °C and 135 °C, as shown in Figure 29. The conversion increases considerably, illustrating the endothermic behaviour which is typical for the reduction process.

Application of the expression for the thermal energy to the optical data yields the results shown in Table 2. Here the thermal energy has been calculated at the pressure at which 10 per cent conversion is observed. As one would expect, the thermal energy is approximately zero at this pressure in all cases.

Reduction of Cu(II)

In both of these studies of reduction, the correlation between Mössbauer and optical results has been fairly good. This opens up the possiblity that pressure induced oxidation state changes may be studied in other transition metal systems for which no Mössbauer nuclide exists. Although this is somewhat of a digression from our examination of the high pressure chemistry of iron compounds, a brief consideration of one other system will serve to establish the greater generality of the effect.

Here we will consider complexes of Cu(II) in which measurable reduction to Cu(I) was observed in five of the six molecules studied [Wang and Drickamer, 1973]. The optical spectrum at 10 kilobars for the diethyldithiocarbamato derivative $Cu(DTC)_2$ is shown in Figure 30. The peak at 22,000 cm^{-1} is assigned to ligand to Cu(II) charge-transfer and the peak near 15,000 cm^{-1} to the *d–d* transition. The high pressure spectrum is shown in Figure 31 where a new peak appears at 18,000 cm^{-1}, assigned to Cu(I) to ligand charge-transfer.

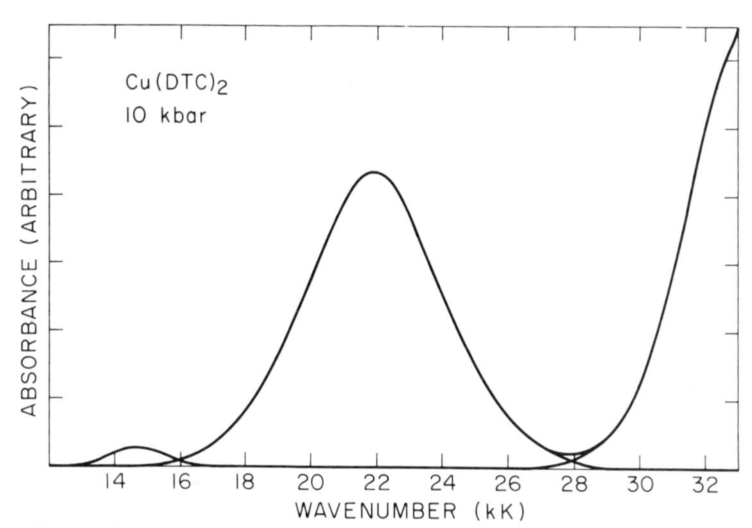

Figure 30. Spectrum of $Cu(DTC)_2$ at 10 kbar. (Reproduced from P. J. Wang and H. G. Drickamer, *J. Chem. Phys.*, **59**, 715 (1973) by permission of the American Institute of Physics).

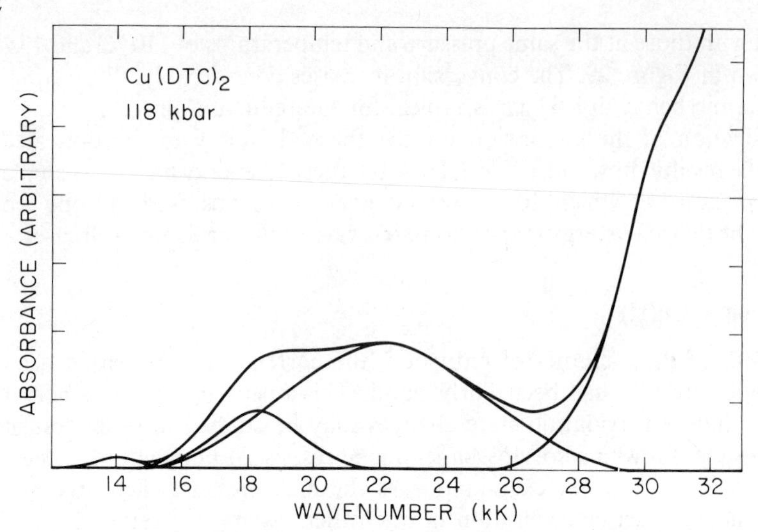

Figure 31. Spectrum of $Cu(DTC)_2$ at 118 kbar. (Reproduced from P. J. Wnag and H. G. Drickamer, *J. Chem. Phys.*, **59**, 715 (1973) by permission of the American Institute of Physics).

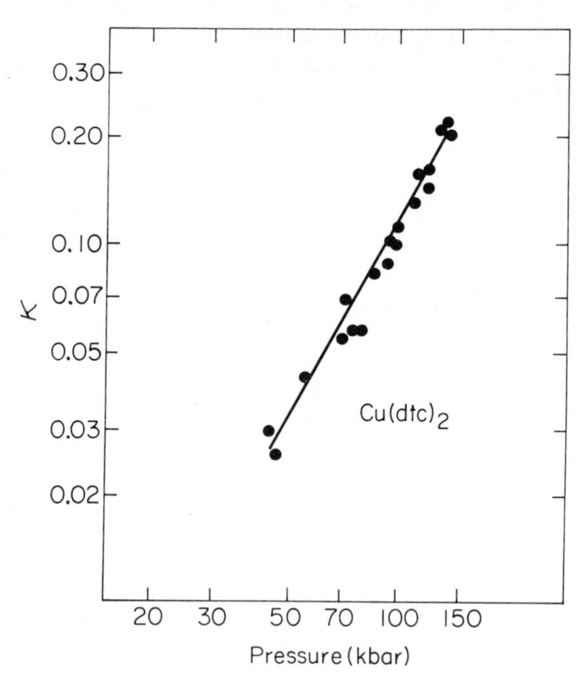

Figure 32. Log K versus log P for $Cu(DTC)_2$. (Reproduced from P. J. Wang and H. G. Drickamer, *J. Chem. Phys.*, **59**, 716 (1973) by permission of the American Institute of Physics).

Estimates of relative peak areas indicate that approximately 16 per cent of the charge-transfer intensity may be attributed to the Cu(I) state. Further evidence for the reduction process is found in the decrease in integrated intensity of the *d–d* excitation peak with pressure. This would be anticipated since no *d–d* transition is possible in the reduced d^{10} configuration of Cu(I). A qualitative estimate of the thermal energy, assuming $\omega = 1.1\omega'$, indicated that E_{th} would vanish at a pressure of 120 kilobars. Thus, the postulation of thermal electron transfer is not unreasonable. It is interesting to note that the equilibrium constant expression $K = AP^M$ used for the ferric/ferrous conversion applies equally well for the cupric/cuprous case, as shown in Figure 32 for the $Cu(DTC)_2$ derivative. A good linear log–log plot is obtained over the range 50–150 kilobars.

SUMMARY

In summary, iron compounds exhibit a number of interesting phenomena under high pressure. Both decreases and increases of spin state of Fe(II) are observed as is the reduction of Fe(III) to Fe(II). It has been possible to analyse the results in the context of a theory of pressure-induced electronic transitions. The essential feature is that relative energy level shifts may lead to the establishment of a new ground state at high pressure. The simultaneous coexistence of converted and unconverted sites over a wide pressure range has been justified on the basis of repulsive interaction between sites. Furthermore, the probability of occurrence of a pressure induced electronic transition depends on the thermal energy associated with the process. This thermal energy has been related to the optical transition data.

To be sure, the materials considered in this discussion of high pressure solid state iron chemistry exhibit fairly covalent bonding and are reasonably compressible, in contrast to the predominantly ionic bonding and low compressibilities found in mineral structures. However, the effect of these differences on the high pressure behaviour of minerals should be largely one of degree, not of kind. Thus, in the first place, it is expected that all high-spin Fe(II) would transform to the low-spin state long before the core is reached. Such a transition must be considered in analyzing volume discontinuity data. The second Fe(II) spin transition, from low- to high-multiplicity, is most likely of lesser geological significance because of the decreased role of back donation in minerals. Finally, the reduction process should play a significant role, particularly in the analysis of electrical conductivity data. It seems clear that ferric sites in mineral structures should transform to nominal ferrous ones at sufficiently high temperatures and pressures. However, in light of the data presented on copper and recent results showing pressure induced reduction of Mn(III) in $K_3[Mn(CN)_6]$ [Ahsbahs *et al.*, 1974], the phenomenon appears to have broader implications beyond iron chemistry. This is a fertile area for additional research.

REFERENCES

Ahsbahs, H., Dehnicke, G., Dehnicke, K., and Hellner, E., *Paper presented at the European High Pressure Conference*, Marburg, Germany. March 19–21, 1974.
Bargeron, C. B., and Drickamer, H. G. (1971). *J. Chem. Phys.*, **55**, 3471.
Bargeron, C. B., Avinor, M., and Drickamer, H. G. (1972). *Inorg. Chem.*, **10**, 1338.
Burns, R. G. (1972). *Can. J. Spectroscopy*, **17**, 51.
Drickamer, H. G., Frank, C. W., and Slichter, C. P. (1972). *Proc. Nat. Acad. Sci.*, **69**, 933.
Fisher, D. C., and Drickamer, H. G. (1971). *J. Chem. Phys.*, **54**, 4825.
Frank, C. W., and Drickamer, H. G. (1972). *J. Chem. Phys.*, **56**, 3551.
Grenoble, D. C., and Drickamer, H. G. (1971). *Proc. Nat. Acad. Sci.*, **68**, 549.
Jorgensen, C. K. (1959). *Molec. Phys.*, **2**, 309.
Konig, E., and Madeja, K. J. (1967a). *Inorg. Chem.*, **6**, 48.
Konig, E., and Madeja, K. J. (1967b). *Spectrochim. Acta.*, **23A**, 45.
Konig, E., and Madeja, K. J. (1968). *J. Am. Chem. Soc.*, **90**, 1146.
Mulliken, R. S. (1952a). *J. Am. Chem. Soc.*, **74**, 811.
Mulliken, R. S. (1952b). *J. Phys. Chem.*, **56**, 801.
Slichter, C. P., and Drickamer, H. G. (1972). *J. Chem. Phys.*, **56**, 2142.
Wang, P. J., and Drickamer H. G. (1973). *J. Chem. Phys.*, **59**, 713.

Behaviour of Iron Compounds at High Pressure, and the Stability of Fe_2O *in Planetary Mantles*

R. G. J. Strens
School of Physics, The University,
Newcastle upon Tyne NE1 7RU
England

INTRODUCTION

The possible occurrence in planetary interiors of such exotic compounds as Fe_2O has been discussed in recent works by Dubrovskii and Pan 'kov (1972), Soroktin (in Bullen, 1973) and Bullen (1973). Studies of the optical and Mössbauer spectra of transition metal compounds at high pressure have demonstrated the reduction of Fe^{3+} to Fe^{2+} (Drickamer and Frank, 1973; Huggins, this volume), Cu^{2+} to Cu^{+} (Wang and Drickamer, 1973), and Mn^{3+} to Mn^{2+} (Gibbons *et al.*, 1974). In addition, high pressure optical spectra of ferromagnesian silicates and magnesiowüstites obtained by Mao (1973, and this volume) can be interpreted as showing that the configuration $Fe^{+}O^{-}$ becomes increasingly stable relative to $Fe^{2+}O^{2-}$ with increasing pressure. Continuation of this process might lead to the formation of $Fe^{0}O^{0}$ (covalently bonded FeO) at pressures in the megabar range. The sum of the Pauling (1960) octahedral covalent radii for Fe and O is only 1.92 Å, compared with the sum of octahedral ionic radii of 2.16 Å, implying a volume reduction of 30 per cent.

Our ideas about the high pressure chemistry of iron are thus in a state of flux. This paper attempts to deal semiquantitatively with two aspects of the subject which seem to be amenable to simple calculation, and which are also of geophysical interest, namely the possible stability of Fe_2O and the oxidation state of iron in the mantle. The spin-pairing transition was considered by Strens (1969), and his calculations of transition pressure and slope are also revised and corrected, and the behaviour of gillespite discussed.

STABILITY OF Fe_2O

At surface pressure and temperature, the stable assemblage of bulk composition Fe_2O is $Fe_3O_4 + 5\ \alpha$-Fe (magnetite + iron). This is replaced at about 830 K by $Fe_{1-x}O + \alpha$-Fe (wüstite + iron), the wüstite reacting to form stoichiometric FeO at high pressure (Katsura *et al.*, 1967). Ferrous ions in FeO are octahedrally coordinated by O^{2-} at 2.1615 Å, and possess the high-spin electronic configuration $(Ar)t_{2g}^4e_g^2$. A transition to the low-spin form $(Ar)t_{2g}^6e_g^0$ is predicted to occur at about 340 kb (Strens, 1969) if it is not pre-empted by some other reaction or electronic rearrangement. Metallic iron also undergoes phase transitions with increasing temperature (α or ε to γ) and pressure (α or γ to ε), the stable form in the lower mantle probably being ε (Andrews, 1973). As pressure and temperature increase along the geothermobar, the sequence of stable assemblages of composition Fe_2O will thus be as follows, unless Fe_2O or some other compound becomes stable:

Magnetite	+ α-iron (bcc)	$T = 300$ K, $P = 0$
Wüstite	+ α-iron	$T > 830$ K
Wüstite	+ γ-iron (fcc)	$T > 1000$ K
FeO(hs)	+ γ-iron	$P > 65$ kb
FeO(ls)	+ γ-iron	$P > 520$ kb
FeO(ls)	+ ε-iron (hcp)	$P > 650$ kb

The pressures and temperatures quoted are approximate, and the last two stages may occur in reverse order. Recent work by Mao (1974) suggests that stoichiometric FeO may again be replaced by wüstite above about 200 kb, and that high-pressure forms of Fe_2O_3 and Fe_3O_4 may occur between about 200–300 kb and the high- to low-spin transition pressure.

The stability of Fe_2O will be determined mainly by the Fe–O distance and structure type assumed. Plots of metal-oxygen distance (r) against the logarithm of the formal charge (z) for both isonuclear and isoelectronic series are linear for $z \leqslant 4$, and suggest an octahedral Fe^+–O distance of about 2.40 Å, only 1 per cent less than the octahedral Na^+–O distance of 2.42 Å found by Shannon and Prewitt (1969), and 11 per cent longer than the Fe^{2+}–O distance in stoichiometric FeO. Since Na_2O crystallises in the antifluorite structure, this structure will be assumed for Fe_2O also. Both Na^+ and Fe^+ would be tetrahedrally coordinated by oxygen, and both the r–z plots referred to above and the r–s curves of Brown and Shannon (1973) indicate a tetrahedral M–O distance some 0.13 Å shorter than the octahedral distance, i.e. 2.27 Å for Fe^+ and 2.30 Å for Na^+. The observed Na^+–O distance in Na_2O is 2.40 Å, suggesting that O–O repulsion or some other factor prevents the Na–O bonds reaching their preferred length. If the same effect occurs for Fe^+, the corresponding distance should be about 2.37 Å. For the purposes of calculation a minimum estimate of 2.27 Å and a maximum value of 2.40 Å will be assumed for the tetrahedral Fe^+O^{2-} bond length, the smaller value conferring greater stability.

The following highly simplified treatment of the energetics of the reaction:

$$Fe_2O \rightleftharpoons FeO + Fe \tag{1}$$

neglects the transitions between the various forms of iron, assumes a stoichiometric FeO, and uses relatively crude treatments of lattice energy.

Table 1. Zero-pressure values of interatomic distance, molar volume, crystal field splitting, and bulk modulus.

	r(Å)	Δ(eV)	V_o(cm^3)	K_o(Mb)
FeO(hs)	2.1615[a]	1.13	12.17	1.44[b]
FeO(ls)	2.01[c]	1.63	9.78	1.80[b]
$Fe^{+}O^{-}$	(2.04)[d]	—	10.23	(1.8)[e]
$Fe^{0}O^{0}$	1.92[f]	—	8.53	(1.8)[e]
Fe_2O	2.27[g]	0.382	21.70	0.603[b]
Fe_2O	2.40[h]	0.297	25.64	0.510[b]
α-Fe	—	—	7.093[i]	1.712[i]
ε-Fe	—	—	6.713[i]	1.827[i]
α-Fe_2O_3	—	—	30.28[j]	2.16[b]
Fe_3O_4	—	—	44.53[j]	1.83[b]

[a]Katsura *et al.*,(1967); [b]bulk moduli calculated by method of Anderson and Anderson (1970); [c]Shannon and Prewitt, (1969); [d]Average of high-spin FeO and Fe^0O^0 values; [e]assumed equal to K_0 for low-spin FeO; [f]Sum of Pauling's, (1960) octahedral covalent radii; [g]lower estimate (see text); [h]higher estimate (see text); [i]Andrews, (1973); [j]Clark, (1966).

The difference between the lattice energies, i.e. the heats of formation of FeO and Fe_2O from the gaseous ions, is given by the approximate relation:

$$E_1 = 12.60\left[\frac{A_1 z_1^2}{r_1} - \frac{A_2 z_2^2}{r_2}\right]\text{ eV} \tag{2}$$

where the Madelung constants are $A_1 = 1.74756$, $A_2 = 5.03878$, $z_1 = 2$, $z_2 = 1$. The Born exponent is taken as 8 for both Fe^{+}–O^{2-} and Fe^{2+}–O^{2-} bonds, and r_1, r_2 are taken from Table 1.

To this must be added the difference in energies required to form the gaseous ions from their elements in the standard state:

$$E_2 = p_2 - (p_1 + s) = 3.95\text{ eV} \tag{3}$$

where $p_1 = 7.90$ and $p_2 = 16.18$ eV are the first and second ionization potentials of iron, and $s = 12.23$ eV is its heat of sublimation (Franklin *et al.*, 1969). Other terms, including the dissociation energy of O_2 and the first and second electron affinities of O, cancel out.

The remaining important energy terms are the changes in crystal field stabilization and electron pairing energies. The crystal field splitting parameter Δ_0 in FeO with the NaCl structure is taken as 1.13 eV at $r_0 = 2.16$ Å, and is assumed to vary as the inverse fifth power of the metal oxygen distance. The tetrahedral splitting parameter is assumed to be 4/9 of the octahedral value at the same interatomic distance, and to be unaffected by the oxidation state of the cation. The last assumption may seem surprising, but it appears to be approximately true that Δ *at a given r* is insensitive to the charge on the transition–metal ion. On these assumptions, the crystal field contribution to the energy of reaction (1) in the case of high-spin FeO is given by:

$$E_3^{hs} = \frac{2\,\Delta_0}{5}\left(\frac{r_0}{r_1}\right)^5 - \frac{12\,\Delta_0}{5}\frac{4}{9}\left(\frac{r_0}{r_2}\right)^5 \tag{4}$$

and by

$$E_3^{ls} = \frac{12\,\Delta_0}{5}\left(\frac{r_0}{r_1}\right)^5 - \frac{12\,\Delta_0}{5}\frac{4}{9}\left(\frac{r_0}{r_2}\right)^5 \tag{5}$$

for the low-spin case. The spin-pairing energy in regular octahedral coordination if given by:

$$E_4 = 2\,\Delta_0\left(\frac{r_0}{r_p}\right)^5 - 4\,K \tag{6a}$$

where r_p is the Fe–O distance at the pressure considered, and $K \simeq 0.74$ eV (Strens, 1969). A more general formulation, applying to distorted polyhedra, is:

$$E_4 = \varepsilon_{14} + \varepsilon_{23} - 4\,K \tag{6b}$$

where the ε_{ij} are the energy differences between the d-orbital energy levels indicated by the subscripts, e.g. ε_{04} is the energy difference between the lowest- and highest-lying 3*d*-levels. The results of the calculation are summarised in Table 2, which shows Fe_2O to be unstable by a margin of 16.3 to 18.0 eV at zero pressure, in agreement with the observation that Fe^+ is below detection limits in both crystalline oxides and silicate melts. Stability of Fe_2O at high pressure then requires:

$$\int \Delta V \mathrm{d}P > 16\,\mathrm{cm}^3\,\mathrm{Mb} \tag{7}$$

This condition is unlikely to be met. Even with an Fe^+–O^{2-} distance of 2.27 Å, which is believed to be the minimum value which can reasonably be assumed,

Table 2. Stability of Fe_2O relative to FeO + α-Fe at zero pressure (energies in eV, negative values favour Fe_2O).

	$r_2 = 2.27$ Å	$r_2 = 2.40$ Å
E_1	3.95	3.95
E_2	12.80	14.32
E_3	−0.45	−0.26
$\sum$	16.30	18.01

Spin-pairing energies for the FeO phases at $P = 0$ and $P = 340$ kb (positive values favour low-spin state)

Spin-state	P (kb)	r_1 (Å)	E_4 (eV)
high	0	2.1615	−0.70
high	340	2.03	0.00
low	0	2.01	+0.15
low	340	1.92	+1.13

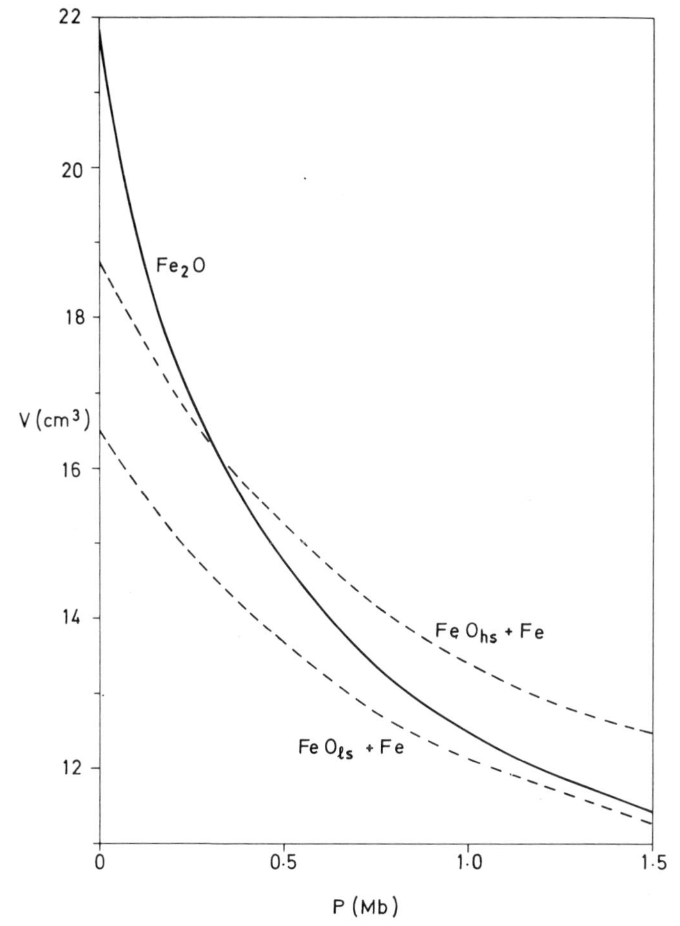

Figure 1. Volume–pressure relations for isochemical assemblages of composition Fe_2O calculated from data in Table 1 using Birch's (1952) equation of state. Only the low-volume ($r_2 = 2.27$ Å) Fe_2O is shown.

the zero-pressure volume of Fe_2O is above that of isochemical mixtures of Fe + FeO. When the P–V curves are calculated using compressibilities derived from the Anderson–Anderson (1970) and the Birch (1952) equations, the data of Figure 1 are obtained. The volume of Fe_2O is less than that of high-spin FeO above 0.3 Mb and $\Delta V \mathrm{d}P$ will favour Fe_2O above about 0.5 Mb. However, high-spin FeO should have transformed either to low-spin FeO or to some other form at that pressure, in which case Fe_2O will become increasingly unstable relative to some form of FeO + Fe to indefinitely high pressures.

Temperature is not expected to have a significant effect on the stability relations at pressures above a few tens of kilobars, where $T\Delta S$ is usually small compared with $P\Delta V$ for (P,T) on the geothermobar.

CHANGES IN OXIDATION STATE WITH PRESSURE

Accepting that there is a general trend towards reduced forms of iron at high pressure, two main mechanisms of reduction may be envisaged, the first represented by the reaction:

$$2\,Fe^{3+} + Fe \rightleftharpoons 3\,Fe^{2+} \tag{8}$$

This should determine the Fe^{3+}/Fe^{2+} ratio in systems containing metallic iron, and would presumably have applied during the early stages of the evolution of planetary mantles. It has long been known that silicate melts held in iron crucibles contain significant amounts of ferric iron, implying that the free energy change of reaction (8) is only a few times RT under these conditions. For small concentrations of Fe^{3+}, the Fe^{3+}/Fe^{2+} ratio will change *e*-fold for each change of RT in this energy. We may thus define an *e*-folding pressure P_e as that which alters the free energy of reaction (8) by RT. For a model temperature of 1800 K, this requires:

$$RT = \int_0^P \Delta V \mathrm{d}P \simeq 150 \text{ kb cm}^3$$

Since 1 kb $\text{cm}^3 \simeq 12\,R$. From the P–V curves of Figure 2, it can be seen that if iron in other compounds behaves like iron in hematite and high-spin FeO,

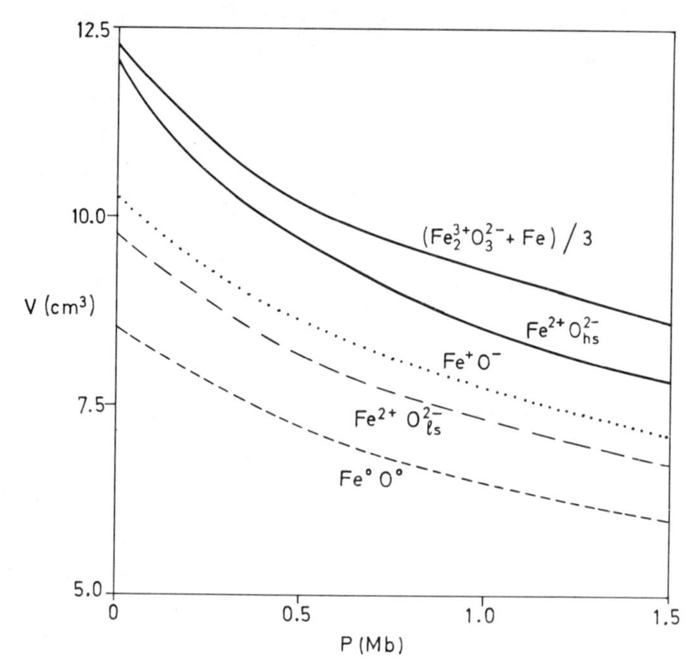

Figure 2. Volume–pressure relations calculated for isochemical assemblages of composition FeO, using data from Table 1 and the Birch equation of state.

then P_e is several hundred kilobars. Replacement of high-spin FeO by either low-spin FeO or Fe^+O^- would reduce P_e to a few tens of kilobars. We might thus expect the Fe^{3+}/Fe^{2+} ratio in the upper mantle to be that determined during the early evolution by reaction (8), varying only slowly with pressure to a depth at which a high-pressure form of FeO appears. The content of Fe^{3+} would then decrease very rapidly with depth. The situation would be complicated by the incorporation of both Fe^{2+} and Fe^{3+} into mineral solid solutions, preventing accurate calculation.

The second reduction mechanism is less well understood, but may be written:

$$Fe^{3+} \rightleftharpoons Fe^{2+} + \square \quad (9a)$$

$$\text{or} \quad Fe^{3+}O^{2-} \rightleftharpoons Fe^{2+}O^- \quad (9b)$$

where the symbol □ represents a 'hole', or missing electron in an orbital or band which would otherwise be occupied. Like unpaired electrons, holes can diffuse through the crystal structure until they are trapped or destroyed by reaction. They are then no longer available for recombination with the reduced metal ions, which can then survive quenching to surface pressures. This may be shown by rhodonite ($MnSiO_3$) shocked to several hundred kilobars for a fraction of a microsecond (Gibbons *et al.* 1974), in which the small amount of Mn^{3+} originally present is reduced. Apparently, the holes formed in this material were removed or destroyed before recombination could occur, despite the extremely short reaction time available before the sample returned to surface temperature and pressure.

Further examples of this mechanism are discussed by Drickamer and Frank (1973) in their book, and by Frank and Drickamer, and Huggins (this volume).

The evidence for further reduction of Fe^{2+} to Fe^+ by the process:

$$Fe^{2+} \rightleftharpoons Fe^+ + \square \quad (10)$$

is not yet conclusive, but the optical spectra of magnesiowüstite and of some ferromagnesian silicates reported by Mao (1973, and this volume) show that the absorption edge at about 4 eV shifts to lower energies at a rate of about 0.0028 eV/kb in the pressure range 25–310 kb. As this absorption edge is usually attributed to the ligand–metal charge-transfer:

$$Fe^{2+} \rightarrow O^{2-} = Fe^+O^- \quad (11)$$

it appears that the stability of Fe^+O^- is increasing relative to $Fe^{2+}O^{2-}$ at about this rate. If the 'compound' symbolised as Fe^+O^- has an interatomic spacing between the sum of the octahedral ionic and covalent radii (figure 2 above), the *e*-folding pressure is probably a few tens of kilobars at mantle temperatures, and the observed shift of 0.0028 eV/kb is comparable with $P\Delta V$.

There is at present no direct evidence for the final stage of the reduction:

$$Fe^+ \rightleftharpoons Fe^0 + \square \quad (12)$$

other than the fact that the Pauling covalent radii of Fe and O suggest a volume some 30 per cent below that of high-spin FeO at zero pressure. Thus, covalent FeO should be stabilized relative to 'ionic' FeO by some 0.007 eV/kb initially, which is a very significant change in stability over the 1.35 Mb pressure range occurring in the mantle. Recent comparisons of observed and calculated density gradients in the lower-mantle (Miller, 1972; Holly, 1974) show density increasing faster than expected from the presumed properties of such assemblages as $MgSiO_3$ (perovskite) + MgO or SiO_2 (PbO_2-II) + MgO. One explanation is that the iron content increases with depth, but increasing covalency of the Fe–O bond might produce a similar effect at constant composition.

The changes envisaged occur at pressures beyond the reach of static high-pressure experiments, but should be fast enough to be capable of investigation by shock-wave methods.

SPIN-PAIRING TRANSITIONS

My previous review of spin-pairing in iron minerals (Strens, 1969) gave an estimate of:

$$P_t = 337 + 0.127T\,(^\circ\text{C})$$

for the transition pressure in stoichiometric FeO. Since then, new estimates of the ionic radius of low-spin Fe^{2+} have appeared and R. G. Burns (pers. comm.) has pointed out an error in the calculation of the electronic entropy.

It was originally estimated that spin-pairing would occur in high-spin FeO when the Fe–O distance had been reduced by compression to 2.03 Å, at which point the Fe–O distance would reduce discontinuously to 1.92 Å in the low-spin form. The assumptions and experimental data used still appear reliable, and the calculation of the transition pressure at 300 K stands. The Shannon and Prewitt (1969) estimate of the radius of low-spin Fe^{2+} converts to a zero-pressure Fe–O distance of 2.01 Å, which would be reduced to about 1.93 Å by compression to the transition pressure if $K_0 \simeq 1.8$ Mb. The difference is within the error of the assumptions made, and no revision of the original value of $\Delta V = 1.77$ cm^3/mole is needed.

The calculation of electronic entropy in the 1969 paper was incorrect, the entropy tending to a maximum of:

$$\Delta S_{\text{el}} \rightarrow R\ln 3 = 2.2 \text{ cal deg mole}^{-1}$$

rather than the value of 3.8 originally given. There is no change in the estimate of $S_v = 1.6$ e.u. The revised entropy change of the spin-pairing transition in stoichiometric FeO then becomes:

$$\Delta S = \Delta S_{\text{el}} + \Delta S_v = 3.8 \text{ cal deg mole}^{-1}$$

The revised equation of the reaction is then:

$$P_t = 337 + 0.09T\,(^\circ\text{C})$$

which cuts the geothermobar at a pressure of about 517 kb, equivalent to a depth of about 1300 km.

The Gillespite I–II Transition

Recent work by Abu-Eid *et al.* (1973), and by Bassett and Hazen and Abu-Eid (this volume) has shown the 26 kb transition in gillespite discovered by Strens (1966) to be displacive rather than electronic. It appears that rotation of the silicate sheets causes the $[BaO_8]$ polyhedron to become a square antiprism, with the $[FeO_4]$ group simultaneously folding along one O–Fe–O diagonal to form a very flattened tetrahedron.

Although crystal field calculations yield relatively poor absolute values of *d*-orbital energies in gillespite (Wood, 1971), they should give good estimates of changes in the relative energies of the *d*-orbital energy levels during a minor change in the geometry of the $[FeO_4]$ group. Calculations by M. Dempsey using the method of Wood and Strens (1972) are given in Table 3, and suggest

Table 3. Changes in *d*-orbital energy levels between gillespite-I at zero pressure and gillespite-II at 26 kb. Energies in cm^{-1}

z^2	− 196
$x^2 - y^2$	− 54
xy	− 122
xz	+ 32
yz	− 12
H_{cf}	− 196
$(\varepsilon_{14} + \varepsilon_{23})$	− 156

Calculated by the method of Wood and Strens (1972) from structure data by Hazen and Burnham (1974). 1 eV = 8067 cm^{-1}. H_{cf} is the crystal-field stabilization energy.

that gillespite II at 26 kb is some 156 cm^{-1} further removed from spin-pairing than is gillespite I at zero pressure. Spin-pairing would require a 2.5 per cent contraction of the Fe–O bonds, equivalent to a pressure increment of about 45 kb. The calculation thus suggests a transition pressure to spin-paired 'gillespite III' of at least 70 kb, if there is no further distortion of the $[FeO_4]$ group. Abu-Eid (pers. comm.) has taken gillespite to 80 kb without finding a transition.

The new interpretation of the 26 kb transition does not affect the estimates of spin-pairing pressures in other iron minerals, as these were based on the properties of phenanthroline derivatives.

REFERENCES

Abu-Eid, R. M., Mao, H. K. and Burns, R. G. (1973). *Carnegie Inst. Washington Yearbook*, **72**, 564.

Anderson, D. L. and Anderson, O. L. (1970). *Jour. Geophys. Res.*, **75**, 3494.

Andrews, D. J. (1973). *Jour. Phys. Chem. Solids*, **34**, 825.

Birch, F. (1952). *Jour. Geophys. Res.*, **57**, 227.
Brown, I. D., and Shannon, R. D. (1973). *Acta. Cryst.*, **A29**, 266.
Bullen, K. E. (1973). *Nature*, **243**, 68.
Clark, S. P. (1966). *Handbook of Physical Constants*, Geol. Soc. Amer. Memoir 97.
Drickamer, H.G. and Frank, C.W. (1973). *Electronic Transitions and the High Pressure Chemistry and Physics of Solids*, Chapman and Hall, London.
Dubrovskii V. A. and Pan'kov V. L. (1972). *Izv. Akad. Nauk SSSR, Fiz. Zemli*, **7**, 48 (CA 77, 129025f).
Franklin, J. L., Dillard, J. G., Rosenstock H. M., Herron, J. T., Draxl, K. and Field, F. H. (1969). Ionization Potentials, Appearance Potentials and Heats of Formation of Gaseous Positive Ions. U.S. Department of Commerce Document NSDRN–NBS 26.
Gibbons, R. V., Ahrens, T. H. and Rossman, G. R. (1974). *Amer. Mineral*, **59**, 177.
Hazen, R. M. and Burnham, C. W. (1974). *Amer. Mineral.*, **59**, 1166.
Holly, D. J. (1974). *M.Sc. Thesis*, University of Newcastle upon Tyne.
Katsura, T., Iwasaki, B., Kimura, S. and Akimoto, S. (1967). *Journ. Chem. Phys.*, **47**, 4559.
Mao, H. K. (1973). *Carnegie Inst. Washington Yearbook*, **72**, 552.
Mao, H. K. (1974). *Carnegie Inst. Washington Yearbook*, **73**, 510.
Miller, G. H. (1972). *M.Sc. Thesis*, University of Newcastle upon Tyne.
Pauling, L. (1960). *The Nature of the Chemical Bond*, 3rd Edn., Cornell, Ithaca, New York.
Shannon, R. D. and Prewitt, C. T. (1969). *Acta Cryst.*, **B25**, 925.
Strens, R. G. J. (1966). *Chem. Commun.*, 777.
Strens, R. G. J. (1969). In: Runcorn, S. K. (Ed), *The Application of Modern Physics to the Earth and Planetary Interiors*, Wiley, London and New York.
Wang, P. J. and Drickamer, H. G. (1973). *Journ. Chem. Phys.*, **59**, 713.
Wood, B. J. (1971). *Ph.D. Thesis*, University of Newcastle upon Tyne.
Wood, B. J. and Strens, R. G. J. (1972). *Mineral. Mag.*, **38**, 909.

Partitioning of Transition Metals in Mineral Structures of the Mantle

R. G. Burns
Department of Earth and Planetary Sciences
Massachusetts Institute of Technology
Cambridge, Massachusetts 02139, USA

INTRODUCTION

The mineralogy and chemical composition of the interiors of the Earth and Moon are currently the focus of active research and debate. The crystal structures of minerals in the mantle not only profoundly affect the dynamic and static physical properties of the Earth's interior, but they also influence the fractionation of elements between the crust and upper and lower mantles. Nearer the surfaces, partial fusion of mantle minerals leads to chemical differentiation at various depths, thereby influencing the compositions of mare basalts and magma extruding onto the earth's surface.

Certain transition elements, because of crystal field effects, are particularly sensitive to changing environments in crystal structures. Some cations are susceptible to pressure-induced variations of oxidation state and coordination symmetry, while others may undergo rearrangements of their electronic configurations at high pressures. These changes lead to partitioning of the elements between different sites within and between coexisting minerals. Crystal field theory has been used to interpret the range of transition pressures observed for the olivine–spinel transformation in various transition metal-bearing silicates and germanates (Syono *et al.*, 1970; Mao and Bell, 1972). This paper discusses some of the factors which affect the crystal chemistry of the transition elements at high pressures in the interiors of the Earth and Moon, and demonstrates how certain elements may serve as geochemical indicators of the evolution of the mantle.

Following a review of structure-types of minerals proposed for the Earth's mantle, the paper describes experimental results on the effect of pressure on coordination numbers, energy levels, ionic covalent bonding, oxidation states, and electronic configurations of the transition elements. These data form the basis for interpreting the distributions of iron, chromium, titanium, and nickel in mantle minerals and the oxidation state of titanium in mare basalts.

STRUCTURE-TYPES OF MINERALS IN THE MANTLE

A growing body of evidence is showing that common rock-forming minerals in the Earth's crust and lower mantle are transformed to denser polymorphs at high pressures (Ringwood, 1966, 1969, 1970, 1973). Many of these phase transformations are believed to take place in the transition zone between the upper and lower mantles, and to be responsible for the sharp increases of seismic velocity data originating at different depths in the mantle. The more significant structure-types of the mantle minerals are summarized in Table 1. These structure-types show a general rise of coordination number of the cation sites in the denser polymorphs. For example, the six-coordinate sites accommodating Mg^{2+} and Fe^{2+} ions in the olivine and spinel structures become 7-, 8-, and 9-coordinated in the denser strontium plumbate, calcium ferrite, and potassium nickel fluoride structure-types, respectively. The eight-coordinate sites of pyroxenes and garnets increase to 12-fold coordination in the perovskite

Table 1. Mineral structure-types in the mantle. (Based on data from Ringwood, 1966, 1973).

	Phase	Structure-type	Coordination numbers of cation sites
	UPPER MANTLE		
	Olivine: $(Mg,Fe)_2SiO_4$	Olivine	6–6–4
	Orthopyroxene: $(Mg,Fe)_2Si_2O_6$	Pyroxene	6–6–4
	Omphacite: $(Ca,Fe,Na)(Mg,Al,Cr)Si_2O_6$	Pyroxene	8–6–4
	Pyrope: $(Mg,Fe,Ca)_3(Al,Cr)_2Si_3O_{12}$	Garnet	8–6–4
↑ TRANSITION ZONE	**350–400 km**		
	Spinel-like phase: $(Mg,Fe)_2SiO_4$	Spinel β-form	6–6–6–4
		Spinel γ-form	6–4
	Complex garnet solid-solution: $(Ca,Na,Fe,Mg)_3(Mg,Si,Al,Cr)_2(SiO_4)_3$	Garnet	8–6–4
	Jadeite: $NaAlSi_2O_6$	Pyroxene	8–6–4
	650–700 km		
	$(Mg,Fe)_2SiO_4$	Strontium plumbate	7–6
	$(Mg,Fe)SiO_3$–$(Al,Cr)_2O_3$	Ilmenite	6–6
	$CaSiO_3$	Perovskite	12–6
	$NaAlSiO_4$	Calcium ferrite	8–6
↓	**1000 km**		
	$(Mg,Fe)_2SiO_4$	Calcium ferrite or Potassium nickel fluoride	8–6 or 9–6
	$(Ca,Mg,Fe)SiO_3$	Perovskite	12–6
	$(Mg,Fe)(Al,Cr)_2O_4$	Calcium ferrite	8–6
	$(Mg,Fe)O + SiO_2$	Periclase + rutile or Caesium chloride + rutile	6–6 + 6 or 8–6 + 6
	LOWER MANTLE		

structure-type, while tetrahedrally coordinated silicon in crustal minerals becomes octahedrally coordinated in the garnet, ilmenite, perovskite, hollandite (derived from feldspars), and rutile (stishovite) structure-types postulated to occur at greater depths in the mantle. The configurations of the oxygen coordination polyhedra about many of these sites are contained in Figures 1(a) and 1(b) discussed later.

The so-called 'post-spinel phase' is particularly topical because the P,T conditions to study it border on the realm of current experimental attainability. The transition: spinel → Sr_2PbO_4 structure-type for $(Mg,Fe)_2SiO_4$ compositions was deduced from studies of analogous compounds of germanium (Wadsley *et al.*, 1968; Morimoto *et al.*, 1972; Ringwood, 1973). Recently, disproportionation of Fe_2SiO_4, Mg_2SiO_4, and $(Mg,Fe)_2SiO_4$ compositions to mixtures of oxides with the periclase and stishovite structures was demonstrated experimentally (Bassett and Ming, 1972; Kumazawa *et al.*, 1973). Although the P,T calibration data are crude at these extreme experimental conditions, there is a suggestion that the disproportionations occur at higher pressures and temperatures than those corresponding to depths around 650 km. in the mantle. Furthermore, thermodynamic arguments (Ringwood, 1973) favour a structural transformation of $(Mg,Fe)_2SiO_4$ compositions to the Sr_2PbO_4 structure-type rather than disproportionation to oxide mixtures. In this paper, we shall assume that the Sr_2PbO_4-structure and its denser polymorphs influence element fractionations at the transition zone-lower mantle boundary. Disproportionation of these compounds to mixtures of simple oxides such as (Mg,Fe)O and SiO_2 probably occurs deeper in the lower mantle. In this region, the metal ions may exist in 8-coordination in the (Mg,Fe)O phase possessing the caesium chloride structure-type.

CRYSTAL FIELD ENERGIES

The characteristic features of all first series transition elements are the incomplete filling and differential splitting of $3d$-orbital energy levels when the cations are in non-spherical environments in crystal structures. According to crystal field theory (Burns, 1970), the *relative* energies of the five $3d$-orbitals are controlled by the symmetry and distortion of the ligands (e.g. oxygen atoms) surrounding the central transition metal ion in a coordination site. The *values* of the energy differences (Δ) of the split $3d$-orbital levels are influenced by such factors as cation valence (e.g. $\Delta(M^{3+}) > \Delta(M^{2+})$), coordination symmetry (e.g. octahedral $\Delta_0 >$ cubic $\Delta_0 >>$ tetrahedral Δ_t), nature of the ligands (i.e. degree of covalent bonding), and metal-ligand interatomic distances, R (i.e. $\Delta \propto 1/R^5$). Octahedral coordination is the most common ligand symmetry for chemical compounds of the transition elements, and oxygen ions predominate around the octahedral sites of most rock-forming minerals containing these cations. In such octahedral sites, the five $3d$-orbitals are split into a group of three relatively stable t_{2g} orbitals and two less-stable e_g orbitals, the energy separation (crystal field splitting) between the t_{2g} and e_g levels being denoted by Δ_0. Each electron in a t_{2g} orbital stabilizes the cation by $2/5\Delta_0$, whereas each electron in an e_g orbital destabilizes it by $3/5\Delta_0$. For example, the Cr^{3+} ion with three $3d$ electrons occupying singly each t_{2g} orbital acquires a very high crystal field stabilization energy (CFSE) of $6/5\Delta_0$ in octahedral coordination (approximately 60 kcal/Cr^{3+} in oxide structures). In tetrahedral environments, the relative energies of the t_2 and e orbital levels (subscript g omitted because a tetrahedron lacks an inversion centre) are reversed relative to octahedral coordination, so that the third $3d$ electron of Cr^{3+} must occupy a less stable t_2 orbital and contribute a destabilization energy to tetrahedrally coordinated Cr^{3+} ions. Moreover, the small tetrahedral crystal field splitting, Δ_t, between the more stable e orbital and less stable t_2 orbital levels is such that if Cr^{3+} ions were present in tetrahedral sites, the acquired CFSE, $4/5\Delta_t$, would amount to less than 30 per cent of the CFSE of octahedrally coordinated Cr^{3+} in sites with identical ligands and metal-ligand distances. Thus, Cr^{3+} ions have a high octahedral site preference energy, and discriminate against tetrahedral sites in geochemical media in general, and in mineral structures in particular. Similar factors apply to the Ni^{2+} ion.

The energy levels of the t_{2g} and e_g orbital groups are split into additional levels when the oxygen ions constituting the sixfold coordination site are distorted from octahedral symmetry. As a result, at least one of each of the t_{2g} and e_g orbital levels is further stabilized relative to the remainder. Cations such as Ti^{3+} and Fe^{2+} with one and six $3d$-electrons, respectively, receive a CFSE of $2/5\Delta_0$ plus an additional stabilization energy resulting from the single or sixth $3d$-electron occupying the most stable orbital of the t_{2g} group. The Cr^{2+} ion, $3d^4$, on the other hand, is stabilized in a distorted six-coordinate site relative to a regular octahedral environment because its fourth $3d$-electron occupies the more stable orbital of the e_g group. Ions such as Cr^{3+} and Ni^{2+} attain no additional stabilization energy in a distorted site, while the Fe^{3+} and Mn^{2+} ions with $3d^5$ configurations receive zero CFSE in any site in oxides.

The t_{2g}–e_g orbital splittings may be obtained from absorption spectral measurements of compounds and minerals containing the transition-metal ions. With Cr^{3+}, for example, values of Δ_0 may be estimated from the position of the band at the lowest energy in the absorption spectra, which corresponds to the ${}^4A_{2g} \rightarrow {}^4T_{2g}$ transition in octahedrally coordinated Cr^{3+} ions. For Cr_2O_3 with the larger metal-oxygen distance (Cr–O = 2.00 Å), $\Delta_0 = 16{,}670$ cm^{-1} which is smaller than the value for Cr^{3+} in Al_2O_3 having shorter metal-oxygen distances (Al–O = 1.91 Å, $\Delta_0 = 18{,}000$ cm^{-1}). Spectral energies of Cr^{3+} ions in several oxide minerals are summarized by Burns (1975). The data show a trend of increasing Δ_0 with decreasing metal-oxygen distance, R, in approximate compliance with the relationship $\Delta \propto 1/R^5$ derived for a point charge model. Thus, Cr^{3+} acquires a higher CFSE in a 'compressed' site, such as an aluminium site, compared to a magnesium site in an oxide structure. The data (Burns, 1974) show that CFSE of Cr^{3+} in octahedral sites of major minerals occurring in the crust and upper mantle increases in the sequence olivine-pyroxene-garnet-spinel.

Similar features apply to the spectra of other transition metal ions in oxide structures (Burns, 1970; Burns and Vaughan, 1975). Computations of the energy levels and stabilization energies of Fe^{2+}, Ti^{3+}, and Ni^{2+} are complicated by the occurrences of the cations in several sites with different coordination symmetries and distortions (Wood and Strens, 1972; Runciman *et al.*, 1973a, b, 1974; Wood, 1974; Burns, 1974). Nevertheless, there is a general increase of Δ_0 with decreasing metal-oxygen distance (Faye, 1972) and greater stability of Fe^{2+} and Ti^{3+} in the more distorted coordination sites of oxide and silicate minerals (Burns, 1970; Burns and Vaughan, 1974).

PRESSURE-INDUCED EFFECTS ON CRYSTAL CHEMISTRY

(a) Changes of Coordination Number

A guiding principle of crystal chemistry is that the coordination number of a cation depends on the radius ratio: R_C/R_A, where R_C and R_A are the ionic radii of cation and anion, respectively. Octahedrally coordinated cations are predicted for $0.414 < R_C/R_A < 0.732$, while 4-fold (tetrahedral) and 8-fold (cubic) coordinations are favoured for radius ratios below 0.414 and above 0.732, respectively. Octahedral ionic radii (Shannon and Prewitt, 1969) of cations of Ti, Cr, Mn, Fe, Co, Ni, and Cu, together with their radius ratios in sixfold coordination with oxygen ($O^{2-} = 1.40$ Å) are summarized in Table 2. The divalent cations, having radius ratios of 0.52–0.59, lie towards the upper limit (0.732) predicted for octahedral coordination, whereas the trivalent cations with $R_C/R_A = 0.38$–0.48 straddle the lower limit (0.414).

A dominant theme pervading the phase transformations observed or predicted in the mantle is that the coordination numbers of many cations tend to increase in the denser polymorphs. This is a direct consequence of the greater compressibility of the highly polarizable coordinating oxygen anions, compared to the cations, resulting in an increase of radius ratios for most cations. The divalent transition metal ions lying nearer to the upper limit of radius ratios for octahedral coordination are more susceptible to pressure-induced increases of coordination number than the trivalent cations. Thus, ions such as Cr^{2+}, Fc^{2+}, and Ni^{2+} are predicted to acquire coordination numbers greater than six at high pressures, whereas Ti^{4+}, Ti^{3+}, and Cr^{3+} are expected to remain in octahedral coordination in the lower mantle.

(b) Shifts of Energy Levels and Changes of CFSE

According to point charge calculations, $\Delta \propto 1/R^5$, so that crystal field splittings and stabilization energies should increase with decreasing metal-oxygen distances. Rising pressures also produce shorter interatomic distances, so that Δ_0 and CFSE values are expected to increase at high pressures if the

Table 2. Ionic radii and radius ratios of certain transition metal ions octahedrally coordinated by oxygen.

Cation M^{2+} / M^{3+} / M^{4+}	Ti^{4+}	Ti^{3+}	Cr^{3+}	Cr^{2+}	Mn^{3+}	Mn^{2+}	Fe^{3+}	Fe^{2+}	Co^{3+}	Co^{2+}	Ni^{2+}	Cu^{2+}	Mg^{2+}	Al^{3+}	Si^{4+}
Ionic Radius (Å) (octahedral)	0.605	0.67	0.615	0.82	0.65	0.82	0.645	0.77	0.525[a]	0.735	0.70	0.73	0.72	0.53	0.40
Radius Ratio (octahedral)	0.43	0.48	0.44	0.59	0.47	0.59	0.46	0.55	0.38[a]	0.53	0.50	0.52	0.52	0.38	0.29

[a]Low-spin configuration in oxide structures.

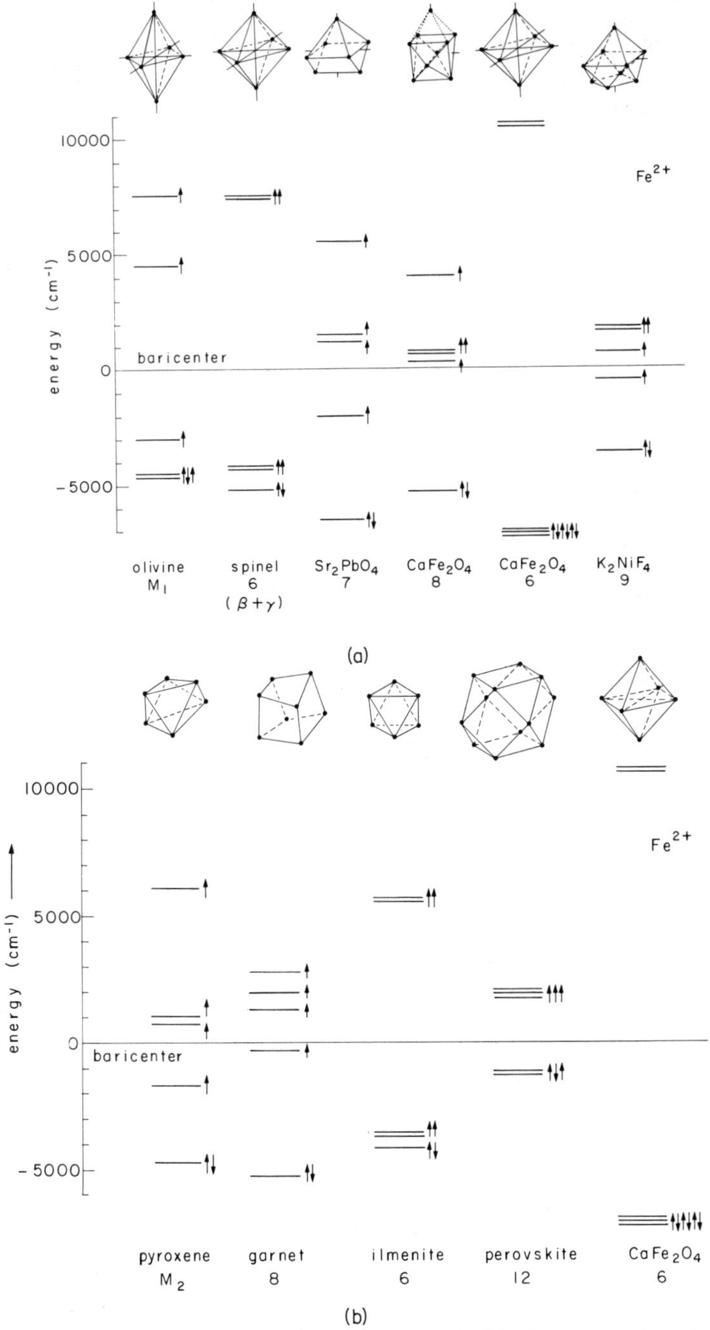

Figure 1. Electronic configurations of Fe^{2+} ions in mineral structures of the mantle:
(a) AB_2O_4 phases, (b) ABO_3 phases (data based on Gaffney, 1972; Runciman *et al.*, 1973a; Runciman and Sengupta, 1974; Burns, 1970).

coordination symmetry is unchanged. Spectral data for minerals (Bell and Mao, 1969; Mao and Bell, 1971, 1972, 1973; Abu-Eid *et al.*, 1973; Abu-Eid, this volume; Sung *et al.*, 1974) and doped MgO and Al_2O_3 (Minomura and Drickamer, 1961; Stephens and Drickamer, 1961) show that most crystal field bands for transition metal ions, including Ti^{3+}, Cr^{3+}, Fe^{2+}, and Ni^{2+}, move to higher energies with rising pressure. Drickamer and Frank (1973) have demonstrated from plots of Δ_p/Δ_1 versus $(R_1/R_p)^5$, where Δ_p, Δ_1, and R_p, R_1 are the crystal field splittings and interatomic distances at pressure p and one atmosphere, respectively, that there is a close correlation between Δ and R^{-5}. The agreement is better for transition metals in the Al_2O_3 structure than for doped MgO crystals. Therefore, Δ_0 and CFSE's increase with rising pressure. As a result, the strong octahedral site preference energy of Cr^{3+} should be enhanced in the mantle. However, the Ni^{2+} ion, also having a high preference for octahedral sites at atmospheric pressures, probably undergoes an increase of coordination number as a result of the radius ratio effect discussed in the previous section, and may discriminate against octahedral sites of minerals in the lower mantle.

The electronic energy levels of the Fe^{2+} ion in each coordination polyhedron of the various structure-types proposed for the mantle have been calculated by Gaffney (1972). These relative energies, together with those estimated for certain cation sites in the olivine, spinel, pyroxene, and garnet structures, are shown schematically in Figures 1(a) and (b).

(c) Changes of Bond-type

Interpretations of the crystal chemistry of transition metal ions by crystal field theory are based on a predominantly ionic model of the chemical bond. As interatomic distances decrease at high pressures, the covalent character of metal–oxygen bonds is expected to increase significantly. Therefore, it is necessary to examine the validity of crystal field theory at high pressures. One measure of the change of the ionic-covalent character of bonds in transition metal compounds is the Racah B parameter, which may be evaluated from absorption spectra.

Calculations based on the two spectral bands of Cr^{3+} for ruby (Drickamer and Frank, 1973) and for chrome-garnets (Abu-Eid, this volume) show that the Racah B parameter decreases by less than three per cent per 100 kb, suggesting that the degree of covalent character of chromium-oxygen bonds increases by less than 20 per cent towards the transition zone-lower mantle boundary. Similar estimates may be obtained for iron–oxygen bonds from the pressure variations of isomer shifts in the Mössbauer spectra of Fe^{2+} silicates (Huggins, this volume). These results suggest that crystal field theory is applicable to transition metal geochemistry throughout the transition zone.

(d) Pressure-induced Reduction

A growing body of spectral evidence has demonstrated that several transition

Table 3. Observed and predicted oxidation states of transition metals in the earth, moon, and meteorites.

Cation		Ti	Cr	Mn	Fe	Co	Ni	Cu
Oxidation States	M(IV)	Ti^{4+} (c, u, m, me)		Mn^{4+} (c)				
		·········→						
	M(III)	Ti^{3+} (m, me, ?c, ?u)	Cr^{3+} (c, m, me)	Mn^{3+} (c)	Fe^{3+} (c, me)	Co^{3+} (c)	Ni^{3+} (?c)	
			- - - →	——→ - - - →	——→			
	M(II)		Cr^{2+} (m, me, ?u)	Mn^{2+} (c, m, me)	Fe^{2+} (c, m, me, u)	Co^{2+} (c, m, me)	Ni^{2+} (c, m, me)	Cu^{2+} (c)
					·········→	·········→		——→
	M(I)				Fe^{+} (?1)	Co^{+} (?1)	Ni^{+} (?1)	Cu^{+} (c, m, me)
	M(0)		Cr (me)		Fe (c, m, me, co)	Co (c, m, me, co)	Ni (c, m, me, co)	Cu (c, m, me)

Legend: c = crust; u = upper mantle; l = lower mantle; co = core; m = moon; me = meteorites
——→ Pressure-induced reduction observed
- - - → Shock-induced reduction observed
·········→ Predicted reduction at high pressures

metal ions undergo reversible pressure-induced reduction to lower oxidation states at very high pressures. Originally, Drickamer and coworkers (for example, Drickamer *et al.*, 1970) demonstrated by Mössbauer spectroscopy that in most synthetic iron compounds there is reduction of Fe^{3+} to Fe^{2+} ions at high pressures. Higher oxidation states of iron, such as ferrates IV and VI, also reduce reversibly with pressure. There is generally some hysteresis, because a fraction of Fe^{2+} ions formed at high pressures remain in iron (III) compounds returned to one atmosphere. This is attributed to locked-in strain. However, efforts to remove strain by powdering the sample generally returns it to an an Fe^{2+}-free state. Analogous high pressure Mössbauer spectral studies of silicate minerals (Burns *et al.*, 1972a, b; Huggins, this volume) have demonstrated that reversible reduction of Fe^{3+} to Fe^{2+} ions occurs at high pressures and temperatures, leading to the suggestion that negligible amounts of ferric iron occur in the lower mantle (Tossell *et al.*, 1972).

Spectral measurements have also demonstrated that reduction of Mn^{3+} to Mn^{2+} (Ahsbahs *et al.*, 1974) and Cu^{2+} to Cu^{+} ions (Wang and Drickamer, 1973) also occur at high pressures. Furthermore, Gibbons *et al.* (1973) observed shock-produced reduction of Mn^{3+} to Mn^{2+} ions in rhodonite, a pyroxenoid mineral.

The available data summarized in Table 3 show a trend towards the stabilization of progressively lower oxidation states at high pressures across the first transition series. These observations indicate that higher oxidation states characteristic of the earth's surface (e.g. Ti^{4+}, Cr^{3+}, Fe^{3+}, Ni^{2+}) may be unstable at the high P,T conditions of the lower mantle. It is also suggested that exotic oxidation states, such as Ti^{3+}, Ti^{2+}, Cr^{2+}, Fe^{1+}, Ni^{1+}, and perhaps Fe^{0}, Ni^{0}, may be prevalent towards the core-mantle boundary.

CRYSTAL CHEMISTRY OF CERTAIN TRANSITION ELEMENTS IN THE MANTLE

(a) Iron

The state of iron in the earth's interior has long been a topic of debate in the earth sciences. Experimental evidence derived from high pressure Mössbauer spectroscopy described earlier suggests that insignificant amounts of Fe^{3+} ions exist at high pressures and temperatures, and that Fe^{2+} ions predominate in the lower mantle. Furthermore, the Fe^{2+} ions are predicted to occur in the low-spin state at great depths (Fyfe, 1960; Strens, 1966), leading to substantial enrichments of iron with increasing depths in the mantle (Strens, 1969; Burns, 1969; Gaffney and Anderson, 1973).

Gaffney (1972) has calculated the CFSE of Fe^{2+} in the various structure-types of the transition zone and lower mantle. At the 1050 km discontinuity, the stabilization energies for high-spin Fe^{2+} decrease in the order $Sr_2PbO_4 - 7$ $> K_2NiF_4 - 9 >$ ilmenite $> CaFe_2O_8 - 8 >$ perovskite. This order suggests that Fe^{2+} ions are strongly enriched in the Sr_2PbO_4 structure-type in the transition zone and in the K_2NiF_4 structure-type of the lower mantle. Gaffney (1972) has further suggested that if Fe^{2+} ions are spin-paired in the lower mantle, they would be strongly enriched in the octahedral site of the $CaFe_2O_4$ structure-type.

Recent spectral data for ferromagnesian silicates (Runciman *et al.*, 1973a, b, 1974; Runciman and Sengupta, 1974; Burns, 1974) enable the CFSE of Fe^{2+} to be estimated more accurately in upper mantle minerals. The calculations show that the CFSE of high-spin Fe^{2+} decrease in the order:

$$\text{garnet} - 8 > \text{pyroxene-M2} > \text{pyroxene-M1} \approx \text{olivine-M1 and M2.}$$

This correlates with the observed enrichment of iron in garnet phases of upper mantle rocks such as kimberlites. These results for iron enrichments observed in the upper mantle compared with those predicted for the lower mantle phases indicate a reversal of iron partitioning from ABO_3 structure-types of the upper mantle, to AB_2O_4 structure-types of the lower mantle, in accord with the calculations of Ahrens (1972).

Finally, it is worth noting that if the lower mantle contains dense oxide phases, such as MgO with the caesium chloride structure, the Fe^{1+} ion, if present, would acquire a high stabilization energy in the 8-coordinate site because of its unique $3d^7$ electronic configuration.

(b) Chromium

The high CFSE acquired by Cr^{3+} ions in octahedral sites, coupled with the enhanced stabilization energy obtained in a compressed site, indicate that Cr^{3+} should continue to have a high octahedral site preference energy in the lower mantle. The coordination number of Cr^{3+} is unlikely to change with pressure because its radius ratio (Table 2 above) lies near the lower limit of the range predicted for octahedrally coordinated cations at low pressures. Although the radius ratio increases with pressure due to the greater compressibility of the O^{2-} ion, it is considered unlikely that higher coordination numbers will be induced for Cr^{3+} at pressures encountered in the lower mantle. Thus, Cr^{3+} ions, if they occur as such in the lower mantle, are expected to favour octahedral sites in the perovskite, K_2NiF_4 and $CaFe_2O_4$ (or periclase) structure-types. In the transition zone, Cr^{3+} ions may be accomodated and stabilized in octahedral sites of the Sr_2PbO_4, $CaFe_2O_4$, ilmenite, perovskite, and garnet structures. In the upper mantle, Cr^{3+} ions will be partitioned between the olivine, pyroxene, garnet, and remnant spinel phases, in concentrations related to the CFSE acquired in each structure. The CFSE data compiled by Burns (1975) suggest that the order of Cr^{3+} enrichment in the upper mantle is:

$$\text{spinel} > \text{garnet} > \text{pyroxene} \approx \text{olivine}$$

Note that the high octahedral site preference energy of Cr^{3+} also mitigates against high concentrations of chromium in fusion products of pyrolite. Melts

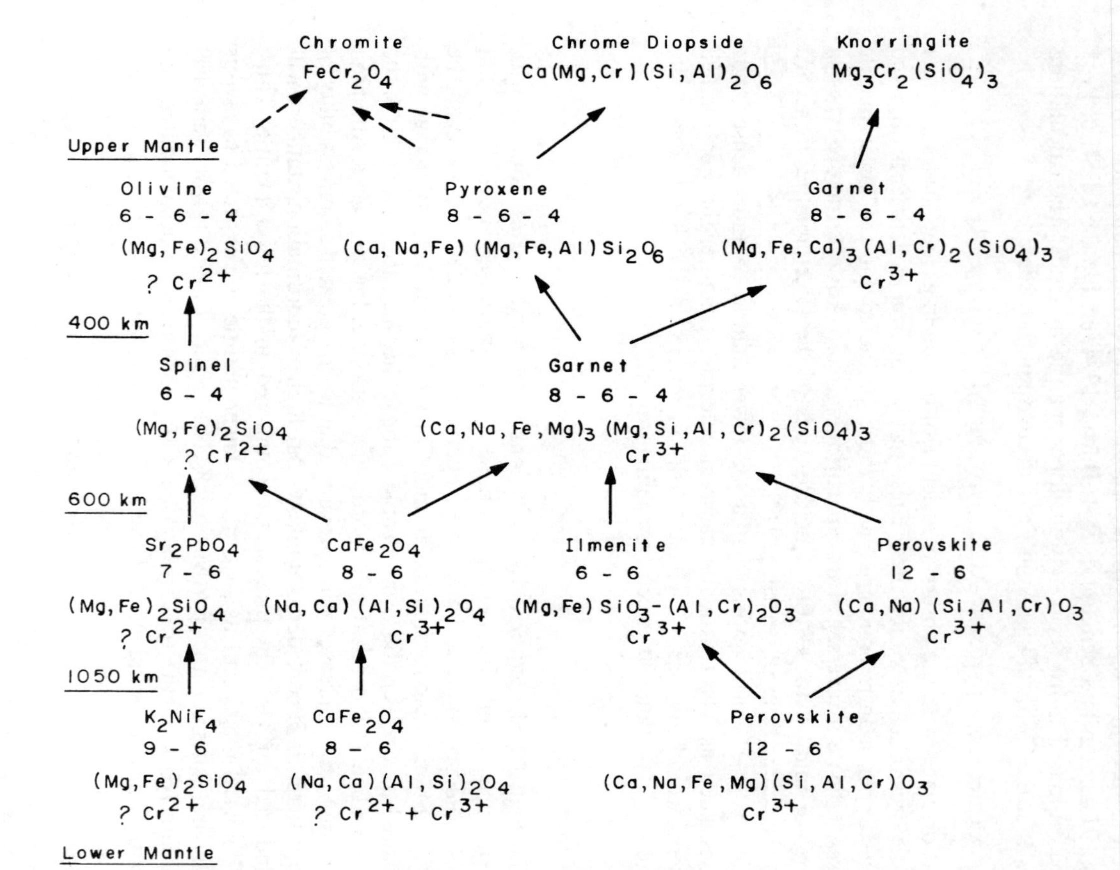

Figure 2. Partitioning scheme for chromium in structure-types in the mantle.

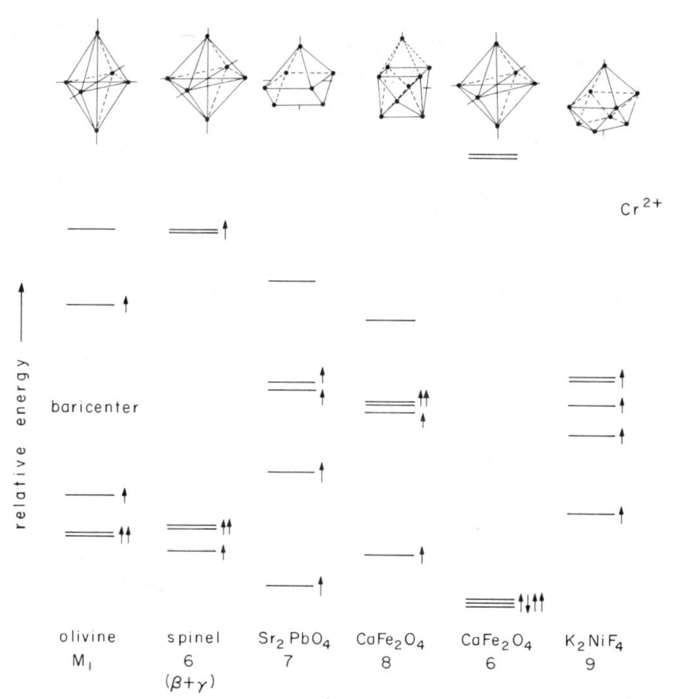

Figure 3. Electronic configurations of Cr^{2+} ions in mineral structures of the mantle relative energy levels based on Gaffney. (1972).

formed from silicates generate an array of octahedral and tetrahedral sites, so that the equilibrium encountered by Cr^{3+}

$$[Cr^{3+} \text{ octahedral}] \text{ crystal} \underset{\text{cryst'n}}{\overset{\text{fusion}}{\rightleftarrows}} [Cr^{3+} \text{ octahedral}] \text{ melt} + [Cr^{3+} \text{ tetrahedral}] \text{ melt}$$

lies strongly to the left. This explains why Cr^{3+} ions are observed to be concentrated in minerals in residual lherzolites (Jackson and Wright, 1970; Carter, 1970; Burns, 1973).

In the fractionation scheme for chromium outlined in Figure 2, it is suggested that Cr^{2+} ions also may occur in the lower mantle. They are predicted to be formed by pressure-induced reduction of Cr^{3+} ions, by analogy with processes observed in compounds and minerals containing Fe^{3+}, Mn^{3+}, and Cu^{2+} ions. If similar behaviour occurs in chromium, the Cr^{2+} ions derived by pressure-induced reduction of Cr^{3+} would be stabilized in the distorted 7-coordinate site of the Sr_2PbO_4 structure-type, and might occur in the low-spin state in the $CaFe_2O_4$ structure-type in the lower mantle, by analogy with low-spin Fe^{2+} (Gaffney, 1972). The electronic configurations of Cr^{2+} in the various structure-types are shown in Figure 3.

Fractionation of Cr^{2+} in the transition zone could eventually lead to the presence of divalent chromium in olivines in the upper mantle (Meyer and

Boyd, 1972; Sobolev, 1972), where it is predicted to be stabilized and enriched in the M1 site of the olivine structure (Burns, 1970). However, since most of the pressure-induced reductions of Fe(III) compounds are reversible with some hysteresis (Drickamer *et al.*, 1970), it is proposed that pressure-released oxidation of most of the Cr^{2+} ions to Cr^{3+} ions would occur in the upper mantle, leading to exsolution of chromite crystallites frequently observed in olivines.

The discussion has centred on chromium coordinated to oxygen in structures derived from olivine, pyroxene, garnet, and spinel. Other structure-types in the mantle that may be relevant to chromium include the rutile and hollandite structures, representing dense polymorphs of silica (stishovite) and feldspars, respectively. The hollandite structure is related to rutile, and both contain octahedral sites. The existence of synthetic Cr^{3+}-rutile structure-types might also accommodate chromium in octahedral sites in the mantle. Octahedral sites also occur in periclase which may be formed by disproportionation of Mg_2SiO_4. Chromium also forms nitride and sulfide phases such as carlsbergite (CrN), daubréelite ($FeCr_2S_4$), and brezinaite (Cr_3S_4), all found in meteorites. Carlsbergite has the rocksalt structure with metals in octahedral coordination. Thiospinels, such as daubréelite, also transform at high pressures to denser structure-types (Vaughan *et al.*, 1971) including a cation-defect NiAs structure (cf. Cr_3S_4) containing chromium in octahedral sites. Thus, octahedral sites favourable for the occurrence and stability of chromium are common to a variety of minerals that might occur in the mantle.

(c) Titanium

Because the Ti^{4+} ion contains no 3*d*-electrons, it acquires zero CFSE in mineral structures. Therefore, its crystal chemistry is controlled by size criteria. The radius ratio of Ti^{4+} lies near the lower limit for octahedral coordination (Table 2 above), so that it is expected to occur in sixfold coordination throughout the mantle and to be distributed between the octahedral sites of the pyroxene, garnet, ilmenite, perovskite, Sr_2PbO_4, $CaFe_2O_4$, and K_2NiF_4 structure-types.

A fraction of the Ti^{4+} ions is expected to undergo pressure-induced reduction to Ti^{3+} ions in the mantle. These Ti^{3+} ions, having a radius ratio well within the limits for octahedral coordination (Table 2 above) and acquiring a moderate CFSE, are expected to behave like Cr^{3+} ions in the lower mantle, and to favour octahedral sites in the perovskite, K_2NiF_4, and $CaFe_2O_4$ structure-types. In the upper mantle, there is evidence to suggest that Ti^{3+} ions are present in the geikielite-ilmenite (S. E. Haggerty, pers. comm.) and garnet (Burns and Burns, 1971; Burns, 1972) phases.

Titanium, like chromium, also forms meteoritic nitride and sulphide phases such as osbornite (TiN) and heiderite (Fe,Cr) $(Ti,Fe)_2S_4$, which are isostructural with carlesbergite (CrN) and brezinaite (Cr_3S_4). Thus, a wide range of octahedral sites are available for Ti, either as Ti^{3+} or Ti^{4+}, throughout the mantle.

The moon rocks have been shown to contain substantial proportions of trivalent Ti, particularly the lunar samples returned from the Apollo 11 and 17 missions, in which up to 30 per cent of the Ti is present as Ti^{3+} ions in pyroxene phenocrysts and orange glass spherules from the mare basalts (Vaughan and Burns, 1973; Sung *et al.*, 1974). The high Ti^{3+}/Ti^{4+} ratios cannot be explained by low oxygen fugacities alone (Mao *et al.*, 1973), suggesting that other factors contribute to the Ti^{3+}/Ti^{4+} ratios in lunar materials. It has been proposed (Sung *et al.*, 1974) that the high Ti^{3+}/Ti^{4+} ratios in lunar materials may be the result of pressure-induced reduction of Ti^{4+} to Ti^{3+} in the lunar interior, or be caused by shock-induced reduction of Ti^{4+} ions by impacting meteorites or comets plunging into basaltic magma or rock on the Moon's surface.

(d) Nickel

Prediction of the distribution of nickel in the mantle produces an enigma. On the one hand, the large octahedral site preference of the Ni^{2+} ion indicates that it will favour octahedral coordination wherever possible. However, the ionic radius of Ni^{2+} is such that it cannot enter very compressed six-coordinate sites and may be partitioned into sites of higher coordination number in the dense structure-types of the lower mantle. Furthermore, the fact that copper, the neighbouring element in the periodic table, undergoes pressure-induced reduction of Cu^{2+} to Cu^{+}, suggests that some Ni^{2+} ions also may be reduced to Ni^{+} ions under the very high pressures of the lower mantle.

In the upper mantle, Ni^{2+} is observed to be strongly enriched in the six-coordinated sites of the olivine structures. Transformation of olivine via the spinel phases to the Sr_2PbO_4 structure-type in the transition zone probably leads to partitioning of the Ni^{2+} ions into the 7-coordinate site. In this site the Ni^{2+} ion would acquire a large CFSE if it occurred in the low-spin state (i.e. highest $3d$-orbital energy level unoccupied). In the lower mantle, both Ni^{2+} (low-spin $3d^8$) and the larger Ni^{+} ($3d^9$) ion would be more stable in the 9-coordinate site of the K_2NiF_4 structure compared to the $CaFe_2O_4 - 8$ and perovskite-12 coordinate sites.

The strong octahedral site preference energy of Ni^{2+} would be retained if the spinel, Sr_2PbO_4, $CaFe_2O_4$, or K_2NiF_4 structure-types disproportionated to a mixture of oxides containing (Mg,Fe)O with the periclase structure. However, the coordination number of the (Mg,Fe)O phase is predicted to be 8-fold under high pressures in the lower mantle, leading to a smaller CFSE for Ni^{2+} compared to an octahedral site.

SUMMARY

Certain transition elements, notably Ti, Cr, Fe, and Ni, serve as geochemical indicators of the evolution of the mantle. Their distributions are profoundly influenced by structure-types of the host minerals occurring in the transition zone and lower mantle, which include Sr_2PbO_4, K_2NiF_4, $CaFe_2O_4$, ilmenite, perovskite, rutile, and hollandite structures. Analysis of orbital energy levels enables fractionation patterns of transition elements in the mantle to be explained on the basis of crystal field stabilization energies acquired in the coordination sites of each structure-type.

Recent high-pressure spectral studies of minerals, together with data obtained for lunar minerals, show that oxidation states (e.g. Ti^{3+}, Cr^{2+}, Fe^{2+}) predominate under conditions of low oxygen fugacity and very high pressures and temperatures found in the earth's interior. Radius ratio criteria suggest that trivalent cations (Ti^{3+}, Cr^{3+}) remain six-coordinated with oxygen in the mantle, whereas coordination numbers of divalent cations increase in the Sr_2PbO_4, K_2NiF_4, $CaFe_2O_4$, and perovskite structure-types. These factors cause distinctive partitioning behaviours for each element. For example, Cr^{3+} is expected to be distributed between several phases in the mantle because each structure-type also contains a six-coordinated site and Cr^{3+} has a high octahedral site preference energy. In the upper mantle, stabilization energies derived from spectral data suggest that the partitioning of Cr^{3+} follows the sequence:

$$\text{spinel} > \text{garnet} > \text{pyroxene or olivine.}$$

High-spin Cr^{2+} ions, which may result from pressure-induced reduction of Cr^{3+}, are stabilized in distorted coordination sites of the Sr_2PbO_4 and olivine structures, while low-spin Cr^{2+} may occur in octahedral sites of the $CaFe_2O_4$ structure in the lower mantle. It is proposed that certain chromites are formed in the upper mantle by pressure-released oxidation of Cr^{2+} ions in phases (olivine, spinel) derived from the Sr_2PbO_4 structure-type.

Titanium is also predicted to undergo pressure-induced reduction to Ti^{3+} ions in the mantle, and to parallel Cr^{3+} ions in its distribution. The abundant Ti^{3+} ions observed in pyroxene phenocrysts and orange glass spherules in mare basalts may be indicative of great depths of partial fusion in the Moon's interior, or be the result of shock-produced reduction of Ti^{4+} to Ti^{3+} ions by impacting planetary objects.

The CFSE of Fe^{2+} conforms wih the observed iron enrichment in garnets relative to pyroxenes and olivines in the upper mantle. However, in the transition zone and lower mantle, Fe^{2+} ions are predicted to be enriched in phases with the Sr_2PbO_4, K_2NiF_4, and $CaFe_2O_4$ structures. Therefore, an increased partitioning of iron into the AB_2O_4 phases relative to ABO_3 phases may occur in the transition zone.

Nickel also has a high octahedral site preference energy, which may be annulled by pressure-induced increase of coordination number of Ni^{2+}. In the transition zone, low-spin Ni^{2+} ions are predicted to be stabilized and enriched in the seven-coordinated site of the Sr_2PbO_4 structure-type. Pressure-released phase transformations of the Sr_2PbO_4 structure-type via 'spinel' intermediaries, lead to the observed enrichment of nickel in olivines of the upper mantle.

ACKNOWLEDGEMENTS

Spectral and high pressure crystal chemical studies of minerals containing transition metal ions are supported by grants from the National Science Foundation (grant no. GA-28906X) and the National Aeronautics and Space Administration (grant no. NGR-22-009-551). Special thanks are due to Mrs. Virginia Mee Burns for bibliographic research and to Ms. Roxanne Regan for preparation of the manuscript.

REFERENCES

Abu-Eid, R. M., Mao, H. K., and Burns, R. G. (1973). *Ann. Rept. Geophys. Lab., Yearbook 72*, 564–567.

Ahrens, T. J. (1972). *Phys. Earth Planet. Interiors*, **5**, 267–281.

Ahsbahs, H., Dehnicke, G., Dehnicke, K., and Hellner, E. (1974). *Conf. High Pressure Research, Abstr.*, p. 26 (Marsburg, Germany, March 1974).

Bassett, W. A. and Ming, L-C. (1972). *Phys. Earth Planet. Interiors*, **6**, 154–160.

Bell, P. M. and Mao, H. K. (1969). *Ann. Rept. Geophys. Lab., Yearbook* 68, 253–256.

Burns, R. G. (1969). In: *The Application of Modern Physics to the Earth and Planetary Interiors* (Ed.: S. K. Runcorn; Publ.: J. Wiley and Sons), pp. 197–211.

Burns, R. G. (1970). *Mineralogical Applications of Crystal Field Theory* (Cambridge University Press).

Burns, R. G. (1972), *Canad. Journ. Spectr.*, **17**, 51–59.

Burns, R. G. (1973). *Geochim. Cosmochim. Acta*, **37**, 2395–2403.

Burns, R. G. (1974). *Amer. Mineral.*, **59**, 625–629.

Burns, R. G. (1975). In: *Chromium: Its Physicochemical Behavior and Petrologic Significance* (Ed.: T. N. Irvine) *Geochim. Cosmochim. Acta*, **39**, 857–864.

Burns, R. G. and Burns, V. M. (1971). *Geol. Soc. Amer., Ann. Meet.*, Washington, Abstr., **3**, 519–520.

Burns, R. G. and Vaughan, D. J. (1975). In: *Infrared and Raman Spectroscopy of Lunar and Terrestrial Minerals* (Ed.: C. Karr, Jr.; Publ.: Academic Press) 39–72.

Burns, R. G., Huggins, F. E., and Drickamer, H. G. (1972a). *Proc. 24th Intern. Geol. Congr., Sect. 14*, 113–123.

Burns, R. G., Tossell, J. A., and Vaughan, D. J. (1972b). *Nature*, **240**, 33–35.

Carter, J. L. (1970). *Bull. Geol. Soc. Amer.*, **81**, 2021–2034.

Drickamer, H. G. and Frank, C. W. (1973). *Electronic Transitions and the High Pressure Chemistry and Physics of Solids* (Chapman and Hall).

Drickamer, H. G., Bastron, V. C., Fisher, D. C., and Grenoble, D. C. (1970). *Journ. Solid State Chem.*, **2**, 94–104.

Faye, G. D. (1972). *Canad. Mineral.*, **11**, 473–487.

Fyfe, W. S. (1960). *Geochim. Cosmochim. Acta*, **19**, 141–143.

Gaffney, E. S. (1972). *Phys. Earth Planet. Interiors*, **6**, 385–390.

Gaffney, E. S. and Anderson, D. L. (1973). *Journ. Geophys. Res.*, **78**, 7005–7014.

Gibbons, R. V., Ahrens, T. J., and Rossman, G. R. (1974). *Amer. Mineral.*, **59**, 177–182.

Jackson, E. D. and Wright, T. L. (1970). *Journ. Petrol.*, **11**, 405–432.

Kumazawa, M. Sawamoto, H., Ohtani, E., and Masaki, K. (1974). *Nature*, **247**, 356–358.

Mao, H. K. and Bell, P. M. (1971). *Ann. Rept. Geophys. Lab., Yearbook 70*, 207–215.

Mao, H. K. and Bell, P. M. (1972). *Ann. Rept. Geophys. Lab., Yearbook 71*, 520–531.

Mao, H. K. and Bell, P. M. (1973). In: *Analytical Methods Developed for Application to Lunar Samples Analysis* (ASTM STP 539, Amer. Soc. Test. Mat.) pp. 100–119.

Mao, H. K., Virgo, D., and Bell, P. M. (1973). *Proc. 4th Lunar Sci. Conf., Geochim. Cosmochim. Acta, Suppl. 4*, vol. 1, pp. 397–412.

Meyer, H. O. A. and Boyd, F. R. (1972). *Geochim. Cosmochim. Acta*, **36**, 1255–1273.

Minomura, S. and Drickamer, H. G. (1961). *Journ. Chem. Phys.*, **35**, 903–907.

Morimoto, N., Tokonami, M., Koto, K., and Nakajima, S. (1972). *Amer. Mineral.*, **57**, 62–75.

Ringwood, A. E. (1966). In: *The Chemical Composition and Origin of the Earth* (Ed.: P. M. Hurley; Publ.: MIT Press) pp. 357–399.

Ringwood, A. E. (1969). *Earth Planet. Sci. Lett.*, **5**, 401–412.

Ringwood, A. E. (1970). *Phys. Earth Planet. Interiors*, **3**, 109–155.

Ringwood, A. E. (1973). *Fortschr. Miner.*, **50**, 113–139.

Runciman, W. A. and Sengupta, D. (1974). *Amer. Mineral.*, **59**, 563–566.

Runciman, W. A., Sengupta, D., and Marshall, M. (1973a). *Amer. Mineral.*, **59**, 444–450.

Runciman, W. A., Sengupta, D., and Gourley, J. T. (1973b). *Amer. Mineral.*, **59**, 451–456.

Runciman, W. A., Sengupta, D., and Gourley, J. T. (1974). *Amer. Mineral.*, **59**, 630–631.

Shannon, R. D. and Prewitt, C. T. (1969). *Acta Cryst.*, **B25**, 925–946.

Sobolev, N. V. (1972) *Proc. 24th Intern. Geol. Congr., Section 2*, 297–302.

Stephens, D. R. and Drickamer, H. G. (1961). *Journ. Chem. Phys.*, **35**, 427–429.

Strens, R. G. J. (1966). *Chem. Comm.*, p. 777.

Strens, R. G. J. (1969). In: *The Application of Modern Physics to the Earth and Planetary Interiors* (Ed.: S. K. Runcorn; Publ.: J. Wiley and Sons) pp. 213–220.

Sung, C. M. Abu-Eid, R. M., and Burns, R. G. (1974). *Proc. 5th Lunar Sci. Conf., Geochim. Cosmochim. Acta*, Suppl. 5, **1**, 717–726. M.I.T. Press.

Syono, Y., Tokonami, M., and Matsui, Y. (1971). *Phys. Earth. Planet. Interiors*, **4**, 347–352.

Tossell, J. A., Vaughan, D. J., Burns, R. G., and Huggins, F. E. (1972). *EOS (Trans. Amer. Geophys. Union)*, **53**, 1130.

Vaughan, D. J. and Burns, R. G. (1973). *EOS (Trans. Amer. Geophys. Union)*, **54**, 618–620.

Vaughan, D. J., Burns, R. G., and Burns, V. M. (1971). *Geochim. Cosmochim. Acta*, **35**, 365–381.

Wadsley, A. D., Reid, A. F., and Ringwood, A. E. (1968). *Acta Cryst.*, **B24**, 740–742.

Wang, P. J. and Drickamer, H. G. (1973). *Journ. Chem. Phys.*, **59**, 713–717.

Wood, B. J. (1974). *Amer. Mineral.*, **59**, 244–248.

Wood, B. J. and Strens, R. G. J. (1972). *Mineral. Mag.*, **38**, 911–917.

Charge-Transfer Processes at High Pressure

H. K. Mao
Geophysical Laboratory, Carnegie Institution of Washington, Washington, D.C. 20008, USA

INTRODUCTION

The efficiency of radiative heat transfer in the earth's mantle is strongly dependent upon the absorption properties of its mineral constituents in the visible and near-infrared regions (Clark, 1957; Lawson and Jamieson, 1958). Petrological and cosmochemical evidence suggests that the earth's mantle consists essentially of ferromagnesian silicates (Birch, 1952; Clark and Ringwood, 1964), which exist as olivine and pyroxenes in the upper mantle; spinel, β-phase, and stishovite in the transition zone (Ringwood, 1956; Ringwood and Major, 1966); and magnesiowüstite and stishovite in the lower mantle (Mao and Bell, 1971a; Bassett and Ming, 1972; Kumazawa *et al.*, 1974). Thus it is critical to know the absorption properties of these phases under the appropriate pressure-temperature conditions in order to understand heat transfer in the earth. The present investigation of ferromagnesium silicates and a series of other minerals containing minor constituents of the mantle, such as Al_2O_3, CaO, Fe_2O_3, and TiO_2 was undertaken to explore these properties at high pressure.

FERROMAGNESIAN SILICATES

Optical spectra of ferromagnesian silicates were studied in a diamond-windowed pressure-cell at room temperature to 300 kbar with the technique described by Mao and Bell (1973). Samples included the complete series of olivine, spinel, and magnesiowüstite in their stable or metastable conditions with composition ranging from their Mg to Fe end members. Greatly enhanced absorption was observed in fayalite and Fe_2SiO_4 spinel at high pressures (Mao and Bell, 1972a). As shown in Figures 1 and 2, the apparent absorption edge at the violet end of the spectrum increases in intensity with pressure and sweeps through the visible region into the near-infrared as pressure is raised above 100 kbar. Above 200 kbar, fayalite becomes completely opaque. The

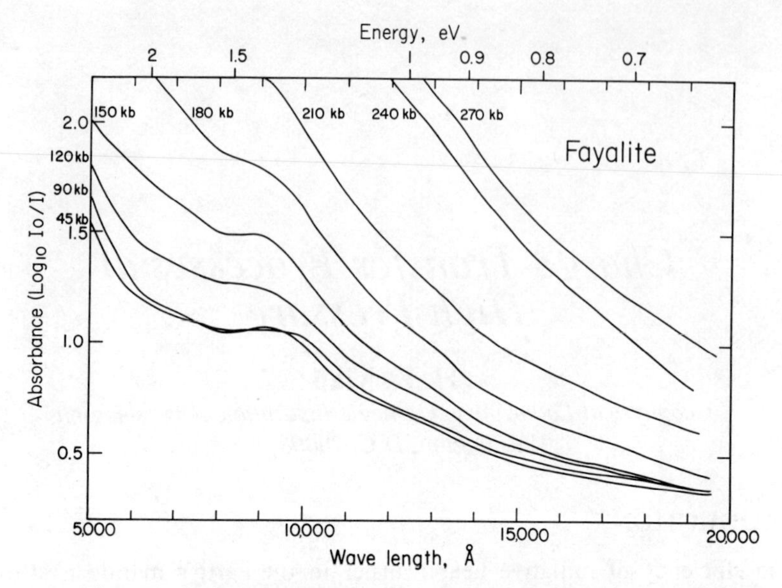

Figure 1. Absorbance of polycrystalline fayalite in the pressure range from 45 to 270 kbar at room temperature. Fayalite is metastable under these conditions.

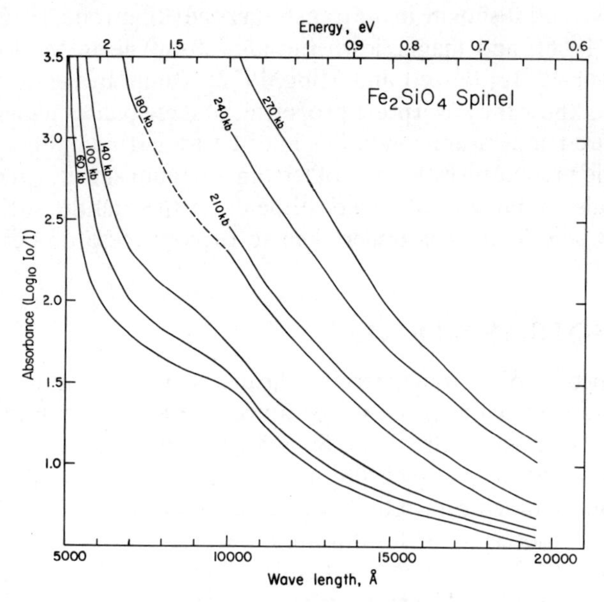

Figure 2. Absorbance of polycrystalline spinel phase of Fe_2SiO_4 in the pressure range from 60 to 270 kbar at room temperature.

Figure 3. Photomicrograph of fayalite taken through a diamond window of the pressure cell. The octagonal outline is the shape of the anvil face of the diamond. The pressure distribution inside the octagon is nearly parabolic, with the maximum at the centre and 1 bar at the edge. The dimension of the octagon is 280 μm. The sample is compressed by the two opposite diamond anvils into a thin foil, approximately 5 μm thick, covering the whole octagon area. The pressure at the centre of the cell is 270 kbar, and the fayalite is opaque. The fayalite is transparent at the rim where the pressure is low.

effect can be observed visually in the diamond pressure cell (Figure 3) and is completely reversible when pressure is released. A similar pressure effect was also observed in a series of natural and synthetic olivines and $(Fe,Mg)_2SiO_4$ spinels (Mao, 1973), but the intensity of the absorption increases with increasing fayalite content as well as increasing pressure, as shown in the series of photomicrographs in Figure 4. The absorption effect was not observable in olivine

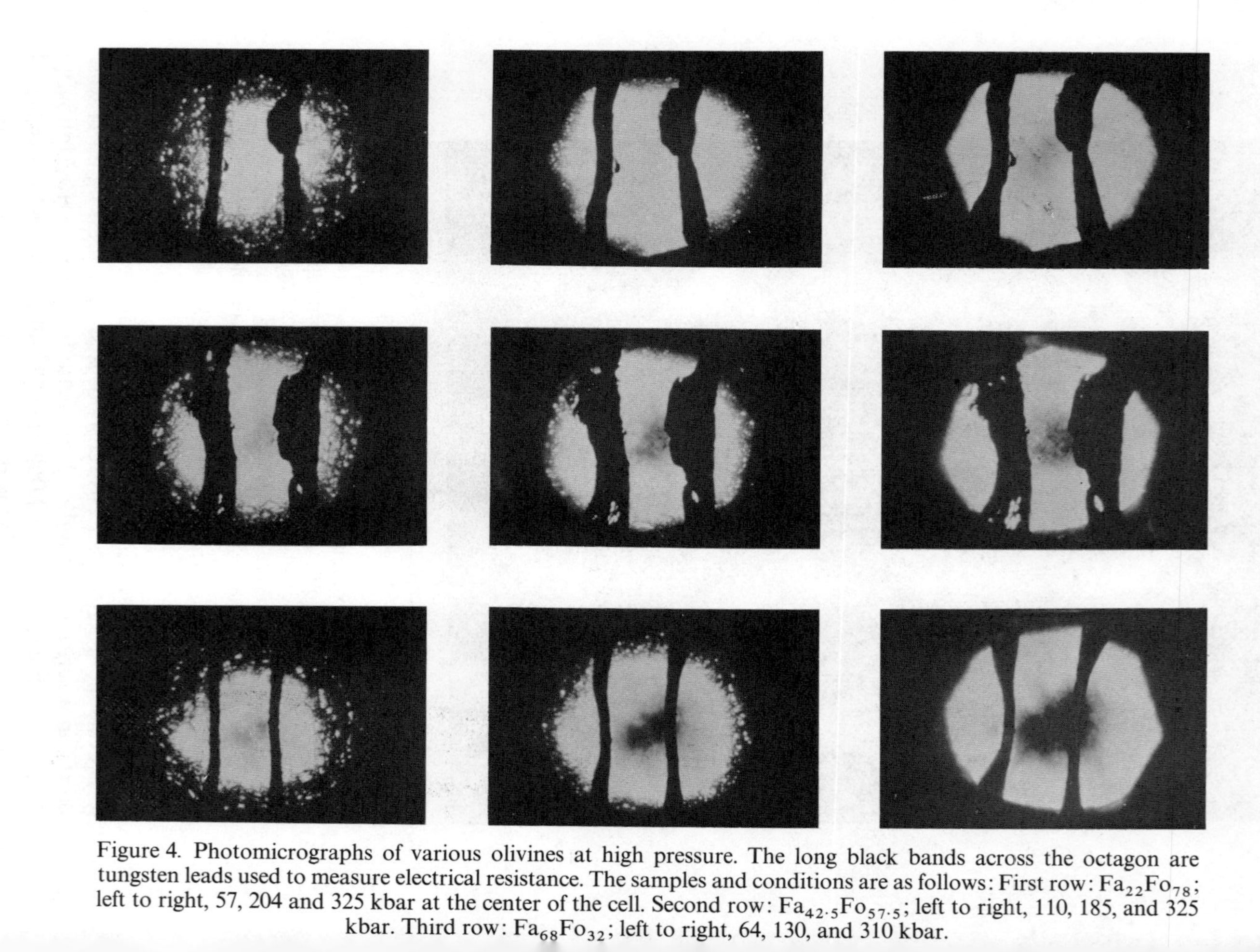

Figure 4. Photomicrographs of various olivines at high pressure. The long black bands across the octagon are tungsten leads used to measure electrical resistance. The samples and conditions are as follows: First row: $Fa_{22}Fo_{78}$; left to right, 57, 204 and 325 kbar at the center of the cell. Second row: $Fa_{42.5}Fo_{57.5}$; left to right, 110, 185, and 325 kbar. Third row: $Fa_{68}Fo_{32}$; left to right, 64, 130, and 310 kbar.

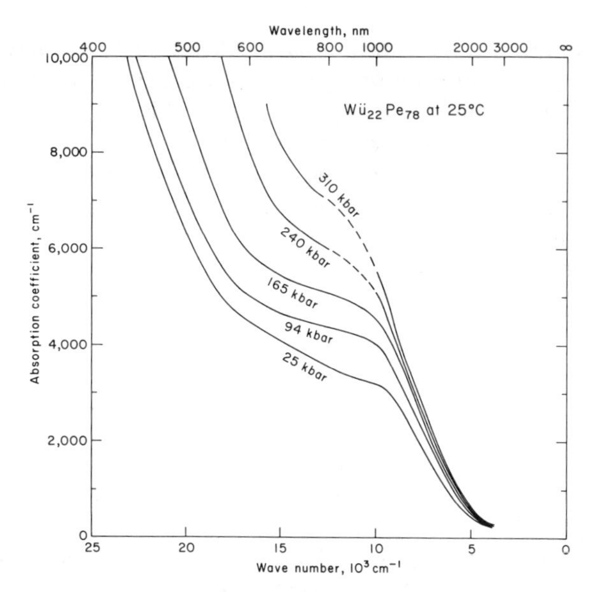

Figure 5. Optical absorption spectra of $Wü_{22}Pe_{78}$ magnesiowüstite at 60 to 310 kbar and room temperature.

with Fa < 10 mole per cent up to 300 kbar at room temperature. Olivine in the mantle with composition $Fa_{20}Fo_{80}$ has an upper stability limit of 130 kbar. The pressure alone is insufficient to induce strong optical absorption.

Olivines at 100 kbar have been heated in the diamond cell with a laser technique described by Bassett and Ming (1972). The optical absorption also increases with temperature and is reversible up to 500 °C. The absorptions induced by temperature and pressure are additive, but temperature is the dominant factor at this pressure.

A pressure effect similar to that observed in olivine and spinel occurs in magnesiowüstite (Mao, 1973). As shown in Figure 5, the apparent absorption edge of wüstite $(Wü)_{22}$-periclase $(Pe)_{78}$ moves through the visible region with increasing pressure, and virtually complete absorption results at 300 kbar. The effect is also sensitive to iron content. The observed absorption is stronger in magnesiowüstite than in olivine with the same Fe:Mg ratio. A considerable darkening effect caused by absorption is observed at 300 kbar in $Wü_9Pe_{91}$, the composition predicted for magnesiowüstite in the lower mantle.

INTERPRETATION

The absorption effects should be important in the lower mantle where the pressure is above 300 kbar and magnesiowüstite and stishovite may be stable. With the strong absorption in the visible and near-infrared regions, magnesiowüstite with less than 10 per cent Wü would be sufficiently opaque to block all radiative heat transfer.

The absorption spectra of transition elements originate in *d*-electron transitions within a transition-metal ion (crystal-field transition) or between neighbouring ions (charge-transfer transition). Considerable effort has been devoted to studies of the effect of pressure and temperature on the crystal-field spectra of ferromagnesian silicates (Shankland, 1970; Mao and Bell, 1971b, 1972b; Fukao *et al.*, 1968), but the effect of pressure on the spectra of olivine, spinel, and magnesiowüstite is apparently not related to *d*–*d* transitions. Instead, the effect appears to involve a metal–ligand charge-transfer process.

High pressure is known to affect charge-transfer processes (Drickamer and Frank, 1973) but the dependence has not been well characterized. In general, high pressures cause increased overlapping of electronic orbitals of neighbouring ions, thus increasing the probability and intensity of charge-transfer absorptions. The maximum of the metal–ligand charge-transfer absorption is usually too far in the ultraviolet region or too intense to be observed in the diamond cell. Only an apparent absorption edge can be observed, as in olivine, spinel and magnesiowüstite. With the edge moving toward the infrared with pressure, it was not clear whether pressure was shifting the charge-transfer band to lower energy, increasing its intensity, or both.

By contrast, metal–metal charge-transfer usually occurs in the visible region with a well defined absorption maximum, making it easier to characterize

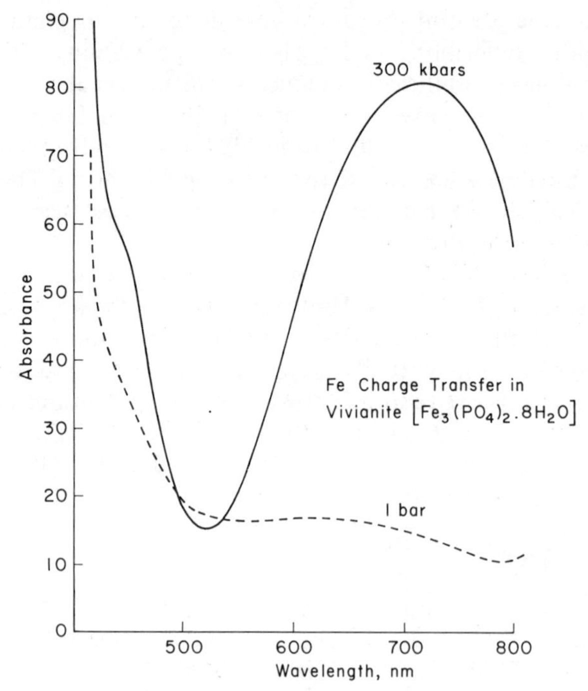

Figure 6. Absorbance of polycrystalline vivianite at 1 bar, 300 kbar, and room temperature.

the effect of pressure. Vivianite, which has a known Fe^{2+}–Fe^{3+} charge-transfer band (Robbins and Strens, 1972), was studied at high pressure (Mao and Bell, 1974). As shown in Figure 6, the Fe^{2+}–Fe^{3+} charge-transfer band at 650 nm is intensified 50 times and shifted to 700 nm at 300 kbar. The intensification and shifting are similar to the pressure-induced shift of the charge-transfer edge of olivine, spinel, and magnesiowüstite. However, since the Fe^{2+}–Fe^{3+} band is at the red end of the visible region, the pressure-induced colour change is from very light blue to deep green, whereas the shift of the violet edge in olivine causes colour change from yellow to brown to black.

PRELIMINARY RESULTS ON OTHER MINERALS

In an attempt to characterize the pressure effects, groups of minerals that may be stable in the mantle, minerals with iron in various oxidation states and crystallographic sites, and minerals with transition elements other than iron were studied up to 300 kbar. Most of the samples were well documented and characterized by X-ray diffraction, electron microprobe analysis, Mössbauer spectroscopy, electron-spin resonance, crystal-field spectroscopy, or wet chemical analysis. The results are summarized in Table 1. In general, the reversible pressure effect is observed in minerals with strong metal–ligand charge-transfer bands in the ultraviolet region or metal–metal charge-transfer bands in the red region. The intensity of the effect varies with crystal structure. Within a group of minerals, the effect increases with increasing content of iron or other transition element.

A pressure effect similar to that in olivine is observed in the enstatite-orthoferrosilite series, but it is weaker and occurs at higher pressure in the pyroxenes. The metal–ligand charge-transfer edge shifts to lower energy with increasing pressure, and the sample darkens. The effect also depends on the iron content. For clinopyroxene with low iron content, including chrome diopside (Mao *et al.*, 1972), ferric diopside (Bell and Mao, 1972), and jadeite, the effect is not observed at 300 kbar.

For pyroxenes and amphiboles with a known Fe^{2+}–Fe^{3+} charge-transfer band, such as hedenbergite, actinolite, and ferrotremolite (Burns, 1970), a pressure effect similar to that in vivianite is observed. The metal–metal charge-transfer band at 600–700 nm greatly intensified with pressure, and the sample became green. The Fe^{2+}–Fe^{3+} charge-transfer band of tourmaline and the Ti^{3+}–Ti^{4+} charge-transfer band of the Allende fassaite at 600–700 nm show similar effects. For anthophyllite and gedrite with low iron content, the effect is not observed.

In a red chromite with Fe^{3+} in octahedral sites and Fe^{2+} in tetrahedral sites, the darkening effect caused by pressure-induced shifting of the charge-transfer edge is very strong. After the same chromite was reduced to a chrome spinel with all ferric iron and most of the ferrous iron intentionally removed by diffusion, the pressure-induced darkening effect was not observed. The spinel remained red even after it was made iron-free because of the crystal–field

Table 1. List of preliminary observations of pressure-induced absorption in the range 1 bar-300 kbar. Type 1 pressure effect is the moving of the apparent absorption edge at the violet end of the spectrum through the visible region into near-infrared. Type 2 pressure-effect is the intensification of the metal-metal charge-transfer band near 700 nm.

Substance		Pressure-Induced Absorptions of Type 1	Type 2
Olivines:	Fa_{100}–Fo_{90}	s to vw[a]	
	Fo > 90	n	
Orthopyroxenes:	Fs–En	s to n	
Clinopyroxenes:	Chrome diopside; Ferridiopside, jadeite	n	
	Hedenbergite, Ti-fassaite		s
Amphiboles:	Actinolite, ferrotremolite		s
	Anthophyllite, gedrite	n	
Spinels:	Chromite	vs	
	Reduced chrome spinel	n	
	Fe_2SiO_4 spinel	s	
Magnesiowüstite:	$Wü_{100}$–$Wü_9$	vs to w	
	Periclase	n	
Garnets:	Almandite, spessartite	vw	
	Andradite	w	
	Ti-andradite	m	
	Kimzeyite	m to s	
FeOOH:	Akaganéite, lepidocrocite, goethite	vs	
Glasses:	$CaO–MgO–Al_2O_3–SiO_2–FeO–TiO_2$		
	Oxidized	s	
	Reduced	s	
	$CaO–MgO–Al_2O_3–SiO_2–FeO$		
	Oxidized	s	
	Reduced	w	
Miscellaneous:	Vivianite, tourmaline		s
	Wulfenite	vs	
	Malachite, chrysocolla		s
	Stishovite, quartz, corundum; Calcite, alkali halides	n	

[a]v, very; s, strong; m, moderate; w, weak; n, no observable effect.

absorption of Cr^{3+} ions. Evidently the chromic ion is not affected by pressure to 300 kbar.

A weak pressure shift of the apparent absorption edge is observed in almandite and spessartite with Fe^{2+} and Mn^{2+} in cubic sites, and andradite, titanium andradite, and kimzeyite with Fe^{3+} in octahedral and tetrahedral sites. The effect is stronger with Fe^{3+} in a tetrahedral site than in an octahedral site.

The lepidocrocite, akaganéite, and goethite polymorphs of ferric oxide hydroxide all show very strong absorption effects at high pressure. Strong effects are also observed in oxidized basaltic glass and lunar orange glass. The effect is weak in a reduced green basaltic glass.

The effect is not observed in stishovite, quartz, corundum, alkali halides, and calcite. Pressure intensifies the green colour in the copper minerals malachite and chrysocolla. A very strong pressure-induced shifting of the absorption edge is observed in wulfenite. All results generally support the interpretation of pressure intensifying the charge-transfer processes.

CONCLUSION

High pressure intensifies the charge-transfer absorption in ferromagnesian silicates and shifts it to lower energy. In the upper mantle where olivine is stable, the pressure alone is insufficient to cause a strong absorption, and temperature plays the dominant role in controlling optical absorption. However, at depths greater than 900 km, olivine is not stable and the pressure-induced absorption is significant in most lower mantle minerals. The charge-transfer band covers the visible and near-infrared regions and completely blocks radiative heat transfer. At such pressures, crystal-field absorption becomes relatively unimportant and charge-transfer absorption prevails.

REFERENCES

Bassett, W. A., and L. Ming, *Phys. Earth Planet. Interiors*, **6**, 154–160, 1972.

Bell, P. M. and H. K. Mao, *Carnegie Institution of Washington Year Book 71*, 531–534, 1972.

Birch, F., *J. Geophys. Res.*, **57**, 227–286, 1952.

Burns, R. G., *Mineralogical Applications of Crystal-Field Theory*, Cambridge University Press, Cambridge, 1970.

Clark, S. P., Jr., *Am. Miner.*, **42**, 733–742, 1957.

Clark, S. P., Jr., and A. E. Ringwood, *Rev. Geophys.*, **2**, 35–88, 1964.

Drickamer, H. G., and C. W. Frank, *Electronic Transitions and the High Pressure Chemistry and Physics of Solids*, Chapman and Hall, London, 1973.

Fukao, Y., H. Mizutani, and S. Uyeda, *Phys. Earth Planet. Interiors*, **1**, 57–62, 1968.

Kumazawa, M., H. Sawaimoto, E. Ohtani, and K. Masaki, *Nature, Lond.*, **247**, 356–358, 1974.

Lawson, A. W., and J. C. Jamieson, *J. Geol.*, **66**, 540–551, 1958.

Mao, H. K., *Carnegie Institution of Washington Year Book 72*, 552–557, 1973.

Mao, H. K., and P. M. Bell, *Carnegie Institution of Washington Year Book 70*, 176–178, 1971a.

Mao, H. K., and P. M. Bell, *Carnegie Institution of Washington Year Book 70*, 207–215, 1971b.

Mao, H. K., and P. M. Bell, *Science*, **176**, 403–406, 1972a.

Mao, H. K., and P. M. Bell, *Carnegie Institution of Washington Year Book 71*, 524–527, 1972b.

Mao, H. K., and P. M. Bell, *Am. Soc. Test. Mater. Spec. Tech. Publ.*, **539**, 100–119, 1973.

Mao, H. K., and P. M. Bell, *Carnegie Institution of Washington Year Book 73*, in the press, 1974.

Mao, H. K., P. M. Bell and J. S. Dickey, Jr., *Carnegie Institution of Washington Year Book 72*, 538–541, 1972.

Ringwood, A. E., *Nature*, **178**, 1303–1304, 1956.

Ringwood, A. E., and A. Major, *Earth Planet. Sci. Letters*, **1**, 351–357, 1966.

Robbins, D. W., and R. G. J. Strens, *Mineralog. Mag.*, **38**, 551–563, 1972.

Shankland, T. J., *J. Geophys Res.*, **75**, 409–414, 1970.

Intervalence-Transfer Absorption in Some Silicate, Oxide and Phosphate Minerals

G. Smith
Department of Solid State Physics.
Australian National University,
Canberra ACT, Australia 2600.

and

R. G. J. Strens
School of Physics, The University,
Newcastle upon Tyne NE1 7RU, U.K.

Intervalence-transfer absorption is one of two main types of charge-transfer absorption. The electron transfer occurs beween neighbouring cations (metal–metal charge-transfer) rather than between anion and cation (ligand–metal charge-transfer). Recent reviews of the subject include those by Robin and Day (1967), Allen and Hush (1967), and Day (1970) with contributions to the theory by Hush (1967, 1968), Robbins and Strens (1968), Drickamer and Frank (1973), and Mayoh and Day (1974) among others.

Metal–metal charge-transfer may usefully be divided into *homonuclear* (e.g. $Fe^{2+} \rightarrow Fe^{3+}$) and *heteronuclear* (e.g. $Fe^{2+} \rightarrow Ti^{4+}$) types, with further subdivison into *symmetric* ($\Delta z = 1$, e.g. $Fe^{2+} \rightarrow Fe^{3+}$ and $Ti^{3+} \rightarrow Ti^{4+}$) and *asymmetric* ($\Delta z \neq 1$, e.g. $Mn^{2+} \rightarrow Mn^{4+}$) types. We are concerned only with symmetric homo- and hetero-nuclear transfers in this paper, although the asymmetric type may well occur in minerals. Charge-transfer between ions of the same charge and variable-valence e.g. Fe^{2+}, $Fe^{2+} \rightarrow Fe^{3+}$, Fe^{+} has not been observed in minerals, and should be extremely weak because of the very small dipole moment expected in such cases.

A simple and useful approach to intervalence transfer is that of Robin and Day (1967), who represent the ground-state wavefunction of a donor-acceptor system as:

$$\psi_0 = (1 - \alpha^2)^{1/2}\psi'_0 + \alpha\psi_x$$

where ψ'_0 is the zero-order wave function, ψ_x the excited state wavefunction, and α is a delocalization factor. When two potentially exchanging ions occupy crystallographically identical sites, e.g. the octahedral Fe^{2+} and Fe^{3+} in magnetite above 120 K, and when their separation is also small, we find $\alpha \rightarrow 1/\sqrt{2}$, and the system is best represented as having an oscillating valence:

$$Fe^{3+} \longleftarrow e \longrightarrow Fe^{3+}$$

In magnetite this oscillation is fast enough ($\tau < 10^{-8}$ s) for the octahedral Fe^{2+} and Fe^{3+} to be indistinguishable by the Mössbauer method.

At the other extreme, as $\alpha \rightarrow 0$, the spectrum of a sample containing donor and acceptor ions is simply the sum of the spectra of the two components. We are concerned in this paper with the intermediate case, $0 < \alpha < 1/\sqrt{2}$, in which localized donor and acceptor ions may be distinguished by Mössbauer or other methods, but sufficient delocalization occurs for bands to be present which do not occur in the spectra of minerals containing only the donor or acceptor ions. In silicates and oxides, α is commonly less than 0.1 per cent, and the charge-transfer process may be written:

$$Fe_A^{2+} + Fe_B^{3+} \xrightarrow{h\nu} Fe_A^{3+} + Fe_B^{2+}$$

The electron transfer is between ions in coordination polyhedra which share edges or faces (the inner-sphere complexes of Allen and Hush, 1967) and apparently occurs *via* ligand orbitals, usually of π type, rather than by direct overlap of metal orbitals (Robbins and Strens, 1968; Smith, 1973; Mayoh and Day, 1974).

Application of the Mulliken approximation is also illuminating:

$$h\nu = (I - A) + (C_i - C_f)$$

where I is the ionization potential of the donor, A the electron affinity of the acceptor, and C_i, C_f are the initial and final Coulomb energies of interaction. For homonuclear transfers between ions in similar sites, both $(I - A)$ and $(C_i - C_f)$ are expected to be small compared with the observed transition energies of 1 to 3 eV, implying that the energy is absorbed into the phonon system of the crystal. This in turn implies that there will be a thermally assisted electron hopping process, which typically has an activation energy about one-quarter that of the optical transition.

Although the optical, magnetic and electronic properties of these donor-acceptor systems may be 'understood' in terms of these simple ideas, it cannot be said that a satisfactory theory exists. We suspect that the search for representations of the behaviour of charge-transfer systems may have been too narrowly based, and that a general theory of ion-pair spectra and related magnetic and electronic properties should be sought.

This unified theory might cover metal–metal and ligand–metal charge-transfer, the spectra of exchange-coupled ion-pairs (Ferguson, Guggenheim and Tanabe, 1967), and such phenomena as photomagnetism (Rijnierse and Enz, 1975), and relate these to our ideas about polarons and hopping conduction (Austin and Mott, 1969), and magnetic interactions in solids.

At a less celestial level, criteria are needed for the identification and assignment of charge-transfer bands in mineral spectra. We have concentrated in this paper on the temperature-dependence of charge-transfer bands in a number of silicate, oxide and phosphate minerals, but also review the pressure-, composition- and polarization-dependence of these bands, and tabulate oscillator strengths, energies and band widths. We hope that these data will be useful both to theoreticians wishing to test models, and to spectroscopists needing to assign spectra.

CLASSIFICATION OF *d–d* AND CHARGE-TRANSFER BANDS IN MINERALS

In general, a mineral spectrum consists of a monotonically varying background absorption, increasing with photon energy, and representing the low-energy wing of intense ligand–metal charge-transfer bands in the near ultraviolet. On this are superimposed a number of discrete absorption bands of *d–d* or charge-transfer origin. In rare cases *F*-centres or vibrational bands contribute to the spectrum. The first problem confronting the experimentalist is then the identification of the band as *d–d* or charge-transfer in origin, and the identification of the absorbing species. Full assignment requires the specification of the ground and excited states involved, this being done in a manner which is consistent with the energy, intensity, width, polarization, and pressure-, temperature- and composition-dependence of the band.

Apart from charge-transfer bands, the mineralogist is mainly concerned with transitions between the 3*d* electron orbitals of first-row transition metals with maximum spin-multiplicity (weak-field case). Such transitions are theoretically forbidden by the Laporte selection rule:

$$\Delta L = \pm 1$$

but become partially allowed in practice through mixing of the *d*-states with nearby states of *p*-character, for example those involved in the ligand-metal charge-transfer. The *d–d* transitions are therefore very much weaker than the ligand–metal charge-transfer bands, and somewhat weaker than the metal–metal charge-transfer bands mentioned above. They are nonetheless often strong enough to give intense colours in thin sections of normal (30 μm) thickness.

Two further selection rules are important. Spin-allowed transitions are those for which:

$$\Delta S = 0$$

If $\Delta S \neq 0$, the transitions are said to be spin-forbidden, with integrated intensities some 100 to 1000 times lower. Symmetry-allowed transitions are those for which the dipole moment operator in the expression:

$$\langle \psi_0 | \mu | \psi_x \rangle$$

is finite, i.e. $\mu \neq 0$. When $\mu = 0$, as for a transition metal ion located on a centre

Table 1. Classification of charge-transfer and *d–d* bands.

Type	Label	ΔL	ΔS	μ	f
Ligand⟶metal charge-transfer	LM	+1	0	large	$1-10^{-1}$
Metal⟶metal charge-transfer	MM	0^a	0	large	10^{-2}–10^{-4}
Normal *d–d*	N	0	0	$\neq 0$	10^{-4}
Vibronic *d–d*	V	0	0	0	10^{-5}
Spin-forbidden normal	FN	0	$\neq 0$	$\neq 0$	10^{-7}
Spin-forbidden vibronic	FV	0	$\neq 0$	0	10^{-8}

[a]In cases considered here

of symmetry, or for directions normal to a mirror plane or parallel to a rotation axis of order > 1, the transition is symmetry-forbidden. We therefore find four classes of *d–d* transition (Table 1), and two classes of charge-transfer transition each with its own characteristic pressure and temperature-dependence, intensity and width.

DEFINITIONS

The integrated absorption (area under band) assuming Gaussian band shape is given by:

$$A = (\pi/2)^{1/2} w\alpha \tag{1}$$

where w is the full width at half height, and α the absorption coefficient to base e.

The oscillator strength f is a measure of the probability that a photon will be absorbed in an encounter with an absorbing species (ion or ion-pair), and for *d–d* bands approximating Gaussian form is given by:

$$f = 1.877 \times 10^{-12} AV/n \tag{2}$$

where V is the molar volume in cm^3, and n is the number of absorbing ions per formula unit. The expression becomes more complicated for charge-transfer bands, in which n is replaced by a term involving the product of the donor and acceptor concentrations $[c]$ and the number of donor–acceptor contacts n_{da}. Because charge-transfer transitions are polarized along the donor–acceptor vector, it is convenient to express their oscillator strength as the value which would be observed were the electric vector of the incident light to lie along this direction. Equation (2) then becomes:

$$f_{\parallel} = 1.877 \times 10^{-12} \frac{A_{\parallel} V}{[c] n_{da}} \tag{3}$$

where $A_{\parallel}$ is the sum of the A values for the three principal vibration directions. The equations in Robbins and Strens (1968) differ only in that the first incorporates the $(\pi/2)^{1/2}$ term of (1) above in the constant, and the second is expressed in terms of optical density.

EXPERIMENTAL

Sections of known orientation were cleaved or cut, ground to the thickness required, polished, and analysed using an electron probe microanalyser. Partial analyses appear in Table 2.

Table 2. Partial analyses of mineral samples.

	SiO_2	Al_2O_3	MgO	FeO	Fe_2O_3	TiO_2	Notes
Biotite B9	35.1	16.2	7.8	17.1	5.5	2.8	r
Chlorite	40.5	3.4	35.8	5.3	n.d.	0.2	p
Cordierite	45.5	31.5	2.5	12.0	4.2	0.7	p,s
Tourmaline	—	—	—	14.0	1.0	—	r
Vivianite	—	—	—	36.9	4.5	—	s
Glaucophane 945	—	—	—	2.0	3.4	—	d
Glaucophane 1044	—	—	—	4.9	3.2	—	d
Glaucophane 615	—	—	—	14.3	1.6	—	d
Kyanite	36.3	60.1	—	—	0.154	0.015	p
Corundum	—	98.6	—	—	1.58	0.021	p

[r]Analyst D. W. Robbins (Robbins and Strens, 1972)
[p]Electron probe analysis by G. Smith. (1973), total iron stated as FeO or Fe_2O_3
[s]FeO and Fe_2O_3 determined by J. H. Scoon
[d]J. E. Dixon, pers. comm. 1966.

Spectra were measured using an Optica CF4 spectrophotometer with mirror optics, which had been modified to accept high-pressure (0–50 kb) and variable temperature (77–700 K) sample holders. The polarizer was a Nicol prism mounted in the common beam, making it unnecessary to use a matched pair of polarizers.

The spectra were converted from the original linear wavelength scale to a wavenumber scale, and analysed into their component bands using a Dupont curve resolver. This instrument permits the fitting of symmetrical or skewed Gaussians or Lorentzians, or of mixed Gaussian–Lorentzian curves.

Most bands were best represented by skewed Gaussians. The spectra were resolved using the minimum number of bands required to give a satisfactory fit (necessarily a subjective judgement), and any two components were considered unresolvable if separated by less than about four-fifths of the width of the narrower band. Band widths, positions and intensities were unconstrained.

While this procedure may result in failure to identify weak contributions to the spectra, it is very unlikely to generate spurious bands. The widths, positions and intensities of the resolved bands vary in accuracy from spectrum to spectrum, and no general estimate of accuracy can be given.

RESULTS AND ASSIGNMENTS

Ferromagnesian Silicates and Vivianite

Ferrous–ferric interactions are important in the spectra of many ferromagnesian minerals, and especially of those containing brucite-like sheets, ribbons or fragments, e.g. mica and chlorite, amphiboles and tourmaline.

Charge-transfer in ferromagnesian minerals usually involves ions in edge-sharing octahedra, although cordierite is an exception. Transfer between Fe ions in corner-sharing polyhedra is not at present known. Electron-transfer is normally into a single acceptor level, although two levels may be present in biotite and chlorite, and possibly in other minerals. Faye and Nickel (1970) suggest that because Fe^{3+} has a symmetrical $^6A_1$ ground state, its t_{2g} levels are unlikely to be split by as much as 3000–4000 cm^{-1}. The splitting of the acceptor orbitals depends on the symmetry of the ligand field, and not on that of the transition metal ion. The t_{2g} splittings for Fe^{3+} will be larger than those of Fe^{2+} in the approximate ratio $\Delta(Fe^{3+})/\Delta(Fe^{2+})$, or about 3:2, and splittings of several thousand wavenumbers will certainly occur, and can be calculated by the method of Wood and Strens (1972).

Biotite. As Robbins and Strens (1972) have given a detailed account of the spectra of trioctahedral micas, we shall consider only developments since their paper appeared. These mainly concern the temperature-dependence of the intensities of the various absorption bands, and the phenomena of substitutional broadening and intensification. Gorbatschev (1972) has provided further qualitative data on the link between Ti content and colour in biotites, but we do not agree with his conclusion that Ti^{3+} is an important colouring agent.

Figure 1 shows that the 13650 cm^{-1} band, which is unequivocally assigned to $Fe^{2+} \rightarrow Fe^{3+}$ charge-transfer, has a strong inverse temperature dependence, the intensity at 587 K being only 28 per cent of that at 94 K. The dependence may be of the form:

$$A(T) = A_0 \exp - (T/\theta)$$

where θ is of the order of the Debye temperature, but further detailed work will be required to establish this.

Curve resolution suggests that there is a second charge-transfer band with inverse temperature dependence at about 16400 cm^{-1} (Table 3a). Its origin is uncertain, and it may represent either electron-transfer into an acceptor orbital lying some 2600 cm^{-1} below that involved in the 13650 cm^{-1} transition, or transfer between different sites or ions.

Robbins and Strens (1972) found that the $O \rightarrow Fe^{2+}$ absorption edge of biotite shifted to lower energies as the TiO_2 content increased, and interpreted this as substitutional broadening of the $O \rightarrow Fe^{2+}$ band by Ti. Unpublished data on the optical properties of mineral solid solutions obtained by Freer at Newcastle suggest an alternative explanation. It appears that the replacement of an electropositive ion by a more electronegative ion, e.g. of Mg by Al in orthopyroxene, or of Mg by Ti in biotite, increases the covalency of the Fe^{2+}–O bond, and shifts the $O \rightarrow Fe^{2+}$ absorption edge to lower energies. Pressure has a similar effect (Mao, this volume), the covalency presumably increasing as the interatomic distance is reduced, and the $O \rightarrow Fe^{2+}$ absorption edge moving into the visible.

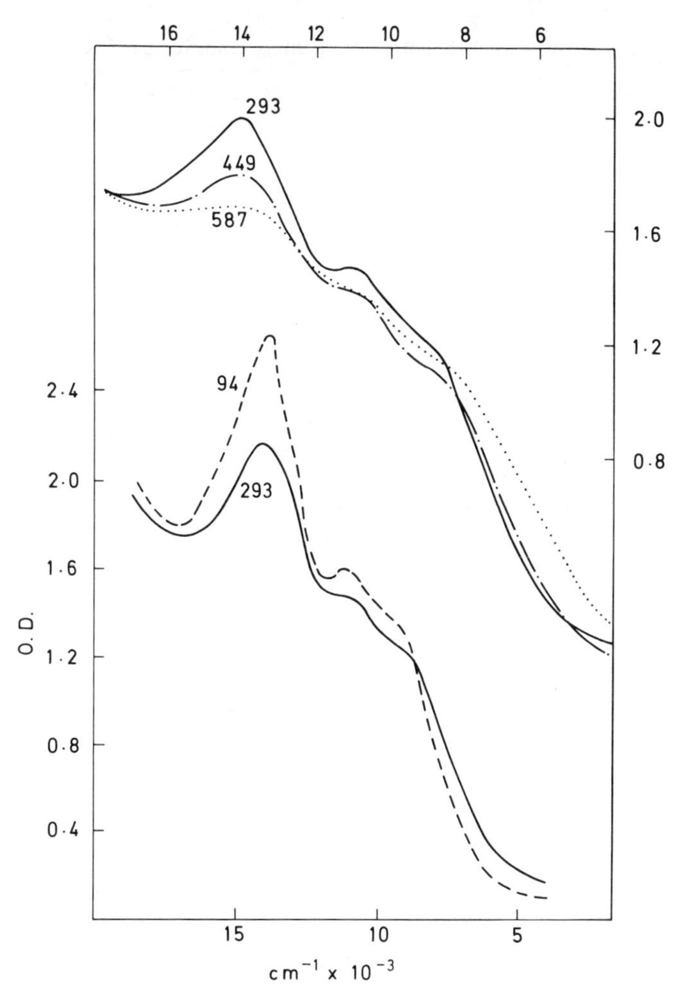

Figure 1. The ω $(\beta + \gamma)$ spectra of biotite B9 of Robbins and Strens (1972). Note the strong inverse temperature dependence of the intensity of the 13650 cm^{-1} charge-transfer band, and the relative insensitivity of the other components of the spectrum to temperature. Thickness 18 μm. Temperatures are absolute (K).

Table 3 (a). Unpolarized ω-spectrum of biotite B9 at 293 K.

Band	$\nu(cm^{-1})$	$\alpha(cm^{-1})$	$w(cm^{-1})$	$10^{-4}A$
I[a]	16400	171	2275	49
II	13650	1382	3150	546
III	10900	631	1900	150
IV	8750	819	2775	285

Table 3 (b). **E** ∥ b (β) spectrum of chlorite at 293 K.

Band	$\nu(cm^{-1})$	$\alpha(cm^{-1})$	$w(cm^{-1})$	$10^{-4}A$
I[a]	16700	11.4	3375	4.8
II	13800	44.2	3575	19.8
III	10900	26.0	1700	5.5
IV	9100	22.4	2100	5.9

[a]Existence of band not certain, intensity very sensitive to method of resolution.

It seems that shifts of the absorption edge in aluminous orthopyroxenes, and in titanian biotites and tourmalines may originate in a reduction in the energy rather than a broadening of the O → Fe^{2+} band. Mechanical broadening of the kind envisaged by Robbins and Strens will occur, but its magnitude may be small compared with the effect described above.

Freer's observation also suggests an additional mechanism of substitutional intensification. Any band which depends for its strength on intensity stealing from an underlying charge-transfer absorption will inevitably intensify if

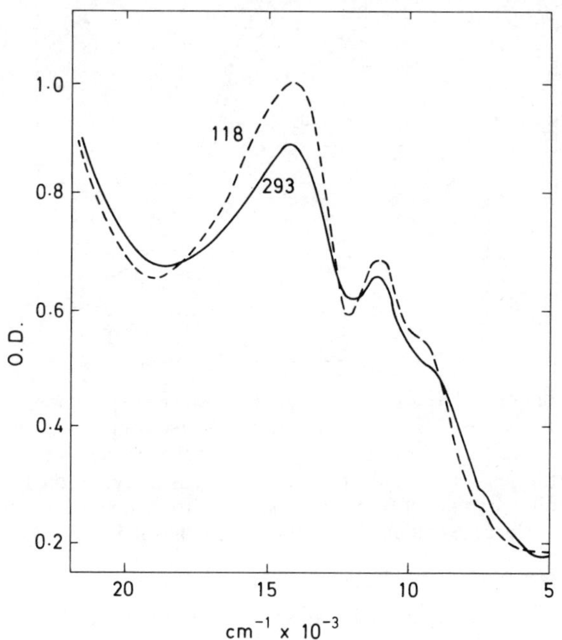

Figure 2. **E**∥$b(\beta)$ absorption spectrum of a green chlorite (thickness 237 μm), showing the strong inverse temperature dependence of the main charge-transfer band. The **E**∥$a(\gamma)$ spectrum is indistinguishable from that illustrated, contrasting with biotite, in which the charge-transfer absorption is polarized **E**∥b > **E**∥a. Temperatures marked are absolute (K).

that absorption is increased by the substitutional or pressure effects mentioned above. This may be important in titanian tourmalines (see below, and Faye *et al.*, 1974).

Chlorite. The chlorite spectrum (Figure 2) closely resembles that of biotite, and both the number of bands and their energies and widths are similar (Table 3b above). The 13800 cm^{-1} and 16700 cm^{-1} bands exhibit inverse temperature dependence, and are assigned to $Fe^{2+} \rightarrow Fe^{3+}$ charge-transfer, the assignment being definite for the former, and probable for the latter. The bands at 10900 and 9100 cm^{-1} are assigned to $t_{2g} \rightarrow e_g$ transitions of Fe^{2+} predominantly located in the 'talc' layer of the chlorite structure (Smith, 1973). It is interesting that the *d–d* bands are 3 to 6 times as strong in biotite as in chlorite, and this may be a measure of the 'substitutional intensification' in the chemically much more complex biotite solid solution.

Amphiboles: We have made no detailed experimental study of amphibole spectra, and consider them only in relation to other minerals. Polarized single crystal spectra of glaucophane were measured by Bancroft and Burns (1969). who also determined the Fe^{2+} and Fe^{3+} site populations by Mössbauer and infrared methods. The spectra show two strong bands at 16300 cm^{-1} (β) and 18100 cm^{-1} ($\gamma > \beta$) which merge to form a single composite band in our diffuse reflectance spectra (Figure 3). From the polarization dependence found by Bancroft and Burns, the composition-dependence shown in Figure 3, and

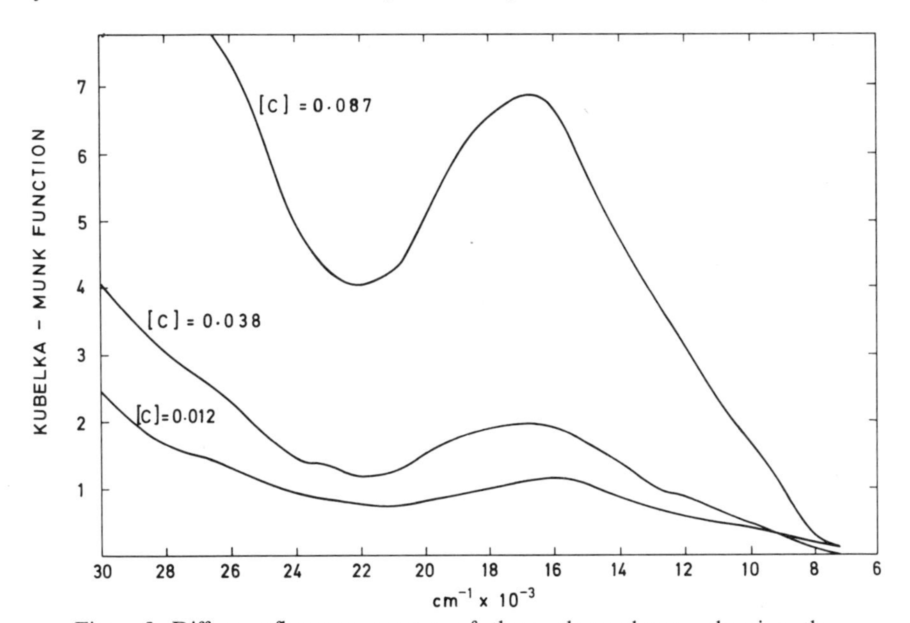

Figure 3. Diffuse reflectance spectra of three glaucophanes, showing the dependence of the intensity of the 17000 cm^{-1} composite charge-transfer band on the product of donor (Fe^{2+}) and acceptor (Fe^{3+}) concentrations.

the width and energy of the absorption, they can be unequivocally assigned to M(3)–M(2) and M(1)–M(2) charge-transfer, and assigned the oscillator strengths in Table 8 below.

Manning and Nickel (1969) and Faye and Nickel (1970) have measured the spectra of riebeckite, holmquistite, hornblende, barkevikite and ferrohastingsite, all of which have strong charge-transfer absorption between 15000 and 18000 cm^{-1}.

Tourmaline. This mineral crystallizes in space group $R3m$ with the general formula $Na(Mg,Fe,Mn,Li,Al)_3\ (Al,Fe)_6\ (Si_6O_{18})\ (BO_3)_3(OH,F)_4$. Most of the Mg and Al occupy octahedral sites 9(b) and 18(c) respectively, although there is some (Mg,Al) disorder between these sites. In most tourmalines, $Fe^{2+} \gg Fe^{3+}$, and both ferrous and ferric iron are concentrated in the large b-sites. The b- and c-site octahedra share edges, so that b–b, b–$c \neq c$–b, and c–c charge-transfer transitions are possible, in addition to d–d transitions of Fe^{2+} and Fe^{3+} in both sites. The three b-site ions in each formula unit are arranged in a brucite-like fragment, the hydroxyl stretching vibrations of which exhibit a fine structure similar to that described in amphiboles by Burns and Strens (1966), and discussed by Strens (1974).

Interpretation of the tourmaline spectra is made difficult by this structural and chemical complexity, and by the existence of several mutually contradictory assignments of the main absorption bands (Table 4). Different authors (and sometimes the same authors at different times) have considered that the two main bands at about 13600 (band I) and 8900 to 9500 cm^{-1} (band II) originate in d–d transitions of Fe^{2+} in the b-site (Wilkins *et al.*, 1969), that they represent d–d transitions of Fe^{2+} in the c- and b-sites respectively (Burns, 1972), that they are both charge-transfer bands (Smith, 1973), and that the former is a charge-transfer band and the latter a d–d transition of b-site Fe^{2+} (Robbins and Strens, 1968). Faye *et al.* (1974) have reviewed much of the earlier work, and we shall consider only the most recent attempts at assignment by them and by Burns (1972), and Burns and Simon (1973).

Figure 4 shows the $\varepsilon(\mathbf{E} \parallel c)$ and $\omega\ (\mathbf{E} \perp c)$ spectra of a blue tourmaline containing 14.02 per cent FeO and 1.00 per cent Fe_2O_3, equivalent to 2.05 Fe^{2+} and 0.13 Fe^{3+} per formula unit, with < 0.2 per cent TiO_2 and MnO. Mössbauer work by Bancroft (personal communication, 1966) shows that about 10 per cent of the Fe^{2+} is in the c-site, but he did not resolve separate b- and c-site Fe^{3+} doublets.

Table 4. Spectra of blue tourmaline at 293 K.

	Band	$\nu(cm^{-1})$	$\alpha(cm^{-1})$	$w(cm^{-1})$	$10^{-4}A$
$\mathbf{E}\perp(\omega)$	I	13600	453	3200	181
	II	8900	384	2650	127
$\mathbf{E}\parallel c(\varepsilon)$	I	13600	29	2600	9.4
	II	9500	42	2600	13.7

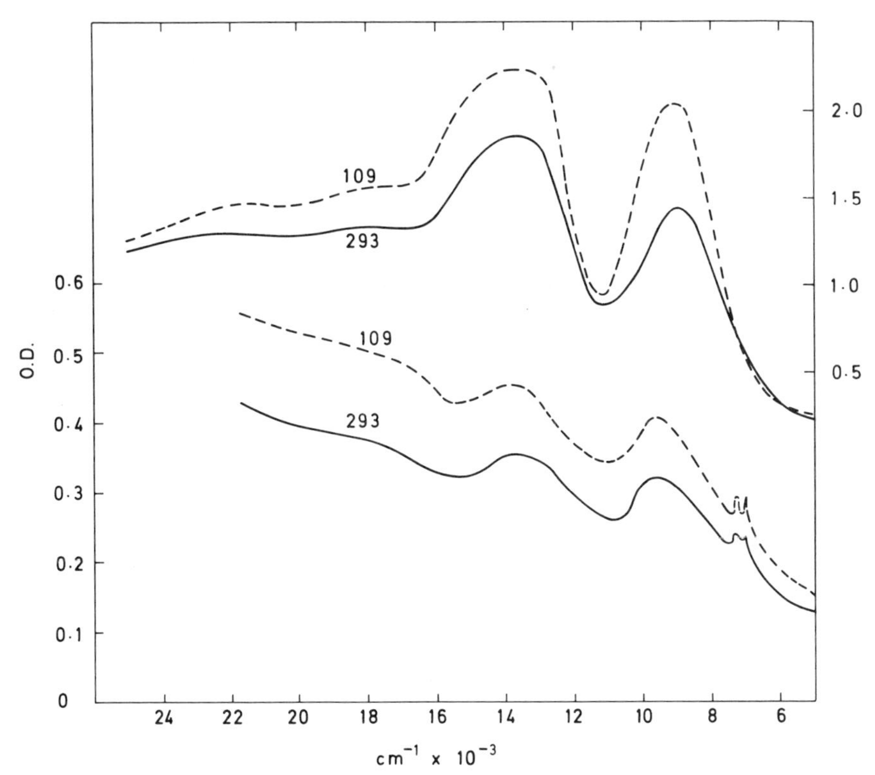

Figure 4. $\mathbf{E} \perp c(\omega)$ spectrum of a blue tourmaline (top) showing the distinct inverse temperature dependence of the main bands at 8900–9500 and 13600 cm^{-1}, which are nonetheless assigned to *d–d* transitions of Fe^{2+} in the *b*-sites. The $Fe^{2+} \rightarrow Fe^{3+}$ charge-transfer band at 18500 cm^{-1} is weak in this sample, which contains little ferric iron. Thickness 60 μm, temperatures marked are absolute (K). The $\mathbf{E} \parallel c(\varepsilon)$ spectrum is shown below.

Both spectra are characterized by two strong bands at 13600 and 8900 to 9500 cm^{-1} the intensities of which exhibit strong inverse temperature dependence. Two weak, broad bands occur at about 23000 and 18500 cm^{-1} in both spectra, and the fundamental and first overtone of the hydroxyl stretching vibration are seen in the ε spectrum at about 3500 and 7000 cm^{-1} respectively.

Faye *et al.* (1974) have published spectra of a number of tourmalines, including elbaites, dravite and schorl. Apart from bands at about 23000 and 18500 cm^{-1}, which they assign to $Fe^{2+} \rightarrow Ti^{4+}$ and $Fe^{2+} \rightarrow Fe^{3+}$ charge–transfer, they claim to have located six major bands and many minor ones between 6000 and 30000 cm^{-1} by curve resolution. For example, in the Villeneuve schorl (their Figure 6b) they split band I into two components at 14500 and 13200 cm^{-1}, band II into components at 9500 and 7900 cm^{-1}, and find less intense bands at 6000 and 16000 cm^{-1}, plus the bands at 23000 and 18500 cm^{-1}, and several very weak absorptions. Five bands are shown at < 20000 cm^{-1} in the resolved spectrum of a blue elbaite (their Figure 3) and four in that of a green elbaite (their Figure 4).

The spectra of a mineral of the chemical and structural complexity of tourmaline will inevitably contain many bands of diverse origin, and we do not doubt the presence of absorptions other than those at 23000, 18500, 13600 and 8900 to 9500 cm^{-1}. However, we do doubt the validity of the resolution procedures used by Faye *et al.*, the band positions and intensities which they derive, and some of the assignments based on those results.

Consider first the problem of resolving two bands of comparable width and intensity, e.g. the two 'components' of bands I and II in Figure 6(b) of Faye *et al.* As the bands approach to within w of each other (where w is the full width at half height of the narrower band), the intensity minimum between the bands becomes a maximum, and resolution becomes impossible unless constraints are placed on the shape and the width, energy or area of one component. The analogy with the Rayleigh resolution criterion used in physical optics is close, although the two situations are not formally identical.

Since the great majority of bands in mineral spectra are skewed Gaussians (the skewness varying from band to band), and the widths, energies and intensities vary from sample to sample, it will not normally be possible to apply the constraints needed to permit resolution of bands less than w apart. The easy (but incorrect) assumption of a symmetrical Gaussian band shape, with a band width remaining constant from sample to sample (and usually taken equal to that of the narrowest band observed) inevitably generates spurious bands. For example, fitting a symmetrical Gaussian to a skewed band generates a spurious satellite lying on the 'tail' of the main band (compare 14000 cm^{-1} band of Faye *et al.*, Figure 3), while any attempt to fit a narrow band on a wider one produces at least one (in general two) satellites of the main band (compare 14000 cm^{-1} band of Faye *et al.*, Figure 4c). These problems of resolution are similar to those encountered in infrared work on amphiboles, and reviewed by Strens (1974).

Because the widths of spin-allowed *d–d* bands at 300K are seldom less than $v/6$ (where v is the wavenumber in cm^{-1} of the band), and range up to $v/3$, while widths of charge-transfer bands are usually in the range $0.23 < w/v < 0.38$ we consider that Faye *et al.* were not justified in splitting band I into two components at 14500 and 13200 cm^{-1} ($v/6 = 2200$ cm^{-1}, $\Delta v = 1300$ cm^{-1}), and we think it improbable that band II needs to be split into components at 9500 and 7900 cm^{-1}. We also suspect that many of the weaker bands shown in their Figures 3, 4 and 6 are artefacts of the resolution procedure.

The assignment by Burns (1972) of bands I and II to *d–d* transitions of Fe^{2+} in the *c*- and *b*-sites respectively is based mainly on the argument that transition energies in the *c*-site should be higher because of the shorter metal-oxygen distances. This assignment is incompatible with the relatively constant intensity ratio of these bands in tourmalines in which the $b{:}c$ site Fe^{2+} ratio varies over at least the range 15:1 to 8:7 found by Faye *et al.* Neither can it account for the inverse temperature dependence common to both bands, and no evidence has been presented to show that the polarization dependence and energies are those to be expected of Fe^{2+} in the *c*- and *b*-sites.

Burns and Simon (1973) claim to have resolved a band at about 14000 cm^{-1} (the intensity of which is proportional to *c*-site Fe^{2+}) from one at 15000 cm^{-1} which they assign to $Fe^{2+} \rightarrow Fe^{3+}$ charge-transfer. In the absence of full details, we merely comment that it is difficult to see how two bands separated by only 1000 cm^{-1} can be resolved without constraining both shape and width. Since the width of band I can be expected to increase with *c*-site Fe content, due to substitutional broadening (Robbins and Strens, 1972) the intensity of a resolved component could well increase with *c*-site Fe^{2+} without the transition necessarily originating in that site. Also, the main $Fe^{2+} \rightarrow Fe^{3+}$ charge-transfer band appears to be at 18500 cm^{-1} (see below), although absorption at other energies cannot be excluded.

Because of these doubts about the validity of the methods used, we analyse the tourmaline spectrum in terms of only four main components, at about 23000, 18500, 13600 and 8900 to 9500 cm^{-1}, while recognizing that other absorptions will be present, but not resolvable on our criteria.

In our view, the key to the assignment of bands I and II is the observation, which now seems well established, that the ratio of the intensities of the two bands does not vary greatly from sample to sample, and that both bands are prominent even in tourmalines which contain little *c*-site Fe^{2+}. Their intensity also appears to be insensitive to Fe^{3+} content. This is strong evidence of a common origin as *d–d* transitions of Fe^{2+} in the *b*-site, and we have found excellent agreement between observed and calculated transition energies, polarizations and intensities on this assumption. The inverse temperature dependence is unusual for a *d–d* band, but can also be explained (see below).

Our calculations have been based on an effective *b*-site symmetry of C_{2v} (true symmetry C_s), with the electronic *z*-axis along the (false) diad which lies near the HO–Fe–OH vector, and the electronic *y*-axis normal to the (true) mirror plane. Two strong $t_{2g} \rightarrow e_g$ transitions are predicted for this symmetry, at 14750 ρ (13600 obs.) and 10500 ρ (8900 to 9500 obs.), with an intensity ratio of about 2:1 ($\sim$ 2:1 obs.), and a polarization ratio $\omega:\varepsilon$ of 11:4 (variable 1:1 to 5:1). The energies agree for $\rho \simeq 0.9$, implying that the distances around the Fe^{2+} ions are some 0.04 Å longer than the average *b*-site distances used in the calculation. This assignment requires the ground state to be B_1 (*xz*), as transitions from $A_2(xy)$ are forbidden, and those from $B_2(yz)$ should be both weak and polarized normal to *c*.

The calculations (Table 5) actually show *yz* lying 322 cm^{-1} below *xz*, but this sequence could easily be reversed by small changes in the *x* and *y* splittings caused by ions outside the first coordination shell, which are not included in the calculation. Provided *yz* remains close to *xz* the observed reduction in intensity of both bands on heating from 109 to 293 K can be accounted for by thermal depopulation of the ground state. A Mössbauer study of the dependence of quadrupole splitting of *b*-site Fe^{2+} on temperature should detect any major

Table 5 (a). Calculated d-orbital energy levels in tourmaline.

		b-site	c-site
$x^2 - y^2$	(4)	9334	12082
z^2	(3)	5012	7095
xy	(2)	− 3490	− 4902
xz	(1)	− 5267	− 6478
yz	(0)	− 5589	− 7797

Calculated by M. J. Dempsey from structure data of Donnay and Buerger (1950) without allowance for relaxation of M–O distances around Fe^{2+}. Energies are in cm^{-1}. Calculation modified from Wood and Strens (1972).

Table 5 (b). Energy, polarization and allowability of allowed transitions in tourmaline.

	Transition	$\nu(cm^{-1})$	Component $\parallel c$	Allowability $\parallel c$
b-site	0–3	10601	0	0
(C_{2v})	0–4	14923	0	0
	1–3	10279	0.27	1.5
	1–4	14601	0.27	3.0
c-site	0–3	14892	0.15	0.30
(C_s)	0–4	19879	0.15	0.53

The five d-orbital energy levels are labelled 0 (the ground state) to 4, in order of increasing energy. The calculated energies must be reduced by a relaxation factor ρ before comparison with the spectra. The 'allowability' provides an estimate of relative intensities.

changes in electron population, and enable the xz–yz separation to be determined accurately.

The use of C_{2v} symmetry rather than the true C_s symmetry, and the necessity to assume a reversal of the calculated order of xz and yz energy levels leave even this assignment open to some doubt, but we regard it as by far the most probable of the d–d assignments suggested, as it is not contradicted by any of the available data. A charge-transfer assignment is difficult to reconcile with the polarization dependence (too much intensity for $\mathbf{E} \parallel c$), the apparent lack of dependence on Fe^{3+} contents, and the presence of two bands (although it is possible to postulate two acceptor levels split by 4700 cm^{-1}).

Although the lack of formal symmetry prevents unequivocal calculation of transition energies in the c-site, there is a fair approximation to C_s symmetry. On this assumption, we find two allowed $t_{2g} \rightarrow e_g$ transitions with a polarization ratio $\varepsilon:\omega$ of about 1:5, and an expected intensity about one-third that calculated for the weaker (8900 to 9500 cm^{-1}) b-site band. Thus, we expect the d–d transitions of Fe^{2+} in the c-site to be relatively weak even in the ω-spectra of tourmalines with a high c-site iron content.

There is little doubt that Faye *et al.* have correctly assigned the 18500 cm^{-1}

band in their samples TBl-1, TGr-2 and schorl to $Fe^{2+} \rightarrow Fe^{3+}$ charge-transfer. The relative intensities of the 18500 cm^{-1} band in these samples correlate well with the product of Fe^{2+} and Fe^{3+} concentrations (which varies by a factor of 20 between TGr-2 and schorl), and the intensity increased sharply when the $[Fe^{2+}][Fe^{3+}]$ concentration product was increased by heat treatment of sample TBl-1. The width, energy and oscillator strength of the 18500 cm^{-1} band in schorl are all within the ranges expected for $Fe^{2+} \rightarrow Fe^{3+}$ charge-transfer bands, and the polarization is that expected ($\omega > \varepsilon$) if b–b interactions predominate.

The assignment by Faye *et al.* of the 23000 cm^{-1} transition to $Fe^{2+} \rightarrow Ti^{4+}$ charge-transfer seems less certain, as absorption occurs even in tourmalines with very low Ti^{4+} contents. The association of (Fe^{2+}, Ti^{4+}) as charge-compensated pairs replacing 2Al, which makes Fe–Ti charge-transfer a dominant process in corundum and aluminosilicates containing only 0.01 per cent TiO_2, is less likely to occur in tourmaline, in which many other charge-compensation mechanisms are available, and other strong absorptions occur.

We believe that more serious consideration must be given to the ${}^6A_1 \rightarrow {}^4A_1$, ${}^4E(G)$ transition of Fe^{3+}, which contributes significantly to the absorption in this region in many minerals. In natural corundum with $[Fe^{3+}] \simeq 0.01$ (Ferguson and Fielding, 1972) this field-independent transition is surprisingly broad (about 1500 cm^{-1} at 273 K), with both intensity and width increased considerably by interaction with neighbouring Fe^{3+} ions.

The dependence of intensity on Ti content, suggested by Figures 5 and 10 of Faye *et al.* could well be indirect. Increasing TiO_2 contents are known to broaden the $O \rightarrow Fe^{2+}$ charge-transfer band in biotite (Robbins and Strens, 1972), and Faye *et al.* report a similar effect in tourmaline. Any broadening of the $O \rightarrow Fe^{3+}$ absorption edge in tourmaline, particularly if accompanied by a simultaneous shift of the absorption maximum to lower energies, would cause all bands which depend for their strength on intensity-stealing from the $O \rightarrow Fe^{3+}$ band to increase in strength with increasing Ti content. In both biotite and tourmaline, the substitutional broadening and intensification are highly anisotropic, the maximum effect being for vibration directions in the plane of the 'brucite' layer ($\perp c$ in tourmaline), so that the ω-polarization of the 23000 cm^{-1} band in tourmaline is only weak evidence for a charge-transfer origin. We consider that the assignment of the 23000 cm^{-1} band remains open.

Pyroxenes. We have not studied pyroxenes, and discuss the spectra only for comparison with other minerals. Manning and Nickel (1969) studied the polarized spectra of titanaugite, and suggested that $Fe^{2+} \rightarrow Fe^{3+}$ charge-transfer occurred at about 14000 cm^{-1}. Dowty and Clark (1973) suggest that the broad absorption in terrestrial and lunar titanian pyroxenes at about 19000–23000 cm^{-1} is attributable to $Fe^{2+} \rightarrow Ti^{4+}$ charge-transfer, rather than absorption by Ti^{3+} assumed by other authors, including Manning and Nickel. Both the above charge-transfer assignments appear reasonable, although they cannot yet be regarded as firmly established.

Vivianite. This mineral crystallizes in space group $C2/m$, with a general formula $(Fe^{2+}_{3-x}Fe^{3+}_{x})(PO_4)_2(8-x)H_2O.xOH$, where $0 \leqslant x \leqslant 3$. It has long been quoted as an example of intervalence-transfer absorption (Day, 1970), as it is very easily oxidized by grinding in air, the colour changing from green through blue to yellow as oxidation proceeds towards the ferric end-member. The blue colour of vivianites containing both Fe^{2+} and Fe^{3+} is attributable to a broad and intense absorption centred at about 15500 cm^{-1} and unequivocally assigned to charge-transfer between Fe^{2+} and Fe^{3+} in pairs of edge-sharing octahedra aligned with the Fe–Fe vector along $b = \alpha$, which is the direction of strongest absorption. The Fe–Fe distance is only 2.84 Å. In the ferrous end-member the iron is coordinated by $2\,H_2O + 4\,O^{2-}$ belonging to the phosphate groups.

Our sample could not be heated above room temperature, or cooled to liquid nitrogen temperature, without suffering irreversible changes. The temperature-dependence was therefore studied only over the range 207–293 K, and significant intensification was noted at the lower temperature. Mao (this volume) reports a fiftyfold increase in absorption at 300 kb.

The oscillator strength is uncertain because of the possibility of differences between Fe^{3+} concentrations in the analysed sample and thin section. The value obtained from the analysis (Table 2 above) is $f_{\|} = 192 \times 10^{-4}$, or about 4 times that normally found for $Fe^{2+} \rightarrow Fe^{3+}$ transitions in silicates and oxides.

Cordierite. This mineral crystallizes in space group $Cccm$, with the general formula $(Mg,Fe)_2Al_4Si_5O_{18}$. The (Mg,Fe^{2+}) occupies the 8(g) sites and Al is located in the 8(k) and 8(l) sites: Gibbs (1966) designated these Me, T1 and T5 respectively. The octahedra about Me share edges with T1 tetrahedra (but not with T5 or Me polyhedra), the squares of the direction cosines of the Me–T1 vector being $0.27 \parallel a = \beta;\ 0.73 \parallel b = \gamma;\ 0 \parallel c = \alpha$. Mössbauer work by Duncan and Johnston (1974) on a Madagascan cordierite with Fe as FeO = 2.38 per cent shows about 80 per cent of the Fe^{2+} in the Me site, with the remainder located in a five-coordinate channel site. About half the Fe^{3+} is in T1, with the balance presumably distributed between Me and T5.

The spectra of magnesian cordierite have previously been described by Farrell and Newnham (1967) and Faye *et al.* (1968), the latter having interchanged the $\mathbf{E} \parallel a(\beta)$ and $\mathbf{E} \parallel b(\gamma)$ spectra (Struntz *et al.*, 1970). Our sample was an iron-rich cordierite (sekaninaite) with 12.05 per cent FeO, 4.21 per cent Fe_2O_3, and 0.70 per cent TiO_2, equivalent to $1.18Fe^{2+}$ and $0.37Fe^{3+}$ per formula unit.

Figure 5 shows the β and γ spectra of sekaninaite. The α spectrum (not illustrated) is very simple, consisting of a doublet with components of approximately equal intensity at 8150 and 10100 cm^{-1}, which increase in intensity with increasing temperature. It seems probable that these are vibronically allowed $^5T-{}^5E$ transitions of Fe^{2+} in the Me site. The β and γ spectra show a similar composite band with components at about 8000 (weak) and 10000 cm^{-1}, (Table 6), which exhibit only moderate direct temperature dependence.

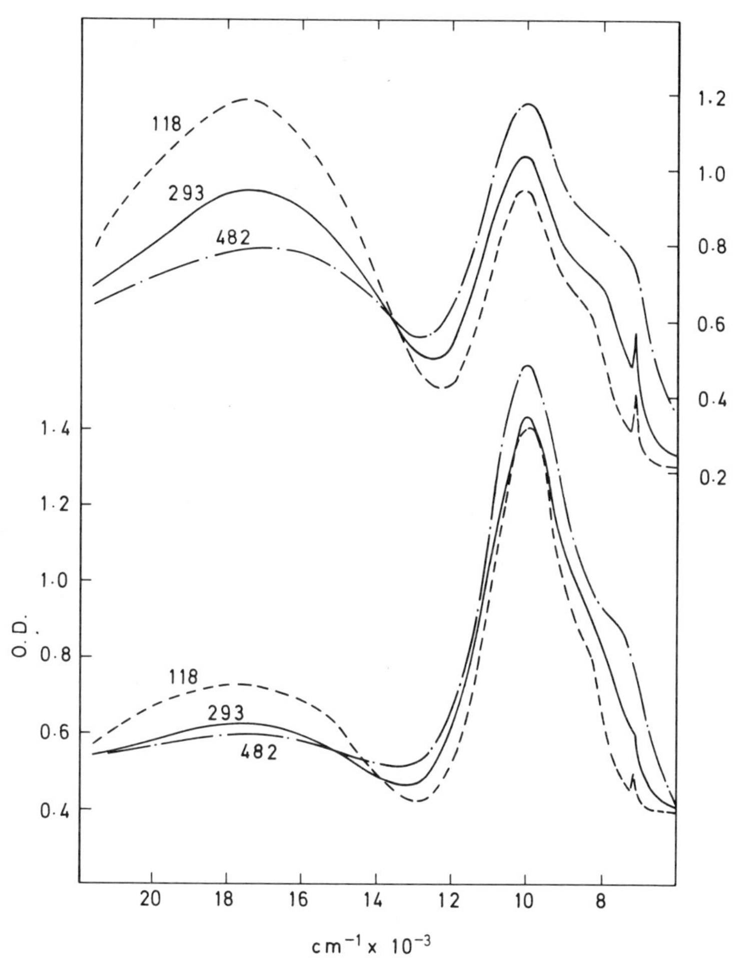

Figure 5. The $\mathbf{E} \| b(\gamma)$ spectrum of an 813 μm section of the iron-rich cordierite sekaninaite (top) showing the direct temperature dependence of the Fe^{2+} *d–d* bands at about 8000 and 10000 cm^{-1}, and the marked inverse temperature dependence of the band at about 17000 cm^{-1} assigned to $Fe^{2+} \rightarrow Fe^{3+}$ charge-transfer between octahedral Fe^{2+} and tetrahedral Fe^{3+}. The $\mathbf{E} \| a(\beta)$ spectrum of the same sample is shown below. Temperatures marked are absolute (K).

These we interpret as *d–d* bands with some vibronic character, i.e. the static and thermally-induced dipole moments are of comparable magnitude.

There is also a broad, strong and possibly composite absorption at about 17000 cm^{-1} in the β and γ spectra, polarized $\gamma > \beta$ with negligible intensity for $\mathbf{E} \| c = \alpha$. This was assigned to Fe^{2+}(Me) → Fe^{3+}(T1) charge-transfer by both Robbins and Strens (1968) and Faye *et al.* (1968), although Farrell and Newnham (1967) had adopted a *d–d* assignment without considering

Table 6. Spectra of sekaninaite (Fe-cordierite) at 293 K.

		$\nu(cm^{-1})$	$\alpha(cm^{-1})$	$w(cm^{-1})$	$10^{-4}A$
$\mathbf{E} \parallel a$ (β)	I	17200	3.8	5000	2.4
	II	9800	27.8	2200	7.7
	III	7900	8.2	1675	1.7
$\mathbf{E} \parallel b$ (γ)	I	21600	4.5	4300	2.4
	II	17000	15.9	6400	12.7
	III	10050	20.1	2200	5.5
	IV	8000	9.3	1750	2.1
$\mathbf{E} \parallel c$ (α)	III	10100	51.2	2175	14.0
	IV	8150	41.5	1850	9.6

the possibility of charge-transfer. We retain the charge-transfer assignment, as the observed polarization ratio of 1:3:0 is that expected for Me–T1 charge-transfer, the energy is too high for a *d–d* transition of Fe^{2+} in a relatively undistorted octahedron, and the inverse temperature dependence is characteristic of metal–metal charge-transfer.

Because the Me site lies near the channel through the cordierite structure, the contributions of ions outside the first coordination sphere to the crystal field splittings of Fe^{2+} in this site are unlikely to cancel, and the simplified (Wood and Strens, 1972) method of calculating *d–d* transition energies is likely to give poor results. This would apply with even greater force to any attempt to calculate possible contributions from Fe^{2+} in the channel sites.

We have disregarded the suggestion by Duncan and Johnston (1974) that the charge-transfer occurs between ions in the Me and channel sites, as there is no strong evidence for this, and no shared edge (which appears to be necessary for observable $Fe^{2+} \rightarrow Fe^{3+}$ charge-transfer to occur). The magnetic interactions mentioned by these authors are commonly observed in minerals containing transition metal ions with non-zero spin, and are only seldom accompanied by observable charge-transfer absorption. Nevertheless, an investigation of the contributions (if any) of channel ions to the cordierite spectrum would be interesting and useful.

The sharp bands at about 7000 cm^{-1} (polarized $\gamma > \beta$) are part of the vibrational spectrum of water molecules in the channels described by Farrell and Newnham (1967), and discussed by Strens (1974).

CORUNDUM AND THE ALUMINOSILICATES

In most respects, charge-transfer in these minerals resembles that in ferromagnesian minerals. Both $Fe^{2+} \rightarrow Ti^{4+}$ and $Fe^{2+} \rightarrow Fe^{3+}$ charge-transfer bands are important in the spectra, the former usually predominating despite the fact that the (Fe^{2+}, Ti^{4+}) concentration product is normally 10 to 100 times less than the (Fe^{2+}, Fe^{3+}) product. The unexpectedly high intensity of the

$Fe^{2+} \rightarrow Ti^{4+}$ bands is attributable to the presence of charge-compensated Fe^{2+}–Ti^{4+} pairs replacing 2 Al^{3+}, while Fe^{2+} and Fe^{3+} are randomly distributed.

Face-sharing octahedra are rare in mineral structures, and charge-transfer between such octahedra in corundum is of particular interest, although its occurrence is to be expected, as the presence of a shared face necessarily implies that of three shared edges. Charge-transfer between corner-sharing polyhedra has not been observed.

Corundum. The coloured varieties of this mineral (often called sapphire) crystallise in space group $R\bar{3}c$. The $[AlO_6]$ octahedra have C_3 symmetry, and are arranged in face-sharing pairs with their Al–Al vectors (2.65 Å) lying along the c-axis of the crystal. Individual octahedra also share edges with three neighbours at similar z-coordinates, the resulting Al–Al vectors (2.79 Å) being directed at about 80° to c. More distant neighbours are important in considering the spectra of pairs of magnetic ions (Ferguson and Fielding, 1972), but probably give rise to no observable charge-transfer bands.

Recent studies of the spectra of synthetic and natural corundum include those of Townsend (1968). Lehmann and Harder (1970), Ferguson and Fielding (1971, 1972) and Eigenmann *et al.* (1972). We have examined the spectra of the Khao Ploi Waen corundum described by Lehmann and Harder (1970), finding five main bands (Table 7a). The reported positions of these bands vary considerably from author to author, probably reflecting different methods of resolving the spectra, and also real differences between samples. For example, minerals rich in Fe^{3+} have spin-forbidden bands of Fe^{3+} which nearly coincide with two of the charge-transfer bands listed in Table 7(a) so that what at first appears to be the same band may have different origins in different samples.

Table 7 (a). Spectra of Khao Ploi Waen corundum at 293 K.

		$\nu(cm^{-1})$	$\alpha(cm^{-1})$	$w(cm^{-1})$	$10^{-4}A$
$\mathbf{E}\perp c(\omega)$	I	17000	7.2	5170	4.7
	—	14100	4.1	2890	1.5
	II	11150	20.3	3450	8.8
$\mathbf{E}\parallel c(\varepsilon)$	III	12900	5.8	6100	4.4
	IV	9700	1.7	2450	0.5

Table 7 (b). Spectra of blue kyanite at 293 K

		$\nu(cm^{-1})$	$\alpha(cm^{-1})$	$w(cm^{-1})$	$10^{-4}A$
γ-spectrum	I	16300	4.1	5800	3.0
	II	11500	1.1	5400	0.8
β-spectrum	I	16300	2.1	4850	1.3
	II	10800	1.2	6000	0.9

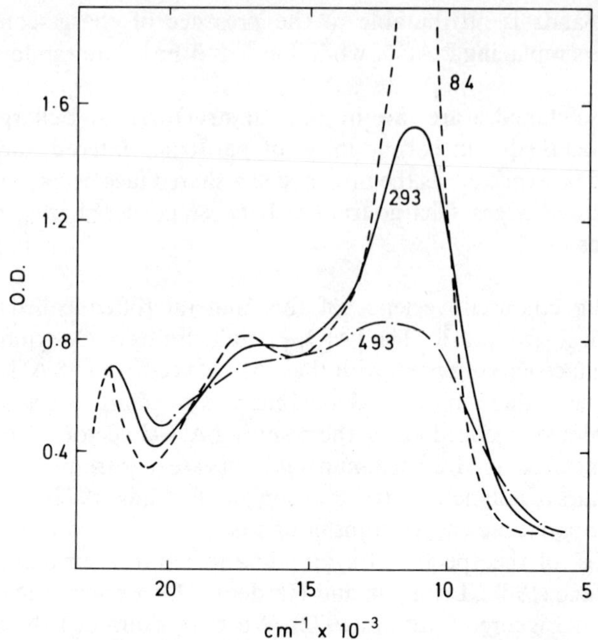

Figure 6. The $\mathbf{E}\perp c(\omega)$ spectra of a 1.41 mm section of corundum from Khao Ploi Waen (Lehmann and Harder, 1970) measured at 84, 293 and 493 K. Note the strong inverse temperature dependence of the intensities of the 11150 and 17000 cm^{-1} charge-transfer bands.

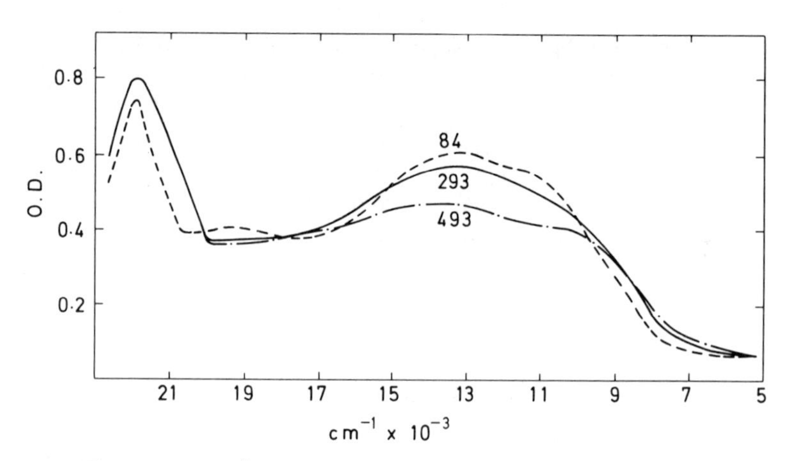

Figure 7. The $\mathbf{E}\parallel c(\varepsilon)$ spectrum of Khao Ploi Waen corundum, showing the strong direct temperature dependence of the ${}^6A_1$–${}^4T_1$, ${}^4E(G)$ band at about 22000 cm^{-1}, and the inverse temperature dependence of the bands at about 9700 and 12900 cm^{-1}. Thickness 1.41 mm.

Our spectra (Figures 6, 7) show a strong inverse temperature dependence of the intensities of bands I, II and III, and a direct dependence similar to that observed by Ferguson and Fielding (1972) for the ${}^6A_1 - {}^4T_1, {}^4E(G)$ band at 22000 cm^{-1}. The response of band IV to temperature is uncertain.

Townsend (1968) found that synthetic sapphire with 0.071 per cent Fe and 0.012 per cent Ti was dark blue, while crystals containing Fe and Ti alone were colourless (Fe) and pink (Ti) respectively. Ferguson and Fielding (1971, 1972) showed that bands I (17000 $cm^{-1}, \perp c$) and II (12900 $cm^{-1}, \parallel c$) could be produced in synthetic sapphire only by the simultaneous addition of Ti and Fe, while band III (11150 $cm^{-1}, \perp c$) occurred in crystals grown from fluoride fluxes, in which ($Fe^{2+}F^-$) was thought to replace (Al^{3+} O^{2-}). In natural sapphires, the corresponding replacement would be (Fe^{2+} OH^-) for (Al^{3+} O^{2-}). These observations led Townsend (1968) and Ferguson and Fielding (1971, 1972) to assign bands I and II to charge-transfer between charge-compensated ($Fe^{2+} - Ti^{4+}$) pairs, and bands III and IV to $Fe^{2+} \rightarrow Fe^{3+}$ charge-transfer. Eigenmann *et al.* (1972) have since advanced what they consider to be conclusive evidence for the presence of charge-compensated pairs in sapphire.

Lehmann and Harder (1970) did not analyse their samples for TiO_2, and did not consider the possibility of either $Fe^{2+} \rightarrow Fe^{3+}$ or $Fe^{2+} \rightarrow Ti^{4+}$ charge-transfer as alternatives to the *d–d* assignments which they advanced. For their Khao Ploi Waen corundum, we found Fe as $Fe_2O_3 = 1.58$ per cent and $TiO_2 = 0.021$ per cent (Table 2 above). If Fe^{2+} contents exceed those of Ti, then these figures would permit reasonable intensities for both Fe–Fe and Fe–Ti charge-transfer.

In view of the unequivocal compositional evidence, and the polarization and temperature dependences, we follow Townsend and Ferguson and Fielding in assigning bands I and II to $Fe^{2+} \rightarrow Ti^{4+}$ charge-transfer between charge-compensated pairs in edge-sharing and face-sharing octahedra respectively, and bands III and IV to the equivalent transitions between randomly distributed Fe^{2+} and Fe^{3+} ions. The band at about 14100 cm^{-1} is probably the ${}^6A_1 - {}^4T_2(G)$ transition of Fe^{3+}, which Lehmann and Harder placed at 14350 cm^{-1}. Their ${}^6A_1 - {}^4T_1(G)$ transition at 9450 cm^{-1} probably contributes significantly to the intensity of band IV in the Khao Ploi Waen corundum, but the work of Ferguson and Fielding definitely indicates that a charge-transfer absorption is present in this region.

Kyanite. This is the densest of the Al_2SiO_5 polymorphs, crystallizing in space group $P\bar{1}$, with a true formula unit $Al_4Si_2O_{10}$. There are four crystallographically inequivalent octahedral Al positions, of which the Al_1 and Al_2 octahedra of Burnham (1963) share edges to form chains running along the *c*-axis. In synthetic Cr-kyanites (Langer, this volume), most of the Cr^{3+} occupies these chain sites. Electron spin-resonance studies of natural kyanites by Troup and Hutton (1964) suggest a 10:40:10:1 distribution of Fe^{3+} between Al_1 (mean Al–O distance 1.907 Å), Al_2(1.918 Å), Al_3(1.916 Å) and Al_4(1.897 Å). We think it probable that the main charge-transfer interaction will be that between transition-metal ions in the chain sites, with a smaller contribution from $Al_{1,2}$–Al_3 interactions. The Al_1–Al_2 charge-transfer would be polarized in the approximate ratio. $\gamma:\beta = 3:1$, with only a small contribution to the α-spectrum. The $Al_{1,2}$–Al_3 charge-transfer would be polarized in the approximate ratio $\gamma:\beta:\alpha = 3:5:2$.

The β and γ spectra of a blue kyanite with $TiO_2 = 0.015$ per cent, Fe as $Fe_2O_3 = 0.154$ per cent, and Cr, V < 0.03 per cent are shown in Figure 8. Both spectra contain a composite band, with components at about 16300 and 11500 cm^{-1} (Table 7b above), of which the former is polarized with $\gamma:\beta \simeq 3:1$, and determines the colour and pleochroism of kyanite, while the latter has intensities more nearly equal. Both bands have marked inverse temperature dependence.

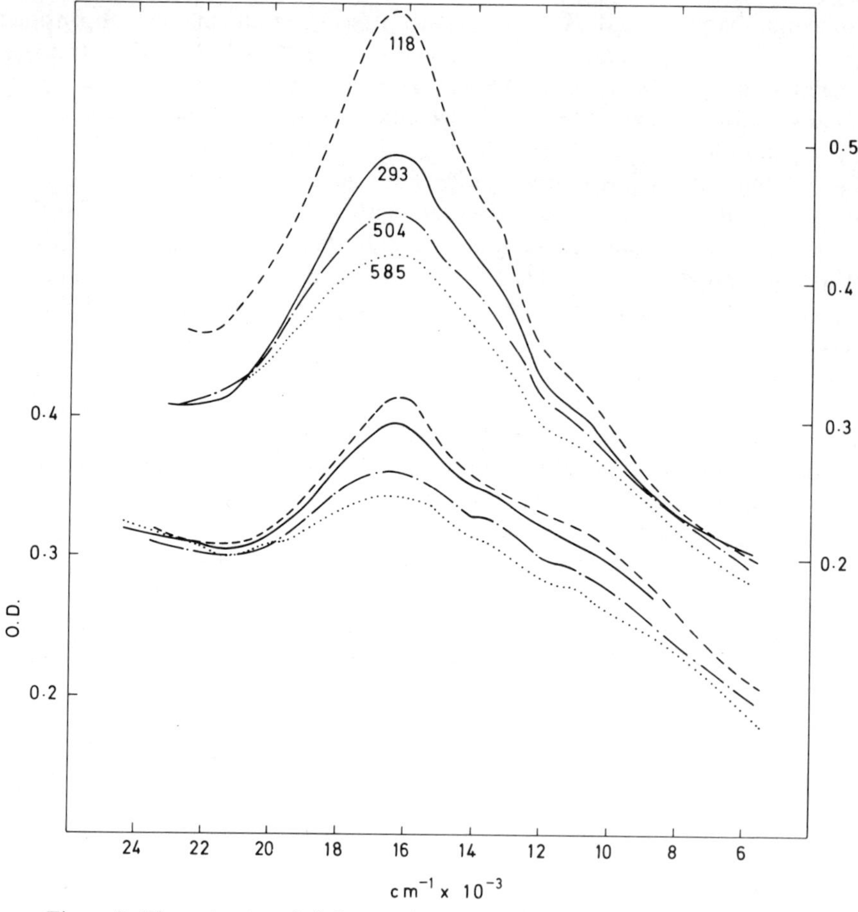

Figure 8. The γ (top) and β (bottom) spectra of a blue kyanite from Pizzo Forno, showing the very strong inverse temperature dependence of the absorption at 16300 cm^{-1} which is mainly responsible for the colour and pleochroism. Thickness 1.245 mm. Temperatures marked are absolute (K).

The spectra illustrated by White and White (1967) and Faye and Nickel (1969) show similar $\gamma:\beta$ intensity ratios, but the latter has rather less intensity in the α-spectrum, with $\gamma:\beta:\alpha$ approximately 8:3:1. Faye and Nickel show the band at 11500 cm^{-1} polarized $\gamma > \beta > \alpha$ with a higher $\gamma:\beta$ intensity ratio than was found for our sample. These differences may reflect ambiguities in the resolution, or be sample-dependent.

There is strong evidence of a link between Ti content and colour. For example, White and White found approximately 10, 20 and 40 ppm Ti in the colourless, pale blue and dark blue zones of an Indian kyanite. Among the many data collected by Rost and Simon (1972) are those for kyanite from Castion/Tessin (Switzerland) which have comparable Fe, Cr and Mn contents, but Ti varying from undetected (colourless) through 0.02 per cent (light blue) to 0.05 per cent (dark blue), and other data in the same paper suggest that colour correlates with Fe and Ti contents. Recent unpublished EDAX data obtained by Freer at Newcastle show an increase in both Fe and Ti on passing from colourless to blue zones, and he also demonstrated bleaching in less than two days at 1250 K in air, implying that the chromophore is destroyed by oxidation. These data are compatible with either Fe–Ti or Ti–Ti charge-transfer assignments, but the latter is eliminated from consideration by the extremely small concentration product, and unreasonably high oscillator strength which this implies. The oscillator strength calculated for the 16300 cm^{-1} band assuming an ordered distribution of Fe^{2+} and Ti^{4+} is $\geqslant 45 \times 10^{-4}$ (Table 8 below), which appears reasonable. The assignment is thus consistent with the intensity, energy and width of the 16300 cm^{-1} band, and its observed temperature-, composition- and polarization-dependence.

The origin of the 11500 cm^{-1} band is less certain, although the polarization- and temperature-dependence strongly suggest that it is also of charge-transfer origin. It may originate in electron transfer to an acceptor orbital lying some 4000 cm^{-1} below that involved in the 16300 cm^{-1} transition, or arise from charge-transfer between Fe^{2+} and Fe^{3+}, which yields an unreasonably high oscillator strength, unless ordering of these ions is assumed.

Bands in the spectra of green kyanite (White and White, 1967) at about 9300 and 22500 cm^{-1} are almost certainly the $^6A_1$–$^4T_1(G)$ and $^6A_1$–$^4A_1$, $^4E(G)$ transitions of Fe^{3+}, but in kyanite the narrowness of the 22500 cm^{-1} band suggests an absence of the pair effects which are so important in corundum (Ferguson and Fielding, 1972).

Andalusite. The structure is composed of chains of edge-sharing [AlO_6] octahedra running along the *c*-axis, and linked together by [AlO_5] and [SiO_4] polyhedra. The Al_1–Al_1 vector thus lies along *c*, the Al–Al distances being alternately 2.693 and 2.866 Å (Burnham and Buerger, 1961). Little is known of the distribution of iron and titanium in the structure, but they are probably concentrated in the Al_1 octahedra.

Faye and Harris (1969) attributed the pleochroism of andalusite to a single broad absorption band at about 20800 cm^{-1}, which they assigned to Ti^{3+}–Ti^{4+}

charge-transfer, without considering the possibility of Fe–Ti transfer. They rejected Fe–Fe charge-transfer, as they considered the energy to be too high.

The only work on the composition dependence of the colour of andalusite appears to be that of Okrusch and Evans (1970), who found no obvious correlation with iron content. However, their data do imply some substitution of (Fe,Mg) + Ti for 2 Al. Heating experiments on andalusites by Freer (unpublished) show that the chromophore is oxidized by heating above 1100 K for several days. Iron-poor andalusite becomes colourless, and iron-rich andalusite turns yellow.

Assuming iron and titanium to be present mainly as charge-compensated pairs, while Fe^{2+}–Fe^{3+} and Ti^{3+}–Ti^{4+} pairs occur at random, or nearly so, we find the concentration products for typical andalusites with [Ti] = 0.001 and [Fe] = 0.01 to be:

$$[\mathrm{Fe,Ti}] \leqslant [\mathrm{Ti}] = 10^{-3}; [\mathrm{Fe,Fe}] \leqslant [\tfrac{1}{2}\mathrm{Fe}]^2 = 2.5 \times 10^{-5}; [\mathrm{Ti,Ti}] \leqslant [\tfrac{1}{2}\mathrm{Ti}]^2 = 2.5 \times 10^{-7}$$

For $Fe^{2+} \rightarrow Ti^{4+}$ charge-transfer these figures yield $f_{\parallel} \geqslant 34 \times 10^{-4}$ for the 20800 cm^{-1} band of andalusite, compared with 57 and 45×10^{-4} for corundum and kyanite respectively. The corresponding values for Fe–Fe and Ti–Ti are unreasonably high and impossibly so at 0.14 and 14 respectively (by definition, $f \leqslant 1$). We therefore assign the 20800 cm^{-1} band to $Fe^{2+} \rightarrow Ti^{4+}$ charge-transfer, finding $f_{\parallel} \geqslant 20 \times 10^{-4}$ using Faye and Harris's analysis.

PROPERTIES OF $Fe^{2+} \longrightarrow Fe^{3+}$ AND $Fe^{2+} \longrightarrow Ti^{4+}$ CHARGE-TRANSFER BANDS

The energies, widths and oscillator strengths found for charge-transfer bands in the ferromagnesian silicates, aluminosilicates, corundum and vivianite are summarized in Table 8. For edge-sharing octahedra in silicates and oxides (*i.e.* excluding cordierite and vivianite) the energies range from about 10000 to 23000 cm^{-1}, the widths are typically 0.23 to 0.38 of the energy, and oscillator strengths at 300 K vary by no more than a factor of 2 from the common figure of 40×10^{-4}.

The polarization- and composition-dependences of the charge-transfer bands are now well established, but much detailed work remains to be done before the temperature- and pressure-dependences are properly understood. The present state of knowledge is reviewed below.

Polarization Dependence of Intensity. The dipole moment of an isolated donor–acceptor pair will lie along the interatomic vector, and have a magnitude Δzer, where Δz is the difference in charge, e is the electronic charge, and r is the separation. If both ions have the same charge, $\Delta z = 0$, and no transition will occur. This accounts for the failure to observe charge-transfer between the abundant Fe^{2+} ions in minerals.

The situation will not be very different for ion-pairs in crystals. The orien-

Table 8. Summary of main Fe^{2+}–Fe^{3+} and Fe^{2+}–Ti^{4+} charge-transfer assignments. Room temperature data (293 K nominal).

Phase	Polz.	Pair	Sites	Geom.	M–M(Å)	$\nu(cm^{-1})$	$w(cm^{-1})$	$10^{-4}A_{\parallel}$	n_{da}	[c]	$V(cm^3)$	$10^4 f_{\parallel}$	Notes
Biotite	$\gamma > \beta$	Fe^{2+}–Fe^{3+}	hc,hh	ESO	3.1	13650	3150	1092	18	0.041	148*	42	rs,s
Chlorite	$\gamma \simeq \beta$	Fe^{2+}–Fe^{3+}	talc	ESO	3.1	13800	3575	39	18	$\leqslant 0.0013$	105*	$\geqslant 33$	s
Glaucophane	$\gamma > \beta$	Fe^{2+}–Fe^{3+}	M(1)–M(2)	ESO	3.09	16300	6550	266	4	0.082	270*	41	bb,s
Glaucophane	β	Fe^{2+}–Fe^{3+}	M(3)–M(2)	ESO	3.21	18100	6600	356	2	0.133	270*	68	bb,s
Schorl	$\omega \gg \varepsilon$	Fe^{2+}–Fe^{3+}	bb	ESO	3.185	18500	4200	150	6	0.036	317*	41	f
Vivianite	α	Fe^{2+}–Fe^{3+}	Fe(II)	ESO	2.84	15500	7200	995	2	0.09*	185*	192	s
Cordierite	$\gamma > \beta$	Fe^{2+}–Fe^{3+}	Me—T1	ESOT	2.846	17000	6400	20	4	0.10*	235*	2.2	s
Corundum	ω	Fe^{2+}–Ti^{4+}	Al	ESO	2.79	17000	5170	9.3	6	$\leqslant 0.00013$	25.6	$\geqslant 57$	s,o
Corundum	ω	Fe^{2+}–Fe^{3+}	Al	ESO	2.79	11150	3450	17.5	6	?	25.6	?	s
Corundum	ε	Fe^{2+}–Ti^{4+}	Al	FSO	2.65	12900	6100	4.4	2	$\leqslant 0.00013$	25.6	$\geqslant 81$	s,o
Kyanite	$\gamma > \beta$	Fe^{2+}–Ti^{4+}	$Al_{1,2}$	ESO	2.793	16300	5480*	4.7	2	$\leqslant 0.0005$	88.2	$\geqslant 45$	s,o
Kyanite	$\gamma \simeq \beta$	Fe^{2+}–Fe^{3+}	$Al_{1,2}$(?)	ESO	2.793	11500*	5800	1.8	2	?	88.2	?	s
Andalusite	α	Fe^{2+}–Ti^{4+}	Al_1	ESO	{2.693, 2.866	20800	6500	10	2	$\leqslant 0.0013$	51.5	$\geqslant 20$	f,fh,o

ESO edge-sharing octahedra; FSO face-sharing octahedra; ESOT edge-sharing octahedron and tetrahedron. *denotes uncertain or estimated values. [rs] Robbins and Strens (1972); [s] Smith (1973); [bb] Bancroft and Burns (1969); [f] Faye *et al.* (1974) [fh] Faye and Harris (1969); [o] ordered distribution assumed, [Fe, Ti] $\simeq$ [Ti].

tation of the dipole moment should not be much affected by the presence of ligands, and as the donor-ligand-acceptor overlap will also be greatest for some direction near the donor–acceptor vector, the transition moment will inevitably lie near that vector.

It is not surprising that the polarization of all known metal–metal charge-transfer bands is identical (within experimental error, which may be 5–10 degrees) with the orientation of metal–metal vectors. The sole apparent exception was thought to be the polar structure of tourmaline, in which the 8900–9500 and 13600 cm^{-1} bands have significant intensity in the $\mathbf{E} \parallel c$ spectrum. However, these are now interpreted as *d–d* bands of Fe^{2+} in the *b*-sites.

Composition Dependence of Intensity. Some problems of definition arise when donor and acceptor ions are non-randomly distributed over the sites considered. In the general case of a partially ordered distribuion of donor and acceptor ions over A and B sites, the $[c]\, n_{da}$ term of equation (3) becomes:

$$[d]_A[a]_B n_{AB} + [d]_B[a]_A n_{BA} \qquad (n_{AB} = n_{BA}) \tag{4}$$

whereas only one of these terms is required for an ordered distribution, e.g. glaucophane M(1) or M(3) → M(2). For a truly random (tourmaline *b*-site) or pseudorandom (chlorite, mica) distribution, $[d]_A = [d]_B$, $[a]_A = [a]_B$ and (4) reduces to:

$$[c]n_{da} = 2n[d]\,[a]$$

i.e. n_{da} is twice the number of contacts. This is taken into account in Table 8.

Distance- and Pressure-dependence of Band Parameters. Figure 9 shows the dependence of band energy on donor–acceptor separation. It appears that a 1 per cent change in this separation changes the energies of Fe–Fe and Fe–Ti bands by about 480 and 940 cm^{-1} respectively. Assuming linear compressibilities and thermal expansions of 2×10^{-4}/kb and 2×10^{-5}/K then a 1 kilobar increase in pressure or a 10 K increase in temperature will respectively decrease and increase the band energy by about 10 cm^{-1} (Fe–Fe) and 20 cm^{-1} (Fe–Ti).

Few data exist for band shifts at high pressure. Shankland and Strens (1966, unpublished work) obtained a very uncertain value of about -3 cm^{-1}/kb for the 13600 cm^{-1} band in biotite. Mao (this volume) shows the $Fe^{2+} \rightarrow Fe^{3+}$ band in vivianite shifted from 15500 cm^{-1} at zero pressure to about 14000 cm^{-1} at 300 kb, an average of -5 cm^{-1}/kb over this range. Abu-Eid (this volume) illustrates what appears to be a much larger shift of about -16 cm^{-1}/kb for the $Fe^{2+} \rightarrow Fe^{3+}$ band in blue omphacite. These figures are in reasonable agreement with the estimate, when it is remembered that local and bulk compressibilities may differ considerably.

So far as we are aware, no data exist for Fe–Ti bands at high pressure. Experimental difficulties arise because thick samples are required to obtain measurable absorption, and the diamond anvil cell cannot be used.

Mao (this volume) has reported an extraordinary fiftyfold intensity increase

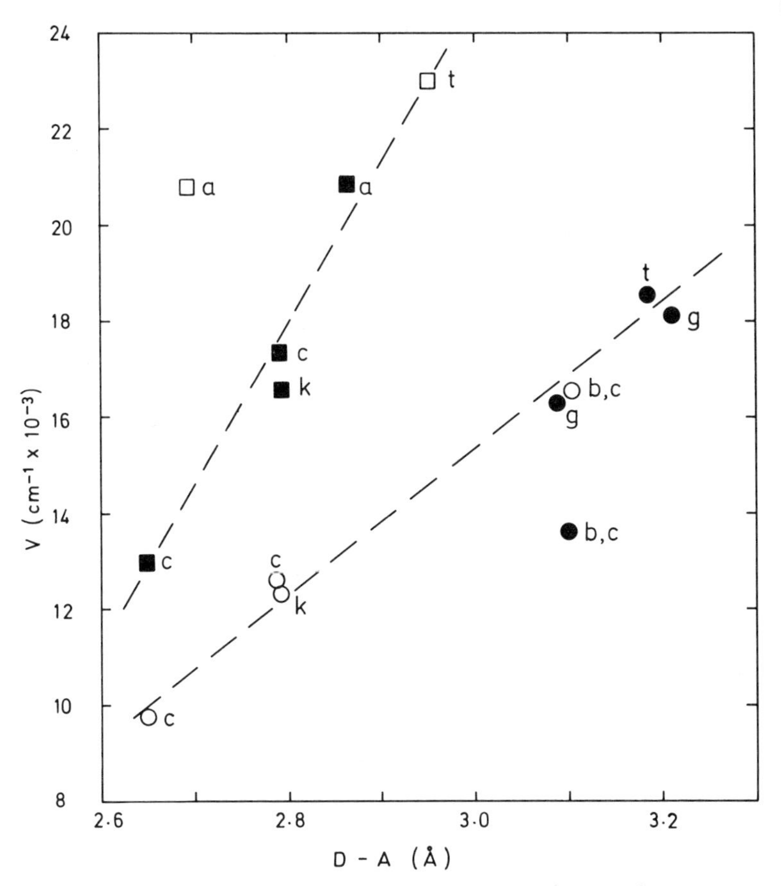

Figure 9. Dependence of the energies of $Fe^{2+} \rightarrow Fe^{3+}$ (●), and $Fe^{2+} \rightarrow Ti^{4+}$ (■) bands on donor-acceptor separation. Filled symbols represent the more certain assignments, open symbols those which are less certain. Only polarized single-crystal spectra of silicate and oxide minerals in which donor and acceptor occupy face- or edge-sharing octahedra are included. *a* andalusite (long and short Al–Al distances plotted); *b,c* biotite and chlorite (bands at about 13650 and 16700 cm^{-1}); *c* corundum; *g* glaucophane (both M(1)–M(2) and M(3)–M(2) bands plotted); *k* kyanite; *t* tourmaline (with *b–b* and *b–c* separations used for Fe–Fe and Fe–Ti bands respectively).

for the Fe^{2+}–Fe^{3+} band in vivianite at 300 kb, while the biotite and omphacite spectra taken at pressures of 40 to 50 kb show only slight intensification. If it is assumed that donor-ligand and ligand-acceptor overlaps increase exponentially with decreasing separation (which should be approximately true), then a pressure dependence of the form:

$$A(P) = A(0) \exp k(P/3K_0)$$

is expected, where k is a constant, and K_0 is the bulk modulus at zero pressure. This could account for the experimental observations if $k \simeq 80$ and $K_0 \simeq 2$Mb.

Temperature-dependence of Band Parameters. The intensities of all intervalence-transfer bands examined by us and by Loh (1972) exhibit inverse temperature-dependence, but not all bands with inverse temperature-dependence are of charge-transfer origin. Thermal depopulation of ground states lying within a few times kT of the first excited state will occasionally result in *d–d* bands with such dependence, e.g. in tourmaline. Nonetheless, inverse temperature-dependence is strong evidence for a charge-transfer assignment.

The origin of this behaviour remains uncertain, but it is probably not due to a reduction in overlap as interatomic distances increase with temperature, as was suggested by Ferguson and Fielding (1972). The pressures of 40–50 kb applied to biotite and omphacite in the experiments mentioned above should have reduced the interatomic distances by about the amount of the increase caused by heating from -200 to $+200$ °C in our experiments, yet the heating apparently caused much larger changes in intensity than the compression. We consider that the answer is more likely to lie in the effect of temperature on magnetic interactions, or unspecified phonon-coupling effects. It is interesting that the spectra of exchange-coupled ion-pairs reported by Ferguson *et al.* (1967) have a very similar temperature-dependence to that of the charge-transfer bands.

The widths of both charge-transfer and *d–d* bands are thought to increase as $T^{1/2}$. The data of Figure 9 imply that band energies should increase by 1–2 cm^{-1}/K using a linear thermal expansion of $2 \times 10^{-5}/K$. Our data show some suggestion of a systematic increase in energy with temperature, but the broadening of the bands with increasing temperature makes this difficult to follow. If such an increase does occur, then it contrasts with the behaviour of *d–d* bands which normally move to lower energies with increasing temperature.

CONCLUSIONS

The literature of mineral spectra now provides many examples of the danger of basing assignments on inadequate data, and a much more thorough and systematic approach to the problems of assignment seems to be required.

For *d–d* bands, recent modifications by M. J. Dempsey (Newcastle) of the method of Wood and Strens (1972) now enable energy, width, polarization-dependence and an indication of the relative intensity to be calculated, and compared with observation. Usually (but not invariably) the *d–d* bands will shift to higher and lower energies with increasing pressure and temperature respectively, and their intensities will increase or remain constant with increasing temperature, and be proportional to the concentration of the absorbing ion.

By contrast, intervalence-transfer (metal–metal charge-transfer) bands will normally shift to lower energies as pressure increases, and their intensities will increase and decrease as pressure and temperature are increased. Their polarization depends on the orientation of the donor–acceptor vector rather than site symmetry, and their intensity depends on the product of donor and acceptor concentrations, rather than the content of a single ion. Although their energies, widths and intensities cannot at present be calculated from structure data, all these parameters are normally higher than those for *d–d* bands of Fe^{2+}, and may therefore be useful in band assignment.

ACKNOWLEDGEMENTS

We thank Professor G. Lehmann for providing the sample of corundum from Khao Ploi Waen used in our work and that of Lehmann and Harder (1970), Dr. J. E. Dixon for three analysed glaucophane samples and Mr. J. H. Scoon of the Department of Mineralogy and Petrology at Cambridge for determining the FeO and Fe_2O_3 contents of the vivianite and sekaninaite. One of us (G. S.) thanks the Science Research Council for a studentship, and the Royal Society and the Japan Academy for supporting his work at The Institute for Solid State Physics, University of Tokyo.

REFERENCES

Allen, G. C. and Hush, N. S. (1967). In: F. A. Cotton (Ed.), *Progr. Inorg. Chem.*, **8**, 357. Wiley, New York.
Austin, I. G. and Mott, N. F. (1969). *Adv. Phys.*, **18**, 41.
Bancroft, G. M. and Burns, R. G. (1969). In J. J. Papike (Ed.), *Mineral. Soc. Amer., Special Paper 2*, 137.
Burnham, C. W. (1962). *Z. Kristallogr.*, **118**, 337.
Burnham, C. W. and Buerger, M. J. (1961). *Z. Kristallogr.*, **115**, 269.
Burns, R. G. (1972). *Canad. Journ. Spectroscopy*, **17**, 51.
Burns, R. G. and Strens, R. G. J. (1966). *Science*, **135**, 890.
Burns, R. G. and Simon, H. F. (1973). Abstracts, *Geol. Soc. Amer. Dallas Meeting*, **5**(7), 563.
Day, P. (1970). *Endeavour*, **29**, 45.
Donnay, G. and Buerger, M. J. (1950). *Acta Crystallogr.*, **3**, 379.
Drickamer, H. G. and Frank, C. W. (1973). *Electronic Transitions and the High Pressure Chemistry and Physics of Solids*. Chapman and Hall, London, Halsted-Wiley, New York.
Dowty, E. and Clark, J. R. (1973). *Amer. Mineral.*, **58**, 230.
Duncan, J. F. and Johnston, J. H. (1974). *Austr. Journ. Chem.*, **27**, 249.
Eigenmann, K., Kurtz, K. and Gunthard, H. H. (1972). *Helv. Phys. Acta*, **45**, 452.
Farrell, E. F. and Newnham, R. E. (1967). *Amer. Mineral.*, **52**, 380.
Faye, G. H. (1968). *Canad. Mineral.*, **9**, 403.
Faye, G. H. and Harris, D. C. (1969). *Canad. Mineral.*, **10**, 47.
Faye, G. H., Manning, P. G. and Nickel, E. H. (1968). *Amer. Mineral.*, **53**, 1174.
Faye, G. H., Manning, P. G., Gosselin, J. R. and Tremblay, R. J. (1974). *Canad. Mineral.*, **12**, 370.
Faye, G. H. and Nickel, E. H. (1969). *Canad. Mineral.*, **10**, 35
Faye, G. H. and Nickel, E. H. (1970). *Canad. Mineral.*, **10**, 616.
Ferguson, J. and Fielding, P. E. (1971). *Chem. Phys. Letters*, **10**, 262.
Ferguson, J. and Fielding, P. E. (1972). *Austr. Journ. Chem.*, **25**, 1371.
Ferguson, J., Guggenheim, H. J., and Tanabe, Y. (1966). *Journ. Chem. Phys.*, **45**, 1134.
Ferguson, J., Guggenheim, H. J. and Tanabe, Y. (1967). *Phys. Rev.*, **161**, 207.
Gibbs, G. V. (1966). *Amer. Mineral.* **51**, 1068.
Gorbatshev, R. (1972). *Neus Jahrb. Mineral. Abh.*, **118**, 1.
Hush, N. S. (1967), *In* F. A. Cotton (Ed.), *Progr. Inorg. Chem.*, **8**, 391, Wiley New York.
Hush, N. S. (1968). *Electrochim. Acta*, **13**, 1005.

Lehmann, G. and Harder, H. (1970). *Amer. Mineral.*, **55**, 98.
Loh, E. (1972). *Journ. Phys. C*, **5**, 1991.
Mayoh, B. and Day, P. (1974). *Journ. Chem. Soc. Lond., (Dalton)*, 846.
Manning, P. G. and Nickel, E. H. (1969). *Canad. Mineral.*, **10**, 71.
Newnham, R. E. and De Haan, Y. M. (1962). *Z. Kristallogr.*, **117**, 235.
Okrusch, M. and Evans, B. W. (1970). *Lithos*, **3**, 261.
Rijnierse, P. J. and Enz, U. (1975). *Phys. Bull.*, **15**,
Robbins, D. W. and Strens, R. G. J. (1968). *Chem. Commun.*, 508.
Robbins, D. W. and Strens, R. G. J. (1972). *Mineral. Mag.*, **38**, 551.
Robin, M. B. and Day, P. (1967). *Adv. Inorg. Chem. Radiochem.*, **10**, 247.
Rost, F. and Simon, E. (1972). *Neues Jarhb. Mineral. Monat.*, **9**, 383.
Smith, G. (1973). *Ph.D. Thesis*, University of Newcastle upon Tyne.
Strens, R. G. J. (1974). In: V. C. Farmer (Ed.) *The Infrared Spectra of Minerals*, Mineral. Soc. G. B., London.
Strunz, H., Tennyson, Ch., and Uebel, P. J. (1970). *Minerals Sci. Engng.*, **2**, 3.
Townsend, M. G. (1968). *Solid State Commun.*, **6**, 81.
Townsend, M. G. (1970). *Journ. Phys. Chem. Solids*, **31**, 2481.
Troup, G. J. and Hutton, D. R. (1964). *Brit. Journ. Appl. Phys.*, **15**, 1493.
White, E. W. and White, W. B. (1967). *Science*, **158**, 915.
Wilkins, R. W. T., Farrell, E. F., and Naiman, C. S. (1969). *Journ. Phys. Chem. Solids*, **30**, 43.
Wood, B. J. and Strens, R. G. J. (1972). *Mineral. Mag.*, **38**, 909.

Mossbauer Studies of Iron Minerals Under Pressures of up to 200 Kilobars

F. E. Huggins
Geophysical Laboratory
2801 Upton Street NW
Washington DC 20008 U.S.A.

INTRODUCTION

It is a fortunate coincidence that iron is not only the most convenient Mössbauer element, but also the element which is probably most critical in determining the geophysical and geochemical properties of the Earth's interior. For, although iron oxides probably only amount to 10 per cent by weight of the Earth's mantle (Ringwood, 1970; Press, 1970), their significance is considerably greater due to a number of their properties which are not shown by any other oxides of major abundance and which may either hinder or assist modelling of the Earth's interior. Such properties of iron oxides include mean atomic weights $\neq 20$–21, variable valency, electrical and magnetic phenomena, and complexities due to the ligand-field splitting of the 3*d*-electron energy levels of the iron cations.

The last three properties result directly from the fact that iron cations have an incomplete 3*d*-electron shell and therefore exhibit a much richer variety of electronic phenomena than the other major elements, which, as ions, have either completely filled or empty electron shells and electronic structures similar to inert gas atoms.

Electronic phenomena of importance to studies of the Earth's interior include redox reactions with volatiles which can determine the oxygen fugacity of the environment; the electronic and magnetic behaviour of iron cations which determine the electrical and magnetic properties of rocks and make inversion studies of such properties of the Earth particularly useful; the possibility of spin-state transitions in iron which may lead to seismic discontinuities and the formation of separate phases for magnesium and low-spin ferrous silicates in the lower mantle; the optical properties of ferrous cations in silicates which affect radiative heat transfer in the Earth; and the ligand-field stabilization energies of ferrous cations which contribute to the distribution and ordering of iron and magnesium between different sites in minerals.

The potential of high pressure Mössbauer studies of minerals for investigations of the Earth's interior based on previous studies of iron compounds has been discussed by Burns *et al.* (1972a). In order to realize the full potential of such studies, however, the simultaneous application of high pressures and high temperatures appears to be necessary. In this paper, results obtained from Mössbauer investigations of minerals under high pressures, at ambient temperatures for the most part, are presented. Such results provide information on the electronic and crystallochemical properties of iron as a function of pressure.

THE MÖSSBAUER EFFECT

The Mössbauer effect is the name given to the phenomenon of recoilless resonant emission and absorption of γ-rays which was first discovered in 1957 (Mössbauer, 1958). Although it is a nuclear property, the phenomenon is sensitive to the chemical and bonding environment around nuclei exhibiting this effect.

Of the stable isotopes of iron, only ^{57}Fe exhibits an observable Mössbauer effect and iron Mössbauer studies relate to this isotope which constitutes only 2.2 per cent of the natural abundance of iron. The energy levels involved in the Mössbauer effect and basic experimentation are described in Figure 1. The energy spectrum is produced by means of the Doppler effect of the moving ^{57}Co source which perturbs the energy of the γ-rays emitted by the source. Doppler velocities of the order of 1 cm/sec (approximately 10^{-10} times γ-ray energy) are sufficient to obtain complete resonance with any iron species. To obtain resonance, the γ-ray must be absorbed and emitted without recoil (recoil energy for an unbound ^{57}Fe atom is approximately 10^{-4} times γ-ray energy) and the probability of achieving this is great only when the nucleus is in a solid. The probability of the recoilless process, f, is given by (Wertheim, 1964):

$$f = e^{-4\pi\langle x^2\rangle/\lambda^2} \tag{1}$$

where λ is the wavelength of the γ-ray and $\langle x^2\rangle$ is the component of the mean square vibrational amplitude of the nucleus emitting or absorbing the γ-ray in the direction of the γ-ray. Using the Debye model for the solid state, it can be shown (Wertheim, 1964) that:

$$\text{for } T \ll \theta_D: f = \exp\left[-\frac{E_R}{k\theta_D}\left(\frac{3}{2} + \frac{\pi^2 T^2}{\theta_D^2}\right)\right] \tag{2}$$

where E_R is the recoil energy of the free atom, θ_D is the Debye temperature for the solid and T is the temperature. Although equation (2) is an approximation it does demonstrate how the recoil-free fraction, f, depends on E_R, θ_D and T. Both increasing T and E_R decrease f whereas increasing θ_D increases f.

The two principal parameters obtained from a Mössbauer spectrum are the isomer shift (IS) and the quadrupole splitting (QS). These are described in Figure 1 below. The isomer shift is the centre of energy of the components in

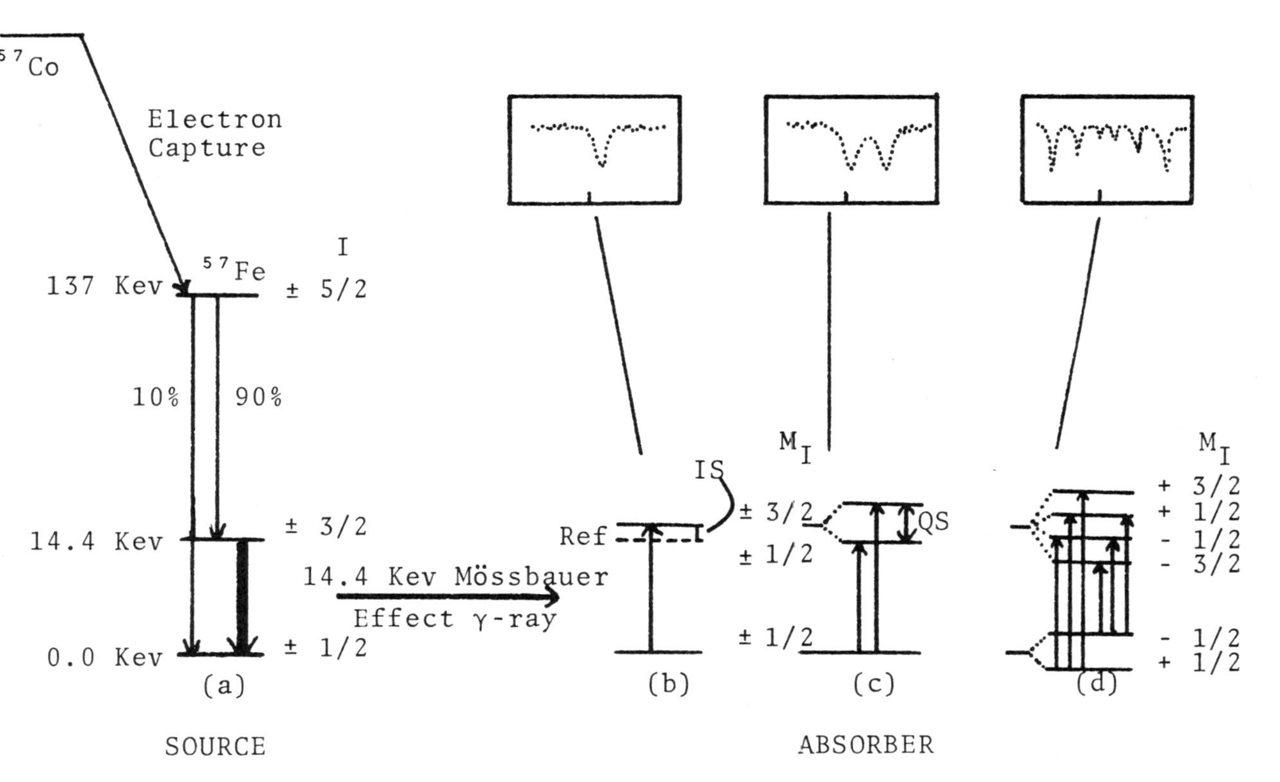

Figure 1. Basic Mössbauer experimentation
(a) Production of ^{57}Fe γ-rays by electron capture by ^{57}Co source. The source is oscillating to give rise to the Doppler shift and a range in energy about 14.4 Kev.
(b) Absorption of γ-rays by ^{57}Fe in sample absorber at energy, IS, relative to some standard, where IS is the isomer shift.
(c) Splitting of energy levels by QS, the quadrupole splitting.
(d) Effect of magnetic field on the energy levels is to remove all remaining degeneracies and to give rise to a six-peak spectrum.

the spectrum of one type of iron species and is a measure of the difference in chemical environments about the ^{57}Fe nucleus. Values are generally quoted relative to some standard (iron metal in this study). Variations in the value of isomer shift arise from differences in the electron density at the nucleus, according to the equation:

$$\mathrm{IS} = \frac{4\pi}{5} Ze^2R^2\left(\frac{\delta R}{R}\right)[|\psi(0)|^2_{\mathrm{abs}} - |\psi(0)|^2_{\mathrm{ref}}] \quad (3)$$

where $\delta R = R_{ex} - R_{gd}$ is the difference in radii of the excited and ground states and $|\psi(0)|^2$ is the electron density at the nucleus, for absorber or reference. Only *s*-electrons have a finite probability of being at the nucleus and chemical effects are observed in highly ionic situations due to the interaction of the cation 3*s* and 3*d* electrons since these electrons have similar radial maxima and shield each other from the nucleus. In more covalent situations, a molecular-orbital approach to the problem would describe the variation of isomer shift as due to interactions involving the metal and ligand orbitals. Nevertheless, since δR is negative, decreasing isomer shift values reflect increasing *s* electron density at the nucleus.

The quadrupole splitting is obtained from a Mössbauer spectrum as the energy separation of the spectral components due to one species of iron atom. The quadrupole splitting arises from an asymmetric charge distribution about the ^{57}Fe nucleus which produces an electric field gradient (EFG) at the nucleus. The interaction of the EFG with the nuclear quadrupole moment of the excited state results in a quadrupole-split spectrum (Figure 1c above) and is described by the equation:

$$\mathrm{QS} = \frac{eQV_{zz}}{2}(1 + \eta^2/3)^{1/2} \quad (4)$$

where Q is the nuclear quadrupole moment, and η, defined as follows:

$$\eta = \frac{V_{xx} - V_{yy}}{V_{zz}}$$

is the asymmetry parameter and V_{xx}, V_{yy}, V_{zz} are principal components of the diagonalized EFG tensor. In general, there will be two contributions to the components of the EFG tensor, one describing the asymmetric nature of the ligands and the other describing the contribution due to a non-spherical distribution of electrons over the five 3*d*-orbitals.

From the values of isomer shift and quadrupole splitting, different chemical and bonding properties of iron cations in silicates including valence state, spin state and crystallographic coordination may be easily determined. Ferrous-ferric ratios can also be estimated from the Mössbauer spectrum, if certain assumptions are made concerning the values of the recoil-free fractions for the different kinds of iron cations.

The interaction of the nuclear magnetic dipole moment with a magnetic field lifts the remaining degeneracies of the nuclear states (Figure 1d) and the selection rules permit six transitions. In an iron-containing solid which is randomly oriented, these transitions characteristically have intensities 3:2:1:1:2:3. The magnetic field may result from the coupling of magnetic moments within the solid or may be applied externally.

More complete descriptions of the Mössbauer effect may be found in the literature (for example: Wertheim, 1964; Danon, 1968; Bancroft, 1974).

EXPERIMENTAL

For this study, it was necessary to adapt the Mössbauer experiment for study of samples under high pressure. The normal Mössbauer experiment is performed with transmission geometry, i.e. the γ-ray detector, the sample absorber and the γ-ray source are in a straight line with the sample presenting as large an area and as small a thickness as practical to the γ-ray flux. To apply pressures of up to 200 kb, an unsupported Bridgeman anvil arrangement and press was used similar to that described by Debrunner *et al.* (1966). The samples were prepared in the same way as Debrunner *et al.* (1966) and Burns *et al.* (1972b) have outlined. Pressure calibration was based on the calibration curve published by Debrunner *et al.* (1966) and cross-correlated using the variation of the quadrupole splitting of hematite with pressure reported by Vaughan and Drickamer (1967).

Unlike the normal experiment, the sample area exposed to the γ-ray flux is not large and to ensure that as much of the flux as possible interacts with the sample, the source of γ-rays is mounted on the end of a screwdriver-like rod which can fit between the tapers of the opposed anvils in order to get within 2 cm of the sample. Due to the small sample aperture of the pressure cell and the low strength of the source confined to the small area at the tip of the screwdriver, count rates are very low and spectra can take up to two weeks to acquire a sufficient number of counts, approximately $10–20 \times 10^3$ counts/channel.

In order to obtain as large an absorption as possible, some mineral samples were synthesized from mixes containing enriched hematite, which had a 90 per cent $^{57}Fe_2O_3$ component. The syntheses were achieved by standard hydrothermal techniques (Huebner, 1971) and product verification was carried out using X-ray diffraction, Mössbauer and optical techniques.

Mössbauer spectra were accumulated in 512 channels of a Nuclear Data multichannel analyser. The Mössbauer drive and ancillary electronics were supplied by Austin Science Associates. Fitting of the spectra was carried out on an IBM 370 series computer using a modified version of the program written by Stone (1967). This program fits the spectral envelope to component Lorentzian peaks using minimization of chi-square as the statistical fit criterion. Convergence was most rapid using the stage-constraint procedures described by Stone.

Table 1. Isomer shift and quadrupole splitting data for minerals investigated as a function of pressure in this study.

FERRIC				FERROUS			
Phase	*P*(kb)	IS	QS	Phase	*P*(kb)	IS	QS
	0	0.39	[a]		0	1.31	3.51
	30	0.39		Almandine[b]	55	1.25	3.52
	55	0.38			100	1.24	3.55
Hematite[d]	80	0.38			150	1.24	3.55
	110	0.37					
	150	0.37					
					0	1.14	2.82
					50	1.10	2.75
	0	0.39	0.62	Fayalite[c]	105	1.06	2.78
$Fe^{3+}_{50}Al_{50}$	57	0.37	0.78		151	1.07	2.73
Garnet[d]	108	0.36	0.76				
	145	0.37	0.76				
	177	0.36	0.76				
					0	1.00	2.62
				Fe^{2+} in Ferri-	57	0.93	2.65
				phlogopite[d,e]	107	0.94	2.69
	0	0.22	0.62		175	0.92	2.72
	25	0.26	1.31				
Ferri-	60	0.29	1.49				
microcline[d]	90	0.30	1.52				
	110	0.31	1.52				
	145	0.31	1.52				
	175	0.31	1.52				
	0	0.19	0.57				
Ferri-	57	0.21	0.62				
phlogopite[d]	107	0.23	0.64				
	175	0.24	0.64				

[a]Values used as pressure calibrant
[b]$(Fe_{0.76}Mg_{0.18}Mn_{0.05})_3Al_2Si_3O_{12}$ from Arizona
[c]From Rockport, Massachusetts
[d]Synthetic
[e]$Fe^{2+}/Fe^{2+}+Mg \approx 0.03$
$Fe^{2+}/Fe^{2+}+Fe^{3+} \approx 0.10$

RESULTS

It is more convenient to describe the results of the Mössbauer experiments on minerals under high pressure in terms of the changes in the individual parameters with pressure, rather than mineral by mineral. Data are listed in Table 1.

(a) Isomer Shift Trends with Pressure

The trends of isomer shift with pressure for a number of ferrous and ferric minerals are shown in Figures 2 and 3. Except for Fe^{3+} in tetrahedral co-ordination, there is a general decrease in isomer shift, indicating an increase in electron density at the nucleus according to equation (3) above. The most

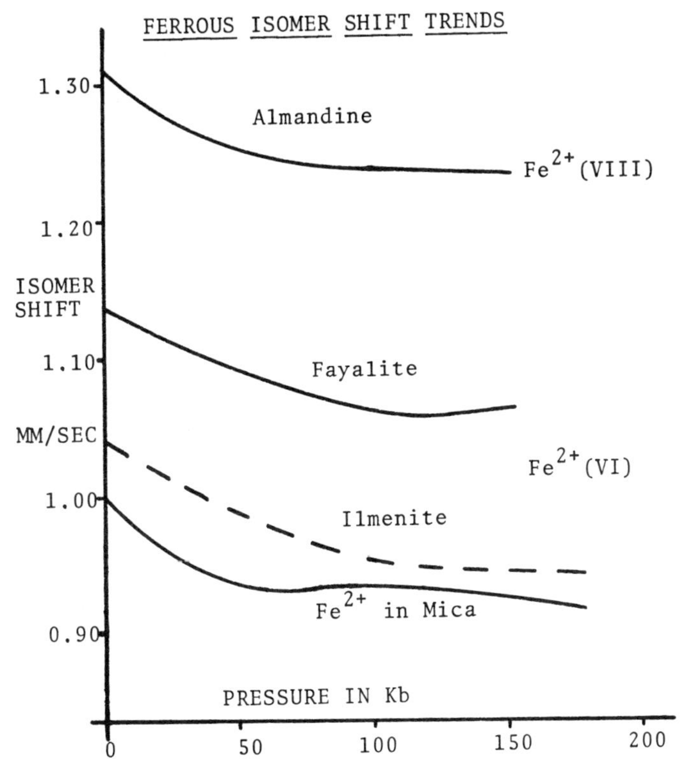

Figure 2. Isomer shift trends with pressure: Ferrous minerals. Ilmenite data taken from Vaughan and Drickamer (1967).

reasonable explanation for this trend is an increase in the contribution of the 3*s*-electrons to the electron density at the nucleus as a result of increasing delocalization of the 3*d*-electrons towards the anions. This is equivalent to describing an increase in covalency of the iron-oxygen bonds due to increased overlap of the metal 3*d* and ligand electron clouds with compression. The increased sharing of the 3*d*-electrons between cation and ligands decreases the shielding of cation 3*s*-electrons from the cation nucleus, resulting in the increase in electron density at the nucleus.

Initially, at least, the decrease in isomer shift appears to be in the order $Fe^{3+}(IV) < Fe^{3+}(VI) < Fe^{2+}(VI) < Fe^{2+}(VIII)$ which is firstly the order of increasing ionic character of the metal–ligand bonds as indicated by the initial isomer shift values and secondly, the expected order of increasing site compressibility. Covalency and site compressibility apparently determine changes in isomer shift. The levelling off at high pressure of the ferrous isomer shift trends suggests that some other process may also come into play which decreases the electron density at the nucleus and counteracts the effect of the spreading *d*-orbitals. One possibility appears to be an increase in the contribution of electron density from the ligand orbitals to the 3*d*-orbitals resulting

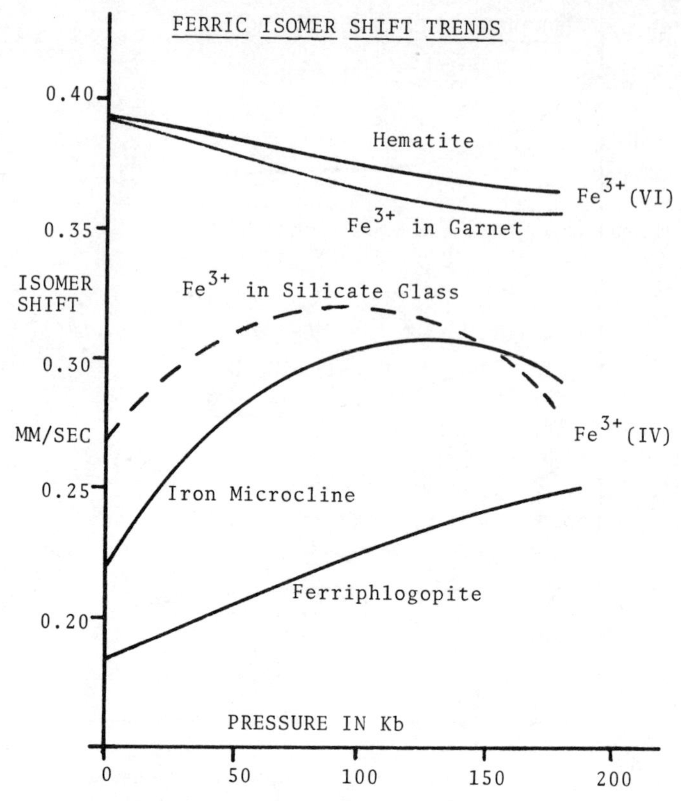

Figure 3. Isomer shift trends with pressure: Ferric minerals. Ferric silicate glass data taken from Lewis and Drickamer (1968).

in increased shielding of the 3*s*-electrons. This would be consistent with the increase in intensity and red shift of ferrous-oxygen metal-ligand charge-transfer bands found by Mao and Bell (1972) and by Mao (1973) in the optical spectra of certain ferrous silicates and oxides.

Ferric in tetrahedral sites in silicates is anomalous in that the *s*-electron density at the nucleus decreases with pressure over the first 100 kb. The maximum in the value of the isomer shift in two of the three systems gives credence to the suggestion made by Vaughan and Drickamer (1967) that the trend comes about from two competing factors. These factors are, firstly, a charge-transfer process (ligand to metal) which increases the 3*d*-electron density on the iron, resulting in a decrease of the 3*s*-electron density at the nucleus and, secondly, a delocalization of the 3*d*-electrons resulting in an increase of 3*s*-electron density at the nucleus. However, because the volume change of this site with pressure is expected to be small, effects involving *d*-electrons, which contribute only in a secondary manner to the electron density at the nucleus, should also be small and because of the large change in isomer shift with pressure, primary effects involving *s*-electrons would appear to be indicated. This suggests that there is considerable cation *s*-electron character in the bonding of the tetrahedral ferric site, which becomes delocalized from the cation with increasing pressure resulting in a decrease in electron density at the nucleus and the observed increase in isomer shift.

Both the delocalization of cation electron clouds and the transfer of ligand charge to the cation result from increased overlap of metal and ligand orbitals, or alternatively, from increased covalency of the iron–oxygen bonds, and indicate that the ionic model becomes less appropriate at high pressure. Molecular orbital calculations using the method originated by Johnson (1973) and recently applied by Tossell *et al.* (1973) to mineralogical systems may enable considerably more quantitative deductions to be made about the change in isomer shift with pressure due to changes in bonding.

Compared to the majority of phases which Drickamer and his coworkers have studied, the isomer shift trends with pressure are relatively small, which is consistent with the expected compressibility differences (silicates being less compressible than most of their phases).

(b) Quadrupole Splitting Trends with Pressure

The quadrupole splitting arises from the interaction of the electric field gradient (EFG) at the nucleus and the nuclear quadrupole moment according

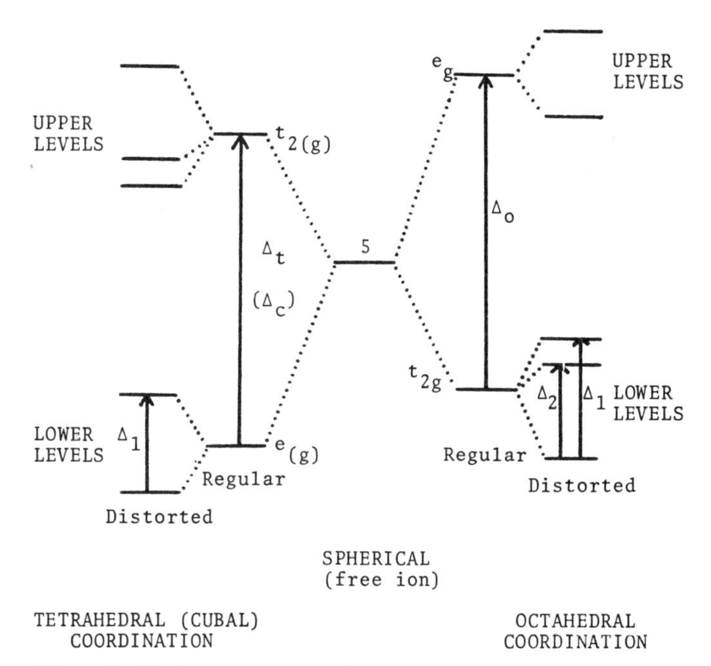

Figure 4. 3*d*-electron energy levels for regular and distorted octahedral and tetrahedral (cubal) sites.

to equation (4). The nuclear quadrupole moment, a nuclear property, will be little affected by pressure and to explain quadrupole splittings with pressure, only the EFG need be considered.

In a site which is only slightly distorted from regular tetrahedral, octahedral or cubal coordination, there is a large distinction between high-spin ferrous and ferric iron in terms of the valence electron and lattice contributions to the EFG and thus to the quadrupole splitting. For high-spin ferric iron, which contains one electron in each of the five $3d$-orbitals, resulting in a spherical electron distribution about the nucleus, the net valence contribution to the EFG is zero and the quadrupole splitting is only a function of the asymmetry in the ligand positions about the ^{57}Fe cation. On the other hand, for high-spin ferrous which contains an additional electron in the lowest set of $3d$-electron energy levels (t_{2g} derived levels for octahedral, e derived levels for tetrahedral, and e_g derived levels for cubal—see Figure 4), the electron contribution outweighs the ligand contribution and so ferrous quadrupole splittings are largely related to the valence electron contribution to the EFG. Due to this difference, ferrous and ferric quadrupole splitting trends will be treated separately.

(*i*) *Ferric Quadrupole Splitting Trends*

The ferric quadrupole splitting trends are shown in Figure 5 and all samples show a large relative increase over the first 50–100 kb and then level off as pressure is further increased. Except in certain high symmetry cases, the quadrupole splitting cannot be related unambiguously to specific changes in the positions of the ligands. It is possible, however, to estimate the quadrupole splitting trend if the compression were to occur at constant distortion (i.e. angles and ratios of bond lengths remain constant—the structure is identical except for a scale factor).

The ligand contribution to the components of the EFG tensor is of the form (Travis, 1971):

$$V_{ij} = \sum_k q_k(3\, x_{ki} x_{kj} - r_k^2 \delta_{ij}) r_k^{-5}$$

where q_k is the charge on the kth ligand
x_{ki} is the ith coordinate of the kth ligand
r_k is the ferric-kth ligand distance
and δ_{ij} is Kronecker's delta: $\delta_{ij} = 1$ if $i = j$
$= 0$ if $i \neq j$

At constant distortion, it follows that if $\langle r \rangle$ is the average bond length:

$$r_k = a_k \langle r \rangle$$
$$x_{ki} = b_{ki} \langle r \rangle$$

where a_k and b_{ki} are constants independent of the compression which only changes the value of $\langle r \rangle$:

Then: $V_{ij} = \sum_k q_k(3\, b_{ki} b_{kj} - a_k^2 \delta_{ij}) a_k^{-5} \langle r \rangle^{-3}$

or: $V_{ij} = A_{ij} \langle r \rangle^{-3}$

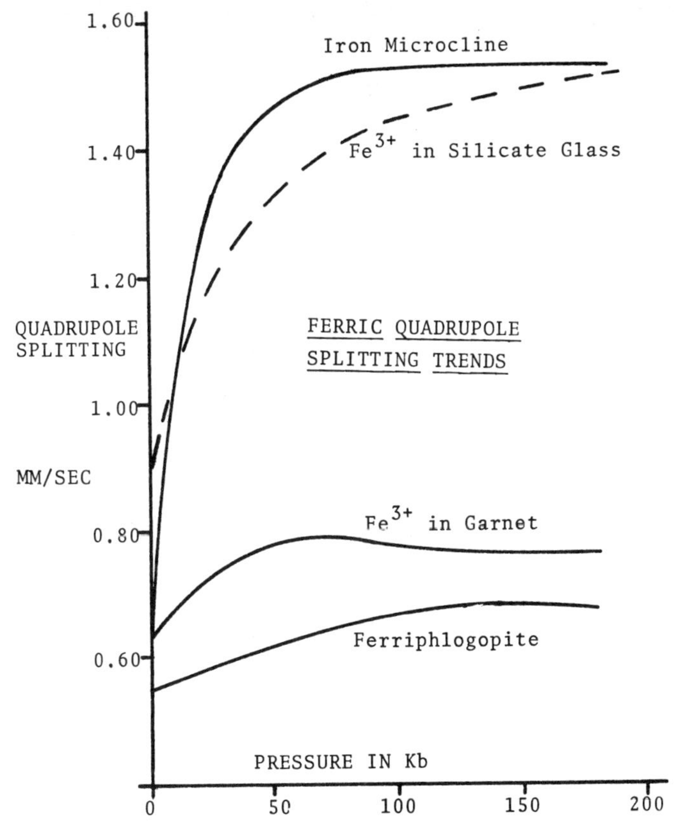

Figure 5. Quadrupole splitting trends with pressure: Ferric minerals. Ferric silicate glass data taken from Lewis and Drickamer (1968).

where A_{ij} is also independent of compression. The asymmetry parameter is defined as:

$$\eta = \frac{V_{xx} - V_{yy}}{V_{zz}} = \frac{\langle r \rangle^{-3} A_{xx} - \langle r \rangle^{-3} A_{yy}}{\langle r \rangle^{-3} A_{zz}}$$

Thus, $\eta = \dfrac{A_{xx} - A_{yy}}{A_{zz}}$ at constant distortion

and is independent of compression. Therefore, equation (4) becomes at constant compression:

$$\mathrm{QS} = \frac{eQA_{zz}}{2}(1 + \eta^2/3)^{1/2} \langle r \rangle^{-3}$$

or

$$\mathrm{QS} = k\langle r \rangle^{-3} = k' V^{-1}$$

where $k(k')$ is independent of compression. The pressure dependence of the QS is therefore inversely proportional to the change in volume with pressure.

$$\ln \mathrm{QS} = \ln k' - \ln V$$

Now

$$\left(\frac{\partial \ln \mathrm{QS}}{\partial P}\right)_{T,\mathrm{dist}} = -\left(\frac{\partial \ln \mathrm{V}}{\partial P}\right)_{T}$$

and so

$$\frac{1}{\mathrm{QS}}\left(\frac{\partial \mathrm{QS}}{\partial P}\right)_{T,\mathrm{dist}} = -\frac{1}{V}\left(\frac{\partial V}{\partial P}\right)_{T} = \beta$$

where β is the compressibility of the solid.

The order of site compressibility can be predicted by analogy with measurements of the differential thermal expansion of polyhedra. Sites which have large thermal expansions should also be highly compressible.

In addition, Anderson and Anderson (1970) have shown how the bulk moduli of oxides of a given structure-type obey the relationship:

$$KV_0 = \mathrm{const}$$

where K is the bulk modulus and V_0 is the molar volume. The constant was shown, using the Born ionic model, to be given by:

$$\mathrm{const} = \frac{AZ_cZ_ae^2(n-1)}{9\,r_0}$$

where A is the Madelung constant; Z_c, Z_a are cation and anion charges; n is the repulsive exponent; and r_0 is a distance characteristic of the oxide. Anderson and Anderson were able to show that $A(n-1)/9r_0$ was approximately 0.157 ± 0.005 for most oxides regardless of structure-type. Therefore, for any oxide:

$$KV_0 = 0.157\ Z_cZ_ae^2$$

Applying this equation to individual polyhedral components in silicates suggests that:

$$\text{for } Z_a = -2\text{: } K_{\mathrm{pol}} \propto Z_c/V_{\mathrm{pol}} \propto 1/\beta_{\mathrm{pol}}$$

This proportionality results in the expected order of site compressibilities.

Although specific compressibility data for phases of compositions found in this study are not available, approximate values (Voigt-Reuss-Hill averages) can be estimated from the compilation of Simmons and Wang (1971). For the garnet, which has a composition midway between grossular and andradite, a maximum value of 0.7 Mb^{-1} for the compressibility can be assigned and the quadrupole splitting would be expected to increase approximately 7 per cent in 100 kb. The observed increase is initially much greater than this suggesting that the structure distorts on compression in such a way as to increase the EFG at the ferric site. At higher pressures, the quadrupole splitting change is less than predicted, indicating that the site distortions are reducing the EFG.

For both ferrimicrocline and ferriphlogopite, the values of compressibility may be estimated from the aluminum analogue. Values of about 2.0 Mb^{-1} are appropriate for both phases, enabling predictions of a 20 per cent increase in quadrupole splitting over 100 kb to be made. The change in ferriphlogopite quadrupole splitting is of this magnitude, whereas the change in ferrimicroline is considerably greater. The agreement for ferriphlogopite must be fortuitous because the linear compressibilities in the mica sheet and perpendicular to the sheet are greatly different indicating that the constant distortion criterion cannot be satisfied.

The quadrupole splitting trend for ferrimicrocline shows a huge increase in the first 50 kb, suggesting an equally impressive rearrangement of the atoms around the ferric cation or possibly a phase change. Whatever the transition may be, the final high pressure state must have a very asymmetric charge distribution around the ferric cation.

(*ii*) *Ferrous Quadrupole Splitting Trends*

The quadrupole splitting of ferrous compounds depends on both the valence electron contribution and the ligand contribution. These contributions tend to be opposite in sign and the valence contribution to be larger, unless the site is considerably removed from the cubic symmetry sites.

Ingalls (1964) developed a model for the quadrupole splittings of ferrous sites which depends on the covalency of the bonds (α^2), the spin–orbit coupling parameter (λ_0), temperature (T) and the splittings (Δ_1, Δ_2—see figure 4 above) of the lower $3d$ energy levels. Parametrically the equation may be written as:

$$\mathrm{QS} = \mathrm{QS}_0 \alpha^2 F(\Delta_1, \Delta_2, \alpha^2, \lambda_0, \mathrm{T}) \tag{5}$$

where QS_0 is the quadrupole splitting at absolute zero for an ideal ionic ferrous site, and F is the difference between the lattice and valence contributions and is known as the reduction function. For the purposes of this study:

$$\mathrm{QS}_0 \alpha^2 = \mathrm{QS}'_0$$

where QS'_0 is the quadrupole splitting at absolute zero for the ferrous compound in question and following the suggestion made by Bancroft (1974), the role of α^2 and λ_0 in determining the reduction function will be ignored. These simplifications lead to an easily manipulated form of equation (5).

$$\mathrm{QS} = \mathrm{QS}_0 \mathrm{F}(\Delta_1, \Delta_2, T)$$

Since ferric quadrupole splittings are usually independent of temperature, the lattice contribution to the ferrous quadrupole splittings may also be assumed to be temperature independent. Therefore, the temperature dependence of ferrous quadrupole splittings comes solely from the valence contribution and is due to the Boltzmann distribution of the sixth $3d$-electron over the two or three lower energy levels. For tetrahedral or cubal sites, the explicit expression for the valence contribution is:

$$F_{\mathrm{val}}(\Delta_1, T) = \frac{1 - \exp(-\Delta_1/kT)}{1 + \exp(-\Delta_1/kT)} \tag{6a}$$

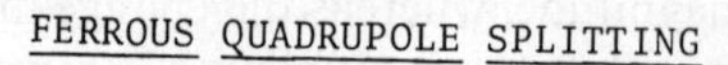

INGALLS' MODEL: QS = QS(0)α^2F(Δ_1,Δ_2,α^2,λ_0,T)

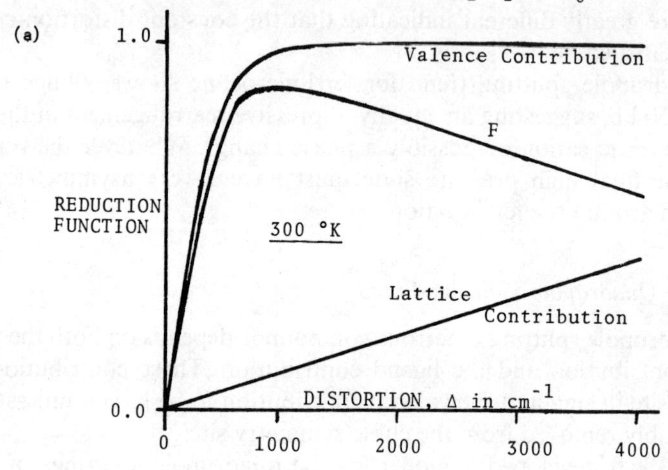

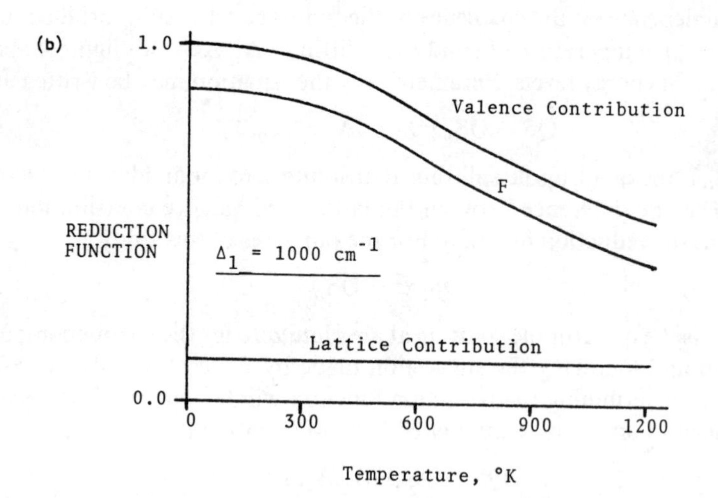

Figure 6. Ingalls' model for ferrous quadrupole splittings.
(a) Variation of quadrupole splitting with distortion at constant temperature.
(b) Variation of quadrupole splitting with temperature at constant distortion.

For octahedral sites:

$$F_{\text{val}}(\Delta_1, \Delta_2, T) = \frac{[1 + \exp(2x) + \exp(2y) - \exp(x) - \exp(y) - \exp(x+y)]^{1/2}}{1 + \exp(x) + \exp(y)} \tag{6b}$$

where $x = -\Delta_1/kT$ and $y = -\Delta_2/kT$.
Various aspects of these equations are plotted in Figure 6.

Since Δ_1 and Δ_2 are not generally expected to change significantly compared to T, the temperature dependence of the quadrupole splitting is often fitted to equations (6a) or (6b) to evaluate Δ_1 and Δ_2, assuming that these energy level splittings are constant over the complete temperature range. Such an assumption negates, for the most part, any sophistication of this approach by attempting to allow for the effect of λ_0 and α^2 on the reduction function.

The pressure dependence of the quadrupole splitting at constant temperature comes about from changes in Δ_1 and Δ_2 with pressure:

$$\text{QS}(P) = \text{QS}(T)F(\Delta_1(P), \Delta_2(P), T)$$

Before the lattice contribution to the quadrupole splitting may be ignored, it must be demonstrated that the values of Δ_1 and Δ_2 are small and that at least one of these values lies well to the low distortion side of the maximum in Figure 6(a). If both values lie to the high distortion side, then the change in quadrupole splitting with pressure will be due in large part to variations in the ligand geometry. For this purpose, values of Δ_1 and Δ_2 were estimated using equation (6a) or (6b) as appropriate from the temperature variation of quadrupole splittings published in the literature. Data taken from Lyubutin *et al.* (1971b), Eibschutz and Ganiel (1967) and Annersten (1974) were used to estimate values for Δ_1 and Δ_2 for almandine, fayalite and Fe^{2+} in ferriphlogopite, respectively, in order that the pressure dependence of the quadrupole splittings of these phases, which are shown in Figure 7, may be explained.

For almandine, a value of between 1050 and 1400 cm^{-1} for Δ_1 was determined. The range of values is to allow for a lattice contribution of up to 2 mm/sec to the quadrupole splitting. However, since the original value of the quadrupole splitting is so large (within 8 per cent of the largest ferrous quadrupole splitting ever recorded) the lattice contribution must be small (< 0.30 mm/sec) and the value of 1100 ± 50 cm^{-1} was assigned to Δ_1. This value is just to the high distortion side of the maxima in Figure 6(a) and suggests that the observed increase in quadrupole splitting of almandine with pressure comes about from a slight decrease in the lattice contribution to the EFG, presumably from a decrease in the site distortion.

For fayalite, values of 620 ± 20 cm^{-1} for Δ_2 and 1400 ± 200 cm^{-1} for Δ_1 for one octahedral site and 710 ± 20 cm^{-1} for Δ_2 and 1500 ± 200 cm^{-1} for Δ_1 for the other site, assuming a lattice contribution of up to 0.5 mm/sec, were estimated from the data of Eibschutz and Ganiel (1967). Although pressure did not resolve the two contributions any better than the spectrum at ambient conditions, the decrease in separation of the superimposed peaks suggested

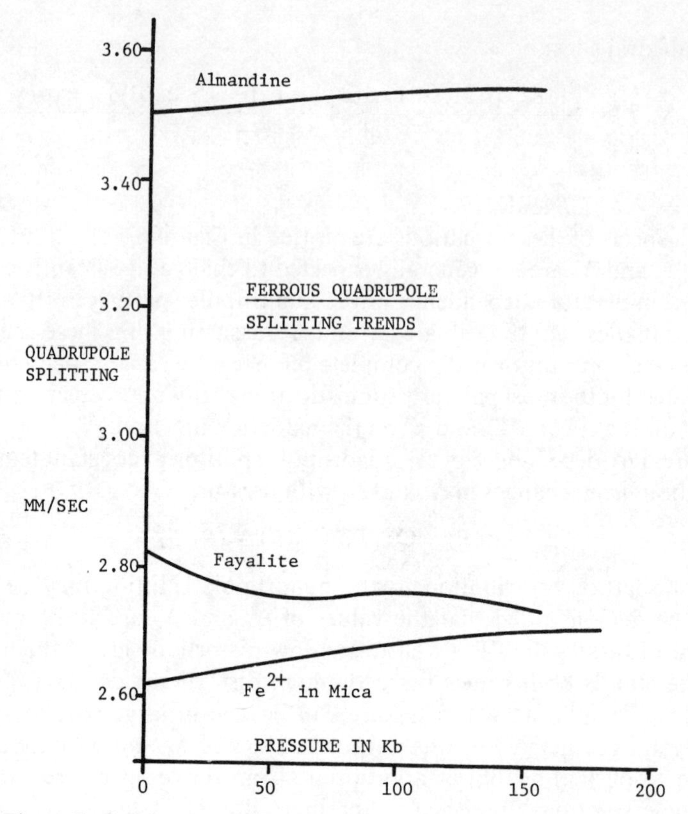

Figure 7. Quadrupole splitting trends with pressure: Ferrous minerals.

that the values of Δ_1 and Δ_2 for both sites were decreasing. The value of Δ_2 for both sites, which lies on the low distortion side of the maxima, must decrease by about 10 per cent over 150 kb to explain the change in quadrupole splitting with pressure, but Δ_1, which has much less influence on the quadrupole splitting, cannot be evaluated as accurately. Although the data are not conclusive, the decrease in the energy level splittings suggests a change in site distortion towards a more regular site, which is consistent with the results obtained by Abu-Eid (this volume) from changes in the M1 absorption spectral peaks with pressure.

For ferrous in ferriphlogopite, an increase in quadrupole splitting with pressure is observed. The data presented by Annersten (1974) suggest that both of the splittings of the ferrous t_{2g} level in phlogopite are less than 500 cm^{-1} and decrease with temperature. The pressure trend may therefore be explained by an increase in these splittings with pressure resulting from an increase in site distortion. Since micas are much more compressible perpendicular to the sheets in the structure, the octahedral sites, which are already compressed along a trigonal axis approximately perpendicular to the sheets (Hazen and Burnham, 1973), would appear to become even more compressed in this direction at high pressure.

(c) Changes in Fe^{2+}/Fe^{3+} Ratio with Pressure

The main emphasis of the research carried out by Drickamer and his group at the University of Illinois concerned the changes in the Mössbauer Fe^{2+}/Fe^{3+} ratio as a result of pressure. Examples of changes in the valence state ratio from zero to 4 or 5 may be found in their studies (Drickamer *et al.*, 1970; Drickamer and Frank, 1973). This phenomenon is explained as an electron transfer from the ligands to the central ferric cation leading to a new electronic ground state involving a ferrous cation and a radical anion species. Since the iron is monitored by Mössbauer spectroscopy and the formation of ferrous at the expense of ferric is observed directly, the phenomenon has been described as 'pressure-induced reduction' although 'pressure-induced electron transfer' would be a more accurate description. For, although in one sense this phenomenon is a redox reaction, in that there is a change of valence state, however since it takes place *within* an individual phase and, except for pressure and temperature, appears independent of conditions external to the phase (Drickamer and Frank, 1973), it is not a true redox reaction in the samc sense that a disproportionation reaction is not.

The systematic study of Frank and Drickamer (1973) showed that the extent of the charge-transfer process was related to the electron-donor properties of the surrounding ligands and that the degree of conversion of ferric to ferrous can be correlated with the change in isomer shift, which also depends on the electron-donating properties of the ligands. More detailed theoretical aspects of this electronic transition are presented by Slichter and Drickamer (1972) and by Drickamer, Frank and Slichter (1972).

The investigations carried out at Illinois were mostly of no direct mineralogical interest and the ferric phases which were of interest, such as hematite and ferric-containing silicate glass, showed no signs of the formation of ferrous by the charge-transfer process.

Except for magnesioriebeckite (Burns *et al.*, 1972b) only very small increases in Fe^{2+}/Fe^{3+} ratio have been observed in silicate minerals which have been studied at room temperature. Since most of these silicates already contained a minor amount of ferrous in addition to ferric at ambient conditions, two further factors must be discussed before the changes in valence state ratio can be ascribed to the charge-transfer process.

The first of these factors is the possible change in the recoil-free fraction, f, of the different iron species with pressure. Equations (1) and (2) above are expressions for f and since the parameters $\langle x^2 \rangle$ and θ_D depend on the vibrational spectrum of the solid, changes in f with pressure can be expected, which will reflect changes in the vibrational frequencies of the solid, especially those related to the cation sites. Whipple (1973) has demonstrated that within the same mineral phase, the ferrous recoil-free fraction may be smaller than the ferric recoil-free fraction by as much as 25 per cent. For example, he found that $f(Fe^{2+},VIII)/f(Fe^{3+},VI) \approx 0.77$ for the garnet structure at 25 °C. Other investigators have also shown that the coordination number and valence state determine the recoil-free fraction at 25 °C (Lyubutin *et al.*, 1970, 1971a, 1971b; Sawatzky *et al.*, 1969). With increasing pressure, absolute values of recoil-free fractions increase towards unity since Debye temperatures increase with pressure and so ratios of recoil-free fractions will also tend to unity, which will lead to an increase in the observed Fe^{2+}/Fe^{3+} ratio with pressure. In principle, an increase in the ratio of up to 25 per cent might be expected. Except for magnesioriebeckite, all observed changes in the ratio of valence states could conceivably be explained in this manner for minerals containing both oxidation states.

The second factor is related to thick absorber phenomena. When the amount of ^{57}Fe in a Mössbauer sample is too great, the area under the peak no longer bears a linear relationship to the amount of ^{57}Fe in the phase, but due to self-absorption effects underestimates the amount of ^{57}Fe. Concomitant with the non-linear absorption are line broadening, inadequacy of the Lorentzian line shape as an approximation to the experimental line shape, and extreme baseline curvature in computer fits of the data. For these reasons, thick absorber phenomena are best avoided. However, since the high pressure geometry is no longer suitable for the absorber to approximate a thin sheet perpendicular to the γ-ray flux, dilution of the sample in boron is necessary to avoid thick absorber problems. Dilution of the sample in boron until ^{57}Fe constitutes only about 1/2 per cent by weight of the mixture coupled with very careful collimation has been found to result in adequate spectra (Huggins, 1974). Since the ferrous absorptions are generally small in these studies of the pressure-induced charge-transfer process in ferric silicates, they approach thin absorber behaviour more closely than the ferric absorptions, if the dilution is insufficient, so that once again the ferrous content is increased relative to the ferric content. This property is especially noticeable when comparing ratios obtained from normal geometry experiments with those from high pressure geometry experiments at 1 bar where the dilution is insufficient.

Due to the uncertainty introduced by the possible effects of thick absorbers and the pressure-induced changes in recoil-free fractions, most minerals containing both ferrous and ferric absorptions do not show sufficient change in valence state ratio with pressures up to 200 kb at room temperature for pressure-induced charge-transfer to be considered important.

The effect of temperature has been shown to be most beneficial in assisting the charge-transfer process and increasing the amount of ferrous formed at the expense of ferric (Drickamer *et al.*, 1970). For this reason, experiments were carried out at pressures of up to 150 kb at temperatures up to 90 °C on a synthetic garnet of composition intermediate between uvarovite and andradite. With increasing pressure, a slight asymmetry of the ferric absorption at high velocity was observed. On raising the temperature, this asymmetry increased until at 90 °C and 110 kb peaks could be fitted which amounted to about 15 per cent of the absorption (figure 8). This doublet is obviously neither due to thick absorber complications nor to the change in recoil-free fraction since no peak was present originally. The most reasonable explanation appears to be that

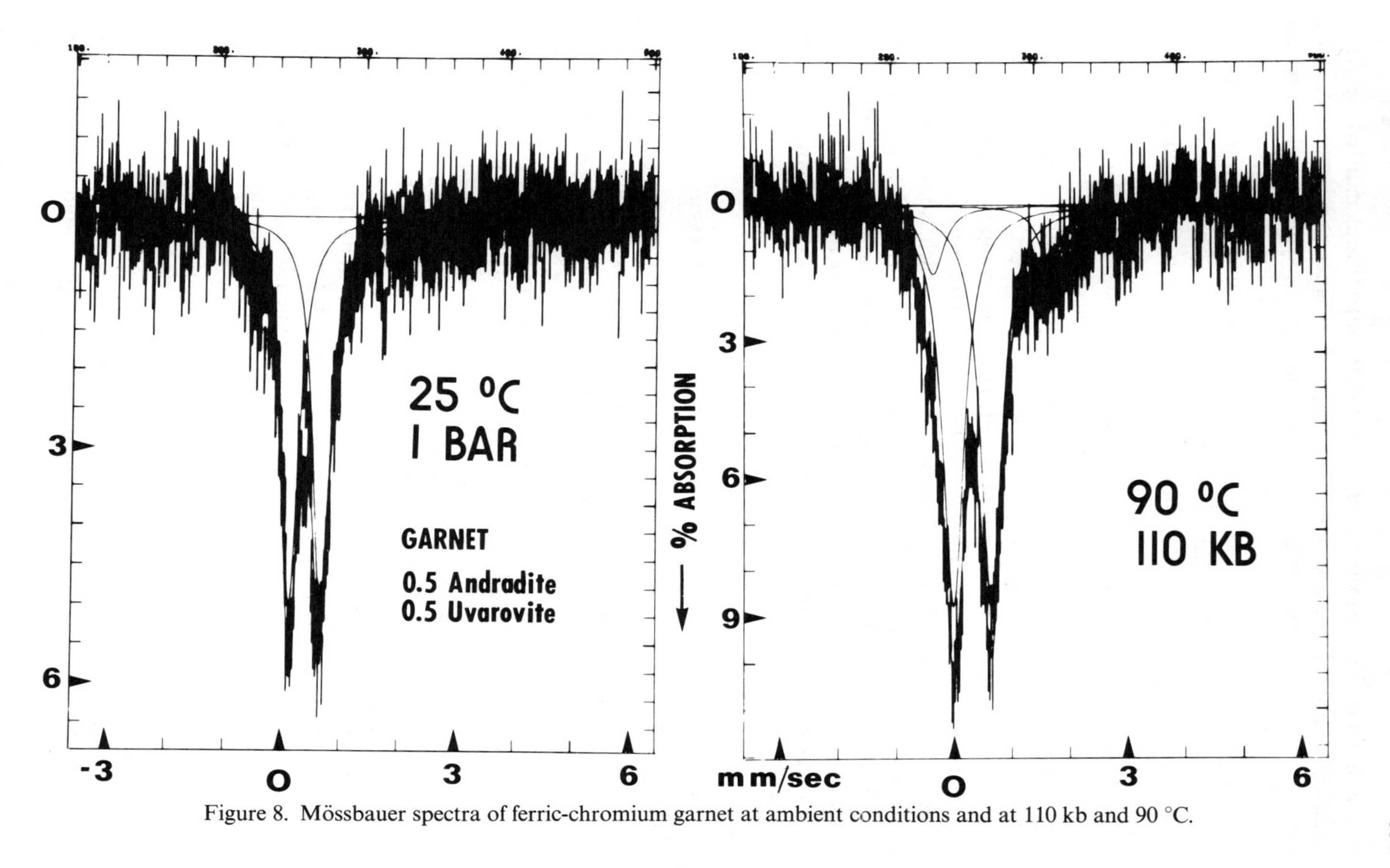

Figure 8. Mössbauer spectra of ferric-chromium garnet at ambient conditions and at 110 kb and 90 °C.

this is ferrous (VI) formed by pressure (and temperature) induced electron-transfer from the ligands to ferric (VI). The Mössbauer parameters (IS ~ 0.8, QS ~ 2.1) are consistent with ferrous (VI) in a small, symmetrical site which is the case for this octahedral site in garnet.

Temperature appears, therefore, to play a critical role in addition to pressure in controlling the electron transfer process in minerals. Thus, at pressures and temperatures appropriate to mantle conditions, the new electronic ground state may be extensive in nominally ferric minerals.

Attempts to use the valence state ratio-isomer shift correlation for ferric silicates, according to the scheme of Frank and Drickamer (1972), are obviously not appropriate, suggesting that such correlations must be restricted to given ligand types. The reasons why the electron-transfer occurs in ferric silicates to such a small extent compared to the organo-ferric compounds must include differences in compressibilities and the poor electron-donor ability of the oxygen ligands in silicates. Magnesioriebeckite, which shows the largest pressure-induced change in Fe^{2+}/Fe^{3+} ratio of any silicate at room temperature, contains some ferric in the hydroxylated sites of the amphibole structure. Such sites are expected to be more compressible than the non-hydroxylated ferric sites most commonly found in silicates and may explain why the largest change in valence state ratio is seen with this phase.

(d) Changes in Magnetism of Oxides and Silicates with Pressure.

Only very small changes have so far been observed in the magnetic properties of oxides at pressures up to 200 kb at 25 °C and no pressure-induced magnetic effects have been encountered in similar experiments with iron silicates, which remain paramagnetic.

With hematite, a reduction in the magnetic field strength of 1 per cent and a change in the sign of the quadrupole splitting at about 30 kb have been observed (Vaughan and Drickamer, 1967) with increasing pressure. The change in the magnetic field strength is also of this magnitude in magnetite and in addition the tetrahedral and octahedral magnetic subpatterns may lose some resolution with increasing pressure.

Compared to the changes in magnetism of iron sulphides with pressure, such changes are minimal. Vaughan and Tossell (1973) reported that the application of 50 kb completely eliminates the magnetism in chalcopyrite and pyrrhotite and Kasper and Drickamer (1968) reported that the application of 36 kb eliminates the magnetism in troilite suggesting that sulphides become paramagnetic at high pressure. The difference between sulphides and oxides must be a function of compressibility and electronic differences, which permit the oxides to retain magnetic coupling between iron sites in oxides while quenching the magnetic coupling between iron sites in sulphides. The change in spin state of the iron cations postulated by Vaughan and Tossell as one possible mechanism to explain the paramagnetism of the sulphides at high pressure would not occur in oxides until considerably higher pressures are attained and would appear to be the most likely reason for the difference.

DISCUSSION

(a) The High Pressure Electronic Transition in Ferric Silicates

The evidence for the new electronic ground state at high pressure in ferric silicates is rather meagre at present, being limited to two phases. However, since both temperature and pressure appear to augment the electron transfer from ligand to metal, by extrapolating to mantle conditions, it is not inconceivable that the new ground state may be extensive for ferric phases in the Earth. For this reason some further discussion of the transition, in particular concerning its possible effects on various aspects of mantle mineralogy, is presented here.

Since oxygen anions usually constitute about 90 per cent by volume of silicate and oxide minerals, any means of reducing the size of the oxygen anions will be favoured by pressure and the formation of O^- at the expense of O^{2-} by means of electron transfer from the oxygen anions to the reducible ferric cation satisfies this condition. As long as the decrease in anion volume exceeds the increase in cation volume, the transfer will be favoured at high pressure. Taking a most naive approach and assuming that O^- has the same ionic radius as F^-, it will be approximately 1.59 $Å^3$ smaller in volume than O^{2-} and this volume change exceeds the increase in volume of the iron cation, which is approximately 0.88 $Å^3$ (based on Shannon and Prewitt (1969) radii). Evidence for radical anions such as O^- could in principle be investigated by electron

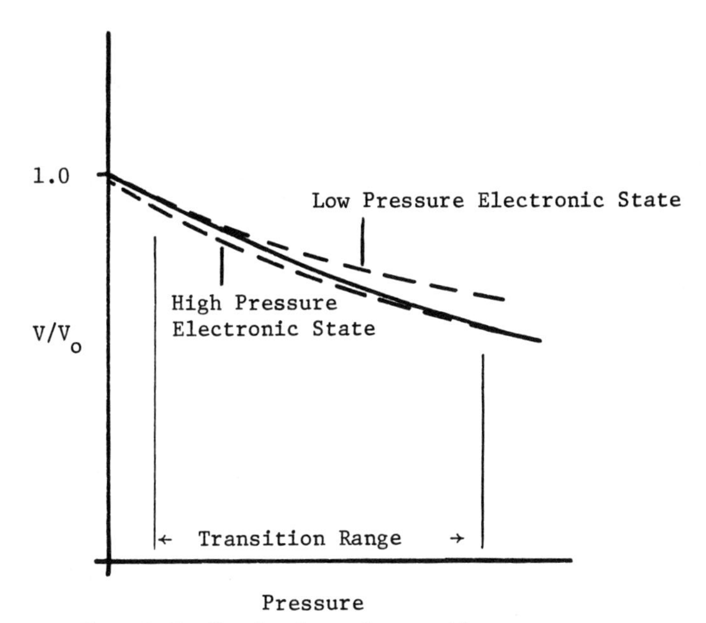

Figure 9. Predicted volume change with pressure for the ligand-to-metal electron transfer transition in ferric compounds.

spin resonance (ESR) at high pressure, but such experiments would be extremely difficult. ESR experiments at ambient conditions on certain phases have shown the presence of small amounts of radical anions. Vinokurov *et al.* (1971) showed that O^- is present in zircon, and Low and Zeira (1972) have shown CO_2^- to be present in $CaCO_3$.

The new electronic ground state that results at high pressure and temperature can be expected to have somewhat different physical properties than the low pressure ground state since a less highly charged pair of ions have been formed. Extending the scheme of Anderson and Anderson (1970) suggests that the bulk modulus of the high pressure state may be less than that of the low pressure phase. Therefore, it is conceivable that the change in electronic state may occur with minimal volume change and that the decrease in bulk modulus may be responsible for the transition. Figure 9 shows schematically what would happen to the volume, if this was the case. Such deviations in $\Delta V/V$ versus pressure may well be lost within the accuracy of even the most precise PV measurements, especially if the original ferric concentration is small.

The charge-transfer process will tend to greatly decrease the amount of ferric iron in the mantle and would appear to invalidate one of the arguments proposed by Ringwood (1966) concerning non-equilibrium between the core and mantle on the basis of the predicted Fe^{2+}/Fe^{3+} ratio in the lower mantle. However, the question to be considered now would be whether equilibrium is possible between iron metal and the oxygen radical anion.

Any electronic rearrangement may well affect the electrical and magnetic properties of the material and this new electronic ground state for ferric phases could result in such changes. However, such investigations have not been reported for materials in which this transition extensively occurs as a result of pressure.

Other than iron, the change in electronic ground state has been observed in cupric systems (Wang and Drickamer, 1973) and the question is raised as to how general this electronic transition may be. The transition requires a fairly oxidizing cation to be reduced to a fairly stable cation and that the ligand-metal charge-transfer band in the optical absorption spectrum to be of relatively low energy. Of the transition metal oxidation states which occur in nature and satisfy this condition, ferric is obviously the best. Although Ti^{4+} is more stable to reduction than ferric, it may also exhibit this property to some extent since it too has a relatively low energy charge-transfer band and the volume change $Ti^{4+} \rightarrow Ti^{3+}$ should not be as large as $Fe^{3+} \rightarrow Fe^{2+}$. Of the less common oxidation states, V^{4+} and Mn^{3+} also appear to have potential in this respect, whereas Cr^{3+}, due to the reducing nature of Cr^{2+}, may well be stable with respect to this transition.

(b) Covalency Changes in the Earth

The covalency of iron-oxygen bonds in minerals are closely allied with structural details, in particular, with bond lengths and coordination number.

In general, due to compression, shorter bond lengths and increases in radius ratio (r_c/r_a) result if the structure remains the same. In addition, with increasing pressure, phase changes occur, resulting in more efficient packing and, if the coordination number increases, longer bond lengths. Since covalency of the bond varies inversely with the bond length, compression results in increasing covalency, which is indicated by the Mössbauer experiments. However, such increases are offset by phase changes to structures containing higher coordination numbers and longer bond lengths. Changes in spin-state from high-spin to low-spin in the transition cations due to compression will result in shorter bond lengths and increased covalency. The electron-transfer phenomenon, on the other hand, by decreasing the effective nuclear charge on neighbouring ions, will result in a decrease in covalency and an increase in radius ratio.

Similar arguments can be advanced for the effect of changes in these phenomena with depth on bulk modulus and Table 2 represents a summary of such arguments. Examples of these changes are also shown in Figure 10. From such arguments, it appears that although covalency increases with pressure, there are a number of factors which will decrease the covalency of metal-oxygen bonds, the most important of these being phase changes involving increases in coordination number with depth. From Table 2 and Figure 10(a), (b), (c) it is apparent that covalency mirrors neither bulk modulus nor radius ratio.

A manifestation of the covalency of ferrous-oxygen bonds in silicates is the pressure-induced shifts and increase in intensity of the ligand → metal charge-transfer bands in $(Mg,Fe)_2SiO_4$ polymorphs and in ferropericlase (Mao and Bell, 1972; Mao, 1973). Such changes are consistent with increasing overlap of cation and anion electron clouds or increasing covalency with compression. It is these features, rather than crystal-field absorptions, which play a dominant role in determining radiative transfer in minerals under mantle conditions.

The ionic model becomes less appropriate with depth in the Earth since the emerging ligand → metal charge-transfer features cannot be treated adequately by crystal-field theory and the structure postulated for high pressure phases

Table 2. Effect of changes in pressure, temperature, ccoordination number, spin multiplicity and cation oxidation state on radius ratio, covalency of metal–oxygen bonds and bulk modulus.

Expected change in property with increasing depth in Earth		Effect of change in property with depth on		
Property	Change	Radius Ratio	Bond Covalency	Bulk Modulus
Pressure	Increase	Increase	Increase	Increase
Temperature	Increase	Decrease	Increase	Decrease
Coordination No.	Increase	Increase	Decrease	Increase
Spin Multiplicity	Decrease	Decrease	Increase	Increase
Cation Oxidation State	Decrease	Increase	Decrease	Decrease?

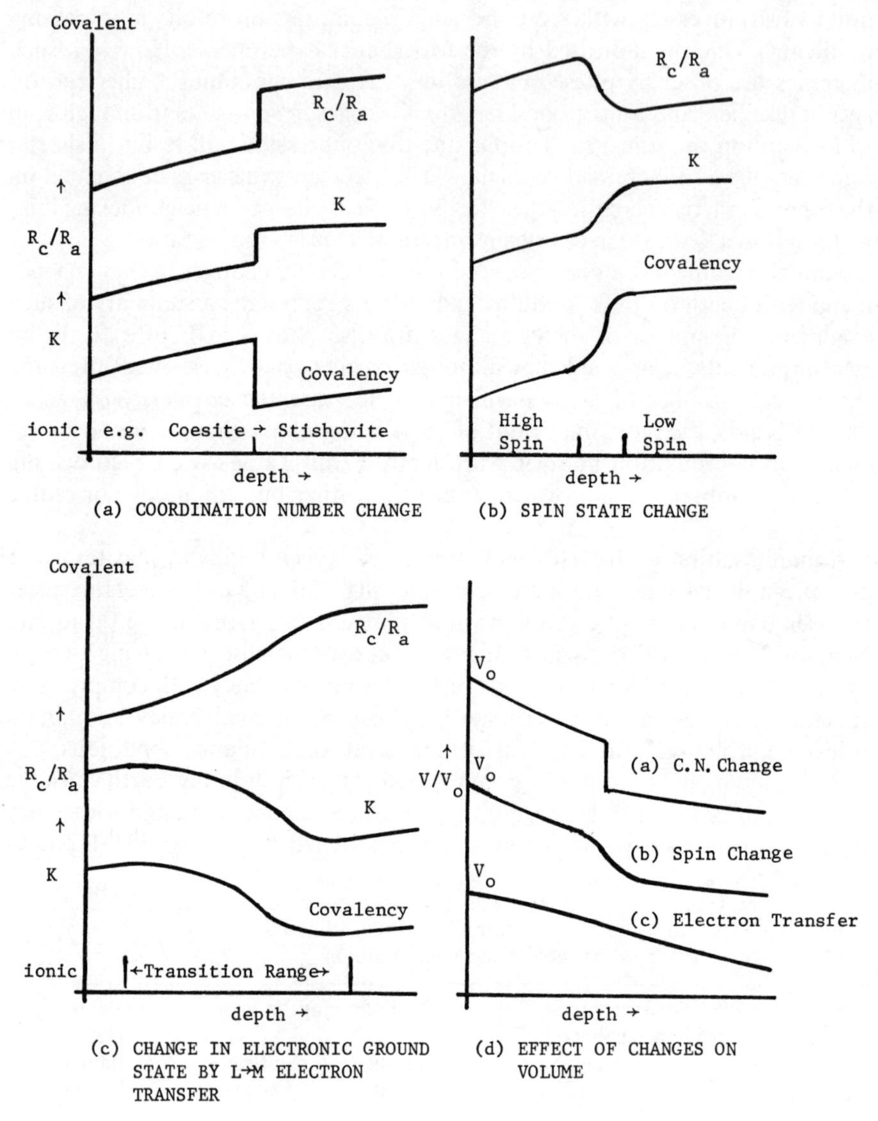

Figure 10. Expected changes with depth in Earth of radius ratio, R_c/R_a, bulk modulus, K, and bond covalency due to changes in: (a) coordination number (b) spin-state (c) ferric electronic ground state (d) expected changes in volume, V, with depth caused by (a), (b) and (c).

(Ringwood, 1970) increasingly violate Pauling's rules as the coordination number increases. Indeed, it is a necessary condition that as the coordination numbers of the cations increase, intimate contact between polyhedra also increases. For instance, all possible structures for MO_2 containing four co-ordinate cations involve only corner sharing between polyhedra, whereas the rutile MO_2 structure containing 6CN cations involves some edge sharing of polyhedra and the fluorite MO_2 structure containing 8CN cations requires that all polyhedral edges be shared. One consequence of increased polyhedral contact is to decrease the distance between neighbouring cations in the structure. This would suggest that metal-metal charge-transfer phenomena should become increasingly prevalent with depth as the coordination number increases.

For these reasons, interpretations of and speculations on the nature of mantle phases under conditions of pressure and temperature found in the mantle must attempt to break away from the conceptually simple theories based on the ionic model in order to consider covalent effects and non-localized phenomena such as the various charge-transfer processes alluded to in this study. Two more complex theories which have but little been applied to questions of the electronic properties of mantle materials are molecular orbital theory and band theory. The more quantitative aspects of band theory appear, at the present, to be inapplicable to even the simplest of silicate structures, whereas molecular orbital theory still retains aspects of the localized treatment, in that for the most part only one site in the structure is considered. If some compromise could be made between these two theories in that if molecular orbital theory could be extended to include all of the atoms in a unit cell or if band theory could be adapted to include some description of localized phenomena, then progress in the theoretical treatment of the electronic states in silicates may be possible.

(c) The Effect of Pressure on Ferrous Crystal-field Stabilization Energies

The ferrous crystal-field stabilization energy arises from the difference in energy of the lowest $3d$-level which is occupied by the sixth ferrous $3d$-electron and that of the baricentre of energy of all five $3d$-electron energy levels. Such values are of interest because they are important energy and entropy factors in considering the ordering of ferrous cations over the different sites in a crystal structure or the distribution of ferrous between minerals. For example, Burns (1969) attempted to rationalize cation distributions in minerals on the basis of crystal-field stabilization energies and site size only and agreement between the predicted and observed distributions was quite satisfactory.

In the course of this study, it has been shown that Mössbauer studies, as a function of temperature, can lead to estimates of the relative energy of the lower $3d$-levels which contain the extra electron. Therefore, if such measurements are coupled with optical absorption measurements which, for ferrous, lead to estimates of the relative energy of the very lowest level and the upper energy levels, values for all five energy levels, for Δ_0 and for the CFSE may be obtained. Such measurements can also be extended as a function of pressure, although with less certainty, to yield values of Δ_0 and CFSE as a function of pressure. Due to differences in site compressibility and in changes in site distortion with pressure, the variation with pressure of Δ_0 and CFSE for different ferrous sites in minerals may not be similar so that the distribution of ferrous between such sites may therefore be quite pressure sensitive.

From the variation of the olivine Mössbauer spectrum and absorption spectrum with pressure, there appears to be no large difference in Δ_0 and CFSE for the two sites with pressure which suggests that the ferrous distribution over the M1 and M2 sites in olivine will not be very pressure sensitive. Since the distribution is also not very temperature sensitive, the near random distribution of Mg and Fe over the two sites will probably persist over the whole P,T stability field of the olivine structure.

Ahrens (1972, 1973) has considered the distribution of ferrous iron between a pyralspite garnet, olivine and spinel and its geophysical significance. From the expected compressibility of the Fe^{2+}(VIII) site in garnet as opposed to that of Fe^{2+}(VI) sites in the $(Mg,Fe)_2SiO_4$ polymorphs and from the fact that the lower energy level splitting does not decrease greatly with pressure based on Mössbauer data, the CFSE of Fe^{2+} is predicted to increase faster in garnet than in these other mantle phases. For this reason, as pressure increases, the distribution of ferrous into garnet relative to olivine and spinel will be favoured. Since increasing temperature randomizes the distribution of Fe^{2+} and Mg between phases, the calculated distribution coefficients of Ahrens (1972) for various geotherms within the Earth based on thermodynamic data, appear to be consistent with these qualitative CFSE arguments for the distribution of ferrous between garnet and olivine.

SUMMARY

From Mössbauer measurements of minerals under high pressures and the trends of Mössbauer parameters with pressure, information has been obtained which is relevant to the electronic properties and the crystal chemistry of iron in oxides and silicates.

Trends of isomer shifts with pressure indicate increasingly complex covalent interactions in iron-oxygen bonds and suggest that theories of bonding for minerals based on the ionic model are less appropriate still at high pressure.

Ferrous quadrupole splitting trends with temperature and pressure coupled with optical absorption data lead to a complete energy level scheme for the $3d$-electron levels. Such knowledge is useful for understanding the distribution of ferrous cations over different sites in minerals as a function of pressure.

Only traces of the pressure-induced charge–transfer transition have been observed in ferric minerals. Such a transition would lead to a new electronic ground state for such minerals which may well have intriguing geophysical and geochemical implications.

ACKNOWLEDGEMENTS

The author would like to acknowledge the advice and support of Dr. R. G. Burns during the course of this project and Drs. D. J. Vaughan, J. A. Tossell, R. M. Abu-Eid and E. R. Whipple for useful discussions of aspects of this study. Ms. Michelle Huggins provided invaluable assistance with the preparation of the manuscript.

This project is supported by a grant from the National Science Foundation (U.S.A.) to R. G. Burns of M.I.T., where this work was done.

REFERENCES

Abu-Eid, R. M. (1974). *Ph.D. Thesis*, Massachusetts Institute of Technology, Cambridge, Massachusetts, U.S.A.

Ahrens, T. J. (1972). *Phys. Earth Planet. Ints.*, **5**, 267–281.

Ahrens, T. J. (1973). *Phys. Earth Planet. Ints.*, **7**, 167–186.

Anderson, D. L. and Anderson, O. L. (1970). *J. Geophys. Res.*, **75**, 3494–3500.

Annersten, H. (1974). *Amer. Mineral.*, **59**, 143–151.

Bancroft, G. M. (1974). *Mössbauer Spectroscopy: An Introduction for Inorganic Chemists and Geochemists*. McGraw-Hill, Maidenhead, Berkshire, England.

Burns, R. G. (1969). *Chem. Geol.*, **5**, 275–283.

Burns, R. G., Huggins, F. E. and Drickamer, H. G. (1972a). *24th International Geological Congress, Section Reports.*, **14**, 113–123.

Burns, R. G., Tossell, J. A. and Vaughan, D. J. (1972b). *Nature Phys. Sci.*, **240**, 33–35.

Danon, J. (1968). *Lectures on the Mössbauer Effect*. Gordon and Breach, Inc., New York.

Debrunner, P., Vaughan, R. W., Champion, A. R., Cohen, J., Moyzis, J. and Drickamer, H. G. (1966). *Rev. Sci. Instr.*, **37**, 1310–1315.

Drickamer, H. G., Bastron, V. C., Fisher, D. C. and Grenoble, D. C. (1970). *J. Sol. State Chem.*, **2**, 94–104.

Drickamer, H. G. and Frank, C. W. (1973). *Electronic Transitions and the High Pressure Chemistry and Physics of Solids*. Chapman and Hall, London.

Drickamer, H. G., Frank, C. W. and Slichter, C. P. (1972). *Proc. Nat. Acad. Sci.* **69**, 933–937.

Eibschutz, M. and Ganiel, U. (1967). *Sol. State Comm.*, **5**, 267–270.

Frank, C. W. and Drickamer, H. G. (1972). *J. Chem. Phys.* **56**, 3551–3565.

Hazen, R. M. and Burnham, C. W. (1973). *Amer. Mineral.*, **58**, 889–900.

Huebner, J. S. (1971). Chapter 5 in *Research Techniques for High Pressure and High Temperature*, edited by G. C. Ulmer. Springer–Verlag, New York.

Huggins, F. E. (1974). *Ph.D. Thesis*, Massachusetts Institute of Technology, Cambridge, Massachusetts, U. S. A.

Ingalls, R. (1964). *Phys. Rev.*, **133A**, 787–795.

Johnson, K. H. (1973). *Adv. Quant. Chem.*, **7**, 143–185.

Kaspar, H. and Drickamer, H. G. (1968). *Proc. Nat. Acad. Sci.*, **60**, 773–775.

Lewis, Jr., G. K. and Drickamer, H. G. (1968). *J. Chem. Phys.*, **49**, 3785–3789.

Low, W. and Zeira, S. (1972). *Amer. Mineral.*, **51**, 1115–1124.

Lyubutin, I. S. and Dodokin, A. P. (1971a). *Sov. Phys. Cryst.* (trans.), **15**, 936–938.

Lyubutin, I. S. and Dodokin, A. P. (1971b). *Sov. Phys. Cryst.* (trans.), **15**, 1091–1092.

Lyubutin, I. S., Dodokin, A. P. and Belyaev, L. M. (1970). *Sov. Phys. Sol. State* (trans.), **12**, 1100–1102.

Mao, H. K. (1973). *Ann. Rep. Dir. Geophys. Lab.*, **72**, 552–564.

Mao, H. K. and Bell, P. M. (1972). *Ann. Rep. Dir Geophys. Lab.*, **71**, 520–524.

Mössbauer, R. L. (1958). *Z. Physik*, **151**, 124–143.

Press, F. (1970). *J. Geophys. Res.*, **75**, 6575–6581.
Ringwood, A. E. (1966). *Geochim. et Cosmochim. Acta*, **30**, 41–104.
Ringwood, A. E. (1970). *Phys. Earth Planet. Ints.*, **3**, 109–155.
Sawatzky, G. A., Van der Woude, F. and Morrish, A. H. (1969). *Phys. Rev.*, **183**, 383–386.
Shannon, R. D. and Prewitt, C. T. (1969). *Acta Cryst.*, **B25**, 925–946.
Simmons, M. G. and Wang, H. (1971). *Single Crystal Elastic Constants and Calculated Aggregate Properties: A Handbook*. M.I.T. Press, Cambridge, Massachusetts.
Slichter, C. P. and Drickamer, H. G. (1972). *J. Chem. Phys.*, **56**, 2142–2160.
Stone, A. J. (1967). Appendix to Bancroft et al. *J. Chem. Soc.*, **A**, 1966–1971.
Tossell, J. A., Vaughan, D. J. and Johnson, K. H. (1973). *Nature Phys. Sci.*, **244**, 42–45.
Travis, J. C. (1971). Chapter 4 in *An Introduction to Mössbauer Spectroscopy*, ed. L. May. Plenum Press, New York.
Vaughan, D. J. and Tossell, J. A. (1973). *Science*, **179**, 375–377.
Vaughan, R. W. and Drickamer, H. G. (1967). *J. Chem. Phys.*, **47**, 1530–1536.
Vinokurov, V. M., Gainullina, N. M., Evgrafova, L. A., Nizamutdinov, N. M. and Suslina, A. M. (1971). *Sov. Phys. Cryst.* (trans.), **16**, 262–265.
Wang, P. J. and Drickamer, H. G. (1973). *J. Chem. Phys.*, **59**, 713–717.
Wertheim, G. K. (1964). *Mössbauer Effect: Principles and Applications*, Academic Press, New York.
Whipple, E. R. (1973). *Ph.D. Thesis*, Massachusetts Institute of Technology, Cambridge, Massachusetts, U.S.A.

Absorption Spectra of Transition Metal-Bearing Minerals at High Pressures

R. M. Abu-Eid
Department of Earth and Planetary Sciences
Massachusetts Institute of Technology
Cambridge, MA 02139 *U.S.A.*

INTRODUCTION

Metals of the first transition series often occur in natural minerals, and the physical and chemical properties of many minerals depend on the nature and state of transition metal ions which are present.

The electronic absorption spectral technique is one of the most powerful tools for studying the crystal chemistry of transition metal ions. In the last decade it has been used successfully in mineralogy and crystal chemistry for such studies as: the determination of cation valence state (Burns, 1965, 1969; Faye *et al.*, 1968), the interpretation of cation site symmetries (Burns, 1969, 1970; Manning, 1967a, b), the explanation of the cause of colour and pleochroism of minerals (Burns, 1966; Manning, 1969; Hush, 1967), the site preferences of cations (Burns and Strens, 1967; Burns and Fyfe, 1967; Keester and White, 1966), and the estimation of the degree of covalency of the metal-ligand bonds (Manning, 1970; Moore and White, 1972). Spectra obtained at elevated pressures should also provide valuable information about the Earth's interior, where electronic transitions such as spin-pairing (Fyfe, 1960; Strens, 1969; Burns, 1969; Gaffney, 1973) or pressure-induced reduction (Drickamer *et al.*, 1969; Burns *et al.*, 1972b) of transition metal ions to lower oxidation states are expected to occur. High pressure spectral techniques enable such transitions to be easily identified, and phase transformations in mineral phases may also be detected (Abu-Eid *et al.*, 1973; Hazen and Abu-Eid, 1974; Abu-Eid and Hazen, 1974; Suchan and Drickamer, 1959).

Besides the geochemical and mineralogical applications, optical absorption spectroscopy has also been used to study radiative conductivity and the variation of thermal conductivity within the Earth (Pitt and Tozer, 1970a, b; Fukao *et al.*, 1968; Shankland, 1970, 1972).

Since the volumes of the polyhedra containing transition metal ions are closely related to the spectral parameter 10Dq, variation of this parameter with pressure can provide information on site compressibilities (Tischer and Drickamer, 1962).

Many assignments of charge-transfer and crystal field bands have been the subject of controversy in the mineralogical literature (Dowty and Clark, 1973a, b; Burns and Huggins, 1973; Burns *et al.*, 1972a, b, c; Burns *et al.*, 1973). Since crystal field and charge-transfer bands are expected to show opposite energy shifts with pressure, high pressure spectra can be used to distinguish them.

In this study, results of high pressure absorption spectral measurements are presented for lunar pyroxene, blue omphacite, fayalite, andradite, uvarovite, piemontite, crocoite, and vanadinite minerals. These minerals display many of the high pressure effects discussed above.

THEORETICAL BASIS

The spectra of minerals containing transition metal ions are dominated in general by two types of feature. Firstly, there are crystal field bands which are related to electronic transitions between the *d*-levels of transition metal ions and which appear, in general, in the energy region 4,000–30,000 cm^{-1}; these bands are characterized by their relatively low intensities and narrow widths. Secondly, there are charge-transfer bands which are of two categories: first, ligand → metal or metal → ligand, and second, metal → metal charge-transfer transitions (Phillips and Williams, 1966). The first category of charge-transfer band arises due to electron transfer from energy levels of highly ligand character to levels of mostly metal character and *vice versa* (McClure, 1959; Lever, 1968). These, in general, are very intense and usually occur at energies higher than those of crystal field bands above 25,000 cm^{-1}. The second category of charge-transfer transition is known as metal → metal charge-transfer which arises from transfer of electrons between the *d*-levels of adjacent metal ions having different oxidation states, e.g. $Fe^{2+} \rightarrow Fe^{3+}$ (Hush, 1967).

(i) Crystal Field Theory

In the crystal field formulation, the negative and positive ions are represented as point charges and the electrostatic potential, V_1, associated with an electron at a point, P, at a distance, d, from ligand (1) and r, from the central metal ion is given by

$$V_1 = +\frac{Ze^2}{d} \tag{1}$$

where Z is the charge on the ligand (Figure 1).

Considering the particular case of an octahedral array of six negative point charges each at distance, R, from the central ion, the total potential, V, at P, arising from the six ligands has been calculated by Dunn *et al.* (1965) and Hutchings (1964), and is expressed as:

$$V = \frac{6\,Ze^2}{R} + \frac{35\,Ze^2}{4\,R^5}\left(x^4 + y^4 + z^4 - \tfrac{3}{5}r^4\right) \tag{2}$$

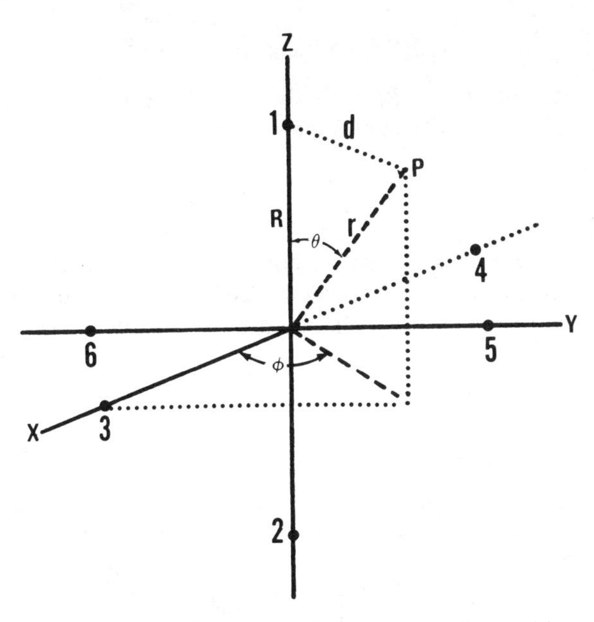

Figure 1. Octahedral array of ligands and the position of an electron at point P described in spherical coordinates in three dimensional space.

The first term, $6\,Ze^2/R$, is usually neglected in crystal field approximation since it is spherically symmetrical and does not participate in the splittings of *d*-orbitals. The quantity, $35\,\mathrm{Ze}^2/4\,\mathrm{R}^5$, is usually abbreviated to D; it follows then that the potential at any point at $\leqslant R$ distance from the central ion can be expressed as:

$$V = D(x^4 + y^4 + z^4 - \tfrac{3}{5}r^4) \tag{3}$$

After establishing the potential, V, the perturbations on the wave functions of *d*-electrons are taken into account. This is done by integrating the radial and angular wave function pairs to obtain integrals of the form:

$$\int_0^\infty \int_0^\pi \int_0^{2\pi} R_{nl}(r)\Theta_{lm}(\theta)\Phi_m(\phi)\, V R_{nl'}(r)\Theta_{l'm'}(\theta)\Phi_{m'}(\phi)r^2 \sin\theta\, \mathrm{d}r\, \mathrm{d}\theta\, \mathrm{d}\phi$$

(Dunn *et al.*, 1965), where $R_{nl}(r)$ is the radial part of a d wave function and $\Theta_{lm}(\theta)\Phi_m(\phi)$ are the angular parts or spherical harmonics. The secular determinant is constructed by solving the above integrals.

For a d^1 electron, a square matrix 5×5 is established from the five $\mathrm{M_L}$ values. Solution of the part of the integral dependent on Φ will reduce the 25 elements of the secular determinant to seven. However, solutions of the Θ integral for different values of m and m' need tedious mathematical treatments and have been given in many references (Dunn *et al.*, 1965; Figgis, 1966; Hutchings, 1964); thus, for example, for $m = m' = 0$, solution of Θ_{lm} and Φ_m integrals is given as:

$$<0|(x^4+y^4+z^4)|0> = \tfrac{5}{7}r^4 \tag{4}$$

The radial part integral $\int_0^\infty R_{nl}^2(r)r^4r^2\mathrm{d}r$ is insoluble (Dunn *et al.*, 1965; Figgis, 1966), and it is replaced by the parameter q where:

$$q = \frac{2}{105}\int_0^\infty R_{nl}(r)r^4r^2\mathrm{d}r \tag{5}$$

This method of calculation permits the construction of energy level diagrams for transition metal ions of various d-electronic configurations; these are frequently called Orgel energy level diagrams (Orgel, 1966). The splitting energies between these levels are expressed as multiples of Dq and the relations among the Dq values for octahedral, tetrahedral, and cube symmetries are:

$$\mathrm{Dq\ (oct.)} = -\frac{9}{4}\mathrm{Dq\ (tet.)} = -\frac{9}{8}\mathrm{Dq\ (cube)}$$

From the above outline of the theoretical aspect of crystal field theory, we should keep in mind the important equivalence:

$$\mathrm{Dq} \equiv \frac{35\,Ze^2}{4\,R^5}\left[\frac{2}{105}\int_0^\infty R_{nl}^2(r)r^4r^2\,\mathrm{d}r\right] \tag{6}$$

From this relation, it is evident that Dq is inversely proportional to the fifth power of the average metal-ligand bond distance, R.

(ii) Ligand Field Theory (LFT) and Racah Parameters

The major modifications of crystal field theory that take into account mixing of cation and ligand orbitals involve using the parameters of $3d$ interelectronic interaction as variables rather than constants (equal to the values for the free ion). Of these parameters, the most significant are the interelectronic repulsion parameters which are frequently known as the Racah parameters B and C. These are related to the covalency of the metal-ligand bond, and they are derived from calculations of the radial wave functions of the electrons involved in the bonding between transition metal ions and the surrounding ligands (Racah, 1942a, b, 1943, and 1949). The B and C energy values are inversely proportional to the spatial extent of the ligand and cation orbitals, hence the larger their value the more ionic is the metal-ligand bond.

The perturbation of the d-wave functions of the central ion, using ligand field models, is mathematically expressed in a secular determinant or energy matrix, the elements of which are given in terms of the crystal field parameter Dq and Racah parameters B and C.

Tanabe and Sugano have derived such secular determinants and from the solutions, they obtained the energies of electronic transitions from the ground state to the higher energy states. This has been done in detail in two classic articles by Tanabe and Sugano (1954a, b) who have also constructed energy level diagrams which depend upon both interelectronic repulsion parameters and the crystal field. In these diagrams, the energies of the levels of a d^n system are plotted in B units as vertical coordinates, and the horizontal coordinate is Dq/B. These diagrams are useful for recognizing the type of electronic transitions, forbidden or allowed, and for rough estimation of the energy of each transition in various d^n systems, i.e. d^1–d^9.

Ligand field theory is a modification of crystal field theory which takes into account the various degrees of overlap between the d-orbitals of the transition metal ion and the ligand orbitals. When there is mixing between the metal and ligand electrons, LFT will have essentially a molecular orbital formulation; however, in the case of small to moderate amounts of overlap it will be basically an adjusted crystal field model (Cotton and Wilkinson, 1966).

For the purposes of this study, more emphasis is placed on crystal field and ligand field models since the degree of covalency of the metal-ligand bond in most mineral phases under investigation is small to moderate.

For qualitative spectroscopic studies, crystal field theory is quite successful; however, for quantitative work it is inadequate, primarily because it does not take into account the possibility of some covalent character of the metal-ligand bond.

(iii) Molecular Orbital Theory and Charge-transfer Spectra

Crystal field models cannot explain all the observed spectral features, and there is a great deal of experimental evidence demonstrating limits to its usefulness (McClure, 1959; Cotton, 1971), which are mainly due to the neglect of the overlap of metal and ligand orbitals.

In the molecular orbital model, ligand orbitals of appropriate symmetries are mixed with the d-orbitals. Consider a strong octahedral crystal field in which the d-orbitals are split into e_g and t_{2g}. Of the p-orbitals of each ligand one is directed along the bond and gives rise to a σ-orbital when combined with a cation orbital. The other two p-orbitals, which are perpendicular to the bond, give rise to π-orbitals when combined with cation orbitals. These two types of orbitals, σ and π, may also be classified into e_g and t_{2g}.

To form molecular orbitals, the atomic orbitals of the ligand and cation orbitals should have similar symmetry, e.g. p-orbitals of e_g symmetry will combine with e_g d-orbitals and p-orbitals of t_{2g} symmetry will combine with t_{2g} d-orbitals. (McClure, 1959; Lever, 1968). Since π-orbitals are not directed along the bond whereas σ-orbitals are so directed, then the former, π, will generally form weaker bonding and antibonding orbitals than do σ.

In both crystal field theory and molecular orbital theory, the difference in energy between e_g and t_{2g} orbitals is a result of the same geometrical factor, the octahedral ligand field, which gives rise to an electrostatic effect in the former and an antibonding effect of σ-ligand orbitals in the latter (McClure, 1959; Cotton, 1971).

To understand the origin of the charge-transfer bands and how they are related to crystal field transitions, Figure 2 is drawn schematically to illustrate both types of transition for an octahedral complex. In this diagram, the energy

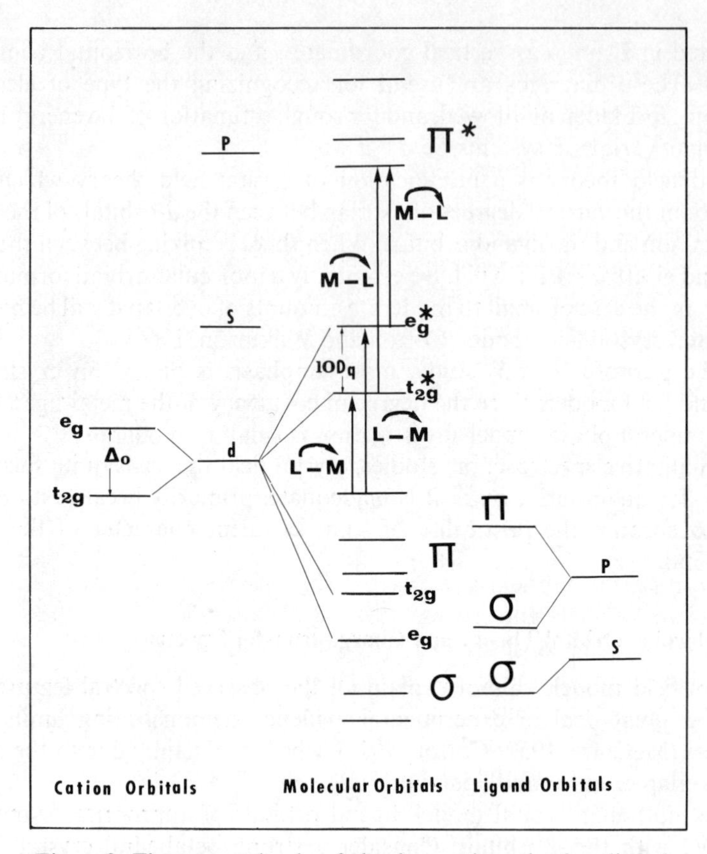

Figure 2. The energy levels of the free cation, the free ligands, and the molecular unit (ML_6), drawn schematically, based on the molecular orbital model.

levels of the free cation and the free ligand are shown in both sides. The *d*-orbitals are split into two sets of levels e_g and t_{2g}. The central portion of the diagram shows the energy levels of the molecular units (ML_6). Molecular orbitals marked by an asterisk are those of antibonding character, whereas others are either bonding or non-bonding. The levels below t_{2g}^* are filled with electrons and the levels above e_g^* are empty; however, t_{2g}^* and e_g^* could be filled or empty or partly filled depending on the number of *d*-electrons of the transition metal ion. The e_g^* and t_{2g}^* levels have mostly metallic character, whereas the levels immediately above and below them have mostly ligand character. Transfer of electrons from the ligand orbital to either t_{2g}^* or e_g^* will give rise to ligand-to-metal ($L \rightarrow M$) charge-transfer bands and from t_{2g}^* or e_g^* to the above levels will produce metal-to-ligand ($M \rightarrow L$) charge-transfer bands (Phillips and Williams, 1966).

A further type of electron transfer is that between the t_{2g}^* and e_g^* level of one cation to that of another neighbouring cation of different oxidation state.

This is called a metal-to-metal electron transfer (M $\rightarrow$ M) or intervalence transfer (Hush, 1967). Metal-metal electron transfer can occur either between the energy levels of cations of the same element but of different oxidation states, e.g. $Fe^{2+} \rightarrow Fe^{3+}$, which is called homonuclear intervalence transfer; or between *d*-levels of cations of different elements which is known as heteronuclear intervalence transfer (Hush, 1967).

The intensities and energies of charge-transfer bands arising from metal–metal electron transfer are dependent on the extent of delocalization of electrons from one metal nucleus to another, this delocalization will lead to either direct overlap of the orbitals of the two metal atoms and the overlap integrals in this case are designated as ΔMM, or metal-ligand-metal overlap through σ or π metal-ligand bonding and their integrals are designated ΔMLM (Hush, 1968). The values of the overlap integrals ΔMM or ΔMLM are also dependent on the interatomic distances, the nature of the ligand, and the crystal geometry.

(iv) Effect of Pressure on Crystal Field and Charge-transfer Bands

In this work, crystal field bands are described in terms of the crystal field and ligand field models, and charge-transfer bands are treated using a molecular orbital approach. The most significant spectal parameters are the band energies, intensities and widths, and their variation with pressure.

Of the crystal field transitions, the energies of some depend solely on Dq, some depend on B, and some on both B and Dq. From equivalence (6) the relation between Dq and the metal-anion distance, R, can be given as: $\mathrm{Dq} \propto 1/R^5$. With increasing pressure, assuming a typical hydrostatic condition, R is expected to decrease and then Dq will increase, so we expect to observe a significant increase in the energies of those bands dependent only on Dq. However, bands which are dependent on B and C, are expected to show very small negative energy shift (i.e. red shift).

Increasing covalency causes B to decrease with pressure, and the effect is usually described in terms of the nephelauxetic ratio $\beta = B/B_0$, where B and B_0 are values appropriate to the complex and the free ion respectively (König, 1971; Reiner, 1969; Lever, 1969).

Spectral bands whose energies depend on both B and Dq need careful treatment since B and Dq are expected to show opposite trends with increasing pressure; however, such bands will usually show a blue shift because Dq is more sensitive to the shortening of metal-ligand bond distances than B.

On the other hand, the effect of pressure on charge-transfer bands can be viewed in three ways:

Firstly, assuming that the *d*-orbitals (e_g and t_{2g} in C.F.T. or e_g^* and t_{2g}^* in M.O.T.) are the major orbitals affected by pressure, then the increase in crystal field splitting parameter 10 Dq is due mostly to raising the e_g^* energy level relative to t_{2g}^* (Figure 2 above). As a result of this shift with pressure energies of charge-transfer transitions M $\rightarrow$ L and L $\rightarrow$ M are expected to decrease with pressure.

Secondly, using the band model, the effect of pressure on charge-transfer bands may be related to closing the energy gap between valence and conduction bands (Drickamer and Frank, 1973). Drickamer (1965) has indicated that with increasing pressure the general tendency is to broaden the energy bands and to lower the energy of the conduction band since it is more sensitive to compression than the valence band. The net effect will be a shift of the absorption edge or the charge-transfer band to lower energy. This is often true especially especially for direct transitions ($\Delta K = 0$). This phenomenon is illustrated in Figure 3. Reducing the energy gap with pressure may also lead to overlap of the highest filled band with the empty conduction band, and may lead to a metallic behaviour of the solids. Band theory also predicts decreasing the resistivities of metals with pressure, and hence increasing their electrical conductivities (Drickamer and Frank, 1973); this is due mostly to reducing the amplitude of lattice vibrations.

Thirdly, using the schematic representation of the potential energies of the ground and excited state (Figure 4), Hush (1968) and Drickamer *et al.* (1972)

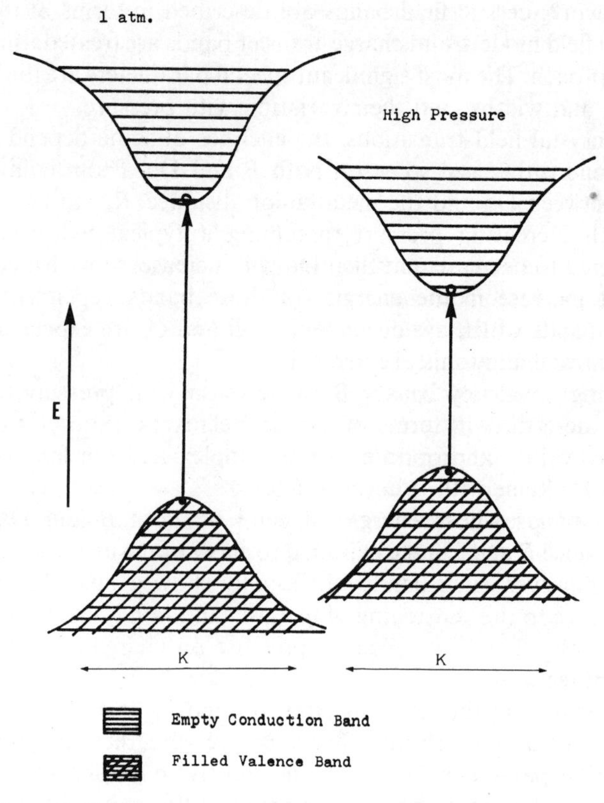

Figure 3. Schematic representation of conduction and valence bands at 1 atm. and high pressure for a direct transition (K is the propagation vector of the wave function).

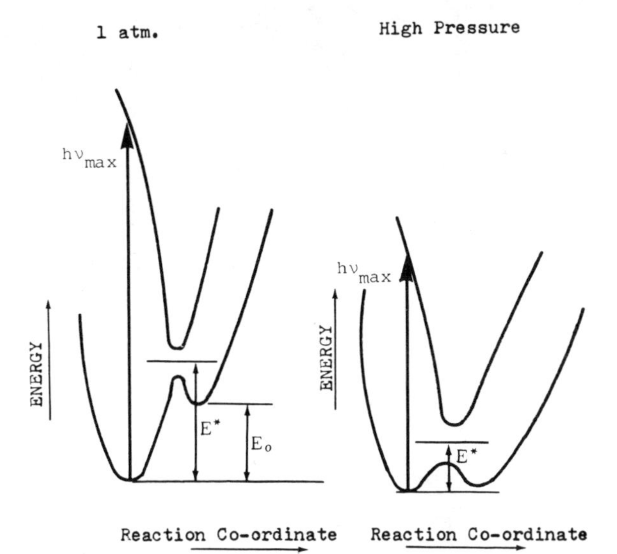

Figure 4. The potential energy wells of the ground and excited state at 1 atm. and high pressure are shown schematically. [Data modified from Hush (1968) and Drickamer *et al.* (1972)].

have expressed the energy of the maximum optical absorption as

$$h\nu_{max} = E_0 + 1/2\Delta\Omega^2\Delta \tag{7}$$

where E_0 is the overall energy difference between the ground and excited state, Δ is the vector difference of equilibrium positions of atoms in the final and initial state of the system, and Ω is the frequency tensor associated with the vibration of all atoms in the medium.

Drickamer *et al.* (1972) have shown that there is a vertical displacement of the potential wells with pressure, i.e. decreasing E_0, Δ is expected to decrease due to the horizontal displacement of the configuration coordinate, and Ω will obviously decrease with pressure due to inhibition of atomic vibration, then the net effect is to decrease the energy of the maximum optical absorption $h\nu_{max}$, i.e. the energy of charge-transfer band.

APPARATUS AND EXPERIMENTAL METHODS

The apparatus used was a standard Cary 17 Spectrophotometer with a newly designed optical attachment, and diamond anvil cells of UP, UXP and PP types (Mao, 1967; Bassett *et al.* 1967; Merrill, 1973; Bassett, 1974). The UP and UXP cells were used for powder diffraction work. The PP type can be mounted on a modified goniometer head, and is used for single crystal X-ray work and for optical absorption studies. In both cases, the gasketing technique developed by Van Valkenburg (1965) is used.

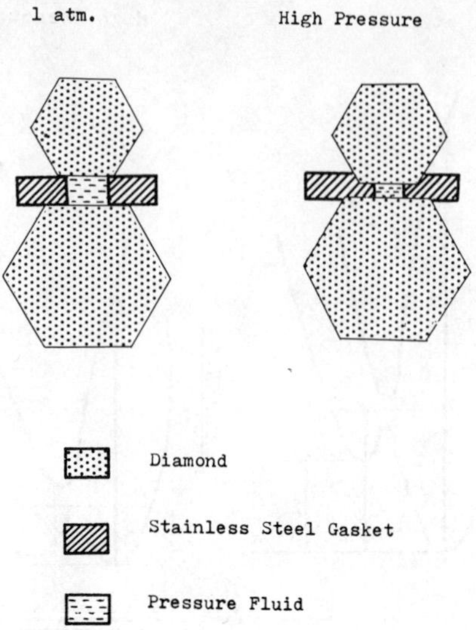

1 atm. High Pressure

Diamond

Stainless Steel Gasket

Pressure Fluid

Figure 5. The gasketing method: A stainless steel gasket between the two diamond anvils at 1 atm. and at high pressure.

The crystal fragment is oriented by means of a spindle and transfer stages (Wilcox, 1959; Chao *et al.*, 1970); it is then polished until flat parallel faces are obtained perpendicular to the desired axis. A stainless steel gasket having a cylindrical hole of about 300 μm at its centre is placed on the flat face of the lower anvil (Figure 5). The hole in the gasket is filled with a liquid of known freezing pressure, and the crystal is dropped in the liquid with the required axis perpendicular to the anvil faces.

The pressure is obtained by driving the two anvil faces against the metal gasket which in turn transfers the pressure to the liquid and then to the sample. The ultimate pressure that can be reached is about 80 kilobars due to the limitation of the gasket yield strength and solidification pressure of the liquid.

The pressure in the single crystal cell is calibrated by employing different liquids of known freezing pressure (Forman *et al.*, 1972). These liquids are:

Material	Phase transition	Pressure (kb)
CCl_4	L–I	1.3
H_2O	L–VI	9.6
C_2H_5Br	L–I	18.3
nC_7H_{16}	L–I	11.4
Isoamyl alcohol	L–S	37
2-Methyl alcohol	L–S	25
Methyl & Ethyl alcohol	L–S	80

Another pressure indicator is the transformation of KCl (I) to KCl (II) at about 19.3 kb.

For powdered specimens, the pressure is calibrated using NaCl as an internal standard. The sample and NaCl, 1:2 weight ratio, are mixed and ground in an agate mortar for a few hours until a very fine powder (grain size $< 5\,\mu m$) is obtained, then the mechanical mixture of the sample and NaCl are mounted on the lower diamond anvil face and the pressure is obtained by driving the two anvils against each other. A lead glass collimator is used to collimate the X-rays to about 100 μm on the central region of the sample, and the pressure is known from determining the diffraction lines of NaCl at 1 atm. and at high pressures and then using either Decker's or Weaver's equation of state for NaCl (Decker, 1966; Weaver *et al.*, 1971).

The Cary 17 spectrophotometer provides a source of monochromatic radiation of a wavelength from 2200–360 nm. The optical system has been modified to enable the measurement of spectra of small crystal fragments ($\sim 50\ \mu m$), or micro-powdered specimens ($\sim 100\ \mu m$) while they are subjected to pressures of up to 250 kilobars.

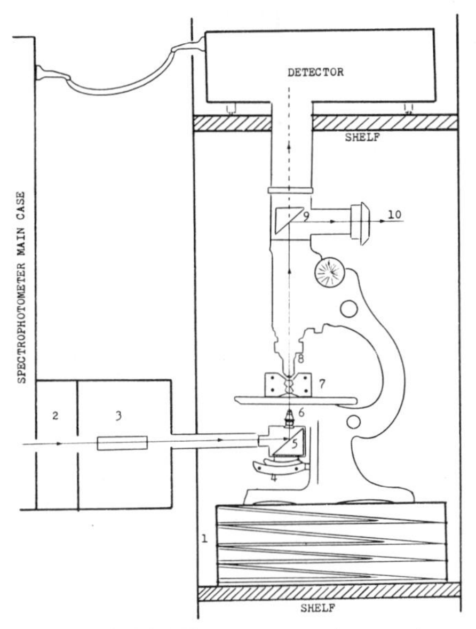

Figure 6 (a). The spectrophotometer optical attachment designed to enable the measurement of spectra of micro-samples ($\geqslant 50\ \mu m$): 1. Jack, 2. Beam divider, 3. Polarizer, 4. X – Y mini-transfer stage, 5. Aluminium mirror, 6. Focusing lens (UMK50), 7. Diamond Cell, 8. UMK50 objective lens, 9. Mirror-shutter combination, 10. lateral eye-piece.

Figure 6 (b). A photograph of the spectrophotometer optical attachment.

To obtain the spectra of such small samples at these elevated pressures, a new optical attachment has been designed and built which is similar to that described by Bell and Mao (1972b). This is illustrated in Figures 6(a) and 6(b). In the new optical system the two light beams, reference and sample, are polarized by means of calcite prisms, and then reflected and focused by means of UMK50 objectives. The focused beam is centred using a mini $x-y$ transfer stage below the reflecting-focusing box. After the sample beam has been focused on the sample in the diamond cell, it diverges and then converges again using another UMK50 objective. The reference beam does the same without going through the sample. The two beams are then either reflected horizontally to the eyepieces, or allowed to travel vertically into the detector.

One of the difficulties encountered is the interference of diamond absorption in the visible region. Figure 7 shows the spectra of a pair of diamond anvils used in the pressure cell. The sharp peaks in the visible region around 415, 403, 394 and 384 nm are related to defect centres in the diamond structure containing trapped nitrogen in various forms (Clark *et al*., 1956; Raal, 1959). To reduce this problem, another pressure cell containing diamond anvils of similar thicknesses as those in the sample beam were employed in the reference beam; the two diamond anvils then absorb similarly and offset the absorption of those used in the sample beam.

Another serious difficulty is the complexity of the natural minerals investigated in this study which contain transition metal ions, often of more than one valence, in multiple distorted or regular sites.

Besides these complications that arise from the diamond absorption and the complex spectra of minerals, the size of the samples used in this investigation presents another problem. Working with small samples (50–100 μm in diameter and about 5 μm thickness) will not permit the resolution of broad weak bands

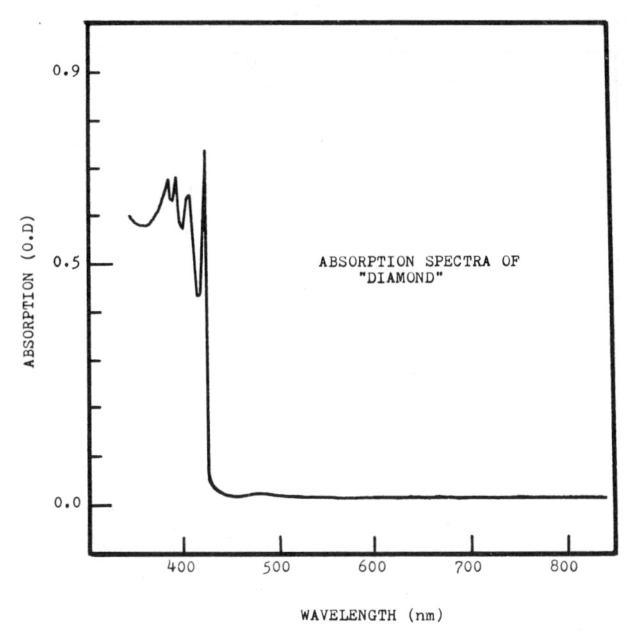

Figure 7. The visible spectra of a pair of diamond anvils used in the pressure cell.

and because of this difficulty, only sharp and intense bands are studied, especially when powdered specimens are being used.

The pressure gradient on the diamond anvil produces a non-hydrostatic pressure condition on the sample; to overcome this problem the light beam is focused on the central region (75–100 μm) where the pressure gradient is minimal. On the other hand, using the single crystal method, this problem is eliminated as long as the fluid pressure medium remains at least partly in the liquid phase.

Extrusion and thinning of the sample with increasing pressure is another problem which exists in every anvil type press; this can be solved by measuring the spectra starting with the highest pressure and then reducing the pressure slowly for lower pressure measurements. This procedure is useful only when there are no irreversible processes taking place.

RESULTS

The effect of pressure has been measured on the absorption spectra of Ti^{3+}, Fe^{2+}, Fe^{3+}, Mn^{3+}, Cr^{3+}, Cr^{6+}, and V^{5+} cations contained in lunar pyroxenes, fayalite, andradite and blue-omphacite, piemontite, uvarovite (synthetic), crocoite, and vanadinite, respectively.

(i) Ti^{3+} in Lunar Pyroxene

Figure 8 shows the polarized spectra of an Apollo 17 lunar pyroxene sample

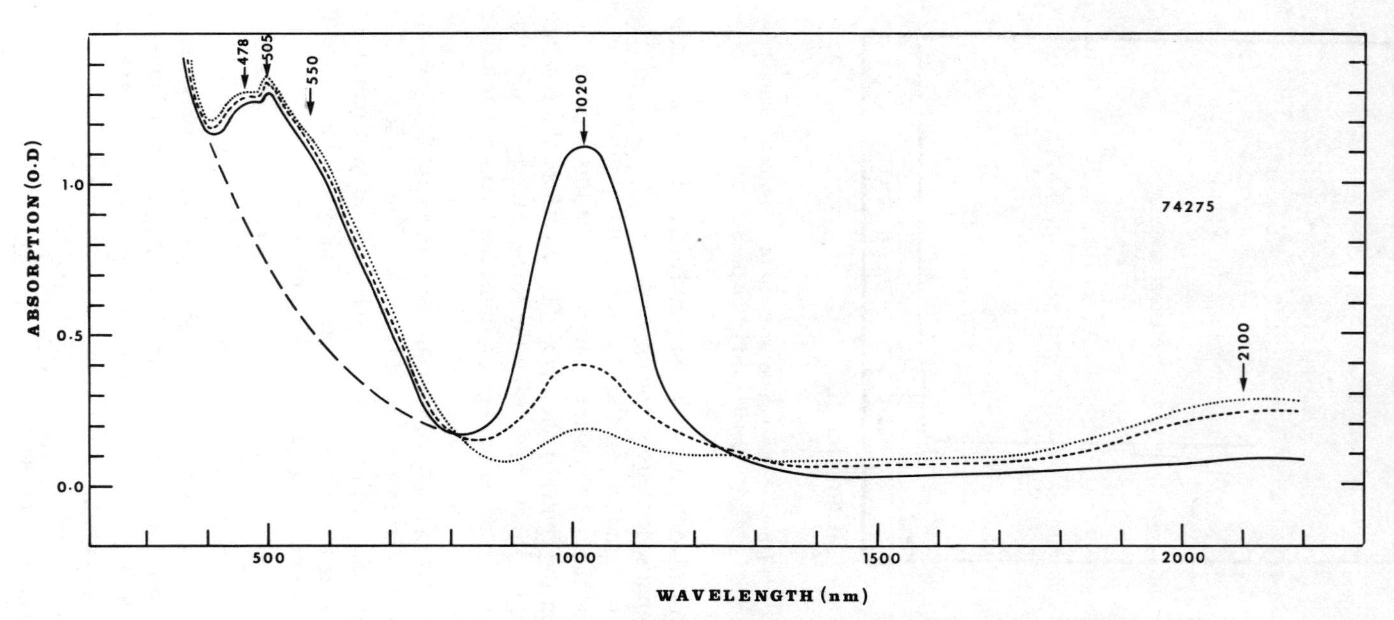

Figure 8. Polarized absorption spectra of lunar pyroxene single crystals in rock 74275. . . . . . *Z* spectrum; – – – – – *Y* spectrum; ———— *X* spectrum.

74275,85 (Sung *et al.*, 1974) containing Ti as TiO_2 6.05 per cent, Cr_2O_3 0.64 per cent, FeO 10.03 per cent.

The broad bands in the infrared region around 1020 nm and 2100 nm are due to spin-allowed transitions in Fe^{2+}. The sharp peak at 505 nm and the broad shoulder around 550 nm are related to spin-forbidden transitions in Fe^{2+}. However, there is a debate on the assignment of the band at 478 nm. Burns *et al.* (1972a, b, c, 1973) have assigned it as a crystal field transition in Ti^{3+}, whereas Dowty and Clark (1973) and Bell and Mao (1972 a) have assigned this band as a charge-transfer band, i.e. due to electron transfer between either Fe^{2+} and Ti^{4+} or Fe^{2+} and Fe^{3+}.

Since the spin-allowed transitions of Ti^{3+} ($^2T_{2g} \rightarrow {}^2E_g$) and Fe^{2+} ($^5T_{2g} \rightarrow {}^5E_g$) are only dependent on 10 Dq, then their energies should be sensitive to variations of metal-ligand bond distances. With increasing pressure, and then shortening the interatomic bond distances, we expect these bands to shift to higher energies if they are indeed crystal field transitions. However, charge–transfer bands are expected to show red shift with pressure. Figure 9 shows the spectra of a lunar pyroxene single crystal which was mounted in a pressure fluid using the gasketing method described earlier in this article. At elevated pressure, both bands at 1020 and 478 nm (Fe^{2+} and Ti^{3+} spin-allowed transitions) shift to higher energies (Figure 9); this trend in energy shift is consistent with the earlier prediction and the assignment of these bands (Burns *et al.*, 1972a, b, 1973) as spin-allowed transitions in Fe^{2+} and Ti^{3+}.

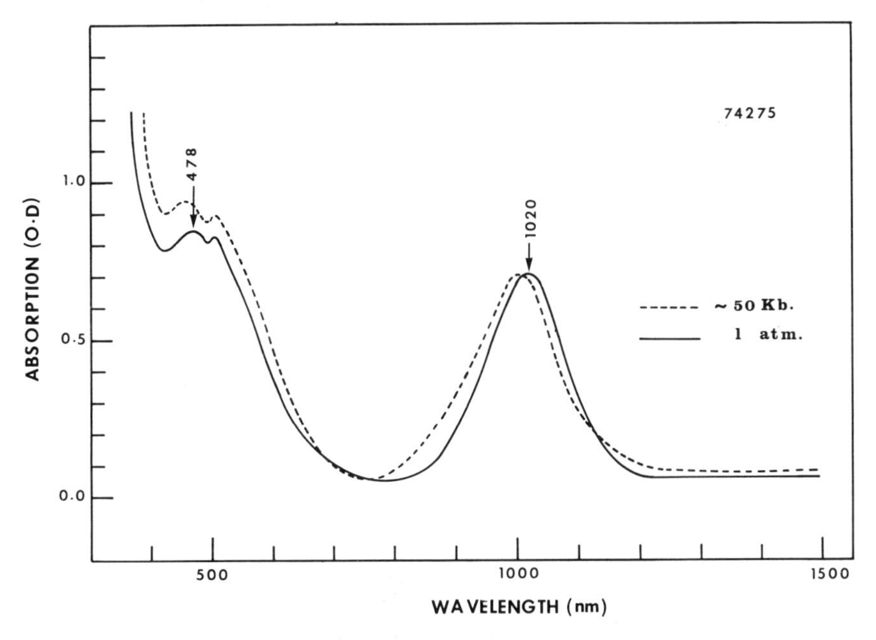

Figure 9. Pressure induced shift of obsorption bands in the α-spectra of pyroxene 74275.

On the other hand, the weak bands at 550 and 505 nm show negligible energy shift with pressure. This observation may support their assignment as spin-forbidden transitions in Fe^{2+} (these two transitions may be related to ${}^5T_{2g} \rightarrow {}^3T_{1g}$, ${}^5T_{2g} \rightarrow {}^3T_{2g}$, or ${}^5T_{2g} \rightarrow {}^1A_{1g}$). Since Fe^{2+} spin-forbidden bands depend mostly on B, and to a lesser extent on Dq, and since both of these parameter have opposite trends of energy shifts with pressure, such bands should not shift significantly with pressure.

(ii) Fe^{2+} in Fayalite

The α-spectra at 1 atm. and at pressure around 40 kb. are shown in Figure 10. The two bands at energies of 7350 cm^{-1} and 11100 cm^{-1} are related to Fe^{2+} spin-allowed transitions, ${}^5T_{2g} \rightarrow {}^5E_g$, in the distorted M_1 site (Burns, 1970). With increasing pressure, the band at 7350 cm^{-1} shifts slowly to higher energy, whereas the band at 11100 cm^{-1} shifts slowly to lower energy. Constructing energy level diagrams for the measured spectra at 1 atm. and 40 kb predicts a small increase in 10 Dq with a decrease in the energy separation between e_g levels and between t_{2g} levels (Figure 11). This suggests that a more regular octahedral site is obtained at higher pressures. At 1 atm. 10 Dq or Δ_0 has been evaluated accurately as 8489 cm^{-1} using the energy separation (Δ_1 and Δ_2) between t_{2g} levels obtained from high temperature Mössbauer spectra (Huggins, 1974; Huggins and Abu-Eid, 1974). The estimated change in Δ_0 in 40 kilobars for the Fe^{2+} ion in the M_1 site is about 100 cm^{-1}.

The sharp peaks in the visible region are assigned to Fe^{2+} spin-forbidden

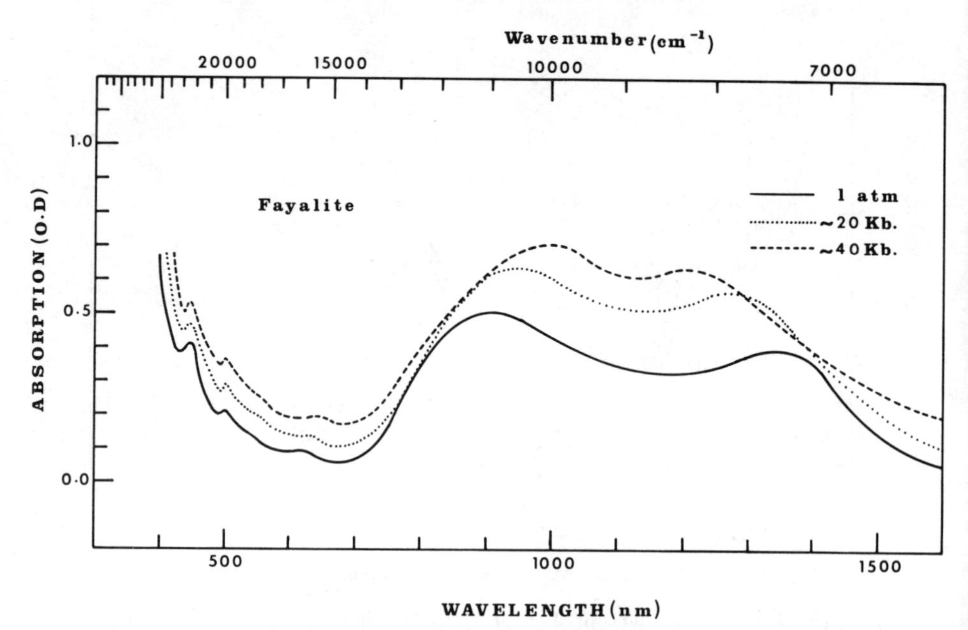

Figure 10. α-Spectra of a fayalite single crystal at 1 atm., 20 and 40 kb.

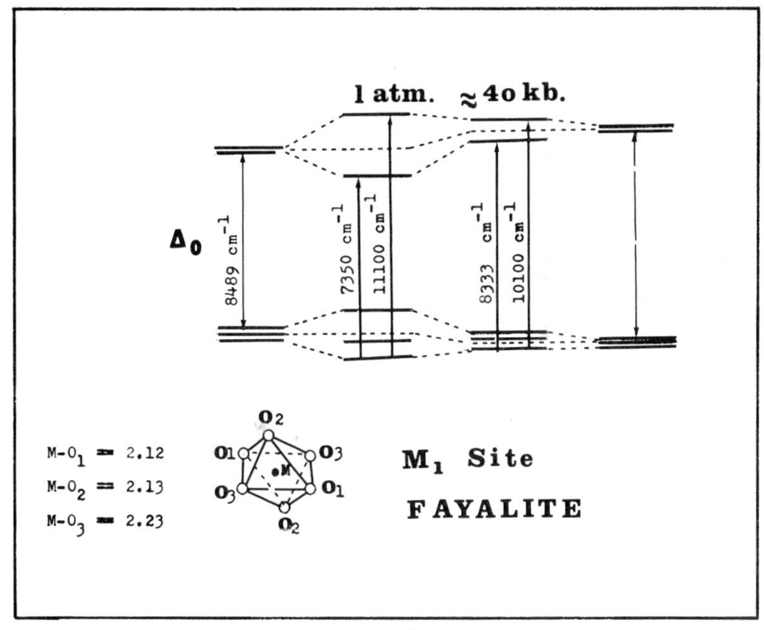

Figure 11. Energy level diagram for Fe^{2+} in the M1 site of fayalite, constructed from the spectra measured at 1 atm. and 40 kb.

transitions; they show negligible energy shifts with increasing pressure like those of Fe^{2+} spin-forbidden transitions in lunar pyroxenes. However, the band around 615 nm shifts to lower energy, suggesting that it may be a charge-transfer band which arises from electron transfer between Fe^{2+} and Fe^{3+} cations (Fe^{3+} exists in trace amounts in olivine). The effect of pressure on the charge-transfer transitions $Fe^{2+} \rightarrow Fe^{3+}$ will be demonstrated in detail with the spectra of the blue omphacite shown later in this paper.

(iii) Fe^{3+} in Andradite

The spectrum of andradite garnet has been measured at 1 atm. (Figure 12). The sharp intense band at 440 nm was assigned to the spin-forbidden transition in Fe^{3+} in the octahedral site $[{}^6A_{1g} \rightarrow {}^4A_{1g}, {}^4E_g]$ by Manning (1972). The energy of this transition is $10B + 5C$. With increasing pressure up to 100 kb, no significant shift has been observed. Since this transition is independent of 10 Dq and depends only on B and C, the negligible energy shift implies that changes in B and C with pressure are very small; alternatively, it may also be related to the incompressible nature of the andradite garnet.

(iv) Mn^{3+} in Piemontite

Piemontite is a manganiferous epidote $[Ca_2(Al,Fe,Mn).AlOH.AlO.SiO_4.Si_2O_7]$ which contains Mn^{3+} in a very distorted octahedral site. Because of

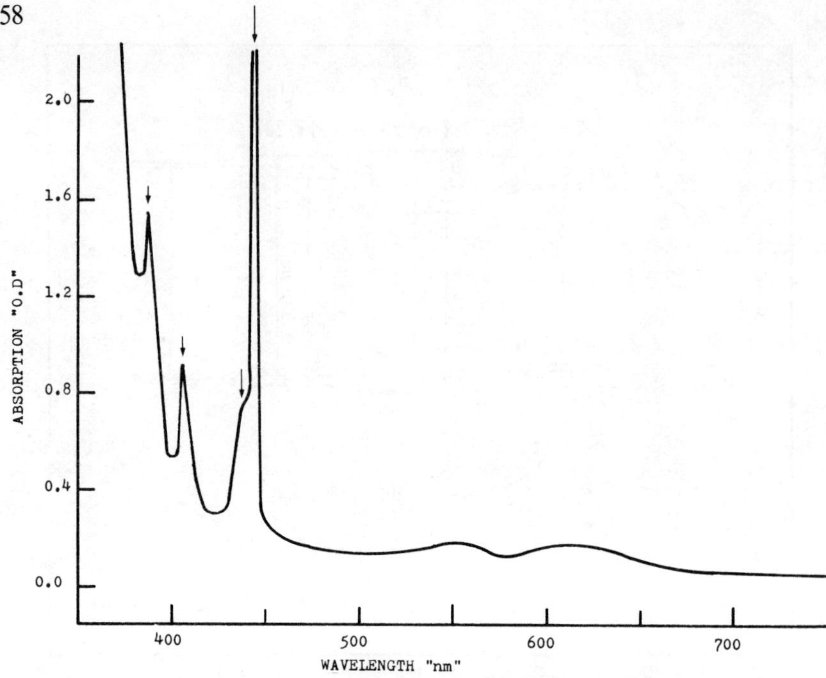

Figure 12. Absorption spectra of Fe^{3+} in andradite garnet, Harvard collection No. 98087. The chemical formula: $Ca_{2.947}Na_{0.036}Mg_{0.007}Fe^{3+}{}_{2.007}Al^{3+}{}_{0.018}$-$Si_{2.979}O_{12}$ (Data from E. R. Whipple, personal communication).

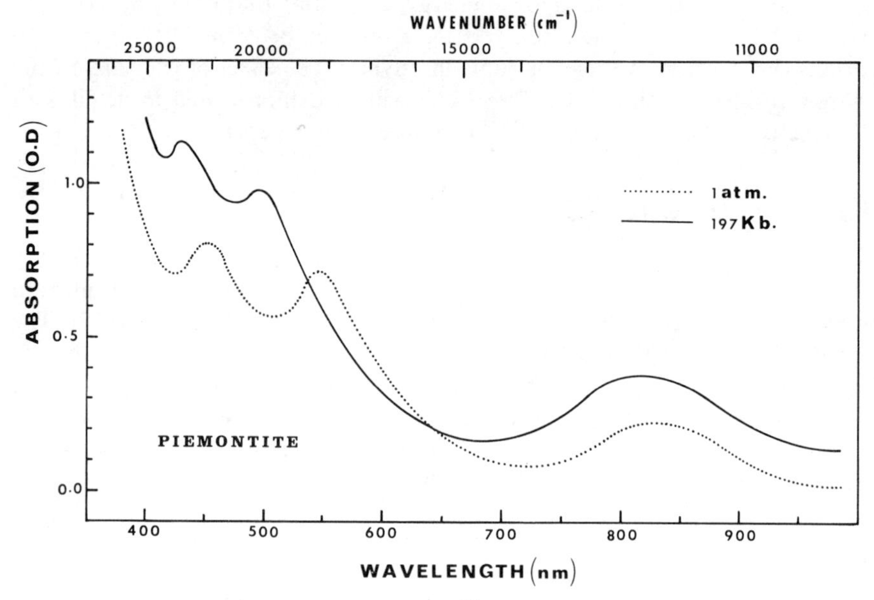

Figure 13. Absorption spectra of Mn^{3+} in a powdered sample of piemontite at 1 atm. and 197 kb.

this distortion, the transition, $^5E_g \rightarrow {}^5T_{2g}$, splits into three, due to the removal of the degeneracy of the lower t_{2g} levels and of the higher energy e_g levels, and three bands are observed in the spectrum at energies of 12,000 cm^{-1}, 18,170 cm^{-1}, and 22,000 cm^{-1} (Burns, 1970). The spectra of piemontite (powder) at 1 atm. and at around 197 kb are shown in Figure 13. With increasing pressure, the band at 18,170 cm^{-1} shifts rapidly to higher energy with increases in width and intensity; however, the band at 22,000 cm^{-1} shifts to higher energy at a slower rate, while the band at 12,000 cm^{-1} shows a very slow blue shift. The magnitude of energy shifts of the high energy bands with pressure is shown in Figure 15; and the mechanism of these transitions at 1 bar and 197 kb is illustrated, using the energy level diagram (Figure 14). From this figure it is obvious that the splitting of the upper two levels has increased slightly, whereas the splitting of the lower three t_{2g} levels has decreased. The estimated change in Δ_0 is about 1,642 cm^{-1} in 197 kb.

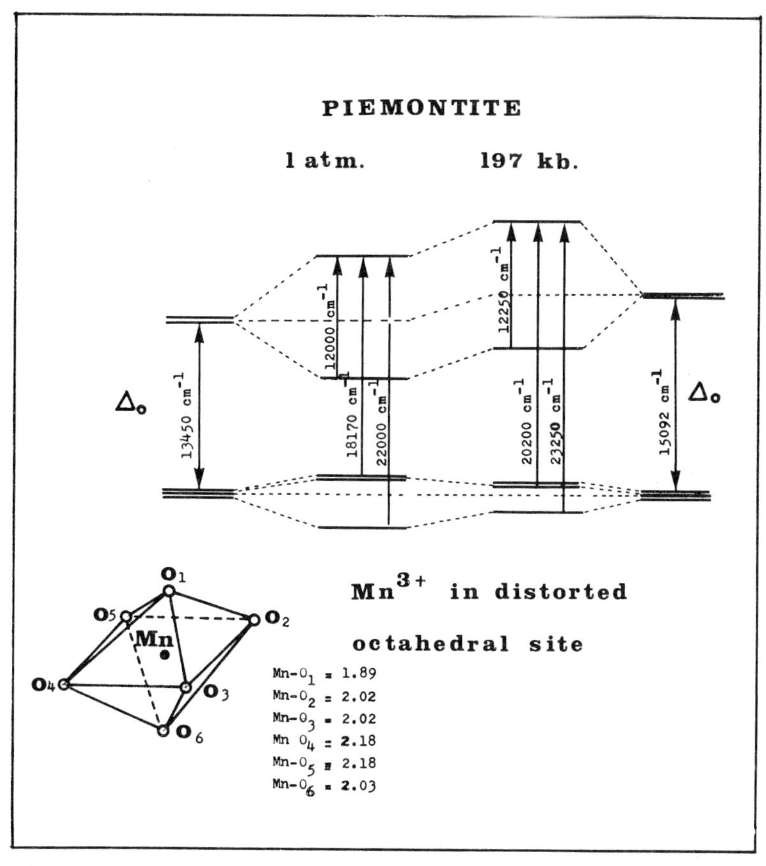

Figure 14. Energy level diagrams of Mn^{3+} in piemontite, constructed from the spectra measured at 1 atm. and 197 kb.

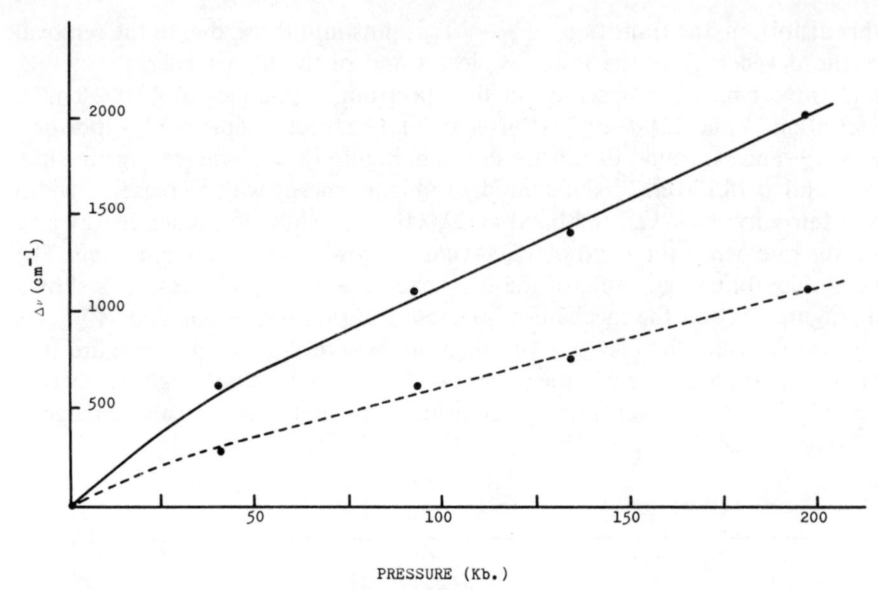

Figure 15. The pressure-induced shift of Mn^{3+} absorption bands in piemontite:
———————— energy shift of the band at 18,170 cm^{-1}.
– – – – – – – – energy shift of the band at 22,000 cm^{-1}.

(v) Cr^{3+} in Synthetic Uvarovite

The end member of the garnet series, uvarovite [$Ca_3Cr_2Si_3O_{12}$], was synthesized by Huggins (1974) from a mixture of CaO, SiO_2 (silica glass), and $(NH_4)_2Cr_2O_7$ kept for 5 days at 1 kb and 675 °C. The spectra of the powdered specimen mixed with NaCl (as internal standard for pressure calibration) were measured at 1 atm., 95 and 177 kb, and are shown in Figure 16.

The two bands in the 1 atm. spectrum at energies 16,667 cm^{-1} (ν_1) and 22,727 cm^{-1} (ν_2) are assigned to electronic transitions from the ground state, $^4A_{2g}(F)$, to the higher energy states, $^4T_{2g}(F)$ and $^4T_{1g}(F)$, respectively. The energies of ν_1 and ν_2 are given as:

$$\nu_1 = 10\ \mathrm{Dq}$$
$$\nu_2 = 15\ \mathrm{Dq} + 7.5B - 6B(1 + \mu)^{1/2}$$

where

$$\mu = \left[\frac{(10\ \mathrm{Dq} - 9B)}{12B}\right]^2$$

The above equations were derived from Tanabe and Sugano's (1954a) secular determinants and were used by Parsons and Drickamer (1958), Reiner (1969), Poole (1964) to calculate the values of B and 10 Dq for Cr^{3+} complexes.

From the first transition, ν_1, 10 Dq can be obtained directly from the energy

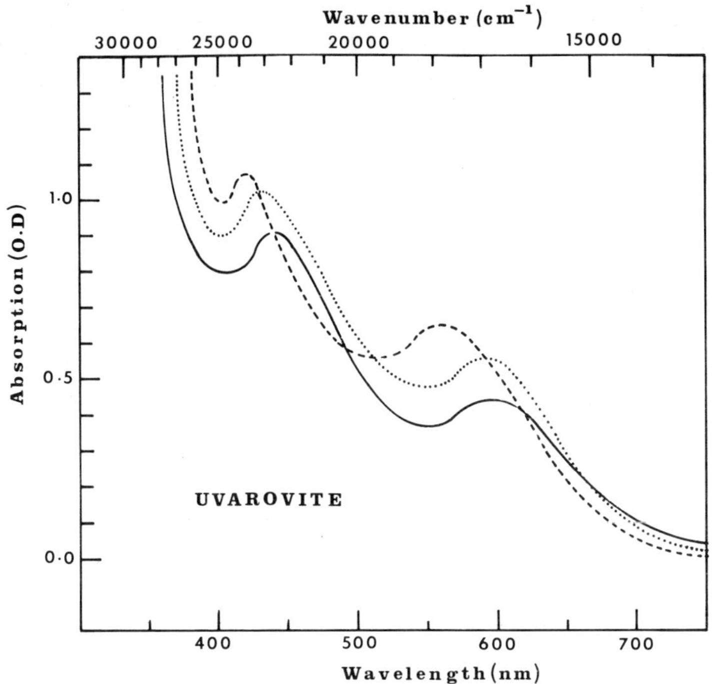

Figure 16. Unpolarized spectra of synthetic uvarovite ($Ca_3Cr_2Si_3O_{12}$) measured at 1 atm., 95 and 177 kb.

of the band maximum and B can be determined from the two transitions, ν_1 and ν_2, by means of the relationship

$$B = \frac{1}{3}\left[\frac{(2\,\nu_1 - \nu_2)(\nu_2 - \nu_1)}{9\,\nu_1 - 5\,\nu_2}\right]$$

The following values of B have been calculated at three different pressures for Cr^{3+} in uvarovite:

Pressure	B(cm^{-1})	ν_1(cm^{-1})	ν_2(cm^{-1})
1 atm.	589	16,667	22,727
95 kb	582	17,007	23,041
177 kb	577	17,668	23,697

It is evident that there is a consistent decrease in the value of B with increasing pressure (about 0.067 cm^{-1}/kb). Further, considering the negligible energy shift of Fe^{2+} spin-forbidden transitions which are dependent mostly on B and the undetectable change of pressure in the energy of Fe^{3+} transitions in andradite which are only dependent on B and C, these results may lead us to believe that up to 200 kb the ionic model in silicate minerals still holds, and only a very slight increase in the degree of covalency of the metal-ligand bond is expected. For uvarovite and Fe^{2+} phases there is also a consistent increase in 10 Dq which is related to shortening the metal–ligand distances and hance, decreasing the volume of the polyhedron.

(vi) $Fe^{2+} \rightarrow Fe^{3+}$ Charge-Transfer in Blue Omphacite

Besides crystal field bands, the effect of pressure on charge-transfer bands has also been studied. The blue omphacite is a silicate mineral which exhibits metal–metal charge-transfer absorption due to an electron transition between energy levels of adjacent Fe^{2+} and Fe^{3+} cations. This mineral was found by Curtis (1974) in a group of peralkaline intrusions located in central Labrador; it contains about 3.5 per cent TiO_2, 10.5 per cent Fe_2O_3, and 5 per cent FeO. The electron microprobe analysis (Curtis, 1974) is given in Table 1.

Omphacite (Clark and Papike, 1966a, b) is the only pyroxene which has two different kinds of octahedral layers. There are two kinds of M1 site alternating along the octahedral chain, usually occupied by Mg^{2+} and Al^{3+}. However, in some cases, Fe^{2+} substitutes for Mg^{2+}, and Fe^{3+} for Al^{3+}, as in blue omphacite. The Fe^{2+} and Fe^{3+} octahedra share edges which may facilitate electron transfer between cations.

Figure 17(a) shows the infrared and visible spectra of the blue omphacite measured at 1 atm. The two bands in the infrared region at wavelengths 1150 and 930 nm are assigned to Fe^{2+} spin-allowed transitions in the M1 site, and the band at 440 nm is assigned to the Fe^{3+} spin-forbidden transition, ${}^6A_{1g} \rightarrow {}^4A_{1g}, {}^4E_g$ (similar to that observed in andradite). The broad intense band at 665 nm is related to electron transfer between ferrous and ferric ions, i.e. $Fe^{2+} \rightarrow Fe^{3+}$ charge-transfer.

Table 1. Electron microprobe analysis of blue omphacite (Data from Curtis, 1974)

Oxide	wt%	Ion		Sum
SiO_2	53.1	Si	1.999	2.00
		Al	0.001	
Al_2O_3	7.33			
		Al	0.324	
TiO_2	3.48	Ti	0.099	
FeO	10.5	Fe^{2+}	0.330	
Fe_2O_3	4.93	Fe^{3+}	0.140	0.99
MnO	0.53	Mn	0.016	
MgO	1.49	Mg	0.084	
CaO	8.78	Ca	0.355	
Na_2O	8.96	Na	0.642	1.00
K_2O	0.10	K	0.004	
Cr_2O_3	0.00			
		$Al/(Al + Fe^{3+}) = 0.70$		
	99.2	$Na/(Na + Ca) = 0.64$		

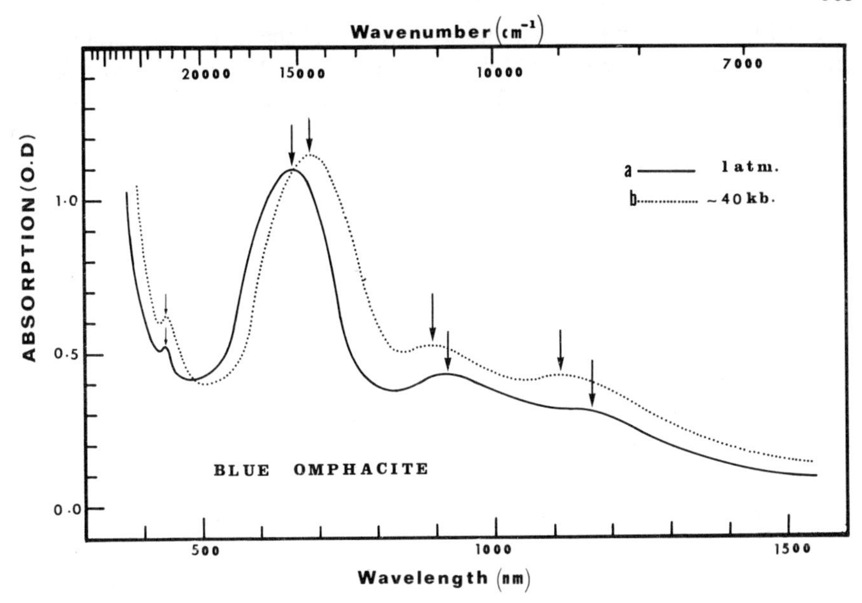

Figure 17. Absorption spectra of the blue-omphacite single crystal at 1 atm., and at pressure around 40 kb. Note the shift of the $Fe^{2+} \rightarrow Fe^{3+}$ charge-transfer band to lower energy at high pressure.

To examine the validity of these assignments, high pressure spectral measurements have been carried out on single crystals of blue omphacite. The spectrum measured at about 40 kb is shown in Figure 17(b). The two bands in the I.R. region have shifted to higher energies. This shift is in agreement with the prediction that they are crystal field dependent transitions of ferrous ions. The band at 440 nm shows negligible energy shifts with increasing pressure; this trend also confirms its assignment as a spin-forbidden transition in Fe^{3+}, since the energy of this transition is crystal field independent, and depends only on Racah parameters, B and C. Since the variations with pressure of B and C. Since the variations with pressure of B and C are very small, only a very slight energy shift is expected for this kind of transition, which is consistent with observation.

On the other hand, the broad intense band at 665 nm shifts to lower energy with increasing pressure while its intensity is increasing significantly. The red shift of this band is a strong evidence for its assignment as a charge-transfer transition. Furthermore, it is of interest to notice in Figure 17(a) that there are no absorption features in the visible region around 478 nm. Although this omphacite contains significant amounts of titanium (Table 1 above), the absence of absorption peaks around 478 nm could be an indication that most of titanium occurs as Ti^{4+}, since there are no crystal field transitions for this cation. Nevertheless, there may be some contribution of absorption, under the band at 665 nm, due to electron transfer between Fe^{2+} and Ti^{4+} ions.

In all silicate minerals so far studied, the tail of the ligand to metal charge-transfer transition is always observed in the spectrum. This tail is also known as the 'absorption edge'. For every case, there is a small shift of this tail to lower energy as pressure increases. Since it is not possible to monitor the band maximum of this absorption, we are not able to determine accurately the magnitude of this energy shift.

(vii) Ligand → Metal Charge-Transfer in Crocoite and Vanadinite

Crocoite [$PbCrO_4$] contains Cr^{6+} in a distorted tetrahedral site. The chromium–oxygen interatomic distances are: $Cr–O_1 = 1.61$ Å, $Cr–O_2 = 1.67$ Å, $Cr–O_3 = 1.66$ Å, and $Cr–O_4 = 1.67$ Å (Quareni and DePieri, 1965). Crocoite shows a very intense absorption in the region between 20,000–25,000 cm^{-1}, Figure 18a. Because of the huge optical density of this band, the absorption maximum could not be distinguished until a very thin crystal fragment (about 25 μm in diameter) was used. The low energy absorption edge measured at 1 atm. (Figure 18(a)) starts at a wavelength around 550 nm, and the broad intense absorption (500–400 nm) is assigned as a ligand to metal charge-transfer. It is believed that the splitting of the band is caused by the distortion of the tetrahedra.

The molecular orbital energy level diagram for the chromate ion in a tetrahedral site (Figure 19) is constructed schematically, based on Wolfsberg and Helmholz (1952) and Johnson (1973) who reported energy levels for the iso-

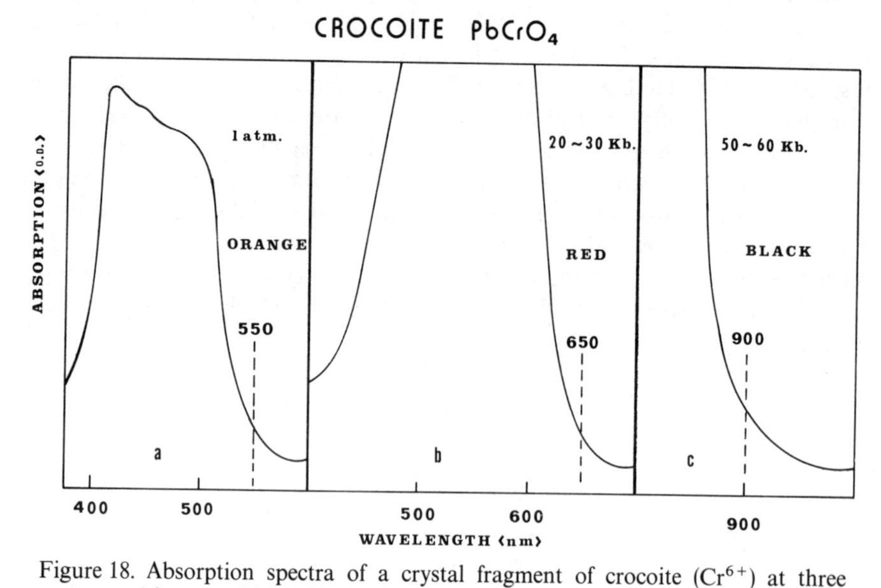

Figure 18. Absorption spectra of a crystal fragment of crocoite (Cr^{6+}) at three pressures: (a) 1 atm., (b) 20–30 kb, (c) 50–60 kb.
Note that the consistent shift of the absorption edge to lower energy with increasing pressure.

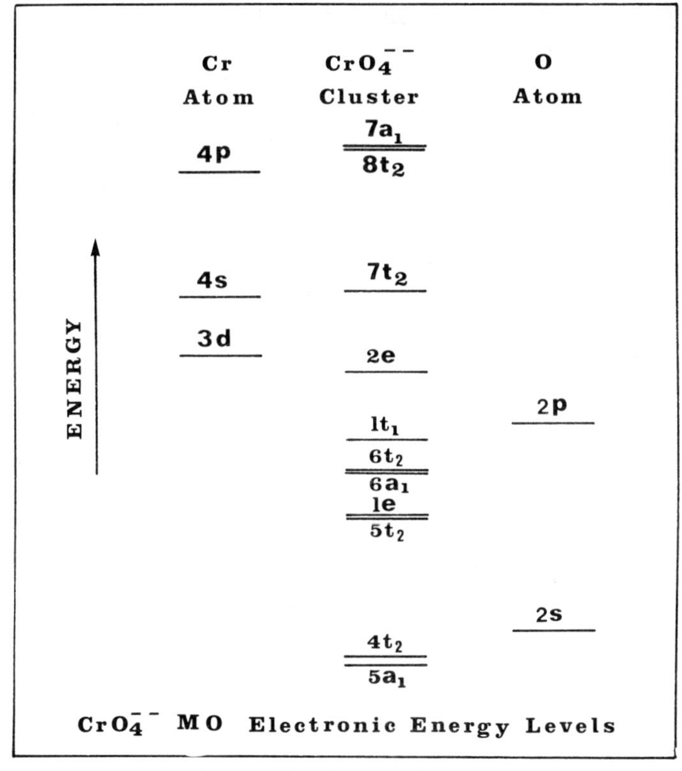

Figure 19. The molecular orbital energy level diagram for the chromate ion (Cr^{6+} in tetrahedral site), constructed schematically after Helmholz (1952); and the calculated energy levels of MnO_4^- (data from Johnson, 1973).

electronic ion, MnO_4^-. In this figure, the levels below $2e$ are filled with electrons, and the levels above $1t_1$ are empty (no *d*-electron for Cr^{6+} ion); the broad band in the visible spectra of crocoite is due to electron transfer from the $1t_1$ to the $2e$ level.

With increasing pressure, the energy separation between the two levels, $7t_2$ and $2e$, will increase and then the energy gap between $1t_1$ and $2e$ levels will decrease; so we expect to observe a red shift of the charge-transfer band as pressure increases.

At around 20–30 kb, the band increases in width and intensity and the absorption edge shifts to about 650 nm (Figure 18b above). This shift is remarkably rapid (about 3,000 cm^{-1} in 25 kb) and accompanied by colour change from orange to red.

Increasing the pressure further, to about 50–60 kb, causes the absorption edge to shift more towards the infrared region to about 900 nm (Figure 18c above), i.e. a negative energy shift of about 7500 cm^{-1} in 55 kb. A colour change, red to black, is also observed around this pressure, and the high energy limb of the band could not be reached due to the huge increase in the band width and intensity.

The same phenomenon has been observed for vanadinite, $[Pb_5(VO_4)_3Cl]$, which contains V^{5+} in tetrahedral sites. The spectral band in this case arises due to electron transfer from the oxygen ligands to the vanadium cations. It occurs at an energy higher than that observed in the spectra of crocoite, which may be due to longer ligand–cation bond distances. The vanadium–oxygen interatomic distances (Trotter and Barnes, 1958) are: $V–O_1 = 1.76$ Å, $V–O_2 = 1.76$ Å, $V–O_3 = 1.72$ Å, and $V–O_4 = 1.72$ Å.

With increasing pressure, the band increases in width and intensity and shifts rapidly to lower energy. The low energy shift of the absorption edge produces a change in colour from orange to red and then to black. The colour changes, in this case, occur at higher pressures than in crocoite, and the 'phase' boundaries are sharper.

The rapid red shift of the absorption edge and the associate change in colour is not fully understood yet, but it may be related to either pressure induced reduction of Cr^{6+} or V^{5+} cations to lower oxidation states, or may be due to changes in the structures of crocoite and vanadinite, or closing the energy gap between conduction and valence bands.

DISCUSSION

(i) Energy Shift of Spectral Bands With Pressure

It was previously believed that crystal field bands would show a shift to high energy with increasing pressure. This prediction was based on the assumption that energies of most crystal field transitions are dependent on the crystal field splitting parameter, Δ_0, or 10 Dq.

From the experimental results presented earlier in this article, it is evident that spin-allowed transitions, dependent mainly on 10 Dq, shift significantly to higher energies as pressure increases. These transitions include those of Ti^{3+} $({}^2T_{2g} \rightarrow {}^2E_g)$, Fe^{2+} $({}^5T_{2g} \rightarrow {}^5E_g)$, Mn^{3+} $({}^5E_g \rightarrow {}^5T_{2g})$, and Cr^{3+} $({}^4A_{1g} \rightarrow {}^4T_{2g})$. The shift to high energy of each transition at elevated pressures is an indication of a consistent increase in 10 Dq value; however, the magnitude of this increase depends on the site compressibility of each cation, the regularity or distortion of the site, and the structure of the mineral phase.

In interpreting the high pressure spectra, some attention should be paid to the possibility of changing the site symmetry of the central ion, i.e. being more regular or more distorted with increasing pressure. In the case of olivine, one of the spectral bands for the M1 site shifts to higher energy with pressure, and the other to lower energy. However, in spite of the opposite energy shift of both bands, there appear to be a net increase in 10 Dq. The decrease in separation of these two bands with pressure indicates that the M1 site is getting more regular.

The change of the polyhedral distortions towards more regular octahedra in fayalite have also been indicated using the high pressure Mössbauer technique (Huggins, 1974). A change in the site symmetries with pressure has also been indicated by Abu-Eid *et al.* (1973) for the ferrous site in gillespite. This mineral contains Fe^{2+} ions in a square planar site, and the site symmetry has been found to change dramatically to a flattened tetrahedron at pressure around 26 kb. The crystal field bands moved discontinuously to lower energies at 26 kb, resulting in a sharp colour change from red to blue with increasing pressure (Abu-Eid *et al.*, 1973; Abu-Eid and Burns, 1975). The change in site symmetry has been confirmed by means of high pressure X-ray single crystal work on gillespite by Hazen and Abu-Eid (1974) and Hazen and Burnham (1974).

Absorption bands dependent only on Racah parameters are expected to shift slightly to lower energies since both B and C are expected to decrease with metal–ligand distance (Minomura and Drickamer, 1961). Zahner and Drickamer (1961) have shown that there is a decrease of about 20 cm^{-1} in the value of B for Ni^{2+} in MgO at 140 kb. In the present study, a decrease of 6 cm^{-1} is found for Cr^{3+} ions in uvarovite as the pressure increases to about 100 kb. A further piece of evidence for the slight changes of B values in silicate minerals is the undetectable energy shift of Fe^{3+} bands in andradite, which are directly related to B and C, and the negligible shift of the spin-forbidden bands of Fe^{2+} ions and olivine and pyroxene which are dependent to a large extent on B and C values. From the above results, it will be safe to assume that the energies of crystal field bands which depend only on B and C, are expected at most to decrease slightly.

The slow red shift of the absorption edge, which has been related to L → M charge-transfer, in all the investigated silicate minerals, is consistent with the prediction made using molecular orbital and band theory models. However, the huge low energy shift of the absorption edge, observed in the spectra of crocoite and vanadinite, may be due to other processes which will be discussed later in this section.

M → M charge-transfer bands that arise from an electron hopping between Fe^{2+} and Fe^{3+} ions shift slowly to lower energies with an increase in their widths and intensities as pressure increases. These trends have been observed in the spectra of fayalite (Figure 10 above) and blue omphacite (Figure 17b above).

Suchan and Drickamer (1959) and Suchan *et al.* (1959) have reported the red shift of the absorption edges in many halide compounds, and related this shift to closing the gap between conduction and valence bands. They also found that the pressure required to eliminate the gap for all iodide compounds is below 30,000 atm. Furthermore, Mao and Bell (1972a) have found that, above 100 kb, the absorption edge of olivine shifts abruptly with pressure into the lower energy infrared region with simultaneous exponential increase in electrical conductivity. Although the abrupt shift has not been explained yet, this effect should be taken into consideration when discussing the mechanism of radiation and absorption of heat in the earth's interior.

(ii) The Degree of Covalency and Racah Parameter B at Elevated Pressures

The degree of covalency of the metal–ligand bond in transition metal complexes is indicated by obtaining the ratio of the interelectronic repulsion parameter B in the complex to that of the free cation, B_0. In the free metal ion, the interelectronic repulsion integrals are approximately proportional to the average reciprocal radius, $\langle r^{-1} \rangle_{3d}$, of the partly filled $3d$-shell (König, 1971). B and C values in the transition metal compounds have been found to be lower than their values in the free transition metal ion. This decrease in B is attributed to an expanded radial distribution of d-electrons in the complex (König, 1971). The covalency parameter, $\beta = B/B_0$, known as the nephelauxetic parameter, is of two types. Firstly, a parameter related to central field covalency which arises from screening the nuclear charge of the cation by ligands (Lever, 1968; König, 1971) which is designated as β_{35}^{*}. Secondly, a parameter related to covalency which arises due to delocalization of d electrons onto the ligands, and designated as β_{33} and β_{55}. Fortunately, the B value for Cr^{3+} in uvarovite, which is determined from the energy of the second band at 22,727 cm^{-1}, is related to the first type of covalency, i.e. it depends on both orbitals, e_g and t_{2g}, and is not restricted only to one type of orbital (Reiner, 1969). The B_0 value for the Cr^{3+} free ion is 918 cm^{-1} (Lever, 1968), and the nepheleuxetic ratio, β_{35}, for Cr^{3+} is calculated at three different pressures using the determined B values which are presented in the **RESULTS** section. They are:

P	$\beta_{35} = B/918$
1 atm.	0.641612
95 kb	0.633987
177 kb	0.628540

The spectrum of natural uvarovite, containing about 10 per cent Cr^{3+}, was measured by Manning (1967c). The two bands related to the spin-allowed transitions in Cr^{3+} occur at energies (ν_1) 16,600 cm^{-1} and (ν_2) 23,100 cm^{-1}, and the B and β values estimated from these transitions are 645.5 cm^{-1} and 0.703 cm^{-1}, respectively. It is evident then that the B value is considerably larger for Cr^{3+} in natural uvarovite than in synthetic uvarovite. As natural uvarovite definitely contains a significant amount of Al^{3+}, replacing Al^{3+} by Cr^{3+} will enlarge the site and thus decrease Δ_0. Reiner (1969) reported that Δ and B values for $Cr_xAl_{2-x}O_3$ mixed crystal series vary with the concentration of Cr^{3+}. He found that both Δ and B decrease with increasing x.

The decrease in Δ_0 is easily explained, since the Cr^{3+} site is getting larger with increasing x and the decrease in B may be due to delocalization of d-

*The repulsion parameters containing e_g or (γ_3) orbitals, only, are designated as β_{33}, and those with t_{2g} or (γ_5) are given as β_{55}; whereas integrals containing both e_g and t_{2g} orbitals are designated as β_{35} (Reiner, 1969).

electrons onto the ligands; this also may be due to the higher electronegativity of Cr^{3+} relative to Al^{3+}.

In the light of the above discussion, we could conclude that B or the degree of covalency of the cation-anion bond is affected by two factors. First, the size of the central ion and, second, the type of next nearest-neighbour cations. Nonetheless, the increase in the degree of covalency in Cr^{3+} in uvarovite up to 177 kb is very small, and the postulated ionic model for silicates at such depths may be justified. The negligible change in B values with pressure is also indicated from the Fe^{3+} spectra in andradite garnet, as discussed previously, the sharp absorption band which occurs at 22,676 cm^{-1} is equivalent to 32.5 B (considering $C/B = 4.5$). Any significant decrease in B with pressure will cause this band to show red shift; unfortunately no significant shift (< 100 cm^{-1}) has been detected which may indicate that the change in B value at 177 kb is less than 3 cm^{-1}.

(iii) Site Compressibilities from 10Dq Values

It was mentioned (equation (6)) that 10 Dq is related to the average metal-to-ligand distance, R, by

$$10\ \mathrm{Dq} \propto R^{-5}.$$

Consider

$$|10\ \mathrm{Dq}| \equiv E \text{ and } R \equiv (V)^{1/3},$$

where E is the energy and V is the volume of the polyhedron. It follows then that

$$E \propto V^{-5/3}. \tag{7}$$

Differentiating with respect to V will give

$$\mathrm{d}E/\mathrm{d}V \equiv -5/3\,V^{-8/3},$$

and then using relation (7):

$$\mathrm{d}E/\mathrm{d}P = 5/3(E/K). \tag{8}$$

From (8), we may be able to determine the values of K at various pressures when 10 Dq is known.

In piemontite 10 Dq increases by about 1,650 cm^{-1} in 197 kb. This increase is due mostly to shortening metal-ligand bond distances, which decrease the volume of Mn^{3+} polyhedron by about 7 per cent. The estimated bulk modulus of the polyhedron is about 2.8 Mb. Comparing this value with the bulk modulus of the mineral which is about 1.074 Mb (Simmons and Wang, 1971), we may conclude that the octahedral site is rather incompressible, suggesting that the calcium polyhedra take up most of the change in volume.

Similar calculations could be done to estimate the bulk moduli of the transition metal ions, especially those having crystal field transitions dependent only on 10 Dq, e.g. Fe^{2+}, Cr^{3+}, Ti^{3+}, and Mn^{3+}. Equation (8) is applicable only when 10 Dq is known, where its value will be equivalent to E, and in most cases the energy of the individual spectral band does not represent E. For this reason, when there is a distorted site that gives rise to more than one crystal field band related to 10 Dq, an energy level diagram has to be constructed to evaluate the crystal field parameter, 10 Dq. Further, it should also be noticed that K, given in equation (8), is the bulk modulus of the site rather than that of the bulk mineral.

Calculating the bulk modulus of a site at various pressures could be very useful for extending the bulk modulus-volume relationships. Anderson (1969) and Anderson and Anderson (1970) have shown that for compounds of constant mean atomic weight the relationship $KV = \text{const.}$ holds. It would be useful to examine this relation for each type of polyhedra in a given phase as well. Obtaining the differential polyhedral compressibilities would enable us to investigate the interrelation of K and V for each type of polyhedra, and hence, to obtain expressions relating the bulk elastic constants to the polyhedral compressibilities. Such relationships may be then used to predict properties and structures of mineral phases postulated to exist at great depths in the earth.

Studying the effect of pressure on the transition metal ion sites in silicates may also be valuable for understanding the mechanism of phase transformations and the ordering phenomena of cations in silicate structures as a function of pressure.

Tischer and Drickamer (1962) investigated the local symmetries and compressibilities of some transition elements in several silicate and phosphate glasses. They found that the tetrahedral sites are less compressible than the bulk glass, and that Co^{2+} and Ni^{2+} tend to transfer from the tetrahedral to the octahedral site. Performing such investigations on silicate minerals containing transition metal ions in various sites will be useful in estimating the enrichment and stability of these cations at various depths in the earth.

(iv) Pressure Induced Reduction

Pressure induced reduction of ferric ions in many organic and inorganic complexes has been reported (Drickamer and Frank, 1973) to occur over a wide range of pressures, depending on the type of the ligands surrounding the metal ion and the nature of the complex. Recently, Wang and Drickamer (1973) have reported pressure induced reduction in copper complexes, reduction of Cu^{2+} to Cu^{+} at pressure around 118 kb. Further, reduction of ferric ions in amphiboles (magnesiorebickite) have been demonstrated by Burns *et al.* (1972) using high pressure Mössbauer techniques. Furthermore, Gibbons *et al.* (1973), from their shock pressure experiments on Mn^{3+} in rhodonite, observed irreversible pressure induced reduction of Mn^{3+} in a sample which has been shocked to 496 kb. They related the reduction of the trivalent manganese ions to the presence of water in the rhodonite structure.

The main evidence of pressure induced reduction is from high pressure optical absorption and Mössbauer studies (Drickamer *et al.*, 1972; Drickamer and Frank, 1973).

The process of reduction involves an electron transfer from an orbital of mostly ligand character to another orbital of primarily metal character. This process has been defined earlier as ligand-to-metal charge-transfer. The reduction process could be approached theoretically by performing an analysis which relates the parameters of the L → M charge-transfer bands (their widths, intensities, energies), to the probability of thermal treansfer (Drickamer *et al.*, 1972; Drickamer and Frank, 1973).

The analogy with the thermal transfer may be expressed in the relation given by Drickamer *et al.* (1972):

$$E_{th} = h\nu_{max} - [(\delta E)^2_{1/2}(\omega/\omega')^2 1/16KT \ln 2] \quad (9)$$

where E_{th} is the difference in energy between the bottom of the ground state and excited state potential wells; ν_{max} is the frequency of the maximum absorption peak; $(\delta E)_{1/2}$ is the half width of the peak; and ω and ω' are the force constants of the ground and excited states, respectively. The reduction process may take place when E_{th} approaches to zero and then the excited state becomes more stable at higher pressures.

Considering the relation given above (9), E_{th} will decrease if $h\nu_{max}$ decreases and $(\delta E)_{1/2}$ increases, assuming that ω and ω' are constants.

In this study, no evidence has been observed for reduction of Ti^{3+}, Fe^{3+}, Mn^{3+}, and Cr^{3+} in silicate minerals up to 200 kb. This conclusion is made since no new crystal field transitions that may be related to lower oxidation states have been observed, and the tails of the L → M charge-transfer bands shift slowly to lower energies.

On the other hand, the spectra of crocoite (Cr^{6+}) and vanadinite (V^{5+}) show unusual changes as pressure increases. The L → M charge-transfer bands of O–Cr and O–V shift rapidly to lower energies with increasing pressure, with a huge increase in their widths and intensities. The rapid low energy shift and increase in half width of the charge-transfer band could be a manifestation of a large decrease in the thermal energy gap (Drickamer *et al.*, 1972) between the ground state (Cr^{6+}) and the excited state (Cr of lower oxidation state and oxidized ligands). As a consequence of the consistent decrease in this energy gap, the electronic configuration of Cr of lower oxidation state and oxidized ligands could well be more stable than Cr^{6+} and regular ligands at high pressure.

Reducing Cr^{6+} (d^0) and V^{5+} (d^0) to lower oxidation states will introduce one or more electrons into the *d*-levels of these cations; if this is the case, we would expect crystal field transitions to be observed in the high pressure spectra. Unfortunately, the increase in width and intensity and the rapid red shift of the charge-transfer bands probably mask any possible crystal field transitions in the visible-near infrared region.

Alternatively, the rapid low energy shift of the absorption edge may be due to closing the gap between conduction and valence bands which may lead to a more metallic character of these minerals at elevated pressures. In addition, significant changes in the structures of these minerals with pressure could be partly responsible for this rapid spectral shift.

Mao and Bell (1972) have reported that a rapid low energy shift of the absorption edges in olivine and spinel is prominent at pressures above 100 kb. So far no reasonable explanation has been given for these phenomena in olivine and spinel. As we mentioned earlier, the absorption edge is related to transfer of electrons from the filled orbitals of mostly ligand character to the empty or partly filled metal orbitals (Figure 2 above). With increasing pressure, the energy gap between metal and ligand orbitals will be narrowed slightly, and then the charge-transfer transition (L $\rightarrow$ M) will gradually decrease in energy.

On the other hand, at sufficient pressures, when 10 Dq increases significantly and the energy gap between conduction and valence bands decreases (i.e. lowering the energies of the highest energy levels which are related to the conduction band) at this point, electron-transfer may take place from the metal orbital, e_g, to the above ligand orbitals, and M–L charge transfer is observed. Then the rapid shift of the absorption edge above 100 kb could be related to the coupled processes of L–M and M–L charge-transfer transitions. Having the two types of electron transfer, (L–M) and (M–L), taking place simultaneously may lead to reluctance of the process of reducing Fe^{2+} to lower oxidation state, e.g. Fe^{1+}. On the other hand, in the case of Cr^{6+} and V^{5+} ions in crocoite and vanadinite, the upper *d*-levels, t_2, are empty and then the process of M–L electron transfer should not be expected to take place. Hence, the excited state (lower oxidation state) may be stable at elevated pressures since it does not have reverse effect, i.e. M–L.

The results of the high pressure optical absorption measurements indicate that most crystal field bands shift to higher energies, whereas charge-transfer bands shift to lower energies. Further, crystal field bands dependent mostly on Racah parameters B and C do not shift significantly, due to negligible changes in the energy values of these parameters with increasing pressure.

Up to about 200 kb, pressure induced reduction has not been observed in Mn^{3+}, Fe^{2+}, Ti^{3+}, and Cr^{3+} cations. However, Cr^{6+} and V^{5+} may have been reduced to lower oxidation states at moderate pressures (30–60 kb). The abrupt red shift of the absorption edge at elevated pressure will block the radiative heat in the upper part of the transition zone.

ACKNOWLEDGEMENTS

The author is grateful for the helpful advice, criticism, and encouragements of Professor Roger G. Burns, and Dr. Frank E. Huggins for the useful discussion and critical review of this manuscript. He would also like to thank Ms. Roxanne Regan for the helpful assistance and preparation of the manuscript. This research was supported by grants from the National Science Foundation (Grant no. GA–40910) and the National Aeronautics and Space Administration (Grant no. NGR–22–009–551).

REFERENCES

Abu-Eid, R. M., H. K., and Burns, R. G. (1973). *Ann. Rept. Geophys. Lab., Year Book 72–73*, p. 564.

Abu-Eid, R. M. (1975). *Ph.D. Thesis*, MIT, Cambridge, Mass.

Abu-Eid, R. M. and Burns, R. G. (1975). Submitted to *Chem. Geol.*

Anderson, D. L. and Nafe, J. E. (1965). *J.G.R.*, **70**, 3951.

Anderson, D. L. and Anderson, D. L. (1970). *J.G.R.*, **75**, 3494.

Anderson, D. L. (1969). *J.G.R.*, **74**, 3857.

Ballhausen, C. J. (1954). *Dan. Man. Fys. Medd. 29*, no. 4.

Ballhausen, C. J. (1962) *An Introduction to Ligand Field Theory*, McGraw-Hill, New York.

Bassett, W. A., Takahashi, T., and Stook, P. W. (1967). *Rev. Sci. Instr.*, **38**, 37.

Bassett, W. A. and Ming, L. (1972). *Phys. Earth Planet. Interiors*, **6**, 154.

Bassett, W. A. (1974). Personal communication.

Bell, P. M. and Mao, H. K. (1972a). *Ann. Rept. Geophys. Lab., Year Book 72–73*, 480.

Bell, P. M. and Mao, H. K. (1972b). *Ann. Rept. Geophys. Lab., Year Book 72–73*, 608.

Burns, R. G. (1965). *Ph.D. Dissertation*, University of California, Berkeley, California.

Burns, R. G. (1966). *Mineral. Mag.*, **35**, 715.

Burns, R. G. and Fyfe, W. S. (1967). In: *Researches in Geochemistry*, Vol. 2, (ed.: P. H. Abelson), J. Wiley and Son, New York, p. 259.

Burns, R. G. and Strens, R. G. J. (1967). *Mineral. Mag.*, **35**, 547.

Burns, R. G. (1969). In: *The Application of Modern Physics to the Earth and Planetary Interiors*, (ed.: S. K. Runcorn), J. Wiley and Sons, New York.

Burns, R. G. (1970). *Mineralogical Applications of Crystal Field Theory*, Cambridge University Press.

Burns, R. G., Abu-Eid, R. M., and Huggins, F. E. (1972a). *Geochim. Cosmochim. Acta, Suppl. 3*, vol. 1, p. 533, MIT Press.

Burns, R. G., Tossell, J. A., and Vaughan, D. J. (1972b). *Nature*, **240**, 33.

Burns, R. G., Huggins, F. E., and Abu-Eid, R. M. (1972c). *The Moon*, **4**, 93.

Burns, R. G. and Huggins, F. E. (1973). *Amer. Mineral.*, **58**, 955.

Burns, R. G., Vaughan, D. J., Abu-Eid, R. M., Witner, M. and Morawski, A. (1973). *Proc. 4th Lunar Sci. Conf., Geochim. Cosmochim. Acta, Suppl. 4*, Vol. 1, 983.

Chao, E. C. T. and Minken, J. (1970). *Amer. Mineral.*, **55**, 1416.

Clark, J. R. and Papike, J. J. (1966a). *Science*, **154**, 127.

Clark, J. R. and Papike, J. J. (1966b). *Prog. 1966 Annual Meetings, G.S.A.*, p. 40.

Clark, C. D., Ditchbure, R. W., and Dyer, H. B. (1956). *Proc. R. Soc. London*, **A234**, p. 363.

Collotti, G., Conti, L., and Zocchi, M. (1959). *Acta Cryst.*, **12**, 416.

Cotton, F. A. and Wilkinson, G. (1966). *Advanced Inorganic Chemistry*, John Wiley and Sons, New York.

Cotton, F. A. (1971). *Chemical Applications of Group Theory*, Jojn Wiley and Sons, New York.

Curtis, L. (1974). Personal Communication.

Decker, D. L. (1966). *J. App. Phys.*, **37**, 5012.

Dollase, W. A. (1969). *Amer. Mineral.*, **54**, 710.

Dowty, E. and Clark, J. R. (1973a). *Amer. Mineral.*, **58**, 962.

Dowty, E. and Clark, J. R. (1973b). *Amer. Mineral.*, **58**, 230.

Drickamer, H. G. (1965). *Solid State Phys.*, **17**, 1–133.

Drickamer, H. G., Lewis, Jr. G. K., and Fung, S. C. (1969). *Science*, **163**, 885.

Drickamer, H. G., Frank, C. W., and Slichter, C. P. (1972). *Proc. Nat. Acad. Sci. USA*, **69**, 933.

Drickamer, H. G. and Frank, C. W. (1973). *Electronic Transitions and the High Pressure Chemistry and Physics of Solids*, Chapman and Hall, London.

Dunn, T. M., McClure, D. S., and Pearson, R. G. (1965). *Some Aspects of Crystal Field Theory*, Harper and Row, New York.

Faye, G. H., Manning, P. G., and Nickel, E. H. (1968). *Amer. Mineral.*, **53**, 1174.
Figgis, B. N. (1966). *Introduction to Ligand Fields*, John Wiley and Sons, New York.
Forman, R., Piermarini, G., Barrett, J. D., and Block, S. (1972). *Science*, **176**, 284.
Fukao, Y., Mizutani, H., and Uyeda, S. (1968). *Phys. Earth Planet. Interiors*, **1**, 57.
Fyfe, W. S. (1960). *Geochim. Cosmochim. Acta*, **19**, 141.
Gaffney, E. S. (1972a). *Ph.D. Thesis*, Cal. Inst. Tech., California.
Gaffney, E. S. and Anderson, D. L. (1973). *J.G.R.*, **78**, 7005.
Gaffney, E. S. (1972b). *Phys. Earth Planet. Interiors*, **6**, 385.
Gibbons, R. V., Ahrens, T. J., and Rossman, G. R. (1974). *Amer. Mineral.*, **59**, 177.
Griffith, J. S. (1964). *The Theory of Transition Metal Ions*, 1st Edition, Cambridge University Press, London.
Hazen, R. M. and Abu-Eid, R. M. (1974). *EOS, Trans. Amer. Geophys. Union*, **55**, 463.
Hazen, R. M. and Burnham, C. (1974). *Amer. Mineral.*, **59**, 1166.
Huggins, F. E. (1974) Ph.D. Thesis, MIT, Cambridge, Mass.
Huggins, F. E. and Abu-Eid, R. M. (1974). *GSA Ann. Meeting, Abstr.*, **6**, 803.
Hush, N. S. (1967). In: *Progress in Inorganic Chemistry*, **8**, (ed.: F. A. Cotton), Interscience Publishers, New York, p. 404.
Hush, N. S. (1968). *Electrochemica Acta.*, **13**, 1005.
Hutchings, M. T. (1964). *Solid State Phys.*, **16**, 227.
Ilse, F. E. and Hartmann, H. Z. (1951a). *Physik. Ch.*, **197**, 239.
Ilse, F. E. and Hartmann, H. Z. (1951b). *Naturforschg.*, **69**, 751.
Johnson, K. H. (1973). *Advances in Quantum Chemistry*, **7**, 143.
Jorgenson, C. K. (1966). *Structure and Bonding*, **1**, 3.
Keester, K. L. and White, W. B. (1966). *I.M.A. Symposia*, p. 22.
König, E. (1971). *Structure and Bonding*, **9**, 175.
Lever, A. B. P. (1968). *Inorganic Electronic Spectroscopy*, Elsevier Pub. Co., Amsterdam.
Manning, P. G. (1967a). *Canad. Jour. Earth Sci.*, **4**, 1039.
Manning, P. G. (1967b). *Canad. Mineral.*, **9**, 723.
Manning, P. G. (1967c). *Canad. Mineral.*, **9**, 237.
Manning, P. G. and Nickel, E. H. (1969). *Canad. Mineral.*, **10**, 71.
Manning, P. G. (1970). *Canad. Mineral.*, **10**, 677.
Manning, P. G. (1972). *Canad. Mineral.*, **11**, 826.
Mao, H. K. (1967). *Ph.D. Thesis*, University of Rochester, Rochester, New York.
Mao, H. K. and Bell, P. M. (1972a). *Science*, **176**, 403.
Mao, H. K. and Bell, P. M. (1972b). *Ann. Rept. Geophys. Lab., Year Book 71*, p. 520.
Mao, H. K. (1973a). *Ann. Rept. Geophys. Lab., Year Book 72*, p. 557.
Mao, H. K. (1973b). *Ann. Rept. Geophys. Lab., Year Book 72*, p. 552.
Mao, H. K. and Bell, P. M. (1973. In: *ASTM Spec. Tech. Publ. 539*, (ed.: D. A. Flong), p. 100.
McClure, D. S. (1959). *Solid State Phys.*, **9**, 399.
Merrill, L. (1973). Ph.D. Thesis, *University of Rochester*, Rochester, New York.
Minomura, S. and Drickamer, H. G. (1961). *Journ. Chem. Phys.*, **35**, 923.
Moore, R. K. and White, W. B. (1972). *Canad. Mineral.*, **11**, 791.
Oelkrug, D. (1971). *Structure and Bonding*, **9**, 1.
Orgel, L. (1966). *An Introduction to Transition-metal Chemistry: Ligand-field Theory*, 2nd edition, Methuen, London.
Parson, R. W. and Drickamer, H. G. (1958). *J. Chem. Phys.*, **29**, 930.
Phillips, G. S. P. and Williams, R. J. P. (1966). *Inorganic Chemistry*, Vol. 2, Oxford University Press.
Pitt, G. D. and Tozer, D. C. (1970a). *Phys. Earth Planet. Interiors*, **2**, 189.
Pitt, G. D. and Tozer, D. C. (1970b). *Phys. Earth Planet. Interiors*, **2**, 179.
Poole, C. P. Jr. (1964). *J. Phys. Chem. Solids*, **25**, 1169.
Quareni, S. and DePieri, R. (1965). *Acta Cryst.*, **19**, 287.
Raal, F. A. (1959). *Proc. Phys. Soc. London*, **74**, 647.

Racah, G. (1942a). *Phys. Review*, **61**, 186.
Racah, G. (1942b). *Phys. Review*, **62**, 438.
Racah, G. (1943). *Phys. Review*, **63**, 367.
Racah, G. (1949). *Phys. Review*, **76**, 1352.
Reiner, D. (1969). *Structure and Bonding*, **6**, 30.
Ringwood, A. E. (1962). *Geochim. Cosmochim. Acta*, **26**, 457.
Ringwood, A. E. (1970). *Phys. Earth Planet. Interiors*, **3**, 109.
Shankland, T. J. (1970). *J.G.R.*, **75**, 409.
Shankland, T. J. (1972). *J.G.R.*, **77**, 3750.
Simmons, G. and Wang, H. (1971). *Single Crystal Elastic Constants and Calculated Aggregate Properties, A Handbook*, MIT Press.
Strens, R. G. J. (1969). In: *The Application of Modern Physics to the Earth and Planetary Interiors*, J. Wiley and Sons, New York, p. 213.
Suchan, H. L., Balchar, A. S., and Drickamer, H. G. (1959). *J. Phys. Chem. Solids*, **10**, 343.
Suchan, H. L. and Drickamer, H. G. (1959). *J. Phys. Chem. Solids*, **11**, 111.
Suchan, H. L., Wiederhorn, S., and Drickamer, H. G. (1959c). *Journ. Chem. Phys.*, **31**, 355.
Sung, C.-M., Abu-Eid, R. M., and Burns, R. G. (1974). *Proc. 5th Lunar Sci. Conf., Geochim. Cosmochim. Acta*, Suppl. 5, vol. 1, 717.
Tanabe, Y. and Sugano, S. (1954a). *Journ. Phys. Soc. Japan*, **9**, 753.
Tanabe, Y. and Sugano, S. (1954b). *Journ. Phys. Soc. Japan*, **9**, 766.
Tischer, R. E. and Drickamer, H. G. (1962). *J. Chem. Phys.*, **37**, 1554.
Tossell, J. A., Vaughan, D. J., and Johnson, K. H. (1974). *Amer. Mineral.*, **59**, 319.
Trotter, J. and Barnes, W. H. (1958). *Canad. Mineral.*, **6**, 161.
Van Valkenberg, A. (1965). Conference International Sur-les-Hautes, Pression, Le Crensot Saone-et-Loive, France.
Wang, P. J. and Drickamer, H. G. (1973). *J. Chem. Phys.*, **59**, 713.
Weaver, J. S., Takahashi, T., and Bassett, W. A. (1971). In: *Accurate Characterization of* the High Pressure Environment, (ed.: Lloyd), *Nat'l. Bur. St. Spec. Publ. 326*, p. 189.
Wilcox, R. E. (1959). *Amer. Mineral.*, **44**, 1272.
Wood, B. J. and Strens, R. G. J. (1972). *Mineral. Mag.*, **38**, 909.
Wolfsberg, M. and Helmholz, L. (1952). *J. Chem. Phys.*, **20**, 837.
Zahner, J. C. and Drickamer, H. G. (1959). *J. Phys. Chem. Solids*, **11**, 9296.
Zahner, J. C. and Drickamer, H. G. (1961). *J. Phys. Chem. Solids*, **35**, 1483.

A Model for the Investigation of Hydroxyl Spectra of Amphiboles

A. D. Law
University Chemical Laboratory,
University of Kent,
Canterbury, Kent CT2 7NH, U.K.

INTRODUCTION

It has come to be realized that the infrared technique for obtaining partial site populations for amphiboles, by studying the hydroxyl stretching peaks in the region of 3800–3500 cm^{-1}, is fraught with practical and theoretical difficulties. Among the practical difficulties the chief is often the prevention of oxidation during sample preparation. Theoretical problems include the question of a proper peak profile (a skew Gaussian may be more appropriate than the normal Gaussian commonly used); the effect of different coordination on the transition moment of the O–H vibration (and hence on the intensity, as opposed to the frequency, of the corresponding absorption); and the necessity at present of assuming the random distribution of O(3) sites not occupied by hydroxyl groups. A review of these and many other aspects of the method is given by Strens (1974).

Another difficulty is that while it is realised that the ordering of cations in the M(1) and M(3) sites will affect the spectra, the precise effect—and hence the interpretation of real spectra in terms of ordering—is not well understood. Strens (1966) developed criteria for the detection of two types of ordering, but the present study has shown that these empirical criteria need to be reconsidered.

In the discussion that follows many theoretical problems, and in particular the ones mentioned above, will be set aside. For details of the amphibole structure the reader is referred to the standard texts. For details of the infrared technique see for example Bancroft and Burns (1969); these authors give fuller details of the assignments of peaks, and of peak nomenclature followed here (see also Table 1).

TOTAL SITE OCCUPANCIES AND THE RANDOM DISTRIBUTION

For an amphibole where the M(1) and M(3) sites are occupied by combi-

Table 1. Assignment of major peaks.

Combinations

Peak	Cations in M(1) M(1) M(3)
A	3 Mg
B	$2\ Mg + 1\ Fe^{2+}$
C	$1\ Mg + 2\ Fe^{2+}$
D	$3\ Fe^{2+}$

Permutations

Cation in M(1)	M(1)	M(3)	M(1). M(3) equivalent	M(1). M(3) not equivalent
Mg	Mg	Mg	A	A
Mg	Mg	Fe	B	B'
Mg	Fe	Mg	B	B''
Fe	Mg	Mg	B	B''
Fe	Fe	Mg	C	C'
Fe	Mg	Fe	C	C''
Mg	Fe	Fe	C	C''
Fe	Fe	Fe	D	D

nations of the major ions Mg and Fe^{2+}, and some minor ions are also present, the overall cation population of 2 M(1) + M(3) is given by

$$T_m = 3\,A_0 + 2\,B_0 + C_0 + \text{terms from minor peaks}$$
$$T_f = B_0 + 2\,C_0 + 3\,D_0 + \text{terms from minor peaks}$$

where T_m and T_f denote respectively total Mg and total Fe^{2+}, the total of other ions, denoted T_a, is determined from minor peaks. (For more detail, see Burns and Prentice, 1968.) For the present purpose the nature of these other (minor) ions is unimportant; typically they will be trivalent iron or aluminium.

If M(1) and M(3) sites are equivalent then these ions may be expected to be randomly distributed: there will be on average, in each M(1) and each M(3) site, $T_m/3$ Mg ions, $T_f/3$ Fe^{2+}, and $T_a/3$ other ions. These quantities will be denoted μ, ϕ and α respectively. On a statistical basis such a random distribution would give rise to peaks in the infrared spectrum with intensities given for the major peaks by:

$$A_r = \mu^3$$
$$B_r = 3\,\mu^2\phi$$
$$C_r = 3\,\mu\phi^2$$
$$D_r = \phi^3$$

and for the minor peaks by similar expressions. The statistical weightings of 3 for B_r and C_r arise from the assumption that the M(1) and M(3) sites are identical (see Table 1).

Table 2.

Notation

$T_m T_f T_a$	Total Mg, Fe^{2+}, other ions in 2M(1) + M(3) (Total = 3).
$\mu\ \phi\ \alpha$	Fractional site occupancies of Mg, Fe^{2+}, other ions (Total = 1 for each site).
P	Site Fraction (defined as $1 - \alpha$).

Subscripts 1, 3 denote M(1) or M(3), e.g. ϕ_1 is the fraction of M(1) sites occupied by Fe^{2+}.

Relationships

$T_f = 2\phi_1 + \phi_3 \qquad T_m = 2\mu_1 + \mu_3$

If $\alpha_1 = \alpha_3$ (random distribution of minor ions):

$T_a = 3\alpha \qquad \phi_1 + \mu_1 = \phi_3 + \mu_3 = P$

If $\alpha_1 \neq \alpha_3$:

$T_a = 2\alpha_1 + \alpha_3 \qquad \phi_1 + \mu_1 = P_1 \qquad \phi_3 + \mu_3 = P_3$

Peak intensities

A_0	Normalised observed intensity of peak A.
A_r	Reference value of intensity (random distribution).
A_c	Calculated value of intensity.

Corresponding notation for all peaks.

It has been pointed out by Krzanowski and Newman (1972) that such a statistically random distribution will include combinations of cations which are implausible for valency reasons (in this case, more than one trivalent ion). However, most amphibole spectra show few trivalent ions in M(1) and M(3), and in any case the random distribution is used only as a reference point (as denoted by the subscript r applied to the peak intensities).

Ordering of cations can be detected by observing the deviation of the pattern of peaks from the random model. Strens (1966) defined two types of ordering which may occur. These are *segregation* (for which the word *ordering* has previously been used; the latter term is here used to describe any non-random arrangement), and *clustering*. By *segregation* is meant the tendency for specific ions to enter specific sites preferentially; by *clustering*, an enhanced incidence of combinations of three like ions in M(1)M(1)M(3) groups. The criteria for ordering given by Strens were (with some changes of notation):

(a) a ratio of A_0/A_r or D_0/D_r greater than unity indicates Mg or Fe^{2+} clustering, respectively;

(b) a ratio ρ defined as $(B_0/B_r)/(C_0/C_r)$ greater than unity indicates that Mg is segregated into M(1), and a value less than unity indicates that Fe″ enters M(1). This arises because there are two M(1) sites to each M(3) so that (for example) segregation of Mg into M(1) relatively enhances peak B. M(3) sites show the opposite segregation.

CATION SEGREGATION

Burns and Law (1970) note that the effect of being able to distinguish M(1) and M(3) sites is to split the *B* and *C* peaks into two components (Table 1 above). This effect was invoked as a partial explanation of broadening ('permutation broadening') of these lines in anthophyllite spectra. Strens (1974) comments on this in more detail and notes that the broadening of these peaks could be used to give some indication of the 'sense' of segregation (i.e. which ion is segregated into which site) even if the individual components cannot be resolved.

The spectra of many amphiboles show little or no such line broadening, however, suggesting that the components overlap almost exactly. The factors which distinguish M(1) and M(3) sites need not necessarily affect the strength of the O–H bond. Further, other causes may be partly or wholly responsible where broadening of *B* and *C* peaks is observed. It is probable that in most cases information about cation ordering must be obtained from the relative intensities of the major peaks alone (the resolution of minor peaks is much less accurate).

We may define separate site populations for M(1) and M(3) sites and use these to calculate intensities for the major peaks. The simplest model assumes that minor ions are still randomly distributed; and the *site fraction P* is then defined as the fraction of M(1) and M(3) sites occupied by Mg and Fe^{2+} together. The parameters defined, and relationships between them, are given in Table 2 above. The expressions for major peak intensities are:

$$
\begin{aligned}
A_c &= \mu_1^2\mu_3 \\
B_c &= \mu_1^2\phi_3 + 2\,\mu_1\phi_1\mu_3 \\
C_c &= \phi_1^2\mu_3 + 2\,\mu_1\phi_1\phi_3 \\
D_c &= \phi_1^2\phi_3
\end{aligned}
$$

The intensities calculated for all peaks from any model must sum to unity, and for the present model this can easily be demonstrated algebraically. Additionally, here the sum of the major peak intensities is P^3. These relationships provide a check on the validity of the formulation of the model.

The relationships between the parameters allow the use of a single parameter to define the extent of segregation. Here ϕ_3 is used. This simplifies the arithmetic a little and facilitates comparison with results from Mössbauer spectroscopy. Calculations were performed using computer programs written in Algol for the Oxford University ICL 1906A machine. The values of T_f, T_m and T_a impose limits on ϕ_3 (see Appendix) and numerical tables of results were obtained for discrete values of ϕ_3 between these limits. Results were also output in graphical form.

Figure 1 shows the graphical results of a calculation based on the measured peak intensities in the spectrum of an actinolite, USNM 44973. Detailed data were obtained by Greaves (1970). A report of the Mössbauer and infrared study of the specimen is given by Burns and Greaves (1971); X-ray results have

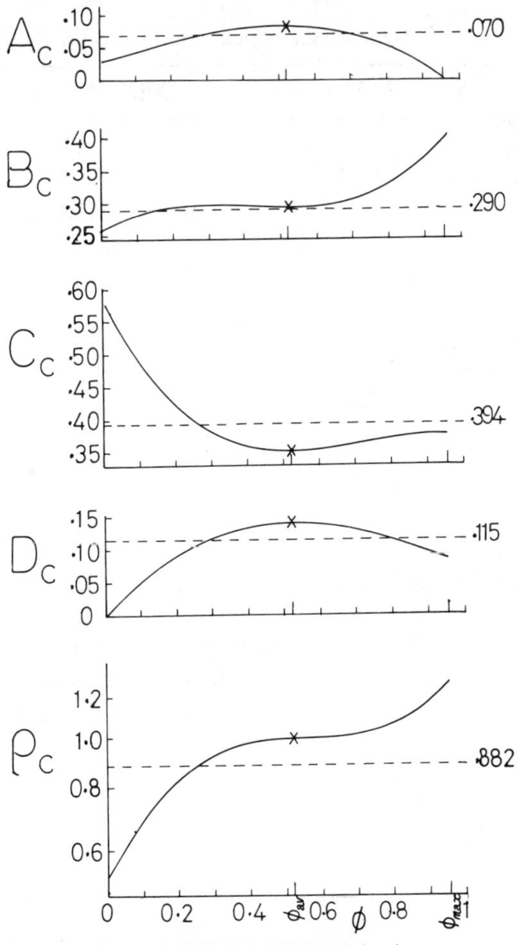

Figure 1. USNM 44973. Major ion segregation only. For symbols, etc., see Figure 3.

been obtained by Mitchell, Bloss and Gibbs (1971) whose analysis data are also quoted here.

The salient features of the results are:

(a) The functions A_c and D_c show maxima at the random distribution point ($\phi_3 = T_f/3$);

(b) B_c and C_c also show stationary values at this point;

(c) ρ rises with ϕ_3, with an inflexion through unity at the random distribution point.

Thus the intensities of peaks A and D cannot exceed A_r and D_r if segregation is the only form of ordering shown; to this extent Strens' criterion for clustering holds for this specimen. The value of ρ would indicate the correct sense of segregation but would be rather insensitive as it remains close to unity for some distance either side of the random distribution point.

Table 3. Site populations for USNM 44973

	M(1) per site	M(2) per site	M(3) per site
1. Infrared	0.72–0.60 Fe^{2+} ($+Fe^{3+}$)	0.47 Fe^{2+} ($+Fe^{3+}$)	0.10–0.35 Fe^{2+} ($+Fe^{3+}$)
2. X-ray	0.61 Fe	0.46 Fe^{2+} + 0.16 Fe^{3+}	0.58 Fe
3. Mössbauer	0.76 Fe^{2+}	0.29 Fe^{2+}	0.40 Fe^{2+}

Sources
1. This study, data for i.r. spectrum from Greaves (1970). See Burns and Greaves (1971).
2. Mitchell, Bloss and Gibbs (1971).
3. Burns and Greaves (1971).

Total Fe **determined in** 2M(1) + M(3)
1. 1.55 Fe^{2+} + 0.14 Fe^{3+} = 1.69 Fe
2. 1.84 Fe
3. 1.86 Fe^{2+}, no Fe^{3+} resolved.

In Figure 1 the observed peak intensities are indicated on the diagrams. There is relatively narrow region, $0.10 < \phi_3 < 0.35$, within which the observed values of all five functions are matched by the model. Table 3 shows site populations for M(1) and M(3) derived from this bracket, and compares these with results from other techniques. Agreement is only fair: but the Mössbauer and X-ray measurements also disagree; there is an inconsistency in the X-ray refinement which assigns more iron to sites in the structure than is given in the chemical formula; and the total iron determined in 2M(1) + M(3) is lower for the infrared measurement than for the other measurements, which may indicate that sample oxidation has occurred. In any case the actinolite spectrum is used here arbitrarily to give a numerical illustration. The model does not yet depend for its validity on its success in estimating site populations in agreement with other methods, given the many other uncertainties in the infrared technique. It is encouraging that for such a basic model even moderate agreement is obtained. At the same time, when corrections for the several likely sources of error can be applied to infrared peak intensities, calculations can be based on this model using the corrected intensities. Thus individual M(1) and M(3) populations can be determined from the hydroxyl spectra.

Figure 2 shows the results for a low iron specimen (Klein 6). Detailed data have been supplied by Greaves (personal communication). The different composition (noted in the caption) results in changes in the forms of the functions (see Appendix). Of these the most important is that ρ no longer has any value below unity for 'real' values of ϕ_3. Thus Strens' criterion for segregation is invalid for this specimen.

For Klein 6 the value determined for ϕ_3 is very precise—it is 0.44 ± 0.02. Agreement with rather uncertain Mössbauer data (Burns and Greaves, 1971) is hard to judge.

The model as so far outlined considers only the effect of major ion segregation, and a poor value for ϕ_3 with a wide bracket might be expected. To improve the model, it can easily be adapted to allow for segregation of minor ions. Separate calculations need to be performed for each value of the minor ion

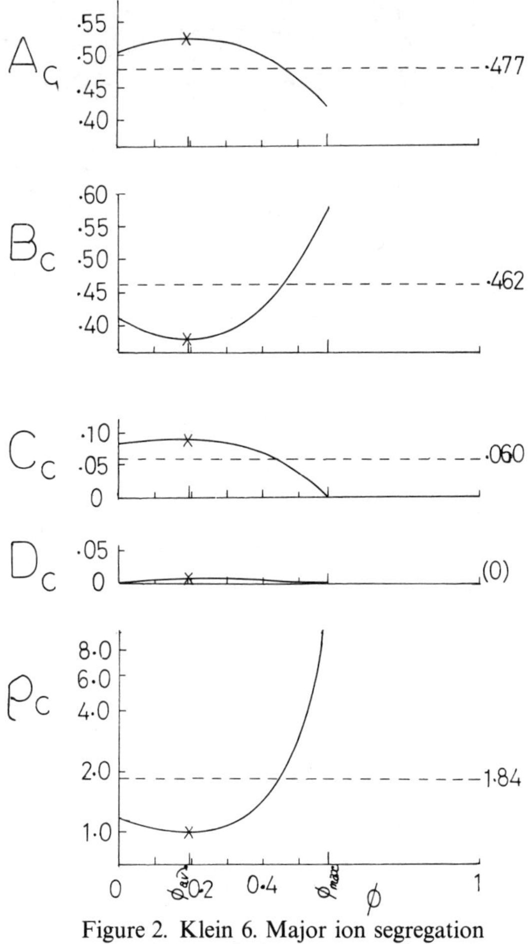

Figure 2. Klein 6. Major ion segregation only. For symbols, etc., see Figure 3.

site occupancies. The expressions for the major peaks are unaltered, but the relationship between the parameters are changed slightly since α_1 is no longer equal to α_3 (Table 2 above). The sum of all peaks is still unity; but there is no longer a simple expression for the sum of the major peaks alone.

Only one set of results is presented. Figure 3 shows the case where all minor ions enter M(3), for USNM 44973. The main change again concerns ρ: the two stationary values have clearly separated. Strens' criterion would here determine the wrong sense of segregation unless segregation were quite considerable.

The bracket for ϕ_3, by comparison with the observed intensities, is $0.25 < \phi_3 < 0.40$ (B_c lies close to B_0 for a wide range of ϕ_3 and is disregarded). Thus the refinement of the model has narrowed the bracket on ϕ_3, and the determined value is different. Trial and error, to find the narrowest bracket, would give an estimate of the best value for minor ion segregation: the use of computer graphics considerably reduces the amount of work involved.

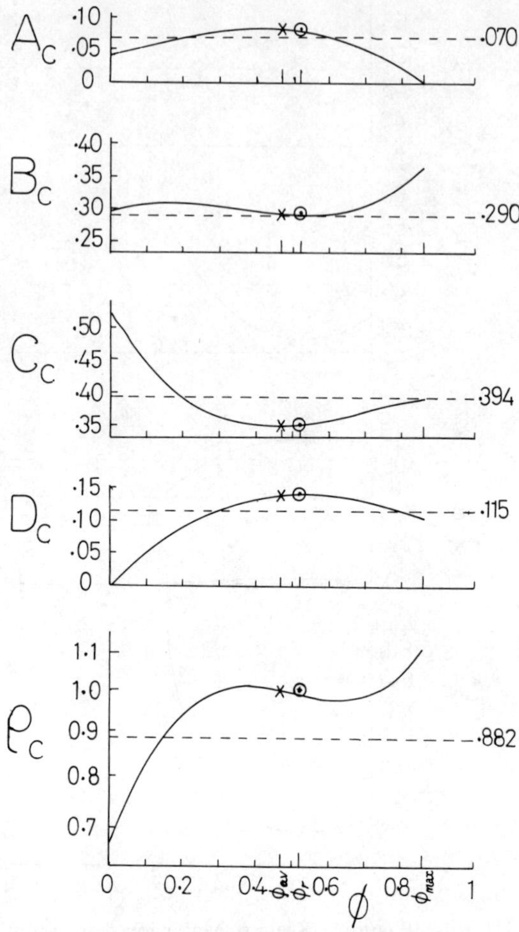

Figure 3. USNM 44973. Major ion segregation for $p_1 = 0.93$, $p_3 = 1.00$. Symbols, x indicates ϕ_{av}, the value of ϕ in the absence of major ion segregation; ⊙ indicates ϕ_r, the value of ϕ for the random distribution (Figure 3 only; for Figures 1 and 2 $\phi_r = \phi_{av}$). The maximum 'real' value of ϕ is ϕ_{max}. The observed values of the functions, indicated by the dashed lines, are given at the right hand side of the plots.

THE PROBLEM OF CLUSTERING

The refinement of the model to include clustering is a qualitatively different problem. For cation segregation it is correct to consider M(1) and M(3) sites, in effect, separately. Clustering is the effect of the occupancy of one site on

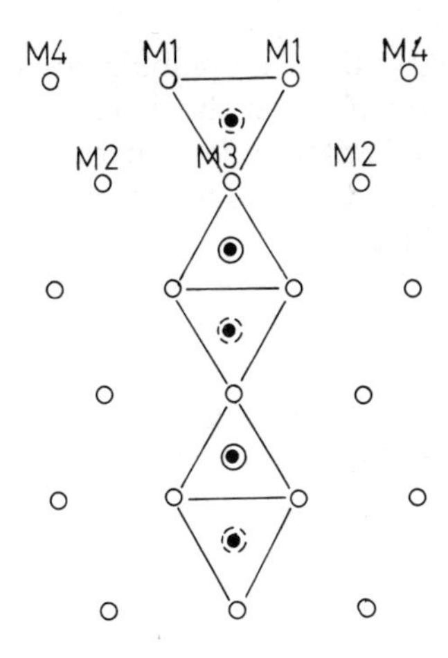

Figure 4. Part of the cation strip of amphiboles, showing OH groups. (100) projection.

another and must be considered in terms of the strip of alternating M(3) and M(1) sites with each M(3) and each pair of M(1) sites related to two OH groups on opposite sides of the cation strip (see Figure 4). Individual M(1)M(1)M(3)OH groups cannot in fact be isolated, and the simple concept of clustering implies that they can. Also, the effect of (say) the presence of Fe^{2+} at a M(3) site might be to decrease rather than to increase the probability of Fe^{2+} also occupying any of the four adjacent M(1) sites. A more general concept, expressing the effect of M(1) and of M(3) occupancies on each other, may prove to be more useful in formulating the model. It is hoped to consider this in a subsequent study.

If we accept for the moment the concept of clustering, we can consider qualitatively the effect of clustering of Fe^{2+}, for example. The most obvious effect is to increase the intensity of peak *D*. However, in the sites not occupied by clusters of Fe^{2+}, the proportion of Mg will be increased. Hence the relative intensity of peak *A* and probably to a lesser extent that of *B* will be increased, and that of *C* will be decreased. Thus clustering of Mg would also be indicated by Strens' criterion, without any actual tendency of Mg to cluster. In addition, peak *B* will be enhanced relative to *C*, and incorrect conclusions about segregation would again be drawn on the basis of Strens' criterion.

CONCLUSION

This study shows that the empirical criteria of Strens (1966) for the detection of ordering in the M(1) and M(3) sites of amphiboles are not in fact reliable in the majority of cases. The concept of cation clustering requires further thought. It is suggested that the only useful criterion for ordering processes is the actual determination of the extent of ordering. The type of model presented will allow such determinations to be made.

ACKNOWLEDGEMENTS

This work was done in the Department of Geology and Mineralogy, Oxford, during the tenure of a Burdett-Coutts studentship. Dr. E. J. W. Whittaker's critical discussion has been of great help.

APPENDIX

Properties of the Stationary Values of A, B, C, D

One stationary value for all functions is at the random distribution point, $\phi_3 = T_f/3$. The second stationary value (S.S.V.) varies. The value of ϕ_3 is 'real' if it lies within the limits of variation[a] of ϕ_3.

Function:	A	B	C	D
1. At $\phi_3 = T_f/3$:	Maximum	Maximum if $T_f > 2P$ Minimum if $T_f < 2P$	Maximum if $T_f < P$ Minimum if $T_f > P$	Maximum
2. Value of ϕ_3 at second stationary point:	$T_f - 2P$	$T_f - \frac{4}{3}P$	$T_f - \frac{2}{3}P$	T_f
3. SSV $\geqslant$ 0 if[a]:	$T_f \geqslant 2P$	$T_f \geqslant \frac{4}{3}P$	$T_f \geqslant \frac{2}{3}P$	$T_f \geqslant 0$ (always)
4. SSV $\leqslant$ P if[a]:	$T_f \leqslant 3P$ (always)	$T_f \leqslant \frac{7}{3}P$	$T_f \leqslant \frac{5}{3}P$	$T_f \leqslant P$

[a] If $T_f \leqslant P$ the limits of ϕ_3 are $0 \leqslant \phi_3 \leqslant T_f$. SSV's of A, B, C are never real. SSV of D is at $\phi_3 = T_f$, the upper limit.

If $P \leqslant T_f \leqslant 2P$ then $0 \leqslant \phi_3 \leqslant P$. SSV's real if both conditions in lines 3, 4 are fulfilled.

If $2P \leqslant T_f \leqslant 3P$ then $T_f - 2P \leqslant \phi_3 \leqslant P$. SSV of A is at $\phi_3 = T_f - 2P$, the lower limit. No other SSV's are real.

REFERENCES

Bancroft, G. M., and R. G. Burns (1969). *Min. Soc. Amer. Sp. Pap.*, **2**, 137–148.
Burns, R. G., and C. Greaves (1971). *Amer. Min.*, **56**, 2010–2033.
Burns, R. G., and A. D. Law (1970). *Nature*, **226**, 73–75.
Burns, R. G., and F. J. Prentice (1968). *Amer. Min.*, **53**, 770–76.
Greaves, C. (1970). *Thesis, B.A. Part II*, Oxford University.
Krzanowski, W. J., and A. C. D. Newman (1972). *Min. Mag.*, **38**, 926–35.
Mitchell, J. T., F. D. Bloss, and G. V. Gibbs (1971). *Zeits. Krist.*, **133**, 273–300.
Strens, R. G. J. (1966). *Chem. Comm.*, p. 519.
Strens, R. G. J. (1974). In, V. C. Farmer, Ed., *The Infrared Spectra of Minerals*, Mineralogical Society monograph 4, 305–330.

Index